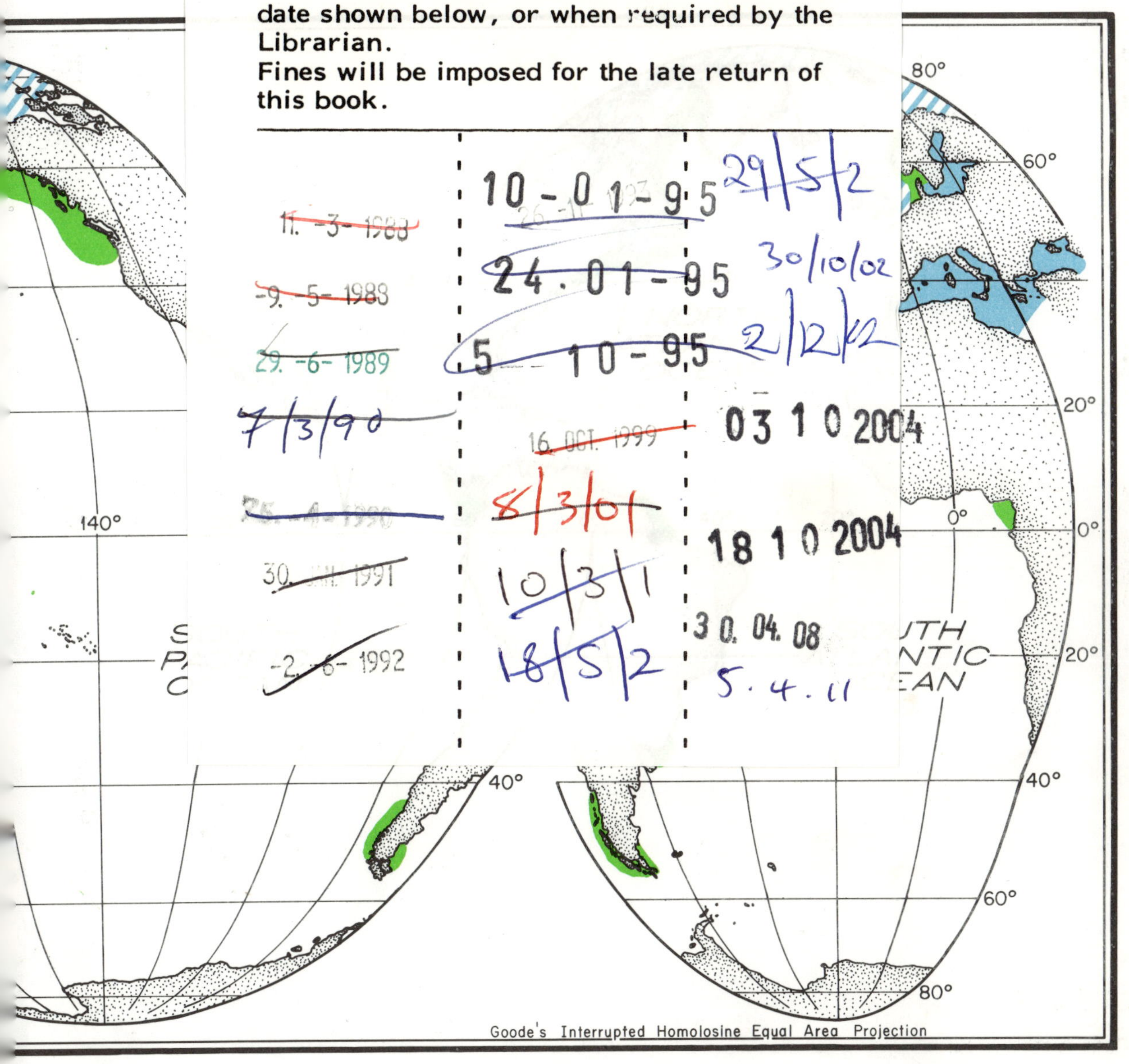
140°
80°
60°
20°
0°
40°
Goode's Interrupted Homolosine Equal Area Projection

ECOSYSTEMS OF THE WORLD 26

ESTUARIES AND ENCLOSED SEAS

ECOSYSTEMS OF THE WORLD

Editor in Chief:

David W. Goodall

CSIRO Division of Land Resources Management, Wembley, W.A. (Australia)

ECOSYSTEMS OF THE WORLD 26

ESTUARIES AND ENCLOSED SEAS

Edited by

Bostwick H. Ketchum†

Woods Hole Oceanographic Institution
Woods Hole, Mass. 02543 (U.S.A.)

ELSEVIER SCIENTIFIC PUBLISHING COMPANY

Amsterdam — Oxford — New York 1983

ELSEVIER SCIENTIFIC PUBLISHING COMPANY
1 Molenwerf
P.O. Box 211, 1000 AE Amsterdam, The Netherlands

Distributors for the United States and Canada:

ELSEVIER SCIENCE PUBLISHING COMPANY INC.
52 Vanderbilt Avenue
New York, N.Y. 10017, U.S.A.

Library of Congress Cataloging in Publication Data
Main entry under title:

Estuaries and enclosed seas.

(Ecosystems of the world ; 26)
Includes bibliographical references and indexes.
1. Estuarine ecology. 2. Marine ecology. I. Ketchum, Bostwick H., 1912- . II. Title: Enclosed seas.
QH541.5.E8E84 574.5'26365 81-9853
ISBN 0-444-41921-7 AACR2

ISBN 0-444-41921-7 (Vol. 26)
ISBN 0-444-41702-8 (Series)

Printed in The Netherlands

This volume is gratefully dedicated to my wife

EMILY WESTBROOK (BROOKIE) KETCHUM

whose love, devotion and patience have supported
and sustained me throughout my professional career

PREFACE

Estuaries constitute a small part of the area and an even smaller part of the volume of the total marine hydrosphere. They play, however, a vital role in the life history and development of many marine populations. Several benthic species live out their entire life cycle within the brackish water of the estuaries; other coastal species find estuaries essential for some part of their life history. These brackish water areas are essential as breeding grounds or nursery areas for many species; they serve as migration routes for the anadromous species which breed in fresh water but spend most of their life in the sea or for the catadromous species which breed in the sea and develop and grow in fresh waters.

A gradient of salinity from the fresh waters of the rivers to the coastal waters at the mouth is characteristic of estuaries. In well-mixed estuaries, the vertical distribution of salinity is uniform but in a stratified estuary, the surface waters may be considerably fresher than the waters at the bottom. At any given place within the estuary, the salinity can vary widely on several time scales. In the middle of the estuarine zone, where the horizontal gradient of salinity is steepest, there can be large changes of salinity as the tide ebbs and flows. As the river flow varies seasonally the fresher water moves seaward during flood periods while the sea water moves landward during periods of low river flow. During dry years the estuary will become saltier while during wet years it becomes fresher. The geographical extent and the magnitude of these changes are largely determined by the unique geomorphology of each estuary, but all estuaries are characterized by an extremely variable set of environmental conditions which impose unusual stresses on the populations that inhabit them.

Section I of this volume consists of seven chapters which discuss these variations and their effects on the resident populations of marine or brackish-water species. Examples are drawn from estuaries throughout the world wherever the appropriate studies have been conducted. The various chapters discuss estuarine characteristics, physics of estuarine circulation, responses to estuarine stress, the phytoplankton of estuaries, the zooplankton of estuaries, estuarine benthos and estuarine fishes.

The characteristics of ten enclosed seas are discussed in Section II of this volume. Each of these enclosed seas is connected with the open ocean in a way which modifies and controls the circulation within the sea and produces an environment and an ecosystem which is unique and different from the adjacent coastal waters. For several of the enclosed seas there is only a single connection which is narrow and shallow; these are the mediterranean-type seas. For others there are multiple connections separated by islands or island arcs. The enclosed seas are characteristically found in the Northern Hemisphere. Their geographic range extends from the tropics to the Arctic.

The enclosed seas discussed include the European Mediterranean Sea, the Black Sea, the Red Sea, the Baltic Sea, the Bering Sea, the Okhotsk Sea, the Japan Sea, the East China Sea, the Gulf of St. Lawrence and the Gulf of California. Most of these seas are brackish and illustrate on a large scale typical estuarine characteristics and processes. In three of the seas, however, the loss of fresh water by evaporation exceeds the addition by precipitation and river drainage. These seas, characterized by high salinities, are the Mediterranean, the Red Sea and the Gulf of California.

Estuaries and enclosed seas have always been important to people as well as to the marine species which they nourish. From the time of earliest history, people have settled on the shores of estuaries and of these enclosed seas partly because of the abundance of living resources for human food which they afford and partly because of the comparative ease of marine transportation. Many of the world's modern major cities are built upon the shores of estuaries or on seaways giving access to marine transportation.

Marine pollution is a comparatively recent problem which has been aggravated by the concentration of human populations along the shores of estuaries and of enclosed seas. Human populations have always been a source of domestic sewage waste, but this can be recycled by the normal marine ecosystem when it is not toxic and is not discharged in excessive amounts that exceed the receiving capacity of the system. Modern technology not only adds toxic industrial wastes to sewage but it also permits the collection and discharge of sewage into local areas of the environment. Sewage treatment technology has not kept pace with the growing demand of expanding population size. Only with a thorough understanding of the dynamics of the estuarine system, physical, chemical and biological, can we hope to evaluate the impact of marine pollution and to preserve, and in some cases to restore, the valuable resources which are now threatened.

It is hoped that this volume will help to guide future studies of the ecosystems of estuaries and enclosed seas. These ecosystems are fascinating and complex; they are also important in maintaining a continuing supply of living marine resources. They are worthy of our best efforts to understand and preserve them.

As Editor I sincerely thank each author for his contribution and I am especially grateful for the cooperative way in which each author has responded to suggestions and editorial comments. To those authors whose manuscripts were completed and submitted on time, I extend special thanks and express my regret that unavoidable delays have postponed the publication of the volume beyond its original scheduled date. I am pleased that the cooperation of the authors has made it possible for the volume to appear complete as it was originally planned.

I also want to thank Dr. David Goodall, Editor in Chief of the series, not only for the opportunity to participate as Editor of this volume but also for his expert editorial advice and for his careful and thoughtful reading of each of the contributions.

Ms. Patricia Anne Pykosz has been of invaluable assistance, not only in typing the voluminous correspondence that is associated with such an effort, but also in typing and retyping revised drafts as they were received from the authors. She also has done an excellent job of keeping the various revisions and drafts in an orderly manner so that the status of each page is readily recognized.

Ms. Ann Martin, who received support from the Pew Memorial Trust, has helped me with the editing of the volume, even to the extent of revising the English presentation of some of the contributions of our foreign authors. She has meticulously read each page for the smoothness of presentation, for punctuation and spelling and has compared the citations in the text with the publications listed in the references. Ann, Pat and my wife, Emily, have all made major contributions in compiling the three indices which appear in the volume; a tedious job which must be carefully and meticulously well done.

I also thank Dr. James Carlton who assisted in checking and completing the Systematic Index. In particular Dr. Carlton provided the authorities for many species' names when this information was omitted by the authors.

Finally, I am grateful for the financial support from the Woods Hole Oceanographic Institution which has made possible the collaboration with Dr. Carlton, Ms. Pykosz and Ms. Martin. The congenial atmosphere and conversations with many members of the scientific staff have contributed greatly in making the development of the entire volume a real pleasure to me.

BOSTWICK H. KETCHUM†
Senior Scientist *Emeritus*
Woods Hole Oceanographic Institution

Buck Ketchum did not live to see this volume, on which he had lavished so much care, in its final form. He died on July 14, 1982, when the volume was in its final stages of proof-reading.

Buck Ketchum was well known and held in high regard throughout the fields of marine science and ecology, and his loss will be mourned by his many colleagues and collaborators in all parts of the world.

This volume distils much of his accumulated knowledge, and its publication shortly after his death will constitute a fitting tribute to him. In the words of a famous epitaph in St. Paul's Cathedral, London:

Si monumentum requieris, circumspice.

DAVID W. GOODALL
Editor-in-Chief

LIST OF CONTRIBUTORS

SAUL ALVAREZ-BORREGO
Centro de Investigacion Cientifica y de Educacion
Superior de Ensenada, B.C.
Espinoza 843
Ensenada, Baja California (Mexico)

ADAM BEN-TUVIA
Department of Zoology
Hebrew University
Jerusalem (Israel)

LLOYD M. DICKIE
Ocean and Aquatic Sciences
Marine Ecology Laboratory
Bedford Institute of Oceanography
Dartmouth, N.S. (Canada)

RICHARD L. HAEDRICH
Memorial University of Newfoundland
St. John's, Nfld. (Canada)

DONALD W. HOOD
P.O. Box 57
Friday Harbor, Wash. 98250 (U.S.A.)

BOSTWICK H. KETCHUM†
Woods Hole Oceanographic Institution
Woods Hole, Mass. 02543 (U.S.A.)

GUNNAR KULLENBERG
Institute of Physical Oceanography
University of Copenhagen
Copenhagen (Denmark)

ARTHUR R. MILLER
Woods Hole Oceanographic Institution
Woods Hole, Mass. 02543 (U.S.A.)

CHARLES B. MILLER
School of Oceanography
Oregon State University
Corvallis, Ore. 97330 (U.S.A.)

SABURO NISHIMURA
Biological Laboratory
Yoshida College
Kyoto University
Yoshida, Sakyo-Ku
Kyoto (Japan)

CHARLES B. OFFICER
Earth Sciences Department
Dartmouth College
Hanover, N.H. 03755 (U.S.A.)

DAVID A. ROSS
Woods Hole Oceanographic Institution
Woods Hole, Mass. 02543 (U.S.A.)

THEODORE J. SMAYDA
Graduate School of Oceanography
University of Rhode Island
Kingston, R.I. 02881 (U.S.A.)

YU.I. SOROKIN
South Department of Institute of Oceanology
U.S.S.R. Academy of Science
Gelendzhik, F. Krasnodar, distr., 353720 (U.S.S.R.)

RONALD W. TRITES
Ocean and Aquatic Sciences
Marine Ecology Laboratory
Bedford Institute of Oceanography
Dartmouth, N.S. (Canada)

WINONA B. VERNBERG
School of Public Health
College of Public Health and Associated Health Programs
University of South Carolina
Columbia, S.C. 29208 (U.S.A.)

W.J. WOLFF
Research Institute for Nature Management
c/o Netherlands Institute for Sea Research
Texel (The Netherlands)

CONTENTS

Section II. Enclosed seas

Chapter 1

ESTUARINE CHARACTERISTICS

BOSTWICK H. KETCHUM†

INTRODUCTION

The estuary, where the flow of the river meets the flood of the tide, is a unique and important part of the aquatic environment. It forms the transition zone between the inland world of fresh water and the sea water lying offshore. As such, it retains some characteristics of both fresh-water and the marine environment but it also has unique properties of its own. Some of these unique properties are common to all estuaries of the world; others are determined by local conditions so that each estuary has its own individual and special properties.

As an ecosystem, the estuary performs several vital functions. Many species live out their entire life cycle in the estuary such as the American oyster (*Crassostrea virginica*), and some species of clams, *Mya arenaria* and *Mercenaria mercenaria*, and the New England (U.S.A) bay scallop (*Argopecten irradians*). For other species such as the anadromous salmon (*Salmo salar*) and the catadromous eel (*Anguilla rostrata*), the estuary forms part of the migration route which these species must pass through in order to reach their breeding grounds. Some fish species spawn in the brackish water of the estuary, but even more importantly the estuary forms important nursery areas for many young fish. There are few truly estuarine fishes and most of the estuarine fish spawners are derived from the coastal waters (see Ch. 7, pp. 183–207). This is in contrast to the many estuarine benthic species (Ch. 6, pp. 151–182). The zooplankton are somewhat intermediate in this regard since some species are predominantly estuarine in their distribution, but many other estuarine zooplankton are derived from the nearby coastal waters (Ch. 5, pp. 103–149). The phytoplankton that are carried into the estuary in the river flow die off rapidly as the salinity increases and most of the phytoplankton of estuaries are commonly also found in the coastal waters offshore (Ch. 4, pp. 65–102). Since most fresh-water species cannot withstand the rigors of the estuarine environment, most species found in estuaries are either endemic or derived from the nearby marine environment.

Conditions in the estuary are more variable than those in either the fresh-water or the coastal marine environment. The variation is primarily in the salinity of the water which increases horizontally from virtually zero in the river water to the coastal water salinity of 30 parts per thousand, more or less. The salinity of the water may vary vertically by several parts per thousand and similar large variations may occur as the tide ebbs and flows. Seasonal changes in river flow introduce another time scale of salinity variation and annual differences occur between wet and dry years. To survive in such a variable environment, organisms have developed various mechanisms to withstand the osmotic stress imposed by the salinity changes as discussed in Ch. 3 (pp. 43–63). Organisms which cannot adapt to these drastic salinity variations simply cannot survive in the estuarine environment.

DEFINITIONS AND CLASSIFICATIONS

Characteristically, the estuary is defined as a body of water in which river water mixes with and measurably dilutes sea water. The definition is either explicitly stated or implied by the treatment of the data by Alexander et al. (1935, 1936), by Rochford (1951), by Ketchum (1950, 1951a, b) and by Tully (1949). An often quoted definition is that given by Pritchard (1952, 1967) and by Cameron and Pritchard (1965). This definition states that "An estuary is a semi-enclosed coastal body of water which has a free connection with the open sea and within which the sea water is measurably diluted with fresh water derived from land drainage" (Pritchard, 1967, p. 3).

This type of definition leaves both the landward and the seaward boundaries of the estuary vague, it excludes some areas that various investigators feel are an essential and integral part of the estuarine system and it precludes certain unique situations which are best evaluated by the use of techniques developed for studies of estuaries.

Ketchum (1951a) has defined the inner end of the estuary as the cross section above which the volume of water involved in raising the level of water from the low tide to the high tide mark is equal to the volume contributed by the river during one half of a tidal cycle. There is no net exchange of water across this boundary during the flood tide; during the ebb there will be a seaward loss of one river flow per tidal cycle. This is a dynamic boundary, not a geographical one, since it implies that the inner boundary of the estuary moves up- and downstream as river flow varies.

In many estuaries, the tide rises and falls far upstream of the area to which sea water penetrates. This is commonly known as the "tidal river" and its characteristics are included in the discussion of estuarine benthos (Ch. 6, p. 173). For example, in the Hudson Estuary, the river is used as a water supply at Poughkeepsie, New York, 120 km north of the entrance to the sea, but the tide rises and falls at Troy, New York, an additional 120 km north of Poughkeepsie. Similarly, in the Delaware Estuary, the water is virtually fresh at Philadelphia, Pennsylvania, while the tide rises and falls as far north as Trenton, New Jersey, an additional 100 km upstream. The definition proposed by Ketchum (1951a) does not apply to the tidal river sections of the Hudson and Delaware systems since, above the observed penetration of salt, the fresh-water volume accumulated is many times greater than the volume introduced by river flow during the time of the flooding tide.

The seaward boundary of the estuary can generally be defined geographically by a line, or a section, between the land masses on each side of the entrance to the estuary from the coastal waters. This can be ambiguous, however, where deltaic structures are formed at the mouth. The seaward limits of the estuary cannot be precisely defined by the salinity distribution, though the change of salinity with distance is generally more gradual in the coastal waters than it is within the estuary itself. All of the coastal water is commonly diluted with fresh water from the many rivers along the coast and from ground-water seepage. The salinity of the coastal water gradually increases offshore until it reaches that characteristic of the deep sea at the break of the continental shelf, generally defined as a depth of one hundred fathoms (183 m). Thus, the coastal waters can be considered an extension of various estuaries discharging into the region and indeed Tully and Barber (1960, 1961) have treated the circulation of the entire northeastern Pacific using fresh water as a tracer, a method commonly used in studies of estuaries, and Ketchum and Keen (1956) similarly studied the exchanges of water on the continental shelf on the East Coast of the United States between Cape Cod and Cape Hatteras.

Large rivers produce plumes of freshened water which float on the more dense coastal sea water and they can be traced for many miles from the geographical mouth of the estuary. Riley (1937) traced the effect of the Mississippi River water as far as 90 km into the Gulf of Mexico and Ryther et al. (1967) indicated that the river plume from the Amazon River formed a lens of appreciably freshened surface water up to 500 km in diameter several hundred kilometers to the north and east of the estuary. Ketchum et al. (1951) used the Hudson River plume and estuarine type calculations to evaluate the exchanges of river water and sea water in an area of about 1500 km^2 in the New York Bight.

McHugh (1967) has defined the offshore estuarine zones of the world as those waters bounded by a surface salinity of 33.5‰ ($g\ kg^{-1}$). His figure is reproduced in Ch. 7 (Fig. 7.1). Extensive river plumes and those areas which have been analyzed using estuarine type calculations are clearly shown in this figure, specifically the northeast Pacific (Tully and Barber, 1960, 1961) and the continental shelf on the East Coast of the United States (Ketchum and Keen, 1956). The chart suggests that many other areas of the world could similarly be analyzed using estuarine methodologies. As we discuss the distribution of organisms within estuaries, it will be clear that many of these are derived from coastal waters, or as McHugh (1967) terms them offshore estuarine zones, and are not necessarily endemic or limited to the estuarine environment *per se*.

In some parts of the world, evaporation from the surface of enclosed bodies of water with a free connection to the sea may exceed the precipitation on the surface and the runoff from land drainage. In the normal type of estuarine circulation, the brackish water leaves the estuary in the surface layer whereas the denser sea water enters the estuary at greater depths. Where evaporation within the enclosed body of water exceeds the sources of fresh water, the flow is the inverse of this, with sea water entering at the surface and the denser water formed within the basin leaving at greater depths. The Laguna Madre on the coast of Texas in the Gulf of Mexico has been treated by Hedgpeth (1947, 1957, 1967) as an inverted estuary. In Section II of this volume, the Mediterranean Sea (Chs. 9, 10) and the Red Sea (Ch. 12) are examples of enclosed seas in which evaporation exceeds precipitation and runoff from the land.

Classification systems

Estuaries have been classified on the basis of the geomorphology, on the basis of the distribution of salinity and density and on the basis of the characteristic type of circulation. These systems are interdependent to a large extent since the geomorphology combined with the tidal range and river flow determine to a large degree the type of salinity distribution and this in turn determines the dominant mechanisms influencing the circulation. For the purposes of Chapter 2 on the physics of estuarine circulation, the quantitative classification system of Hansen and Rattray (1966) has been used (pp. 16, 17). This system interrelates the vertical salinity and vertical velocity structures, conditions which determine the type of circulation.

Geomorphology

Drowned river valleys. During the last ice age, sea level was about 100 m lower than it is today and the major rivers of the world flowed through wide river valleys in the flat coastal plain in areas which are now fully submerged and form the continental shelf. The river flows were probably torrential as the ice sheet melted and these rivers cut trenches and canyons across the shelf and at its edge which are still dominant submarine features today. As sea levels rose, the valleys occupied by these rivers became submerged (drowned) and now form the estuaries of the rivers.

Bar-built estuaries. In other areas of generally flat coastal plain, oceanic currents have transported sand and have built offshore bars enclosing shallow, sometimes extensive, bodies of water, the bar-built estuaries. The entrance to these embayments is generally narrow and shallow so that exchanges with the sea are restricted. Consequently, the tidal range within them is generally decreased and the circulation pattern is more dependent upon the wind than it is upon tidal ebb and flow.

Fjords. In more mountainous coastal regions the glaciers cut deep gorges through which the rivers now flow on their way to the sea. The walls of fjords are steep and precipitous and depths in the center of the fjord may reach hundreds of meters. There are two main types of fjords: those with a relatively shallow sill at the mouth and those in which the connection with the sea extends to depths comparable to those within the fjord. The former type of fjord is commonly found along the coast of Norway (Saelen, 1967), whereas those without the shallow outer sill are more typical of the northwest coast of the North American continent (Pickard, 1961).

Deltaic formations. Rivers which carry large amounts of suspended sedimentary material may deposit this material in the shallow coastal waters outside of the mouth of the river as the water spreads over wider and wider areas and the current flow slows down. These deltas can become very complex with many channels connecting the area of the original river mouth with the coastal waters offshore. In the main channel of these rivers, a salt wedge is commonly formed with very limited mixing between the sea water below and the fresh water at the surface.

Salinity and density distributions

Well-mixed estuaries. In a well-mixed estuary, there is virtually no change in salinity between the surface and the bottom. This type of salinity distribution is frequently found among the drowned river valleys or the bar-built estuaries where the mixing is strong compared to the tidal and river flows. In such well-mixed estuaries, the transport of fresh water seaward and of salt water landward is primarily accomplished by horizontal eddy diffusion and the net flow at all depths may be seaward.

Weakly stratified estuaries. In weakly stratified estuaries, there is a salinity variation of a few parts per thousand between the surface layer and the deeper waters with exchanges of both fresh water and salt across a salinity gradient at mid-depths. Coastal sea water, undiluted with river water, is found above the bottom only near the mouth of the estuary and conversely river water undiluted with sea water is found at the surface only near the river entrance. Between the two ends, there is a gradient of salinity in both the surface and the deeper layers. Depending upon conditions and varying within the same estuary from time to time, there may be a net flux of water seaward in the surface layers and landward in the deeper layers to provide some advective transport of fresh water and salt. This type of salinity distribution is commonly found among the drowned river valley estuaries and, more rarely, in bar-built estuaries.

Strongly stratified estuaries. Where the surface salinity differs from the bottom salinity by several parts per thousand, there is a strong gradient of salinity at mid-depth. Such a strongly stratified estuary typically has a net flux of the surface waters seaward and a net flux of the deep waters landward to transport the river waters out of the estuary and to maintain the salt distribution. Undiluted sea water may penetrate near the bottom to considerable distances along such an estuary, but there is mixing across the interface between the surface and deeper water along the entire length of the estuary. Strongly stratified estuaries are commonly found among drowned river valleys, but once again, the same estuary may vary from a strongly to a weakly stratified condition both in different parts of the same estuary or as the river flow fluctuates. Fjord-type of distribution and the arrested salt wedge are special cases of the highly stratified estuary.

Fjords. The salinity distribution in fjords commonly consists of three layers: a surface brackish-water layer, an intermediate layer of increasing salinity with depth, and the deep water within which the salinity is constant to the bottom. The major part of the circulation is confined to the surface and intermediate layers with the deep water taking an active part only during unusually stormy periods. In fjords with a shallow sill at the mouth, the circulation is largely limited to the upper layers with the deep waters becoming stagnant and isolated from the surface layers by a strong density gradient. Under such circumstances, organic materials accumulate in the deep water and, as these are decomposed, the oxygen of the water is exhausted and hydrogen sulfide is formed. Appropriate storm conditions may flush out this deep water periodically, a process which produces very odorous and obnoxious conditions nearby. In those fjords without a shallow sill, deep water does not become stagnant and develops a circulation which is, however, largely independent and separate from the circulation in the more brackish surface layers. Depending on the depth of the sill at the mouth, all gradations between these two extremes may be found.

Arrested salt wedge. Where the river flow is large, undiluted fresh water may flow out of the estuary above a deeper layer of undiluted sea water which forms the salt wedge. There is virtually no mixing between the two water masses involved (pp. 30, 31). The inner edge of the salt wedge intersects the river bottom and will move up- and downstream as river flows fluctuate. There is little circulation within the salt wedge, but the flow in the upper layer tends to accelerate in a down-estuary direction because of the decrease in depth and consequently of the cross-sectional area. Where the salt wedge structure occurs in rivers forming deltas, the fresh water is drained off through the many channels in the delta and the acceleration in the flow is not so pronounced.

TIDES AND TIDAL CURRENTS

The word estuary is derived from the Latin *aestus*, the tide. The importance of the tide in estuaries was thus early recognized. The ebb and flow of the tides provide much of the energy for the mixing which takes place within estuaries. Other major sources of energy are, of course, the river flow and the winds. Tides and tidal currents are the most predictable of oceanographic phenomena and annual tide tables are prepared showing the predicted times of high and low tide day by day for many locations throughout the world. The tides are predictable because they are produced by astronomical forces which are well known and because the mathematical theory of tides is well developed. It is beyond the scope of this volume to discuss the theory of tides but this is presented in detail elsewhere and summarized in all standard oceanographic texts (see for example Marmer, 1926; Schureman, 1941; Sverdrup et al., 1942; Defant, 1961). Certain characteristics of tides and tidal currents are, however, of great importance in determining estuarine circulation and these characteristics are briefly introduced.

The tides of estuaries, and of other coastal embayments, derive their energy from the ocean tides rather than from local direct action of lunar and solar gravitational forces. The limited depth of the ocean, the presence of continents and depth irregularities greatly modify the shape and the rate of progression of the ocean tides. Some long stretches of coastline may, nevertheless, experience high tide simultaneously, while along other coastal stretches the oceanic tide arrives at different times relative to the astronomical forces.

An example of a simultaneous high tide is shown in Fig. 1.1 for the northeast coast of United States (Redfield, 1958). Between Fire Island inlet on Long Island, New York, and Cape Hatteras, North Carolina, the estimated time of high water at outlying coastal positions is within 10 min of the same time. In other parts of the world, irregularities of the coastline and of the depths over the continental shelf can delay the arrival of high tide through frictional damping of its forward progress.

The character of the tidal wave

In estuaries and embayments, the character of the tidal wave is determined by interference between the wave entering from the coastal waters outside and the reflection from the shores. There is always some attenuation of the wave due to friction with the bottom and reflection from irregularities in the boundaries. The reflection from the location of the head of tide in the embayment produces two waves moving in opposite directions at any given time. Under these circumstances, there may be augmentation or diminution of the tidal amplitude so that the maximum and minimum ranges of tide may occur in any part of the system depending upon the geographical form of the embayment, the damping of the primary and reflected waves by frictional effects and constrictions or expansions of the cross-sectional area at various locations. Redfield (1950) has provided an analysis of tidal phenomena in narrow embayments and has applied the analysis to tides on the continental shelf (1958), to open-ended channels in coastal waters (1953), to the tidal system in Lake Maracaibo, Venezuela (1961) and to the tide in coastal waters (1978, 1980).

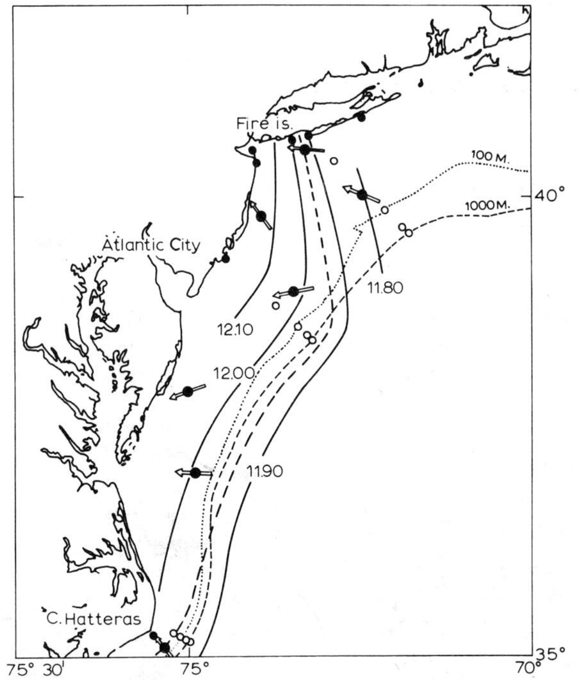

Fig. 1.1. Estimated time of high water, in hours after the moon's transit at Greenwich, on the continental shelf of the United States. Arrows show the direction of observed currents three hours before high water. (After Redfield, 1958.)

The standing wave

In estuaries or embayments where the opening to the sea is not unduly restricted, where the entering wave and its reflection have the same amplitude and period and where the waves are not greatly attenuated, a "standing wave" is produced. In this tidal regime, the tide rises and falls simultaneously in all parts of the estuary, slack water coincides with the time of high or low water and the maximum tidal current flow occurs at mid-flood and mid-ebb.

The progressive tidal wave

In some larger estuaries, or those with a more restricted connection to the ocean and where the entering wave is attenuated by friction with the bottom and reflections from irregularities in the boundaries, the entering wave is larger than the reflected wave except near the head of the embayment where complete reflection may be assumed to occur. The difference appears as a "progressive wave" in which the time of high tide, or of any other stage of the tide, occurs progressively later as the distance between the point of observation and the entrance increases. If there is a full tidal wave or more within the estuary at any given time, it will be low tide at one location at the same time that it is high tide at another, with a corresponding slope to the sea surface. In a progressive wave, the maximum currents are found at the time of high water and of low water, and slack water is observed at the time of mid-flood and mid-ebb.

The mixed tidal wave

In some elongated estuaries, the tide will have the characteristics of a standing wave in one section and of a progressive wave in another. Characteristically, the wave tends to be a standing wave near the mouth, where the entering wave from the coastal waters dominates the system and also near the head of the tide where total reflection of the wave is assumed to occur. Between the entrance and the head of tide, the time of high or low water occurs progressively later as the estuary is ascended, but the characteristics differ from those of a pure progressive wave as described above. The time of maximum current flow and of slack water will both differ from the time of high water or low water by an interval of time greater than zero but less than one-quarter of a tidal period — that is, the characteristics will be intermediate between those of a standing wave and a progressive wave.

The tidal prism or intertidal volume

In an estuary characterized by a standing tidal wave, the tidal prism is the difference between the volume of water contained in the estuary at high tide and at low tide. This volume is simply calculated by multiplying the range of the tide by the average of the surface area at high water and at low water. The volume of water entering the estuary on the flooding tide is equal to this intertidal volume minus the volume introduced by the river in one half of a tidal period. The volume leaving the estuary on the ebb is equal to the volume which entered the estuary on the flooding tide plus the volume introduced by the river in a full tidal cycle.

The tidal prism, or the intertidal volume, of an estuary characterized by a progressive tidal wave cannot be computed simply by multiplying the surface area by the range of tides as can be done for an estuary with a standing wave regime since it is never high tide or low tide at all locations within the estuary at any given time. Ketchum (1952) calculated the intertidal volume for the Delaware Bay and River Estuary which is characterized by a progressive wave regime with about three quarters of a tidal cycle within the system at any given time. The tide is monitored at several stations and the tide tables are published annually by the U.S. Department of Commerce, Coast and Geodetic Survey (currently being published by the U.S.D.C. National Oceanic and Atmospheric Administration). From the available tide stations, nine were selected so that a given stage of the tide occurs an hour later at each successive up-river station. For each of the nine stations, a tide curve was prepared showing the height of the water throughout the lunar tidal cycle of 12.42 h. The results are presented in Fig. 1.2. The slope of the sea surface at each location at various times is shown by each individual curve; the progression of the wave up the estuary and river is shown by the successively later times that any given stage of the tide arrives at each station between the seaward end (Cape Henlopen) and the river end (Trenton) of the estuary.

For each of the 12 lunar tidal hours, the height of tide at each of the nine stations was determined, and the resulting heights were plotted against the length of the estuary. The results are shown in Fig. 1.3 for four of the curves thus obtained. Distances in this figure are expressed in nautical

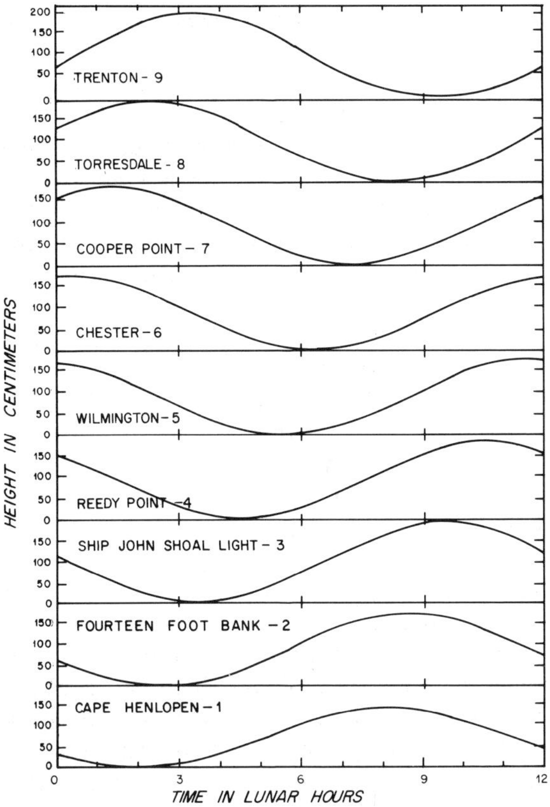

Fig. 1.2. The height of tide at nine gauging stations in the Delaware River Estuary (U.S.A.) throughout a tidal cycle. The location of the stations is as follows: 1, 38°48′N 75°05′W; 2, 39°03′N 75°11′W; 3, 39°18′N 75°23′W; 4, 39°34′N 75°34′W; 5, 39°44′N 75°33′W; 6, 39°51′N 75°21′W; 7, 39°37′N 75°08′W; 8, 40°03′N 74°59′W; 9, 40°12′N 74°46′W. (Ketchum, 1952; data from U.S. Department of Commerce, 1952.)

miles from the seaward end, Cape Henlopen, to the river end, Trenton, 220 km from the entrance. The contour of the sea water surface above mean low water is shown in this figure on an exaggerated vertical scale (about 60 000:1). There is about three quarters of a tidal wave in the estuary at any one time so that when it is high tide halfway up the estuary it is approaching mid-tide at both ends, and conversely when it is high tide at one end it is approaching low tide at the other.

There is always some intertidal volume within the estuary since it is never low tide at all locations at any time. To derive the intertidal volume, it is thus necessary to compute both the maximum and minimum intertidal volume, which can be done from the data by multiplying the height of the tide for each of 12 lunar hours by the appropriate surface area for the segment represented by each of the tidal stations. For the Delaware Estuary, the maximum and minimum enclosed intertidal volumes were accumulated from the river end at Trenton (220 km) to the seaward end at Cape Henlopen with the result shown in Fig. 1.4. The volume entering through any cross section of the estuary on a flooding tide is the difference between the maximum and minimum enclosed intertidal volumes after subtracting the volume added by the river in half a tidal period. Thus, while it is considerably more complicated to compute the flux

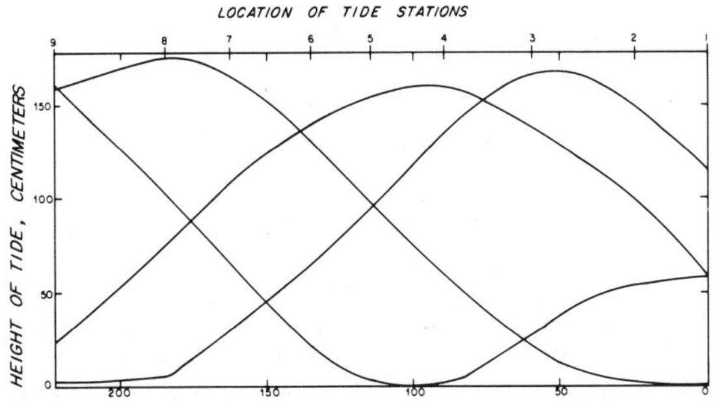

Fig. 1.3. The height of tide at various locations along the Delaware River Estuary (U.S.A.) for four periods of time during the tidal cycle (Ketchum, 1952).

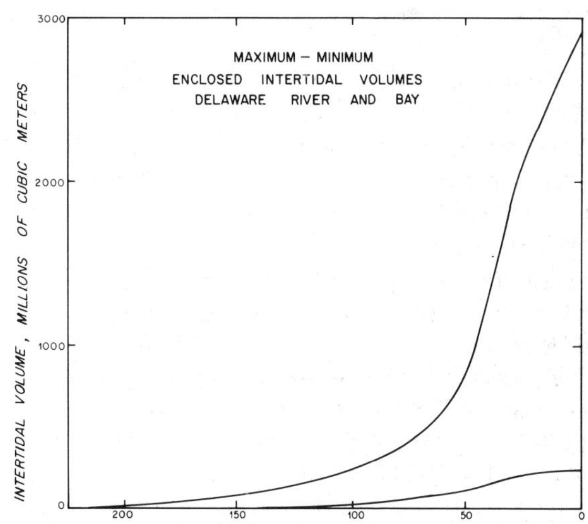

Fig. 1.4. The enclosed intertidal volume for Delaware River and Bay (U.S.A.) accumulated from the river end (220 km) to the ocean, at Cape Henlopen (Ketchum, 1952).

across any given boundary of an estuary resulting from the ebb and the flow of the tides in an estuary with a progressive wave, it is possible to do so when adequate tidal information is available.

Tidal currents

Within the estuary, between confining banks, the tidal currents are reversing with the flood tide currents flowing up-estuary and the ebb tide current flowing in the opposite direction. The magnitude of the tidal currents is generally much greater than that of the residual current which serves to carry the fresh water of the river seaward. It is often desirable to evaluate this residual current, or non-tidal drift, from direct current measurements. This invariably involves the evaluation of the difference between two substantially larger numbers, with consequent increases in uncertainty in the computed small difference.

In offshore coastal waters and in some of the wider embayments, the tidal current is not confined to a definite channel so that during a complete tidal cycle the current will have set in all directions of the compass. This type of current is called a rotary current. In the Northern Hemisphere the rotation is generally in a clockwise direction. These rotary currents are commonly shown on navigation charts as current roses (Fig. 1.5) with the direction at each hour of the tidal cycle being shown by an arrow, the length of the arrow being proportional to the speed of the current.

Rotary currents have no true slack water, such as that found in reversing currents within the estuary. There are minimum and maximum velocities, however, separated from each other by an interval of about three hours so that during a complete tidal cycle there are two periods of minimum and two of maximum flow generally in opposite directions. Two current roses illustrating these characteristics of a rotary current are presented in Fig. 1.5.

Because of the continually changing direction and speed of the current, and because of the variation from minimum to maximum flows, the current rose commonly has the form of an ellipse rather than of a circle. Adjacent coastlines and variable depth contours can also modify the direction and speed of the current flow at various times so that these also influence the shape of the current rose. Off the mouths of some large estuaries,

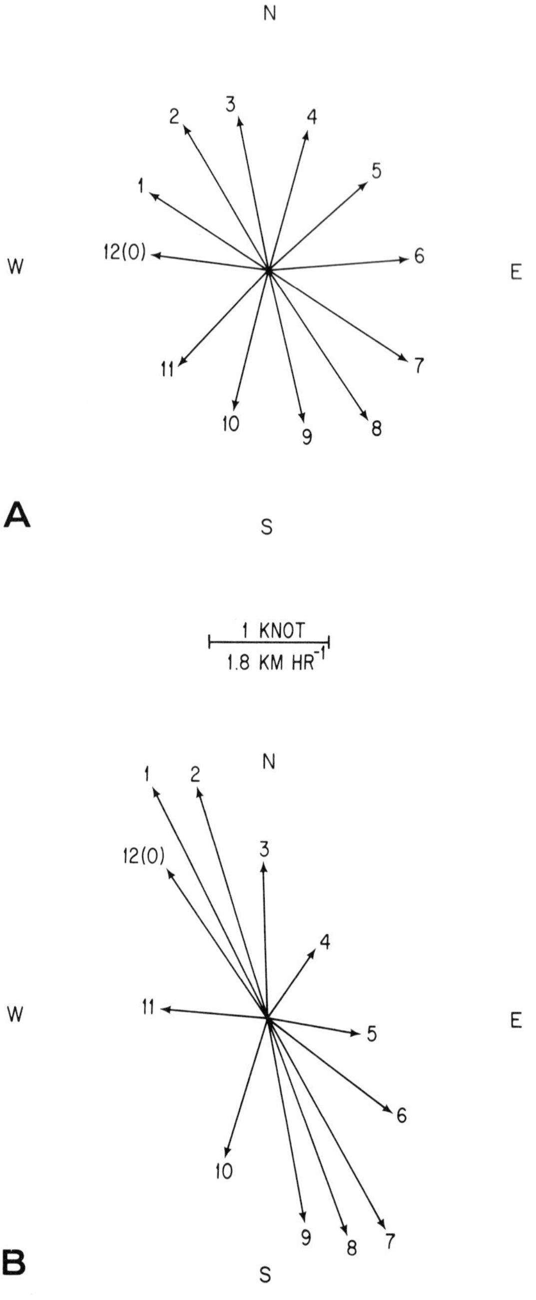

Fig. 1.5. Current roses for two locations on Georges Bank off the East Coast of New England, U.S.A. The numbers at the end of each current arrow are the times in hours after the maximum flood at the reference station, Pollack Rip Channel. A. Eastern edge of Georges Bank, 41°29′N 67°04′W; depth 30 fathoms (54 m). B. Northern edge of Georges Bank, 41°48′N 67°34′W; depth 25 fathoms (45 m). For current rose A, the maximum current is 1.5 times the minimum; for B, the maximum is 3.0 times the minimum. (Data from U.S. Department of Commerce, 1978.)

Chesapeake Bay for example, the elliptical shape is so pronounced that the tidal current essentially becomes a reversing current even in offshore waters.

The Stokes Transport

In an estuary with a progressive tidal wave, where the currents are maximum at high water and low water, there will be a difference in the cross-sectional area as a result of the difference in tidal height. If the current speeds are equal and in opposite directions during the flood and the ebb, there will be a net up-estuary transport of water on the flood as a result of the greater cross-sectional area. This phenomenon was first recognized by Stokes (1847) for progressive gravity waves in deep water, and is known as the Stokes Transport. Pritchard (1957) pointed out that the Stokes Transport had sometimes been ignored in studies of circulation in estuaries. This can cause considerable error in the calculated volume transport and in the exchanges of salt or any other substance, such as a pollutant, dissolved in water. Since the Stokes Transport is up-estuary on the flood, there must be a compensating net non-tidal transport of equal magnitude directed seaward. The water involved in this seaward transport, however, may have very different chemical characteristics to those of the water transported on the flood, so that the distribution of substances in the water cannot be simply derived or predicted. Pritchard (1980) discusses the Stokes Transport and specifies the correct sequence in which the integration of observed tidal velocities must be made in water to obtain a correct tidally averaged transport.

Some comments on the direct measurements of currents in the evaluation of estuarine circulation are appropriate at this point. It is clear from the discussion to this point and from the more detailed discussions in Chapter 2 on the physics of the circulation (pp. 15–41) that considerable information on transport and current velocities can be calculated from the geomorphology of the estuary, the tidal range and the river flow. These are all gross averages, however, and studies of the details of the circulation require direct current measurements. Errors have been made in the past, however, by efforts to oversimplify the system and by lack of adequate data to get a full evaluation of the net transport. Approximations can be made from more limited data, but ideally it is essential to have a number of stations in a cross section and to make observations at several depths for extended periods of time in order to obtain accurate information. Frequently observations are made throughout one or two lunar tidal cycles of 12.42 h each, but in most estuarine systems the height of tide is rarely or never the same in successive lunar tidal cycles and a few centimeters difference produces a large error when multiplied by a large surface area. To take account of the longer period lunar fortnightly tidal cycle, the observations should be continued for a period of 14 days or 28 tidal cycles. It is a major undertaking, therefore, to determine the transport of water in an estuary by direct current measurements and in the author's opinion this has never been adequately and correctly accomplished.

The Kelvin Wave

As a result of the rotation of the earth, currents in the Northern Hemisphere turn to the right; to the left in the Southern Hemisphere. In a channel or a confined embayment, the turning of the current is prevented by the presence of the banks; but the current velocity and the height of the water surface is greater to the right of the current, facing downstream. This phenomenon was first described by Sir William Thomson (later Lord Kelvin) in 1879 and consequently has been called the "Kelvin Wave". In the Northern Hemisphere on the flooding tide, facing up-estuary, the current velocity is greater on the right-hand side of the estuary and the water level slopes downward from right to left. The component of gravity acting down that slope is exactly balanced by the deflecting force of the earth's rotation acting in the opposite direction. On the ebb tide, the increased current velocity and the slope of the sea surface is found on the opposite side of the estuary since the direction of the current flow is the important controlling force.

In very broad reaches of the estuary, the effect of the Kelvin Wave is to produce a cyclonic gyre in the net transport of water and of the components contained therein. This can produce important variations from the average condition in the distributions of salinity, organisms or pollutants. An illustration of the effect of this inequality of flood and ebb current flows on the distribution of salinity in the Chesapeake Bay on the east coast of the United States is shown in Fig. 1.6A (McHugh,

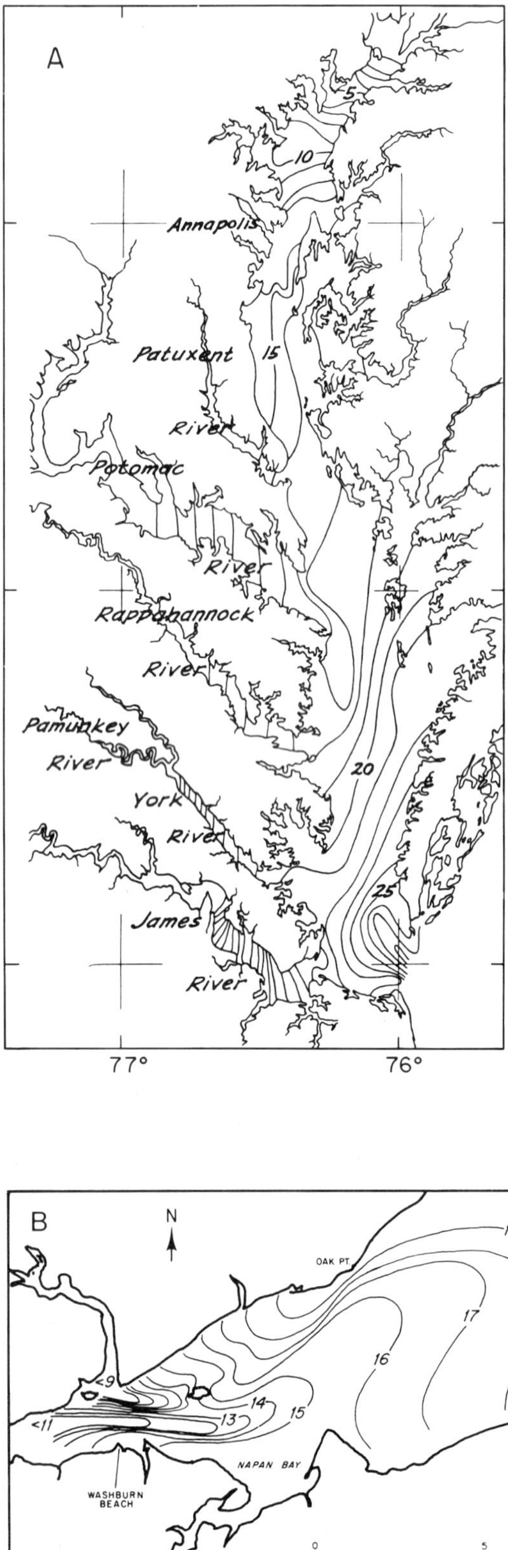

Fig. 1.6. The effect of the Kelvin Wave on the distribution of salinity in (A) the Chesapeake Bay, U.S.A. (after McHugh, 1967), (B) in the Miramichi River Estuary (Canada) on August 30, 1951. Observations were made during the falling tide between 10:12 and 13:15 local time. (After Bousfield, 1955b.)

1967). The higher salinity waters penetrate many kilometers farther into the estuary on the eastern part of the Bay whereas the brackish fresh water extends far to the south on the western part of the Bay. In the Chesapeake, this effect is, no doubt, greatly augmented by the fact that the major rivers all enter the Bay on the western side and there is restricted cross-Bay transport of the fresh water introduced. A similar, nearly synoptic pattern of the surface distribution of salinity in the Miramichi River, New Brunswick, is presented in Fig. 1.6B (Bousfield, 1955b). This distribution is not so greatly affected by the unilateral introduction of river water.

SALINITY VARIATIONS

Variations in salinity occur vertically, horizontally and with time. It is these fluctuations which pose the most extreme stress on organisms living within the estuary, as has been mentioned above.

The steady-state distribution

For some of the calculations discussed in Chapter 2 on the physics of the circulation (pp. 15–41), the salinity distribution in the estuary must be in a steady state in order to make the calculations valid. In the steady state, the salinity at any given point within the estuary will return to the same value at the same stage of each successive tide. The various conditions for and consequences of steady-state distribution are the following:

(1) *River flow:* In order to produce a steady-state distribution, the river flow must remain constant or vary within narrow limits. At times of increasing river flow, the estuary will gradually become fresher; at times of decreasing flow, the estuary will become more saline. The duration of the constant flow of the river must be sufficient to provide the volume of fresh water which has accumulated within the estuary.

(2) *Fresh-water flux:* In any complete cross section of the estuary during a complete tidal cycle, there must move seaward a volume of fresh water equal to the volume introduced by the river in the same period of time.

(3) *Salt transport:* Through any complete cross section of the estuary during a complete tidal cycle there must be no net exchange of salt across the section. There are, nonetheless, exchanges of salt water, but the amount of salt transported into the estuary through the section is exactly balanced by the amount of salt transported out of the section in the more brackish water carrying the river water seaward.

The fluctuations of salinity averaged over each of three complete cross sections of Raritan Bay, New Jersey, for each lunar hour during a tidal cycle, are presented in Fig. 1.7 (Ketchum, 1951b). The total variation in salinity between high and low water was greatest in the sections closest to the mouth of the Raritan River decreasing as the distance from the river increases. In an idealized estuary, these would presumably be smooth curves, and the minor irregularities in the data indicate the magnitude of variations to be expected in a moderately turbulent estuarine area.

Estuarine zones

The estuarine zones of importance to the phytoplankton and benthos are described in Tables 4.2 and 6.8 (pp. 68, 174). Rochford (1951) has defined various estuarine zones on the basis of the horizontal distribution of salinity as shown in Fig. 1.8. Using equilibrium conditions, he defines the gradient zone as a transition between the fresh-water zone and the tidal and marine zones. Under flood conditions, the gradient moves seaward and becomes more gradual. Under drought conditions, the gradient moves landward and becomes somewhat steeper than it is in the equilibrium stage. The mean excursion of the water at the mid-point of the gradient is illustrated in Fig. 1.8 by the horizontal arrow, and the extent of the salinity fluctuation between high and low tide is shown by the vertical arrow. The latter is derived from the excursion and the slope of the gradient.

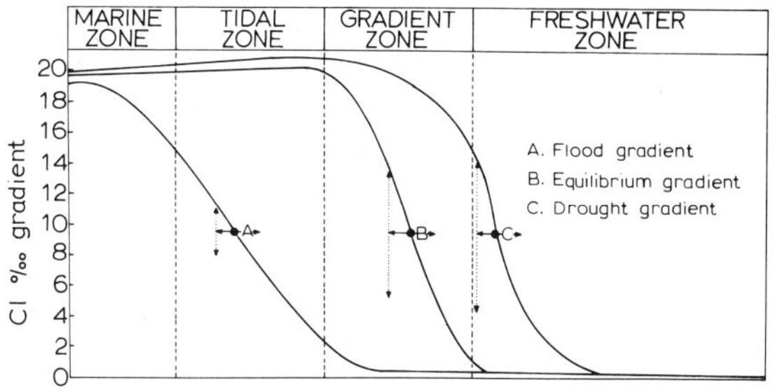

Fig. 1.8. Zonal distribution of chlorinity in a schematic estuarine system in Australia (after Rochford, 1951).

An example of the changes in salinity as related to changes in river flow in the Miramichi Estuary is presented by Bousfield (1955a). His figure is reproduced in Fig. 1.9. The Miramichi is vertically stratified so that both the surface and the bottom (7.3 m) salinities are illustrated. The midpoint of the salinity gradient moved a distance of about 30 km upstream in both surface and bottom waters

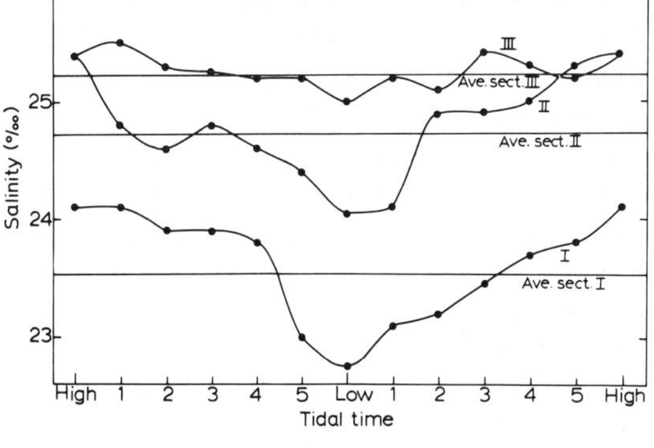

Fig. 1.7. Average salinity variations during a tidal cycle for three cross sections in Raritan Bay (U.S.A.). Section *I* was 3.3 km; Section *II*, 5.4 km; and Section *III*, 8.5 km from the mouth of the Raritan River (Ketchum, 1951b).

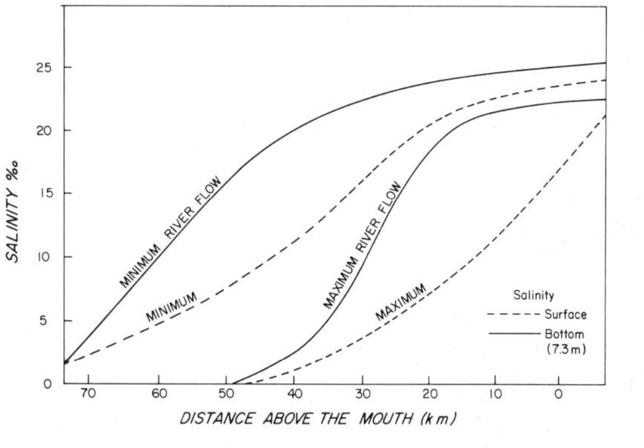

Fig. 1.9. Salinity distributions at the surface and bottom (7.3 m) in the Miramichi Estuary (Canada) during the minimum (56 $m^3\ s^{-1}$) and maximum river flow (140 $m^3\ s^{-1}$) during the summer of 1951. (After Bousfield, 1955a.)

as the river flow dropped from 5000 cfs (140 m^3 s^{-1}) to 2000 cfs (56 m^3 s^{-1}).

POLLUTION

It is impossible to discuss estuarine ecosystems without some consideration of pollution. Because of the convenience of sea transport, many of the major cities of the world are built on the shores of estuaries. The concentration of population and of industries on the banks of estuaries has inevitably led to the discharge of the waste materials into the estuaries so that many are seriously polluted today. There are few unpolluted estuaries left for study and most of these are comparatively small.

As described more fully in Chapter 2 (pp. 33–35), the rate of dilution within the estuary greatly exceeds that in the river and the amount of diluting water available increases as one moves seaward. This is because of the increased volume of sea water which becomes entrained in the brackish water moving seaward to maintain the fresh-water and salt balance. The concentration of pollutants in the water decreases as the volume available for dilution increases.

For an instantaneous discharge of a pollutant into an estuary, there will be a maximum concentration of the pollutant at the point of discharge and the concentration will decrease rapidly both upstream and downstream of this location. During subsequent tidal cycles, the maximum concentration will gradually decrease at the same time that the location of this maximum moves downstream in accord with the net flux of material seaward. If a pollutant were monitored at a location downstream of the point of release, the concentration would gradually increase to a maximum value and then decrease as the material is flushed out of the estuary. O'Connor and Thomann (1971) have presented a one-dimensional model of this process and have computed the expected distribution of a conservative pollutant as a function of both distance downstream from the point of release and as a function of time measured in tidal cycles. Their results are reproduced in Fig. 1.10. It will be noted in the upper figure that the pollutant is also carried upstream from the point of release, though the major transport over the course of time is downstream.

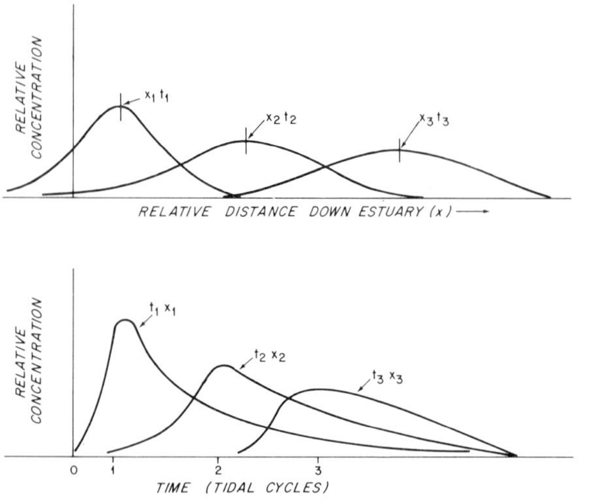

Fig. 1.10. Computed distribution of a conservative pollutant in a hypothetical estuary as a function of distance along the main axis (x) of the estuary and of time (t, tidal cycles). (After O'Connor and Thomann, 1971.)

It is more common for a pollutant to be discharged into an estuary continuously, as from a sewer outfall or an industrial discharge. In such a situation, the pollutant will tend to reach a steady-state distribution, the characteristics of which are controlled by the same processes and mechanisms which determine the distribution of fresh and salt water within the estuary. Seaward of the point of discharge, the pollutant concentration will be proportional to the distribution of fresh water; in the landward direction, the distribution of the pollutant will be proportional to the distribution of salt water (Ketchum, 1955). The distribution of total phosphorus, a conservative pollutant introduced with domestic sewage, over a 65-km stretch of the Hudson River estuary is shown in Fig. 1.11 (Ketchum, 1969). The maximum concentration of total phosphorus (12.3 μg at l^{-1}) was observed at a location just off the southern tip of Manhattan Island. The salinity of this sample (S_0) and fresh water content (F_0) were used to compute the relative upstream distribution of salt and downstream distribution of fresh water from this point of introduction. Samples were taken from surface, mid-depth and near bottom, and all of these data are incorporated in this single figure. The linear relationship between total phosphorus and the relative salt content up-estuary from the point of introduction and with the relative fresh-water con-

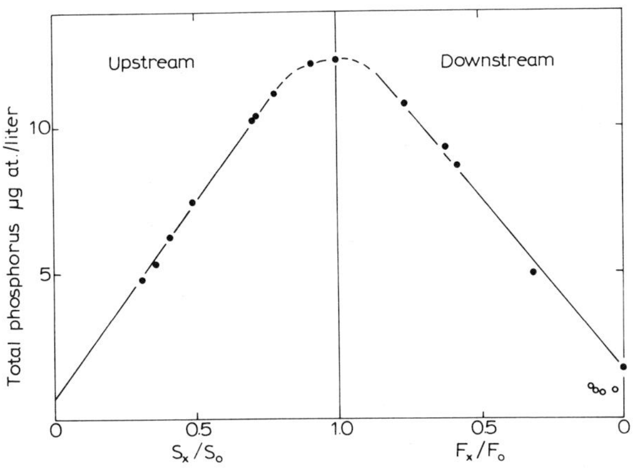

Fig. 1.11. The distribution of total phosphorus in the estuary of the Hudson River (U.S.A.) as a function of the salt content upstream of the point of introduction and of the fresh-water content downstream of this location. The maximum concentration was observed off the southern tip of Manhattan Island (after Ketchum, 1969).

tent down-estuary from that location is seen to apply regardless of the depth at which the water is collected.

The four open circles in the lower right-hand corner of this diagram represent the total phosphorus content of coastal offshore sea water in this area. While these samples contain some fresh water, it is not necessarily all derived from the Hudson Estuary. Ketchum (1967) has discussed the total phosphorus content of the waters off the entrance to New York Harbor and presented the proportions of phosphorus contributed by each of three sources, namely the brackish effluent from the Hudson River Estuary, the surface coastal water and the deeper water over the continental shelf. The maximum concentration of phosphorus within the estuary is about ten times the natural concentration offshore, and this pollution has led to eutrophication and oxygen deficits within the estuary (Ketchum, 1974).

For generations, it has been assumed that the waters are "free" and subject to any use which abutting landowners desire. Thus, there has been inadequate legal control of the discharge of pollutants into our streams and estuaries. To the polluter, this is a great financial advantage since the cost of treatment of his waste is "external" to his balance sheet for plant operation. The costs of the disposal of pollutants in this way, including the loss of natural resources and of amenities, are borne by the population as a whole. Though the aggregate costs may be great, the cost to an individual is small, except for a few fishermen, and only by the concerted action of a large number of people can the true costs be evaluated. It is only in recent decades, with the rise in the concern about environmental quality, that the true costs of pollution of our streams and estuaries have been fully appreciated. This is not to depreciate the valiant efforts made by earlier conservationists. The works of the early conservationists such as President Theodore Roosevelt and Gifford Pinchot, and of the many ecologists who staunchly fought for a quality environment, achieved renowned successes in their time. The park lands and forest preserves of the western parts of the United States are monuments to their efforts.

REFERENCES

Alexander, W.B., Southgate, B.A. and Bassindale, R., 1935. *Survey of the River Tees. Part II. The Estuary: Chemical and Biological.* Department of Scientific and Industrial Research, Water Pollution Research, Technical Paper No. 5. H.M. Stationery Office, London, 171 pp.

Alexander, W.B., Southgate, B.A. and Bassindale, R., 1936. Summary of Tees Estuary Investigations. *J. Mar. Biol. Assoc., N.S.*, 20: 717–724.

Bousfield, E.L., 1955a. Some physical features of the Miramichi Estuary. *J. Fish. Res. Board Can.*, 12(3): 342–361.

Bousfield, E.L., 1955b. Ecological control of the occurrence of barnacles in the Miramichi Estuary. *Nat. Bull. Can.*, No. 137: 69 pp.

Cameron, W.M. and Pritchard, D.W., 1965. Estuaries. In: M.N. Hill (Editor), *The Sea, II.* Wiley and Sons, New York, N.Y., pp. 306–324.

Defant, A., 1961. *Physical Oceanography, II.* Pergamon Press, New York, N.Y., 598 pp.

Hansen, D.V. and Rattray, M., 1966. New dimensions in estuary classification. *Limnol. Oceanogr.*, 11: 319–326.

Hedgpeth, J.W., 1947. The Laguna Madre of Texas. *Trans. 12th North Am. Wildl. Conf.*, pp. 364–380.

Hedgpeth, J.W., 1957. Estuaries and lagoons. II. Biological aspects. In: J.W. Hedgpeth (Editor), *Treatise on Marine Ecology and Paleoecology, I. Geol. Soc. Am., Mem.*, 67(1): 693–729.

Hedgpeth, J.W., 1967. Ecological aspects of the Laguna Madre, a hypersaline estuary. In: G.H. Lauff (Editor), *Estuaries. Am. Assoc. Adv. Sci., Publ.*, 83: 408–419.

Ketchum, B.H., 1950. Hydrographic factors involved in the dispersion of pollutants introduced into tidal waters. *J. Boston Soc. Civil Eng.*, 37(3): 296–314.

Ketchum, B.H., 1951a. The exchanges of fresh and salt waters in tidal estuaries. *J. Mar. Res.*, 10(1): 18–38.

Ketchum, B.H., 1951b. The flushing of tidal estuaries. *Sewage Ind. Wastes*, 23(2): 198–209.

Ketchum, B.H., 1952. *The Distribution of Salinity in the Estuary of the Delaware River.* Woods Hole Oceanographic Institution, Woods Hole, Mass., Technical Report No. 52–10: 60 pp. (unpublished).

Ketchum, B.H., 1955. Distribution of coliform bacteria and other pollutants in tidal estuaries. *Sewage Ind. Wastes*, 27(11): 1288–1296.

Ketchum, B.H., 1967. Phytoplankton nutrients in estuaries. In: G.H. Lauff (Editor), *Estuaries. Am. Assoc. Adv. Sci., Publ.*, 83: 329–335.

Ketchum, B.H., 1969. Eutrophication of estuaries. In: *Eutrophication: Causes, Consequences, Correctives.* National Academy of Sciences, Washington, D.C., pp. 197–208.

Ketchum, B.H., 1974. Population, resources, and pollution, and their impact on the Hudson Estuary. *Ann. N.Y. Acad. Sci.*, 250: 144–156.

Ketchum, B.H. and Keen, D.J., 1956. The accumulation of river water over the continental shelf between Cape Cod and Chesapeake Bay. In: *Papers in Marine Biology and Oceanography.* Pergamon Press, London, pp. 346–357.

Ketchum, B.H., Redfield, A.C. and Ayers, J.P., 1951. The oceanography of the New York Bight. *Pap. Phys. Oceanogr. Meteorol.*, 12(1): 46 pp.

McHugh, J.L., 1967. Estuarine nekton. In: G.H. Lauff (Editor), *Estuaries. Am. Assoc. Adv. Sci., Publ.*, 83: 581–620.

Marmer, H.A., 1926. *The Tide.* Appleton, New York, N.Y., 282 pp.

O'Connor, D.J. and Thomann, D.J., 1971. Water quality models: chemical, physical and biological constituents. In: G.H. Ward, Jr. and W.H. Espey, Jr. (Editors), *Estuarine Modelling: An Assessment.* Water Pollution Control Research Series, Water Quality Office, Environmental Protection Agency, Washington, D.C., pp. 102–168.

Pickard, G.L., 1961. Oceanographic features of inlets in the British Columbia mainland coast. *J. Fish. Res. Can.*, 18: 907–999.

Pritchard, D.W., 1952. Estuarine hydrography. *Adv. Geophys.*, 1: 243–280.

Pritchard, D.W., 1957. Discussion of "On estimating streamflow into tidal estuary" by D.K. Todd and Gueng-Ku Lab. *Trans. Am. Geophys. Union*, 38: 581–584.

Pritchard, D.W., 1967. Observations of circulation in coastal plain estuaries. In: G.H. Lauff (Editor), *Estuaries. Am. Assoc. Adv. Sci., Publ.*, 83: 37–44.

Pritchard, D.W., 1980. A note on the Stokes Transport in tidal estuaries. In B. Patel (Editor), *Management of the Environment.* Wiley Eastern Limited, New Delhi, pp. 217–226.

Redfield, A.C., 1950. The analysis of tidal phenomena in narrow embayments. *Pap. Phys. Oceanogr. Meteorol.*, 11(4): 36 pp.

Redfield, A.C., 1953. Interference phenomena in the tides of the Woods Hole region. *J. Mar. Res.*, 12(1): 121–140.

Redfield, A.C., 1958. The influence of the continental shelf on the tides of the Atlantic Coast of the United States. *J. Mar. Res.*, 17: 432–448.

Redfield, A.C., 1961. The tidal system of Lake Maracaibo, Venezuela. *Limnol. Oceanogr.*, 6(1): 1–12.

Redfield, A.C., 1978. The tide in coastal waters. *J. Mar. Res.*, 36(2): 255–294.

Redfield, A.C., 1980. *The Tides of the Waters of New England and New York.* Woods Hole Oceanographic Institution, Woods Hole, Mass., 108 pp.

Riley, G.A., 1937. The significance of the Mississippi River drainage for biological conditions in the Northern Gulf of Mexico. *J. Mar. Res.*, 1(1): 60–74.

Rochford, D.J., 1951. Studies in Australian estuarine hydrology. I. Introductory and comparative features. *Aust. J. Mar. Freshwater Res.*, 2(1): 116 pp.

Ryther, J.H., Menzel, D.W. and Corwin, N., 1967. Influence of the Amazon River outflow on the ecology of the western tropical Atlantic. I. Hydrography and nutrient chemistry. *J. Mar. Res.*, 25(1): 69–83.

Saelen, O.H., 1967. Some features of the hydrography of Norwegian fjords. In: G.H. Lauff (Editor), *Estuaries. Am. Assoc. Adv. Sci., Publ.*, 83: 63–70.

Schureman, P., 1941. *Manual of Harmonic Analysis and Prediction of Tides.* U.S. Department of Commerce, Coast and Geodetic Survey. U.S. Government Printing Office, Washington, D.C., 317 pp.

Stokes, G.G., 1847. On the theory of oscillatory waves. *Trans. Cambridge Philos. Soc.*, 8: 441–455.

Sverdrup, H.U., Johnson, M.W. and Fleming, R.H., 1942. *The Oceans, Their Physics, Chemistry, and General Biology.* Prentice-Hall, New York, N.Y., 1087 pp.

Thomson, W. (Lord Kelvin), 1879. On gravitational oscillations of rotating water. *Proc. R. Soc. Edinb.*, 10: 92–100.

Tully, J.P., 1949. Oceanography and prediction of pulp mill pollution in Alberni Inlet. *Bull. Fish. Res. Board Can.*, 83: 1–169.

Tully, J.P. and Barber, F.G., 1960. An estuarine analogy in the sub-arctic Pacific Ocean. *J. Fish Res. Board Can.*, 17: 91–112.

Tully, J.P. and Barber, F.G., 1961. An estuarine model of the sub-arctic Pacific Ocean. In: M. Sears (Editor), *Oceanography. Am. Assoc. Adv. Sci. Publ.*, 69: 425–454.

U.S. Department of Commerce, 1952. *Tide Tables, East Coast North and South America (Including Greenland).* Coast and Geodetic Survey, U.S. Government Printing Office, Washington, D.C., 249 pp.

U.S. Department of Commerce, 1978. *Tidal Current Tables, Atlantic Coast of North America.* National Oceanic and Atmospheric Administration, National Ocean Survey, Rockville, Md, 214 pp.

Chapter 2

PHYSICS OF ESTUARINE CIRCULATION

CHARLES B. OFFICER

INTRODUCTION

Our purpose in this chapter is to relate in a narrative style what we know, and in many cases do not know, about the physical processes of circulation and mixing in estuaries. This chapter is not definitive in itself but hopefully may serve the purpose of indicating some of the phenomena that do occur in estuarine waters. These can be quite diverse, and it is important to appreciate this diversity. For those investigating biological, chemical, geological or engineering aspects of estuaries it is essential to have a thorough understanding of the types of physical conditions that may exist and consequently affect the transport, distribution and reaction phenomena that they are investigating.

The first three sections of this chapter are directed toward the physics of estuarine circulation and mixing. The last section covers examples of biological, chemical and geological effects that are dependent on estuarine physics.

Definitions

It is convenient for our purposes of estuarine hydrodynamics to use the definition of an *estuary* given by Pritchard (1952a) and Cameron and Pritchard (1965). This is that an estuary is a semi-enclosed body of water having a free connection with the open sea and within which the sea water is measurably diluted with fresh water derived from land drainage. Traditionally the term estuary has been applied to the lower reaches of a river into which sea water intrudes and mixes with fresh water draining seaward from the land. The term has been extended to include bays, inlets, gulfs and sounds into which several rivers empty and in which the mixing of fresh and salt water occurs.

As has been pointed out and elaborated by Ketchum (1951a, 1951b) and Stommel (1953) as well as others, one can use the fresh-water flow into an estuary as a natural tracer to distinguish various estuarine conditions and to define some of the gross properties. The fresh-water fraction at any given location can be given simply in terms of the observed salinity as referred to the normal ocean salinity of the coastal ocean waters into which the estuary empties. From the observed salinity distribution I define a *well-mixed* condition as one in which there is essentially no variation in the salinity in a vertical column. One can also distinguish a stratified condition with a halocline between the upper and lower portions of the water column, each layer having a nearly constant salinity. I define a *weakly, or partially, stratified* condition as one in which there is a change in salinity of only a few parts per thousand ($<3‰$) from surface to bottom and a *strongly, or highly, stratified* condition as one in which there is a change in salinity of several parts per thousand ($>3‰$) from surface to bottom. One can also distinguish a condition in which there is an interface, or near-interface, between two different water types. I define an *arrested salt wedge* as a situation in which there is a stable salt wedge underlying a strong, fresh-water flow above, and a *fjord-type entrainment* flow as one in which there is a relatively stagnant deep water mass overlain by a thin river runoff flow. When the river is narrow and the flow is great, fresh water may fill the mouth, causing the brackish river water to form a distinguishable *plume* that floats out to sea over the salt water. It is important to appreciate that all of these are descriptive terms only and cannot, in general, be applied to an estuary as a whole; any given estuary may show well-mixed or stratified conditions, for example, as a function of longitudinal distance along the estuary, season of the year, or even in some cases phase of the tidal cycle.

Classification system

I am adopting a classification system which is based on the type of circulation and the related salinity structure of the estuary. The circulation in each estuary must result in the seaward transport of the fresh water being constantly added by the river. As this fresh water moves seaward, it is mixed with more and more sea water so that a landward flux of salt to maintain the salt balance is required. Each of these transports of fresh water and of salt water can be provided by advection or by horizontal diffusion, much of the energy for which is provided by the ebb and flow of the tides. The interaction of these various properties and processes — river flow, tidal currents, advection and diffusion — within the confines of the geomorphology of the estuarine basin produce the salinity distribution within each estuary.

Hansen and Rattray (1966) have presented a quantitative classification system for the description of estuarine conditions for which there are negligible lateral variations. This system is summarized in the stratification–circulation diagrams of Figs. 2.1 and 2.2. In both figures the ordinate is the ratio of the tidal averaged salinity difference between the bottom and the surface, $\bar{s}_b - \bar{s}_s$, to the depth-averaged salinity value $\langle \bar{s} \rangle$ at a given location in an estuary. For a well-mixed estuary,

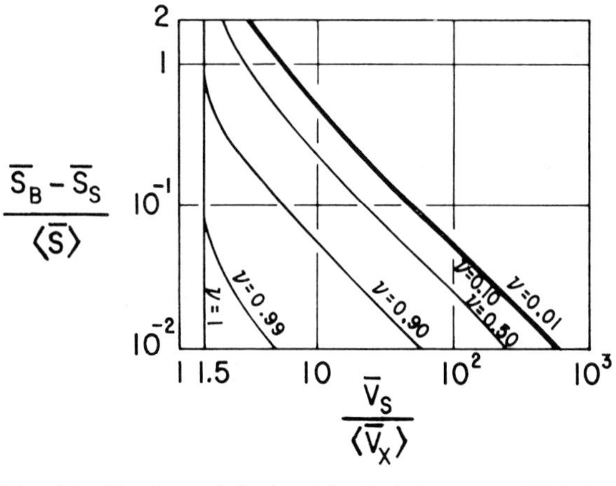

Fig. 2.1. Fraction of horizontal salt balance provided by diffusion (ν) as a function of salinity stratification and convective circulation (from Hansen and Rattray, 1966).

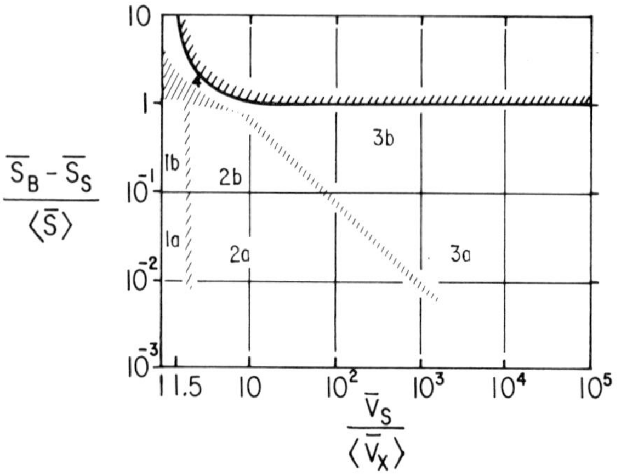

Fig. 2.2. Proposed classification scheme for estuaries (from Hansen and Rattray, 1966).

$\bar{s}_b - \bar{s}_s$ and the ratio will both be small; for a stratified estuary the ratio may approach unity; for a highly stratified estuary or an arrested salt wedge the ratio may be greater than one. The abscissa in each figure is the ratio of the net circulation velocity at the surface averaged over a tidal cycle, $\bar{v}_s$, to the net river runoff flow averaged over the cross section, $\langle \bar{v}_x \rangle$. In the upper part of the estuary, containing only river water, this ratio will equal one. As more and more sea water is entrained, progressively larger volumes of brackish water must move seaward to transport a given volume of fresh water. Thus, the net surface circulation velocity, and the ratio, will increase. The family of curves plotted in Fig. 2.1 represent different values of ν, which is the ratio of the salt flux due to tidal diffusion to the total salt flux due to longitudinal dispersion up the estuary. When $\nu = 1$, the up-estuary longitudinal dispersion flux is due entirely to tidal diffusion. As $\nu \to 0$, the up-estuary longitudinal dispersion flux is due almost entirely to the contribution from the net circulation flux related to the combined effects of the vertical variation of the tidal-averaged salinity and the profiles of net circulation velocity.

The diagram of Fig. 2.1 is divided into various characteristic regions in Fig. 2.2 corresponding to typical estuarine types. In type *1*, *a* and *b*, the net flow is seaward at all depths, and the up-estuary salt transfer is entirely due to tidal diffusion. For type *2*, *a* and *b*, the net flow reverses at depth, and both tidal diffusion and net circulation contribute to the up-estuary salt flux. The distinction between categories *a* and *b* in both cases is that between a well-mixed (*a*) and a stratified (*b*) condition. Fjord entrainment flow is in category *3*, with essentially no flux contribution from tidal diffusion; and type *4* (not in Fig. 2.2) is characteristic of the flow condition associated with an arrested salt wedge.

Dyer (1977) has extended the stratification–circulation diagram of Hansen and Rattray to include lateral variations. Here the ordinate of Fig. 2.2 is replaced by the ratio of the maximum tidal-averaged salinity difference across the estuary at the surface to the salinity value averaged over tide and cross section. For many estuaries the lateral variations in salinity are as large as the vertical variations, indicative of corresponding flux and momentum effects due to lateral circulation. In such instances simplification to two-dimensional, vertical and longitudinal, considerations may be quite inadequate to describe the resultant circulation and mixing.

Motion and continuity equations

The theoretical description of estuarine circulation and mixing is contained in the three equations of motion plus the continuity equations for mass and salt. Various simplifications and reductions in the defining terms of these equations have to be made to obtain solutions. One such simplification that is often used is to reduce the turbulent stress and flux terms to the product of the spatial gradient of velocity or salinity and a coefficient of viscosity (N) or a coefficient of diffusion (K). Conceptually one can object to such simplifications, but within limits they have proven useful. Among other things, however, the coefficients are scale-dependent, and care must be taken in their application. The scale-dependence can be expressed for either type of coefficient as the product of a characteristic velocity and distance, or depth, or alternatively as the product of the square of a characteristic velocity and time.

In addition to the internal, or turbulent, stress terms in the equations of motion there are a number of other possible stress terms. The coastal tides, considered as a driving force at the mouth of the estuary, are important. The tidal motions and resultant tidal mixing are often observable and obvious in estuaries. The wind stress at the ocean surface can also produce observable effects as measured over time sequences longer than a single tidal cycle. Correspondingly there is the bottom frictional stress. When effects within the bottom boundary layer are not of direct interest, this stress can often be approximated as the product of a bottom frictional coefficient and the square of the near-bottom current velocity. The density difference between the fresh water near the estuary head and the more saline brackish water at the estuary mouth is an important longitudinal stress term. This stress, and that related to the difference in elevation of the water surface from near the estuary head to its mouth, provide the driving forces for the density gradient circulation commonly observed in estuaries. For estuaries which have a substantial lateral extent the Coriolis force can also be contributive. In addition there are the inertial terms in the equations of motion related to spatial acceleration changes in the flow dynamics. At places where there are gross changes in the flow geometry, such as at bends or constrictions in an estuary, these terms are of first importance.

For the salt continuity equation there are in addition to the diffusion terms the simple advection, or mass transfer, terms. Appropriate source and sink terms must also be included in the equations of continuity for other observable quantities, as well as the reaction, or decay, terms if the quantity is non-conservative. If, as in the case of dissolved oxygen and biochemical oxygen demand or nitrogen cycle relations, the chemical or biological decay of a given quantity is related to or affects the gain or decay of another quantity, one has a set of coupled continuity equations with various reaction terms for the decay or increase of the quantities considered and their feedforward and feedback contributions.

Numerical modeling

Because of the various complexities enumerated above, and those due to the geometric configuration of a given estuary, it is often difficult to obtain analytic solutions which can be used for anything more than a conceptual idea of what is actually occurring. Consequently numerical modeling has come into vogue. This modeling has proven to be a useful research and engineering tool. At the present, however, our ability to model far exceeds our basic understanding of the phenomena to be modeled; we do not understand enough of many of the processes to be able to express in analytic form what is to be modeled. It is our basic understanding which is the more serious deterrent at the present rather than our modeling capability. Elliott (1977) in his introduction to a symposium on numerical modeling expressed a complementary view, which I quote here:

"Perhaps the most immediate application of estuarine models is to applied problems where a model can assist in the making of management decisions. However, they do have a more fundamental use when they are used to test, by numerical experiment, theoretical ideas which relate to the dynamic and kinematic balances within an estuary. As computers become larger and faster, numerical experimentation may play an important role in our understanding of turbulent systems."

Finally, a note of caution. A numerical model can only be as good as the data used to calibrate *and independently verify it*. Therefore, modelers working with two- or three-dimensional schemes must conduct comprehensive field studies and obtain *more than one set of data*. While much effort has gone into numerical schemes, adequate physical verification has yet to be made.

WELL-MIXED AND STRATIFIED ESTUARINE CONDITIONS

Tidal effects

Tidal effects are of substantial importance for most well-mixed and stratified estuaries. They provide the ultimate driving force for a large part of the turbulence phenomena that are observed; the tidal velocities usually dominate the currents; and tidal mixing, or exchange, is important in the longitudinal dispersion processes.

Typically, tidal velocities range from 50 to 100 cm s^{-1} (about 2 to 4 km h^{-1} or 1 to 2 knots). The net circulation velocity, as averaged over a tidal cycle, at any depth may range from 10 to 20 cm s^{-1} in either a down-estuary or up-estuary direction. The velocity averaged over cross section and tide, or net river flow, may range from 1 to 2 cm s^{-1} for estuaries associated with modest rivers.

A convenient measure of the longitudinal dispersion processes is the longitudinal dispersion coefficient, K_x. This coefficient represents a combination of the processes of longitudinal tidal mixing and net circulation. When tidal mixing is dominant, one may anticipate that K_x would decrease in an up-estuary direction as the amplitude of the tidal motion decreases. Characteristic values of K_x for the upper and middle reaches of an estuary range from 0.1×10^6 to 1×10^6 $cm^2 s^{-1}$, and for the lower reaches as the estuary empties into the ocean they range from 1×10^6 to 10×10^6 $cm^2 s^{-1}$.

For the first several decades of the twentieth century, the accumulation of pollutants within estuaries was computed on the assumption that the volume of water entering the estuary on the flood tide (the tidal prism) was available for dilution on that tidal cycle. This calculation tacitly assumed complete mixing throughout the estuary, and that none of the water leaving the estuary on the ebb tide returned during the next flood tide (Ketchum, 1950, 1951a, b). Both of these assumptions are unrealistic and both give exaggerated estimates of dilution and consequently underestimates of the accumulation.

In modifying the tidal prism concept, Ketchum (1951a, b) assumed that the maximum distance over which mixing could be assumed complete was approximately defined by a segment the length of which corresponded to the excursion of a water particle during the tidal flow. For succeeding segments the high-tide volume of a landward segment was defined as equal to the low-tide volume of the adjacent seaward segment. The segment dimensions were computed as though the entire low-tide volume of the seaward segment moved as a piston to fill the high-tide volume of the landward segment on the flood tide. An exchange ratio was defined as the ratio between the intertidal volume and the high-tide volume of each segment. From this, the salinity distribution was computed sequentially and the results approximated the observed distributions in some estuaries where tidal mixing effects are dominant (Ketchum, 1951a, b; Ketchum and Keen, 1953).

Density gradient circulation

One of the more common, and reasonably well-documented, hydrodynamic effects in well-mixed and stratified estuaries is longitudinal, density gradient circulation and mixing. The driving forces for this type of circulation are the longitudinal surface slope, acting in a down-estuary direction, and the longitudinal density gradient force, acting in an up-estuary direction. The surface slope force is constant as a function of depth, and the density gradient increases essentially linearly as a function of depth. For the condition in which there is a small river runoff flow, which is common for many estuaries, the net effect, then, is that the surface slope force will be dominant in the upper portion of the water column, producing a net circulation flow down-estuary, and that the density gradient force will be dominant in the lower portion of the water column, producing a net circulation flow up-estuary. The theoretical description of this type of circulation has been given in detail by Rattray and Hansen (1962) and Hansen and Rattray (1965), and in a more simplified form by Officer (1976).

Examples of observations of the net longitudinal circulation velocity and salinity, as averaged over a tidal cycle, for two estuaries, the James (Chesapeake Bay, U.S.A.) and the Mersey (England), are shown in Figs. 2.3 and 2.4. In both cases the net circulation velocity is down-estuary in the upper portion of the water column and up-estuary in the lower portion

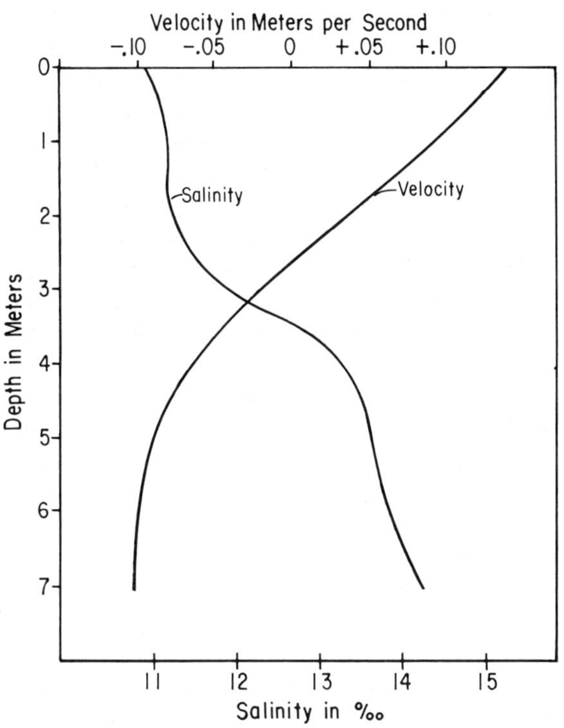

Fig. 2.3. Typical mean salinity and horizontal velocity versus depth curves for the James Estuary, Chesapeake Bay, U.S.A. (from Pritchard, 1954).

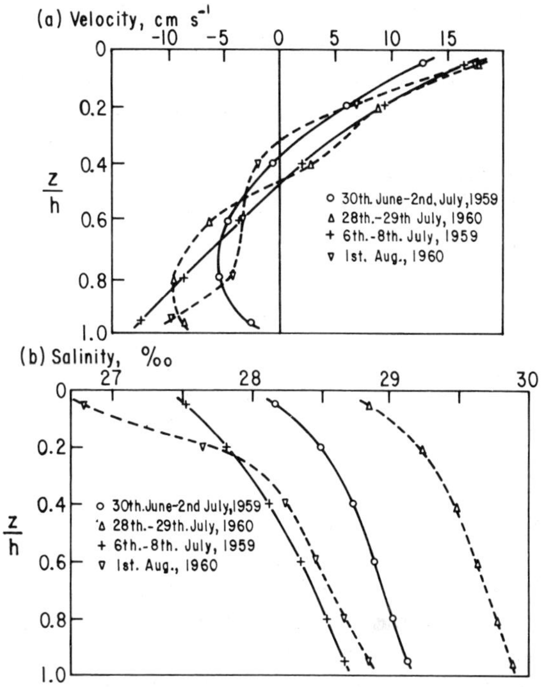

Fig. 2.4. Mean profiles as a function of relative depth (z/h) over a tidal period of (a) longitudinal velocity and (b) salinity for the Mersey Estuary, England (from Bowden, 1963).

of the water column, and the salinity increases as a function of depth with a pronounced halocline near mid-depth for the James Estuary.

Estuarine parameters

Table 2.1 is a summary from Officer (1977) of observed values of some estuarine parameters for a number of estuaries. The vertical salinity difference ($s_b - s_s$) is the difference of salinity between the bottom and the surface; the velocity at the surface and the bottom ($v_s - v_0$ and $v_b - v_0$) is the net circulation velocity minus the cross-sectional averaged, or river flow velocity (positive values are seaward velocities); the longitudinal gradient of salinity ($\partial s/\partial x$) is measured in the central portion of the estuary where there was an observable longitudinal change in salinity and the water depth (h) is the average depth for the same stretch of the estuary. One may see from the $s_b - s_s$ values that a wide range of conditions is covered from nearly well-mixed in the Long Island Sound and the Mersey Estuary, to stratified for the Southampton and

TABLE 2.1

Observed values, averaged over a tidal cycle, of some estuarine parameters (see text for explanation of symbols)

Location	Source[1]	$s_b - s_s$ (‰)	$v_s - v_0$[2] (cm s^{-1})	$v_b - v_0$[2] (cm s^{-1})	v_0[2] (cm s^{-1})	$\partial s/\partial x$ (‰ km^{-1})	h (m)
Long Island Sound (U.S.A.)	a	0.8	22.0	−5.0	0[3]	0.035	31.0
Mersey Estuary (England)	b	0.9	12.0	−9.0	0[3]	0.238	20.4
Southampton Estuary (England)	c	1.46	8.0	−6.0	0[3]	0.222	12.0
James Estuary (Virginia, U.S.A.)	d	3.0	12.0	−10.0	0[3]	0.587	7.0
Columbia River Estuary (U.S.A.)	e	15.0	29.0	−27.0	43.0	1.411	10.6
Vellar Estuary (India)	f	23.5	9.0	−4.0	0[3]	2.014	2.7

[1]Sources: a, Riley (1952, 1955a, b); b, Bowden (1960, 1962, 1963); c, Dyer (1973); d. Pritchard (1952b, 1954); e, O'Brien (1952) and Hansen (1965); f, Dyer and Ramamoothy (1969).
[2]Down-estuary velocities positive, up-estuary velocities negative.
[3]v_0 is one tenth or less of the value for v_s or v_b; nominal value of zero.

James Estuaries, to strongly stratified for the Columbia River and Vellar Estuaries. The longitudinal salinity gradient, $\partial s/\partial x$, increases in a manner similar to the vertical salinity difference, $s_b - s_s$, from one estuary to the next. The values of the bottom velocity, v_b, or in the case of the Columbia River Estuary, $v_b - v_0$, are negative in each case and are a substantial fraction of the corresponding surface velocity.

Table 2.2 is a summary from Officer (1977) of computed values for various coefficients for these same estuaries. The quantities N_z and K_z are the vertical viscosity and diffusion coefficients, k the bottom frictional coefficient, and i the surface slope. We know in a general way that N_z and K_z are dependent on scale, or depth; that in estuaries they will in some manner be related to the magnitude of the tidal action, which is the ultimate driving force for the turbulence phenomena observed there; and that they will decrease with increasing Richardson number, or — what is the same — with an increasing ratio of the increase in potential energy due to vertical mixing to the available kinetic energy of the turbulence. The lowest values of N_z and K_z are for the James and Vellar Estuaries. This is probably due to a combination of a decreased effect of scale, or depth, and an effect of increased stratification, or Richardson number, particularly for the Vellar Estuary. The values for the Columbia River Estuary are high; here there is a strong tidal action which in itself would indicate higher N_z and K_z values and would also tend to offset the stratification effect in the determination of the Richardson number. The Mersey Estuary shows the highest N_z and K_z values, and there is strong tidal action here. The values of the longitudinal surface slope, i, are much the same for four of the estuaries. The value of i is substantially higher for the Columbia River Estuary, related to its higher river runoff. The values of the bottom frictional coefficient, k, fall within the range normally found in other investigations.

To illustrate the kind of variation of K_z that may be expected as related to the mid-depth, halocline

TABLE 2.2

Computed values of some estuarine parameters based upon the data in Table 2.1 (see text for explanation of symbols)

Location	N_z ($cm^2\ s^{-1}$)	K_z ($cm^2\ s^{-1}$)	k	i (cm km^{-1})	K_z/N_z
Long Island Sound (U.S.A.)	8.1	9.5	11.0×10^{-3}	0.03	1.2
Mersey Estuary (England)	41.0	25.0	4.5×10^{-3}	0.16	0.6
Southampton Estuary (England)	12.0	3.3	3.3×10^{-3}	0.12	0.3
James Estuary (Virginia, U.S.A.)	4.4	2.2	0.8×10^{-3}	0.14	0.5
Columbia River Estuary (U.S.A.)	17.0	8.0	0.7×10^{-3}	0.54	0.5
Vellar Estuary (India)	0.9	0.1	6.0×10^{-3}	0.17	0.1

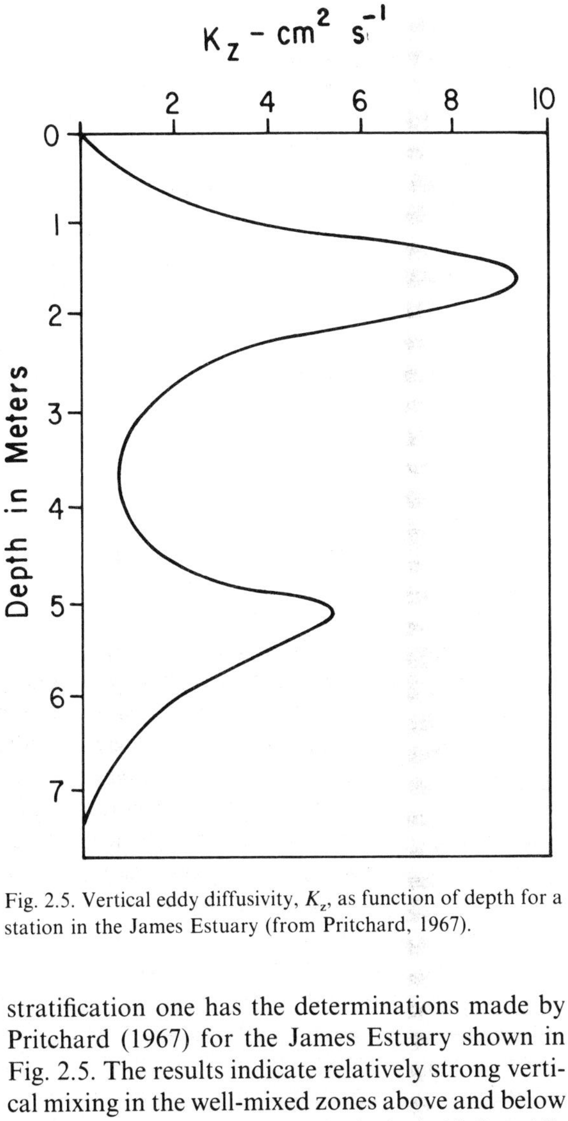

Fig. 2.5. Vertical eddy diffusivity, K_z, as function of depth for a station in the James Estuary (from Pritchard, 1967).

stratification one has the determinations made by Pritchard (1967) for the James Estuary shown in Fig. 2.5. The results indicate relatively strong vertical mixing in the well-mixed zones above and below the halocline, and a region of relatively high stability in the vicinity of the halocline. It is of interest to note that, even in this halocline region of a tidal estuary, the values of K_z are an order of magnitude or more greater than those values commonly determined for the seasonal thermocline region of inland lakes (Hutchinson, 1957; Hesslein and Quay, 1973).

Wind stress effects

Although longitudinal density gradient flow is common, it is not in itself enough for an understanding of estuarine circulation. For larger estuaries, in particular, lateral effects, estuarine geometry and atmospheric and wind stress effects become important. Many of these latter effects only become apparent for measurement sequences which have been taken over time periods that are long compared with the tidal cycle.

Weisberg (1976) made an interesting sequence of measurements of the current velocity 2 m above the bottom in the Providence River Estuary over a period of two months. The Providence River Estuary is geometrically simple with a longitudinal axis bearing 335° from the river mouth toward its head. The measuring station was located in the deepest portion of the channel near the mouth of the estuary. The results clearly showed that the current velocity fluctuations at time scales between the long-term longitudinal density gradient and the tidal oscillations were large and almost exclusively wind-induced. Fig. 2.6 presents the time series of current measurements along the axis of the estuary after low-pass filtering to remove tidal and higher-frequency oscillations. Also included is the time series for wind velocity from a nearby land station similarly oriented and filtered. In this figure a positive value of the wind velocity indicates a wind blowing in the direction 335°. From the current measurements we see that there is a mean up-estuary, near-bottom flow related to the density gradient circulation, and a striking mirror-image correlation of the wind and the current about this mean. The inverse relationship follows from continuity, since the Providence River lacks the storage capacity for sustained non-tidal transport — that is, water being advected landward by wind drag on the sea surface must return seaward at depth and vice versa. One has in this case, then, the tidal oscillations, the four-to-five day wind-induced variations, and the steady-state density gradient circulation.

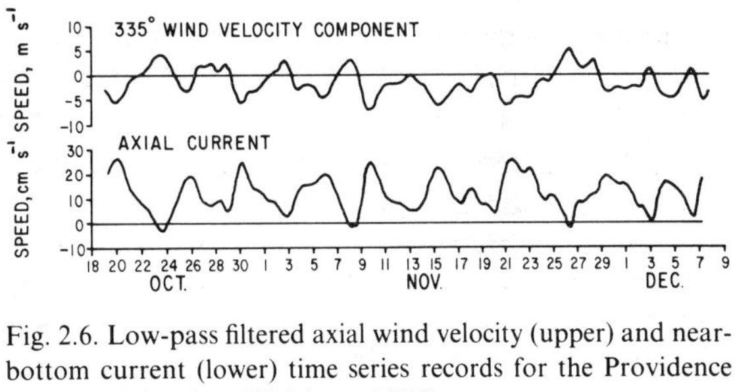

Fig. 2.6. Low-pass filtered axial wind velocity (upper) and near-bottom current (lower) time series records for the Providence River, U.S.A. (from Weisberg, 1976).

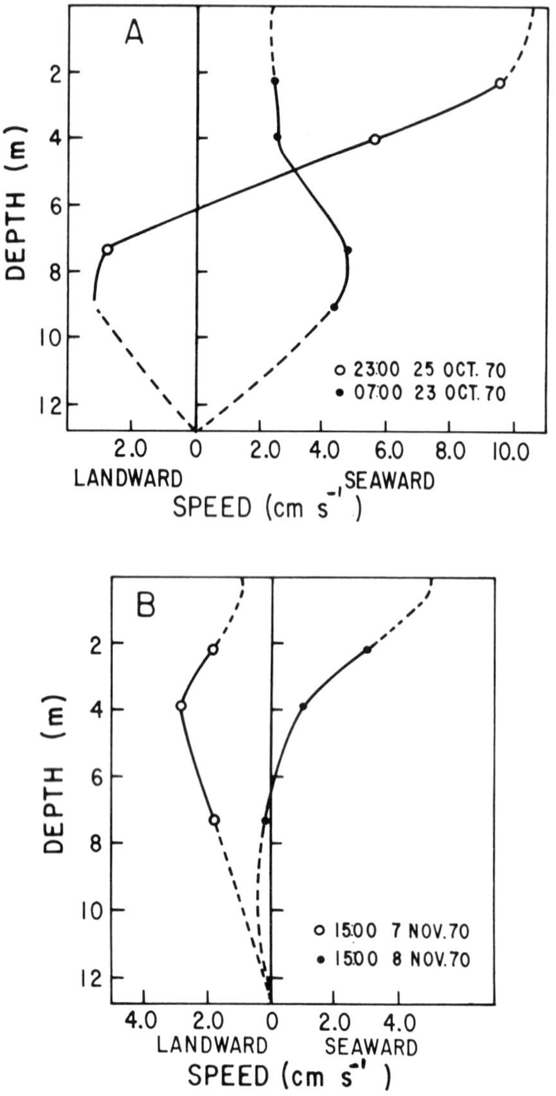

Fig. 2.7. Vertical profiles of the net north–south component of current velocity for various wind conditions for the west passage of Narragansett Bay, Rhode Island, U.S.A. (from Weisberg and Sturges, 1976).

Weisberg and Sturges (1976) made a similar sequence of measurements in the more complex estuarine environment of the west passage of Narragansett Bay. The west passage is oriented approximately north–south with the ocean to the south, and is separated from the east passage of Narragansett Bay by Conanicut Island. Fig. 2.7 shows some of their results for the mean, or tidal-averaged, current as a function of depth. Fig. 2.7A shows two contrasting cases, one for strong winds toward the southwest (25 Oct. '70) and the other for strong winds toward the northwest (23 Oct. '70). The former case results in a two-layer flow with the upper-layer speed greatly enhanced. The latter case results in a net transport to the south at all depths, with water being supplied from the east to the west passage. The flow near the surface, although south, is impeded by the wind relative to the larger speeds in the deeper waters. Fig. 2.7B shows how quickly the west passage responds to a change in wind direction. Within a day the entire net current distribution shifted from northerly to southerly in response to a change in the wind direction.

Long-term variations

Elliott and Hendrix (1976) discussed an interesting experiment made over a ten-day period in the Potomac Estuary. Vertical profiles of the longitudinal current were made at hourly intervals at two mid-channel stations in the estuary. The results from one of these stations are shown in Fig. 2.8. The results have been low-pass filtered to remove the tidal variations, and there is a hiatus in the middle of the measurement sequence. During the study period the Potomac Estuary reversed the direction of its net circulation — that is, following an initial period of normal, density gradient flow the estuary reversed and started to flow landward (negative values) in the surface layers and seaward in the bottom layers. This continued for two days, after which there was a positive seaward flow throughout most of the water column, followed by a three-layer flow pattern in which there was landward flow near mid-depth. Near the end of the study period the pattern returned to the normal, density gradient flow with seaward currents in the surface layers and landward currents at depth. They suggest that the current reversal was associated with a reversal of the horizontal salinity gradient near the mouth of the estuary, and that this salinity reversal was caused by the combined effects of surface water being blown out of the estuary and by higher salinity water, which would be advected into the estuary by a compensating return flow, upwelling within the estuary. Following the relaxation of the wind, the internal density distribution would tend to drive the internal circulation in the opposite sense to that usually associated with a partially mixed estuary.

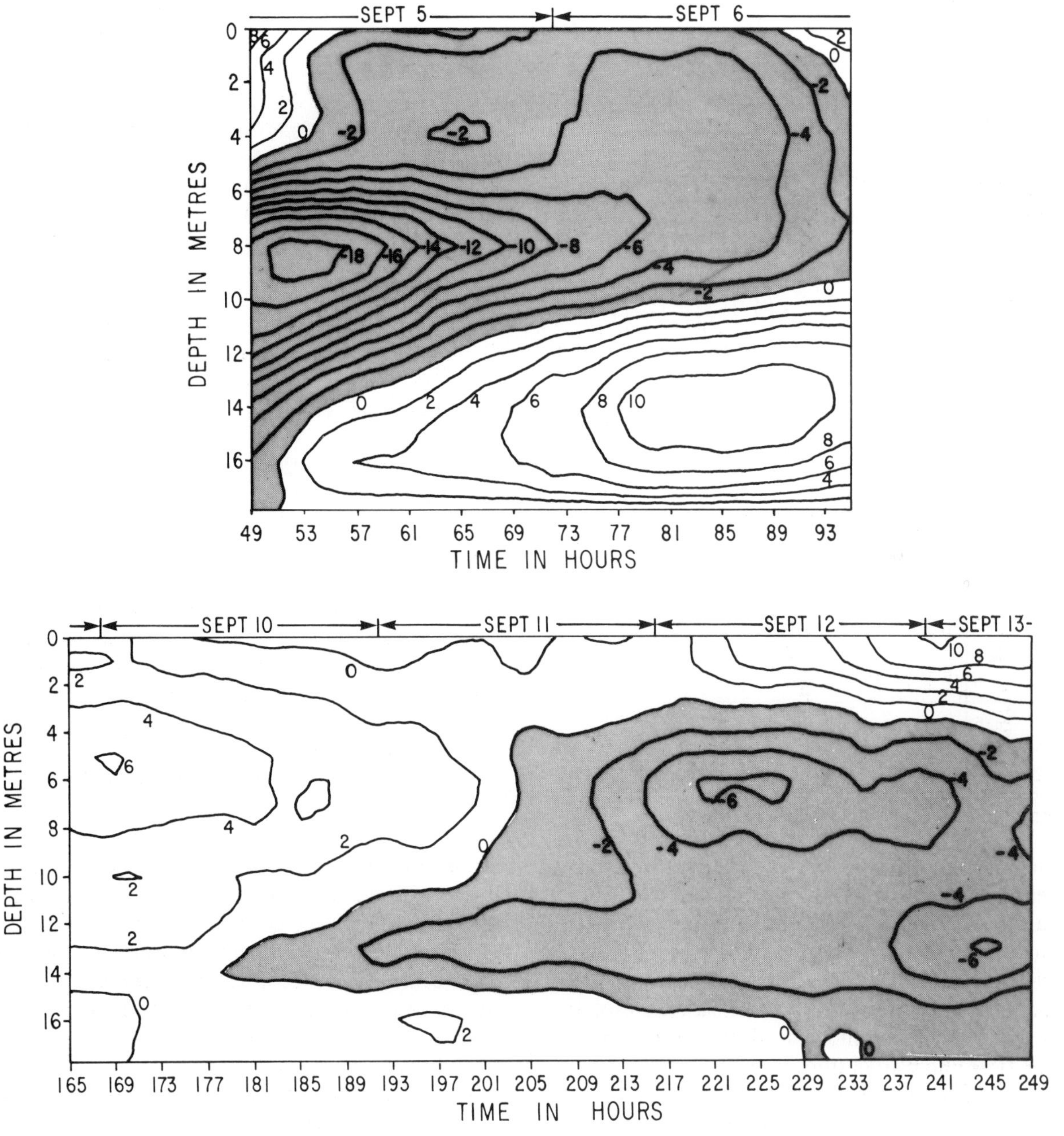

Fig. 2.8. Low-pass filtered residual flow for a station in the Potomac Estuary, Virginia, U.S.A. Depth–time contour plot of the isotachs (cm s^{-1}). Shaded portions denote negative (landward) flow. (Reproduced with permission from A.J. Elliott and T.E. Hendrix; from Elliott and Hendrix, 1976).

Elliott (1976a) continued this type of investigation with a one-year observational study in the lower Potomac Estuary from a reference mooring with current meters at 3, 8 and 12 m in a total water depth of 15 m. An example of the type of data obtained is shown in Fig. 2.9. Particular note should be made of the magnitude of the tidal variations as compared with the non-tidal, net circulation flow. In this example for the near-bottom current meter the average net circulation flow is up-estuary (negative values), as one might expect, but there are periods of flow reversal as well.

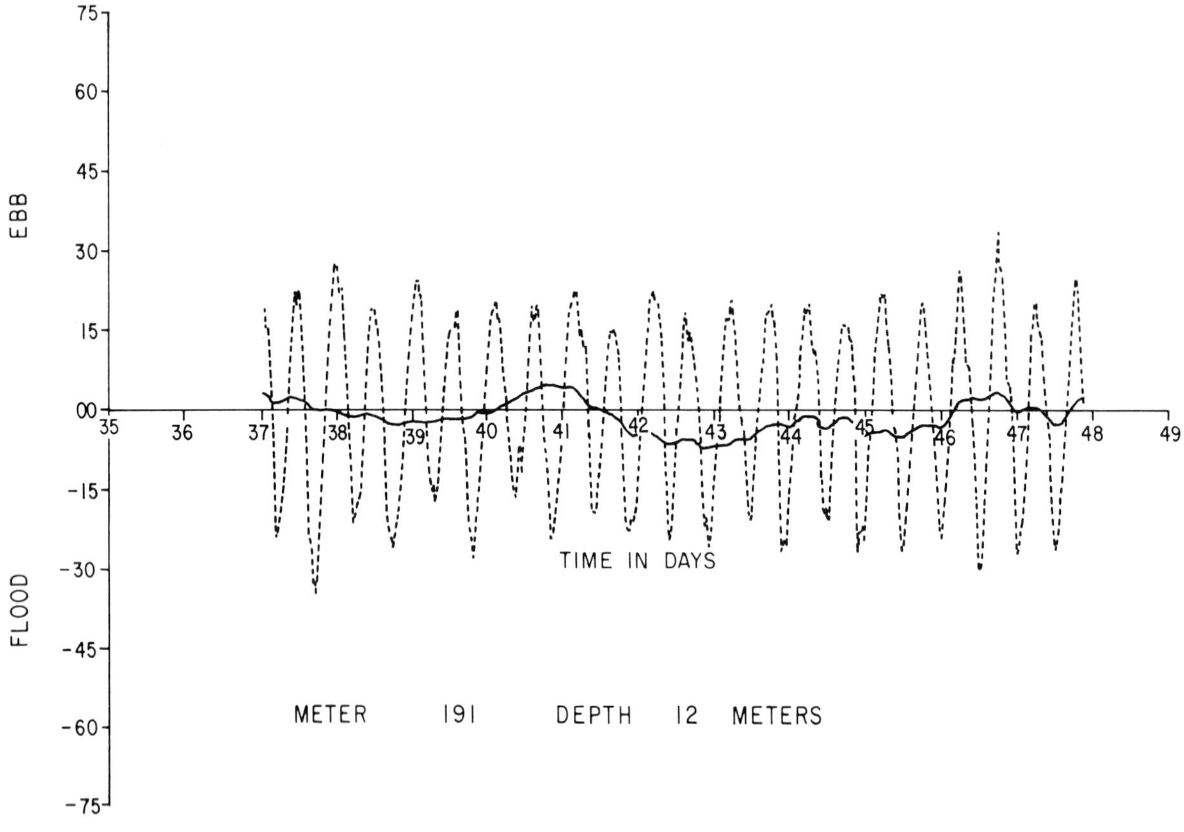

Fig. 2.9. An example of current data (cm s^{-1}) from a reference mooring in the Potomac Estuary, U.S.A. The solid line represents the low-pass filtered, net non-tidal flow; the tidal signal is shown by the dashed line. (Reproduced with permission from A.J. Elliott; from Elliott, 1976a.)

The response from the current meters at the three depths was used to identify six different circulation patterns. The classical estuarine circulation was the most common of the six circulation types, having a mean duration of 2.5 days and occurring for 43% of the total time. The reverse of the classical estuary (surface inflow, outflow at depth) and storage (landward flow at all three levels) were the second most frequent patterns, each occupying about 20% of the total time and having mean durations of 1.6 days. The response of the estuary was also described through the use of empirical orthogonal function analysis. The first mode, which could account for 55% of the variance, was highly correlated with local wind stress and net surface slope. The remainder of the total variance could not be related to local effects and was due in part to signal noise and in part to non-local forcing which could not be isolated.

Wang and Elliott (1978) examined long-period sea level fluctuations in Chesapeake Bay. In their study evidence for non-local forcing in the bay due to exchange with the coastal ocean was obtained. The non-tidal sea level fluctuations in Chesapeake Bay occurred at three dominant periods — 20, 5 and 2.5 days. The 20-day fluctuations originated from changes in sea level at the coast (mouth of the bay) which had been generated earlier by alongshore winds. The amplitude of the fluctuations decreased away from the mouth of the bay, apparently due to bottom friction. The 5-day sea level fluctuations partly originated from sea-level changes at the coast, but in addition, lateral winds provided a local driving force through the Ekman effect. The amplitude of the 5-day fluctuations were nearly uniform throughout the bay. The 2.5-day sea level fluctuations were seiche oscillations forced by longitudinal winds over the bay. The effect of non-local forcing was negligible — that is, the seiche oscillations in the bay did not respond to sea level changes at the coast. The amplitude of the 2.5-day sea level fluctuations increased toward the head of the bay.

Lateral variations

Further complexities in understanding estuarine circulation are often encountered as lateral observations are included in the investigation. One of the more common effects that is observed is illustrated in Fig. 2.10 from Pritchard (1952b) for the James Estuary. The shallow-water regions on either side of the main channel have essentially only a down-estuary net circulation flow at all depths. The figure also shows the lateral variation in the level of no motion, the slope of the surface being interpreted as a Coriolis effect. Nichols and Poor (1967) obtained similar results for the Rappahannock Estuary with the up-estuary net circulation flow being restricted to the main channel.

Elliott (1976b) discussed the net flows in the Patuxent Estuary during a two-day period of response and recovery of the estuary following a period of strong northwesterly, down-estuary winds. The net flows for two sections across the estuary are shown in Fig. 2.11. At section PX-08 the net flow had an estuarine character with a seaward surface flow and a deep return flow. The near-surface waters, however, indicate a considerable cross-channel velocity shear. At section PX-16 the lateral shear is further demonstrated by a reversal of sign. At station PX-16C there was a net seaward flow, while throughout the rest of the section the net flow is directed up-estuary.

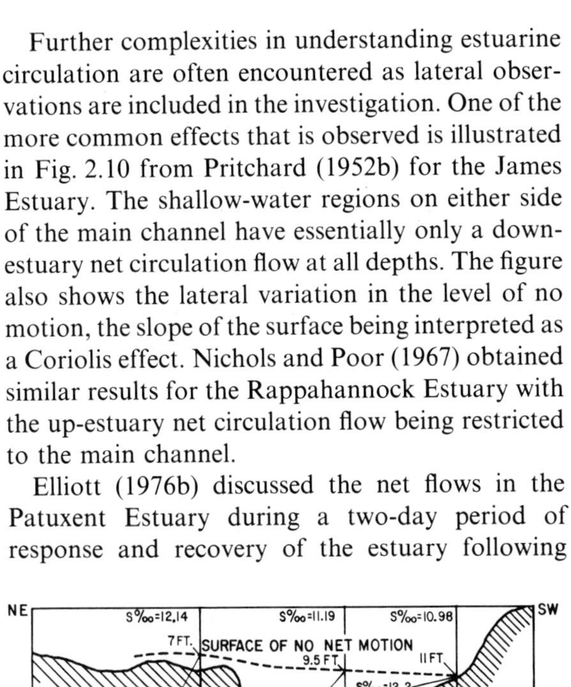

Fig. 2.10. Cross section in the James Estuary showing slope of surface of no net motion (from Pritchard, 1952b).

Dyer (1974, 1977) examined the salt balance in a number of estuaries. His investigations included the contributions related to the vertical and lateral velocity shear and salinity variations as well as the tidal mixing contributions to the determination of the net longitudinal salt flux. His results showed that in a highly stratified estuary the lateral contribution was small, and that in two partially mixed estuaries the mean transverse and vertical flux terms were of similar magnitude. The lateral effects were important in maintaining the salt balance. He also discussed the importance of estuarine geometry — channel curvature and changes in cross-sectional shape — as a control on the lateral variations in the longitudinal velocity as well as in the introduction of transverse velocity components.

Numerical modeling

Numerical modeling has served a useful research purpose in permitting solutions to be obtained to theoretical concepts which are too complex for direct analytic solution. We shall discuss here some

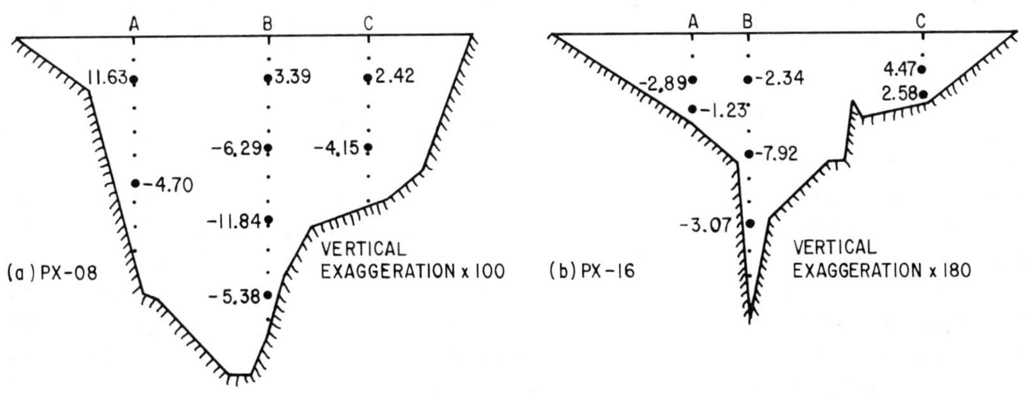

Fig. 2.11. The net flows in cm s^{-1} computed from moored current meters at stations PX-08 and PX-16 in the Patuxent Estuary, Maryland, U.S.A. The sections face up-estuary, and ebb flow is positive (from Elliott, 1976b).

numerical modeling experiments which have been conducted to examine various aspects of circulation and mixing in well-mixed and stratified estuaries. All of these examples are two-dimensional in the longitudinal and vertical directions; all use the momentum, mass and salt continuity equations including, for the most part, the coupling effects and non-linear terms.

Festa and Hansen (1976) modeled the steady-state gravitational circulation in an estuary of uniform geometry. The vertical viscosity and diffusion coefficients, N_z and K_z, were taken constant; and a bottom boundary condition of zero longitudinal circulation velocity was assumed. One of the more interesting results from their investigation is shown in Fig. 2.12. This is a plot of the vertical velocity, positive upward, against longitudinal distance. The vertical velocity is not a directly measurable quantity in an estuary and can usually only be inferred from continuity considerations. From the figure we see that the vertical velocity decreases rapidly to zero up-estuary of the location, $x = 60$ km, and increases to a maximum value down-estuary. All values are positive indicating upward vertical velocities. The location $x = 60$ km is the approximate position in this numerical example at which the horizontal flow changes from an essentially riverine character with seaward flow at all depths and no density gradient contribution to an essentially estuarine character with seaward surface and landward bottom flows.

Blumberg (1977) modeled the real time, or tidal, variations in an estuary using realistic estuarine bathymetry. The coefficients, N_z and K_z, were taken to be variable, dependent on the depth, vertical velocity gradient and the Richardson number; and a quadratic bottom friction law was assumed with a bottom frictional coefficient, k. The external inputs to the model were the salinity and tidal amplitude as a function of time at the ocean boundary, and the fresh-water discharge at the river boundary. An example comparing computed and measured tidal velocities for the Potomac Estuary at three locations up-estuary from its mouth is shown in Fig. 2.13. The numerically computed distributions show the proper phase and in some cases the proper magnitude as compared with the actual tidal velocity observations.

Hamilton (1975) modeled the real time variations in an estuary of varying width and depth but with a rectangular cross section. The N_z and K_z coefficients were taken to be a function of the tidal velocity and the Richardson number, and a quadratic bottom friction law was assumed. The model was applied to the Rotterdam waterway, and the internal dependence coefficients in the N_z and K_z expressions were chosen to give a best fit to the observations. Fig. 2.14 shows the observed and Fig. 2.15 the computed tidal salinity variations at seven locations along the waterway. There is good qualitative agreement both as to the magnitude and shape of the salinity variations.

Bowden and Hamilton (1975; Bowden, 1977;

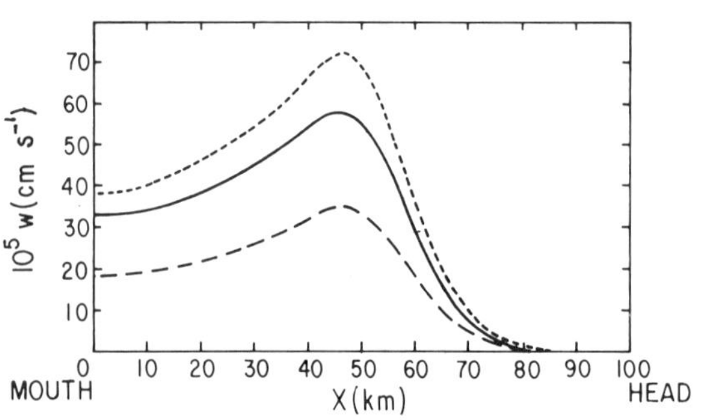

Fig. 2.12. Longitudinal variation of the vertical velocity field, w, for relative depth values $(z/h) = 0.25$ ---, 0.50 —, and 0.75 -- (from Festa and Hansen, 1976).

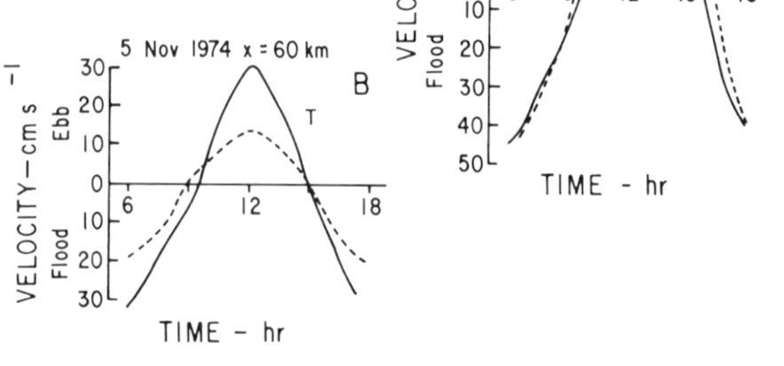

Fig. 2.13. Comparison of observed velocity distribution (—) with computed ones (---) for three different sections of the Potomac Estuary. Depth indicated by T (top) and B (bottom). Distances are measured up-estuary from the mouth. (From Blumberg, 1977).

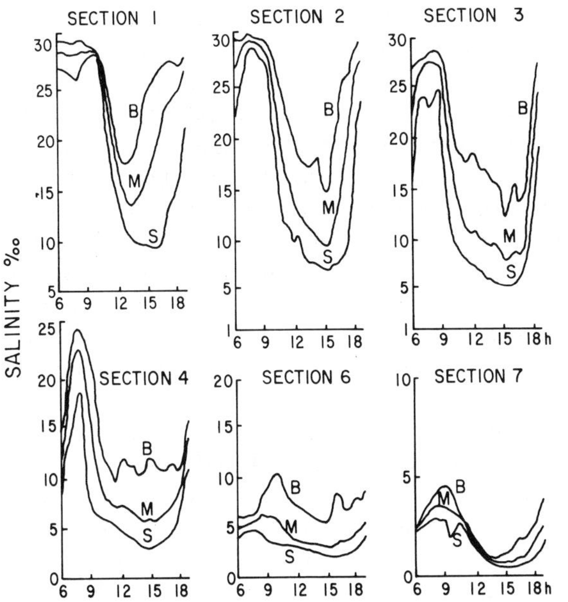

Fig. 2.14. Lateral averaged observed surface (*S*), bottom (*B*), and depth mean (*M*) salinity variations during a tidal cycle at selected sections in the Rotterdam waterway, The Netherlands (from Hamilton, 1975).

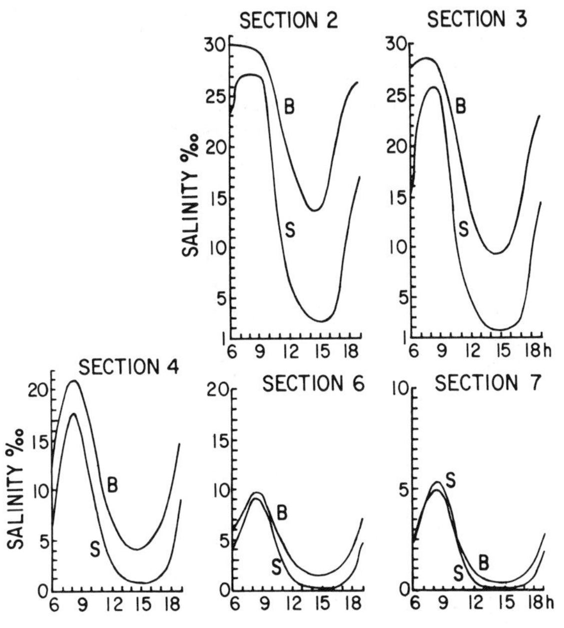

Fig. 2.15. Computed surface (*S*) and bottom (*B*) salinity variations during a tidal cycle at the same sections as in Fig. 2.14 for the Rotterdam waterway (from Hamilton, 1975).

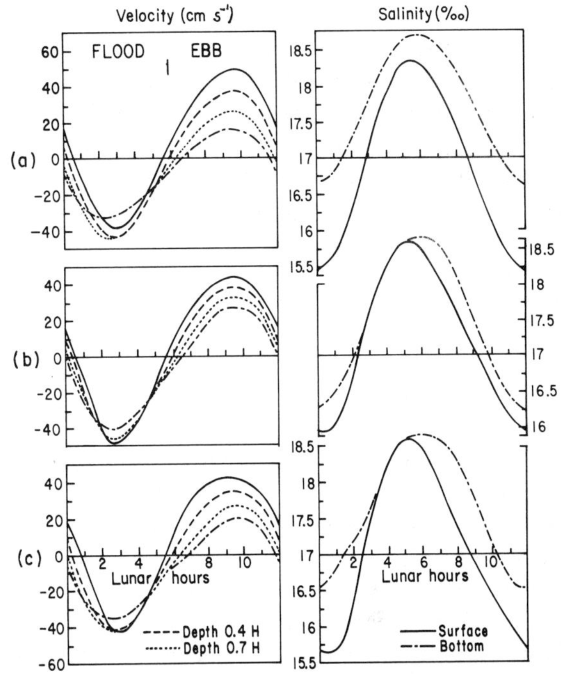

Fig. 2.16. Velocity-time and salinity-time curves for three forms of the eddy coefficients N_z and K_z: (a) constant, (b) tidal variable, (c) dependent on the Richardson number. Curves are given for various relative depth values, z/H, from the surface to the bottom. (From Bowden and Hamilton, 1975.)

Hamilton, 1977) used the same model to carry out some numerical experiments for the simulation of tidal velocity and salinity variations in response to various assumed dependences for the parameters N_z and K_z. The viscosity and diffusion coefficients were taken as (a) constants, (b) functions of the depth and mean tidal velocity, and (c) functions also of the Richardson number. Fig. 2.16 shows the velocity and salinity curves as functions of time for these three cases. The station considered is in a uniform part of the estuary with a width of 1 km and a mean depth of 16 m. In each case the tidal amplitude at the mouth was 1 m, and the river discharge and salinity at the mouth were held constant. The velocity curves have several features in common, such as the ebb lasting longer than the flood, typical of a tidal wave progressing into shallow water. The velocity decreases throughout the depth on the ebb but on the flood it increases to a mid-depth maximum, features to be expected because of the density current effect. The salinity curves also have some features in common, but there are significant differences. When the eddy coefficients are variable instead of constant, the flood and ebb parts of both velocity and salinity curves become more asymmetrical; and vertical mixing is more intense on the flood than on the ebb. These features correspond to observations in several estuaries, including those of the Mersey and the Severn in England, and the Rotterdam waterway in The Netherlands.

FJORD AND SALT WEDGE CIRCULATION, RIVER PLUMES, AND ESTUARY FRONTS

Interface effects

There are a variety of interface, or near-interface, conditions that can occur in estuaries, including fjord and salt wedge circulation, river plumes, and estuary fronts. In some cases the interface between water bodies of differing physical characteristics can be sufficiently sharp to be observed visually or acoustically. In other cases the near-interface conditions are such that there is little exchange between the water bodies, and each retains its identifiable characteristics.

Fjord shallow circulation

Fjords are deep coastal inlets usually with a sill, or sills, at or near their entrance. They are often long with steep sides and of relatively constant width. The most significant influence in many inlets is the fresh-water river runoff. In such cases the circulation is restricted to the upper portion of the water column with a strong outflow in a shallow, surface layer and an inflow immediately beneath. The deeper water behind the sill usually does not participate in this circulation. It is relatively stagnant, replacement occurring during events which often have an annual or longer time scale.

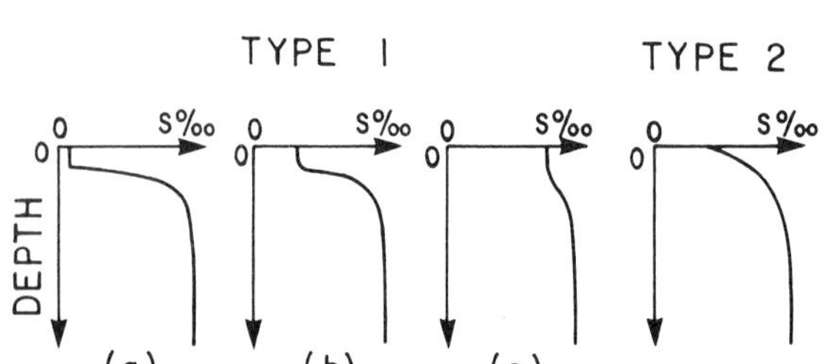

Fig. 2.17. Types of vertical salinity profiles for the upper portion of the water column in British Columbia (Canada) inlets (from Pickard, 1961).

For the upper portion of the water column Pickard (1961) distinguishes two types of salinity–depth profiles for the inlets of British Columbia as shown in Fig. 2.17. In type *1*, associated with high runoff conditions, the salinity profiles vary from *1a* to *1b* to *1c*, or *2* depending on the amount of mixing, from the inlet head to its entrance, showing a strong halocline in the inner to middle reaches of the inlet. For type *2* associated with low runoff conditions, there is no surface layer; and the salinity increases with depth from the surface. An example of fjord salinity profiles from Pickard and Rodgers (1959) is shown in Fig. 2.18. In the upper reaches of the estuary, fresh surface water is separated from the deeper sea water by a strong salinity gradient. This condition gradually changes and in the outer basin the surface layer is only slightly diluted with fresh water and the profile is nearly isohaline.

Rattray (1967) and Winter (1973), as also summarized in Rattray (1977), give a theoretical de-

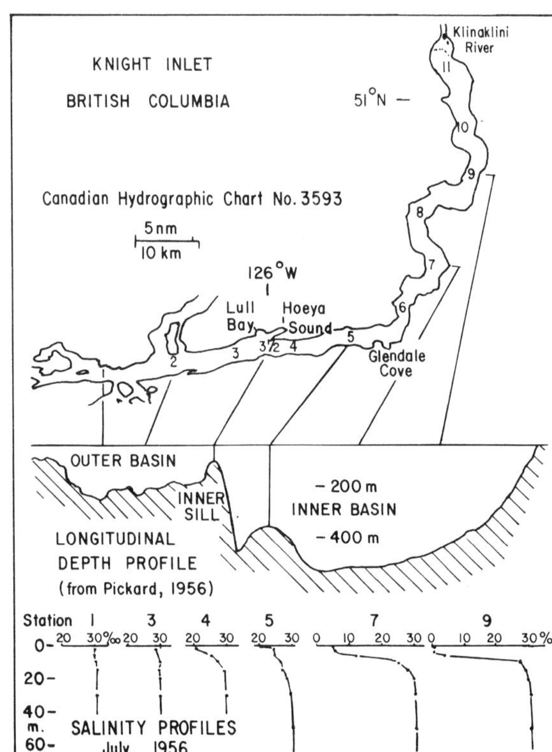

Fig. 2.18. Knight Inlet, British Columbia: plan view and station locations, longitudinal depth profile, and salinity–depth profiles (from Pickard and Rodgers, 1959).

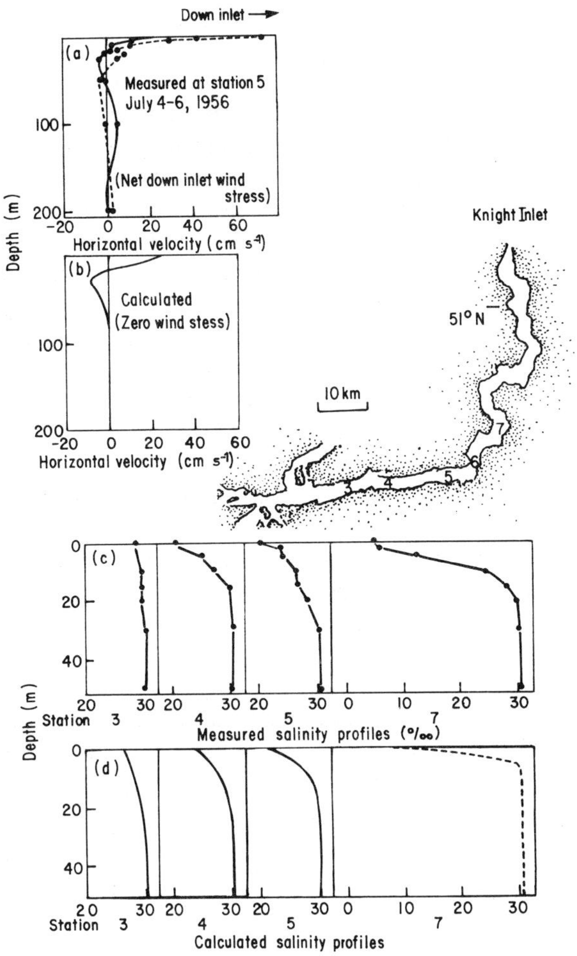

Fig. 2.19. Comparison of (a) measured and (b) calculated horizontal velocity profiles for station 5 in Knight Inlet. Comparison of (c) salinity measurements at the four stations indicated in the map of Knight Inlet with (d) calculated depth profiles of salinity. (From Winter, 1973.)

scription for the near-surface fjord circulation. The dynamic balance in this case is between the horizontal pressure gradient, or surface slope, the longitudinal density gradient terms, the vertical gradient of turbulent stress, or vertical viscosity term, plus the non-linear field accelerations, or inertial terms. Fig. 2.19 shows a comparison of observed and calculated longitudinal velocity and salinity profiles for Knight Inlet; in this comparison the internal solution parameters have been chosen to fit the observations. Nevertheless, the analytic solutions do reproduce the principal features of the observations.

Wind effects can be important for fjord circulation. An example is shown in Fig. 2.20. During

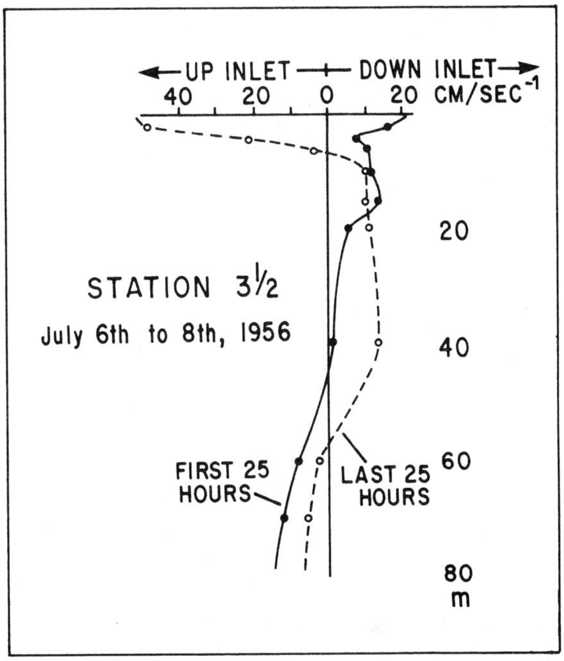

Fig. 2.20. Net current profiles over two 25-h periods at a station in Knight Inlet (from Pickard and Rodgers, 1959).

the first 25-h measurement period there was no wind, and during the second period a wind of 12 knots (22 km h^{-1}) blew in an up-inlet direction. The net circulation velocity distribution for each period was quite different. During the first period there was a down-inlet flow at all depths down to 40 m with an up-inlet flow beneath. During the second period the surface flow was reversed, and the down-inlet flow was constrained to an intermediate level just below the surface.

Fjord deep circulation

The second principal feature of fjord circulation is the intermittent exchange of deep water behind sills. An interesting example is given by Saelen (1967) for Nordfjord, Norway, as shown in Fig. 2.21. The station is located in the deep basin

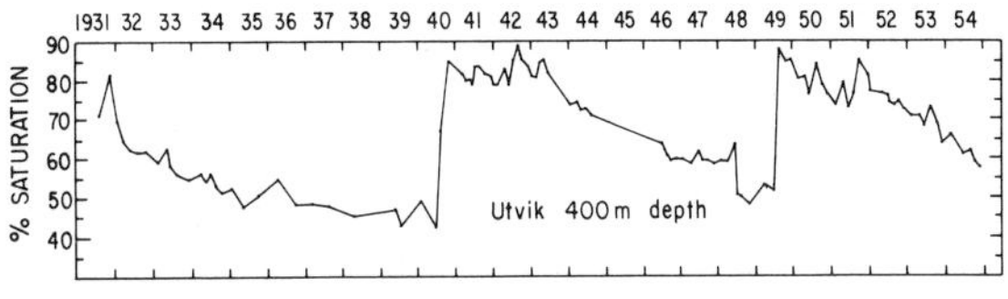

Fig. 2.21. Variation of oxygen (% saturation) at 400 m depth at a station in the inner basin of the Nordfjord, Norway (from Saelen, 1967).

landward of the inner sill. From 1932 to 1940 the percentage of oxygen saturation decreased steadily. In 1940 it increased abruptly to 90% with the same sequence being repeated from 1940 to 1949 and 1949 onward. The conclusion is reached that the deep inflows which occur on an annual basis in the outer basin are not strong enough to carry over to the inner basin save for exceptional years.

Anderson and Devol (1973) investigated the flushing of the deep waters in Saanich Inlet, an intermittently anoxic basin on Vancouver Island (Canada). The intrusion of water over the sill occurred on an annual basis during late summer when dense water reached the sill of Saanich Inlet from nearby Haro Strait. Prior to flushing of the deep water of the basin it is devoid of oxygen and nitrate. An example of the nitrate distribution at a station in the inlet on 6 August 1969 prior to flushing, and on 23 October 1969 after there had been an intrusion of new water, is shown in Fig. 2.22. The intruded water descended and mixed with the deep water of the basin, displacing a residual portion of the deoxygenated layer upward. For the two years investigated Anderson and Devol estimated that 64% and 33% of the basin volume below sill depth was replenished in each respective year.

Cannon (1975) investigated the deep water renewal in Port Susan Estuary, an inlet tributary in Puget Sound (U.S.A.). The primary renewal of bottom water occurs in late summer and early fall. Current meter measurements were made 2 m off the bottom on the sill at the entrance to Port Susan Inlet during the period from 27 July 1970 to 25 September 1970. Renewal occurred during four

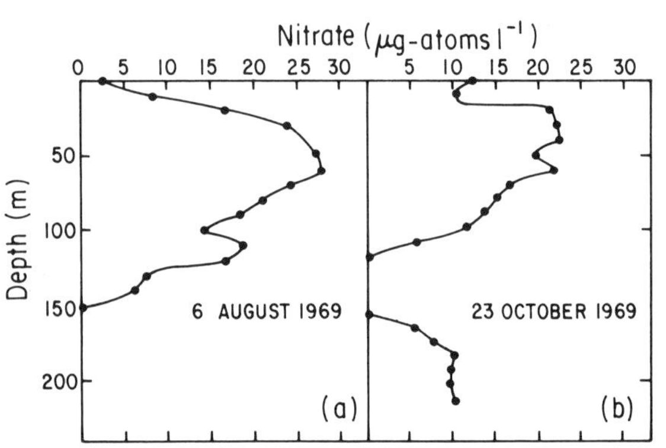

Fig. 2.22. Vertical distribution of nitrate-nitrogen at a station in Saanich Inlet (B.C., Canada) on (a) 6 August 1969 and (b) 23 October 1969 (from Anderson and Devol, 1973).

intervals of about five days' duration each. The inflow during each interval was sufficient to replace all of the water in the inlet below sill depth.

Ebbesmeyer et al. (1975) investigated the intrusion of deep water into Dabob Bay, another inlet tributary to Puget Sound. The intruded water parcels, having vertical dimensions of from 10 to 100 m and horizontal dimensions of from 1 to 10 km, maintained their identities for periods of days to weeks by their distinctive salinities and temperatures. An important aspect in the decay of each parcel was current shear, and periods of maximum shear were due primarily to a second intrusion of new water.

Salt wedge circulation

For an arrested salt wedge there is an interface, or strong halocline, between the overlying fresh and underlying saline waters. Salt wedge circulation is characterized by a strong discharge, or river runoff, in the upper layer and relatively weak tidal currents. The salt wedge is bounded at its upper end by the river bottom and there is little entrainment across the salt wedge interface, so that the net circulation in the lower layer is approximately zero. Theoretical descriptions of salt wedge circulation have been given by Stommel and Farmer (1952) and Rattray and Mitsuda (1974). In this case the longitudinal density gradient is unimportant, and the horizontal pressure gradient is determined by the surface and interface slopes. In the upper layer the flow is accelerated in a down-estuary direction because of the change in depth of the salt wedge interface, and the inertial terms are important in the dynamic balance. The principal frictional stress is that at the interface, given by a quadratic velocity relation or as a boundary layer effect.

One of the more thorough investigations of a salt wedge is that given by Fukushima et al. (1966, 1969) for the Ishikari River (Japan). Using an echo sounder, they obtained high-frequency reflections in the form of a continuous profiling record from the salt wedge interface, or possibly from detritus or bubbles accumulated on the interface, in much the same manner as one obtains a continuous record of the bottom with a conventional echo sounder. An example is shown in Fig. 2.23. The salinity and longitudinal current velocity structure at the time of these measurements is shown in

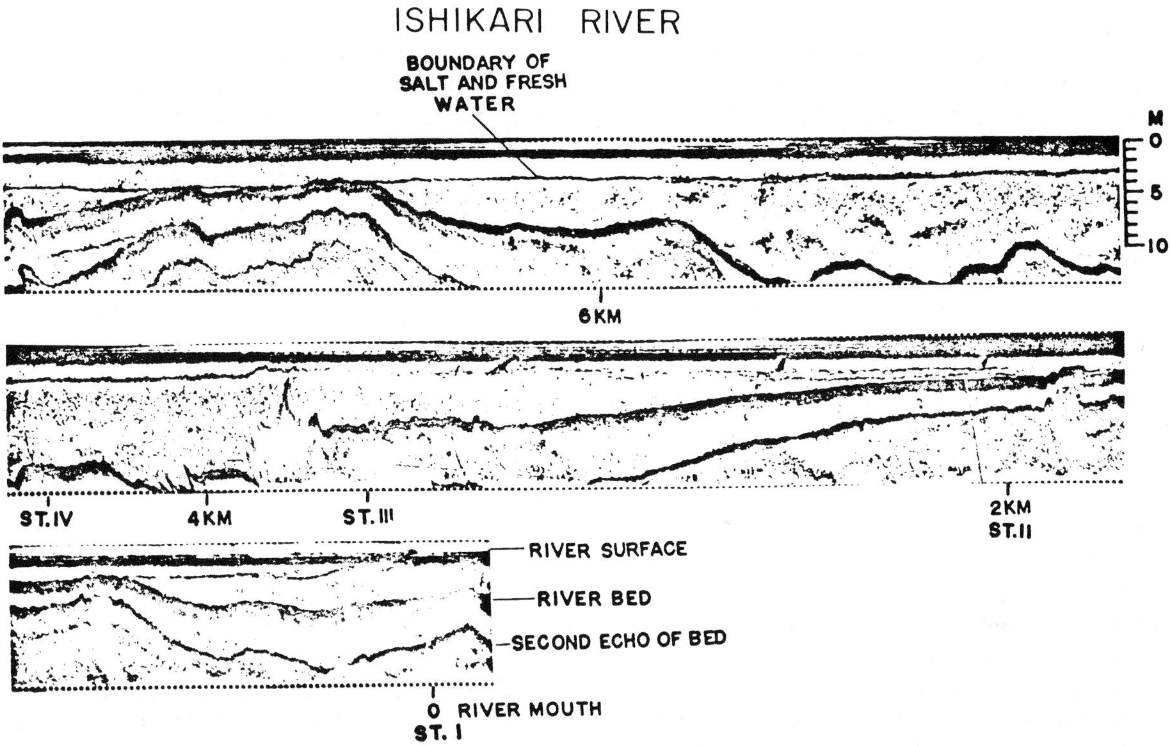

Fig. 2.23. An echo sounder profile of a salt wedge in the estuary of the Ishikari River (Japan) on 22 July 1964 (from Fukushima et al., 1966).

Fig. 2.24. The river discharge rate at this time was at a seasonal low value, and the salt wedge extended for a considerable distance up-river. It is to be noted from Fig. 2.24 that the salt interface is quite sharp, particularly in an up-river direction, with a salinity change of several parts per thousand over a depth interval of a fraction of a meter. The salt wedge interface, as shown on the echo sounder records, is quite sharp up-river and distinct from the bottom echo. The salt wedge return becomes more diffuse immediately downstream from the bottom projection at 3.7 km, but it gradually regains its distinctive characteristics as the river mouth is approached. The authors conclude that this diffuse characteristic on the record is indicative of entrainment mixing effects across the interface associated with flow over the bottom projection.

River plumes and estuary fronts

When the river discharge is large and the estuary or river valley narrow, a distinct plume of brackish water can often be observed beyond the estuary mouth in the neighboring coastal waters. The plumes are shallow pools of lighter water overlying more dense ambient water. There is a steep gradient in salinity between the water bodies, which inhibits vertical mixing; and frequently the lateral boundaries of plumes are formed by sharp fronts.

Garvine (1974) and Garvine and Monk (1974), as also summarized by Garvine (1977), investigated the plume of the Connecticut River in Long Island Sound and its associated transverse fronts. The results from one series of measurements are shown in Figs. 2.25 and 2.26. Fig. 2.25 is a plot of density,[1]

[1] Sigma$_t$(σ_t) is a measure of the density of sea water and is defined as $\sigma_t = (\rho_t - 1)$ 1000 in which ρ_t is the specific gravity of the sample at t°C relative to distilled water at 4°C. In practice, sea water density is not measured directly. It is precisely related to the salinity and temperature of the sample and values for σ_t are derived from tables such as those prepared by Knudsen (1901) or by the U.S. Navy Hydrographic Office (1952). A set of six convenient nomographs to determine σ_t directly for the temperature range of −2°C to 33°C and the salinity range of 0‰ to 42.5‰ have been published by the U.S. Navy Hydrographic Office (undated). These nomographs eliminate the interpolation required by the tables while giving the desired accuracy of 0.01 σ_t units.

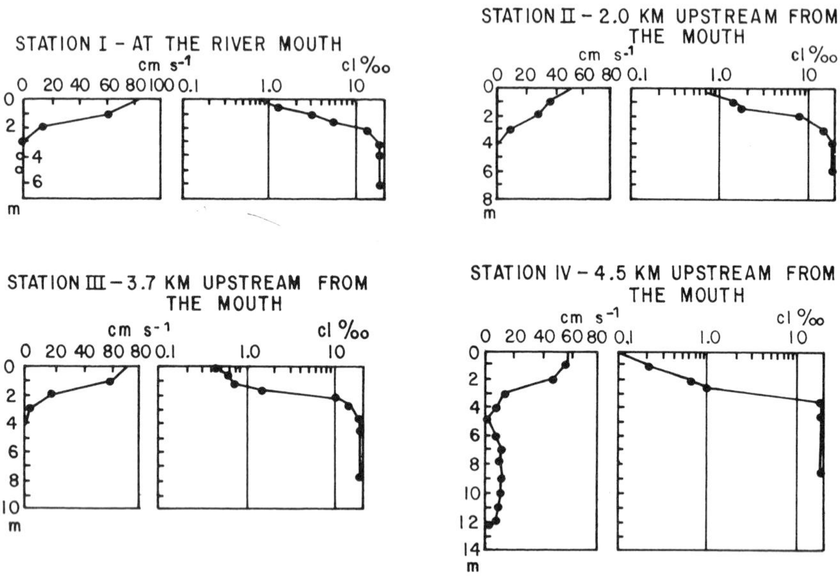

Fig. 2.24. Vertical distributions of velocity (cm s^{-1}) and chlorinity (‰) at selected stations in the Ishikari River on 22 July 1964 (from Fukushima et al., 1966).

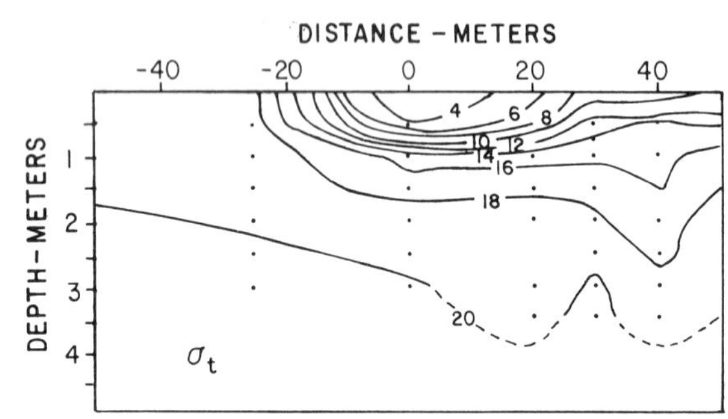

Fig. 2.25. Density (σ_t) section of the Connecticut River (U.S.A.) plume normal to its transverse front on 16 March 1973. Solid dots indicate points where salinity and temperature were measured. (From Garvine and Monk, 1974.)

σ_t, contours along a section normal to the front; here the plume is seen to be a very thin, 1 to 2 m, surficial water layer. Fig. 2.26 is a plot of horizontal velocity contours normal to the front. One of the more prominent features illustrated is the strong convergent velocity into the front from both sides.

In addition, the front forms a distinctive surface feature. The outflowing river water is yellowish brown in color, whereas the sound water is nearly blue green. The offshore boundary between the two water masses, which is exceedingly sharp at the front, presents a feature that may be readily photographed from the air. Furthermore, because of the convergent velocities at the surface into the front, it is also often distinguished by the presence of a foam line or the accumulation of other floating materials.

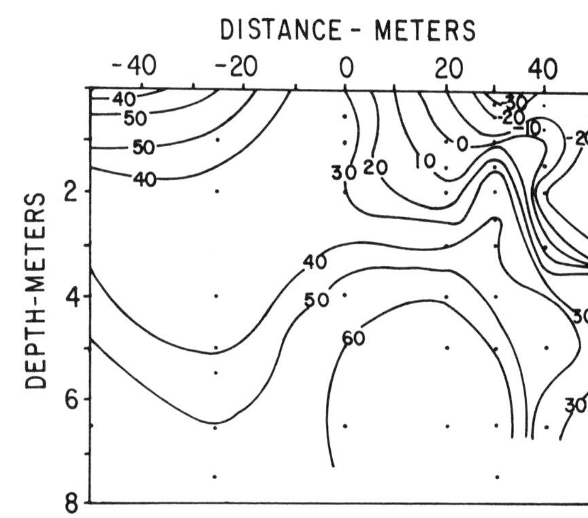

Fig. 2.26. Isotachs of the horizontal velocity field (cm s^{-1}) in the Connecticut River plume normal to the front for 16 March 1973. Solid dots indicate points of measurement. (From Garvine and Monk, 1974).

Another type of front associated with estuaries has been observed in Delaware Bay by Szekielda et al. (1972). Here it is conjectured that the juxtaposition of different density water masses occurs because of the higher tidal velocities and flows for the deeper, central channel in advecting fresher and lighter water down-estuary as compared with the corresponding flows in the adjoining shoal waters. The fronts tend to be more prominent on an ebb tide, becoming less distinct or disappearing on a flood tide. As with the case of river plumes these fronts are also associated with lateral convergence velocities at the surface on both sides, and with the consequent accumulation of floating organic and detrital materials and the formation of foam lines.

BIOLOGICAL, CHEMICAL AND GEOLOGICAL RELATED EFFECTS

Hydrodynamically related effects

Some biological, chemical and geological effects in estuaries are dependent, in part, on the physics of the circulation and mixing for their description. It is not the intention to give a complete summary of such investigations but rather to describe a few selected examples which illustrate some interesting processes. In some cases the phenomena are sufficiently well understood that analytic expressions and corresponding analytic or numerical modeling solutions may be given. In other cases only a qualitative description may be given.

Conservative and non-conservative distributions

One of the simpler examples is that of a conservative, miscible quantity in an estuary. At some distance away from the source and providing that the waters are well mixed in the vertical and lateral directions, the conservative quantity, c, will mix, or dilute, in the same manner as the fresh-water fraction, f, in a down-estuary direction from its source, and will mix in the same manner as the salt, s, in an up-estuary direction. In either case the quantity dc/ds is a constant. This, then, provides a simple means for ascertaining whether a given quantity remains conservative during its passage through an estuary.

An example of the dilution of such a conservative property in relation to salinity is the distribution of silica in the Merrimack River Estuary (Fig. 2.27) as observed by Boyle et al. (1974). The source of silica was from continental weathering up-river

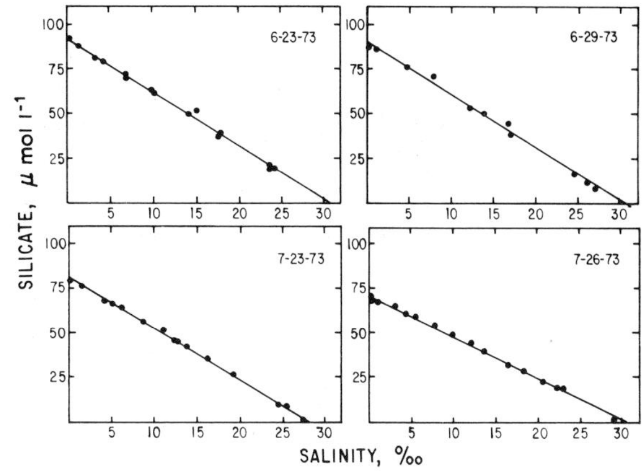

Fig. 2.27. Silica–salinity diagrams for the Merrimack Estuary (U.S.A.) on four different measurement periods (from Boyle et al., 1974).

from the area of investigation and there are no tributary sources. The vertical intercepts on the graphs represent the source concentrations. Another example is shown in Fig. 2.28 for phosphorus in the Charlotte Harbor Estuary on the west coast of Florida, as obtained by Alberts et al. (1970). Here the source is from mining operations for phosphate adjacent to the Peace River. The corresponding available nitrogen for nutrient uptake in phytoplankton growth is small in comparison with the phosphorus so that loss by this process as well as by sedimentation is negligible in comparison with the dilution effects.

Peterson et al. (1975) describe an example in which the estuarine silica distribution is not conservative. This is for the northern reach of San Francisco Bay. The major variations in the dissolved silica distribution are seasonal, and are related

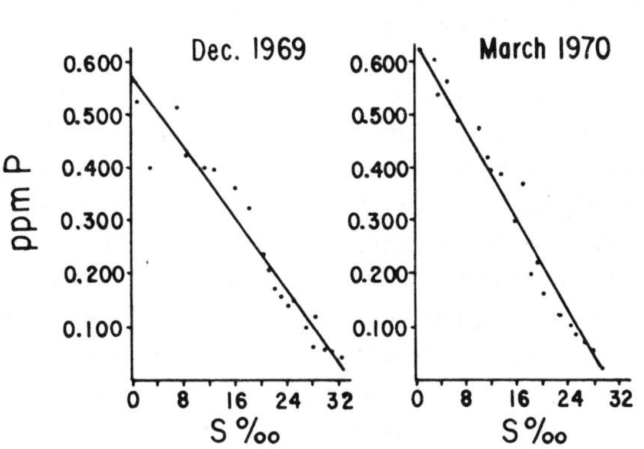

Fig. 2.28. Phosphorus concentrations versus salinity in the Charlotte Harbor Estuary (Florida, U.S.A.) (from Alberts et al., 1970).

to the variations in rates of river supply and silica utilization by phytoplankton. When the rate at which river inflow supplies silica to the bay is large compared with the rate at which silica is used within the estuary, the decrease in dissolved silica as salinity increases is controlled primarily by mixing of river and ocean waters. When the dissolved silica utilization rate increases significantly relative to the supply rate, the dissolved silica concentration in the estuary is considerably less than that predicted by simple mixing, and is often lower in concentration than dissolved silica in the adjacent ocean water. The time of maximum silica utilization within the estuary is during the summer phytoplankton maximum. Fig. 2.29 is an example of the progression from a conservative to a non-conservative silica–salinity relation from winter to summer conditions. Peterson et al. (1978) have used the numerical model of Festa and Hansen (1976), with the addition of the silica continuity equation including a term for phytoplankton silica uptake, to model the silica distribution in San Francisco Bay. Their results adequately describe the observed silica distributions and permit estimates of the uptake rates to be made.

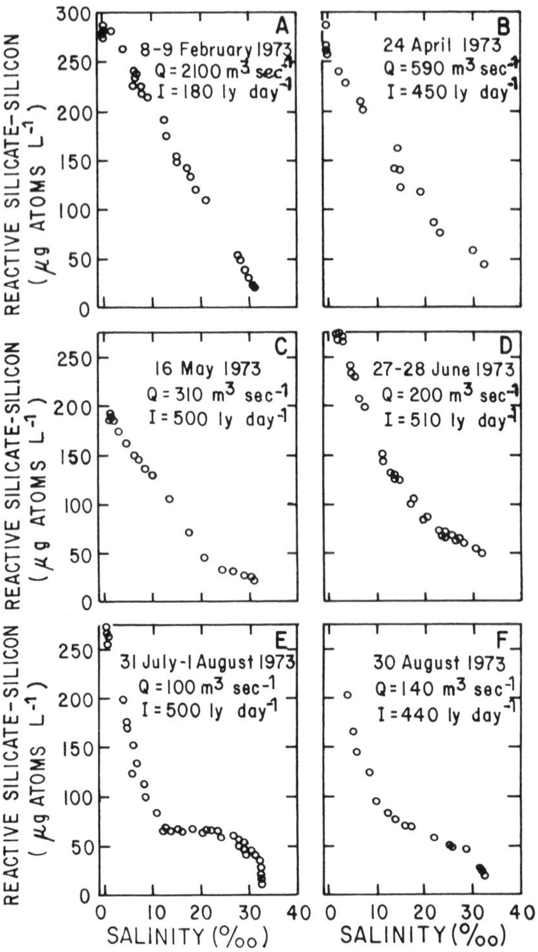

Fig. 2.29. Silica–salinity relations at 2 m depth in the northern reach of San Francisco Bay (Calif., U.S.A.). Q indicates combined mean monthly discharges from the Sacramento and San Joaquin Rivers, and I indicates mean monthly insolation at Richmond. (From Peterson et al., 1975.)

Flushing time

A concept that has proved useful in assessing the relative importance of various possible effects in estuarine waters is that of flushing, or retention, time. Flushing time, τ, is defined as the time necessary to replace any given conservative quantity in a given volume of an estuary, river or coastal region at the rate at which that quantity is being injected into the volume. It is thus a measure of the residence time in a Lagrangian sense of a particle in an estuary, or portion of an estuary. For rivers and lakes τ expresses the effect of the advective river flow, R, through the volume, V, and is given simply by V/R. For estuaries it also includes the important effects of tidal mixing and of density-gradient and wind-induced circulation flows. For coastal waters it includes all three of these effects plus that due to the coastal currents. It is important to emphasize, however, that flushing time is usually of use only as an approximate index.

Flushing time enters estuarine transportation and reaction considerations in two ways — in terms of the effective dilution, and in connection with time-dependent reactions such as oxidation, respiration, photosynthesis and plankton production. The volume-averaged concentration of a conservative quantity, c, is given in terms of τ as $W/(V/\tau)$ where W is the rate at which the quantity is being added. The flushing time enters here in terms of the quantity, V/τ, which we may refer to as the assimilation; and it is a direct measure of the dilution characteristics of the volume. The flushing time also enters in the first order, exponential, reaction terms in the form $k\tau$ where k is the reaction constant. For small $k\tau$ values, little reaction takes place within the defined volume; for large $k\tau$ values, the reaction may be essentially completed within the volume considered.

Ketchum (1951a, b) showed that the fresh-water flow, or river discharge, R, into a tidal river, estuary or coastal region can be used as a natural tracer to determine the flushing time. The flushing time is given by fV/R, where f is the volume-averaged fresh-water fraction; and the assimilation is given by R/f. The quantity, $1/f$, is then a measure of the dilution potential over that due to river advection only.

Values that have been determined for τ, V and $1/f$ for various selected rivers, estuaries and coastal regions are given in Table 2.3 from Officer and Ryther (1977). The substantial increase in the dilution potential from rivers to estuaries and from estuaries to coastal regions should be noted. This succession is much as expected, and gives some indication in gross terms of the increase in the dilution potential over that due to river advection only with the addition of tidal mixing and circulation flows in estuaries, and with the further addition of offshore currents in open coastal waters.

As was indicated previously, hydrodynamic flushing times can be used in comparison with the time constants of other biological and chemical processes to ascertain possible growth or depletion effects under various estuarine conditions. Ketchum (1954) made such a comparison for planktonic populations in a number of estuaries. In essence his comparison was of the net tidal exchange to the open coastal waters with the reproduction rates of plankton to determine steady-state populations.

One may also consider a retention time for chemical constituents, such as phosphorus, iron and aluminum, which undergo various reactions within the defined estuarine volume including incorporation in the sediments. Stumm and Morgan (1970) defined such a retention time, T, as cV/W, similar to the definition of the hydrodynamic flushing time. Depending on the rapidity with which the given chemical constituent is incorporated into the sediment and the various chemical and biological reactions which it may undergo within the water and sediment, T can be smaller or larger than τ.

Still further, one can consider retention, or biodeposition, effects related to the biological uptake of various constituents through rooted plants, phytoplankton or filter-feeding organisms. Hobbie (1976) mentions, for example, that an oyster bed can completely process the whole volume of an estuary in a few days to weeks, times which are comparable to estuarine hydrodynamic flushing times. In such cases, if the oysters retain a significant portion of a given constituent, such filtering processes can be of greater importance than the physical or chemical processes in the ultimate determination of the fate of such constituents introduced into an estuary.

Biological circulation effects

Circulation effects are a determining factor for some biological phenomena. In the first example

TABLE 2.3

Flushing time (τ), volume (V), and dilution potential ($1/f$), for various rivers, estuaries and coastal regions

Region	Source[1]	τ (days)	V (10^6 m^3)	$1/f$
Raritan River (New York, U.S.A.)	a	8	36	2
Houston ship channel (Texas, U.S.A.)	b	56	67	1
Delaware River (U.S.A.)	c	28	1130	2
Boston Harbor (Mass., U.S.A.)	d	2	67	24
Mersey Estuary (England)	e	4	147	7
Raritan Bay (New York, U.S.A.)	a	14	800	12
Passamaquoddy Bay (N.B., Canada)	f	16	4700	41
New York Bight (U.S.A.)	g	8	23 000	63
Bay of Fundy (Canada)	f	76	880 000	52

[1] Sources: a, Ketchum (1951a); b, O'Connor et al. (1977); c, Ketchum (1952); d, Bumpus et al. (1953); e, Hughes (1958); f, Ketchum and Keen (1953); and g, Ketchum et al. (1951).

discussed, the circulation is wind-induced; in the others it is the normal, density gradient estuarine circulation.

Seliger et al. (1970) discussed the mechanisms for the accumulation of high concentrations of the bioluminescent, tropical marine dinoflagellate *Pyrodinium bahamense* in Oyster Bay, Jamaica. Previous to their investigation Carpenter and Seliger (1968) had determined from dye tracer experiments that the flushing time for the bay as a whole was 1 to 1.5 days. In contrast, the observed generation times for the dinoflagellate were 3 to 3.5 days. Thus, there appeared to be a paradox concerning the large accumulations of the dinoflagellate within the bay. The resolution was found in the effect of the wind-induced circulation. During the day the wind generally blows east-northeast across the bay, forcing the brackish surface layer out of the bay. There is a compensating westward flow of the uppermost saline water beneath the brackish surface water, which transports dinoflagellate-rich waters to the shallows in the eastermost part of the bay where the exchange rate is much lower than that for the bay as a whole. Fig. 2.30 shows the sequence of events for the morning of 28 July 1968 for this differential displacement of dinoflagellate-rich, saline water; in the eastern shallow areas the surface populations are further enhanced through photosynthesis and phototaxis.

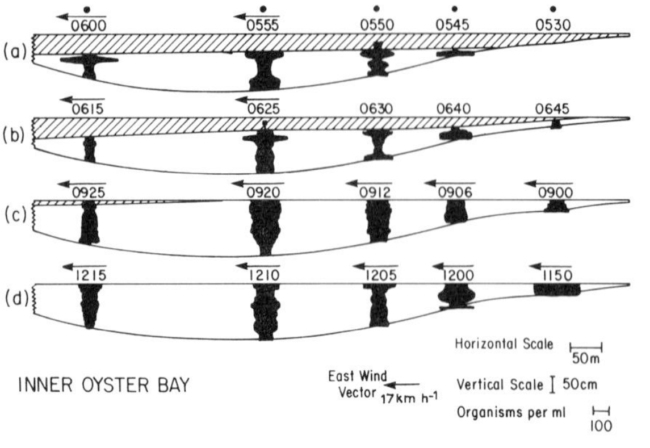

Fig. 2.30. Salinity, wind, and depth distributions of *Pyrodinium bahamense* concentrations along an east–west transect in Oyster Bay, Jamaica, during the morning of 26 July 1968. The brackish water layer ($S = 10–20‰$) is indicated by the cross-hatched areas. Saline waters ($S \geq 30‰$) are the clear areas. The westward pointing arrows are wind velocity vectors; the measured phototactic distributions of *P. bahamense* are given by the solid areas, the width at any depth being proportional to the concentration of organisms. (a) shows the phototactic accumulation of *P. bahamense* directly below the brackish layer. The remaining drawings show the differential displacement of *P. bahamense* into the previously barren eastern shallows and the subsequent uniformity produced by vertical mixing. (From Seliger et al., 1970.)

One of the most interesting investigations of the effects of density-gradient, estuarine circulation on a biological population is that by Tyler and Seliger (1978) of the yearly cycle of the red tide dinoflagellate *Prorocentrum mariae-lebouriae* in Chesapeake Bay. The yearly migration cycle of *P. mariae-lebouriae* covers a distance of 240 km from its bloom area in the upper bay to the bay mouth, following the circulation cycle of the outflowing surface water and the inflowing bottom water. *P. mariae-lebouriae* in surface outflowing waters at the mouth of the bay is recruited in late winter into more dense inflowing coastal waters. Strong stratification produced by late winter or early spring surface runoff results in the development of a stable pycnocline. *P. mariae-lebouriae*, now in northward-flowing bottom waters, is retained in these bottom waters. It accumulates in a subsurface concentration maximum below the pycnocline and is transported northward to reach its bloom area in the upper portion of the bay by late spring. The rapidly decreasing depth of the upper bay causes the pycnocline to rise, mixing the previously light-limited *P. mariae-lebouriae* and its nutrient-rich bottom waters to the surface, where rapid growth ensues. Once in surface waters positive photophobotaxis, combined with wind- and tide-driven surface convergences, produce dense surface patches or red tides. *P. mariae-lebouriae* is effectively retained in the bay until late winter by sequential inoculation into the tributary estuaries on the western shore, which exchange relatively slowly with bay waters. By late winter the annual cycle is complete. *P. mariae-lebouriae* is again in surface waters at the mouth of the bay where it is reintroduced into northward flowing, bottom waters. This sequence of events is illustrated in Fig. 2.31 for the yearly cycle.

On a smaller temporal and spatial scale Bousfield (1955) observed a similar circulation cycle for the transport of larval stages of the brackish-water barnacle *Balanus improvisus* during its 18-day planktonic life period in the Miramichi Estuary (Canada). The first three nauplius stages, occurring

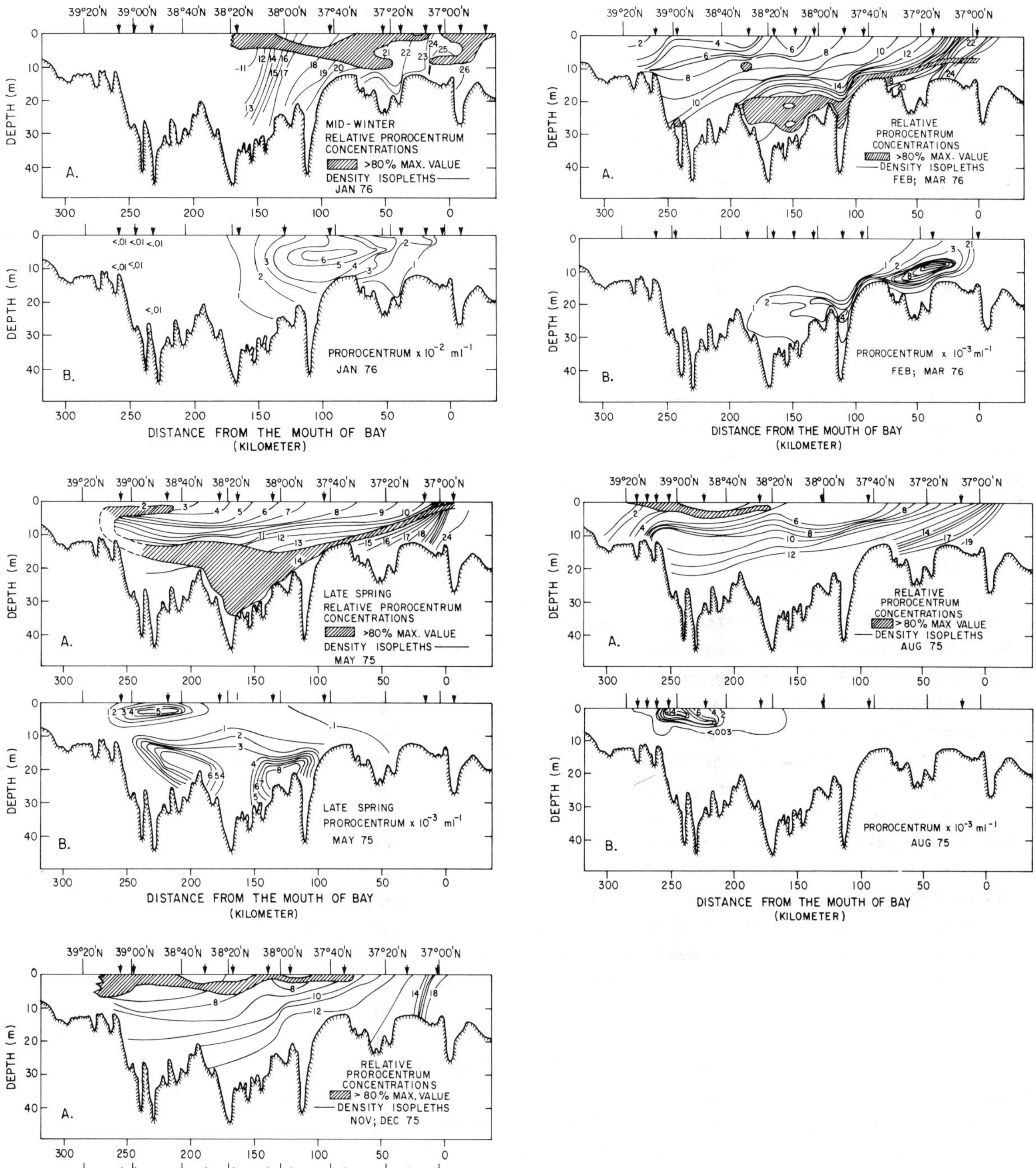

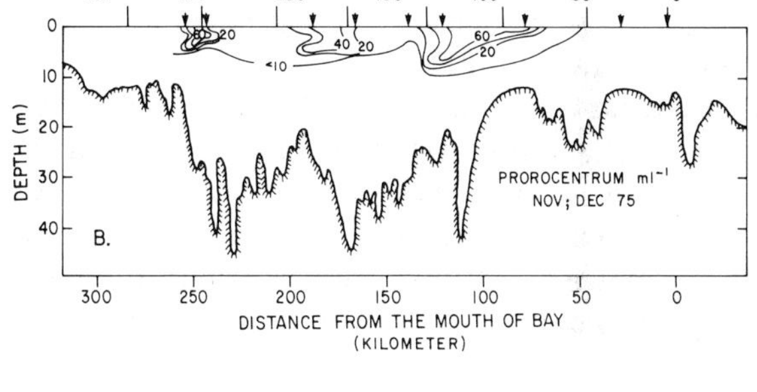

Fig. 2.31. North–south vertical sections along the axis of Chesapeake Bay (U.S.A.). A. Density isopleths (σ_t) and hatched areas indicating location in the vertical water column where *Prorocentrum mariae-lebouriae* is most abundant. B. Optical counts of *P. mariae-lebouriae* corresponding to relative concentrations of A. Data are presented for mid-winter (January 1976); late winter (February, March 1976); late spring (May 1975); summer (August 1975) and late autumn (November, December 1975) conditions. (From Tyler and Seliger, 1978.)

mainly in the surface waters, are transported successively seaward. The fourth and fifth stages, found at mid-depths, are concentrated near the mouth of the estuary. And the sixth stage nauplius and the cyprid, occurring mainly in the bottom waters, are carried progressively landward toward the head of the estuary.

Nutrient traps and turbidity maxima

To conclude, I shall discuss one further estuarine hydrodynamic mechanism which occurs in a number of estuaries and can affect the distribution and transport of certain chemical and geological parameters. For the middle to lower reaches of a well-mixed or stratified estuary, longitudinal density-gradient flow will usually be dominant; and the net bottom current will be in an up-estuary direction. For the upper reaches of an estuary and its associated tributary river, the longitudinal salinity gradient will be negligible, and the flow will be riverine in character and down-estuary at all depths. In addition materials, or their derived products, which are carried seaward by the surface waters may sink during their down-estuary transport to the bottom waters, and be carried back up-estuary. One might anticipate then that the null zone between these two flow conditions could become a high concentration region for constituents or materials that are transported with the water movement, as well as an accumulation zone for the bottom sediment transport. This does, indeed, occur and is referred to as a nutrient trap for the nutrients and as the turbidity maximum for the suspended particulate material.

Redfield (1955; Redfield et al., 1963) discussed such a nutrient trap for the Gulf of Venezuela. Fig. 2.32 is a plot of total inorganic and organic phosphorus along a longitudinal section of the Gulf. The quantity of phosphorus in the deeper water at the head of the Gulf is twice that found in its offshore source. The additional phosphorus is attributed to the sinking of organic material which is carried up-estuary with the bottom water transport.

Postma (1967) and Schubel (1969, 1971) discussed the mechanisms involved in the formation of the turbidity maximum. They attributed the turbidity maximum to a combination of local resuspension of bottom sediments by tidal action in the bottom sediment accumulation zone and a sediment trap. The sediment trap is described in much the same manner as the nutrient trap. Particles entering from the river in the mean down-estuary near-surface flow settle into the near-bottom up-estuary flow in the lower reaches of the estuary, join particles entering from the ocean, and are carried back up-estuary. There the suspended sediment is mixed with the near-surface layer in the vicinity of the null zone, and a continuous exchange of particles takes place. This action forms an effective sorting mechanism; and the size range of particles in the turbidity maximum is narrow, the coarser particles being deposited and the finer ones being swept through. An example of the turbidity maximum and the longitudinal distribution of suspended sediment in an estuary is illustrated in Fig. 2.33, taken from Nichols and Poor (1967) for the Rappahannock Estuary (U.S.A.).

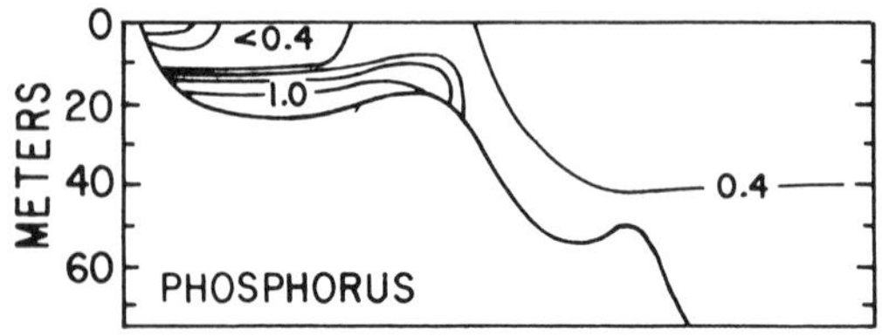

Fig. 2.32. Distribution of total inorganic and organic phosphorus in a section along the axis of the Gulf of Venezuela. Total phosphorus values are in mg-atom m^{-3}. (From Redfield et al., 1963.)

Peterson et al. (1975) described the salient features of the winter turbidity maximum and the summer phytoplankton maximum in the northern reach of San Francisco Bay. An example of their results is shown in Fig. 2.34. The magnitude of the turbidity maximum increases with the increased winter input load of suspended particle; and its longitudinal position varies in response to the tributary river discharge rate. The summer phytoplankton maximum is attributed to the increased residence time in the null zone, which in turn permits increased local growth accumulation of the plankton population.

Festa and Hansen (1978) have used their hydrodynamic numerical model to model the turbidity maximum. Their analysis includes the continuity equation for suspended sediment, with a term representing the settling velocity of the suspended material, and with a river and ocean source of

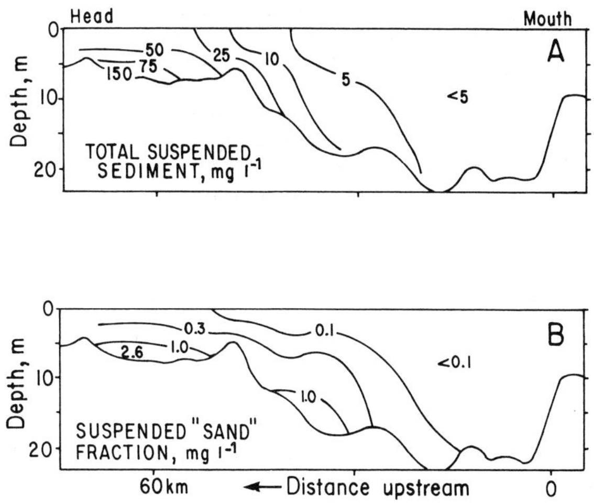

Fig. 2.33. Longitudinal distribution of (A) total suspended sediment and (B) suspended sand fraction greater than 62 μm in the Rappahannock Estuary (Virginia, U.S.A.). Contoured values are in mg l^{-1}. (From Nichols and Poor, 1967.)

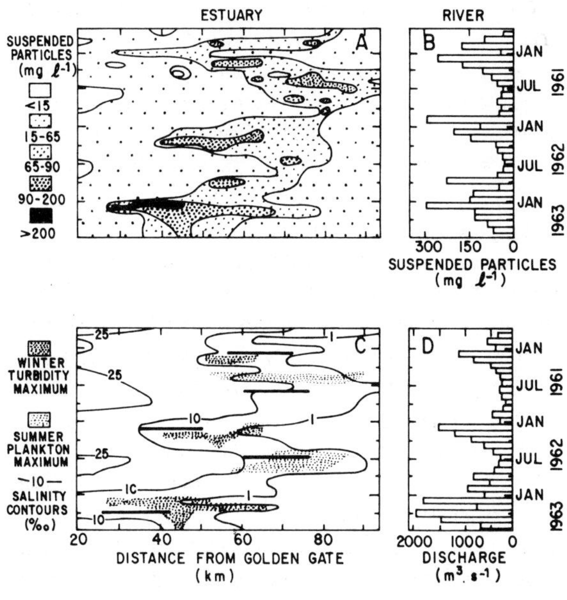

Fig. 2.34. Seasonal distribution of properties in the northern reach of San Francisco Bay, 1961–1963: A. Suspended particle concentration in the estuary. B. Suspended particle concentration in the Sacramento River. C. Salinity at 1 m depth in the estuary, compared with location of suspended particle (turbidity) maximum and approximate location of non-tidal current null zone (solid black lines). D. Combined discharge from the Sacramento and San Joaquin Rivers. (From Peterson et al., 1975.)

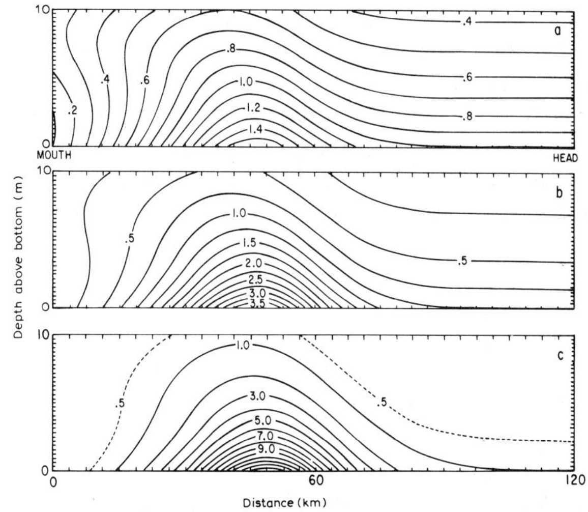

Fig. 2.35. Suspended sediment distributions for particle settling velocities of (a) 1×10^{-3} cm s^{-1}, (b) 2×10^{-3} cm s^{-1}, and (c) 3×10^{-3} cm s^{-1}. The river to ocean source ratio is 10 for each case. (From Festa and Hansen, 1978.)

sediment. An example of their results showing the anticipated variation in the shape and magnitude of the turbidity maximum with changes in the settling velocity is shown in Fig. 2.35.

REFERENCES

Alberts, J., Harriss, R. and Hanke, A., 1970. *Studies on the Geochemistry and Hydrography of the Charlotte Harbor Estuary, Florida.* Progress Report No. 2, Mote Marine Laboratory, Sarasota and Placida, Fla.

Anderson, J.J. and Devol, A.H., 1973. Deep water renewal in Saanich inlet, an intermittently anoxic basin. *Estuarine Coastal Mar. Sci.*, 1: 1–10.

Blumberg, A.F., 1977. A two dimensional numerical model for the simulation of partially mixed estuaries. In: M. Wiley (Editor), *Estuarine Processes, 2.* Academic Press, New York, N.Y., pp. 323–331.

Bousfield, E.L., 1955. Some physical features of the Miramichi estuary. *J. Fish. Res. Board Can.*, 12: 342–361.

Bowden, K.F., 1960. Circulation and mixing in the Mersey estuary. *Int. Assoc. Sci. Hydrol.*, 51: 352–360.

Bowden, K.F., 1962. The mixing processes in a tidal estuary. In: *Advances in Water Pollution Research. Proc. First Int. Conf. Water Pollution Research*, 3: 329–346.

Bowden, K.F., 1963. The mixing processes in a tidal estuary. *Int. J. Air Water Pollut.*, 7: 343–356.

Bowden, K.F., 1977. Turbulent processes in estuaries. In: *Estuaries, Geophysics and the Environment.* National Research Council, Geophysics of Estuaries Panel, National Academy of Sciences, Washington, D.C., pp. 46–56.

Bowden, K.F. and Hamilton, P., 1975. Some experiments with a numerical model of circulation and mixing in a tidal estuary. *Estuarine Coastal Mar. Sci.*, 3: 281–301.

Boyle, E., Collier, R., Dengler, A.T., Edmond, J.M., Ng, A.C. and Stallard, R.F., 1974. On the chemical mass balance in estuaries. *Geochim. Cosmochim. Acta*, 38: 1719–1728.

Bumpus, D.F., Butcher, W.S., Athearn, W.D. and Day, G.C., 1953. *Inshore Survey Project, Boston, Final Harbor Report.* Technical Report, W.H.O.I. Reference No. 53–20, Woods Hole Oceanographic Institution, Woods Hole, Mass.

Cameron, W.M. and Pritchard, D.W., 1965. Estuaries. In: M.N. Hill (Editor), *The Sea, 2.* Wiley and Sons, New York, N.Y., pp. 306–324.

Cannon, G.A., 1975. Observations of bottom water flushing in a fjord-like estuary. *Estuarine Coastal Mar. Sci.*, 3: 95–102.

Carpenter, J.H. and Seliger, H.H., 1968. Studies at Oyster Bay in Jamaica, West Indies, II, Effects of flow patterns and exchange on bioluminescent distributions. *J. Mar. Res.*, 26: 256–272.

Dyer, K.R., 1973. *Estuaries: A Physical Introduction.* Wiley and Sons, London, 140 pp.

Dyer, K.R., 1974. The salt balance in stratified estuaries. *Estuarine Coastal Mar. Sci.*, 2: 273–281.

Dyer, K.R., 1977. Lateral circulation effects in estuaries. In: *Estuaries, Geophysics and the Environment.* National Research Council, Geophysics of Estuaries Panel, National Academy of Sciences, Washington, D.C., pp. 22–29.

Dyer, K.R. and Ramamoothy, K., 1969. Salinity and water circulation in the Vellar estuary. *Limnol. Oceanogr.*, 14: 4–15.

Ebbesmeyer, C.C., Barnes, C.A. and Langley, C.W., 1975. Application of an advective-diffusion equation to a water parcel observed in a fjord. *Estuarine Coastal Mar. Sci.*, 3: 249–268.

Elliott, A.J., 1976a. *A Study of the Effect of Meteorological Forcing on the Circulation in the Potomac Estuary.* Special Report 56, Chesapeake Bay Institute, The Johns Hopkins University, Baltimore, Md.

Elliott, A.J., 1976b. Response of the Patuxent estuary to a winter storm. *Chesapeake Sci.*, 17: 212–216.

Elliott, A.J., 1977. Circulation models. In: M. Wiley (Editor), *Estuarine Processes, 2.* Academic Press, New York, N.Y., pp. 321–322.

Elliott, A.J. and Hendrix, T.E., 1976. *Intensive Observations of the Circulation in the Potomac Estuary.* Special Report 55, Chesapeake Bay Institute, The Johns Hopkins University, Baltimore, Md.

Festa, J.F. and Hansen, D.V., 1976. A two dimensional numerical model of estuarine circulation: The effects of altering depth and river discharge. *Estuarine Coastal Mar. Sci.*, 4: 309–323.

Festa, J.F. and Hansen, D.V., 1978. Turbidity maxima in partially mixed estuaries: A two dimensional numerical model. *Estuarine Coastal Mar. Sci.*, 7: 347–359.

Fukushima, H., Kashiwamura, M. and Yakuwa, I., 1966. Studies on salt wedge by ultrasonic method. In: American Society of Civil Engineers, *Proceedings of the Tenth Conference on Coastal Engineering*, pp. 1435–1447.

Fukushima, H., Yakuwa, I. and Takahashi, S., 1969. Salinity diffusion at the interface of stratified flow in an estuary. In: International Association for Hydraulic Research, *Proceedings of the Thirteenth Conference*, 3: 191–197.

Garvine, R.W., 1974. Physical features of the Connecticut river outflow during high discharge. *J. Geophys. Res.*, 79: 831–846.

Garvine, R.W., 1977. River plumes and estuary fronts. In: *Estuaries, Geophysics and the Environment.* National Research Council, Geophysics of Estuaries Panel, National Academy of Sciences, Washington, D.C., pp. 30–35.

Garvine, R.W. and Monk, J.D., 1974. Frontal structure of a river plume. *J. Geophys. Res.*, 79: 2251–2259.

Hamilton, P., 1975. A numerical model of the vertical circulation of tidal estuaries and its application to the Rotterdam waterway. *Geophys. J. R. Astron. Soc.*, 40: 1–21.

Hamilton, P., 1977. On the numerical formulation of a time dependent multilevel model of an estuary with particular reference to boundary conditions. In: M. Wiley (Editor), *Estuarine Processes, 2.* Academic Press, New York, N.Y., pp. 347–364.

Hansen, D.V., 1965. Currents and mixing in the Columbia river estuary. In: Marine Technology Society and American Society of Limnology and Oceanography, *Joint Conference on Ocean Science and Ocean Engineering*, pp. 943–955.

Hansen, D.V. and Rattray, M., 1965. Gravitational circulation in straits and estuaries. *J. Mar. Res.*, 23: 104–122.

Hansen, D.V. and Rattray, M., 1966. New dimensions in estuary classification. *Limnol. Oceanogr.*, 11: 319–326.

Hesslein, R. and Quay, P., 1973. Vertical eddy diffusion in the thermocline of a small stratified lake. *J. Fish. Res. Board Can.*, 30: 1491–1500.

Hobbie, J.E., 1976. Nutrients in estuaries. *Oceanus*, 19(5): 41–47.

Hughes, P., 1958. Tidal mixing in the narrows of the Mersey estuary. *Geophys. J. R. Astron. Soc.*, 1: 271–283.

Hutchinson, G.E., 1957. *A Treatise on Limnology, 1.* Wiley and Sons, New York, N.Y., 540 pp.

Ketchum, B.H., 1950. Hydrographic factors involved in the dispersion of pollutants introduced into tidal waters. *J. Boston Soc. Civil Eng.*, 37: 296–314.

Ketchum, B.H., 1951a. The exchange of fresh and salt waters in tidal estuaries. *J. Mar. Res.*, 10: 18–37.

Ketchum, B.H., 1951b. The flushing of tidal estuaries. *Sewage Ind. Wastes*, 23: 198–209.

Ketchum, B.H., 1952. *The Distribution of Salinity in the Estuary of the Delaware River.* Technical Report, W.H.O.I. Reference No. 52–103, Woods Hole Oceanographic Institution, Woods Hole, Mass.

Ketchum, B.H., 1954. Relation between circulation and planktonic populations in estuaries. *Ecology*, 35: 191–200.

Ketchum, B.H. and Keen, D.J., 1953. The exchanges of fresh and salt waters in the Bay of Fundy and in Passamaquoddy bay. *J. Fish. Res. Board Can.*, 10: 97–124.

Ketchum, B.H., Redfield, A.C. and Ayers, J.P., 1951. The oceanography of the New York bight. *Pap. Phys. Oceanogr. Meteorol.*, 12(1): 1–46.

Knudsen, M., 1901. *Hydrographical Tables.* G.E.C. Gad, Copenhagen, 63 pp. (also reprinted by Woods Hole Oceanographic Institution, Woods Hole, Massachusetts, U.S.A., 1946.)

Nichols, M. and Poor, G., 1967. Sediment transport in a coastal plain estuary. *Am. Soc. Civil Eng., J. Waterways Harbor Div.*, 93, WW4: 83–95.

O'Brien, M.P., 1952. Salinity currents in estuaries. *Trans. Am. Geophys. Union*, 33: 520–522.

O'Connor, D.J., Thomann, R.V. and DiToro, D.M., 1977. Water quality analyses of estuarine systems. In: *Estuaries, Geophysics and the Environment.* National Research Council, Geophysics of Estuaries Panel, National Academy of Sciences, Washington, D.C., pp. 71–83.

Officer, C.B., 1976. *Physical Oceanography of Estuaries and Associated Coastal Waters.* Wiley and Sons, New York, N.Y., 465 pp.

Officer, C.B., 1977. Longitudinal circulation and mixing relations in estuaries. In: *Estuaries, Geophysics and the Environment.* National Research Council, Geophysics of Estuaries Panel, National Academy of Sciences, Washington, D.C., pp. 13–21.

Officer, C.B. and Ryther, J.H., 1977. Secondary sewage treatment versus ocean outfalls: An assessment. *Science*, 197: 1056–1060.

Peterson, D.H., Conomos, T.J., Broenkow, W.W. and Scrivani, E.P., 1975. Processes controlling the dissolved silica in San Francisco bay. In: L.E. Cronin (Editor), *Estuarine Research, 1.* Academic Press, New York, N.Y., pp. 153–187.

Peterson, D.H., Festa, J.F. and Conomos, T.J., 1978. Numerical simulation of dissolved silica in the San Francisco bay. *Estuarine Coastal Mar. Sci.*, 7: 99–116.

Pickard, G.L., 1961. Oceanographic features of inlets in the British Columbia mainland coast. *J. Fish. Res. Board Can.*, 18: 907–999.

Pickard, G.L. and Rodgers, K., 1959. Current measurements in Knight inlet, British Columbia. *J. Fish. Res. Board Can.*, 16: 635–678.

Postma, H., 1967. Sediment transport and sedimentation in the estuarine environment. In: G.H. Lauff (Editor), *Estuaries. Am. Assoc. Adv. Sci. Publ.*, 83: 158–179.

Pritchard, D.W., 1952a. Estuarine hydrography. *Adv. Geophys.*, 1: 243–280.

Pritchard, D.W., 1952b. Salinity distribution and circulation in the Chesapeake bay estuarine system. *J. Mar. Res.*, 11: 106–123.

Pritchard, D.W., 1954. A study of the salt balance in a coastal plain estuary. *J. Mar. Res.*, 13: 133–144.

Pritchard, D.W., 1967. Observations of circulation in coastal plain estuaries. In: G.H. Lauff (Editor), *Estuaries. Am. Assoc. Adv. Sci. Publ.*, 83: 37–44.

Rattray, M., 1967. Some aspects of the dynamics of circulation in fjords. In: G.H. Lauff (Editor), *Estuaries. Am. Assoc. Adv. Sci. Publ.*, 83: 52–62.

Rattray, M., 1977. Fjord and salt wedge circulation. In: *Estuaries, Geophysics and the Environment.* National Research Council, Geophysics of Estuaries Panel, National Academy of Sciences, Washington, D.C., pp. 36–45.

Rattray, M. and Hansen, D.V., 1962. A similarity solution for circulation in an estuary. *J. Mar. Res.*, 20: 121–133.

Rattray, M. and Mitsuda, E., 1974. Theoretical analysis of conditions in a salt wedge. *Estuarine Coastal Mar. Sci.*, 2: 375–394.

Redfield, A.C., 1955. The hydrography of the Gulf of Venezuela. *Deep-Sea Res.*, 3 (suppl.): 115–133.

Redfield, A.C., Ketchum, B.H. and Richards, F.A., 1963. The influence of organisms on the composition of sea water. In: M.N. Hill (Editor), *The Sea, 2.* Wiley and Sons, New York, N.Y., pp. 26–77.

Riley, G.A., 1952. Hydrography of Long Island and Block Island sounds. *Bull. Bingham Oceanogr. Coll.*, 13: 5–39.

Riley, G.A., 1955a. Oceanography of Long Island sound, I, Introduction. *Bull. Bingham Oceanogr. Coll.*, 15: 9–14.

Riley, G.A., 1955b. Oceanography of Long Island sound, II, Physical oceanography. *Bull. Bingham Oceanogr. Coll.*, 15: 15–46.

Saelen, O.H., 1967. Some features of the hydrography of Norwegian fjords. In: G.H. Lauff (Editor), *Estuaries. Am. Assoc. Adv. Sci. Publ.*, 83: 63–70.

Schubel, J.R., 1969. Size distributions of the suspended particles of the Chesapeake bay turbidity maximum. *Netherlands J. Sea Res.*, 4: 283–309.

Schubel, J.R., 1971. Tidal variation of the size distribution of suspended sediment at a station in the Chesapeake bay turbidity maximum. *Netherlands J. Sea Res.*, 5: 252–266.

Seliger, H.H., Carpenter, J.H., Loftus, M. and McElroy, W.D., 1970. Mechanisms for the accumulation of high concentrations of dinoflagellates in a bioluminescent bay. *Limnol. Oceanogr.*, 15: 234–245.

Stommel, H., 1953. Computation of pollution in a vertically mixed estuary. *Sewage Ind. Wastes*, 25: 1065–1071.

Stommel, H. and Farmer, G.H., 1952. *On the Nature of Estuarine Circulation, Part I.* Technical Report, W.H.O.I. Reference No. 52–88, Woods Hole Oceanographic Institution, Woods Hole, Mass.

Stumm, W. and Morgan, J.J., 1970. *Aquatic Chemistry.* Wiley and Sons, New York, N.Y., 583 pp.

Szekielda, K.-H., Kupferman, S.L., Klemas, V. and Polis, D.F., 1972. Element enrichment in organic films and foam associated with aquatic frontal systems. *J. Geophys. Res.*, 77: 5278–5282.

Tyler, M.A. and Seliger, H.H., 1978. Annual subsurface transport of a red tide dinoflagellate to its bloom area: Water circulation patterns and organism distributions in the Chesapeake bay. *Limnol. Oceanogr.*, 23: 227–246.

U.S. Navy Hydrographic Office, 1952. *Tables for Sea Water Density.* Hydrographic Office Publication No. 615, Washington, D.C., 265 pp.

U.S. Navy Hydrographic Office, undated. *Nomograph for the Determination of Density, 1 to 6.* Hydrographic Office Miscellaneous Publication 15530.

Wang, D.-P., and Elliott, A.J., 1978. Nontidal variability in the Chesapeake bay and Potomac river: Evidence for nonlocal forcing. *J. Phys. Oceanogr.*, 8: 225–232.

Weisberg, R.H., 1976. The nontidal flow in the Providence river of Narragansett bay: A stochastic approach to estuarine circulation. *J. Phys. Oceanogr.*, 6: 721–734.

Weisberg, R.H. and Sturges, W., 1976. Velocity observations in the West passage of Narragansett bay: A partially mixed estuary. *J. Phys. Oceanogr.*, 6: 345–354.

Winter, D.F., 1973. A similarity solution for steady state gravitational circulation in fjords. *Estuarine Coastal Mar. Sci.*, 1: 387–400.

Chapter 3

RESPONSES TO ESTUARINE STRESS

WINONA B. VERNBERG

INTRODUCTION

Estuarine animals are recruited mainly from the sea. Many species that live in the shallow water zone on the continental shelf use the estuary as a nursery ground before migrating to the open sea. While estuarine species are few, those that have become adapted to this fluctuating environment are present in large numbers. Before describing the physiological adaptations of the estuarine organisms, it would be appropriate to look briefly at the estuarine environment.

One of the chief characteristics of an estuary is fluctuation in salinity. Typically the water is fresher at the surface, and more saline on the bottom. Tidal changes also affect salinity by increasing salinity on a flood tide, while decreasing salinity on an ebb tide. It is not unusual for changes in salinity to be sudden and dramatic. For example, Ferguson et al. (1981) reported a change in estuarine salinity from 30 to 10‰ within a 1-h period. In addition, there may be seasonal variations in salinity, so that the water may be nearly fresh during winter and spring rains, but strongly saline in summer. Temperature fluctuations in the estuarine system also tend to be greater than those of either offshore waters or the open ocean, since inflowing river water tends to be colder than sea water in winter and warmer in summer.

Generally estuaries are turbid due to the amounts of silt in the water, and light cannot penetrate to as great a depth as it does in coastal and open-ocean waters. Therefore, estuarine animals cannot depend on visual cues for feeding, environmental selection, or in finding a mate to the extent that open-ocean organisms can. The substratum in an estuary is nearly always covered with silt. Mud is the most common type of bottom, and while the bottom muds are rich in organic detritus (derived primarily from the vegetation along the upper tide levels) conditions are often anoxic a few centimeters beneath the surface of the substrate.

In any environment an animal has two alternative reactions when encountering a stressful situation: it can either migrate to a more favorable environment, or it can remain and adapt. If conditions become too stressful, of course, the animal will die. Environmental stresses need not be only physical in nature, since all organisms must compete intra-specifically and inter-specifically for resources such as space and food. For each species there is a certain sector of the total range of abiotic and biotic factors that is compatible with life. At either end of this gradient, however, there is a point beyond which an organism cannot survive, and beyond that point exists the organism's lethal zone (Fig. 3.1). The adaptations of these tolerance limits are referred to as resistance adaptations. Between

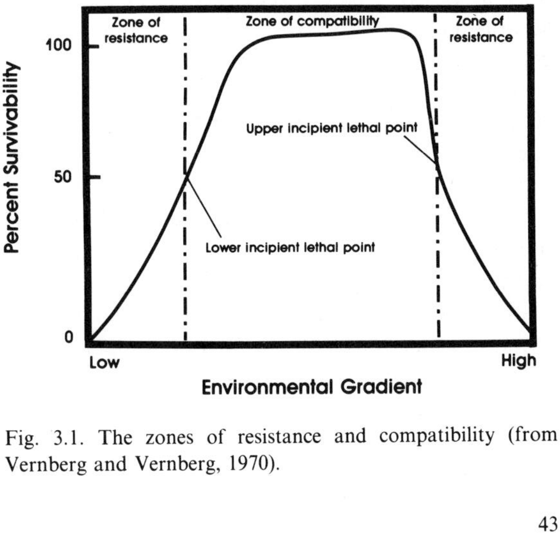

Fig. 3.1. The zones of resistance and compatibility (from Vernberg and Vernberg, 1970).

the upper and lower tolerance limits, the adaptations of organisms are termed capacity adaptations. The dividing line between the zone of resistance and the zone of compatibility is not simply determined since numerous intra- and extra-organismic factors are involved, such as stage in the life cycle, body size, temperature, salinity, oxygen, desiccation and the possible interaction of physical factors.

The response of an animal to any given set of environmental factors may reflect an altered physiological state, which is termed acclimation, and is dependent upon the conditions to which the animal has been exposed. When the environment changes, the animal readjusts by altering its physiological response; however, the range of these environmentally induced responses to changes is determined by the genetic composition of the organism and is not limitless.

RESISTANCE ADAPTATIONS

Salinity

While it is true that estuarine organisms characteristically are euryhaline, a critical salinity boundary, 5 to 8‰, separates fresh-water and marine faunas. This critical salinity principle, first suggested by Remane (1934) and later documented by Khlebovich (1969), is based on a number of physiological parameters including survival, growth, locomotion and osmoregulation. For example, the body fluid concentration of brackish-water animals may be reduced to a salinity of about 5‰ before serious damage results, but below 5‰ distortion of cellular electro-chemical properties occurs and the tissue albumin fraction undergoes a marked change. Above 5 to 8‰ salinity poikilo-osmotic organisms can survive; below this concentration hyperosmotic regulation is required.

There are a number of generalizations that can be made about responses of estuarine organisms to salinity: (1) those organisms living in estuaries subjected to wide salinity fluctuations can withstand a wider range of salinities than species that occur in high-salinity estuaries; (2) intertidal zone animals tend to tolerate wider ranges of salinities than do subtidal and open-ocean organisms; (3) low intertidal species are less tolerant of low salinities than are high intertidal ones; and (4) more sessile animals are likely to be more tolerant of fluctuating salinities than those organisms which are highly mobile and capable of migrating during times of salinity stress. Some estuarine animals, such as one species of fiddler crab (*Uca rapax*) can survive extremely hypersaline as well as hyposaline conditions. This crab is commonly found living on the salt flats of Puerto Rico in salinities as high as 90‰.

Many studies on the lethal limits of estuarine organisms have been conducted in constant salinities, but this may not be the best method to establish mortality patterns. In a study on the oyster drill (*Urosalpinx cinerea*), Zachary and Haven (1973) demonstrated that exposure to constant salinity had a different effect than did exposure to fluctuating low salinities. At any constant low salinity value, the stress on the drill was immediate, and as a result, initial mortalities were very high. In contrast, at similar low fluctuating salinities, the initial 10-day exposure produced little or no mortality; it was only after longer exposure that mortality began to increase. The authors speculated that the initial low-mortality period may reflect the fact that, at low fluctuating salinities, stress is not continuous but is interrupted by periods of higher, less detrimental salinities. Such stress tolerance would be of great adaptive significance to an estuarine organism, since ability to survive in temporarily low but fluctuating salinities could afford the animal a greater chance to survive until salinities returned to normal.

Often there is a difference in the tolerance limits of larval stages compared to those of adults of the same species. For example, adult fiddler crabs (*Uca pugilator*) can survive prolonged periods at 5‰ salinity. Larval stages, on the other hand, cannot tolerate exposure to salinities below 20‰ (Vernberg and Vernberg, 1972b; Vernberg et al., 1974). There is some evidence that salinity tolerance of progeny may depend upon the salinity to which the parent stock is adapted. For example, offspring of the amphipod *Gammarus duelieni*, reared at a salinity of 2‰, tolerate transfer to fresh water better than do offspring from different stock grown at either 30‰ or 40‰ (Kinne, 1953). Tolerance responses of larvae can be influenced also by the salinity in which they are spawned. An illustration of this phenomenon can be found in a recent study by

Newkirk et al. (1977) on the larvae of the American oyster (*Crassostrea virginica*). This species can be found in waters ranging from virtually oceanic to brackish. Tolerance studies of larvae transferred from different habitats to low-salinity water demonstrate that larvae which were spawned in low-salinity waters had the highest survival rates. Through cross-fertilization studies the authors presented evidence suggesting that both a cytoplasmic and a genetic factor are involved in the tolerance limits of the oyster larvae to low-salinity water.

Many adult estuarine organisms have the capacity to acclimate to low salinity, and there are many examples in the literature indicating that animals acclimated to low salinity generally have greater resistance to low salinity than animals acclimated to high salinity (see reviews by Kinne, 1970; and Theede, 1975).

In addition to observed organismic effects, it has also been shown that altered salinity can influence the lethal limits of isolated tissues. The intact stenohalinic bay scallop (*Argopecten irradians*) does not survive well in salinities below 14‰, and pieces of its gill tissue also die at this salinity. In contrast, two euryhaline bivalves, the American oyster (*Crassostrea virginica*) and the ribbed mussel (*Modiolus demissus*), are able to tolerate much lower salinities. The adult oyster dies at 7‰ and gill tissues at 4‰; the ribbed mussel dies at 5‰ and its gill tissues at 2 to 3‰ (Vernberg et al., 1963).

Temperature

Since temperature as well as salinity often fluctuates in estuarine situations, it is not surprising that temperature has an effect on salinity tolerance. As a generalization, it may be stated that tropical species can withstand low salinity best at high temperatures while cold-water species can tolerate low salinities best at low temperatures. Many estuarine organisms are well enough adapted to these fluctuations so that stressful interaction between temperature and salinity is observed only at the extremes. The estuarine intertidal hermit crab (*Clibanarius vittatus*) for example, can tolerate wide ranges of both temperature and salinity (90% survival after 48-h exposure to temperatures from 5 to 35°C and salinities from 2 to 40‰). The response surface curves generated for 48-h exposure to combinations of temperatures and salinity are shown in Fig. 3.2 (A. Young, pers. comm., 1978). Most of the observed mortality proved to be due to temperature (55%) or salinity (25%); temperature–salinity interaction accounted for only a small fraction (18%) of mortality.

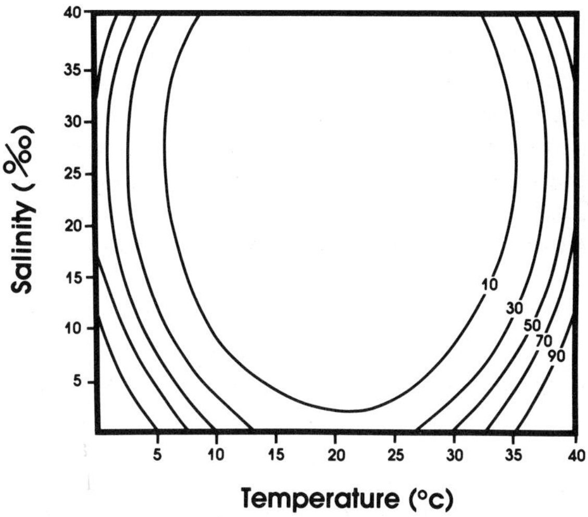

Fig. 3.2. Estimation of percentage mortality in *Clibanarius vittatus* acclimated to 25°C based on fitted response surface to observed mortality after 48 h (from A. Young, pers. comm., 1978).

In some estuarine animals their distribution within a particular habitat can be correlated with temperature–salinity tolerances. In a study on the mystacocarid *Derocheilocaris typica*, Crouse and Found (1975) reported that survival was lowest under conditions of high temperature and low salinity. At any given salinity, highest survival was observed at 7°C, while at any single temperature highest tolerance levels were noted when the animals were maintained at 31‰. Distribution of this species has been reported from intertidal water which ranges in temperature from 1.5°C in Massachusetts to 28°C in Florida and salinities ranging from 11 to 36‰. These observations, plus temperature–salinity studies, led these authors to suggest that the depth at which the mystacocarid population is found **both seasonally and geographically** may reflect a selection process which permits these organisms to avoid temperature and salinity extremes.

Another example of the correlation between thermal–haline tolerance and the ecology of a species can be found in the work of Kenny (1969) on the polychaete *Clymenella torquata*. This species shows a greater tolerance to low salinities at low temperatures and it is abundant in estuarine systems where the lowest salinities typically are found during the winter.

Animals from colder geographical areas are generally less heat-tolerant and more cold-tolerant than animals from warmer regions. Within an estuary, temperature tolerance of a species may be further correlated with its habitat; animals from the upper regions of the intertidal zone are often or usually more thermally resistant than organisms inhabiting the deeper channels. The species composition of an estuary, especially in the temperate zone, may change seasonally, apparently in response to temperature. For example, during the winter northern-affinity crustaceans from offshore waters will often migrate into more southerly located estuaries, but leave when the waters warm up; other species then occupy their vacated niches.

Resistance to thermal extremes has been shown to be influenced both by salinity and by previous thermal history. Acclimation to high salinity can increase the resistance of estuarine animals to temperature extremes. Acclimation to temperature not only increases thermal resistance in a predictable manner, but it also influences the effect of salinity on heat resistance. For example, the beneficial effect of high salinity is reduced when animals are acclimated to higher temperatures. Furthermore, specific ions influence thermal resistance, and variability in the ionic composition of the aquatic medium seems to play a role in thermal resistance over a range of salinities. Calcium is one ion which is known to influence osmoregulation, and it seems to modify the thermal response, although not in a predictable manner (Vernberg and Silverthorn, 1979).

The effect of temperature may be greatly modified by the stage in the life cycle, as can be illustrated by the studies on larvae and adults of fiddler crabs, genus *Uca*. Larvae of tropical fiddler crabs can survive low temperature better than adults (Vernberg, 1969), while larvae of temperate zone species show the opposite response (Vernberg and Vernberg, 1975).

Often the temperature tolerance of larvae can be correlated with time of release into the water column, so that larvae normally released during the winter or spring months are more tolerant to cold temperatures and less tolerant to warm temperatures than are larvae released during the summer months.

Oxygen

Many estuarine habitats experience periods of oxygen deprivation, and in many estuarine systems the oxygen concentration in the lower part of the water column is consistently less than in water above. This is a result of heavier, more saline water from the ocean wedging underneath the less saline surface waters from fresh-water runoff. Sometimes there is little or no mixing between the two layers. The concentration of the bottom layer is effectively sealed off from the atmosphere, and the oxygen that is present is significantly reduced by biological activity. Thus, it is not surprising that many estuarine organisms have the capability to survive low oxygen concentrations.

Burrowing animals can tolerate anoxic conditions for prolonged periods of time as has been documented in many studies. One example of such a study can be found in the work of Thompson and Pritchard (1969) on two burrowing crustaceans, *Callianassa californiensis* and *Upogebia pugetensis*. These two estuarine species were found to be resistant to anoxia for an average time of 138 and 81 h respectively. This prolonged resistance permits these species to survive for a number of tidal changes, during which time the oxygen levels may increase. The interspecific difference in resistance capacity can be correlated with habitat difference. *Upogebia* burrows are found at or slightly below the low tide mark, while those of *Callianassa* are in the intertidal zone where they are exposed longer to anoxic conditions.

There is evidence suggesting that species which are highly mobile are less tolerant to oxygen lack than are less agile species which may be occasionally trapped in low oxygen environments. This is well illustrated in a study by Wieser and Kanwisher (1959). One group of animals with low mobility and no special respiratory organs were compared with three highly agile species of amphipods which are equipped with gills. In laboratory studies these workers found that the nematodes and the mites became quiescent when oxygen was removed, but even after a 12-h exposure these animals recovered when oxygen was made available. This response correlates well with their mode of existence, since these animals are often trapped under large seaweeds at low tide and, during the time span between high and low tide, oxygen tensions drop to very low levels. In contrast, the amphipods quickly became inactive under anoxic stress and died. To insure survival these animals must escape the seaweed habitat when the tide recedes and the oxygen levels drop.

Since the larvae of most species of estuarine organisms are planktonic, they are often less tolerant of oxygen lack than are the adults. Larvae of fiddler crabs (*Uca pugilator*) for example, are sensitive to reduced oxygen levels, while the adults can survive anoxia for long periods of time (Teal and Carey, 1967). The adaptive nature of these responses is obvious in that the larvae live in the well-oxygenated water column while the adults normally experience low oxygen conditions in their burrows in the marshes and sand beaches. Young larvae of the mud-flat snail (*Nassarius obsoletus*) are much more sensitive to lack of oxygen than older ones. The change in sensitivity becomes apparent between seven and nine days after hatching. This stage immediately precedes that at which the larvae begin alternately swimming and creeping over the bottom surface in search of a suitable substratum. Thus, the adaptation to lowered oxygen levels could indicate that these older larvae move downward in the water column to less oxygenated water, and the observed increased tolerance to reduced oxygen levels would be of adaptive significance (Vernberg and Vernberg, 1975).

Subtidally, the lack of oxygen and/or low oxygen concentration can limit the vertical distribution of animals. Numerous investigators have studied the vertical distribution of subtidal meiofaunal assemblages and have determined that about 70% of the fauna are located in the upper 2 cm, with 95% in the upper 5 cm. Elmgren (1975) demonstrated that the meiofaunal communities in the Baltic were distinctly zoned by depth and oxygen content; the deep oxygen-poor zone was extremely impoverished with only a few nematode species present.

Low dissolved-oxygen concentrations can have a profound effect on temperature–salinity tolerance in some organisms. In a study on the sand shrimp, *Crangon septemspinosa* under different temperature–salinity conditions, Haefner (1969, 1970) found that 100% survival occurred over a relatively wide temperature–salinity range, from 4 to 22°C and 20 to 40‰. In comparable mortality studies but under low dissolved oxygen concentrations, temperature–salinity tolerances were greatly narrowed, and high survival was observed only at low temperature and high salinity (Fig. 3.3). Thus,

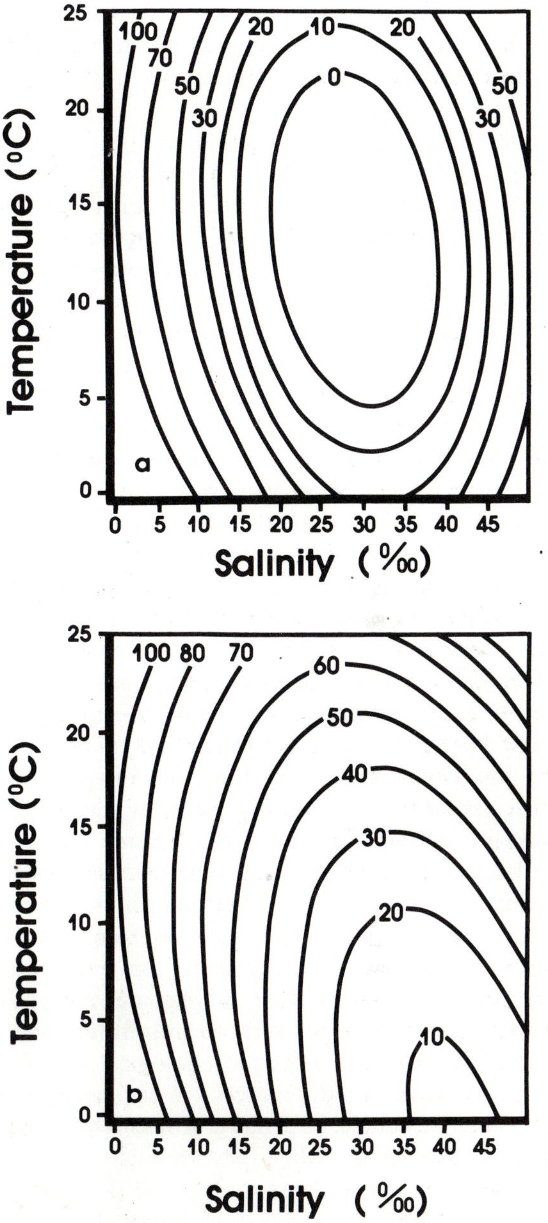

Fig. 3.3. Estimation of percentage mortality of ovigerous *Crangon. septemspinosa*, based on response surface fitted to observed mortality under twelve combinations of salinity and temperature (upper figure) in aerated water, and (lower figure) at low concentrations of dissolved oxygen (from Haefner, 1970).

a low dissolved oxygen content can markedly influence the survival of these estuarine species which require relatively high oxygen levels.

Desiccation

Desiccation may be a major problem for sessile animals living near the high tide mark in the intertidal zone, although it is not a serious problem for most estuarine organisms, nor for mobile intertidal organisms. Many sessile animals are able to tolerate desiccation for long periods of time, and their resistance to desiccation appears to be related to vertical zonation. Brown (1960) analyzed three factors that influenced desiccation in relation to vertical distribution of six species of intertidal gastropods: (1) rate of water loss; (2) amount of water loss a mollusc could survive; and (3) the length of time the animals could tolerate various saturation deficiencies. Another study suggests that the upper level of distribution of intertidal organisms is determined by the interplay between rate of water loss and the time required to regain the water lost when the organism is again covered by the tide (Davie, 1969). Generally, small organisms are less resistant to water loss because of the greater ratios of surface area to volume. Some slower-moving animals must also be able to resist desiccation when the tide recedes and they are exposed to air. Certain intertidal-zone fish, for example, can survive out of water for periods of up to several days (Gibson, 1969). Their survival depends on keeping the body, particularly the respiratory surfaces, wet, and some have morphological modifications that aid in this process. Two species of *Adamia*, for example, have fine grooves in their skin which hold moisture when they are out of water. Other species rely on behavioral patterns, such as burrowing in the moist substrate. Survival time in some of these species is greatly reduced in completely dry air, but is unaffected when relative humidity is 35% or higher (Gibson, 1969).

STRATEGIES FOR CHEMICAL REGULATION AND EXCRETION

The body fluids of many animals living in the open oceans have the same osmotic content as the surrounding sea water, and, since this environment remains relatively constant, these animals do not have to face the problem of water balance. Estuarine animals, on the other hand, must have some physiological means to adjust to alternating high and low salinities. Otherwise, their tissues and cells would tend to lose ions and/or gain water when they encounter low-salinity water, or become very dehydrated through osmotic loss in hypersaline water.

One solution to flooding and loss of electrolytes is morphological, for as a general rule estuarine organisms tend to have body surfaces that show a marked decrease in permeability over those of oceanic ones (Fig. 3.4). Indeed, it has been demonstrated that crabs and polychaetes from fresh water lose less salt and less water through their body covering than estuarine species, which in turn are more resistant than marine species (Lockwood, 1964; Oglesby, 1965; Herreid, 1969; Smith, 1970a, b). To achieve this impermeability estuarine animals may have increased calcium deposits in the exoskeleton or an increased number of mucous glands, or they may show morphological adaptations such as hard-shell structures to protect and isolate the respiratory surface from adverse environmental conditions. Recently evidence has

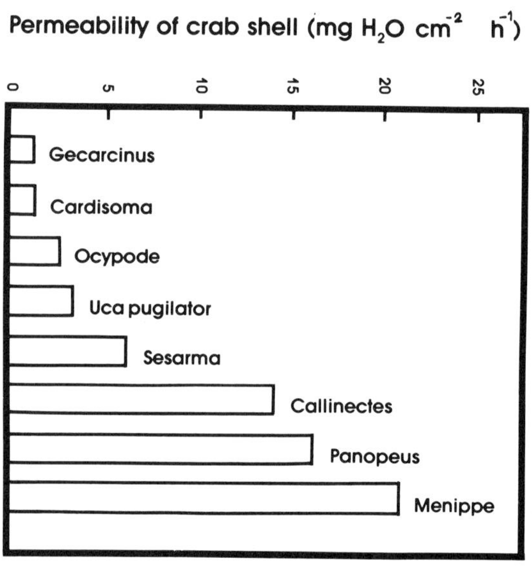

Fig. 3.4. Permeability of crab shells to water in air. Land crabs are on the left, aquatic species on the right. An average of six determinations was made for each species. The average standard deviation among the measurements for each species was ±30% of the mean value. (From Herreid, 1969.)

been presented that an individual animal may adaptively vary its own apparent water permeability or water exchange rate in response to changing salinity (Smith, 1976). The short-term changes in apparent water permeability appear to be related to circulatory adjustments. Cornell, as quoted in Smith (1976), studied a non-osmoregulating spider crab, *Libinia emarginata*, which can tolerate 80% sea water. At this salinity Cornell was able to demonstrate a reduction in apparent water permeability and at the same time an increase in urinary output. Smith has emphasized that it is important to use the term "apparent water permeability" since changes in permeability *per se* may not be involved.

There is evidence from both laboratory and field studies that salinity can influence body shape in such a way as to minimize interaction with low-salinity waters. When the brackish-water hydroid *Cordylophora* was reared in salinities ranging from almost full strength sea water to fresh water, it became apparent that the gross form of the entire colony was a function of salinity. With decreasing salinity the colonies were shorter and their cells were long and narrow. As a result much less surface area was available to interact with the ambient environment (Kinne, 1958). Remane and Schlieper (1971) have documented differences in the appearance of brackish-water forms compared to forms from highly saline regions. For example, the bivalve *Cardium edule* taken from low-salinity waters was smaller and had a reduced number of ribs in comparison to specimens taken from more saline waters. In another species, *Mya arenaria*, the shells of specimens in low-salinity water tended to have a more rounded appearance in comparison to those from higher-salinity waters.

In addition to the reduction in exchange between the organism and the ambient environment by means of changes in the body wall or body shape, many estuarine animals have met the challenge of salinity fluctuation by regulating either their volume or their osmotic or ionic composition. A good illustration of volume regulation can be found in Oglesby's studies (see 1975 review) on soft-bodied estuarine worms. The amount of water that a worm takes up when transferred from one salinity to another depends on two factors: (1) magnitude of the transfer; and (2) species of worm involved. Stenohaline species will osmotically take up much more water than euryhaline ones, which can live in highly diluted seawater for long periods. When a worm is transferred to a lower salinity, there is an initial osmotic inflow that apparently occurs through the body wall. The uptake is very rapid during the first few hours, followed by a plateau. Following this plateau period, one of two types of responses may occur depending upon the species. In one there is no subsequent loss of water and the worms remain swollen until they die or are transferred to a higher salinity. This response is true not only of stenohaline worms but also of species where the salinity transfer is beyond tolerance limits. In the second type of response, a subsequent loss of weight follows the plateau period, and the weight of the worms approaches its initial weight. Typically, the more euryhaline a species is the less water it is likely to gain initially and the more water it is likely to lose subsequently. It is generally agreed that the process of weight loss involves an active metabolic response by the worms.

Some animals are unable to regulate the osmotic concentration of their extracellular body fluid in relation to the external environment and they are said to be osmoconformers. In contrast, the osmoconcentration of the extracellular fluid of an osmoregulator remains relatively constant despite external fluctuating environmental osmotic conditions. There are, of course, many intergrades between these two artificial categories, and many species demonstrate a response somewhere in between. Different stages in the life cycle of a species also may vary in osmoregulatory ability. The generalized comparison of principal adaptations of osmoregulating and osmoconforming organisms is shown in Fig. 3.5.

Osmoregulating organisms may differ in the method of maintaining their osmotic concentration. When exposed to low salinity some polychaetes are capable of active uptake of Na^+ and Cl^-, and the gills of the Chinese wool-handed crab (*Eriocheir sinensis*) can take up sodium from an external medium with a sodium concentration of 8 mmol, even when the internal sodium concentration is 300 mmol. Sharks maintain a relatively high blood osmoconcentration by having a relatively high concentration of urea. Changes in the amount of free amino acids in cells may be used to maintain osmoconcentration, while in other species organic substances, such as glucose, may be used to

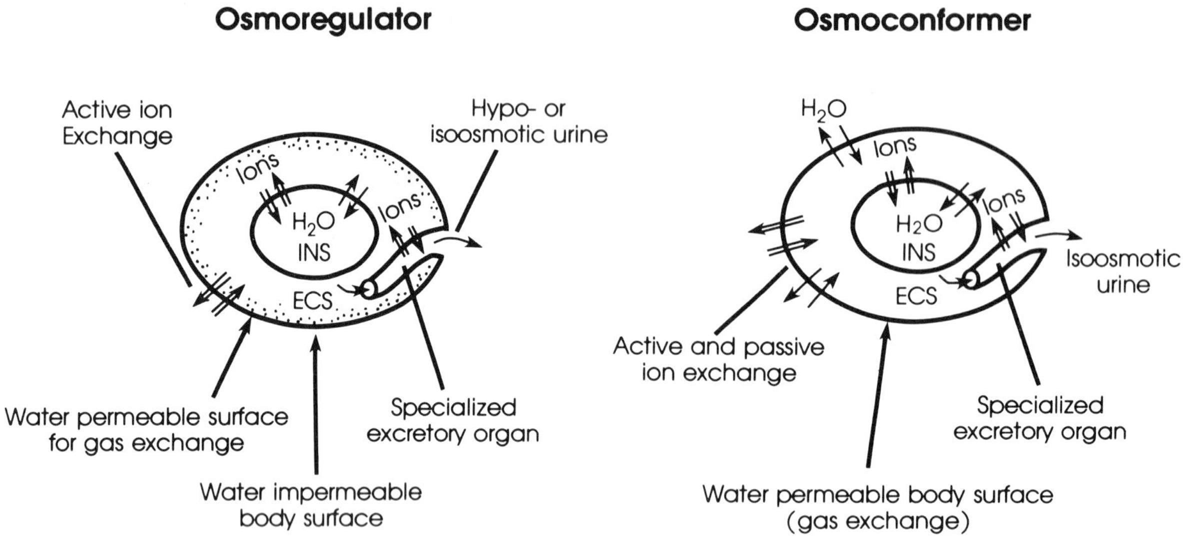

Fig. 3.5. Generalized comparison of principal adaptations of osmoregulating and osmoconforming organisms. *ECS* = extracellular space; *INS* = intracellular space. (From Vernberg and Vernberg, 1970.)

increase or maintain osmoconcentration. The role of amino acids in controlling the intracellular osmotic pressure in estuarine invertebrates has recently been comprehensively reviewed by Bishop (1976).

It should be noted that not all euryhalinic species osmoregulate their body fluids to any extent, but instead appear to tolerate tissue dilution and may depend more on cellular osmoregulating abilities.

While osmotic regulation is important, the selective regulation of specific ions or ionic regulation is essential, since protoplasmic integrity cannot be maintained in absence or excess of certain ions. For example, calcium added to low-salinity water that is ordinarily lethal may enable animals to survive.

Most estuarine organisms produce an isosmotic urine. In high-salinity waters urine formation would not be expected to create an ionic unbalance, but in low-salinity water the problem is greatly magnified, for isosmotic urine depletes electrolytes from the body. The strategy usually used is to actively accumulate one ion and excrete another, using special cells to trap the ions from the ambient fluid. In fish and some of the higher invertebrates, the kidney or renal organ is of great value in both volume regulation and salt retention. There are some exceptions, but generally animals living in regions where they are exposed to low salinity tend to have more renal units, a greater total surface for glomerular filtration and longer and more highly differentiated renal tubules that function for salt absorption than do animals living in constant, high-salinity environments. The gills also play an important role in salt excretion and/or uptake.

Intertidal animals are capable also of hypo-osmoregulating. These animals are able to maintain their body fluid at a lower osmotic concentration than the surrounding water when salinities are exceptionally high. When an organism is trapped in a hyperosmotic situation, it is subject to dehydration through osmotic loss of water. The water lost osmotically will tend to be recovered by drinking sea water and excreting the salts extra-renally. The osmolarity of the tissues are adjusted to match the environment by formation or retention of organic molecules such as free amino acids.

Most euryhaline animals do use a number of strategies to survive fluctuating salinities, as illustrated in the study by Fyhn (1976) on the euryhaline barnacle *Balanus improvisus*. This species succeeds in waters which range from totally fresh to hypersaline. At higher salinities animals osmoconform, but in more dilute waters they osmoregulate. As the hemolymph osmolality increases so does the free amino-acid concentration, particularly proline. The relationship between muscle tissue water content and hemolymph osmolality seems to reflect changes in the concentration of intra-cellular solutes. This phenomenon indicates cell volume regulation, and appears to be dependent upon adjustments of the intracellular concentration of amino acids.

FEEDING

Zooplankton are often assumed to be the primary herbivores in pelagic systems, serving to transfer energy from phytoplankton to secondary consumers. This assumption is not always valid for the estuarine system, however. It has been estimated that the zooplankton consume approximately 50 to 60% of the net phytoplankton produced in Long Island Sound, New York (Riley, 1967), but in the vicinity of Cape Hatteras, North Carolina, zooplankton proved to be relatively unimportant in food chains, grazing only 2 to 9% of the available phytoplankton (Williams et al., 1968). Further south in Lake Pontchartrain, Louisiana, Darnell (1961) found that phytoplankton was utilized directly by only a few of the consumer species, and in the Cochin backwater, a tropical estuary in India, phytoplankton production far exceeded the rate of consumption by the zooplankton herbivores (Qasim, 1970). In warm-water estuaries then, the zooplankton are not the primary herbivores, and instead of phytoplankton serving as the most important primary producers, marsh grasses, sea grasses, reeds and other marsh vegetation fill this role. Much of this plant material becomes a part of the sediment and estuarine food chains tend to be based either on organic detritus derived from these various plants or on the benthic and epiphytic microflora. Thus, benthic herbivores rather than zooplankton represent the critical herbivore link in many estuarine food webs.

In the estuarine intertidal zone many of the sessile animals and those with limited locomotor ability tend to be filter feeders, but the particular modes of feeding are by no means identical. In bivalves, for example, the feeding organs are the mantle, the gills and the mouth palps, while barnacles use long-haired thoracic appendages as a net to sweep the water for food. As a general rule, filter feeders do not discriminate beyond particle size. They do filter some material that has no energetic value and they must filter a large volume of water to be able to concentrate molecules against a gradient from a dilute medium. Active, mobile species frequently are detritus feeders, scraping material from sand grains, or eating fresh and decaying flesh of dead marine organisms. Others are predatory, feeding primarily on living prey. More sluggish and/or burrowing forms tend to be omnivorous, feeding on mud, sand or bottom deposits of organic materials consisting of feces and micro-organisms. These animals swallow everything indiscriminately by means of extroversible gullets, tentacles, or other "shoveling" devices. Despite the fact that deposit feeders do suck up material indiscriminately from the surface of the sediment, there is evidence in molluscs that the material carried to the digestive tract is carefully sorted in the mantle cavity (Hylleberg and Gallucci, 1975).

The feeding habits of subtidal benthic estuarine organisms can be correlated with the hydrodynamic characteristics of an estuary. Animals living in areas where currents are slow and where the mean diameter of the sediments is less than 0.09 mm are generally detritus feeders. In areas where the mean grain size is in excess of 0.09 mm, filter feeding is apt to be the dominant form of feeding behavior, and if the grain size exceeds 0.15 mm, then predation is common (Carriker, 1967).

Estuarine species, like most animals, have varied diets even when food is superabundant, and if one food source is limited or destroyed, they are able to utilize another. The diversity of food materials can be illustrated by a study of the food consumed by adult blue catfish in Lake Pontchartrain, Louisiana. In this study Darnell (1967) found that the diet of these fish consisted of 12% molluscs, 17% macrocrustaceans, 30% fishes, 26% vegetation and 15% miscellaneous material. The type of food eaten seemed to depend more or less on availability.

One of the most successful estuarine fish is the striped mullet (*Mugil cephalus*), and the feeding habits throughout its life cycle can serve as an example of estuarine nutrition. This species feeds on mosquito larvae, copepods and other zooplankton until it attains a length of about 30 mm. Then it either feeds on detritus by sucking up the surface layers of mud, or browses on micro-algae attached to submerged surfaces. Sediment-feeding fish may take up small mouthfuls at random or may skim along the bottom, lips barely touching the sediment, and suck up the top layer. Both types of feeders strain a small quantity of the sediment through the pharyngeal filter and expel undesired material. Browsing animals nibble the attached algae, digest suitable food, and expel the remainder. When offered a choice, the mullet showed a definite preference for live material over plant detritus (Odum, 1970). An algal diet has a higher caloric content than does detritus, and laboratory experiments on the bay scallop (*Argopecten irradians*) showed greatest growth rate on a diet of mixed phytoplankton, although slow but continuous growth was noted in those scallops fed a diet of fresh detritus (Kirby-Smith, 1976).

The stimulus that initiates feeding behavior varies widely. In sessile animals, for example, feeding may be stimulated by water movement, and animals living in the surf zone or in areas exposed to strong bottom currents require a much faster flow of water to induce feeding behavior than do sessile organisms living in sheltered, silted areas (Crisp, 1965). In other species the presence of food particles in the water is sufficient stimulus to initiate feeding activity. In the intertidal zone visual cues are important, and, although estuarine waters tend to be murky, there is evidence that some visual cues can be important in feeding behavior in the water column also. When striped mullet (*Mugil cephalus*) were fed in isolation, feeding began only after some delay. When isolated fish viewed through a plastic divider a group of fish that were actively feeding, then feeding was greatly facilitated in the experimental fish. One interesting aspect of this study was that bluefish (*Pomatomus saltatrix*), which are natural predators of mullet, were effective in initiating feeding behavior in mullet as long as the fish were separated by a plastic divider. Without such separation and sharing a common environment, mullet show a well-documented avoidance reaction to bluefish, suggesting that in nature the mullet must rely on additional recognition factors other than vision for discriminating and avoiding potential predatory species (Olla and Samet, 1974). It has also been noted that fish of the same species which are allowed to interact freely will have a much higher feeding rate than will animals which are fed in isolation (Darnell and Wissing, 1975).

Environmental factors can have a profound effect on feeding patterns. Many animals, for example, will feed only in a certain thermal range, and most have an optimum temperature for feeding which will change with latitude. In the bivalve *Mytilus californianus* the pumping rate is the same at 6.5°C from latitude 48°N as it is in a population at 12°C at 38°N or in a population from latitude 34°N at 14°C. The temperature at which feeding is blocked off also is a function of latitude (Mangum, 1969). When the feeding responses of Arctic nereid worms were measured, 80% of the species responded at 5°C and 20% of them were still feeding at 2.5°C. Among the temperate-zone species, however, the feeding response was observed only at or above 15 to 20°C; although these worms could tolerate 5 to 10°C they would not feed at these low temperatures.

Salinity can also influence feeding behavior. The stomachs of the subtidal oysters *Crassostrea virginica* were found to contain food irrespective of tide or time of day, indicating that these oysters were feeding almost continuously (Loosanoff and Nomejko, 1946). The decisive factor, however, may be a salinity gradient, for oysters living in parts of Delaware Bay where the salinity was relatively constant fed during both low and high tide, but in regions with a pronounced salinity gradient they fed only at high tide when the salinity was relatively high (Korringa, 1952).

RESPIRATORY ADAPTATIONS

The success of an animal living in the fluctuating estuarine environment depends on proper functioning of its metabolic machinery, and those species living in the intertidal zone have a double challenge since they must be able to survive both aquatic and terrestrial life. Oxygen levels, of course, are quite different in the two environments. The oxygen content of an aquatic environment may vary markedly, from supersaturation to anoxia, whereas oxygen levels in the air are relatively constant. Temperature fluctuations, on the other hand, are more extreme on land than in water. Furthermore, intertidal-zone animals may experience greater desiccation which tends to dry out their respiratory membranes.

Crustaceans provide a good example of morphophysiological adaptations to living in the intertidal zone. Both the number and total volume of gills of crabs living on land are less than those of aquatic species. The gill area per unit of body weight is reduced in terrestrial and semi-terrestrial animals, and this trend correlates well with the increased availability of air in the intertidal zone. The structure of the gills also has undergone modification with adaptation to a terrestrial existence. In the ocean, gill leaves are held apart by water currents; without the water currents the leaves would tend to stick together. In many intertidal crabs the subunits of the gills are highly sclerotized so that the gills are supported and held rigid. There is also a tendency for an increase in respiratory rate with the approach of terrestrialism which can be noted at whole body, gill tissue, and enzyme level (Vernberg and Vernberg, 1972a).

Metabolically, animals associated with an estuary show some degree of adaptation to environmental change. Although numerous endogenous and exogenous factors affect estuarine organisms, this discussion will be limited to the influence of salinity, temperature, and dissolved oxygen.

It is difficult to generalize about the respiratory response of estuarine animals to different salinities, since a wide diversity of metabolic responses to variable salinities has been reported. Four general types of metabolic response have been described by Kinne (1967):

(1) The rate of oxygen consumption is not influenced by salinity changes. The crab *Eriocheir* and the fish *Fundulus heteroclitus* follow this pattern.

(2) The metabolic rates increase in reduced salinity and decrease in high salinity. Examples of this type of response can be found in the polychaete *Nereis diversicolor* and the crab *Carcinus maenus.*

(3) In some animals, such as the ghost crab (*Ocypode quadrata*) and the shrimp *Metapenaeus*, metabolic rates increase in both low and high salinities.

(4) In still other species oxygen uptake rates decrease whenever the animals are moved to a new salinity. The anthozoan *Metridium marginatum* and the bivalve *Mytilus edulis* are examples of this type of response.

Some caution must be exercised in interpreting salinity response in intact organisms since a change in salinity may alter the locomotor activity of an organism. Thus, observed changes in metabolic rate may reflect behavioral changes rather than the effect of salinity on basic metabolic processes.

While the metabolic response of intact estuarine organisms to changing salinities often cannot be clearly linked to habitat preferences, results of studies on metabolic response to temperature show a very good correlation. Since estuarine animals are poikilothermic, their metabolic rate will in general vary directly with temperatures — low temperature will lower the rate of oxygen uptake and increased temperature will raise the rate. Usually there will be a range of temperature where an organism is thermally insensitive; this range appears to be the range that coincides with the animal's period of annual activity. For example, in temperate-zone animals their temperature range of insensitivity frequently is between 15 and 25°C.

It is generally accepted that animals experiencing fluctuating thermal environments tend to acclimate metabolically to temperature. Moreover, the temperature range over which thermal acclimation is expressed may be correlated with the ecology and biogeography of the population being studied. Thus, temperate-zone fiddler crabs are more metabolically labile at lower temperatures than tropical species, which demonstrate a greater degree of lability at higher temperatures (Fig. 3.6). Much of the literature on thermal acclimation has been summarized in reviews by Vernberg and Vernberg (1972a), Prosser (1973), Wieser (1973), Newell (1975), Pandian (1975), and Hoss and Peters (1976).

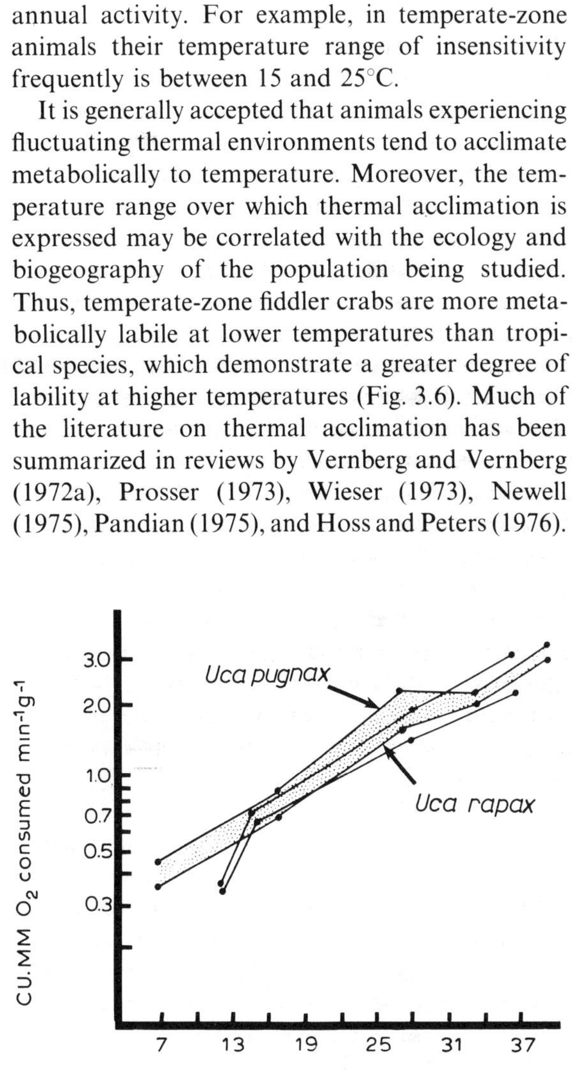

Fig. 3.6. Genotypic limits of oxygen consumption of tropical and temperate zone fiddler crabs (*Uca rapax*, tropical, *U. pugnax*, temperate). (From Vernberg and Vernberg, 1972a.)

Fluctuations in ambient oxygen may greatly influence the metabolic machinery of estuarine animals. Not all species have solved this problem, and hence those that have not are restricted to highly oxygenated waters. Others, however, have evolved adaptive mechanisms that enable them to live under hypoxic conditions. Some fishes, such as menhaden (*Brevoortia tyrannus*) and mackerel (*Scomber scomber*), require high oxygen levels; in contrast, bottom-dwelling species such as the toadfish (*Opsanus tau*) can survive prolonged periods of anoxia. Generally, oxygen consumption rates of species living in oxygen-poor environments are lower than the rates of species living in habitats where oxygen levels are high.

The response of animals to low oxygen tensions varies from species to species. Some species, especially those that live in oxygen-deficient habitats, are able to regulate their metabolic rate down to very low levels. In contrast, the metabolic rate of those that cannot survive in poorly oxygenated waters tend to conform to ambient oxygen levels; that is, as oxygen levels drop, respiration rates decrease. These two categories, which may respectively be termed "regulators" and "conformers", do not, however, fit all species. Lasserre (1976) refers to these remaining species as "complex metabolic types" and has noted that they appear to be adapted to fluctuating or high oxygen level habits. The various responses of representative species are illustrated in Fig. 3.7.

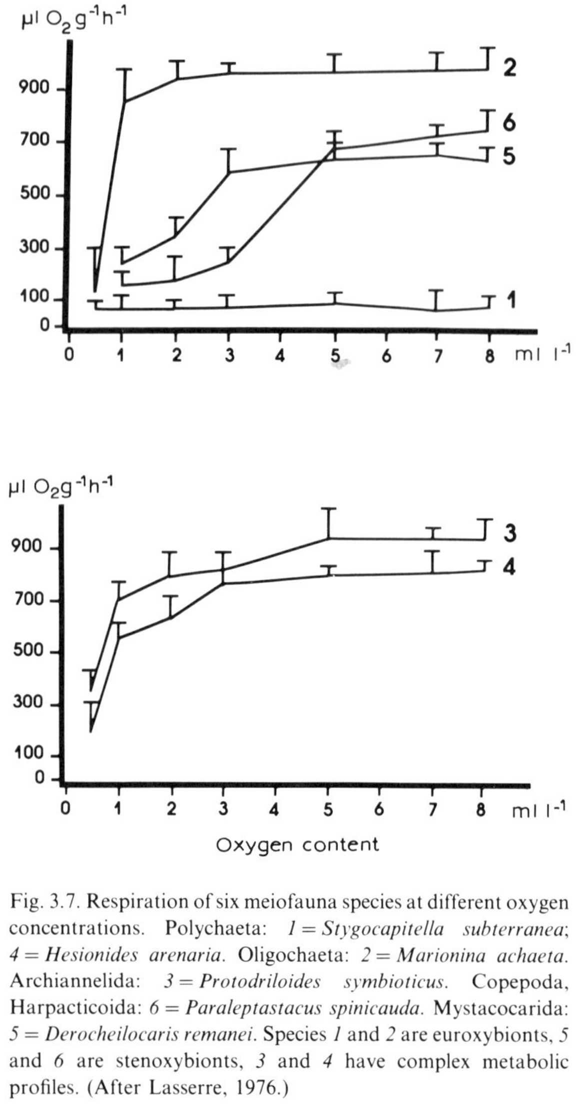

Fig. 3.7. Respiration of six meiofauna species at different oxygen concentrations. Polychaeta: *1* = *Stygocapitella subterranea*; *4* = *Hesionides arenaria*. Oligochaeta: *2* = *Marionina achaeta*. Archiannelida: *3* = *Protodriloides symbioticus*. Copepoda, Harpacticoida: *6* = *Paraleptastacus spinicauda*. Mystacocarida: *5* = *Derocheilocaris remanei*. Species *1* and *2* are euroxybionts, *5* and *6* are stenoxybionts, *3* and *4* have complex metabolic profiles. (After Lasserre, 1976.)

Mechanisms for maintaining respiratory independence during hypoxia vary among species, as can be seen in a study on the bivalves *Arctica islandica* and *Mytilus edulis* (Taylor and Brand, 1975). Both of these species are able to regulate their oxygen uptake over a wide range of oxygen tensions, but *A. islandica* relies on an increase in ventilation rate, while *M. edulis* increases the level of oxygen utilization. The mechanism used can be correlated with the habitats of these two species. *A. islandica* lives subtidally, so an increase in ventilation rate can be brought about by increasing the duration of the pumping periods. This method is not available to *M. edulis* since it is an intertidal species, and hence must rely on alternate respiratory responses to maintain respiratory independence at low oxygen tensions (i.e. sharp increase in oxygen utilization).

In many organisms the ability to provide oxygen for cellular metabolism is enhanced by respiratory pigments. These molecules, which are distributed in various parts of an organism, function to combine reversibly with oxygen. Normally these pigments form a loose chemical combination with oxygen as they pass near a respiratory surface where the external oxygen tension is higher than that of the body fluids. When a region of low tension is reached such as that found in tissues and cells, oxygen is released. The capacity for respiratory pigments to combine with oxygen varies intra- and inter-specifically, but there are certain phylogenetic trends. Sea water saturated with air at 20°C contains 0.538% by volume of oxygen. Invertebrates

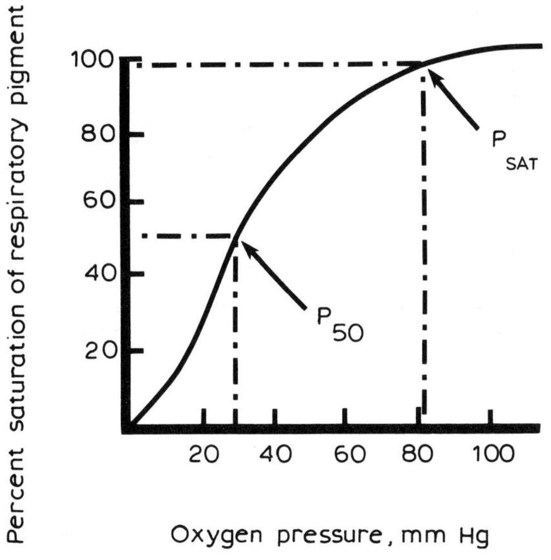

Fig. 3.8. Oxygen-equilibrium curve. P_{50}, the oxygen pressure at which 50% of the respiratory pigment is saturated. P_{sat}, the oxygen pressure at which 95% of the respiratory pigment is saturated.

with respiratory pigments have concentrations ranging from 0.5 to 10%, and in poikilothermic vertebrates the range is from 5 to 12%.

Even when a respiratory pigment is present in an organism, its functional significance to the organism must be demonstrated. One useful measurement of function is to characterize the oxygen-equilibrium curve (Fig. 3.8). Essentially this is done by measuring and plotting the percent of saturation of the respiratory pigment at different oxygen pressures. The pressures required for 95% saturation (the loading tension, P_{sat}) and 50% saturation (the unloading tension, P_{50}) are of particular importance for comparative purposes. For example, the P_{50} values for respiratory pigments of a number of birds and terrestrial animals range from 30 to 74 mbar, while for various aquatic crustaceans these values range from 9 to 20 mbar.

Environmental factors are known to influence the position of oxygen equilibrium curves. Recent studies by Mangum and her coworkers demonstrate the adaptive aspects of environmental salinity on respiratory pigment (in this instance, hemocyanin) function in the blue crab, *Callinectes sapidus* (Weiland and Mangum, 1975; Mangum and Towle, 1977). The blue crab can survive in waters ranging from fresh to hypersaline, and the young crabs migrate from the open oceans into the estuaries to feed and grow in low-salinity waters. As the young crabs proceed upstream, and salinity changes from 35 to 1‰, total blood salts are reduced by approximately 37%, and the oxygen affinity of hemocyanin is lowered by removal of salts from the blood. This would appear to be highly disadvantageous to the crab when considered in the context of its physiological needs. Swimming upstream the crab accumulates acidic metabolites in the blood stream, and, once the crab has reached its destination, it must be able to grow and mature at reduced oxygen levels. A complex mechanism compensates for the lower oxygen affinity of hemocyanin as the salinity and the blood salts decrease during the migration upstream. As the osmotic concentration of the crab's blood decreases by 37%, the fluids inside its cells reach osmotic equilibrium by reducing the size of the free amino-acid pool. This reduction in the amino-acid pool involves an increase in deamination, and as a result a blue crab has four times as much ammonia in its blood at 5‰ salinity as it does at 35‰. Since the ammonia molecules bind a hydrogen ion, lowering the concentration of free H^+, blood pH increases as the crab migrates upstream. Thus, the oxygen affinity of blue crab hemocyanin is lowered by the loss of salts from the blood, but it is raised by the removal of hydrogen ions. Consequently, the effectiveness of the respiratory pigment remains unchanged over a wide range of salinities due to the opposite effects of H^+ and salts in the blood of the crabs.

POPULATION CONTINUITY

In many species reproduction is a major physiological phenomenon dominating all other processes, for, although reproduction is not necessary for an individual to survive, the continuity of the species depends upon the ability to reproduce itself. Both sexual and asexual modes of reproduction have evolved in the animal kingdom, but exclusively asexual reproduction is not widespread. In sexually reproducing higher animals, reproduction is a cyclical physiological process that can be divided into three phases, all of which must function to ensure continuity: gonadal development, spawning and fertilization, and development and growth. Generally, reproductive cycles consist of a vegetative or resting stage with little gonadal activity, and a reproductive stage initiated by gametogenesis followed by the release of gametes. Patterns of reproductive periodicity may be annual, semiannual, or continuous throughout the year.

In estuarine intertidal animals, gamete or larval release must be synchronized with the tidal cycle. More mobile species move down to the water to release their larvae; sessile organisms must time their release of reproductive products to coincide with periods when the adults are submerged. Subtidal animals, particularly sessile ones, face somewhat different problems since they are always covered with water. When and where their larvae are released depends upon larval tolerances to estuarine conditions. Other estuarine species at breeding time will migrate long distances offshore into the deeper, more stable oceanic waters. The principal environmental factors that initiate and influence the reproductive cycles of estuarine organisms are temperature, salinity, photoperiod, and lunar phase.

Temperature

The temperature range necessary for reproduction is generally much narrower than the temperature extremes within which other bodily processes can function. Thus the boring gastropod *Urosalpinx cinerea* is capable of locomotor activity at 10°C and will drill oysters at 15°C, but oviposition does not take place until temperatures reach 20°C (Stauber, 1950). This difference between organismic and reproductive thermal limits has ecological significance, for an organism living in a habitat with a narrow temperature range might have to migrate to a thermally different habitat to reproduce.

Changes in water temperature often initiate the breeding response. Most often it is rising temperatures which induce spawning, but in some species, such as the barnacles *Balanus balanoides* and *B. balanus*, gamete release is induced by falling temperatures (Crisp, 1957).

In the temperate zone, temperature often interacts with other environmental factors to influence the reproductive cycle. For example, for the bay scallop (*Argopecten irradians*) the cycle is closely correlated with seasonal changes in temperature, day length and food abundance (Sastry, 1975). This scallop has a very predictable cycle. In early spring when temperatures and day length are increasing but food abundance is still low, the primary germ cells and oogonia begin development. Gonadal growth begins when the temperatures reach approximately 20°C and when the day length and abundance of food are at their maxima, usually in May and June. Gonadal growth continues during the summer when food is most abundant. Around mid-September, when temperature, food abundance and day length are all decreasing, maturation occurs, followed by the breeding period. Breeding ends about mid-November when the temperatures have reached 17 to 18°C. The scallop then enters a resting stage during the winter when the temperatures, food abundance, and day length are all minimal. Laboratory studies showed three sequential events in the reproductive cycle of the scallop which can be affected differentially by temperature (Sastry, 1975). Initiation of gonadal growth in the vegetative and resting stages required both suitable temperatures (20–30°C) and an adequate food supply. Inadequate food supply and higher temperature, however, decreased gonadal index and caused resorption of oogonia and early oocytes. After the gonads had accumulated certain minimum reserves, the food supply was not critical, and the development of the gonads to maturation occurred when temperatures reached 20°C or higher.

The rate of larval development will increase as temperatures rise to a certain point. As might be expected, the range of temperatures tolerable to a developing larva varies with the species and can be correlated with the thermal characteristics of the habitat (Vernberg and Vernberg, 1972a; Vernberg and Vernberg, 1975). Not all stages in the development of an estuarine organism are equally sensitive to temperature; in many the cleavage stages are more sensitive to temperature fluctuations than are the larval stages.

Salinity

There is much variation in the ability of estuarine animals to reproduce and grow in the fluctuating salinities which are characteristic of their environment. Reduced salinity decreases the reproductive potential of many estuarine animals which characteristically migrate to higher-salinity waters to breed and spawn. Salinities may also differentially influence the various stages of the reproductive process in many organisms, as can be seen in the studies by Bayne (1965) on the bivalve *Mytilus edulis*. Fertilization will occur in salinities from 15 to 40‰, but successful development to the trochophore stage occurs only in salinities of 30 to 40‰.

Estuarine animals often live in low-salinity waters where they are unable to reproduce. Thus, repopulation depends upon the invasion of larvae or adult migratory populations from higher-salinity waters. Indeed, larval susceptibility to low salinity can be a major limiting factor in the distribution of an animal. Populations of the polychaete *Nereis diversicolor* from Sweden can survive in salinities down to 6‰, but successful cleavage and early development of the larvae are limited to salinities of about 8 to 27‰. There can be marked differences in the salinity ranges suitable for populations of these polychaetes, as Smith (1977) has shown. For example, Finnish populations can develop in salinities sufficiently low to inhibit fertilization in the Swedish animals. At the same time, a salinity which is optimal for the Swedish population is lethal to the Finnish larvae.

There are a few species, however, that can breed over a very wide range of salinities. One of the most successful is the brachyuran crab *Halicarcinus australis*, which is found in the Swan Estuary in Western Australia (Lucas and Hodgkin, 1970). Salinities in the area range from essentially fresh water in winter to 36‰ during the summer. Ovigerous females are found throughout the year over the entire salinity range, although the peak breeding season occurs from October to January when salinities are relatively high.

Then, of course, some species seek out reduced salinity waters for breeding. Salmon, for example, leave open ocean waters and move inland to fresh water in order to spawn. Panikkar and Aiyar (1939) reported a number of estuarine fish in India which have their peak of breeding activity during the monsoon season when salinities are at the lowest levels. Apparently the lowered salinity serves to stimulate reproductive activity, since even continuous breeders in the estuary spawn at a greater rate after rain.

Photoperiod

It has been demonstrated that photoperiod has a profound effect on the reproductive cycles of many animals, but the exact manner in which light initiates these effects is not known. In a discussion of photoperiod effects on marine animals, Segal (1970) pointed out that it is not known if light acts directly on the gametes, reproductive structures, neurons serving the oviduct musculature, or an integrative center removed from the reproductive elements. Many animals require alternating periods of light and dark to initiate spawning; when they are kept under constant conditions of light or dark, they will not spawn. For example, it has been known for some time that ascidians spawn in response to light following darkness (Whittingham, 1967). As pointed out earlier, photoperiod often interacts with other factors in the environment to initiate the reproductive process.

Lunar phases

Reproductive cycles of many estuarine animals tend to coincide precisely with various phases of the moon. One of the most spectacular examples of lunar periodicity has been described in the intertidal insect *Clunio marinus*, which lives in tangles of algae in the intertidal zone (Fig. 3.9). The imagos are short-lived and must copulate and deposit eggs in a few hours after leaving the pupa. The eggs, which are deposited in strings, are glued to the substratum. Because all of these activities must take place when the habitat area is relatively dry, the semi-lunar periodicity of hatching and emergence during the neap tides in spring is not surprising. Synchronization of emergence with these low-water spring tides guarantees a concentration of the short-lived adults in a dry environment, permitting the adults to reproduce and deposit their eggs (Neumann, 1975). Epidemic swarming has been linked to lunar periodicity in a number of species. The crab *Cardisoma guanhumi* offers a good example of this phenomenon (Gifford, 1962). This crab lives intertidally and along drainage ditches in the south Florida coast, and it must return to the sea to spawn. The spawning season extends from late June to early December, and spawning occurs in sharp peaks near the time of full moon, and in lesser peaks at new moon periods. During these periods

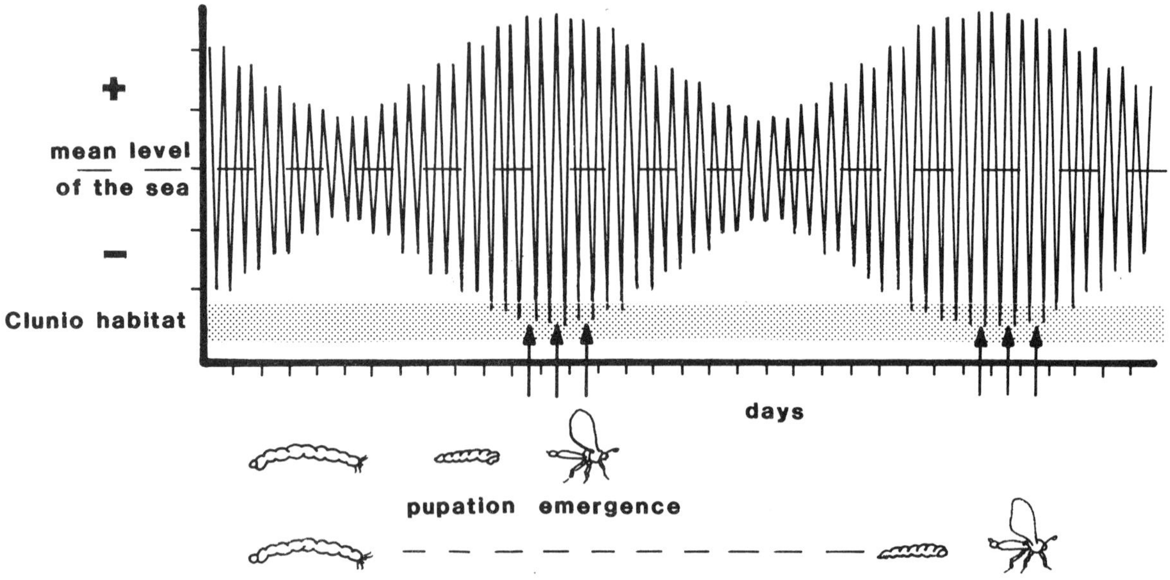

Fig. 3.9. Relationship of emergence time to water level and phase of the moon during one month. The gray strip indicates the level of the *Clunio* habitat being exposed during extreme low waters. (From Neumann, 1975.)

hundreds of egg-bearing (ovigerous) females appear, moving toward the sea; the highest concentration of crabs appears between one night before and the night after the full moon. The spawning period then stops abruptly.

PERCEPTION OF THE ENVIRONMENT

In the estuarine environment the most commonly used sensory modalities are those for photoreception, phonoreception, chemoreception, and those largely undefined organs for the perception of gravity and pressure. The sensory modalities which are most important to an organism in the perception of its physical and biological environment, and the range of stimuli that will cause a receptor to respond, can be correlated with an animal's way of life. For example, light perception is less developed and less important to an animal living in the bottom of a muddy estuary than it is to a fiddler crab living in the well-lighted and unobstructed environment of the intertidal zone where it uses vision in exploration of the environment and in the avoidance of predators.

Often larval and adult stages will not utilize the same modalities to carry out life's functions. For example, at the time of settlement, barnacles orient upward toward the light and have only a very weak orientation to water currents. After settlement, however, adult barnacles are oriented by water current rather than light, and they change from a predominantly vertical to a nearly horizontal position (Crisp and Stubbings, 1957).

Many sessile organisms do not have special photoreceptor organs but they do have dermal light receptors. Although these receptors have not been identified, it is probable that the light response results from accumulation of small amounts of photosensitive substances in the cytoplasm. In some sessile organisms, a dermal light sense will permit the animal to orient toward light by bending the body, and is also responsible for the shadow reflexes observed in animals such as barnacles. Although some can detect either a decrease or increase in light intensity, most respond more vigorously to decreased illumination. Such a response functions as a protective mechanism and has great adaptive value to the animals, allowing them to retract or retreat when shadow suggests the presence of a predator.

In general, larvae of most species that have been studied are photopositive when first released, starting life by swimming toward light, and become photonegative with age. Thus, the young larvae are brought to the phytoplankton-rich surface waters where they grow and develop. When they are ready to settle, they become photonegative and retreat to the bottom.

Among free-swimming animals, placement in a favorable position in the water column is of utmost importance for survival. Positioning does not occur by chance, for each species occupies a well-defined area in the water column. A number of environmental factors are thought to be used by the animals in the selection of their position in the water column. Response to light is generally considered to be an important factor, as are changes in hydrostatic pressure. Knight-Jones and Morgan (1966) have postulated that, since an organism in the water column can adapt to gradual change in pressure up to a certain point, when pressure builds up beyond this point the animal responds by compensatory swimming. Thus, pressure could aid in setting the limits of vertical migration.

Other animals are able to utilize tidal salinity changes in seeking a suitable environment, employing both rheostatic and chemoreceptive modalities. While reduced salinities can prevent some animals from carrying out normal vertical migrations, changing salinities associated with the ebb and flow of the tide can be utilized to advantage by others, such as the European eel, *Anguilla vulgaris* (Creutzberg, 1961). The eels hatch in the tropical waters of the Sargasso Sea and slowly migrate to inland waters on the coast of Europe and Africa, reaching areas as far east as Sicily in the autumn of their third year. They then metamorphose from willow-leaf-shaped larvae into transparent elvers, and become distributed along the European and North African coast. The elvers are able to move with the tides into the estuaries in these areas by staying in the surface waters during the incoming flood tides and then descending into bottom water during ebb tides, thereby avoiding being washed back out to sea. Once they are in the estuary with the ebb tide flowing over them, the elvers can discriminate between the overhead water masses. When a water mass containing inland water passes over them, they orient themselves by heading into the current and then make their way into inland waters to complete their development. The animals apparently are able to distinguish the specific odor of inland water by chemoreception, for they showed no response when exposed in laboratory experiments to tap water. Inland water, however, proved to be a very strong attractant.

Some estuarine animals also can use chemoreception to detect a change in salinity. The behavior of the lugworm (*Arenicola marina*) was studied in response to dilution of water overlying its burrow. When salinities became intolerable, the lugworms became inactive, compressing themselves at the bottom of the burrow. However, the worms sampled the overlying water about once every hour, and when salinity returned to its initial high value, normal activities resumed (Shumway and Davenport, 1977). Thus, the burrowing behavior of *A. marina* provides adequate protection against a changing external environment, and the worm is able through chemical perception of the environment to determine when it has returned to normal.

Once intertidal-zone larvae reach the settling stage in their development, they must choose the proper substrate. Failure to do so would certainly lessen greatly their chances for survival. In many species, larvae at this stage of development alternate between crawling and swimming, which suggests that the larvae are searching for the proper substrate. Settlement appears to be initiated primarily through tactile or chemosensory perception, and is presumed to be under nervous control. Scheltema (1974) reviewed the literature on larval settlement in marine invertebrates, and noted that there are six kinds of biological interactions involved in settlement: "(1) responses to submerged filmed surfaces, (2) responses to microorganisms in bottom sediments, (3) responses to alga substrates, (4) responses to wood, (5) intraspecific responses, or gregariousness, and (6) interspecific responses." As Wilson (1952) concluded, "it is evident that larvae do not settle just anywhere and that many of them have some power of selection, whether by touch, smell, or by other means."

Sound reception is common among both invertebrates and vertebrates in the estuarine environment. Among invertebrates, sound reception is primarily a response to sonic vibrations carried through the substrate rather than an airborne component. Among fiddler crabs (genus *Uca*) acoustical as well as visual signals are used during courtship (Fig. 3.10). The display of the males, which involves waving the large major cheliped, is especially conspicuous and each species has a characteristic display pattern (see review by Crane, 1975). Sound production is also an important component of courtship behavior and is again

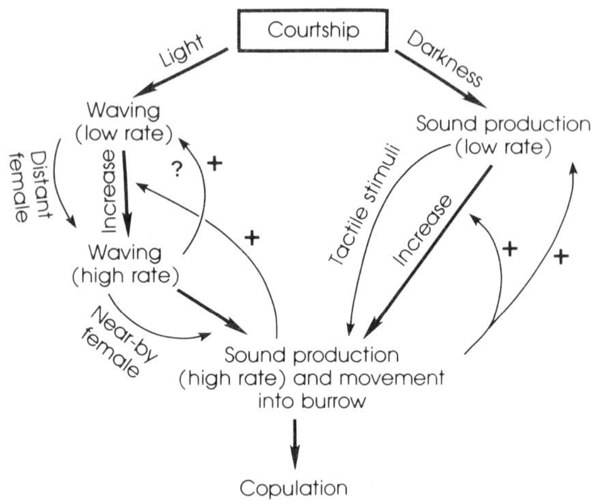

Fig. 3.10. Courtship in *Uca pugilator* during the day (left) and at night (right). Normal sequences of response by male to female are indicated by the thick arrows. Arrows to left of normal sequence indicate effect of stimuli provided by female upon the male. Arrows to right indicate effects of signals produced by male upon behavior of adjacent males. (From Salmon and Atsaides, 1969.)

distinctive for each species (Salmon and Atsaides, 1969). In *Uca pugilator*, when females are absent the male crabs wave during the day and produce sound at night. If a female is nearby during the day, the male waves more rapidly, and, if she approaches the male's burrow, the waving is followed by sound production. At night the male produces sounds at low rates by rapping his large cheliped against the substrate, but increases the sound rate when touched by a female. Females have been observed in the field to orient and move toward males at night in response to sound from distances as great as 25 cm.

For animals living subtidally in murky estuarine waters visual display is of limited value. Instead, some fish rely primarily on hearing for communication.

Basically fish produce sounds in two ways, either by stridulation by bony elements or by movement of the swim bladder (Fine et al., 1977). Sound production in fish is not finely organized, and only major mood responses can be obtained. The male toadfish (*Opsanus tau*) for example, has a well-defined spawning ritual. He first seeks a suitable nesting site, such as a tin can or a large empty shell, and stands guard. When he is ready to spawn, he emits a characteristic boat whistle call, which in turn acts as a stimulus to attract females that are ready to lay eggs. After the female lays a clutch of eggs, she leaves the nest, but the male remains until the young are free-swimming. If other fish approach the nest while the male toadfish is standing guard, he will emit a series of aggressive grunt sounds, presumably to cause the intruders to leave the nest area (Winn, 1964).

In summary then, the adaptation of estuarine organisms reflects their modes of existence. As a group they can be characterized broadly as being more euryhaline and eurythermal than open-ocean forms. Many have evolved considerable tolerance to oxygen deprivation, while ability to tolerate desiccation is limited primarily to intertidal zone species. As a group, these organisms have evolved a wide range of physiological adaptations that permit them to utilize the fluctuations of the estuarine environment to their advantage. Thus, while an estuary is by many standards a stressful environment, the normal fluctuations present few problems to the animals that live there.

ACKNOWLEDGEMENTS

I gratefully acknowledge the assistance of Ms. Donna Mirkes and Dr. Alan Young for critical reading of the manuscript, and Ms. Kathi Hicks for editorial assistance and typing the manuscript.

REFERENCES

Bayne, B.L., 1965. Growth and the delay of metamorphosis of larvae of *Mytilus edulis* L. *Ophelia*, 2: 1–47.

Bishop, S.H., 1976. Nitrogen metabolism and excretion: Regulation of intracellular amino acid concentrations. In: M. Wiley (Editor), *Estuarine Processes*. Academic Press, New York, N.Y., pp. 414–431.

Brown, A.C., 1960. Desiccation as a factor influencing the vertical distribution of some South African gastropoda from intertidal rocky shores. *Port. Acta Biol., Ser. B*, 7, 1: 11–23.

Carriker, M.R., 1967. Ecology of estuarine benthic invertebrates: A perspective. In: G.H. Lauff (Editor), *Estuaries. Am. Assoc. Adv. Sci. Publ.*, 83: 442–487.

Crane, J., 1975. *Fiddler Crabs of the World (Ocypodiae: Genus* Uca). Princeton University Press, Princeton, N.J., 738 pp.

Creutzberg, F., 1961. On the orientation of migrating elvers (*Anguilla vulgaris turt.*) in a tidal area. *Neth. J. Sea Res.*, 1(3): 257–338.

Crisp, D.J., 1957. Effect of low temperature on the breeding of marine animals. *Nature*, 179: 1138–1139.
Crisp, D.J., 1965. Surface chemistry, a factor in the settlement of marine invertebrate larvae. *Botanica Gothoburgensia III. Proceedings of the 5th Marine Biological Symposium, Göteborg*, pp. 51–65.
Crisp, D.J. and Stubbings, H.G., 1957. The orientation of barnacles to water currents. *J. Anim. Ecol.*, 26: 179–196.
Crouse, M.G. and Found, B.W., 1975. Preliminary observations on the salinity and temperature preferences of *Derocheilocaris typica*, Pennak and Zinn, 1943. *Cah. Biol. Mar.*, 16: 751–762.
Darnell, R.M., 1961. Trophic spectrum of an estuarine community, based on studies of Lake Pontchartrain, Louisiana. *Ecology*, 42: 553–568.
Darnell, R.M., 1967. Organic detritus in relation to the estuarine ecosystem. In: G.H. Lauff (Editor), *Estuaries. Am. Assoc. Adv. Sci., Publ.*, 83: 376–382.
Darnell, R.M. and Wissing, T.E., 1975. Nitrogen turnover and food relationships of the pinfish *Lagodon rhomboides* in a North Carolina estuary. In: F.J. Vernberg (Editor), *Physiological Ecology of Estuarine Organisms*. The Belle W. Baruch Library in Marine Science, No. 3: 81–111.
Davie, P.S., 1969. Physiological Ecology of *Patella*. III. Desiccation Effects. *J. Mar. Biol. Assoc. U.K.*, 49: 291–304.
Elmgren, R., 1975. Benthic meiofauna as indicator of oxygen conditions in the Northern Baltic proper. *Proc. 3rd Baltic Symp. Mar. Biol. Merentutkimuslaitoksen Julk. Havsforskningsinst. Skr.*, 239: 265–271.
Ferguson, R.L., Thayer, G.W. and Rice, T.R., 1981. Marine primary producers. In: F.J. Vernberg and W.B. Vernberg (Editors), *Functional Adaptations of Marine Organisms*. Academic Press, New York, 2: 9–69.
Fine, M.L., Winn, H.E. and Olla, B.L., 1977. Communication in fishes. In: T.A. Seboek (Editor), *How Animals Communicate*. Indiana University Press, Bloomington, Ind., pp. 472–518.
Fyhn, H.J., 1976. Holeuryhalinity and its mechanisms in a cirriped crustacean, *Balanus improvisus*. *Comp. Biochem. Physiol.*, 53A: 19–30.
Gibson, R.N., 1969. The biology and behaviour of littoral fish. *Oceanogr. Mar. Biol. Annu. Rev.*, 7: 367–410.
Gifford, C.A., 1962. Some observations on the general biology of the land crab, *Cardisoma guanhumi* (Latreille) in South Florida. *Biol. Bull.*, 123: 207–223.
Haefner, P.A., Jr., 1969. Temperature and salinity tolerance of the sand shrimp, *Crangon septemspinosa* Say. *Physiol. Zool.*, 42: 388–397.
Haefner, P.A., Jr., 1970. The effect of low dissolved oxygen concentrations on temperature-salinity tolerance of the sand shrimp, *Crangon septemspinosa* Say. *Physiol. Zool.*, 43: 30–37.
Herreid, C.F., II, 1969. Integument permeability of crabs and adaptation to the land. *Comp. Biochem. Physiol.*, 29: 423–429.
Hoss, D.E. and Peters, D.S., 1976. Respiratory adaptations: Fishes. In: M. Wiley (Editor), *Estuarine Processes*. Academic Press, New York, N.Y., pp. 335–346.
Hylleberg, J. and Gallucci, V.F., 1975. Selectivity in feeding by the deposit-feeding bivalve *Macoma nasuta*. *Mar. Biol.*, 32: 167–178.
Kenny, R., 1969. The effects of temperature, salinity and substrate on distribution of *Clymenella torquata* (Leidy) Polychaeta. *Ecology*, 50: 624–631.
Khlebovich, V.V., 1969. Aspects of animal evolution related to critical salinity and internal state. *Mar. Biol.*, 2: 338–345.
Kinne, O., 1953. Zur Biologie und Physiologie von *Gammarus duebeni* Lillj. *Z. Wiss. Zool.*, 157: 427–491.
Kinne, O., 1958. Adaptation to salinity variations — some facts and problems. In: C.L. Prosser (Editor), *Physiological Adaptation*. American Physiology Society, Washington, D.C., pp. 92–105.
Kinne, O., 1967. Physiology of estuarine organisms with special reference to salinity and temperature. In: G.H. Lauff (Editor), *Estuaries. Am. Assoc. Adv. Sci. Publ.*, 83: 525–540.
Kinne, O., 1970. Temperature. Animals. Invertebrates. In: O. Kinne (Editor), *Marine Ecology*, I, Part I. Wiley–Interscience, New York, N.Y., pp. 407–514.
Kirby-Smith, W.W., 1976. The detritus problem and the feeding and digestion of an estuarine organism. In: M. Wiley (Editor), *Estuarine Processes*. Academic Press, New York, N.Y., pp. 469–480.
Knight-Jones, E.W. and Morgan, E., 1966. Responses of marine animals to changes in hydrostatic pressure. *Oceanogr. Mar. Biol. Annu. Rev.*, 4: 267–299.
Korringa, P., 1952. Recent advances in oyster biology. *Q. Rev. Biol.*, 27: 226–308, 339–365.
Lasserre, P., 1976. The benthic boundary layer. In: I.N. McCave (Editor), *Metabolic Activities of Benthic Macrofauna and Meiofauna: Recent Advances and Review of Suitable Methods of Analysis*. Plenum Press, New York, N.Y., pp. 95–142.
Lockwood, A.P.M., 1964. *Animal Body Fluids and Their Regulation*. Harvard University Press, Cambridge, Mass., 171 pp.
Loosanoff, V.L. and Nomejko, C.A., 1946. Feeding of oysters in relation to tidal stages and its periods of light and darkness. *Biol. Bull.*, 90: 244–264.
Lucas, J.S. and Hodgkin, E.P., 1970. Growth and reproduction on *Halicarcinus australis* (Haswell) (Crustacea, Brachyura) in the Swan Estuary, Western Australia. I. Crab instars. *Aust. J. Mar. Freshwater Res.*, 21: 149–162.
Mangum, C., 1969. Low temperature blockage of the feeding response in boreal and temperate zone polychaetes. *Chesapeake Sci.*, 10: 64–65.
Mangum, C. and Towle, D., 1977. Physiological adaptation to unstable environments. *Am. Sci.*, 65: 67–75.
Neumann, D., 1975. Lunar and tidal rhythms in the development and reproduction of an intertidal organism. In: F.J. Vernberg (Editor), *Physiological Adaptation to the Environment*. Intext Educational Publishers, New York, N.Y., pp. 451–463.
Newell, R.C., 1975. Factors controlling metabolic capacity adaptation in marine invertebrates. In: F.J. Vernberg (Editor), *Physiological Ecology of Estuarine Organisms*. The Belle W. Baruch Library in Marine Science, No. 3: 111–127.
Newkirk, G.F., Waugh, D.L. and Haley, L.E., 1977. Genetics of larval tolerance to reduced salinities in two populations of oysters, *Crassostrea virginica*. *J. Fish Res. Board Can.*, 34: 384–387.
Odum, W.E., 1970. Utilization of the direct grazing and plant detritus food chains by the striped mullet *Mugil cephalus*.

In: J.H. Steele (Editor), *Marine Food Chains*. University of California Press, Berkeley, Los Angeles, Calif., pp. 222–240.

Oglesby, L.C., 1965. Steady-state parameters of water and chloride regulation in estuarine nereid polychaetes. *Comp. Biochem. Physiol.*, 14: 621–640.

Oglesby, L.C., 1975. An analysis of water-content regulation in selected worms. In: F.J. Vernberg (Editor), *Physiological Ecology of Estuarine Organisms*. The Belle W. Baruch Library in Marine Science, No. 3: 181–204.

Olla, B.I. and Samet, C., 1974. Fish-to-fish attraction and the facilitation of feeding behavior as mediated by visual stimuli in striped mullet, *Mugil cephalus*. *J. Fish. Res. Board Can.*, 31: 1621–1630.

Pandian, T.J., 1975. Mechanisms on heterotrophy. In: O. Kinne (Editor), *Marine Ecology, 2*. Wiley, New York, N.Y., pp. 61–249.

Panikkar, N.K. and Aiyar, R.G., 1939. Observations on breeding in brackish-water animals of Madras. *Proc. Indian Acad. Sci.*, B9: 343–364.

Prosser, C.L., 1973. Physiological adaptations in animals. In: F.J. Vernberg (Editor), *Physiological Adaptation to the Environment*. Intext Education Publishers, New York, N.Y., pp. 19–36.

Qasim, S.Z., 1970. Some problems related to the food chain in a tropical estuary. In: J. Steele (Editor), *Marine Food Chains*. University of California Press, Berkeley, Los Angeles, Calif., pp. 46–51.

Remane, A., 1934. Die Brackwasserfauna. *Zool. Anz.*, 7: 34–74.

Remane, A. and Schlieper, C., 1971. *Biology of Brackish Water*. Wiley and Sons, New York, N.Y., 372 pp.

Riley, G.A., 1967. The plankton of estuarines. In: G.H. Lauff (Editor), *Estuaries*. *Am. Assoc. Adv. Sci. Publ.*, 83: 316–326.

Salmon, M. and Atsaides, S.P., 1969. Sensitivity to substrate vibration in the fiddler crab, *Uca pugilator* Bosc. *Anim. Behav.*, 17: 68–76.

Sastry, A.N., 1975. Physiology and ecology of reproduction in marine invertebrates. In: F.J. Vernberg (Editor), *Physiological Ecology of Estuarine Organisms*. The Belle W. Baruch Library in Marine Science, No. 3: 279–300.

Scheltema, R.S., 1974. Biological interactions determining larval settlement of marine invertebrates. *Thalassia Jugosl.*, 10: 263–296.

Segal, E., 1970. Light. In: O. Kinne (Editor), *Marine Ecology*. Wiley–Interscience, New York, N.Y., pp. 159–211.

Shumway, S.E. and Davenport, J., 1977. Some aspects of the physiology of *Arenicola marina* (Polychaeta) exposed to fluctuating salinities. *J. Mar. Biol. Assoc. U.K.*, 57: 907–924.

Smith, R.I., 1970a. The apparent water permeability of *Carcinus maenas* (Crustacea, Brachyura, Portunidae) as a function of salinity. *Biol. Bull.*, 139: 351–362.

Smith, R.I., 1970b. Chloride regulation of low salinities by *Nereis diversicolor* (Annelida, Polychaeta). *J. Exp. Biol.*, 43: 75–92.

Smith, R.I., 1976. Apparent water-permeability variation and water exchange in crustaceans and annelids. In: P.S. Davies (Editor), *Perspectives in Experimental Biology, I. Zoology*. Pergamon Press, Oxford, New York, N.Y., pp. 45–60.

Smith, R.I., 1977. Physiological and reproductive adaptation of *Nereis diversicolor* to life in the Baltic Sea and adjacent waters. In: D.J. Reish and K. Fauchald (Editors), *Essays on Polychaetous Annelids*. Special publication of the Allan Hancock Foundation, University of Southern California, Los Angeles, Calif., pp. 25–34.

Stauber, L.A., 1950. The problem of physiological species with special reference to oysters and oyster drills. *Ecology*, 31: 107–118.

Taylor, A.C. and Brand, A.R., 1975. A comparative study of the respiratory responses of the bivalves *Arctica islandica* (L.) and *Mytilus edulis* (L.) to declining oxygen tension. *Proc. R. Soc. Lond. B.*, 190: 443–456.

Teal, J.M. and Carey, F.G., 1967. The metabolism of marsh crabs under conditions of reduced oxygen pressure. *Physiol. Zool.*, 40: 83–91.

Theede, H., 1975. Aspects of individual adaptation to salinity in marine invertebrates. In: F.J. Vernberg (Editor), *Physiological Ecology of Estuarine Organisms*. The Belle W. Baruch Library in Marine Science, No. 3: 213–226.

Thompson, R.K. and Pritchard, A.W., 1969. Respiratory adaptations of two burrowing crustaceans, *Callianassa californiensis* and *Upogebia pugettensis* (Decapoda, Thalassinidae). *Biol. Bull.*, 136(2): 274–287.

Vernberg, F.J. and Silverthorn, S.V., 1979. Temperature and osmoregulation in aquatic species. In: R. Gilles (Editor), *Mechanisms of Osmoregulation in Animals*. Wiley and Sons, New York, N.Y., pp. 537–562.

Vernberg, F.J. and Vernberg, W.B., 1970. *The Animal and the Environment*. Holt, Rinehart and Winston, New York, N.Y.

Vernberg, F.J. and Vernberg, W.B., 1975. Adaptations to extreme environments. In: F.J. Vernberg (Editor), *Physiological Ecology of Estuarine Organisms*. The Belle W. Baruch Library in Marine Science, No. 3: 165–180.

Vernberg, F.J., Schlieper, C. and Schneider, D.E., 1963. The influence of temperature and salinity on ciliary activity of excised gill tissue of molluscs from North Carolina. *Comp. Biochem. Physiol.*, 8: 271–285.

Vernberg, W.B., 1969. Adaptations of host and symbionts in the intertidal zone. *Am. Zool.*, 9: 357–365.

Vernberg, W.B. and Vernberg, F.J., 1972a. *Environmental Physiology of Marine Animals*. Springer-Verlag, New York, N.Y., 346 pp.

Vernberg, W.B. and Vernberg, F.J., 1972b. The synergistic effects of temperature, salinity, and mercury on the survival and metabolism of the adult fiddler crab, *Uca pugilator*. *Fish. Bull.*, 70: 415–420.

Vernberg, W.B., CeDoursey, P.J. and O'Hara, J., 1974. Multiple environmental factor effects on physiology and behavior of the fiddler crab, *Uca pugilator*. In: F.J. Vernberg and W.B. Vernberg (Editors), *Pollution and Physiology of Marine Organisms*. Academic Press, New York, N.Y., pp. 381–426.

Weiland, A.L. and Mangum, C.P., 1975. The influence of environmental salinity on hemocyanin in the blue crab, *Callinectes sapidus*. *J. Exp. Zool.*, 193: 265–274.

Whittingham, D.G., 1967. Light-induction of shedding of gametes in *Ciona intestinalis* and *Molgula manhattenis*. *Biol. Bull.*, 132: 292–298.

Wieser, W. (Editor), 1973. *Effects of Temperature on Ectothermic Organisms*, Springer-Verlag, Heidelberg, 298 pp.

Wieser, W. and Kanwisher, J., 1959. Respiration and anaerobic survival in some sea weed inhabiting invertebrates. *Biol. Bull.*, 117: 594–600.

Williams, R.B., Murdock, M.B. and Thomas, L.K., 1968. Standing crop and importance of zooplankton of shallow estuaries. *Chesapeake Sci.*, 9: 42–51.

Wilson, D.P., 1952. The influence of the nature of the substratum on the metamorphosis of the larvae of marine animals, especially the larvae of *Ophelia bicornis* Savigny. *Ann. Inst. Oceanogr., Monaco*, 27: 49–156.

Winn, H.E., 1964. Observations on the reproduction habitats of darters (Pisces–Percidae). *Am. Midl. Nat.*, 59: 190–212.

Zachary, A. and Haven, D.S., 1973. Survival and activity of the oyster drill *Urosalpinx cinera* under conditions of fluctuating salinity. *Mar. Biol.*, 22: 45–52.

Chapter 4

THE PHYTOPLANKTON OF ESTUARIES

THEODORE J. SMAYDA

INTRODUCTION

The phytoplankton of estuaries have the same environmental requirements for survival and growth as those in other aquatic environments, such as fresh water or the open sea. Essential nutrients required for metabolism and growth include phosphorus, nitrogen, silicon (for diatoms), various vitamins, and iron, zinc and other trace metals. Adequate light to allow for photosynthesis is also required, and the water quality must be suitable to permit the occurrence, normal growth pattern and completion of species' life cycles. Sea salt composition, ionic ratios and concentrations must also be appropriate.

A salient difference between the estuarine environment and those of fresh water and open ocean is the much greater environmental variability in estuaries. This is a consequence of the continuous formation of these waters through the admixture of differing volumes of fresh water and saltier oceanic waters. Accordingly, the associated fluctuations in salinity are great (cf. Chapters 1 and 2) which directly influence species composition. Phytoplankton composition and dynamics are also affected by the changing environmental conditions accompanying the variable admixture of fresh and salt water with their differing turbidities, dissolved organic substances and nutrients. The external supplies of nutrients into estuaries are particularly variable. Pollution represents one source, both from direct discharges and from non-point sources including agricultural fertilizer runoff and storm overflow from streets. The influx of deeper, nutrient-rich oceanic waters mixed with estuarine waters represents an additional nutrient source. The contributions from each of these sources are seasonally variable; even more frequently they are also influenced by meteorological conditions and variable nutrient levels in these point and non-point sources.

A phytoplankton flora specifically adapted to the estuarine habitat has evolved. Estuaries range in size from large coastal embayments (Smayda, 1973a) to tiny polls (Braarud and Føyn, 1958), and exhibit a wide range of physiographic features (Castañares and Phleger, 1969), characteristics which further influence estuarine phytoplankton characteristics. Those estuarine conditions significant to phytoplankton growth, and the apparent adaptations of the phytoplankton to these conditions, will be the focus of this chapter.

ESTUARINE HABITAT FEATURES

Rochford's (1959) generalized scheme of estuarine habitat zonation and associated properties is given in Fig. 4.1. Although prepared to represent an idealized Australian estuarine system, it is useful generally. The main axis of an estuary can be subdivided into zones, although similar onshore/offshore zonations may occur within the arms of an estuary. The levels and ranges of salinity, nutrients and turbidity differ between zones. Thus, the fresh-water zone is usually the most turbid and richest in nutrients, and the marine zone least so. The turbidity is elevated through the riverine discharge of sedimentary material, through the scouring and roiling of bottom sediments into suspension, and through chemical precipitation. Fresh-water runoff also introduces nutrients. There is a region intermediate between these fresh-water and marine zones where tidal mixing and hydro-

	ZONE		FRESHWATER	GRADIENT	TIDAL	MARINE
A	Shape					

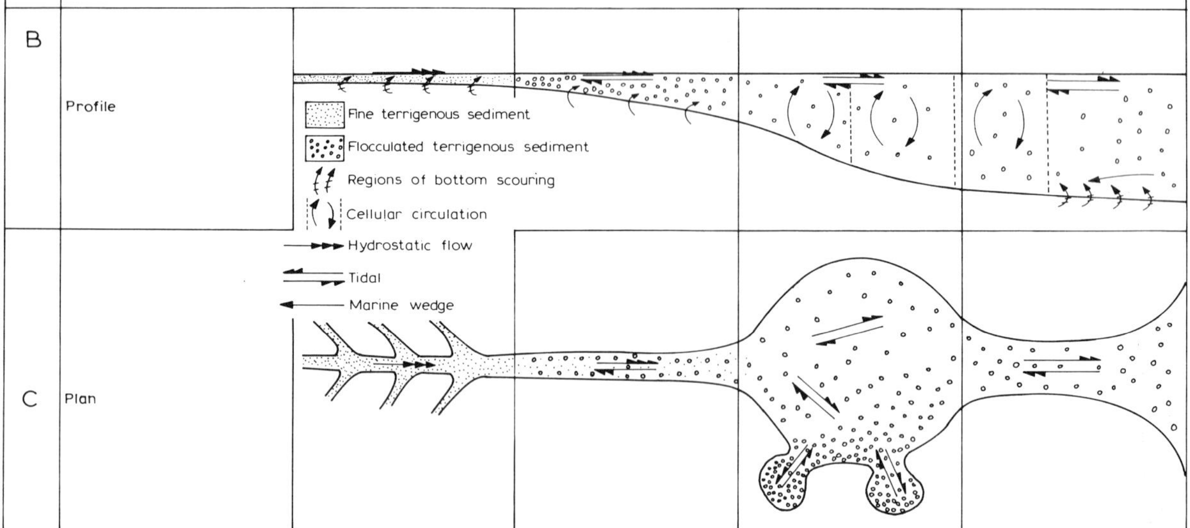

	ZONE		FRESHWATER	GRADIENT	TIDAL	MARINE
D	Average dimensions					
	Depth		3 – 6 m.	6 – 10 m.	10 – 20 m.	20 – 25 m.
	Width		up to 20 m.	up to 30 m.	~ 1000 – 3000 m.	~ 1000 – 2000 m.
	Length		50 – 100 km.	10 – 30 km.	10 – 15 km.	5 – 10 km.
E	Hydrological properties					
	Chlorinity	annual range	< 0.10 ‰	0 – 19.00 ‰	19 – 20.50 ‰	19.50 – 19.80 ‰
		"tidal" range	< 0.10 ‰	3 – 5 ‰	< 0.50 ‰	< 0.20 ‰
	Temperature	annual range	10 – 15 °C.	10 °C.	7 °C.	5 °C.
		"tidal" range	1 – 2 °C.	1 – 2 °C.	< 15 °C.	< 1 °C.
	Phosphate Phosphorus	concentration	20 μg l⁻¹	10 μg l⁻¹	< 5 μg l⁻¹	5 – 10 μg l⁻¹
		% "organic"/inorganic	80 %	50 %	20 %	30 %
	Nitrate Nitrogen	concentration	500 μg/l.	100 μg/l.	< 25 μg/l.	< 25 μg/l.
F	Circulation pattern		Hydrostatic unidirectional flow of homogeneous water	Hydrostatic and tidal oscillating axial flow of stratified water	Tidal oscillating diffuse flow of homogeneous water. Some vertical cellular flow	Weak oscillating flow of homogeneous water over wedge intrusion of marine waters
G	Tide	range	Absent	0.5 – 1.0 m.	1 – 2 m.	1 – 2 m.
		velocity	up to 150 cm s⁻¹	up to 90 cm s⁻¹	up to 45 cm s⁻¹	up to 45 cm s⁻¹

Fig. 4.1. Diagram showing the values and ranges of properties within each of the four zones of an idealized Australian estuarine system (from Rochford, 1959).

graphic diffusion are the predominant processes. Topographic features influence the extent of this zone, which is characterized by its own levels of and ranges in salinity, nutrients and turbidity.

A gradient in environmental properties characterizes estuaries, rather than distinct boundaries between different estuarine zones. It is this estuarine feature, together with seasonal and daily fluctuations in the values and ranges of environmental properties along this gradient, which influences the composition and dynamics of estuarine phytoplankton communities. Moreover, this gradient is variable rather than fixed. These features and the dynamic nature of estuarine zonal gradients are evident from the distributions of salinity in the Ythan Estuary (Scotland) (Fig. 4.2). The salinity gradient and absolute levels of salt content at given locations within this estuary exhibit significant seasonal and tidally induced variations, particularly in the upper estuarine regions. Thus, at a station 7 km from the mouth, the surface salinity varied from 0 to ~10‰, from ~15 to >30‰ at the 5-km station, and was always >33‰ at the 2-km station. It is such features, together with the seasonal and daily fluctuations and ranges in environmental properties along this gradient, which

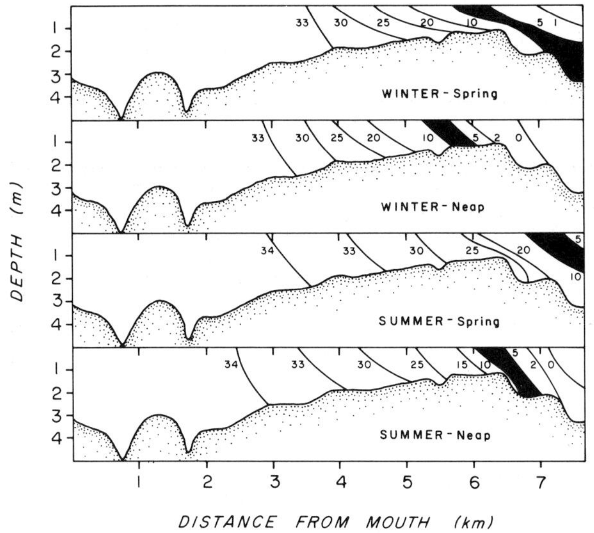

Fig. 4.2. Seasonal distribution of salinity in Ythan Estuary (Scotland) at high water, spring tides and neap tides (from Leach, 1971).

determine the composition and dynamics of estuarine phytoplankton communities.

Fresh-water inflow is the most important determinant of estuarine characteristics. Total salinity and ionic composition, properties which have osmotic effects, transparency and ice formation are influenced by the volume of fresh-water influx, as is the accretion of nutrients and humic substances (= terrestrial litter breakdown products) which influence productivity and chemical water quality. The volume of fresh-water inflow also influences the circulation patterns developed to maintain hydrostatic balance; these patterns also influence the residence time of estuarine properties.

Salinity

Local conditions of rainfall, evaporation, runoff and the degree of exchange with coastal waters strongly influence the annual mean, ranges and gradient in salinity characteristic of a given estuary (see Ch. 2). This effect is detectable even in the large brackish, inland seas (Fig. 4.3). The mean salinity (~6‰) in the Baltic Sea (Chapter 13), for example, is about one third that in the Black Sea (Chapter 11). The wide range in salinity that is characteristic of estuaries also occurs in these large seas. Thus, in the Baltic Sea salinity ranges from about 0 to 10‰, and from about 4 to 22‰ in the Black Sea. Such differences in salinity between estuaries, or within an estuary along the salt gradient, are accompanied by changes in salt composition. The ionic composition of oceanic water compared to that of less saline waters, including Lake Superior, is presented in Table 4.1. The expected reduction in sodium and chloride ions accompanies a reduction in total salt, whereas the divalent calcium and magnesium cations and the sulphate and carbonate anions increase. Since these differences in absolute concentrations and in the ratios of

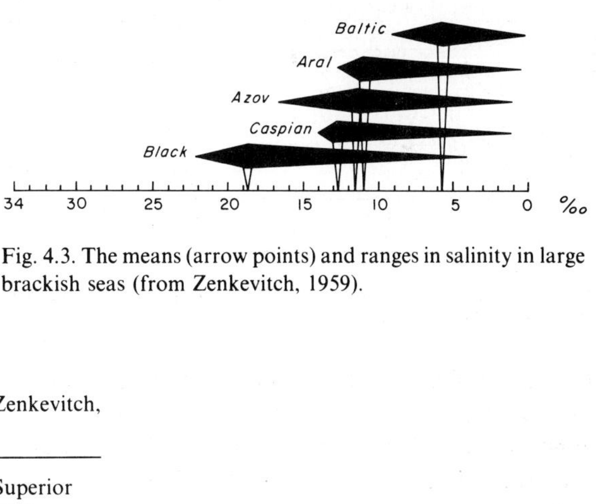

Fig. 4.3. The means (arrow points) and ranges in salinity in large brackish seas (from Zenkevitch, 1959).

TABLE 4.1

Percentage of composition of the dissolved salts in various seas (after Zenkevitch, 1963)

	Ocean (35‰)	Caspian Sea (11–13‰)	Aral Sea (~10‰)	Lake Superior (~0.06‰)
Sodium	30.59	24.82	21.30	?
Potassium	1.106	0.66	0.79	5.52
Calcium	1.197	2.70	5.00	22.42
Magnesium	3.725	5.70	5.41	5.35
Chlorine	55.292	41.73	33.93	1.89
Bromine	0.188	0.06	0.03	?
Sulphates	7.692	23.49	31.29	3.62
Carbonates	0.207	0.86	1.75	47.42

monovalent:divalent cations and anions affect osmotic properties, the selection and distribution of species within an estuary are influenced by both the total salt content and its ionic compositon. This suggests that there are biological differences based on salinity alone between estuaries. Numerous attempts, accordingly, have been made to classify marine waters and zones within an estuary based on salinity. The "Venice System" is a recent example (Table 4.2; see also Table 6.8, p. 174).

According to this scheme, the mean salinity of the ocean is about 35‰ and ranges from about 30 to 34‰ in most coastal waters. Fresh water is always less than 0.5‰; such water masses or estuarine zones are termed **limnetic.** Thus, the salinity range in an estuary potentially ranges from 0.5 to 35‰ (see Fig. 4.2). Water of less than 35‰ salinity represents diluted sea water and, *sensu stricto*, might be termed brackish. However, in the "Venice System" 30‰ is set as the upper level of diluted sea water; and estuaries or zones within an estuary between 0.5 to 30‰ are classified as **mixohaline**, a definition of the more ambiguous term "brackish". Within this grouping, oligohaline, mesohaline and polyhaline zones, or basic estuarine types, are recognized.

Although this salinity-based classification is generally helpful, the actual zonation and distribution of estuarine species correspond only approximately. The most appropriate classification of estuarine waters would be one which has a biological basis — that is, based on the regulation of species occurrence and distribution by salinity (= osmotic conditions). However, an effective classification has yet to be devised because of inadequate observations and the complicating factor that local conditions significantly influence species occurrence. A helpful subdivision of the "Venice System" based on biological observations exists for the Baltic Sea (see legend to Table 4.2 for source):

mixomesohaline 18 to 5‰ → α mesohaline ~18 to ~10‰
→ β mesohaline ~10 to ~5‰

mixooligohaline 5 to 0.5‰ → α oligohaline ~5 to ~3‰
→ β oligohaline ~3 to ~0.5‰

Thus, α oligohaline species are those which predominate in mixohaline waters from ~5 to ~3‰.

TABLE 4.2

The Venice System[1] for the classification of marine waters according to salinity (see also Table 6.8)

Zone	Salinity (‰)
Hyperhaline	>40
Euhaline	30–40
Mixohaline	0.5–30(40)
Mixoeuhaline	>30 but <adjacent sea
Mixopolyhaline	18–30
Mixomesohaline	5–18
Mixooligohaline	0.5–5
Limnetic (fresh-water)	<0.5

[1] A classification recommended at the Symposium on the Classification of Brackish Waters, Venice 8–14 April 1958. It was published, without identifiable authorship, as a supplement included in the proceedings of this symposium appearing in *Arch. Oceanogr. Limnol.*, XI (Suppl.).

Light

The compensation depth, the depth at which the rates of phytoplankton photosynthesis and respiration are equal, generally occurs at the 1% isolume depth. In the euphotic zone, above the compensation depth, photosynthesis exceeds respiration and therefore phytoplankton growth can occur; growth cannot occur below depths where respiration exceeds photosynthesis. Marine waters are not optically similar, and a scheme of optical classification of sea water has been proposed (see Jerlov, 1970; Fig. 4.4). In the clearest of oceanic waters (Type I), 1% of the incident radiation is detectable at about 105 m, and at about 30 and 6 m in the clearest (Type 1) and most turbid (Type 9) coastal waters, respectively. In estuaries the compensation depth can be less than 1 m. The greater attenuation of light in estuarine waters results from its scattering and absorption by the high concentrations of suspended particles introduced by river runoff and sedimentary material roiled from the sea floor by current and tidal flow (see Fig. 4.1). Large quantities of dissolved organic matter known as *Gelbstoff* (yellow substance) produced during the decomposition of terrestrial debris are introduced by river runoff and also strongly absorb light. An inverse relationship exists between the amount of

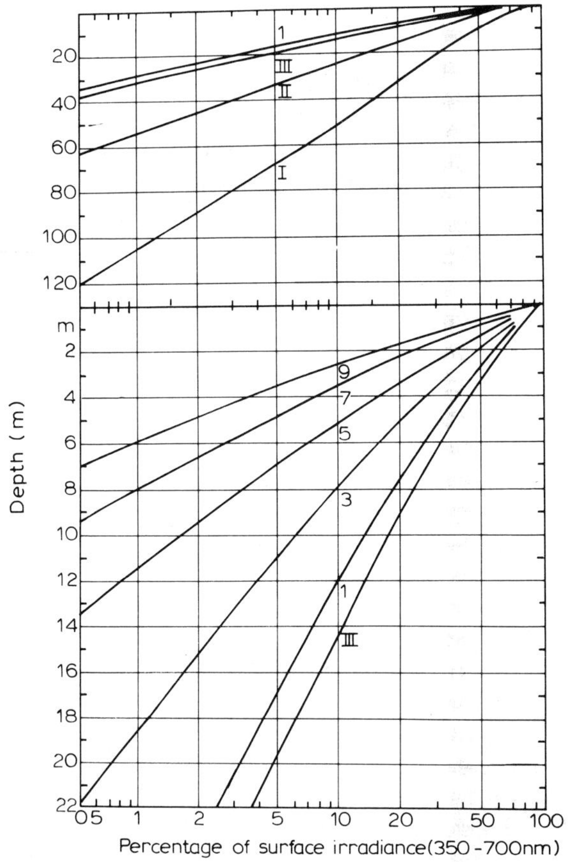

Fig. 4.4. Percentage of surface downward irradiance in photosynthetically available wavelengths (350–700 nm) as function of depth (m) in various oceanic (*I. II, III*) and coastal (*1–9*) water types (from Jerlov, 1970).

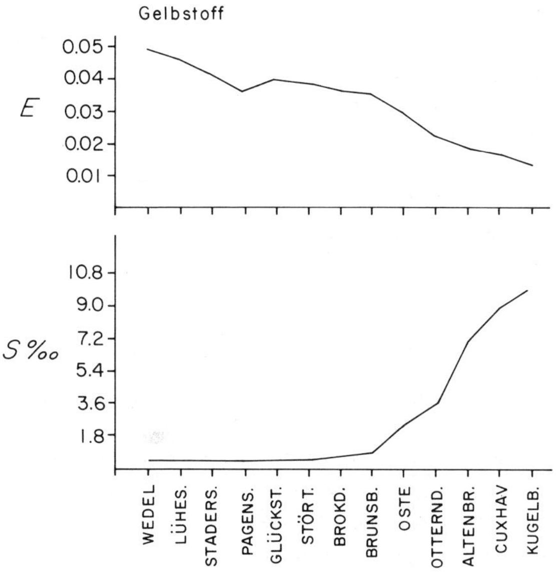

Fig. 4.5. The relationship between the extinction (*E*) of light due to "yellow substance" (*Gelbstoff*) and salinity in the Elbe Estuary from Wedel to Cuxhaven (modified from Schulz, 1961).

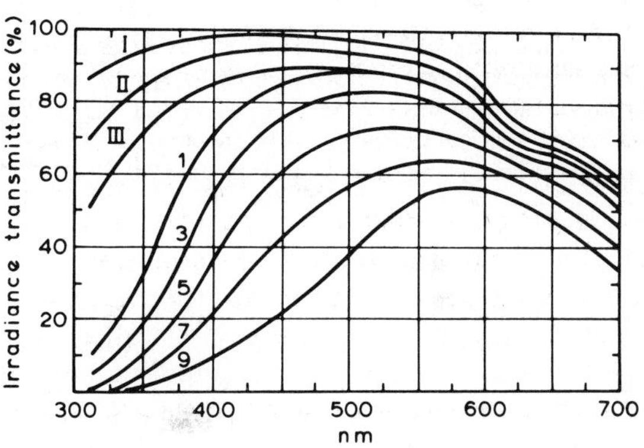

Fig. 4.6. Spectral transmittance (% m^{-1}) of downward irradiance for high solar attitudes in upper layers of oceanic water types *I*, *II*, and *III*, and in various coastal types (from Jerlov, 1970).

Gelbstoff present and estuarine salinity (Fig. 4.5). The amount of suspended material would also tend to increase with decreasing salinity.

These two causes of light attenuation in estuarine waters also influence the spectral transmission. Fig. 4.6 illustrates that in the clearest oceanic waters maximum transmittance occurs in the blue wavelength (400–470 nm) portion of the spectrum. However, with increasing turbidity due to suspended particles and the presence of *Gelbstoff* a dramatic and progressive shift in maximum transmittance from the blue to green (520 nm) and orange (600 nm) portions of the spectrum occurs. It follows that the quantity and quality of light penetrating within a given estuary will vary along the salt gradient. Aarthun (1961) has demonstrated such a condition in a glacier-fed Norwegian fjord, Hardangerfjord (Fig. 4.7). Large quantities of glacier-bed suspensoids are introduced into the fjord during summer melting. This is accompanied by a gradient in euphotic zone depth which increases from 10 to 30 m over a down-fjord distance of about 100 km. The spectral distribution of transmitted light also varies along this gradient (Fig. 4.7C). Obviously, seasonal variations in the incident radiation and the extent to which ice formation occurs also influence the light characteristics within the euphotic zone. Estuarine depth determines whether the sea floor is within the

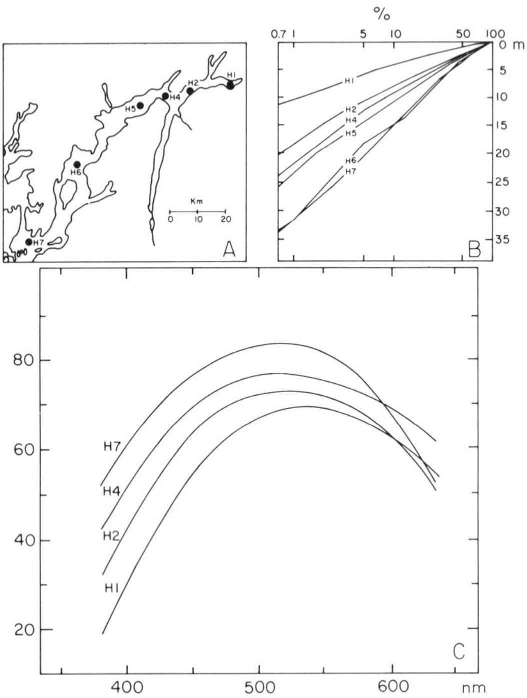

Fig. 4.7. Stations in the Norwegian fjord Hardangerfjord (A) at which the transmission of total daylight with depth as a percent of surface intensity was measured (B) as well as the spectral distribution of light from 375 to 650 nm wavelength at a depth of 1 m (C) during July (modified from Aarthun, 1961).

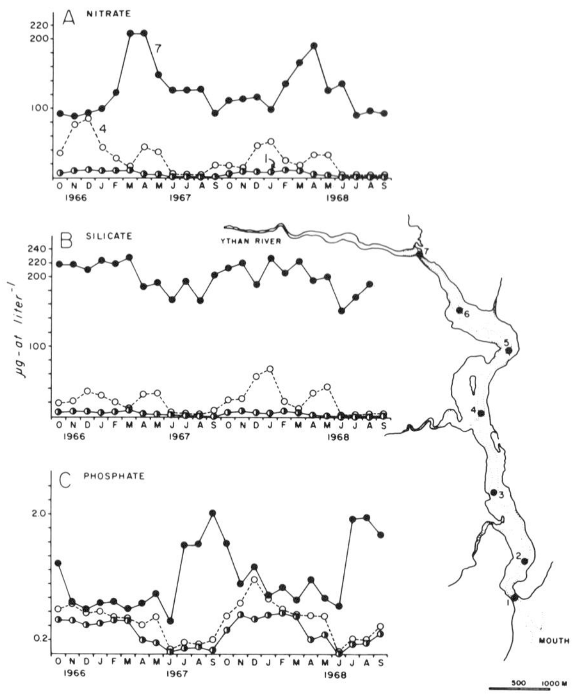

Fig. 4.8. Seasonal distribution of monthly mean concentrations as μg-at l^{-1} in nitrate (A), reactive silicate (B) and reactive inorganic phosphate (C) at Stations *1*, *4* and *7* in Ythan Estuary (modified from Leach, 1971).

euphotic zone. This, in turn, determines whether sea grasses and macroalgae contribute towards primary production.

In summary, considerable variations in both the intensity and quality of photosynthetically available light occur along a given estuarine gradient and between estuaries, which are usually more turbid than coastal and oceanic water masses.

Nutrients

The nutrient content of estuaries is determined by the concentrations in its riverine and coastal water sources, and the proportions of each in the mixture. The seasonal cycles and gradient distributions of the essential phytoplankton nutrients — phosphate, nitrate and silicate — in the Ythan Estuary illustrate some important properties of estuarine nutrient conditions (Figs. 4.8, 4.9). Ythan Estuary, located in northeastern Scotland, is a long (~8 km) and narrow (300–600 m) tidal extension of the Ythan River which drains about 800 km^2 of agricultural land (Leach, 1971). A marked horizontal gradient related to the tidal incursion characterizes the distribution of these nutrients. The concentrations of nitrate and silicate in the fresh-water end of the estuary (station *7*) are about an order of magnitude greater than in marine water. The mean annual concentrations of nitrate increase upstream by about 20-fold from 7 to 130 μg-at l^{-1}, and about 35-fold for reactive silicate from 6 to 200 μg-at l^{-1}; regional differences in phosphate concentration were less marked, ranging from 0.30 to 1.05 μg-at l^{-1}.

The seasonal fluctuations in nutrients in the marine end of Ythan Estuary (station *1*) were related to the biological cycle in the contiguous coastal waters of the North Sea (Leach, 1971). At the head of the estuary (station *7*), seasonal fluctuations in concentrations of nitrate and phosphate were related to rainfall and to the seasonal application of fertilizers to agricultural land in the drainage area. Fresh-water inflow supplied about

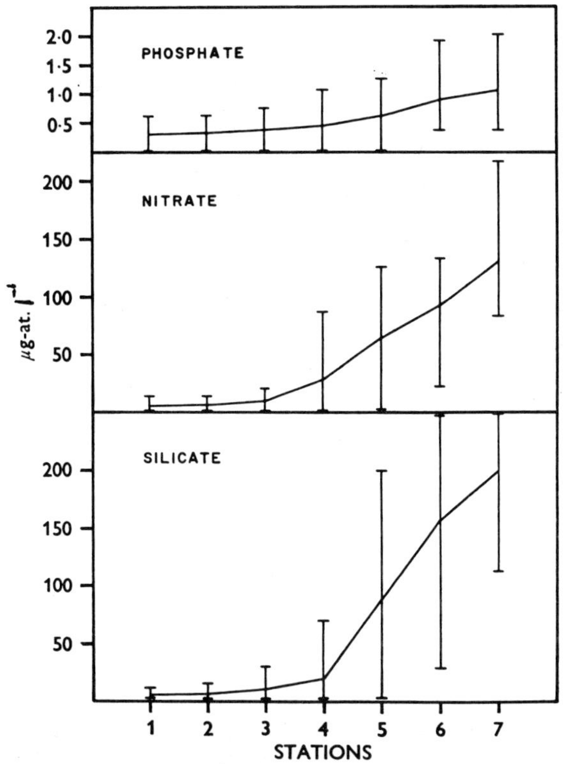

Fig. 4.9. Annual mean concentration and range in phosphate, nitrate and silicate for Ythan Estuary stations at high water (from Leach, 1971).

70% of the nitrate and 80% of the silicate to Ythan Estuary. Marine water influx contributed about 70% of the phosphate. The latter is consistent with observations on several coastal salt ponds and lagoons located in the northeastern United States that such estuarine types are phosphate "sinks" (Smayda, 1973a). That is, tidal mixing transports phosphorus in dissolved inorganic, organic and particulate form. The advected inorganic phosphate is assimilated by the phytoplankton and recycled within the estuary. The particulate fraction either sinks to the bottom or is ingested by benthic organisms and is then excreted or remineralized as phosphate and reassimilated by phytoplankton.

Phytoplankton growth within estuaries is not totally dependent on external sources of nutrients although they can contribute to an elevated productivity in a given estuarine system. Nutrient cycling mechanisms within estuaries also play a major role. Two aspects must be recognized. First, the individual nutrient elements exist as pools, or reservoirs, of differing instantaneous availability; and secondly, the turnover, or replenishment time (R) of the nutrient species comprising these supply pools is a more significant determinant of availability than their measured concentrations. Replenishment time is defined as $R = [S]/S_p$, where $[S]$ is the concentration of the nutrient, and S_p is the daily production rate as phytoplankton particulate content of that nutrient. Thus, replenishment time is the number of days required to replace $[S]$ when it is being removed at the rate S_p (Furnas et al., 1976).

Studies carried out in Chesapeake Bay have helped to clarify both aspects of the nutrient control of productivity. Chesapeake Bay is a highly productive coastal plain estuary, approximately 300 km in length, located on the eastern coast of the United States (Fig. 4.10). It is primarily an estuary of the Susquehanna River, but also receives the inflow from numerous secondary rivers, and exhibits a well-defined two-layer circulation pat-

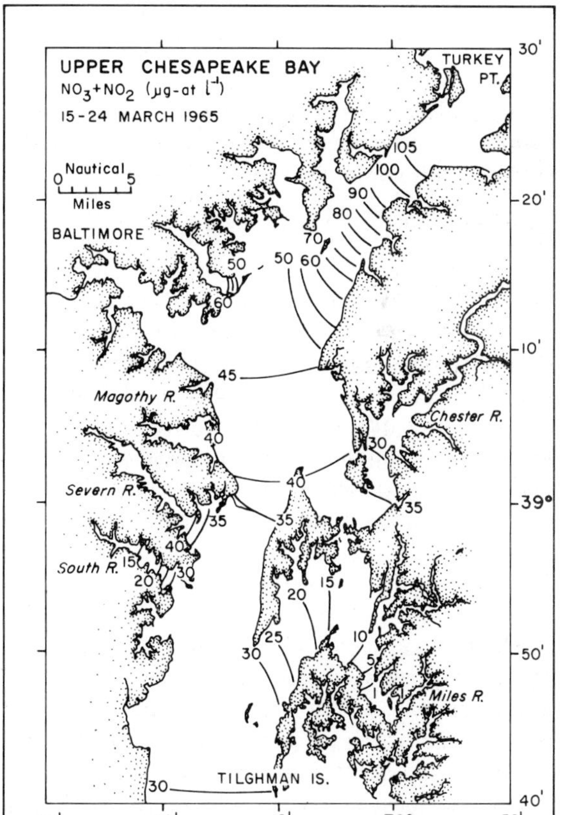

Fig. 4.10. Surface concentrations of nitrate+nitrite in upper Chesapeake Bay during March 1965 (from Carpenter et al., 1969).

tern. Carpenter et al. (1969) have demonstrated that copious winter and spring runoff from the Susquehanna River is the primary source of nitrate within Chesapeake Bay (Fig. 4.10). In contrast, upstream gradients in the nitrate distribution are negative for the secondary rivers. The distributions of phosphorus (Carpenter et al., 1969; Taft et al., 1975) and ammonium (McCarthy et al., 1975, 1977) contrast markedly with that for nitrate. Point sources for these nutrients have not been identified, and seasonal fluctuations and gradients are damped. Phosphorus is present as soluble reactive and dissolved organic phosphorus (both about 0.40 μg-at l^{-1}) and polyphosphate (0.9 μg-at l^{-1}) (Taft et al., 1975). High levels of phosphorus in the polluted headwaters of secondary rivers, such as the Potomac River, are observed. However, little of this enters into Chesapeake Bay proper, possibly because of sedimentation and assimilation by phytoplankton in the upstream segment.

The phosphorus in nutrient pools in Chesapeake Bay generally occurs in soluble reactive form (mostly as inorganic orthophosphate), as organic compounds, and as polyphosphates. Nitrogen in a form usable by phytoplankton occurs as nitrite, nitrate, ammonium, urea and other compounds of dissolved organic nitrogen. It is now well known (Kuenzler and Perras, 1965) that phytoplankton preferentially use inorganic phosphate rather than dissolved organic phosphorus. However, the pool of dissolved organic phosphorus present as phosphomonoesters can be utilized by the phytoplankton upon formation of cellular alkaline phosphatase, triggered by the decreased concentrations of orthophosphate. Similar preferential usage and switching characterize utilization of the nitrogen pool by the phytoplankton. McCarthy et al. (1977) determined the percentage of utilization by phytoplankton populations in Chesapeake Bay relative to availability of the different forms of nitrogen present within the nitrogen pool. Nitrate represented 68% of the average available nitrogen pool, ammonium 15%, urea 7% and nitrite 10%. However, the average percentage contribution of each to the total nitrogen utilized represented 22%, 51%, 20% and 7%, respectively. Thus, preferential utilization of the nitrogen pool components occurs as: ammonium > urea > nitrate > nitrite. Moreover, the uptake of nitrate + nitrite is regulated by the ammonium concentration (Fig. 4.11A). Concentrations of ammonium greater than 1.0 to 1.5 μg-at l^{-1} almost totally suppress nitrate uptake, which is activated at lower ammonium concentrations. In situations where the total nitrogen pool available is less than the immediate phytoplankton demand,

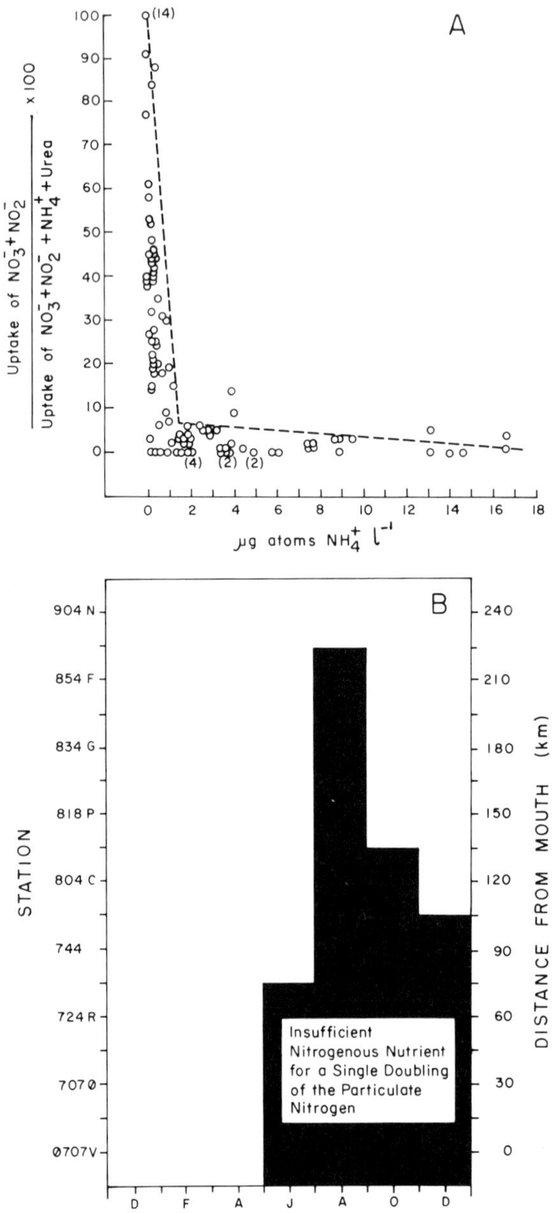

Fig. 4.11. A. Effect of ambient ammonium concentration on phytoplankton selection of nitrate in Chesapeake Bay. B. Regions and times within this bay when available nitrogenous nutrients were insufficient to allow doubling of phytoplankton particulate nitrogen. (Modified from McCarthy et al., 1975.)

utilization of the different nitrogen forms occurs at rates proportional to their availability.

The accretion of nutrients from external sources usually occurs at a rate much slower than the instantaneous demand for nutrients by phytoplankton, so that the nutrient turnover rate regulates nutrient supply and, hence, availability. In Chesapeake Bay, experimentally determined turnover times for the phosphorus pool ranged from several minutes to four days (Taft et al., 1975), and for ammonium from three to twenty hours (McCarthy et al., 1975). In Narragansett Bay the required summer turnover times of nitrate were usually about six hours (Furnas et al., 1976).

Since these turnover rates exceed the rates of advected supply from external sources, turnover must be due to internal estuarine mechanisms. At least three basic internal recycling mechanisms can be distinguished: (1) excretion of remineralized nutrients accompanying grazing by herbivorous zooplankton or benthic organisms, (2) release through sediment roiling and diffusive flux of nutrients from the interstitial water of sediments following microbial remineralization; and (3) kinetic, steady-state exchanges between nutrients present in particulate phase (phytoplankton, bacteria, sedimentary particles) and in the dissolved phase.

The first two mechanisms will be emphasized here since more extensive information is available to evaluate their importance. However, Taft et al. (1975) have evaluated the third mechanism with regard to phosphorus cycling in Chesapeake Bay.

The evidence suggests that the rapid turnover of nutrients resulting from excretion by herbivorous zooplankton and benthic organisms is the most important mechanism, at least in Narragansett Bay and Chesapeake Bay (Carpenter et al., 1969; McCarthy et al., 1975; Furnas et al., 1976). Although there is not a perfect balance between excreted supplies and the amounts required by the phytoplankton, excretion rates have been established primarily for the larger zooplankton. The simultaneous contributions from microzooplankton and the nekton remain to be assessed. The evidence strongly suggests that nutrient regulation of estuarine phytoplankton production is more closely regulated by this internal biocycling mechanism than by supplies from external sources. Obviously, the importance of benthic excretion in this recycling process will depend on such physical factors as water column depth and vertical mixing induced by the tidal currents and the wind. Hale (1975) has shown that, in Narragansett Bay, excretion of ammonium and phosphate is influenced by species composition of the benthic communities, and the rate increases exponentially with temperature.

Carpenter et al. (1969) concluded that the rates of nitrogen and phosphorus regeneration released into the water column from the sediments of Chesapeake Bay were inadequate to provide nutrient requirements of the phytoplankton in the euphotic zone. However, water column depth, the degree of bottom scouring and the extent to which vertical mixing is continuous or seasonal influence the relative importance of this mechanism within a given estuary and between estuaries. Therefore, although this recycling mechanism does not appear to be important in Chesapeake Bay, its importance elsewhere is not precluded.

Despite the continuous supply and turnover of nutrients from external sources and *in situ* recycling, nutrients may limit estuarine phytoplankton growth. Thus, the supply of nitrogen was inadequate over a large portion of Chesapeake Bay from May through December to allow a doubling of phytoplankton particulate nitrogen (Fig. 4.11B). Experimentation has shown that nitrogen and silica may be limiting to phytoplankton growth in Narragansett Bay (Furnas et al., 1976). In the Dutch sector of the estuarine Wadden Sea, which extends several hundred kilometers from Den Helder, The Netherlands, to southern Denmark, reactive silicate alone limits the diatom yield (Van Bennekom et al., 1974). This unusual situation develops because of the large concentrations of nitrogen and phosphorus and low concentrations of silicate present in the Rhine River, which is eutrophic and whose discharge exerts a major influence on the Dutch Wadden Sea.

The nutrient and chemical cycles of estuaries are vulnerable to changes resulting from the riverine discharge of wastes from agricultural and industrial chemical processes and domestic sewage. As previously mentioned, runoff of agricultural fertilizers in the drainage basin of the Ythan Estuary increased the amounts of nitrogen and phosphorus introduced in river runoff. Increased nutrient loading of estuaries will increase the phytoplankton

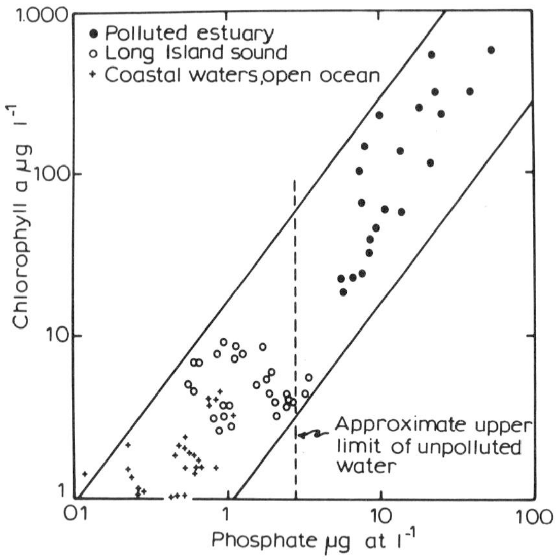

Fig. 4.12. Relationship between inorganic phosphate and chlorophyll content in waters ranging from unpolluted sea water to polluted estuaries (from Ketchum, 1969).

yield, as Ketchum (1969) has shown (Fig. 4.12). Nevertheless, *in situ* nutrient recycling mechanisms exert a major role in regulating the dynamics of nutrient supply to the phytoplankton in estuaries.

PHYTOPLANKTON RESPONSES TO ENVIRONMENTAL CONDITIONS

Salinity

The occurrence and distribution of the various algal groups represented in the phytoplankton are influenced by salinity (Table 4.3). Planktonic blue-green algae and euglenids tend to become relatively more important as habitat salinity decreases, being particularly well adapted to fresh-water habitats. Dinoflagellates exhibit the reverse trend, as do diatoms and coccolithophorids, although these two groups tend to be better represented at intermediate salinity ranges than dinoflagellates. Motile and non-motile planktonic representatives of the green algae (Chlorophyceae) seem to be equally well represented in both limnetic and marine water. Thus, while each of these algal groups is represented in marine, mixohaline and fresh waters, these differences in their general distributional trends reflect differences in their general tolerance to salinity. Interspecific differences in tolerance to salinity likewise occur.

The dissolved salts contributing to salinity have two kinds of effects upon aquatic organisms (Guillard, 1962). The specific chemical nature of the ions in solution and their specific effects on cellular processes represent one mode of action. The occurrence of differences in ionic composition of representative waters of varying salinity was pointed out previously (Table 4.1). The second effect of salt concentration is an osmotic effect which depends on the total salt content and regulates the movement of water into or out of cells.

The estuarine environment is subject to highly fluctuating levels of salinity (Fig. 4.2), and planktonic organisms move up and down the estuarine salinity gradient in consequence of tidal and other current movements. Therefore, the phytoplankton are often under widely fluctuating osmotic stress (cf. Ch. 3, pp. 43–63) which accompanies routine

TABLE 4.3

Relative occurrence of planktonic species from different microalgal groups in waters of differing salinity

Group	Sea water (>18‰)	Mesohaline/ oligohaline (0.5–18‰)	Fresh water (<0.5‰)
Blue-green algae	+	++	+++
Diatoms	+++	++	++
Coccolithophorids	+++	++	+
Dinoflagellates	+++	+	+
Euglenids	+	+	+++
Chlorophytes	+++	+++	+++

+++ greatest frequency; ++ intermediate; + least.

exposure to differing salt concentrations. The osmotic pressure of the cell sap of the planktonic microalgae usually exceeds that of the surrounding medium (Guillard, 1962). Representative fresh-water phytoplankton have excess osmotic pressures ranging from 4 to 8 bar; for marine diatoms 4.5 bar; and blue-green algae 11.5 bar. Brackish-water species also tend to maintain an osmotic value higher than that of the surrounding water. Fresh-water and marine species of phytoplankton thus confront a dual osmotic problem: the former must prevent a dilution of their cellular sap contents, while the marine representatives must prevent loss of water if transported from brackish water to areas of higher salinity. Therefore, species which require low salinity for survival and are displaced to downstream estuarine regions of high salinity require osmotic mechanisms to prevent excess water loss, which would result in plasmolysis and, eventually, death. Conversely, marine species displaced upstream into estuarine regions of very low salinity require mechanisms to minimize the uptake of water, which would lead to increased turgor and cause eventual bursting of the cells. Have estuarine phytoplankters evolved mechanisms to minimize this problem of osmotic stress?

Braarud's (1951, 1961) pioneering experiments on the influence of salinity on cell division of representative species of phytoplankton have revealed a widespread euryhaline nature. Fig. 4.13A illustrates this characteristic for a coccolithophorid (*Syracosphaera*), two species of dinoflagellates (*Amphidinium*, *Exuviaella*) and a cryptomonad flagellate (*Cryptomonas*). Growth occurred over a range from 5 to 10‰ to levels above full-strength sea water (35‰). Differences between species in growth response, salinity optimum and range of tolerance are also evident. The tendency for dinoflagellates to be less important in mixohaline and fresh water (Table 4.3) is also consistent with their growth responses to reduced salinity (Braarud, 1961).

Paasche (1975) examined the effect of salinity on the growth rates of the widely distributed diatoms *Skeletonema costatum* and *Skeletonema subsalsum*; both occur in the brackish Baltic Sea (Fig. 4.13B, C). The euryhaline nature of *Skeletonema costatum*, consistent with its occurrence in the Baltic Sea, is clearly evident: the optimal salinity for cell growth ranges from ~10 to 40‰ salinity, and good growth

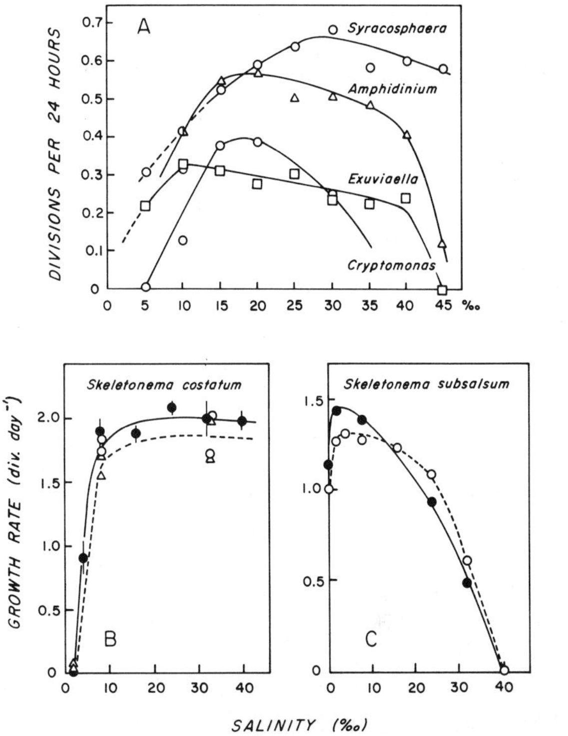

Fig. 4.13. A. Influence of salinity on the rate of cell division for several species of phytoplankton, for *Skeletonema costatum* (B) and for *Skeletonema subsalsum* (C). In the response-curves for *S. costatum*, filled circles represent the Oslofjord clone maintained at 27‰ salinity, open circles response of the same clone maintained at 8‰, and the triangles a Baltic clone maintained at 8‰. For *S. subsalsum*, cells were preconditioned to 8‰ salinity (open circles) and to each experimental salinity (filled circles). (A from Braarud, 1951; B, C from Paasche, 1975).

occurs even at 5‰ salinity. *Skeletonema subsalsum* grew best between 2 to 20‰, 2‰ being optimal, a response consistent with its preference for brackish water in nature. Note that despite the stenohaline distribution of *S. subsalsum* it, too, is characterized by euryhaline tolerance, a general characteristic of the phytoplankton. Paasche also established that both species grew equally well when exposed to different salinities, whether or not these clones were preadapted to, or near, the experimental salinity, or if they were exposed suddenly to higher or lower salinities from the preadapted level of 8‰ (Fig. 4.13B, C). This particular demonstration of euryhaline adaptability is very relevant since estuarine populations are likely to encounter abrupt changes in salinity level.

The general occurrence of euryhalinity amongst

the phytoplankton is due to their ability to osmoregulate, using several different osmoregulatory mechanisms (Guillard, 1962). This is not to say that estuarine phytoplankton have completely overcome the problems associated with osmotic shock or prolonged osmotic stress. Studies of the osmotic and ionic requirements of representative phytoplankton species indicate that their ability to maintain suitable cellular osmotic pressure while under osmotic stress is eventually lowered (Droop, 1958; Schmitz, 1959; Guillard and Myklestad, 1970). Destruction of cells may then occur because of specific cation deficiencies — such as for sodium, potassium, calcium or magnesium — or stressful total cation or total anion concentrations. Guillard and Myklestad (1970) have shown that even small changes in ionic composition (Table 4.1) may markedly influence the rate of growth during osmotic stress. Thus, even if cellular death did not occur, reductions in growth rate when accompanied by significant losses due to sinking, grazing and advection would lead to a population decrease, and possibly to the disappearance of the species in that segment of the estuary.

Changes in cellular osmotic pressure and ionic composition may also affect other cellular activities important to estuarine survival. These include chlorophyll synthesis (McLachlan, 1961) and rates of photosynthesis (Nakanishi and Monsi, 1965; Qasim et al., 1972). Diatom cell size (Chin et al., 1965) and cell wall morphology (Paasche et al., 1975) may also be influenced by salinity.

Attenuation by flushing

Diluted estuarine water continually drifts seaward, carrying away the constant input of fresh water that has entered the estuary from rivers. Since this offshore flow occurs in the upper layers, it usually displaces the euphotic zone populations, whereas more saline water flows into the estuary, usually at depth. This circulation pattern maintains the salt balance and general hydrostatic equilibrium (Chs. 1, 2). The entrained plankton populations are transported with the water and displaced from the estuary by this net seaward movement. In addition to osmotic stress, the indigenous estuarine species thus are subject to washout, another major problem. Stenohaline organisms exhibiting a preference for low salinity (Fig. 4.13C) are more vulnerable to displacement, or washout, from suitable habitats than are euryhaline species. The distribution of two diatom species in the Elbe River estuary illustrates the problem (Fig. 4.14). *Coscinodiscus lacustris* and *Coscinodiscus commutatus* exhibit dissimilar abundance and distributional patterns along the salt gradient. *C. lacustris*, which is found in the less saline portions of the estuary, apparently prefers lower salinity than does *C. commutatus*. How, given offshore displacement, are such discrete distributions established and maintained?

Ketchum (1954) has presented a detailed analysis of this problem (Fig. 4.15). The ability of an indigenous population or species to maintain itself within an estuary, or in a given segment of the estuarine salinity gradient, is determined by its rate of growth relative to the proportion of the population lost through flushing from the estuary, or a given segment, during each tidal cycle. That is, in a locally growing population, which may or may not be augmented from upstream or downstream, the steady-state population level can be determined by the rates of reproduction + recruitment and the loss

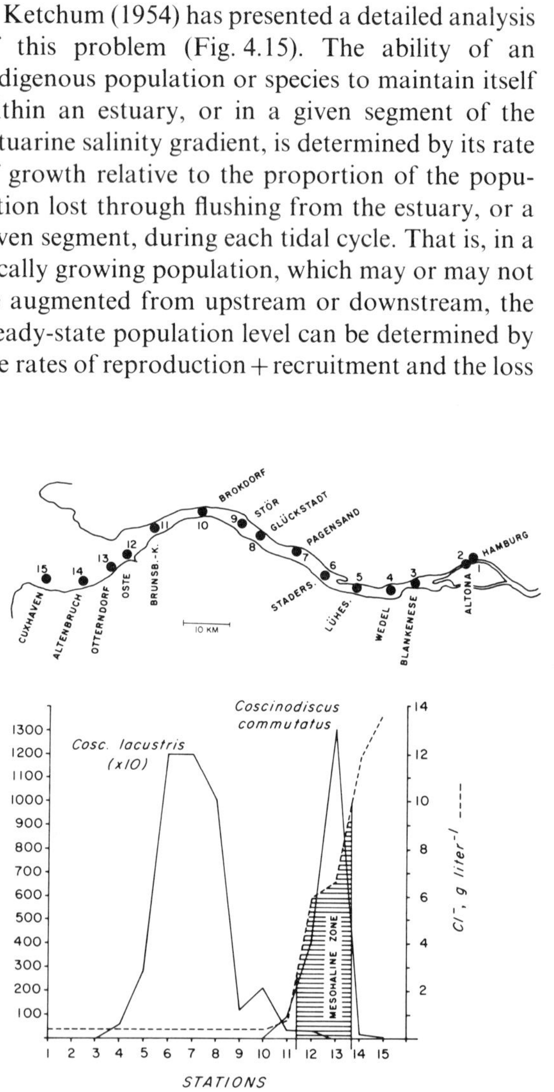

Fig. 4.14. The distribution of chloride ion concentration and of *Coscinodiscus lacustris* and *Coscinodiscus commutatus* during August 1957 in the Elbe Estuary between Hamburg and Cuxhaven, a distance of 105 km (modified from Schulz, 1961).

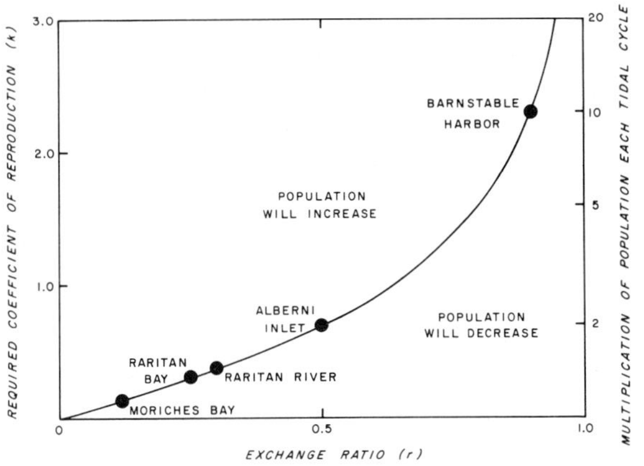

Fig. 4.15. The coefficient of reproduction required to maintain steady state, endemic phytoplankton populations in estuaries as a function of the exchange ratio; 1.44 divisions per tidal cycle correspond to $k = 1.0$ (from Ketchum, 1954).

rate due to circulation. The relative loss rate can be approximated from the exchange ratio (r): the proportion of water moving seaward during each tidal cycle which does not return on the following flood tide (Ketchum, 1954; cf. Ch. 2, p. 18):

$$P_m = P_0 \, (1 - r)^m \tag{1}$$

where m is the number of tidal cycles, r is the exchange ratio, P_0 is the initial population and P_m is the population size after m tidal cycles, assuming no growth. Combining the population growth equation:

$$P_t = P_0 \, e^{kt} \tag{2}$$

where P_t and P_0 represent the population levels after time t and initially, and growing at the rate k, with eq. 1 indicates that the population after m tidal cycles will be:

$$P_m = P_0 \, (1 - r)^m e^{mk} \tag{3}$$

To maintain a steady-state population, $P_m = P_0$, the growth rate required to balance population losses due to washout can be defined in terms of the exchange ratio r:

$$e^{mk} = \frac{1}{(1 - r)^m} \tag{4}$$

or,

$$k = -\ln (1 - r) \tag{5}$$

Therefore, when the rate of reproduction exceeds the rate of loss by flushing, the population will increase, but it will decrease when flushing exceeds reproduction.

Fig. 4.15 illustrates the relationship between growth rates and washout rates. In estuaries characterized by low exchange with offshore waters, relatively low growth rates allow an indigenous population to remain within the estuary. Thus, in Moriches Bay (New York, U.S.A.) a population generation (doubling) time of about three days is sufficient to achieve this, whereas a 4-h generation time is required in Barnstable Harbor (Massachusetts, U.S.A.), which is rapidly flushed. Thus, coastal species rather than estuarine forms predominate in Barnstable Harbor unlike in Moriches Bay (Ketchum, 1954).

The development of estuarine distributions such as those represented by *Coscinodiscus lacustris* and *C. commutatus* (Fig. 4.14) can thus be explained, despite dispersal through circulation. Seasonal and regional differences in occurrence and abundance of estuarine species reflect the balance between their recruitment and dispersal and loss rates accompanying circulation patterns. Thus, for *Coscinodiscus lacustris* the growth rate exceeds flushing in the Elbe Estuary between Blankenese and Otterndorf, and is considerably greater in the region from Staderstand and Pagensand, where it is at a population maximum. Conversely, flushing exceeds growth rate in the estuarine segment where *C. lacustris* is absent. Similar application of Ketchum's model can explain the different distributional pattern exhibited by *Coscinodiscus commutatus*.

Bakker and De Pauw (1974) have made some very important related observations on the phytoplankton composition in the tidally mixed Westerschelde Estuary and in the stagnant, man-made, brackish Lake Veere, both located in southwestern Netherlands. Significant differences in phytoplankton composition, despite the same salinity range, characterize these two regions. The dominant species in the Westerschelde are numerous large, heavily silicified diatoms, including *Coscinodiscus commutatus*, *Actinocyclus ehrenbergii*, and *Biddulphia sinensis*, whereas the dominant species in Lake Veere, which is flushed only twice a year, are small-celled microflagellates and diatoms such as *Skeletonema costatum*, *Chaetoceros muelleri* and *Detonula confervacea*. Bakker and De Pauw (1974) have suggested that the degree of turbulence is also a significant factor in determining the species composition along an estuarine salinity gradient. Tidal turbulence helps to keep in suspension large-celled, heavily silicified species (*Actinocyclus*, *Coscinodiscus*, etc.) which are transported up and down the estuary. Uniform euryhaline conditions, continuous resuspension to allow for photosynthesis and growth, and relatively long residence times are required to allow mass occurrences such as illustrated for *Coscinodiscus lacustris* and *C. commutatus* (Fig. 4.14). The absence of tidal or frequent wind-induced turbulence, as in Lake Veere, leads to water-mass stratification, the sinking of larger species kept in suspension through turbulence, and a longer estuarine residence time of the species present, generally motile species, small slowly sinking diatoms and other nannoplankters.

The effect of turbulent shear on cellular function and integrity during mixing also is a selective factor. It is generally believed that flagellates are more sensitive to turbulent motion than diatoms. Hulburt (1956) has applied such an explanation to observed phytoplankton succession in a coastal salt pond. He reasoned that the reduced tidal flushing and associated turbulence during summer favored the predominance of an indigenous microflagellate community, including the occurrence of a distinct "river community". During fall, winter and spring, when water column stratification was less pronounced and coastal water contributed 83% of total pond volume, diatoms predominated. Differences between the pond and coastal populations were then negligible.

COMPOSITION OF ESTUARINE PHYTOPLANKTON

The floristic composition of the phytoplankton communities found in estuarine waters varies in response to several factors. At a very general level, species selection depends on biogeographic factors. Tropical species, for example, cannot be expected to occur in boreal regions. Within the biogeographic range of a given species, regional variations in its occurrence and relative importance commonly are observed. The presence or absence, and the abundance of a given species of phytoplankton within a given estuary, are influenced by such locally characteristic environmental conditions as temperature, light, nutrients, salinity, and other water quality characteristics. The phytoplankton species composition reported for representative estuarine types will be described to demonstrate both the diversity and some general similarities encountered in estuarine phytoplankton composition. The taxonomy of phytoplankton presently is undergoing considerable re-examination aided by the use of transmission and scanning electron microscopy. I have not attempted to incorporate these taxonomic revisions with regard to changes in species' names. Rather, I have retained the species' names applied by the investigators.

Large brackish seas

The taxonomic character of the phytoplankton assemblages found in large brackish seas differs from that of coastal and oceanic marine communities, presumably as a consequence of differences in salinity. Table 4.4 compares the phytoplankton species representation in the different algal groups found in the inland seas of the U.S.S.R. Several general trends are evident. The total number of phytoplankton species decreases with decreasing salinity. This is due primarily to a decrease in the number of species of planktonic diatoms and, secondarily, of dinoflagellates. Conversely, both the numbers and proportions of planktonic blue-green and green algal species increase with decreasing salinity. In the Aral Sea the most common blue-green algal species are *Chroococcus turgidus*, *Merismopedia glauca*, *Anabaena bergi* and *Lyngbya aestuarii*. *Oocystis socialis* is an important chlorophycean (Zenkevitch, 1963). Dinoflagellates found include *Exuviaella cordata*, *Peridinium trochoideum*, *P. achromaticum*, *P. subsalsum* and *Gonyaulax levanderi*. Diatoms, despite the few species present, dominate the biomass, *Actinocyclus ehrenbergi* var. *crasa* in particular. *Chaetoceros wighami*, *C. subtilis*, *Coscinodiscus grani* var. *aralensis*, *Skeletonema costatum*, *Melosira borreri*, *Thalassiosira decipiens* and *Bacillaria paradoxa* also occur.

In the slightly more saline Caspian Sea blue-green algae predominate, notably *Aphanizomenon*

TABLE 4.4

Numbers of phytoplankton species present in the brackish seas of the U.S.S.R. (modified from Zenkevitch, 1963)

	Kara Sea		Black Sea		Sea of Azov		Caspian Sea		Aral Sea	
Surface salinity:	7–34‰		17–18‰		<12‰		11–13‰		~10‰	
Species:	number	%	number	%	number	%	number	%	number	%
Diatoms	155	56	150	43	41	22	59	31	18	46
Dinoflagellates	84	30	146	42	52	28	28	15	11	30
Coccolithophores	–	–	18	5	–	–	–	–	–	–
Cyanophyceae	6	2	6	2	35	19	54	29	6	15
Chlorophyceae	15	6	13	4	48	26	29	15	4	10
Flagellates[1]	16	8	17	5	12	6	17	9	–	–
Σ species	276		350		190		187		39	

[1] Includes silicoflagellates, euglenids.

flos-aquae, *Nodularia spumigena*, *Anabaena bergi* and *Merismopedia tenuissima*. Chlorophyceans include those reported from the Aral Sea and *Dictyosphaerium ehrenbergianum* var. *subsalsa*. Included in the 59 species of diatoms reported by Zenkevitch in the Caspian Sea are 17 species of *Chaetoceros*, 10 species of *Coscinodiscus* and 6 species of *Thalassiosira*; more than 50% of the pelagic diatom species reported belonged to these three genera. *Skeletonema costatum*, *Cyclotella caspia* and *Actinocyclus ehrenbergi* are also present. Of the dinoflagellates, *Exuviaella cordata* is important in the food web, and *Prorocentrum micans* var. *scutellum* and *Gonyaulax polyedra* are significant.

Karayeva and Makarova (1973) found more diatom species in their recent study of the features and origin of the Caspian Sea diatoms, and stated that only 77 (27%), of the total number of diatom species present were planktonic, and that only one third of the total number were limnetic species.

The phytoplankton composition in the Caspian Sea, the world's largest landlocked body of water (Karayeva and Makarova, 1973), is intermediate between that of the Aral Sea and the Sea of Azov, with which there is great taxonomic affinity. The eutrophic Sea of Azov, an arm of the Black Sea and essentially an estuary of the Don River, is characterized by an extraordinary productivity of the species-rich blue-green algal component, including those species listed for the Caspian Sea. The diatom component includes *Skeletonema costatum*, *Thalassiosira nana*, *Coscinodiscus biconicus*, *Chaetoceros subtile*, *Rhizosolenia calcar-avis*, *Biddulphia mobiliensis*, *Ditylum brightwelli* and *Thalassionema nitzschioides*. These last five species are typical representatives of coastal waters; their occurrence in the Sea of Azov is attributed to introductions from the Black Sea.

Despite the low mean salinity of the Black Sea (Table 4.4; see also Ch. 11, pp. 253–292), the character of the phytoplankton flora is similar to that in marine coastal waters, including the North Sea and Norwegian coastal waters. The Black Sea phytoplankton is characterized by the occurrence of a species-diverse coccolithophorid community in which *Coccolithus huxleyi* dominates and, relative to the less saline seas, an increased importance of diatoms and dinoflagellates and decreased importance of planktonic blue-green algae and chlorophyceans. The similarities between the Black Sea phytoplankton composition and that in the much more saline coastal waters are not enigmatic. The Black Sea may be viewed as a tributary of the Mediterranean Sea from which euryhaline species have been introduced. Thus, representative coastal diatoms include *Skeletonema costatum*, *Thalassiosira radians*, *Cerataulina pelagica*, *Leptocylindrus danicus*, *Thalassionema nitzschioides*, *Rhizosolenia calcar-avis*, *R. fragilissima*. The genus *Coscinodiscus* is also well represented. Among the dinoflagellates are *Prorocentrum micans*, *Exuviaella cordata*, *Ceratium furca*, *C. fusus*, *C. tripos* and many species of *Peridinium*.

The large, brackish Baltic Sea (see Ch. 13, pp. 309–335), slightly larger than the Black Sea, is connected to the North Sea through the Belt Sea, Kattegat and Skagerrak (Fig. 4.16). Over the distance (~1800 km) from the innermost part of the

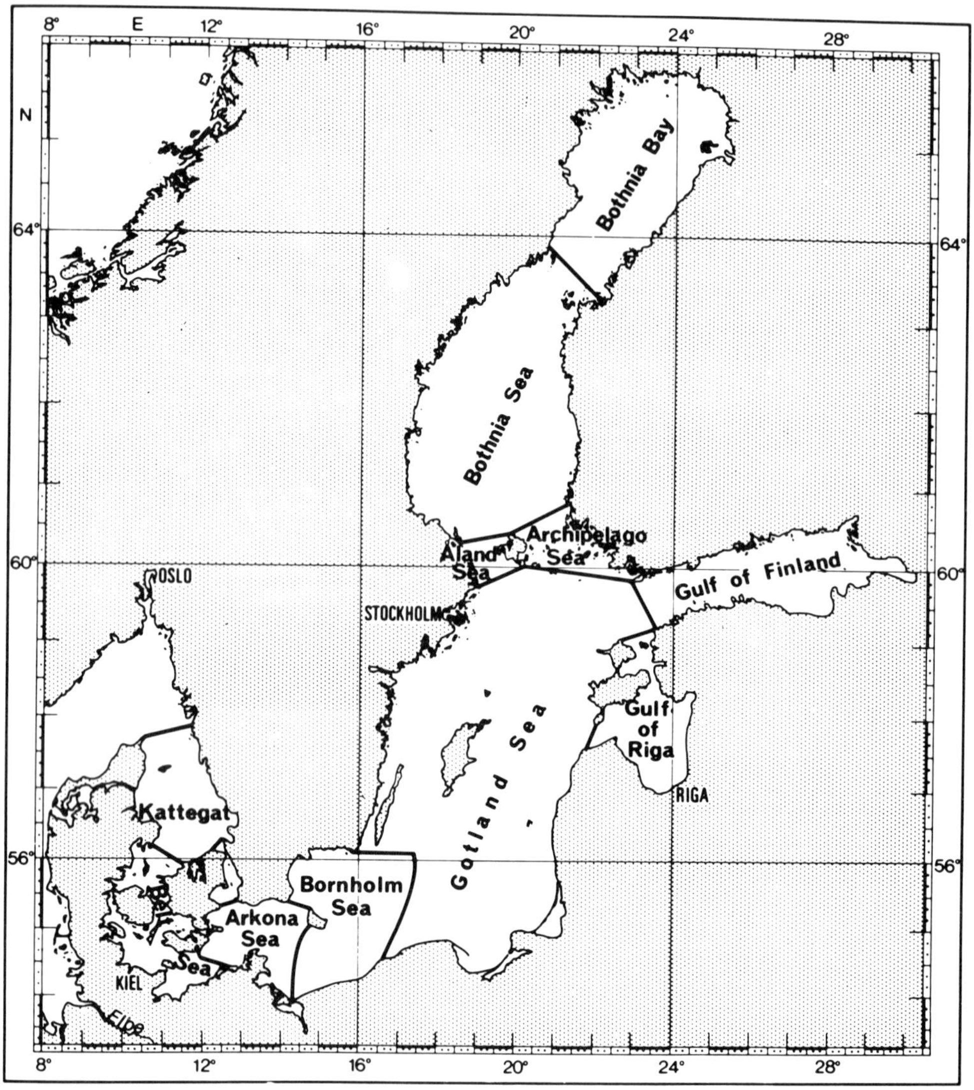

Fig. 4.16. Baltic Sea and its contiguous waters; region north of the Kattegat known as the Skagerrak (from Rheinheimer, 1977; see also Ch. 13, pp. 309–335).

Baltic Sea (Gulf of Bothnia) to the entrance into the Skagerrak the salinity increases from about 0 to 24‰ (Edler, 1977; Lenz, 1977a). Thus, the associated salinity gradient from the Baltic Sea and these outlets to the North Sea represents an estuarine gradient, as illustrated here for the transect between the Bornholm Sea and the Belt Sea (Fig. 4.17). The changing phytoplankton composition along this gradient is particularly interesting. Significant blooms of blue-green and chlorophycean algae occur within the Baltic Sea, particularly at lower salinities — an occurrence consistent with that already described for the large inland seas. The Gulf of Finland (~60°N), where salinity is <5‰, is characterized by a pollution gradient (Bagge and Niemi, 1971; Niemi, 1972; Lehmusluoto and Pesonen, 1973). The dominant blue-green algal species in the eutrophic regions are *Oscillatoria agardhii*, *O. limnetica*, and *Microcystis pulverea* var. *incerta*; important chlorophyceans include species of *Chlamydomonas* and *Scenedesmus*; key diatoms include *Stephanodiscus hantzschii* var. *pusillus*, *Chaetoceros perpusillus* and *Nitzschia acicularis*. The species composition in the oligotrophic and slightly

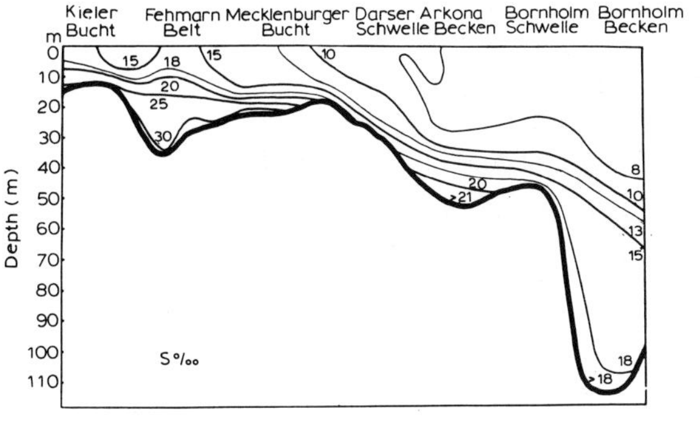

Fig. 4.17. Salinity distribution in a section from the western Baltic Sea to Kiel Bight (Kieler Bucht) located in the Belt Sea (from Remane and Schlieper, 1958).

eutrophic regions differs. *Aphanizomenon flos-aquae*, *Nodularia spumigena* and *Anabaena lemmermannia* dominate the blue-green algal community. A rich chlorophycean flora also occurs, including *Oocystis lacustris, Ankistrodesmus spiralis* and *A. falcatus* var. *mirabilis*. Among the diatoms are *Chaetoceros wighami* and *Skeletonema costatum* and, particularly in more oligotrophic waters, *Achnanthes taeniata*, *Melosira arctica* and *Thalassiosira baltica*. The dinoflagellate *Gonyaulax catenata* and the euglenid *Eutreptia marina* also occur.

The diatoms *Actinocyclus ehrenbergi*, *Coscinodiscus grani*, *Chaetoceros danicum*, *C. gracilis*, *C. holsaticus*, *Navicula grani*, *N. vanhoffeni*, *Nitzschia frigida* and *Skeletonema subsalsum*, and the dinoflagellates *Peridinium achromaticum*, *P. depressum*, *P. pellucida* and *Dinophysis baltica* also occur. The character of the phytoplankton in nearby regions, though still contiguous with the Baltic Sea, changes (Edler, 1977). The blue-green and chlorophycean components become less important, whereas species of diatoms and dinoflagellates characteristic of coastal-water communities appear. These changes suggest that the entrance into the Baltic Sea represents a salinity barrier. Some species which do not usually occur within the Baltic Sea and presumably have a minimal salinity requirement between 7 to 15‰ include (Smayda, 1958; Edler, 1977): *Chaetoceros borealis*, *C. decipiens*, *C. laciniosus*, *C. socialis*, *Guinardia flaccida*, *Lauderia borealis*, *Leptocylindrus danicus*, *Rhizosolenia alata*, *R. fragilissima*, *R. setigera*, *Thalassiosira gravida*, *T. nordenskioeldii* and the dinoflagellates (Nordli, 1953) *Ceratium fusus*, *C. furca* and *C. tripos*. Some species, including *Achnanthes taeniata*, *Chaetoceros holsaticus* and *C. wighami*, are introduced into these waters from the Baltic. The euryhaline (Fig. 4.13) diatom *Skeletonema costatum* is distributed throughout the Baltic Sea "estuarine" system. Edler's (1977) description of the seasonal successional cycle in the Kattegat illustrates the changed character of the phytoplankton, presumably influenced by the more saline conditions. Large diatoms (*Actinocyclus octonarius*, *Coscinodiscus* spp., *Porosira glacialis*) dominated the winter phytoplankton minimum, which were succeeded by others (*Achnanthes taeniata*, *Thalassionema nitzschioides*, *Thalassiosira nordenskioeldii*) transitional into the spring phytoplankton bloom. Diatoms also dominated then, including *Chaetoceros diadema*, *C. holsaticus*, *C. socialis*, *C. teres*, *C. wighamii*, *Detonula confervacea* and *Skeletonema costatum*. This community was replaced during late spring by species which persisted through the summer and early fall: *Rhizosolenia alata*, *R. fragilissima* and the dinoflagellates *Ceratium fusus*, *C. lineatum*, *C. longipes*, *C. tripos* and *Peridinium divergens*. Blue-green algae also occurred during summer, particularly in polluted waters when the temperature reached 17 to 18°C.

The annual primary production in the Kattegat varied regionally from about 75 to 185 g C m^{-2} (Edler, 1977), which compares to 80 to >120 g C m^{-2} in the eutrophic, mixo-oligohaline (see Table 4.7) waters of the Gulf of Finland (Lehmusluoto and Pesonen, 1973). The photosynthesis of diatoms and dinoflagellates is primarily responsible for this production in the Kattegat, and blue-green algae in the Gulf of Finland. The biogeographic data available for the large brackish seas is consistent in showing that such floristic differences are significantly influenced by the salinity. Blue-green algae and chlorophyceans become favored at very low salinity and, accordingly, in the very dilute reaches of an estuary. At higher salinities diatoms, dinoflagellates and coccolithophorids become favored.

Elbe River Estuary

Over a distance of about 100 km, between Hamburg and Cuxhaven, the salinity of the Elbe Estuary increases from <0.5 to 18‰ (Caspers, 1959; Figs. 4.14, 4.18). Schulz (1961) has reported that 700 species and subspecies of phytoplankton and zooplankton occur along this salt gradient. Although the distribution within the estuary of the

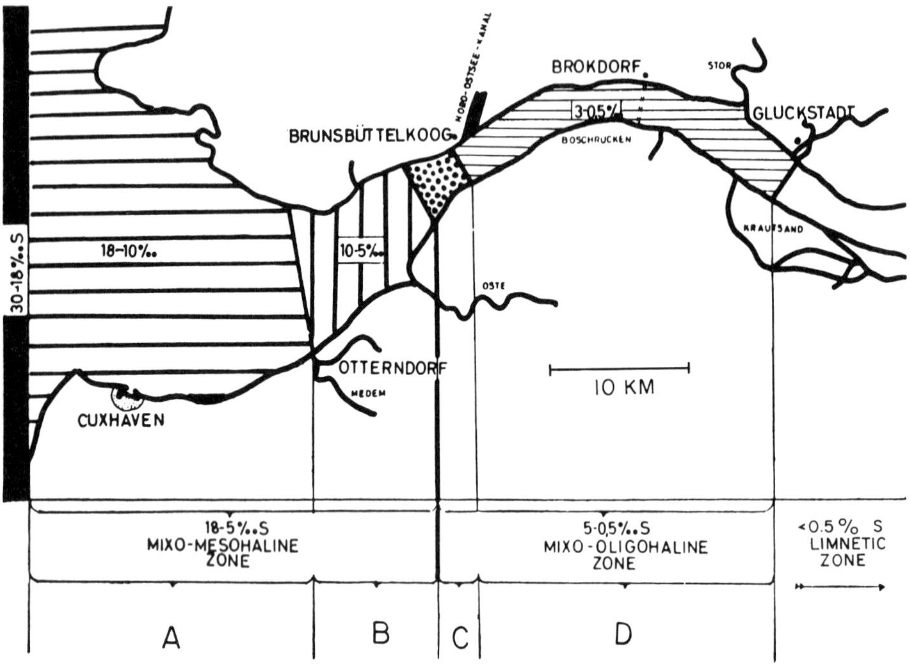

Fig. 4.18. Hydrographic subdivisions of the Elbe Estuary based on salinity distribution. The subzones shown are A, α mesohaline (18–10‰ salinity); B, β mesohaline (10–5‰ salinity); C, α oligohaline (5–3‰ salinity); and D, β oligohaline (3–0.5‰ salinity) (see p. 000). (From Caspers, 1959.)

individual species varies, recognizable communities occur that are characteristic of the different salinity zones. Four zones defined according to the Venice System (Table 4.2) are found (Fig. 4.18). In the limnetic (fresh-water) region of the estuary (<0.5‰) chlorophyceans and diatoms are preeminent. Massive summer blooms $>10^7$ cells l^{-1} occur; the dominant species include the chlorophyceans *Ankistrodesmus falcatus*, *Crucigenia rectangularis*, *Dictyosphericum* sp., *Richteriella botryoides*, *Scenedesmus* sp. and *Tetraedron arthrodesmiformis* and the diatoms *Melosira granulata*, several species of *Cyclotella*, *Stephanodiscus hantzschii* and *Nitzschia actinastroides*.

The transition zone (0.5 to 1‰) into the mixo-oligohaline region (0.5–5‰) is characterized by extensive blooms of the diatom *Actinocyclus normanni*, which is supplemented and eventually outdominated downstream by the diatoms *Coscinodiscus lacustris* (Fig. 4.14), *Cyclotella striata* and *Stephanodiscus lucens*. The chlorophyceans also occur but are reduced in abundance relative to the limnetic zone.

Massive occurrences of *Coscinodiscus commutatus* (Fig. 4.14) characterize the mixo-mesohaline zone (5–18‰; Fig. 4.18), accompanied by the brackish species *Chaetoceros subtilis*, *C. danicus* and several species of *Actinocyclus* and *Melosira*. Chlorophyceans are reduced in species representation and overall abundance; *Dictyosphaerium ehrenbergianum* and *Scenedesmus obliquus* are present. At the mouth of the estuary near Cuxhaven (Figs. 4.14, 4.18), where the salinity is about 16‰, the estuarine phytoplankton community begins to assume a floristic character similar to northern European coastal waters. Chlorophyceans and *Coscinodiscus* are less significant, whereas the diatom genera *Biddulphia*, *Chaetoceros*, *Rhizosolenia*, and *Thalassiosira* each are represented by several or many species. Dinoflagellates, including *Ceratium furca*, *C. fusus* and *C. tripos* and many *Peridinium* species, also occur.

Aside from these noteworthy patterns in phytoplankton community structure, several species including the diatoms *Bacillaria paradoxa* and *Thalassiosira fluviatilis* occur throughout the estuary. The preeminence of chlorophyceans and of *Coscinodiscus* and *Actinocyclus* in the Elbe Estuary is consistent with the general trend for these organisms to become important members of estuarine phytoplankton communities, particularly at lower salinities.

The predominance of blue-green algae and chlorophyceans in the less saline regions of the Elbe Estuary, specifically, and the general character of the phytoplankton flora along the estuarine gradient, are consistent with the observations from large brackish seas. The great importance of the large centric diatoms *Actinocyclus normanni*, *Coscinodiscus lacustris* and *C. commutatus* in the Elbe Estuary is also interesting. These genera are also important in the brackish seas, but generally are unimportant in coastal and oceanic waters. This suggests that among the diatoms they are particularly well adapted to an estuarine environment. Both genera are characterized by large-celled species. Thus, their preponderance within a given estuary can be expected to influence food web processes. The abundant *Actinocyclus ehrenbergii* in the Aral Sea, for example, is grazed by the dominant copepod *Diaptomus salina*, which is foraged on by fish (Zenkevitch, 1963). In contrast, dinoflagellates and chlorophyceans are grazed on by rotifers and certain crustaceans.

This general estuarine character and role of *Actinocyclus* and *Coscinodiscus* is further suggested by their dynamics in a Sierra Leone estuary (8°30′N, 13°W) characterized by exceptional high temperatures (>25°C) and river runoff during the wet season (Bainbridge, 1960). The most salient feature of the phytoplankton community was the dominance of large centric diatoms, usually of minor importance elsewhere, characterized by the following succession: *Coscinodiscus concinnus* (January), *Actinocyclus octonarius* (February–April), *Actinocyclus octonarius* + *Coscinodiscus leptoporus* (May–June), *Actinocyclus octonarius* + *Coscinodiscus jonesianus* (June–July). Bainbridge observed zooplankton to be less abundant in the inner reaches of this estuary, where these large centric diatoms predominated. The primary grazer on these large cells was the clupeoid fish, *Ethmalosa dorsalis*.

Coastal lagoons, salt ponds, oyster polls

Collectively these are minor embayments, often landlocked, connected to a major body of water through a narrow channel or breachway usually less than 100 m wide. The remarkably heterogeneous phytoplankton assemblages and their responses to the locally unique environmental conditions provide important insights into the variety of species and their successional patterns occurring in estuarine environments. Two landlocked fjord arms in Norwegian coastal waters — Nordåsvatn (Braarud and Hope, 1952) and Hunnebunnen (Braarud and Føyn, 1958) — provide examples. Hunnebunnen (1.8 km long, 0.7 km wide, maximum depth 11 m) is characterized by a year-round anoxic layer at depth. In the transition zone between the oxygenated and hydrogen sulphide layers, an abundant Euglenaceae population persists and may actually inhabit the anoxic waters. Following ice melting, a prolific phytoplankton abundance and succession persists from April through October, during which time there is a succession of both brackish species, including the diatoms *Cyclotella caspia*, *Chaetoceros ceratosporum*, *C. danicus*, *C. wighami*, and of marine forms such as *Skeletonema costatum*, *C. socialis*, and the dinoflagellates *Ceratium fusus*, *C. lineatum*, *Exuviaella baltica*, *Dinophysis borealis*. Significant vertical gradients in the distribution of several species occur because of vertical salinity gradients. Such estuarine distributions are in addition to horizontal gradient distributions, such as those described for the Elbe Estuary.

Great Pond, located in Massachusetts on the northeastern coast of the United States, provides a vivid, representative example of very small (2.4 x 0.4 km) and shallow (3 m) estuarine systems in which microflagellates dominate the summer flora and successional pattern (Hulburt, 1956). The spring diatom bloom replaced the dinoflagellate *Heterocapsa triquetra* and consisted of *Leptocylindrus minimus*, *Nitzschia delicatissima* and *Chaetoceros debilis*. *Skeletonema costatum* occurred year-round. The diatom flora was derived from contiguous Vineyard Sound and dominated except during the summer. The composition of the summer flora, which varied with salinity, consisted primarily of microflagellates and progressed from a June abundance of an unidentified chrysophycean to dominance by the chlorophycean *Bipedinomonas pyriformis* and cryptophycean *Rhodomonas minuta* (July), to an August abundance of *Nephrochloris salina* and the diatom *Chaetoceros simplex*. *Thalassiosira nana* and the dinoflagellates *Massartia rotundata* and *Gyrodinium estuariale* also exhibited summer maxima. In the less saline regions two *Chlamydomonas* species predominated. Hulburt concluded that this predominance of microflagellates is favored by the decreased turbulence, minimal advective losses because of weak currents, and overall shallowness.

The altered phytoplankton composition in Great South Bay and Moriches Bay, Long Island, located at the same latitude on the eastern coast of the United States, illustrates the effect of eutrophication on such shallow estuarine systems. Nitrogen fertilization from duck farm wastes has led to an enormous proliferation of the chlorophytes *Nannochloris atomus* and *Stichococcus* cf. *cylindrus*, which have virtually replaced the normal tidal estuarine phytoplankton in these embayments (Ryther, 1954). The diatom flora, including *Thalassiosira nana*, *Leptocylindrus minimus*, *Chaetoceros simplex* and *Skeletonema costatum*, is similar to that found in Great Pond, and characterized by summer populations of 10^7 to 10^9 cells l^{-1}. *Nannochloris* populations exceeded 10^{10} cells l^{-1}.

The phytoplankton observations in these shallow coastal lagoons characterized by restricted exchange with offshore waters illustrate another characteristic of estuarine phytoplankton composition. Reduction to lower estuarine salinity accompanied by elevated nutrient supplies, such as during eutrophication, provides both an osmotic and nutritional stimulus favoring the predominance of chlorophyceans over diatoms and dinoflagellates.

The reader who is interested in the phytoplankton composition in other representative estuarine types may wish to consult the works of Braarud et al. (1958, 1974) and Sakshaug (1972) dealing with Norwegian fjords, and Conover (1956) for Long Island Sound.

PHYTOPLANKTON DYNAMICS IN SOME REPRESENTATIVE ESTUARINE SYSTEMS

Kiel Bight

The Kiel Bight (Kieler Bucht) forms the southwest part of the Belt Sea (Fig. 4.19). Through the Fehmarn Belt (Figs. 4.17, 4.19) it receives the westward flow of Baltic Sea water, which continues through the Belt Sea northwards into the Kattegat and eventually into the North Sea via the Skagerrak (Lenz, 1977a). Although the salinity in the Bight is increased over Baltic Sea levels (Fig. 4.17), it is subject to large annual fluctuations, the extremes ranging from 10 to 30‰.

A series of fjord-like extensions penetrates inland from the Kiel Bight; these include Kiel Fjord, Eckernförde and Flensburg Fjord (Fig. 4.19). Kiel Fjord, the site of classic observations for more than a century, including those by Victor Hensen, Karl Brandt and Hans Lohmann (Gran, 1912; Porep, 1970), has an average depth of 15 m, extends inland 17 km and receives the runoff from a minor river, the Schwentine. It is not a threshold fjord — that is, a sill is not present; thus its bottom water is in direct exchange with the Kiel Bight. A summary of some hydrographic and biological observations carried out in Kiel Fjord and Eckernförde will illustrate some key features of phytoplankton dynamics in this type of estuarine situation. Significant observations are reported in the unpublished dissertations of Smetacek (1975) and Von Bodungen (1975).

Observations made on water samples at 2 m collected over an annual cycle at Kiel Fjord station *2* (54°24′N, 10°12′E), where the depth is 14 m, illustrate the following trends (Table 4.5). Monthly salinity values ranged from 13.60 to 19.82‰. Significant differences in salinity occur between the top and bottom layers throughout Kiel Bight. For example, at the entrance to Eckernförde (station *A*), the mean monthly difference between surface and bottom layer salinity ranges from 1.9 to 8.6‰ (Lenz, 1977a). Thus, strong seasonal halocline stratification characterizes this region, a situation further facilitated by the annual temperature cycle, which ranges over approximately 20°C in the upper layers, but only about 10°C at depth. Extreme surface layer values range from −1.2 to 20°C. The combination of low temperatures and salinity periodically results in ice formation, as during the winter of 1962–1963 when an ice layer 20 to 35 cm in maximum thickness persisted for two months (Hickel, 1969). Despite seasonal stratification and periodic icing, strong wind-induced vertical turbulence also occurs during the winter.

An annual cycle in nutrients of a type characteristic of many coastal waters in the boreal–temperate region occurs (Rheinheimer, 1977). Pronounced seasonal oscillations characterized the ammonia (0.6–39.1 μg-at l^{-1}), nitrate (0.09–12.4) and phos-

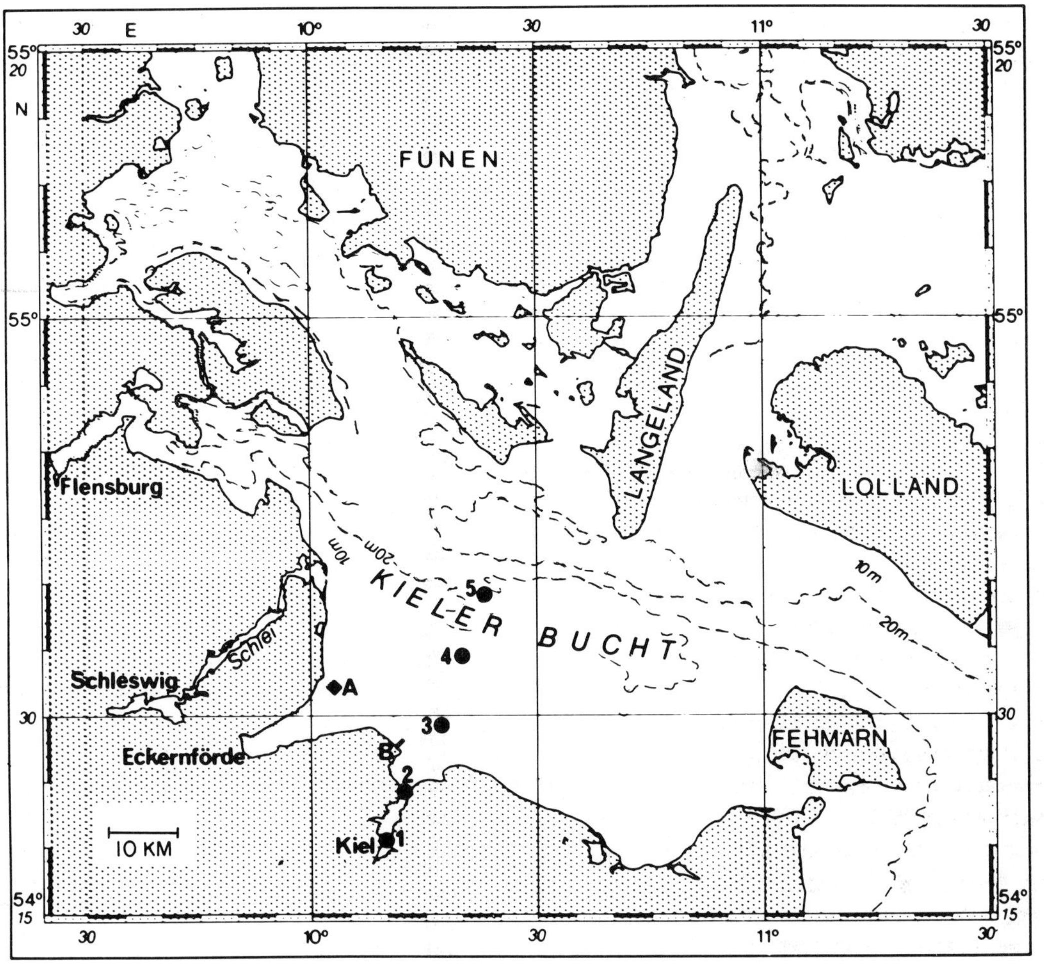

Fig. 4.19. Kiel Bight (Kieler Bucht) and locations of Kiel Fjord and Eckernfjord (Eckernförde) (from Rheinheimer, 1977).

phate (0.3–10.2) concentrations. These nutrients exhibit winter maxima and summer minima. Nutrient concentrations tend to be higher within the fjords' inner reaches and lower offshore in the Bight. Summer ammonia concentrations tend to exceed nitrate levels during the period of maximum stratification. The source of these elevated ammonia levels has been attributed to excretion by herbivorous zooplankton (Rheinheimer, 1977) during their grazing of phytoplankton. Since these animals are ammonotelic, nitrate levels remain low because of limited excretion.

The pronounced decrease in nutrients observed during April is a result of the spring bloom in phytoplankton (Table 4.6), whose continuous growth during the summer keeps nutrients at the observed low levels. The amplitude of the bloom maximum compared to the minimum during the year is about 1800-fold expressed as cell numbers, 19-fold as chlorophyll and 31-fold as carbon. This amplitude in seasonal abundance of phytoplankton standing stock is considerably greater than that characteristic of Cochin Backwater (India), the tropical estuary to be considered later. Diatoms are almost exclusively responsible for the observed variations, with dinoflagellates becoming very important during July.

More extensive studies are being carried out at station *A*, at the entrance to Eckernförde (Fig. 4.19), using a greater sampling frequency. These provide deeper insight into phytoplankton dynamics and its environmental regulation in the Kiel Bight estuarine system. Von Bodungen et al. (1975) have shown that a series of phytoplankton

TABLE 4.5

Temperature, salinity and nutrients (as μg-at l^{-1}) at 2 m depth at Kiel station *2* (from Lenz, 1977a; Rheinheimer, 1977)

	Temperature (°C)	Salinity (‰)	Nutrients		
			NH_4^+	NO_3^-	PO_4^{3-}
January 1974	3.6	14.63	8.2	29.6	1.5
February	3.1	17.17	2.0	13.2	2.8
March	3.8	16.87	4.1	8.7	1.7
April	8.6	14.80	0.6	0.1	0.5
May	10.4	14.99	3.8	1.9	0.9
June	12.8	15.12	4.0	0.3	0.5
July	16.1	16.02	3.1	0.5	1.0
August	17.7	14.65	1.8	0.2	0.3
September	15.9	16.69	2.1	0.4	0.9
October	10.1	18.12	13.5	0.7	0.8
November	7.6	17.27	4.1	5.1	1.5
December	6.7	17.08	4.9	11.5	1.9
January 1975	5.3	19.82	39.1	12.4	1.7
February	3.9	14.49	3.2	10.3	0.9
March	4.2	13.60	1.5	2.7	?10.3

TABLE 4.6

Phytoplankton abundance as numbers (cells l^{-1}), chlorophyll and carbon (μg l^{-1}) at 2 m depth at Kiel station *2* (from Lenz, 1977b, c)

	Cell number	% Diatoms	Chlorophyll	Carbon[1]
January 1974	26	98	0.9	26.8
February	12	100	1.0	32.3
March	668	100	2.7	72.3
April	11 700	100	8.4	?
May	115	99	3.0	59.3
June	21 950	100	6.8	372.5
July	6290	10	7.8	?
August	577	74	4.4	282.0
September	331	95	2.9	113.5
October	315	92	3.7	44.8
November	20	53	1.6	37.8
December	?	?	1.1	25.5
January 1975	113	99	1.1	12.0
February	20	98	1.1	27.0
March	2290	99	16.8	117.5

[1] Converted from ATP data presented in Table 5.13 by Lenz (1977c), assuming a C: ATP ratio of 250:1.

blooms, five in all, occurs over an annual cycle (Fig. 4.20). The first bloom, in March, was caused by the diatoms *Detonula confervacea* and *Skeletonema costatum*; in May the dinoflagellate *Exuviaella baltica* bloomed; the third in July was a blue-green algal bloom characterized by *Anabaena* sp. and *Nodularia* sp. and some diatoms; the fourth bloom in August was dominated by the diatom *Rhizosolenia faeroensis*; and the dinoflagellates *Ceratium fusus* and *C. tripos* produced an October bloom.

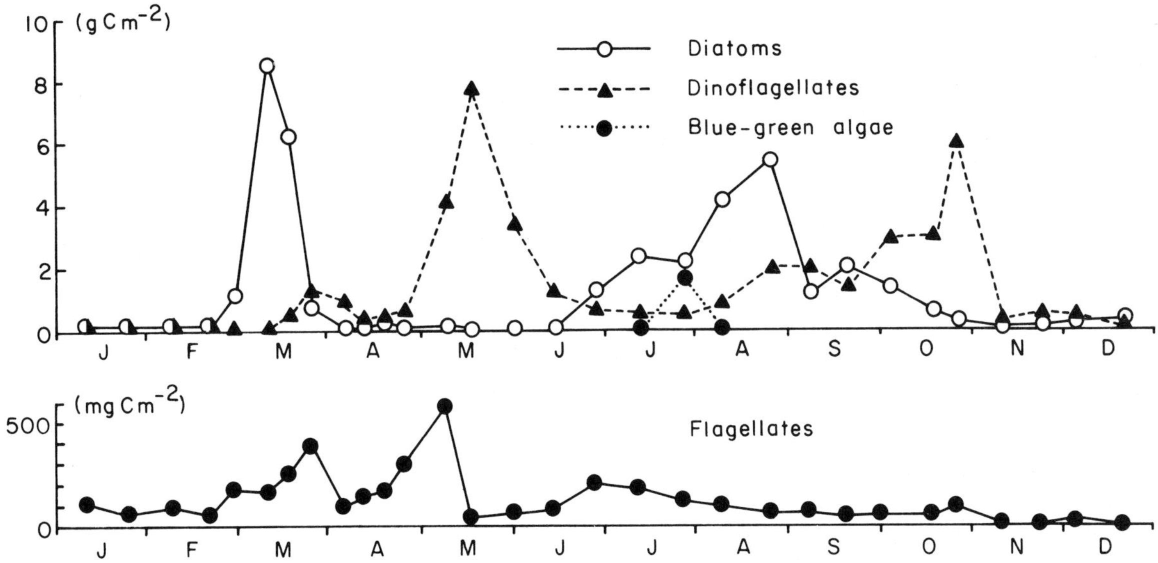

Fig. 4.20. Seasonal cycle in carbon content of various systematic groups represented in the phytoplankton in the 20-m water column at Station *A* located at the entrance to Eckernförde (from Von Bodungen et al., 1975).

Hickel's (1967) study at nearby outer Flensburg Fjord indicates that inception of the spring bloom in Kiel Bight is delayed until light limitation is overcome. The daily compensation depth at this time varies from 2 to 6 m, and the incident light intensity from about 70 to 450 g cal cm^{-2}. The bloom begins in late February or March when vertical mixing of the water column becomes less intense and an increased residence time of the phytoplankton within the euphotic zone accompanies thermohaline stratification of the water column. The subsequent sequence of blooms, like that observed by Von Bodungen et al. (1975), then reflects nutrient availability and, secondarily, zooplankton abundance. The collapse of the March spring bloom in April, 1974 is attributable to nutrient exhaustion (Table 4.5) and not to grazing (Fig. 4.21). Zooplankton abundance remained low during this bloom, and the non-predated diatom population sank to the sea floor where it was visible to divers as a loose, green mat. Remineralization of this decomposing phytoplankton deposit restored nutrients to the water column at depth, and uptake by the dinoflagellate *Exuviaella baltica* led to its

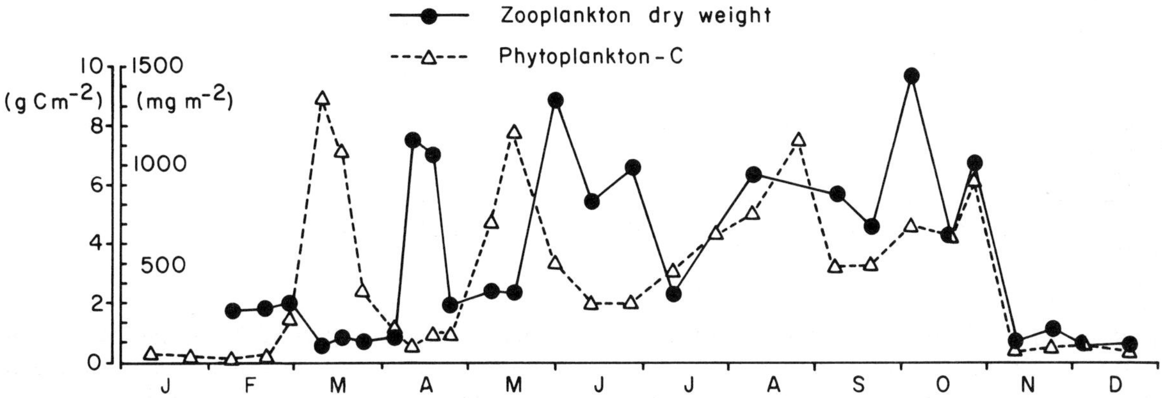

Fig. 4.21. Seasonal cycles in zooplankton biomass (mg m^{-2} dry weight) and phytoplankton carbon (g m^{-2}) in the 20-m water column at Eckernförde Station *A* (from Von Bodungen et al., 1975).

mid-May bloom. The biomass peak of 7 g C m^{-2} approached the spring diatom maximum of 9 g C m^{-2}. Unlike during the March–April bloom, however, grazers became important, and heavy grazing of the *Exuviaella* bloom (Fig. 4.21) contributed to its decline in June. Von Bodungen et al. (1975) reported that the excess phytoplankton biomass — that is, the ungrazed portion — sedimented to the halocline depth and remineralized, releasing nutrients into the water column. This mechanism, together with excretion of nutrients by zooplankton, provided periodic buildups of nutrients leading to the July (<4 g C m^{-2}) and August (7 g C m^{-2}) blooms (Fig. 4.20). The final bloom (6 g C m^{-2}) of *Ceratium*, mostly *C. fusus*, observed also by Hickel (1967), followed extensive wind-induced mixing of the stratified water column which released accumulated bottom-water nutrients into the euphotic zone. The daily compensation depth and incident radiation then varied from 2 to 11 m and ~50 to 350 g cal cm^{-2} (Hickel, 1967). A decreased zooplankton population decimated through grazing by the ctenophores *Bolinopsis infundibulum* and *Pleurobrachia pileus* (Lenz, 1977c) is also favorable to development of this bloom. Termination of this bloom accompanied the occurrence of nutrient-poor, low-saline conditions.

Phytoplankton dynamics in the Kiel Bight estuarine system differs from that occurring in many open coastal waters of boreal–temperate regions, which are usually characterized by two major phytoplankton blooms over an annual cycle. In the brackish Kiel Bight, blooms occur more or less continuously from late February through late October. The initiation of this bloom sequence is delayed until light limitation is overcome, and its eventual termination also occurs because of light limitation. Within this period, nutrient regulation of phytoplankton growth occurs. Blooms and maintenance growth are then dependent on the pulsing of nutrients into the euphotic zone from the sea floor and their recycling within the water column during grazing or microbial activity. The zooplankton and phytoplankton populations are asynchronous in their abundance cycles (Fig. 4.21). There is usually a lag between phytoplankton blooms and zooplankton peaks, the delay interval being more pronounced at lower temperatures. The consequence of this is that regulation of phytoplankton dynamics by the zooplankton through their grazing and nutrient-excretion activities and trophic energy transfer is seasonally variable. Periodically, considerable amounts of ungrazed phytoplankton sink to the sea floor to be consumed by the benthic community, and even contribute to the development of anoxic conditions (Horstmann, 1972).

The Baltic Sea region generally represents an oligotrophic estuarine system characterized by low primary production rates and reduced nutrient recycling (Table 4.7). Annual production in the non-eutrophic portions throughout this region varies from about 15 to 75 g C m^{-2}; it is 60 g C m^{-2} in the central region of the Kiel Bight. However, significant increases in annual production accompany eutrophication; it is 200 g C m^{-2} in heavily polluted inner waters of Kiel Fjord and 158 g C m^{-2} in the slightly eutrophic waters at the entrance to Eckernförde. Data available for the seasonal production cycle (Fig. 4.22) in inner Kiel Fjord demonstrate that the bulk of the annual production occurs during the summer, when nutrient turnover

TABLE 4.7

Annual primary productivity as g C m^{-2} in the Baltic Sea region (see Fig. 4.16 for locations)

Area	Production	Source[2]
Gulf of Finland		1
oligotrophic	15–39	
polluted	193	
Gotland Sea		
western	38–59	2
Danzig	73–118	3
Bornholm Sea	78	3
Arkona Sea	73	3
Kiel Bight		
Kiel Fjord station *1*	200	4
Eckernförde (*A*)	158	4
central region	60	5
Sound[1]		6
upper	183	
mid	148	
lower	73	
Belt Sea	55–66	7
Kattegat	51–82	7

[1] Region between Arkona Sea and Kattegat (Fig. 4.16).
[2] Sources: 1, Lehmusluoto and Pesonen (1973); 2, Schulz and Kaiser (1973); 3, Renk (1974); 4, Probst (1977); 5, Horstmann (1972); 6, Edler (1977); 7, Steemann Nielsen (1964).

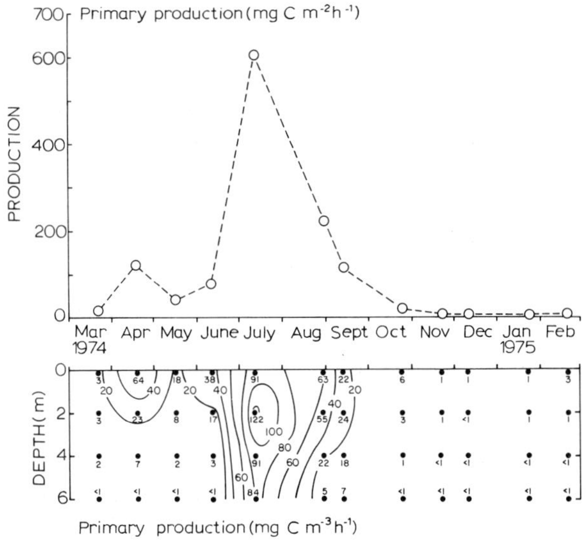

Fig. 4.22. Primary production per unit of sea surface (mg C m^{-2} h^{-1}) and within the water column (mg C m^{-3} h^{-1}) at Station *1* in the inner Kiel Fjord (from Probst, 1977).

must be very rapid to sustain the observed growth. While the elevated production here is attributable to eutrophication from pollution, that at the entrance to Eckernförde is not (Probst, 1977). Nutrients added through pollution are responsible for only about 25 g C m^{-2} of the annual rate of 158. The residual rate is still considerably above Kiel Bight levels of 60 g C m^{-2}. This indicates, therefore, that any mechanism, natural or otherwise, which adds or recycles nutrients rapidly will stimulate primary production in this general estuarine system. Thus, it is basically a nutrient-limited ecosystem. Horstmann's (1972) observations that eutrophication by pollution stimulated production in Kiel Bight by 4%, and by 25% in a more localized area, through increased growth of the existing phytoplankton flora is consistent with this conclusion.

Narragansett Bay

Narragansett Bay (~41°30′N, 71°15′W), located on the northeastern coast of the United States, is a shallow (mean depth of 9 m), well-mixed estuary of 460 km^2, contiguous to the south with Rhode Island Sound (Fig. 4.23). The length of the north–south axis of Narragansett Bay is 42 km and the average width 8 km. Although fed by five rivers,

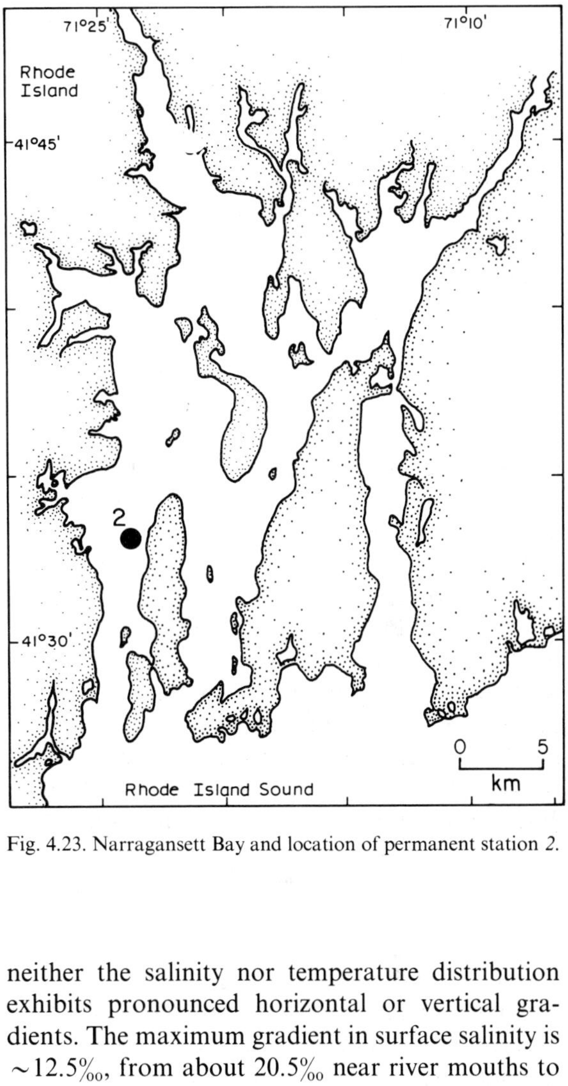

Fig. 4.23. Narragansett Bay and location of permanent station *2*.

neither the salinity nor temperature distribution exhibits pronounced horizontal or vertical gradients. The maximum gradient in surface salinity is ~12.5‰, from about 20.5‰ near river mouths to 32.5‰ at the mouth of the bay (Hicks, 1963). Minimal surface salinity of 24‰ is more representative, however. Tidal currents accompanied by considerable vertical mixing dominate the circulation pattern, and non-tidal flow occurs as a two-layer flow, with flow inwards at the bottom. Estimated flushing time of the bay is about thirty days (Kremer and Nixon, 1978). The extreme temperature range measured at the permanent station (Fig. 4.23) is −1.7 to 24.8°C; winter ice formation is usually not a problem in the shallower, less saline reaches of the bay. Narragansett Bay, because of tidal and wind-induced mixing, is not characterized by either a well-defined halocline or thermocline throughout most of its reaches. In this regard it differs from Kiel Bight and Cochin

Backwater described in this chapter, and from many of the large estuarine systems on the east coast of the United States, including Delaware Bay and Chesapeake Bay (see Fig. 4.10), and Norwegian fjords.

The mean daily light intensity ranges from about 85 to 670 g cal cm^{-2} and the seasonally variable compensation depth in the less turbid lower Narragansett Bay ranges from about 5 to 10.6 m. Representative seasonal trends in nutrient levels at the permanent station are shown in Table 4.8. Ammonia, nitrate, phosphate and silicate are characterized by late fall–early winter maxima, and spring–summer minima. Absolute concentrations range from undetectable levels to maxima of about 15 and 13 μg-at l^{-1} for ammonia and nitrate, respectively, and about 3 μg-at l^{-1} for reactive phosphate. The data selected for presentation in Table 4.8 illustrate an important feature of Narragansett Bay over a 20-year period of investigation, and presumably of estuaries generally. Considerable variation between years occurs in the nutrient, phytoplankton and zooplankton cycles, both in terms of patterns and abundance or concentration, influenced, no doubt, by climatological differences between years. Accordingly, some winters are exceptionally cold, others unusually warm; the same applies to the summer period and to incident light available for photosynthesis. In 1977, for example, surface temperatures in lower Narragansett Bay remained below 0°C from late December (1976) through February; the lowest recorded temperature was −1.7°C and ice persisted throughout this period. In contrast, in 1960 temperatures during this period ranged from 2 to 4.7°C, except during one week (<2°C). Such variations in the physical field obviously influence biological processes and cycles, as already pointed out for Kiel Bight, as well as in biologically essential nutrients. The latter is evident in the cyclical patterns for the various nutrients given in Table 4.8. Maximum silicate concentrations occurred from October 1975 through February 1976, but during the next yearly cycle they were nearly exhausted by December (1976), i.e., three months sooner than during the previous cycle. Similar dynamics in the periods of maximum concentrations are more or less evident for the other nutrients. Such annual differences in nutrient dynamics are accompanied by altered plankton cycles. Phytoplankton abundance expressed as chlorophyll concentration during this 1975–1977 time frame

TABLE 4.8

Maximum monthly surface concentrations of nutrients (μg-at l^{-1}) and chlorophyll (μg l^{-1}) observed from October 1975 through March 1977 at the permanent station in Narragansett Bay (T.J. Smayda, unpublished)

	NH_4^+	NO_3^-	PO_4^{3-}	SiO_3^{2-}	Chl
October 1975	13.5	9.33	2.62	24.43	3.3
November	9.7	7.79	2.28	25.85	6.3
December	7.5	12.78	2.21	32.45	2.1
January 1976	3.8	9.86	1.28	21.80	2.7
February	7.9	5.57	1.30	24.34	15.3
March	2.0	0.95	0.13	2.02	6.1
April	3.0	0.76	0.41	4.95	2.6
May	1.0	0.60	0.34	10.34	5.3
June	1.4	0.60	0.82	23.11	5.9
July	3.2	0.80	1.70	30.26	15.0
August	4.0	1.24	2.27	18.72	16.6
September	5.8	1.85	1.41	23.16	11.3
October	14.9	7.53	2.77	26.13	1.5
November	11.9	12.40	2.07	25.97	31.7
December	3.4	4.35	1.18	1.41	17.9
January 1977	1.1	6.09	0.94	0.82	18.7
February	1.2	1.86	0.38	1.06	9.2
March	2.1	1.33	0.07	2.16	13.1

indicates that the late winter phytoplankton bloom (15.3 $\mu g\ l^{-1}$ chlorophyll) occurred during February in the 1975–1976 season, but three months earlier, in November (31.7 $\mu g\ l^{-1}$ chlorophyll), during the 1976–1977 season. Thus, nutrient levels and cycling during this time of year are markedly influenced by the extent of phytoplankton growth.

The extraordinary year-to-year variation in phytoplankton abundance and seasonal cycling in Narragansett Bay has been described by Smayda (1976). Weekly sampling at the permanent station over a 15-year period has shown that a winter bloom usually begins in December but has started as early as November (1968) and as late as February (1965). While maximum abundance frequently occurs in January, it has occurred as late as March (1964, 1969) or April (1966). Maximum abundance has ranged from 4.5×10^6 (1970) to 42×10^6 cells l^{-1} (1963). Although this winter–spring bloom is a major characteristic of phytoplankton dynamics in Narragansett Bay, in reality a series of secondary pulses occurs year-round (Fig. 4.24, Hulsizer, 1976). This occurrence of nearly continuous growth compares to the five-pulse cycle described by Von Bodungen et al. (1975) for Eckernförde in Kiel Bight (Fig. 4.20), and the classic two-pulse cycle for open coastal waters. Considerable variations in bloom durations, times of occurrence and maximum abundance also characterize the period from late spring through autumn (Smayda, 1976). Maximum abundance has ranged from 7×10^6 cells l^{-1} (1967) to 59×10^6 cells l^{-1} (1975); duration of these seasonal blooms has ranged from a couple of weeks in August, 1969, to almost continuous blooming from July through September, 1968. The most extraordinary finding is that recently during four years (1970, 1973–1975) the annual phytoplankton maximum occurred in mid-August rather than sometime during the winter–spring period. The reasons for this shift are unknown, but it is not attributable either to pollution in these non-polluted waters of lower Narragansett Bay or to changes in species dominance.

Diatoms, particularly *Skeletonema costatum*, are primarily responsible for the observed phytoplankton dynamics (Smayda, 1957, 1973b; Pratt, 1959). *Skeletonema* abundance represents about 80% of total numerical abundance over the annual cycle. *Detonula confervacea* and *Thalassiosira nordenskioeldii* are important during the early stages of the winter–spring bloom. They are gradually replaced by *Leptocylindrus danicus*, *L. minimus*, *Cerataulina pelagica*, *Asterionella japonica* and *Rhizosolenia fragilissima* as important secondary dominant species. Unlike Kiel Bight, neither blue-green algae nor dinoflagellates are important in Narragansett Bay. The unimportance of blue-green algal species probably reflects the much higher salinity in this bay. The absence of dinoflagellates, notably *Ceratium*, is enigmatic. Significant blooms of the chrysophycean flagellate *Olisthodiscus luteus* occur during the summer (Pratt, 1966; Tomas, 1978). Bloom developments of this species, which can form non-toxic red tides, are somewhat variable and tend to show an inverse relationship with the occurrence of *Skeletonema costatum*. Pratt has presented experimental evidence that the interaction between these two species is partially mediated by the secretions of growth-promoting and growth-inhibiting organic substances. Aside from the sporadic summer occurrences of the dinoflagellates *Prorocentrum triangulatum*, *Peridinium trochoideum* and *Massartia rotundata*, diatoms predominate during the summer and autumn as well. Nannoplanktonic cells $<20\ \mu m$ in dimension also play a major role.

The inception of the major phytoplankton bloom, usually during December, coincides with the period of the annual minimum in daylength (9 h light, 15 h dark) and in incident radiation (60–135 g cal cm^{-2} day^{-1}). This situation together with Narragansett Bay being a well-mixed estuary, implicates light as a key factor regulating inception of this bloom. In Kiel Bight watermass stratification, which prolongs the residence time of the phytoplankton within the euphotic zone, is required for initiation of the March bloom there; this requirement, however, is usually unnecessary in Narragansett Bay, despite the incident light conditions and well-mixed water column. The overall shallowness of Narragansett Bay and the fact that the compensation depth includes much of the mean water column depth of 9 m at this time, allows sufficient residence time within the vertically mixed population in the euphotic zone. These conditions eliminate the need for water column stratification and allow this seemingly enigmatic bloom to occur (Smayda, 1957). This is not to say that available light never limits inception of this bloom. During the 1972–1973 winter growth season incident light

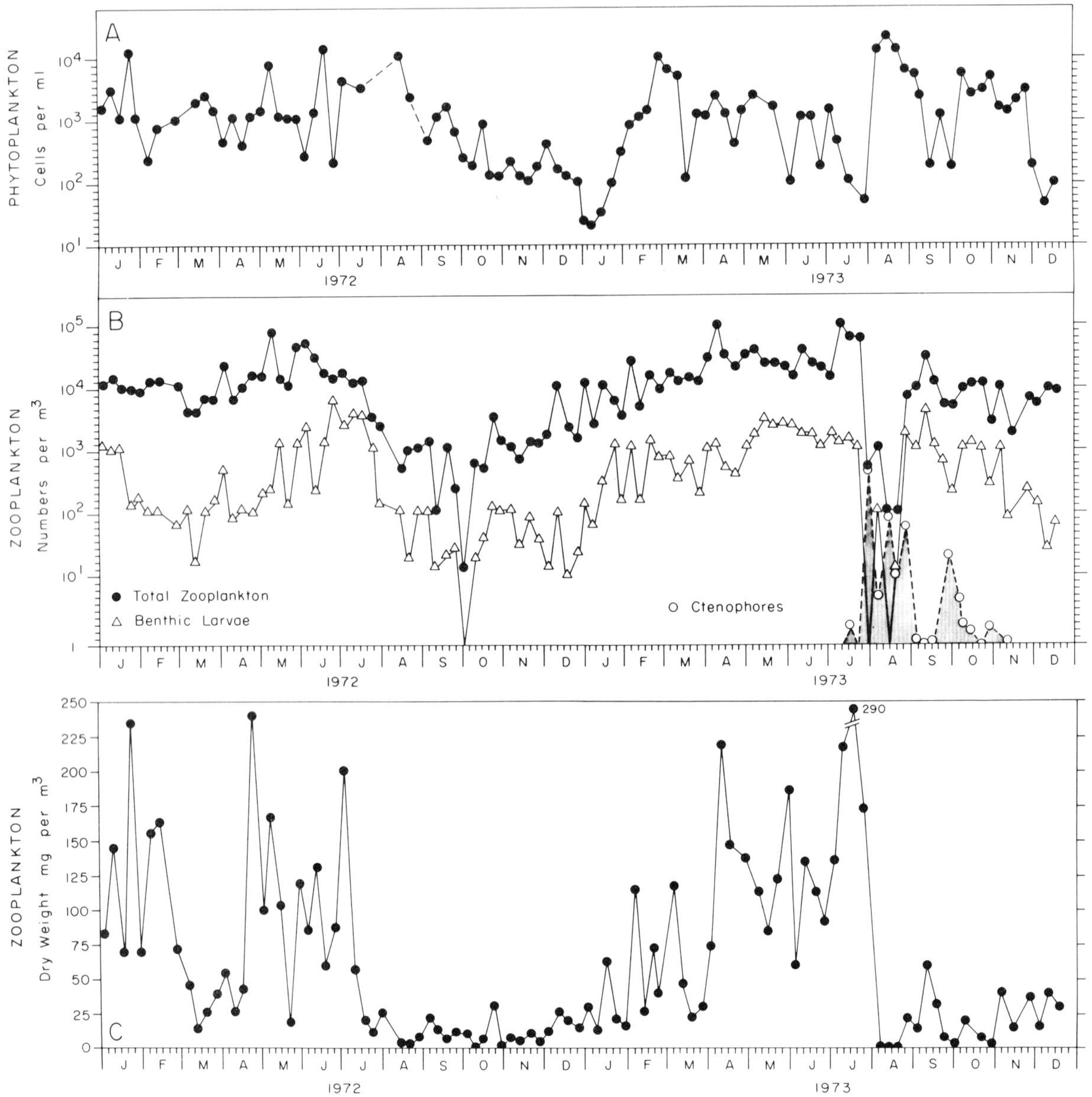

Fig. 4.24. Cycle of total phytoplankton numerical abundance at 0 m (A), cycles of zooplankton numerical abundance (B) and of zooplankton biomass (C) in Narragansett Bay from January 1972 through December 1973. (From Hulsizer, 1976.)

intensity was unusually low, relative to the long-term mean, until mid-January (Fig. 4.24). Inception of the bloom that year, which reached its maximum in late February, commenced shortly after the daily *in situ* intensity at the depth in question reached 40 g cal cm^{-2} (Hitchcock and Smayda, 1977). This is believed to be the critical *in situ* light intensity required to initiate a winter or spring bloom in temperate coastal waters.

Zooplankton levels during this prebloom period were fairly stable (Fig. 4.24), which suggests that their grazing activities were not decisive in regulating initiation of this bloom. Pratt (1965) related annual variations in the date of bloom inception to zooplankton abundance. Experimental observations in this bay suggest that a relaxation in zooplankton grazing pressure may indeed trigger the start of the bloom. Elevated prebloom concentrations (Table 4.8) indicate that nutrients do not regulate bloom inception, although the bloom maximum in abundance may be influenced (Pratt, 1965).

The observations agree with the other observations of year-to-year variability of growth conditions and plankton behavior in Narragansett Bay. Factors regulating bloom inceptions and declines vary between years and seasonally. The dynamics of *Skeletonema costatum*, for example, during the winter and spring, from November through June, 1971–1972, were shown experimentally to be regulated by temperature, light nutrients, grazing, possibly species interactions, and hydrographic disturbances (Smayda, 1973b). Experimental evidence consistently indicates that the availability of nutrients, notably nitrogen, becomes a key factor in regulating phytoplankton dynamics after the winter–spring bloom begins (Martin, 1968, 1970; Smayda, 1973b; Vargo, 1976). The primary sources of the required nitrogen appear to be excretion by herbivorous zooplankton and benthic invertebrates, as well as other benthic processes, as shown by the higher summer levels of ammonia compared to nitrate (Table 4.8).

Phytoplankton dynamics is partly regulated by zooplankton grazing and nutrient excretion. Zooplankton are dominated by the copepods *Acartia clausi* and *A. tonsa*; the latter species predominates during the summer and *A. clausi* is dominant during winter and spring (Martin, 1965; Hulsizer, 1976). Total zooplankton abundance varies seasonally, exhibiting a series of numerical and biomass peaks (Fig. 4.24). Annual variations in the relative importance of each species occur, as well as in rates of mean monthly production (Table 4.9). Thus, these phytoplankton/zooplankton interactions vary seasonally and annually.

Collapse of the winter–spring bloom in May or June appears to result from a combination of intensive copepod grazing and nutrient limitation. Despite considerable grazing, excretion provides less than 10% of the daily nitrogen needs of the large phytoplankton standing stock (Martin, 1968; Smayda, 1973b; Vargo, 1976). Most of the nutrient-limited and surplus phytoplankton stock then (as well as year-round) sinks to the bottom sediments. An extensive benthic bivalve community, including the clam *Mercenaria mercenaria*, which filter-feeds on phytoplankton (Pratt and Campbell, 1956), and other filter-feeding invertebrates, is present in Narragansett Bay. Renewed and extensive phytoplankton growth occurs during the summer (Table 4.8, Fig. 4.24) despite low ambient nitrogen concentrations, partly in response to the recycling of nutrients from the benthic community. About 55% of the total annual phytoplankton production during 1974 occurred during June through August (Table 4.10). Production rates frequently exceeded 2 g C m^{-2} day^{-1} during this period (Fig. 4.25). The community is then dominated by nannophytoplankton <20 μm in size, which account for 90% of the observed production and frequently have carbon doubling times of less than 24 h (Durbin et al., 1975). Larger diatom species also grow rapidly during the summer. Experimental determinations of nutrient uptake rates for these natural populations reveal that nitrogen recycling times of from 2 to 24 h, and from 2.5 to 17 days for silicon, are required to support the observed daily production

TABLE 4.9

Primary production (g C m^{-2}) during 1974 at Station *2* in Narragansett Bay (Furnas et al., 1977)

	Production (g C m^{-2})	Percent of annual total
January–April	52.0 (13.0 g C m^{-2} $month^{-1}$)	17 (4.2% $month^{-1}$)
May	22.6	7
June	42.2	14
July	65.8	21
August	63.1	21
September	25.7	8
October–December	36.6 (12.2 g C m^{-2} $month^{-1}$)	12 (4% $month^{-1}$)
Annual total	308.0	

TABLE 4.10

Mean zooplankton biomass measurements as mg m^{-3} of zooplankton dry weight at Narragansett Bay permanent station (from Hulsizer, 1976)

	1972	1973
January	121.2	28.7
February	130.9	56.7
March	31.0	55.4
April	91.6	114.3
May	102.1	127.4
June	91.7	100.8
July	63.5	163.9
August	5.1	5.6
September	13.9	29.0
October	10.5	9.6
November	6.7	31.6
December	12.0	30.3
Annual average	58.2	66.3

and uptake rates. Although nutrients excreted from the larger-sized ($>153\ \mu m$) zooplankton and benthic communities (see Hale, 1975) fall below required levels, contributions by the micro-zooplankton and nekton have yet to be determined. The significant feature is that summer phytoplankton dynamics in this well-mixed estuary are almost immediately linked to benthic remineralization activities. This differs from the situation described for

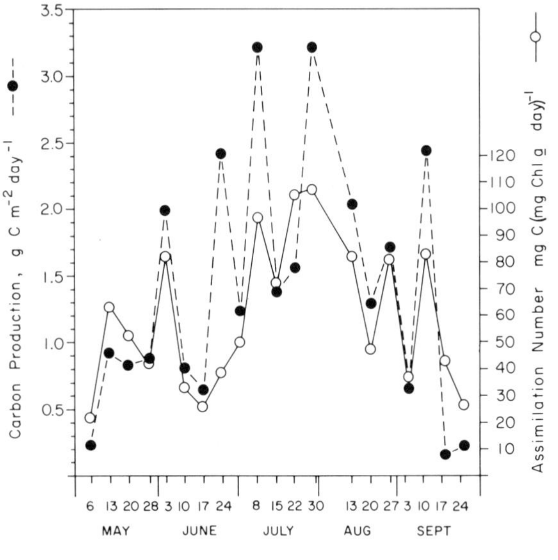

Fig. 4.25. Primary production and assimilation numbers from May to September 1974 at permanent station in lower Narragansett Bay (from Furnas et al., 1976).

Kiel Bight. There, because of summer thermohaline stratification, nutrients accumulate below the euphotic zone and become available for phytoplankton only after the October breakdown of water column stratification leading to the observed *Ceratium* bloom. Zooplankton excretion represents the predominant summer recycling mechanisms in Kiel Bight.

Exclusive of zooplankton regulation of phytoplankton dynamics through nutrient excretion, phytoplankton abundance is also regulated by zooplankton grazing (Martin, 1970; Hulsizer, 1976). A pronounced bloom of *Skeletonema costatum* occurred in August 1973 following decimation of the *Acartia tonsa* population during predation by the ctenophore *Mnemiopsis leidyi* (Fig. 4.24). Durbin and Durbin (1981) have suggested that a late summer decrease of *Acartia tonsa* in 1976 may have been initiated by a decrease in available phytoplankton, and accelerated through extensive concomitant predation by ctenophores and the planktonophagic fish *Brevoortia tyrannus* (menhaden).

A fairly abundant zooplankton population occurs in Narragansett Bay (Table 4.10) relative to other estuaries on the east coast of the United States, and it consists of the fast-growing *Acartia* species (Durbin and Durbin, 1981). The biomass doubling time of *Acartia tonsa*, for example, was generally less than one day at summer temperatures. Nonetheless, the bulk of the annual net phytoplankton production rate of 308 g C m^{-2} (Table 4.9) appears not to be utilized by the zooplankton, but enters the benthic food chain. Based on the mean daily potential production rates reported by Durbin and Durbin for *Acartia clausi* and *A. tonsa*, a combined total production of 25 g C m^{-2} can be calculated for the March to early October growth period considered by them. This represents only about 8% of the annual net phytoplankton production. These *Acartia* species usually contributed about 50% of the total zooplankton biomass during the Durbin and Durbin survey. An unknown proportion of phytoplankton production is also washed out of Narragansett Bay. The mean annual biomass of demersal fish in Narragansett is reported to be 0.16 g C m^{-2} (Oviatt and Nixon, 1973), or about 0.1% of the net annual phytoplankton production. While this relative abundance in demersal fish biomass is comparable to that in other New England coastal waters, Oviatt and Nixon state that it represents only about 3% of the summer biomass levels of the larger fauna.

Despite the large annual phytoplankton productivity in Narragansett Bay (Table 4.9), a commercially important pelagic or demersal fishery does not exist. This demonstrates that elevated primary productivity is not in itself sufficient to support a phytoplankton-based fishery either directly, as in upwelling regions, or through trophic transfer.

In summary, Narragansett Bay is a shallow, highly productive estuary characterized by a modest salinity gradient, and is well mixed throughout the year. Nearly continuous phytoplankton growth occurs year-round aided by a tight coupling to the benthic community, which consumes and remineralizes phytoplankton and releases nutrients into the water column. The phytoplankton and zooplankton cycles are unbalanced in that there is a considerable surplus of phytoplankton production over that grazed within the water column. This appears to support a benthic invertebrate community that is important in terms of biomass.

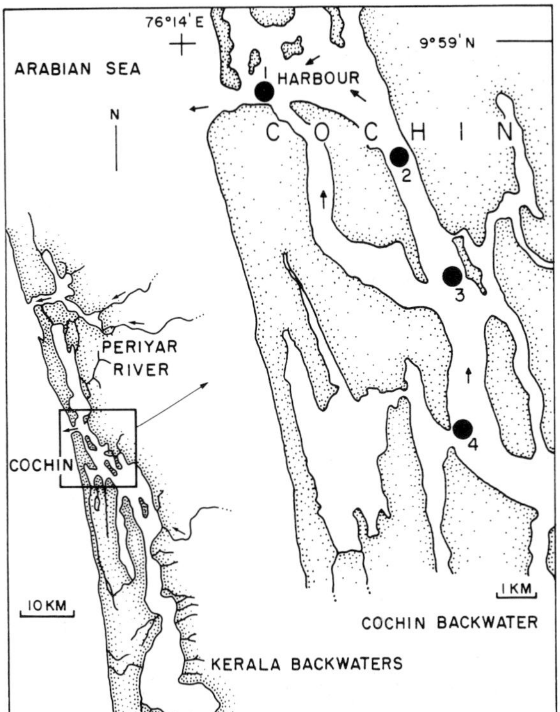

Fig. 4.26. Cochin Backwater, its connection to the Arabian Sea, and stations referred to in text (after Wellershaus, 1974).

Cochin Backwater

Cochin Backwater, one of the largest estuaries in India, is a shallow tropical estuarine system of about 500 km^2 located in the southwest Indian State of Kerala (Fig. 4.26). It is permanently connected to the Arabian Sea by an inlet about 450 m in width, and through a chain of lagoons and swamps receives the runoff from two permanent large rivers (Periyar and Pampa), and from several seasonal tributaries during monsoon months (Qasim et al., 1968; Qasim et al., 1969). The depth near its outlet to the Arabian Sea ranges from 5 to 15 m; the inner reaches are from 2 to 5 m in depth.

The seasonal monsoon rains and extremely high year-round temperatures (>25°C) represent major environmental features of the Cochin Backwater that differ significantly from the other estuaries chosen here for comparison (Fig. 4.27). Climatologically, the annual cycle consists of a dry premonsoon season from January to about May, a wet southwest monsoon period in June and July, followed by a postmonsoon season from about August to December characterized by reduced precipitation (Qasim et al., 1969; Wellershaus, 1971, 1974). The mean annual rainfall is about 3200 mm, of which about 40 to 60% occurs during the two-month southwest monsoon season and 75% from May to September (Fig. 4.27; Qasim et al., 1968). The associated runoff produces a profound effect on environmental conditions. Thus, the Cochin Backwater alternately exhibits two distinct types of estuarine patterns over an annual cycle. During the premonsoonal period it is a well-mixed vertically homogeneous estuary. During the monsoon period salinity falls rapidly from about 31 to 33‰ to <2‰ in the surface waters, to about 10‰ at 1 m and can reach 20‰ at 4 m; bottom (9 m) salinity approximates 32‰ throughout the annual cycle. Following a subsidence in runoff, increased intrusion of Arabian Sea water (34‰) leads to the development of a two-layer circulation pattern and water column stratification throughout postmonsoon season (Fig. 4.27). This trend is reversed when markedly reduced precipitation and runoff during the premonsoonal season from January to May contributes to a breakdown of this two-layered circulation pattern. The estuary is then highly saline

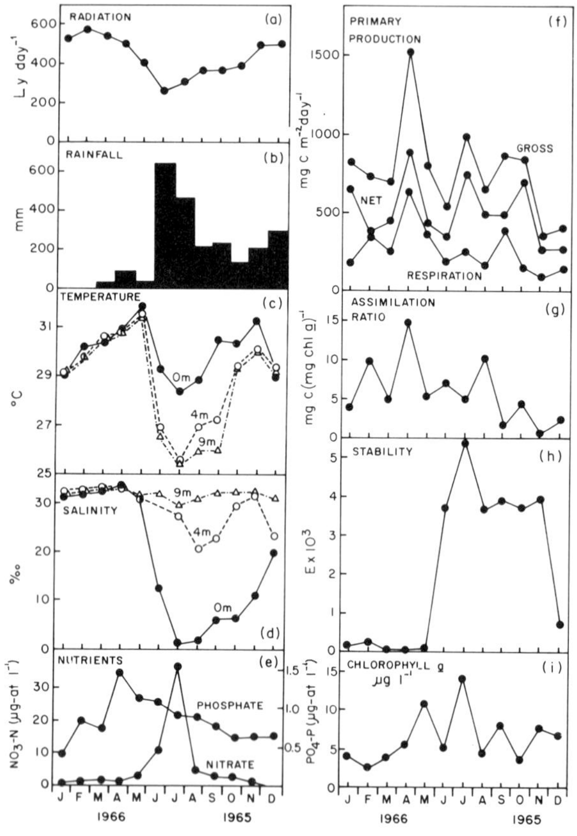

Fig. 4.27. Seasonal changes in various environmental factors, phytoplankton chlorophyll, primary production, respiration and assimilation number at station *2* in Cochin Backwater. The values for nutrients (e), production + respiration (f), assimilation ratio (g) and chlorophyll concentration (i) are for the euphotic zone; vertical stability (h) is that calculated for the 0–4 m stratum. (From Qasim et al., 1969.)

throughout the water column and well mixed through tidal turbulence.

This annual climatological and hydrographic pattern is accompanied by changes in other environmental factors significant to plankton dynamics. Light available for phytoplankton photosynthesis decreases significantly during the monsoon season, both because of a reduction in incident radiation and greatly increased water column turbidity (Fig. 4.27; Qasim et al., 1968). The average daily radiation decreases about two-fold from about 550 g cal cm^{-2} to its annual minimum of 250 g cal cm^{-2} during the monsoon season (Fig. 4.27). Of the incident radiation 80% is attenuated in the upper 1 m, and the high turbidity persistent during the seasons prior to and after the monsoon leads then to a 70% attenuation. The consequence of this is that the euphotic zone is very shallow; the compensation depth for photosynthesis ranges from 2.5 to 4 m. This compensation depth, which is approximately the depth to which 1% of the incident light penetrates, can be compared to that illustrated for representative water types (Fig. 4.4). The Cochin Estuary is obviously much more turbid than even coastal water type 9, which is characterized by 1% penetration at 6 m.

Concurrent with the tremendous volume of fresh water discharged into the Cochin Estuary during the monsoon season, an incursion of cold, saline water from the Arabian Sea occurs. This reflects the occurrence of upwelling along the Cochin coast at this time (Banse, 1959; Sankaranarayanan and Qasim, 1969). This decreases mid- and bottom-water temperatures by about 6°C (Fig. 4.27), though temperatures still remain above 25°C.

The seasonal changes in nutrients are significantly influenced by monsoonal conditions (Sankaranarayanan and Qasim, 1969; Qasim et al., 1969). Nitrate concentrations increased approximately 20-fold to 40 µg-at l^{-1} during the height of the monsoon season (Fig. 4.27); this nutrient seems to be derived from the land through runoff, exclusive of any upwelling contribution. Minimal concentrations occurred during the dry season, which suggests that marine inputs are of little importance. The concentration of silicate was high year-round, with surface levels ranging from about 5 µg-at l^{-1} to monsoon season peaks of 50 to 60 µg-at l^{-1} throughout the estuary. Lowest silicate levels occurred from December to April. Sankaranarayanan and Qasim (1969) concluded that, as for nitrate, fresh-water discharge is a primary source of silicate, its concentrations being inversely related to salinity and increasing with suspended silt loads. The annual cycle in phosphate differs in being more damped; the maximum monsoonal concentrations of 1.5 to 2.5 µg-at l^{-1} are three to five times greater than the minimal levels (Fig. 4.27). Nutrient levels generally do not exhibit a vertical gradient from January through April; the water column is then well mixed. Fresh-water discharge is then minimal, and the Cochin Estuary is primarily an extension of the adjoining coastal waters. Sankaranarayanan and Qasim (1969) considered fresh-water inputs of both inorganic and dissolved organic phosphorus as contributing to its observed annual cycling.

However, differences in vertical gradients relative to those for nitrate and silicate suggest that other enrichment mechanisms also supply this nutrient. For example, from June through December phosphate levels increased with depth while silicate levels decreased, and the nitrate distributions showed both trends. Qasim and Gopinathan (1969) gave evidence that phosphorus concentrations increase during flood tides and decrease during ebb flow. It is unclear to what extent this input represents a release from the bottom sediments during tidally induced roiling, or is introduced directly from Arabian Sea water. In contrast, the concentrations of nitrate and silicate increased during ebb tides, which the authors suggested was consistent with fresh-water inputs.

The increased turbidity and elevated nutrient concentrations during the monsoon period are associated with the presence of large quantities of suspended material consisting of inorganic (seston) and organic (detritus) particles. Qasim and Sankaranarayanan (1972) have established that the phytoplankton abundance represents only from 0.1 to 1% of this detrital load. The primary sources of organic detritus and seston are resuspended bottom sediments and land runoff. Organic detritus is of particular interest because it represents a potential food source to detrital feeders living within the estuary. The detritus in Cochin Estuary largely comprises aggregated fine silts covered with an organic film and fragments of dead algae, rooted plants, animal matter and other materials washed into or produced within the estuary. Contrary to expectations, Qasim and Sankaranarayanan (1972) found that the quantity of detritus sinking to the estuarine floor was low during the monsoon and postmonsoon seasons. Maximum daily depositional rates of 1 kg dry weight m^{-2} occurred during the premonsoon season, even though significant deposition occurred year-round. Despite this significant detrital input, it has low nutritional value, determined to be more or less seasonally constant at about 400 cal per gram dry weight — that is, *c.* 10% that for phytoplankton. While detritus nonetheless represents a source of available carbon in this local food web, evidence suggests that trophodynamics in Cochin Estuary is primarily based on the phytoplankton.

The seasonal chlorophyll distribution (Fig. 4.27) shows a series of peaks, with maximum concentration of about 15 $\mu g\, l^{-1}$ occurring during the monsoon season. This corresponds to about a three-fold increase over minimal levels; additional observations were presented by Qasim and Reddy (1967). The available data suggest that while seasonal variations in phytoplankton abundance occur, possibly in response to nutrient enrichment, there is considerable annual and spatial variability. Moreover, the seasonal cycle in abundance both is somewhat damped relative to those in colder estuaries, and is characterized by relatively high year-round abundance.

Table 4.11 illustrates the annual phytoplankton cycle during two different quantitative sampling periods. During 1972–1973 Gopinathan et al. (1974) observed two maxima, the major bloom during the premonsoon season (January and February) and a monsoon maximum in July and August. The euryhaline (Fig. 4.13) and eurythermal (Smayda, 1973b) diatom *Skeletonema costatum*, as in Kiel Bight and Narragansett Bay, was responsible for the major bloom; during the monsoon maximum the dinoflagellates *Prorocentrum micans* and *Ceratium furca* along with *Skeletonema* were the dominant components. The silicoflagellates *Dictyocha fibula* and *Distephanus speculum* occurred sporadically throughout the year, and chlorophyceans such as *Spirogyra* and the desmids *Euastrum*, *Cosmarium*, *Micrasterias*, and blue-green algae such as *Oscillatoria* sp. and *Merismopedia* sp. occurred during the monsoon and postmonsoon seasons. The nitrogen-fixing, oceanic blue-green algal species *Trichodesmium thiebautii* also occurred. Qasim et al. (1972) compared oscillations in seasonal phytoplankton abundance (generally lower than in the Gopinathan et al. study) to salinity. Diatoms occurred throughout the year, but predominated from November through February. A major *Skeletonema costatum* bloom occurred during November and December when a chlorophycean identified as *Chlorella* dominated the "others" component (Table 4.11). Dinoflagellates attained a maximum in April and May when salinity was still high, but were absent during the monsoonal period. *Coscinodiscus* species exhibited a succession of blooms in the salinity range from 0 to 25‰. Gopinathan's (1972) study of the net phytoplankton revealed that *Coscinodiscus excentricus*, *C. radiatus*, *C. centralis* and *C. jonesianus* were particularly abundant during the mon-

TABLE 4.11

Average monthly surface salinity and phytoplankton abundance (cells l^{-1}) at one location in Cochin Backwater (from Qasim et al., 1972 except Total 2 which represents Station III data in Gopinathan et al., 1974)

Month	Salinity (‰)	Diatoms	Dino-flagellates	Others	Total	Total 2
January	23.80	33 700	500	6600	40 800	899 000
February	28.12	48 900	1200	13 200	63 300	341 000
March	32.75	11 400	2400	2400	16 200	101 000
April	32.02	15 300	14 400	7950	37 650	192 000
May	25.46	17 700	19 500	3300	40 500	136 000
June	0.97	22 500	5400	600	28 500	534 000
July	0.00	18 900	nil	7500	26 400	400 000
August	0.00	13 500	nil	4200	17 700	161 000
September	0.40	4800	600	10 800	16 200	413 000
October	2.50	6900	300	6000	13 200	332 000
November	16.80	64 200	1200	37 200	102 600	203 000
December	29.20	351 600	1200	25 200	378 000	108 000

soon and postmonsoon seasons. George (1958) reported that the chlorophyceans *Spirogyra* sp. and *Wolffia* sp. were common during periods of reduced salinity, and "swarms" of the blue-green alga *Anabaenopsis* occurring during the premonsoon season were accompanied by fish kills. The composition of the phytoplankton flora in Cochin Estuary in response to osmotic conditions is thus consistent with the floristic characteristics of estuarine phytoplankton discussed previously.

The annual production in Cochin Estuary exhibits several peaks, with daily production varying from about 0.2 to 1.5 g C m^{-2} (Fig. 4.27). Production peaks occur during each meteorological season, with the primary peak occurring during the premonsoon period. The annual net production is 124 g C m^{-2}, the amount which is available for herbivorous zooplankton, fish and benthic filter feeders. This relationship with zooplankton has been partially evaluated. Wellershaus (1974) observed considerable variations in the composition and abundance of the zooplankton (Fig. 4.28). During the premonsoon season numerical abundance of zooplankton is high throughout the homogeneous, well-mixed water column, with a March maximum, whereas monsoonal populations are considerably reduced in abundance. Rapid flushing from the estuary then prevents a suitable residence time for population growth, and the reduced salinity represents an additional selective pressure. The reduced flushing and increased salinity occurring during the postmonsoon season lead to a partial recovery and increased abundance of the zooplankton. Expressed on a biomass basis (Fig. 4.28), maximum zooplankton abundance occurs during the premonsoon season. Qasim et al. (1969) have evaluated the general metabolic requirements of this zooplankton biomass on the assumption that its daily respiratory needs for carbon represent 12% of its dry weight. Comparison of this to phytoplankton net production yields daily zooplankton carbon consumption rates amounting to 10% of that produced by the phytoplankton during the monsoon period, 20% during the postmonsoon season, and 46% during the premonsoon months (Fig. 4.28). On an annual basis this represents a consumption of about 30 g C m^{-2} compared to net production of 124 g C m^{-2}. Qasim et al. (1969) have concluded that there is a large surplus of phytoplankton production over that grazed by the zooplankton, and that this may support the appreciable abundance of herbivorous fishes, such as mullet, present in Cochin Backwater. The extent to which nutrient or grazing regulation of seasonal phytoplankton dynamics occurs in Cochin Backwater remains to be established.

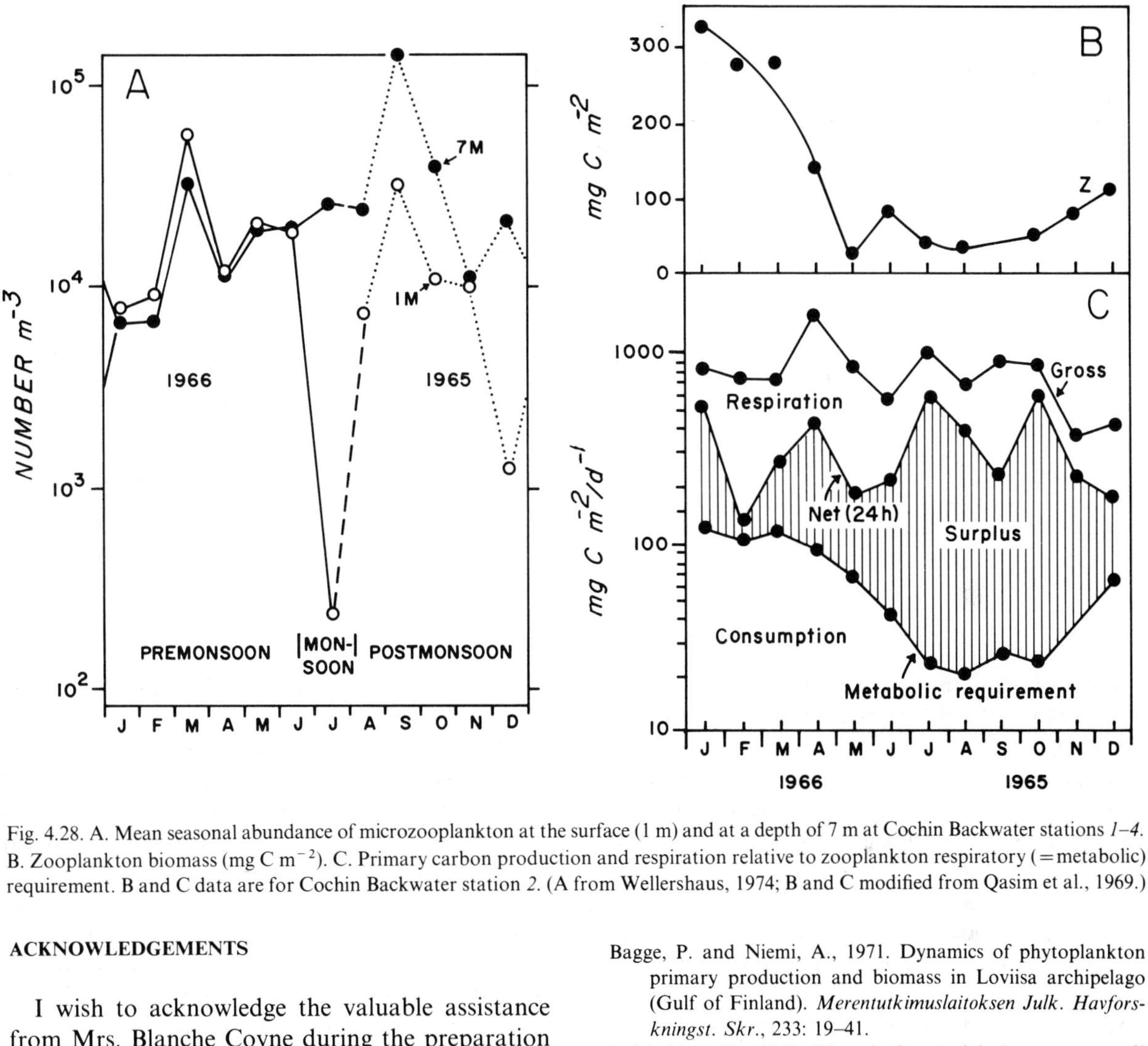

Fig. 4.28. A. Mean seasonal abundance of microzooplankton at the surface (1 m) and at a depth of 7 m at Cochin Backwater stations *1–4*. B. Zooplankton biomass (mg C m^{-2}). C. Primary carbon production and respiration relative to zooplankton respiratory (=metabolic) requirement. B and C data are for Cochin Backwater station *2*. (A from Wellershaus, 1974; B and C modified from Qasim et al., 1969.)

ACKNOWLEDGEMENTS

I wish to acknowledge the valuable assistance from Mrs. Blanche Coyne during the preparation of this chapter. She aided in the collection and preparation of source materials, typed the manuscript, and drafted many of the figures. Many of the Narragansett Bay observations presented here, including previously unpublished material, were aided by support from the U.S. National Science Foundation, specifically grants OCE71-00556 and OCE76-22563.

REFERENCES

Aarthun, K.E., 1961. The natural history of the Hardangerfjord. 2. Submarine daylight in a glacier-fed Norwegian Fjord. *Sarsia*, 1: 7–20.

Bagge, P. and Niemi, A., 1971. Dynamics of phytoplankton primary production and biomass in Loviisa archipelago (Gulf of Finland). *Merentutkimuslaitoksen Julk. Havforskningst. Skr.*, 233: 19–41.

Bainbridge, V., 1960. The plankton of inshore waters off Freetown, Sierra Leone. *Fish. Colonial Off. Publ.*, 13: 1–48.

Bakker, C. and De Pauw, N., 1974. Comparison of brackish-water plankton assemblages of identical salinity ranges in estuarine tidal (Westerschelde) and stagnant (Lake Veere) environment (southwestern Netherlands). I. Phytoplankton. *Hydrobiol. Bull.*, 8: 179–189.

Banse, K., 1959. On upwelling and bottom-trawling off the southwest coast of India. *J. Mar. Biol. Assoc. India*, 1: 33–49.

Braarud, T., 1951. Salinity as an ecological factor in marine phytoplankton. *Physiol. Plant.*, 4: 28–34.

Braarud, T., 1961. Cultivation of marine organisms as a means of understanding environmental influences on populations. In: M. Sears (Editor), *Oceanography*. Am. Assoc. Adv. Sci., Washington, D.C., pp. 271–298.

Braarud, T. and Føyn, B., 1958. Phytoplankton observations in a brackish water locality of south-east Norway. *Nytt Mag. Bot.*, 6: 47–73.

Braarud, T. and Hope, B., 1952. The annual phytoplankton cycle of a landlocked fjord near Bergen (Nordåsvatn). *Rep. Nor. Fish. Mar. Invest.*, 9(16): 1–26.

Braarud, T., Føyn, B. and Hasle, G.R., 1958. The marine and fresh-water phytoplankton of the Dramsfjord and the adjacent part of the Oslofjord March–December 1951. *Hvalrådets Skr.*, 43: 1–102.

Braarud, T., Hofsvang, B.F., Hjelmfoss, P. and Överland, A.A.K., 1974. The natural history of the Hardangerfjord. 10. The phytoplankton in 1955–56. The quantitative phytoplankton cycle in the fjord waters and in the offshore coastal waters. *Sarsia*, 55: 63–98.

Carpenter, J.H., Pritchard, D.W. and Whaley, R.C., 1969. Observations of eutrophication and nutrient cycles in some coastal plain estuaries. In: *Eutrophication: Causes, Consequences, Correctives*. National Academy Sciences, Washington, D.C., pp. 210–221.

Caspers, H., 1959. Die Einteilung der Brackwasser-Region in einem Ästuar. Symposium on the Classification of Brackish Waters. *Arch. Oceanogr. Limnol. (Suppl.)*, 11: 153–170.

Castañares, A.A. and Phleger, F.B., 1969. *Coastal Lagoons, A Symposium*. Universidad Nacional Autónoma de México, Mexico, 686 pp.

Chin, T.G., Chen, C.F., Liu, S.C. and Wu, S.S., 1965. Influence of temperature and salinity on the growth of three species of planktonic diatoms. *Oceanogr. Limnol. Sin.*, 7: 373–384.

Conover, S., 1956. Oceanography of Long Island Sound, 1952–1954. IV. Phytoplankton. *Bull. Bingham Oceanogr. Coll.*, 15: 62–112.

Droop, M.R., 1958. Optimum relative and actual ionic concentrations for growth of some euryhaline algae. *Verh. Int. Ver. Theor. Angew. Limnol.*, 8: 722–730.

Durbin, A.G. and Durbin, E.G., 1981. Standing stock and estimated production rates of phytoplankton and zooplankton in Narragansett Bay, R.I. *Estuaries*, 4: 24–41.

Durbin, E.G., Krawiec, R.W. and Smayda, T.J., 1975. Seasonal studies on the relative importance of different size fractions of phytoplankton in Narragansett Bay (USA). *Mar. Biol.*, 32: 271–287.

Edler, L., 1977 *Phytoplankton and Primary Production in the Sound*. Department of Marine Botany, University of Gothenburg, Gothenburg, 82 pp.

Furnas, M.J., Hitchcock, G.L. and Smayda, T.J., 1976. Nutrient–phytoplankton relationships in Narragansett Bay during the 1974 summer bloom. In: M. Wiley (Editor), *Estuarine Processes, I*. Academic Press, New York, N.Y., pp. 118–133.

George, M.J., 1958. Observations on the plankton of the Cochin Backwaters. *Indian J. Fish.*, 5: 375–401.

Gopinathan, C.P., 1972. Seasonal abundance of phytoplankton in the Cochin Backwater. *J. Mar. Biol. Assoc. India*, 14: 568–577.

Gopinathan, C.P., Ramachandran Nair, P.V. and Kesavan Nair, A.K., 1974. Studies on the phytoplankton of the Cochin Backwater, a tropical estuary. *Indian J. Fish.*, 21: 501–513.

Gran, H.H., 1912. Pelagic plant life. In: J. Murray and J. Hjort (Editors), *The Depths of the Ocean*. MacMillan and Co., London, pp. 307–386.

Guillard, R.R.L., 1962. Salt and osmotic balance. In: R.A. Lewin (Editor), *Physiology and Biochemistry of Algae*. Academic Press, New York, N.Y., pp. 529–540.

Guillard, R.R.L. and Myklestad, S., 1970. Osmotic and ionic requirements of the marine centric diatom *Cyclotella nana*. *Helgol. Wiss. Meeresunters.*, 20: 104–110.

Hale, S.S., 1975. The role of benthic communities in the nitrogen and phosphorus cycles of an estuary. In: F.G. Howell, J.B. Gentry and M.H. Smith (Editors), *Mineral Cycling in Southeastern Ecosystems*, ERDA Symp. Ser. Conf. No. 740513, pp. 291–308.

Hickel, W., 1967. Untersuchungen über die Phytoplankton-Blüte in der westlichen Ostsee. *Helgol. Wiss. Meeresunters.*, 16: 3–66.

Hickel, W., 1969. Planktologische und hydrographisch-chemische Untersuchungen in der Eckernförder Bucht (westliche Ostsee) während und nach der Vereisung im extrem kalten Winter 1962/1963. *Helgol. Wiss. Meeresunters.*, 19: 318–331.

Hicks, S.D., 1963. Physical oceanographic studies of Narragansett Bay, 1957 and 1958. *Spec. Sci. Rep., U.S. Fish. Wildl. Serv.*, 457: 1–30.

Hitchcock, G.L. and Smayda, T.J., 1977. The importance of light in the initiation of the 1972–1973 winter–spring diatom bloom in Narragansett Bay. *Limnol. Oceanogr.*, 22: 126–131.

Horstmann, U., 1972. Über den Einfluss von häuslichem Abwasser auf das Plankton in der Kieler Bucht. *Kieler Meeresforsch.*, 28: 178–198.

Hulburt, E.M., 1956. The phytoplankton of Great Pond, Massachusetts. *Biol. Bull.*, 110: 157–168.

Hulsizer, E.E., 1976. Zooplankton of lower Narragansett Bay, 1972–1973. *Chesapeake Sci.*, 17: 260–270.

Jerlov, N.G., 1970. Light, General Introduction. In: O. Kinne (Editor), *Marine Ecology, I. Environment Factors, Part 1*. Wiley–Interscience, New York, N.Y., 95–102.

Karayeva, N.I. and Makarova, I.V., 1973. Specific features and origin of the Caspian Sea diatom flora. *Mar. Biol.*, 21: 269–275.

Ketchum, B.H., 1954. Relation between circulation and planktonic populations in estuaries. *Ecology*, 35: 191–200.

Ketchum, B.H., 1969. Eutrophication of estuaries. In: *Eutrophication: Causes, Consequences, Corrections*. National Academy Sciences, Washington, D.C., pp. 197–209.

Kremer, J.N.K. and Nixon, S.W., 1978. *A Coastal Marine Ecosystem Simulation and Analysis*. Ecological Studies 24. Springer-Verlag, New York, N.Y., 217 pp.

Kuenzler, E.J. and Perras, J.P., 1965. Phosphatases of marine algae. *Biol. Bull.*, 128: 271–284.

Leach, J.H., 1971. Hydrology of the Ythan Estuary with reference to distribution of major nutrients and detritus. *J. Mar. Biol. Assoc. U.K.*, 51: 137–157.

Lehmusluoto, P.O. and Pesonen, L., 1973. Eutrophication in the Helsinki and Espo sea areas measured as phytoplankton primary production. *Oikos (Suppl.)*, 15: 202–208.

Lenz, J., 1977a. The Baltic Sea as an estuary. In: G. Rheinheimer (Editor), *Microbial Ecology of a Brackish Water Environment*. Ecological Studies 25. Springer Verlag, New York, N.Y., pp. 12–25.

Lenz, J., 1977b. Plankton populations. In: G. Rheinheimer

(Editor), *Microbial Ecology of a Brackish Water Environment*. Ecological Studies 25. Springer Verlag, New York, N.Y., pp. 79–89.

Lenz, J., 1977c. Seston and its main components. In: G. Rheinheimer (Editor), *Microbial Ecology of A Brackish Water Environment*. Ecological Studies 25. Springer Verlag, New York, N.Y., pp. 37–60.

Martin, J.H., 1965. Phytoplankton–zooplankton relationships in Narragansett Bay. *Limnol. Oceanogr.*, 10: 185–191.

Martin, J.H., 1968. Phytoplankton–zooplankton relationships in Narragansett Bay. III. Seasonal changes in zooplankton excretion rates in relation to phytoplankton abundance. *Limnol. Oceanogr.*, 13: 63–71.

Martin, J.H., 1970. Phytoplankton–zooplankton relationships in Narragansett Bay. IV. The seasonal importance of grazing. *Limnol. Oceanogr.*, 15: 413–418.

McCarthy, J.J., Taylor, W.R. and Taft, J.L., 1975. The dynamics of nitrogen and phosphorus cycling in the open waters of the Chesapeake Bay. In: T. Church (Editor), *Marine Chemistry in the Coastal Environment. Am. Chem. Soc. Symp. Ser.*, No. 18: 664–681.

McCarthy, J.J., Taylor, W.R. and Taft, J.L., 1977. Nitrogenous nutrition of the plankton in the Chesapeake Bay. 1. Nutrient availability and phytoplankton preferences. *Limnol. Oceanogr.*, 22: 996–1011.

McLachlan, J., 1961. The effect of salinity on growth and chlorophyll content in representative classes of unicellular marine algae. *Can. J. Microbiol.*, 7: 399–406.

Nakanishi, M. and Monsi, M., 1965. Effect of variation in salinity on photosynthesis of phytoplankton growing in estuaries. *J. Fac. Sci. Univ. Tokyo, III.* 9: 19–42.

Niemi, A., 1972. Observations on phytoplankton in eutrophied and non-eutrophied archipelago waters of the southern coast of Finland. *Mem. Soc. Fauna Flora Fenn.*, 48: 63–74.

Nordli, E., 1953. Salinity and temperature as controlling factors for distribution and mass occurrence of Ceratia. *Blyttia*, 11: 16–18.

Oviatt, C.A. and Nixon, S.W., 1973. The demersal fish of Narragansett Bay: an analysis of community structure, distribution and abundance. *Estuarine Coastal Mar. Sci.*, 1: 361–378.

Paasche, E., 1975. The influence of salinity on the growth of some plankton diatoms from brackish water. *Nor. J. Bot.*, 22: 209–215.

Paasche, E., Johansson, S. and Evensen, D.L., 1975. An effect of osmotic pressure on the valve morphology of the diatom *Skeletonema subsalsum* (A. Cleve) Bethge. *Phycologia*, 14: 205–211.

Porep, R., 1970. Victor Hensen (1835–1924) sein Leben und sein Werk. *Kieler Beitr. Gesch. Mediz. Pharm.*, 9: 1–147.

Pratt, D.M., 1959. The phytoplankton of Narragansett Bay. *Limnol. Oceanogr.*, 4: 425–440.

Pratt, D.M., 1965. The winter–spring diatom flowering in Narragansett Bay. *Limnol. Oceanogr.*, 10: 173–184.

Pratt, D.M., 1966. Competition between *Skeletonema costatum* and *Olisthodiscus luteus* in Narragansett Bay and in culture. *Limnol. Oceanogr.*, 11: 447–455.

Pratt, D.M. and Campbell, D.A., 1956.Environmental factors affecting growth in *Venus mercenaria*. *Limnol. Oceanogr.*, 1: 2–17.

Probst, B., 1977. Primary production. In: G. Rheinheimer (Editor), *Microbial Ecology of a Brackish Water Environment*. Ecological Studies 25. Springer Verlag, New York, N.Y., pp. 71–78.

Qasim, S.Z. and Gopinathan, C.K., 1969. Tidal cycle and the environmental features of Cochin Backwater (a tropical estuary). *Proc. Indian Acad. Sci., Sec. B*, 59: 336–348.

Qasim, S.Z. and Reddy, C.V.G., 1967. The estimation of plant pigments of Cochin Backwater during the monsoon months. *Bull. Mar. Sci.*, 17: 95–110.

Qasim, S.Z. and Sankaranarayanan, V.N., 1972. Organic detritus of a tropical estuary. *Mar. Biol.*, 15: 193–199.

Qasim, S.Z., Bhattathiri, P.M.A. and Abidi, S.A.H., 1968. Solar radiation and its penetration in a tropical estuary. *J. Exp. Mar. Biol. Ecol.*, 2: 87–103.

Qasim, S.Z., Wellershaus, S., Bhattathiri, P.M.A. and Abidi, S.A.H., 1969. Organic production in a tropical estuary. *Proc. Indian Acad. Sci., Sec. B*, 59: 51–94.

Qasim, S.Z., Bhattathiri, P.M.A. and Devassy, V.P., 1972. The influence of salinity on the rate of photosynthesis and abundance of some tropical phytoplankton. *Mar. Biol.*, 12: 200–206.

Remane, E. and Schlieper, C., 1958. Die Biologie des Brackwassers. *Binnengewässer*, 30: 1–348.

Renk, H., 1974. Primary production and chlorophyll content of the Baltic Sea. Part III. Primary production in the southern part of the Baltic. *Pol. Arch. Hydrobiol.*, 21: 191–209.

Rheinheimer, G., 1977. Oxygen and some inorganic nutrients. In: G. Rheinheimer (Editor), *Microbial Ecology of a Brackish Water Environment*. Ecological Studies 25. Springer-Verlag, New York, N.Y., pp. 26–36.

Rochford, D.J., 1959. Classification of Australian estuarine systems. Symposium on the Classification of Brackish Waters. *Arch. Oceanogr. Limnol.*, (*Suppl.*), 11: 171–177.

Ryther, J.H., 1954. The ecology of phytoplankton blooms in Moriches Bay and Great South Bay, Long Island, New York. *Biol. Bull.*, 106: 198–209.

Sakshaug, E., 1972. Phytoplankton investigations in Trondheimsfjord, 1963–1966. *K. Nor. Vidensk. Selsk. Skr.*, 1: 1–56.

Sankaranarayanan, V.N. and Qasim, S.Z., 1969. Nutrients of the Cochin Backwater in relation to environmental characteristics. *Mar. Biol.*, 2: 236–247.

Schmitz, W. 1959. Zur Frage der Klassification der binnenländischen Brackwässer. Symposium on the Classification of Brackish Waters. *Arch. Oceanogr. Limnol.* (*Suppl.*), 11: 179–226.

Schulz, H., 1961. Qualitative und quantitative Plankton-Untersuchungen im Elbe-Ästuar. *Arch. Hydrobiol.* (*Suppl. Elbe-Ästuar*), 26: 5–105.

Schulz, S. and Kaiser, W., 1973. Biological effects of the salt water influx into the Gotland Basin in 1969/70. *Oikos* (*Suppl.*), 15: 21–27.

Smayda, T.J., 1957. Phytoplankton studies in lower Narragansett Bay. *Limnol. Oceanogr.*, 2: 342–359.

Smayda, T.J., 1958. Biogeographical studies of marine phytoplankton. *Oikos*, 9: 158–191.

Smayda, T.J., 1973a. A survey of phytoplankton dynamics in the coastal waters from Cape Hatteras to Nantucket. In: Coastal and Offshore Environmental Inventory, Cape

Hatteras to Nantucket Shoals. *Mar. Publ. Ser., Univ. R.I.*, No. 2: 3-1–3-100.
Smayda, T.J., 1973b. The growth of *Skeletonema costatum* during a winter–spring bloom in Narragansett Bay, Rhode Island. *Nor. J. Bot.*, 20: 219–247.
Smayda, T.J., 1976. Plankton processes in mid-Atlantic near-shore and shelf waters and energy-related activities. In: B. Manowitz (Editor), *Effects of Energy-Related Activities on the Continental Shelf.* Brookhaven National Laboratory, BNL 50484, Upton, N.Y., pp. 70–94.
Smetacek, V., 1975. *Die Sukzession des Phytoplanktons in der westlichen Kieler Bucht.* Dissertation, Christian-Albrechts-Universität, Kiel, Institut für Meereskunde, 151 pp.
Steemann Nielsen, E., 1964. Investigations of the rate of primary production at two Danish light ships in the transition area between the North Sea and the Baltic. *Medd. Dan. Fisk Havunders. N.S.*, 4: 31–77.
Taft, J.L., Taylor, W.R. and McCarthy, J.J., 1975. Uptake and release of phosphorus by phytoplankton in the Chesapeake Bay estuary, USA. *Mar. Biol.*, 33: 21–32.
Tomas, C.R., 1978. *Olisthodiscus luteus* (Chrysophyceae). I. Effects of salinity and temperature on growth, motility and survival. *J. Phycol.*, 14: 309–313.
Vargo, G.A., 1976. *The Influence of Grazing and Nutrient Excretion by Zooplankton on the Growth and Production of the Marine Diatom,* Skeletonema costatum *(Greville) Cleve, in Narragansett Bay.* Thesis, University of Rhode Island, Kingston, R.I., 162 pp.
Van Bennekom, A.J., Krijgsman-van Hartingsveld, E., Van der Veer, G.C.M. and Van Voorst, H.F.J., 1974. The seasonal cycles of reactive silicate and suspended diatoms in the Dutch Wadden Sea. *Neth. J. Sea Res.*, 8: 174–207.
Von Bodungen, B., 1975. *Der Jahresgang der Nährsalze und der Primärproduktion des Planktons in der Kieler Bucht unter Berücksichtigung der Hydrographie.* Dissertation, Christian-Albrechts-Universität, Kiel, Institut für Meereskunde, 116 pp.
Von Bodungen, B., Von Bröckel, K., Smetacek, V. and Zeitzschel, B., 1975. Ecological studies on the plankton in the Kiel Bight. I. Phytoplankton. *Merentutkimuslaitoksen Julk Havforskningsinst. Skr.*, 239: 179–186.
Wellershaus, S., 1971. On the hydrography of the Cochin Backwater, a south Indian estuary. *J. Mar. Biol. Assoc. India*, 14: 487–495.
Wellershaus, S., 1974. Seasonal changes in the zooplankton population in the Cochin Backwater (a South Indian estuary). *Hydrobiol. Bull.*, 8: 213–223.
Zenkevitch, L., 1959. The classification of brackish-water basins as exemplified by the seas of the U.S.S.R. Symposium on the Classification of Brackish Waters. *Arch. Oceanogr. Limnol. (Suppl.)*, 11: 53–61.
Zenkevitch, L., 1963. *Biology of the Seas of the USSR.* Interscience Publishers, New York, N.Y., 955 pp.

Chapter 5

THE ZOOPLANKTON OF ESTUARIES

CHARLES B. MILLER

INTRODUCTION

The term estuary has been applied to such a broad range of marine habitats that some limits must be chosen for useful discussion of the life of the estuarine plankton. McHugh (1967) includes any habitat with a surface layer of substantially reduced salinity, such as the entire subarctic Pacific (see p. 185, Fig. 7.1). That is clearly too broad for present purposes. The Aral and Caspian Seas are estuarine in many of their characteristics (Zenkevitch, 1963), but they are too aberrant in fauna and chemistry to consider in a general review. In some cases hypersaline lagoons have estuary-like properties, but they, too, are estuaries in only a special sense.

Here we will consider those estuaries where a river meets and mixes with the sea. The important distinguishing features of these habitats are salinity gradients in the horizontal and sometimes the vertical dimensions, and horizontal oscillations of the water caused by the tide. Of course, a number of other environmental features, such as plant nutrient concentration gradients, are correlated with the oscillating salinity gradient in these typical estuaries. Because of recent rises in sea level, most rivers enter the sea through drowned and silt-filled river valleys. Therefore, the entire water column in estuaries of this sort is very directly influenced by benthic processes. Many forms, for example, overwinter as dormant eggs in sediment. Net flux of water through the habitat is also an important characteristic of estuaries, and it plays an important role in the lives of planktonic animals.

TYPIFICATION OF THE ZOOPLANKTON OF ESTUARIES

Circulation and zooplankton distribution

The circulation of estuaries of moderate length and not excessively wide connection to the sea at the mouth can be usefully described by several sorts of "null" points. In situations where the water has a strong density stratification in the vertical direction, the net flow near the mouth is upstream near the bottom and downstream near the surface. The upstream limb becomes less strong in the upstream direction, and eventually reverses. The point of this reversal is a "null" point. It is close to the limit of upstream transport of marine sediment, and can be expected to be the approximate average limit of upstream transport of plankton from the near-shore ocean into the river. This turns out to be so, as will be shown below.

In situations where there is small fresh-water flow and enough tidal energy to thoroughly mix the estuary in the vertical direction, there is also an average null point in the circulation. This occurs at about the point where the water present at higher high-water (in a mixed tidal regime) fails to reach the river mouth at a lower low-water, and thus is returned up the river at the next turn of the tide. Below this point there is a great degree of mixing on every tidal cycle with the water of the near-shore ocean, and the plankton can be expected to be largely a reflection of the general neritic assemblage. Above the null point there are opportunities for endemic populations to develop. Since the environmental conditions are more influenced by the river and by effects of the surrounding land, these endemic populations have a different habitat from that downstream of the null point. They are, therefore, different from the downstream populations.

While there are a large number of papers on the zooplankton of estuaries, there are only a few studies of distribution in which the sampling and sample evaluation are adequate to demonstrate the relationships of species presence and abundance to the distribution of physical factors. The general principles will be demonstrated by reference to the distributions in a few cases, primarily in North America and Australia. This limited survey will also introduce many of the phenomena affecting estuarine plankton: seasonality, the role of ctenophore predation, and other matters. A second section will deal with the overall biology of the dominant forms, several genera of copepods.

The Middle St. Lawrence (Canada)

Bousfield et al. (1975) have given an account of the distribution of zooplankton species in the St. Lawrence River estuary in summer. The St. Lawrence is a very large river (mean discharge 10^4 m^3 s^{-1}; little seasonal variation) which drains the eastern portion of North America centered around 45°N (see Ch. 16, pp. 403–425). Its estaury is around 560 km long, but the salinity gradient is principally located between Ile aux Lièvres (69.7°W) and Ile d'Orléans (71°W). The bottom of this section slopes irregularly upward from depths of about 100 m at Ile aux Lièvres to about 25 m at Pte. Argentenoye on Ile d'Orléans. The channel is along the north bank, and the sampling was mostly concentrated in it. Sampling was done with a pump; the water was filtered with 0.158 mm mesh. The animals caught in this way Bousfield et al. term the *medium* zooplankton. All of the results of this study are presented as "average distributional patterns" over the study period, which must essentially mean they represent the patterns at the mean stand of the tide.

Fig. 5.1 gives the salinity and temperature distributions. It is clear from these that the density gradient is predominantly vertical in the lower reaches and predominantly horizontal in the upper

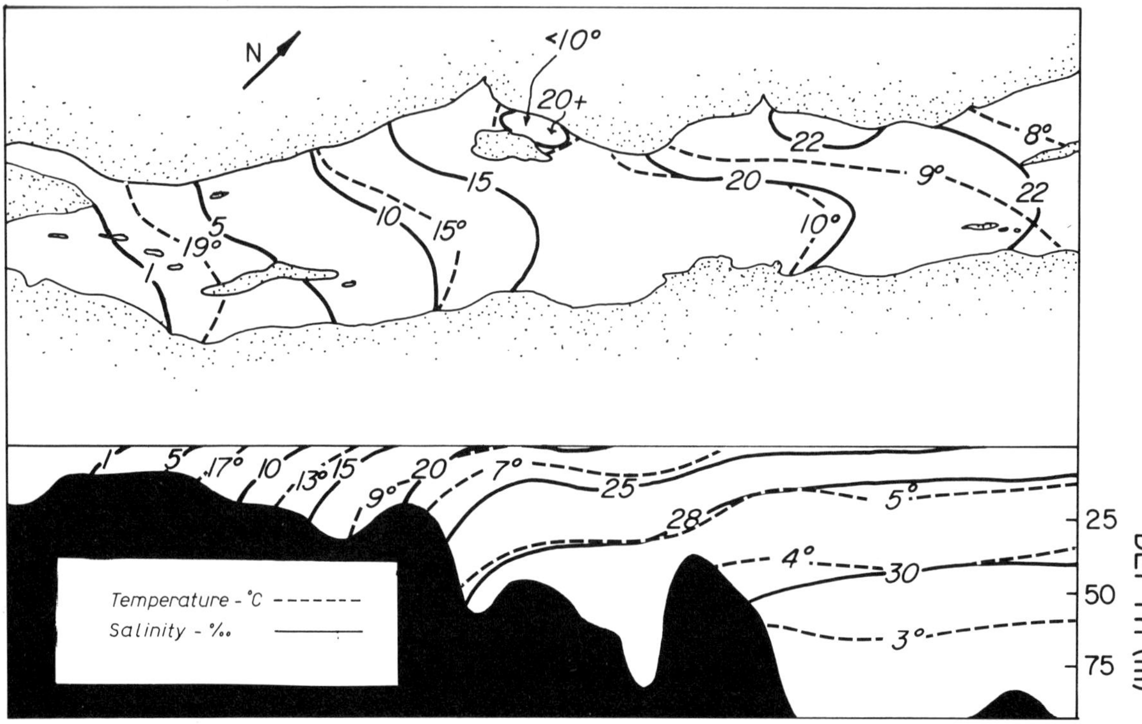

Fig. 5.1. Temperature and salinity distributions in the Middle St. Lawrence Estuary between Ile d'Orléans on the left and Ile aux Lièvres on the right. The surface values are shown in the upper chart. The vertical section below follows the main channel close to the north bank. (Redrawn from Bousfield et al., 1975.)

reaches. The medium zooplankton are mostly calanoid copepods, but mysids, harpacticoid copepods, and cladocerans are also important components. The species found are divided by Bousfield et al. into three groups on the basis of their origins and relationships to salinity:

(1) *Marine coastal species which enter the estuary from the sea.* A subgroup (a) of these are restricted to the deeper more seaward parts of the estuary, and their distribution extends into it in the pattern of the salt wedge. An example is the distribution of *Calanus finmarchicus* shown in Fig. 5.2. A second subgroup (b) has distributions that extend into coastal waters, but they reach their population maxima within the estuary. These forms may be most abundant near the bottom or near the surface. Fig. 5.3 gives the distribution of *Eurytemora herdmanni*, which has a mid-river maximum near the bottom.

(2) *Estuarine endemic species which live only in the brackish water of the upper estuary.* Examples are *Eurytemora affinis* (Fig. 5.4), the harpacticoid *Ectinosoma curticorne*, and the mysid *Neomysis americana*.

(3) *Fresh-water species which extend into brackish water in the upper estuary.* Fig. 5.5 shows the distribution of *Bosmina longirostris*, a fresh-water cladoceran, in the upper estuary.

Table 5.1 (modified from Bousfield et al., 1975) gives the average maximum (with respect to depth) numbers of the principal dominant forms along the estuary, together with a classification by the groups listed above (1a, 1b, 2, and 3). The plankton in each zone (or group) consists of a very small number of species compared to other marine pelagic habitats. The biomass maximum occurs in the weakly brackish water from 1 to 10‰ and is due mostly to the relatively small numbers of *Neomysis americana*, but augmented significantly by *Eurytemora affinis* and *Ectinosoma curticorne*. *Neomysis americana*, like other estuarine mysids, is primarily an epibenthic form, and only a small fraction are typically in the water column.

Acartia longiremis is quite an important component of the plankton of the St. Lawrence Estuary and has its maximum density within the estuary. In contrast, *Acartia clausi* is a very small component and belongs to the coastal marine group of species

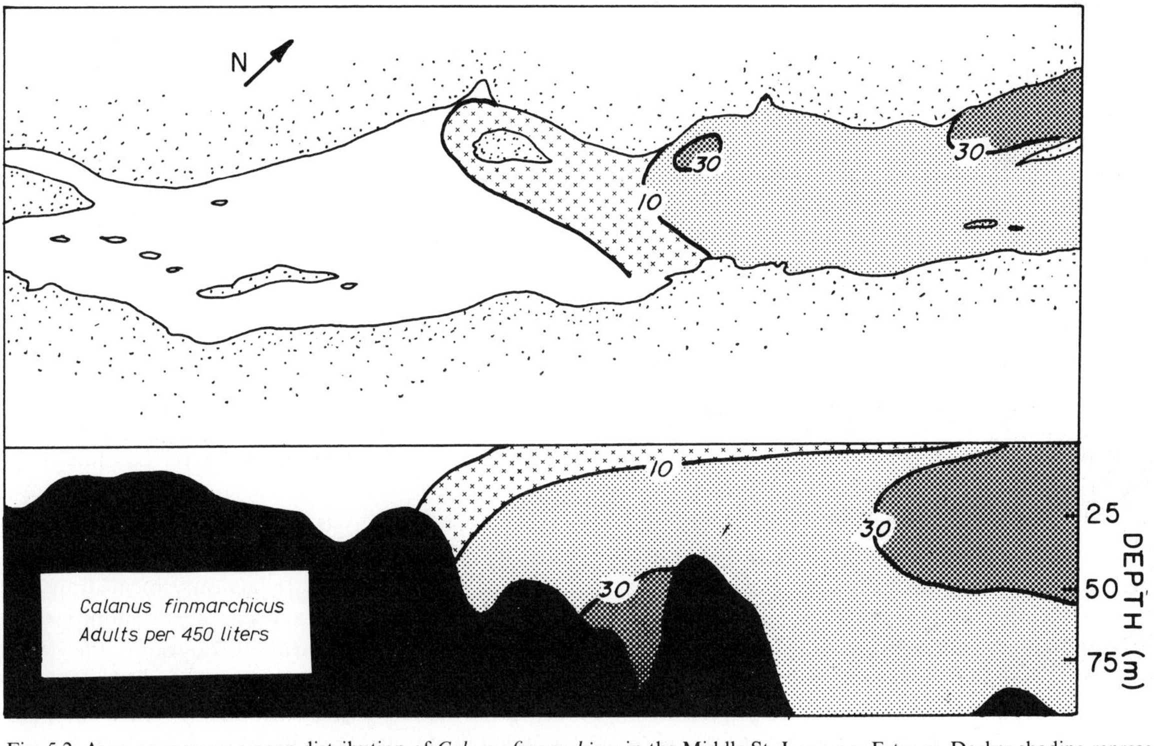

Fig. 5.2. Average summer season distribution of *Calanus finmarchicus* in the Middle St. Lawrence Estuary. Darker shading represents greater numbers. Collection stations as shown in Figs. 5.3 and 5.5. (Redrawn from Bousfield et al., 1975.)

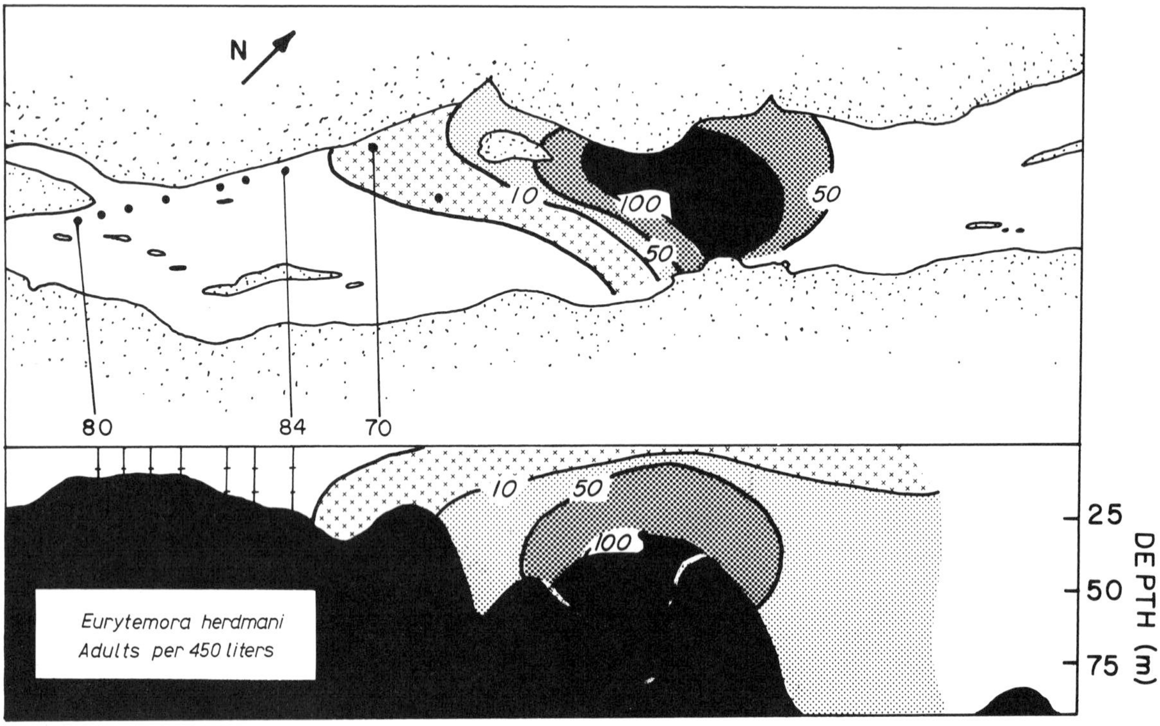

Fig. 5.3. Distribution of *Eurytemora herdmani* in the Middle St. Lawrence Estuary showing collection stations, depths of sampling, and station numbers corresponding to Table 5.1 for the upstream section. (Redrawn from Bousfield et al., 1975.)

approaching the estuary in the salt wedge. This is an expression of a latitudinal gradient in the relative importance of the species in this genus. In New England waters *A. clausi* is an important member of the plankton most of the year, and extends deep into the estuaries. In many places it must be classed as both marine and estuarine endemic.

The null point of the flow and zooplankton distribution patterns is in the neighborhood of Ile aux Coudres. Marine coastal forms do not penetrate upstream much beyond the island. The salt-tolerant, fresh-water species extend downstream almost to this point. Most of the estuarine-endemic species are most dense just below Ile d'Orléans and are nearly absent below the null point. Thus, the brackish region above the null point is a distinctive habitat. One endemic form, however, *Eurytemora herdmani*, is distributed in a zone of the St. Lawrence Estuary centered just downstream of the null point. A population of a species of this genus is found at equivalent parts of a number of estuaries. The mechanisms by which they are retained within the estuary are obscure. The vertical center of distribution for the older life cycle stages is deep in the salt wedge. A balance between upstream and downstream movement is probably achieved by some form of vertical migration: ontogenetic, diel, or possibly without a timing pattern common to the whole population.

Narragansett Bay (U.S.A.)

There are no major rivers between the St. Lawrence and Narragansett Bay. On the whole the plankton of the smaller estuaries of the region (St. John's, Ste. Croix, Penobscot, etc.) has not been studied.

Narragansett Bay is the largest of a sequence of shallow, glacially formed embayments on the south coast of New England. It has only local drainage (the Providence River), and, therefore, salinities are quite high and uniform throughout the year (25–31‰ at mid-bay above Conanicut Island). The bay is divided by islands into north–south passages which connect at about mid-bay. The shoal depths and strong tides produce thorough vertical mixing

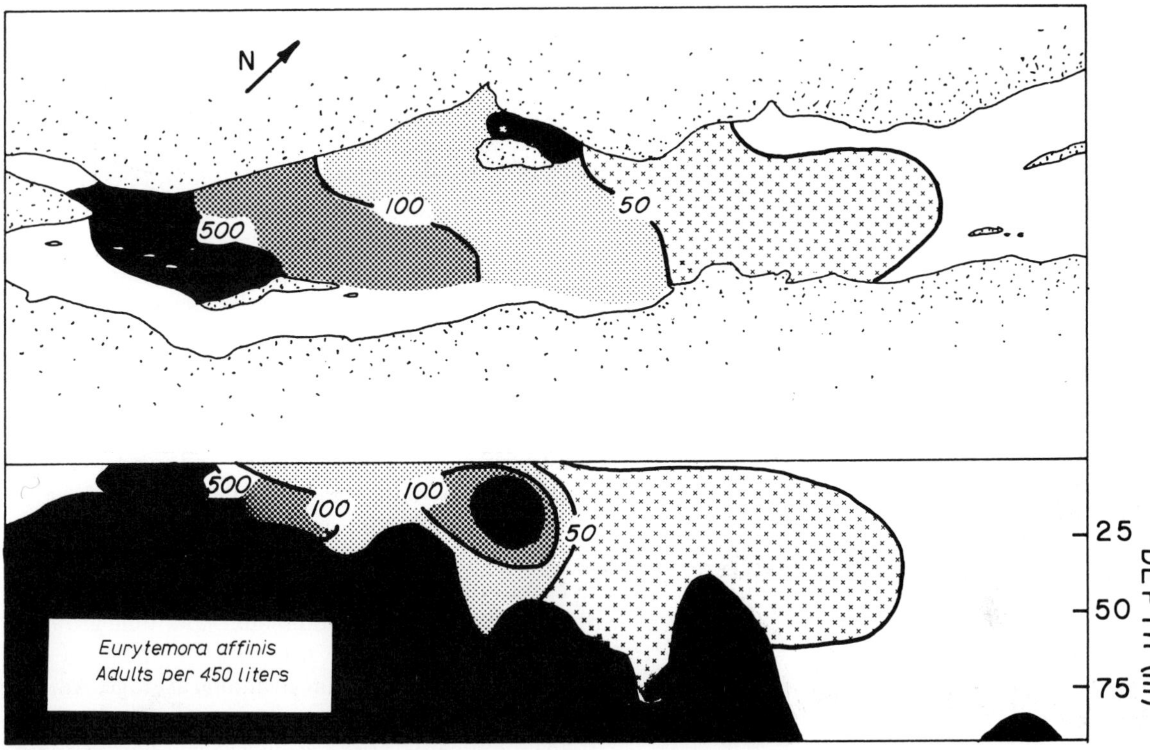

Fig. 5.4. Distribution of *Eurytemora affinis* in the Middle St. Lawrence Estuary. Collection stations are as shown in Figs. 5.3 and 5.5. (Redrawn from Bousfield et al., 1975.)

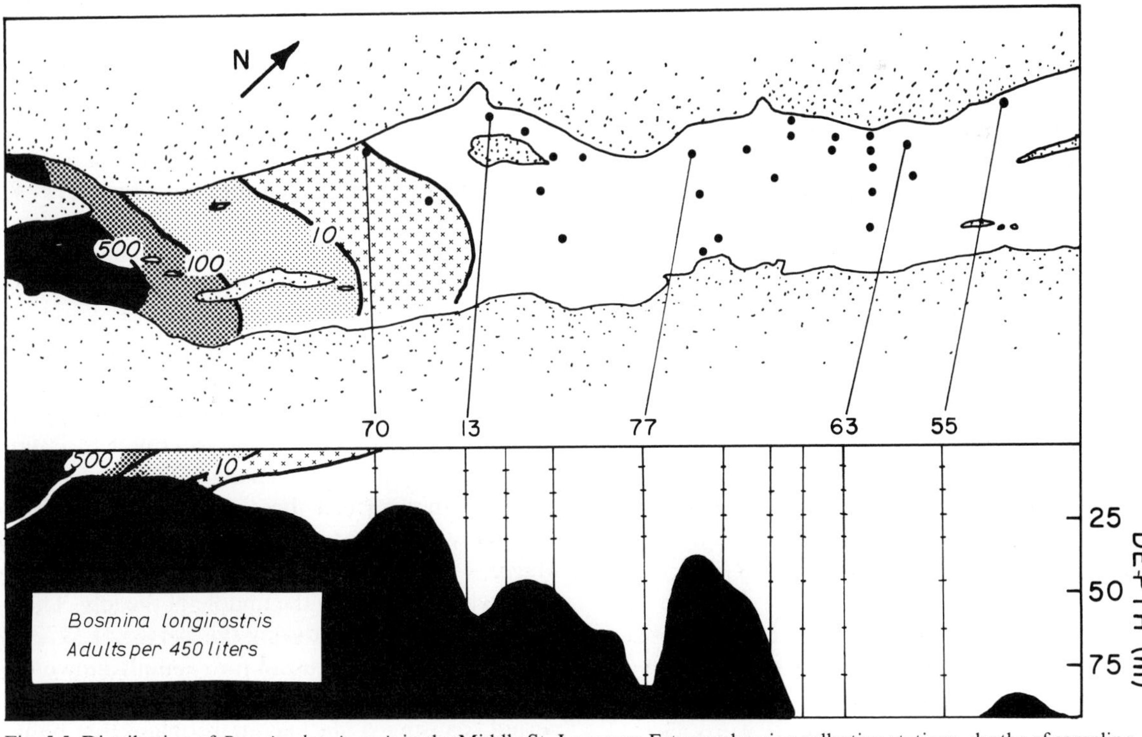

Fig. 5.5. Distribution of *Bosmina longirostris* in the Middle St. Lawrence Estuary showing collection stations, depths of sampling, and station numbers corresponding to Table 5.1 for the downstream section. (Redrawn from Bousfield et al., 1975.)

TABLE 5.1

Average maximum numbers of principal zooplankton species per 450-l sample at principal collecting stations in the Middle St. Lawrence Estuary (simplified from Bousfield et al.,1975); for designations of ecological groups, see the text

Species	Ecological Groups	Station number Ile d'Orléans (downstream →)																		(← upstream) Ile aux Lièvres
		81	80	83	84	70	13	73	76	75	74	77	23	20	51	65	67	64	26	24
Bosmina longirostris	3	700	90	30	8	1	–	–	–	–	–	–	–	–	–	–	–	–	–	–
Ectinosoma curticorne	2	2000	1000	700	250	25	3	1	–	2	–	–	–	–	–	–	–	–	–	–
Neomysis americana	2	25	100	150	9	6	2	1	–	2	4	–	–	–	–	–	–	–	–	–
Eurytemora affinis	2	700	1700	1900	1600	260	250	625	170	135	145	12	29	55	–	5	–	1	4	2
Eurytemora herdmani	1b	–	–	–	–	2	30	45	100	50	2	150	120	100	55	90	30	35	30	25
Acartia clausi	1a	–	–	–	–	1	5	7	5	2	1	5	5	5	8	7	10	14	18	20
Acartia longiremis	1b	–	–	–	–	8	42	84	167	25	7	840	50	84	334	117	125	125	15	84
Calanus finmarchicus	1a	–	–	–	–	–	10	20	25	5	–	35	25	15	20	20	15	13	10	15
Calanus hyperboreus	1a	–	–	–	–	–	1	1	2	1	1	5	4	3	4	4	4	5	3	30
Totals		3425	2900	2800	1910	450	495	950	550	285	165	1075	280	345	435	300	285	295	165	270

all the year. Temperatures vary from January or February lows of 1 to 5°C up to August highs around 23°C (Martin, 1965; Hulsizer, 1976).

Surveys of the zooplankton of Narragansett Bay have been made by Williams (1906), Frolander (1955), Faber (1959), Martin (1965), and Hulsizer (1976). Special studies have been conducted on the role of zooplankton in the control of phytoplankton blooms and species composition (Martin, 1968, 1970; Smayda, 1973) and on the role of ctenophores in controlling the populations of copepods (Kremer, 1976; Kremer and Nixon, 1976). The best of the surveys is that of Hulsizer (1976) for the years 1972 and 1973. She sampled weekly at a station in West Passage off the north end of Conanicut Island. Her data can be compared to those of Martin (1965) for stations at the mouth of the Providence River (head of the bay) and at the mouth of West Passage (seaward end of the bay).

Copepods are the dominant planktonic group in the bay, accounting for 80% or more of the individual organisms on an annual average. Among older copepodite stages that are easily identified, the dominants are *Acartia clausi*, *Acartia tonsa*, *Pseudocalanus minutus*, and *Oithona* spp. Rotifers are abundant in late winter, and cladocerans are abundant in early summer. The two species of *Acartia* alternate seasonally (Fig. 5.6). *Acartia clausi* copepodites IV, V, and VI are present in densities of the order of 5×10^3 m^{-3} from January through June. In July they drop quickly to zero and are rapidly replaced by densities of *A. tonsa* (IV, V, and VI) of the order of 10^2 m^{-3}. *Acartia clausi* reappears and begins to increase again in October, whereas *A. tonsa* disappears in about January. *Pseudocalanus minutus* has the same sort of cycle as *A. clausi* with winter and spring densities around 1400 m^{-3} (Fig. 5.7). *Oithona* spp. have no clear seasonal cycle. Typical densities are 150 m^{-3}.

The seasonal cycle of biomass (Fig. 5.8) is not altogether explained by changes in zooplankton numbers. In particular there is a March or March–April low in biomass. This is also apparent in Martin's (1968) data from the head of the bay although they begin in the middle of the low. There was a drop in salinity during this period of 1972 of about 4‰. It may be usual for rainfall, snowmelt, and, therefore, flushing to reach a peak at this period, causing a decline in zooplankton populations. The effect is less than one order of magni-

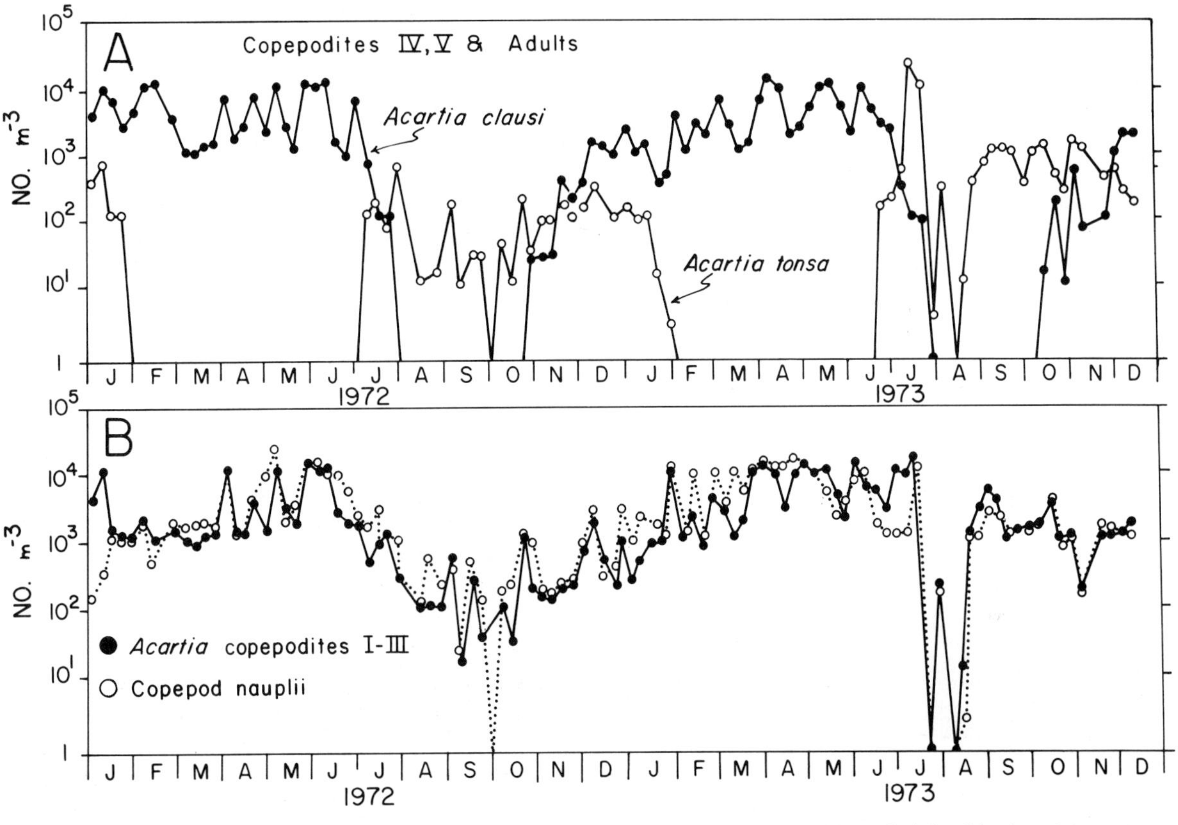

Fig. 5.6. Abundance cycles of *Acartia clausi* and *Acartia tonsa* in central Narragansett Bay. Numerical densities for adults and older copepodites (summed) are shown for each species in part A. Total densities of younger copepodites and of nauplii are shown in part B. The scales are linear between each logarithmic interval, that is, 100 to 200 is represented by the same distance along the ordinate as 800 to 900, etc. (E.E. Hulsizer, pers. comm.) (Redrawn from Hulsizer, 1976.)

tude, so it may be poorly demonstrated in the available graphs of numbers, all of which are on logarithmic scales.

Seasonal alternation of the two *Acartia* species has been noted in a number of estuaries on the east coast, and has been variously attributed to: (1) a shift in the relative competitive coefficients of the two species primarily caused by temperature, but influenced by salinity (Conover, 1956; Jeffries, 1962, 1967); or (2) simple responses of population increase rates to more and less suitable conditions within the seasonal cycle. In some west coast estuaries (see discussion for Yaquina Bay below) two species of *Acartia* can have their seasonal peaks simultaneously and in the same geographic position. These are forms closely related to *A. clausi* and *A. tonsa* respectively, and it seems unlikely, therefore, that competition is a principal causative factor in the east coast case. It is clear that *A. tonsa* lays resting eggs in response to cool temperatures, which is an important factor in removing it from the early winter plankton. This resting stage spends the late winter period in the mud (Zillioux and Gonzalez, 1972). The biology of resting egg production has been well studied elsewhere and is discussed below. The oversummering mechanism of *A. clausi* in the Narragansett Bay area is not apparent at present. Resting eggs have not been reported.

The species composition at Hulsizer's station in mid-bay was more similar to that found by Martin at the mouth than to that found at the head. She points out that for both mid-bay and the mouth: "*Acartia clausi* was more abundant than *Acartia tonsa*; *Pseudocalanus minutus* was present during most of the winter instead of only for a short period, and *Paracalanus parvus* was present in more than trace amounts." In fact, Narragansett Bay has

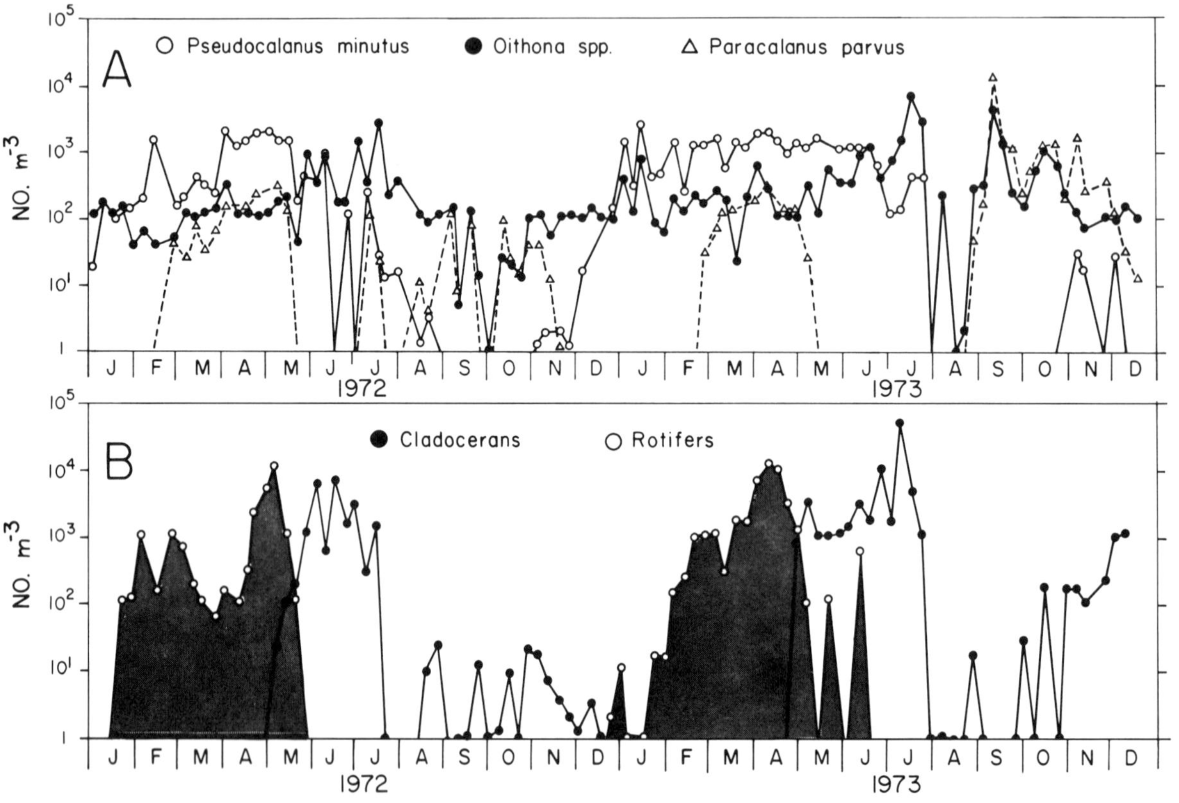

Fig. 5.7. A. Abundance cycles of *Oithona* spp., *Pseudocalanus minutus* and *Paracalanus parvus* in Narragansett Bay. B. Abundance cycles of cladocerans and rotifers. Scales in both parts as in Fig. 5.6. (Redrawn from Hulsizer, 1976.)

no null point above which euryhaline species from the coastal ocean are excluded. It does not extend far enough inland, and it is too well mixed longitudinally for isolated upstream habitats to develop. Martin's (1965) data for 1959 through 1961 show that all of the species important at the mouth are also important, if somewhat reduced in abundance, at the head. *Pseudocalanus minutus*, for example, has a seasonal peak from February through July at the mouth in densities around 1500 m^{-3}. During the same period it is present at the head at densities around 200 m^{-3}. It tends to disappear in late summer at the head, while it is only reduced to about 300 m^{-3} at the mouth. Both *Acartia* species are most abundant at the head, two to ten times more abundant than at the mouth. So there is a gradient in habitat conditions; it does not extend to only those habitat types that support populations of brackish water species found in other east coast estuaries. *Eurytemora affinis*, for example, is not included in any of the published species lists.

Oithona spp. were much more abundant in Martin's samples than in Hulsizer's. Hulsizer attributes this partly to her use of 0.158-mm mesh, whereas Martin used 0.119-mm mesh.

In 1973 Hulsizer found no *A. tonsa* present in mid-bay during part of August. Both total zooplankton numbers and total biomass dropped to very low levels. This was clearly attributable to an increase in the numbers of the ctenophore *Mnemiopsis leidyi* to about 50 m^{-3} (Hulsizer, 1976). Kremer (1976) has shown that this ctenophore increased to high levels in 1971, 1972, and 1974, as well (Kremer has no estimates for 1973). Numbers of zooplankton were not so greatly depleted in 1972 as in 1973. The overall zooplankton abundance only declined in 1972 from winter, spring and early summer levels of 10^4 m^{-3}, to a minimum in summer and fall of 10^3 m^{-3}. This decrease is, of course, substantial, and a summer "bloom" of ctenophores that depletes the zooplankton occurs in most years. It was observed in

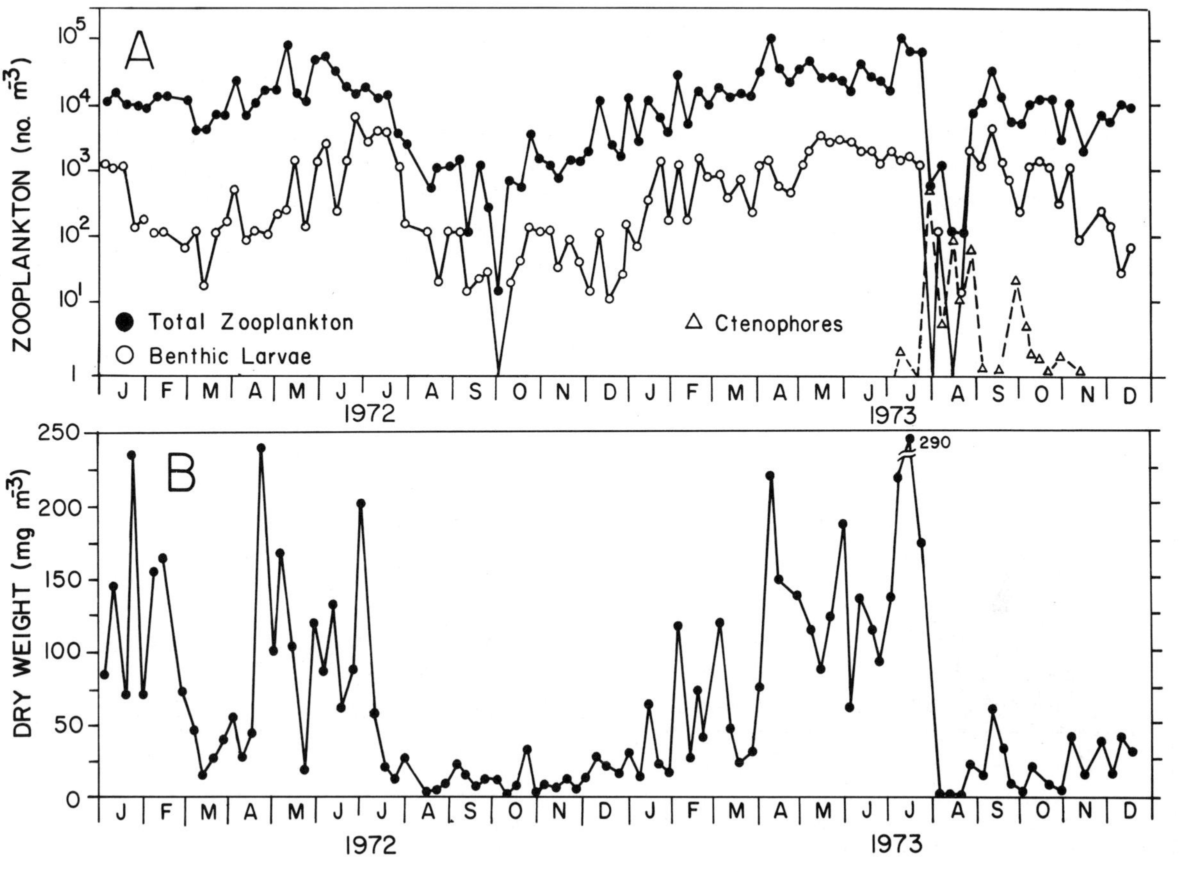

Fig. 5.8. A. Cycles of total holoplanktonic animal abundance (no. m^{-3}) in Narragansett Bay with comparisons to abundance of benthic larvae and ctenophores. Scale as in Fig. 5.6. B. Seasonal cycle of dry weight biomass of zooplankton in Narragansett Bay; linear scale. (Redrawn from Hulsizer, 1976.)

1975, 1976, and 1977 (E. Deason, pers. comm., 1978). Martin's average data for 1959–1962 curiously contain no reference to ctenophores, although his September–October averages for *A. tonsa* are much less than those for August or November. This period, however, was the main seasonal peak for *Oithona* spp. in those years. Possibly there were no ctenophores.

Ctenophore abundance during the population maxima of three years is shown in Fig. 5.9 from Kremer (1976). The data are averages for twenty stations scattered throughout the bay (Kremer and Nixon, 1976). Kremer (1976) has shown that *Mnemiopsis leidyi* is present in "a remnant population (1–2 animals per 10^4 m^3) throughout the winter." She measured fecundity, growth, and maturation rates for the ctenophore, which imply that it can increase in biomass and numbers by four

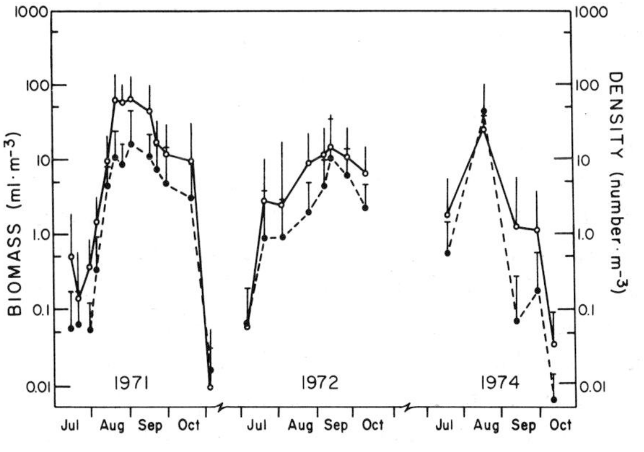

Fig. 5.9. Average biomass of the ctenophore *Mnemiopsis leidyi* in three sampling seasons in Narragansett Bay. Open circles are biomass. Filled circles are numerical density. Bars are positive standard deviations over about twenty stations. (Redrawn from Kremer, 1976.)

orders of magnitude in about sixty days under typical summer conditions of food and temperature. The rapid July increase of copepods in the upper bay in late summer (Martin gives a mean of 9,500 *A. tonsa* m^{-3} in July for the head of the bay) provides a food source. Extremely high ctenophore densities appear there first and spread south (Kremer and Nixon, 1976). The large standing stocks reached by August are responsible for the large drop in overall copepod densities in August through September. Martin and Hulsizer both show that a strong resurgence of *A. tonsa* and the annual increase of *A. clausi* follow the September or October decline of *M. leidyi*. Kremer (1976) states that (1) reduced growth, size and fecundity caused by low food stocks, (2) predation, and (3) possibly progressive infestation by coelenterate parasites account for the decline. Predators include the butterfish (*Peprilus triacanthus*) (Oviatt and Kremer, 1977), and, in 1974 only among years in which observations were made, *Beröe ovata*, another ctenophore of coastal waters. The role of parasitism in the population decline has not been quantitatively evaluated.

The role of zooplankton in control of the phytoplankton populations of Narragansett Bay is complex and continues to be a subject of active study (Smayda and others at the University of Rhode Island). Pratt (1965) and Martin (1965, 1970) concluded that the winter–spring diatom bloom, dominated by *Skeletonema costatum*, was made possible by "release of grazing pressure." However, the data on grazing pressure presented by Martin (1970) indicate that, although it is very modest from August onward, the diatom population usually does not bloom until November–December. Smayda's (1973) data are in good agreement with this. Grazing in both sets of data appears to be very important in removing the plants after the prolonged high in standing stocks during winter. Smayda's (1973) work makes clear that the control of phytoplankton standing stocks and growth is very complex. He states: "the dynamics of *Skeletonema costatum* during the 1971–72 winter–spring bloom in Narragansett Bay could be related to temperature, light, nutrients, grazers, possibly species interactions, and hydrographic disturbances."

Chesapeake Bay and Delaware Bay (U.S.A.)

The east coast of the United States has a broad continental shelf that adjoins a broad, flat coastal plain. Rivers draining the eastern Appalachian Mountains enter this coastal plain along the Fall Line, and almost immediately become estuaries. Delaware Bay and Chesapeake Bay are the principal examples. The bays are shallow: the main channel of the Chesapeake is 6 to 65 km wide and 320 km long, but has a depth of only 20 m. The depth at the mouth is even less, 12 m. Precipitation and snowmelt cycles in the area result in maximum flow in spring and minimum flow in fall. Salinity is correspondingly lowest in spring and highest in fall. Temperature follows the seasonal heating and cooling cycle, being highest in summer and lowest in winter. Stroup and Lynn (1963) give charts showing seasonal average distributions of salinity and temperature in the Chesapeake. Surface distributions are illustrated for spring in Fig. 5.10. The SW to

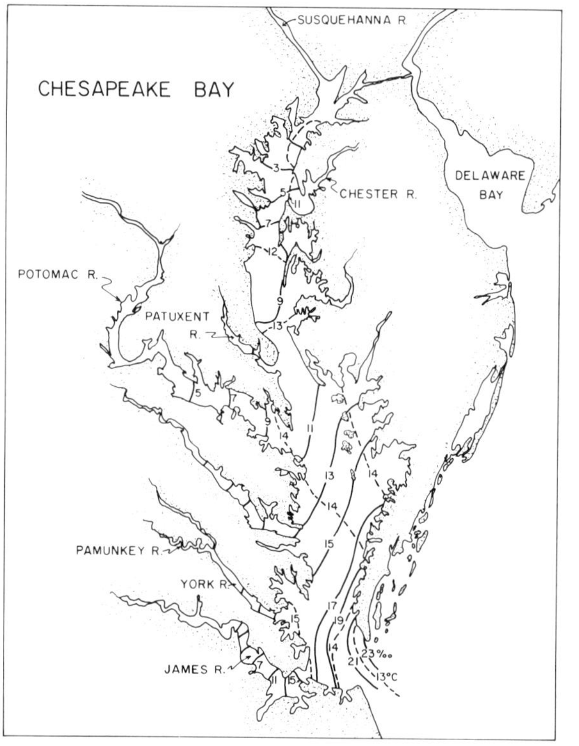

Fig. 5.10. Average spring salinity and temperature distributions in Chesapeake Bay. Solid contours are salinity in parts per thousand. Dashed contours are temperature in degrees centigrade. (Data of Stroup and Lynn, 1963.)

NE trend of the isohaline lines is caused by the greater influx of fresh water along the western shore. The surface temperature tends to be uniform over the bay in all seasons, with a summer high around 26.5°C and a winter minimum around 3.5°C (4°C at mouth, 2°C in upper reaches). Wide expanses of the Chesapeake are frozen in some winters. The vertical distribution of summer salinity and temperature are shown in Fig. 5.11. Summer has the maximum vertical temperature gradient. In other seasons the bay is almost isothermal. The salinity distribution suggests that the physical null point should correspond with roughly the 10‰ isohaline. The distributions established for the oceanic group of zooplankton in Delaware Bay are in agreement, extending to the intersection of this isohaline with the bottom.

The seasonal cycle of oxygen concentration in Chesapeake Bay is important biologically. The great length of the bay means that exchange rates between the ocean and the null point are very low. Therefore, the salt wedge becomes depleted in summer (Fig. 5.12). This probably restricts zooplankton distribution in the main channel, but only Bosch and Taylor (1970) have studied the effects.

Chesapeake Bay zooplankton have been studied by a number of workers, but general distributions in the main bay and the seasonal cycle of species composition and abundance in the bay as a whole are still not well known. This is clearly because of the tremendous length of the estuary. An overall study by Wilson (1932) suffers from too many errors. The dominant species of *Acartia* have been

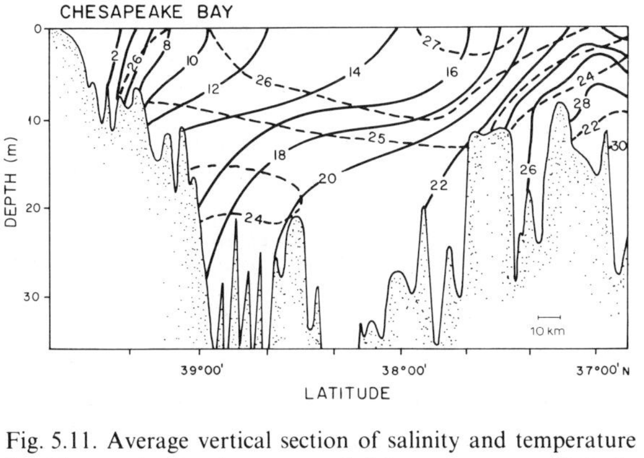

Fig. 5.11. Average vertical section of salinity and temperature for Chesapeake Bay in summer. Solid contours are salinity in parts per thousand. Dashed contours are temperature in degrees centigrade. (Data from Stroup and Lynn, 1963.)

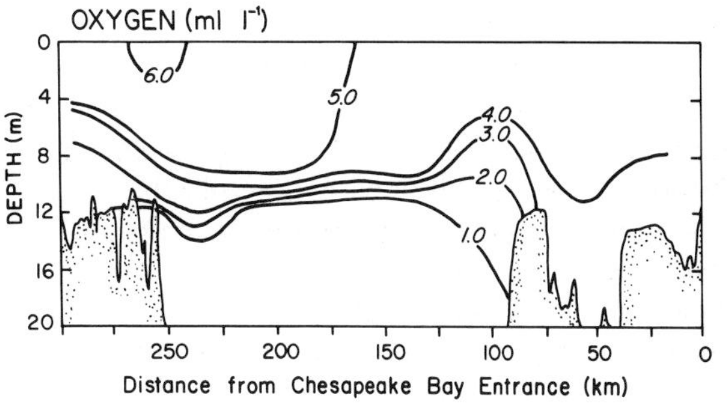

Fig. 5.12. Oxygen concentration on a section down the axis of Chesapeake Bay in September. Values are ml l^{-1}. (Data of Bosch and Taylor, 1970.)

shown (Bowman, 1961) to have been misidentified, and a number of forms are characterized as having seasonal maxima at adjacent stations in opposite seasons. Most of the better work concerns distributions from the main channel up to fresh water in one or another of the subestuaries of the western shore. Seasonal cycles in the Patuxent River have been studied by Herman et al. (1968), and the secondary production of the dominant copepods in that river has been studied by Heinle (1966, 1969a, b, 1974) and Heinle and Flemer (1975). General distributional work has been done in York River by Van Engel and Joseph (1968) and by Burrell (1972). There has been some work by Sage et al. (1976) on the Potomac. Allen et al. (1976) studied the cycle of abundance and production of copepods in Rhode River. One of the subestuaries of the eastern shore, the Chester River, has been cursorily studied by Forns (1972). In addition there is a large literature on the prospective and actual impacts of thermal and other pollution by power plants and sewage treatment in Chesapeake Bay. On the whole this literature is not in the journals, and has not been consulted for this review. Bongers et al. (1975) refer to many of these works.

There are a number of studies of the biology of particular planktonic organisms of the Chesapeake system: the ctenophore *Mnemiopsis leidyi* (Bishop, 1967; Burrell and Van Engel, 1976); the scyphozoan medusa *Chrysaora quinquecirrha*, the sea nettle (Cones and Haven, 1969; Cargo and Schultz, 1966, 1967); copepods (Herman and Mihursky, 1964; Heinle, 1969b, and see above); and the cladoceran *Podon polyphemoides* (Bosch and Taylor, 1970, 1973).

The general distributional and seasonal studies

of the Chesapeake can be compared to those of Cronin et al. (1962) and Deevey (1960a) for the Delaware Bay. This latter bay is much less complexly branched than the Chesapeake, but the basic salinity gradient and seasonality of physical factors are similar. This is fortunate, since these are the only works showing the distribution of the whole plankton community over the whole salinity gradient. Fresh-water, estuarine, and coastal groups can be identified. Cronin et al. (1962) list these as follows:

(1) **Coastal:** copepods — *Centropages typicus*, *C. hamatus*, *Labidocera aestiva*, *Temora longicornis*, *Paracalanus parvus*, *Pseudocalanus minutus*; cladocerans — *Penilia avirostris*, *Evadne nordmanni*. An average (two years of quarterly data) seasonal cycle of distribution and abundance of *C. typicus* is shown in Fig. 5.13. It is present in densities of a few hundreds per cubic meter at most, enters from the ocean, and is usually densest in the deeper layers or salt wedge. Significant densities are generally restricted to salinities greater than about 10‰. This corresponds to the seasonally shifting null point in the Delaware. Densities of *C. typicus* are lowest in summer, highest in spring and fall, a reflection of

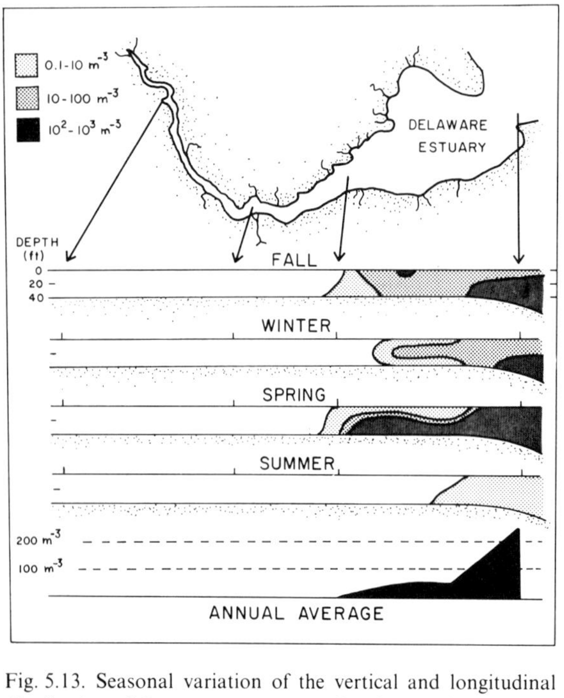

Fig. 5.13. Seasonal variation of the vertical and longitudinal distribution of *Centropages typicus* and *C. hamatus* in Delaware Bay. (Redrawn from Cronin et al., 1962.)

their cycle on the continental shelf in this area (Bigelow and Sears, 1939). *Temora longicornis*, with highest densities in spring, has a pattern similar to that in Fig. 5.13; there are usually less than 100 m^{-3}. *Labidocera aestiva* is almost always less than 100 m^{-3}, usually less than 10 m^{-3}, and disappears in winter. Again, this is true all along the coast, and the species has been shown to overwinter as a resting egg in the sediment (Grice and Gibson, 1975). *Paracalanus parvus* and *Pseudocalanus minutus* are seasonal alternates with southern affinities (maximum in summer and fall) and northern affinities (maximum in winter and spring), respectively. The cladocerans *P. avirostris* (southern) and *E. nordmanni* (northern) alternate similarly. Each is modestly abundant in some years and apparently absent in others. All of these coastal species are listed by Van Engel and Joseph (1968) as the principal downstream forms in Chesapeake Bay. Their data indicate exclusion from upstream areas below 20‰ salinity.

(2) **Estuarine:** copepods — *Acartia tonsa*, *Acartia clausi*, *Eurytemora affinis*, *Scottolana canadensis* (harpacticoid), and *Pseudodiaptomus coronatus*; cladocerans — *Podon polyphemoides*. Both species of *Acartia* have distributions extending over the continental shelf, but each has its maximum abundance in brackish water. *Scottolana canadensis* can produce phenomenal numbers of planktonic nauplii in some branches of the Chesapeake in August to November (see Fig. 5.14). The older stages are mostly benthic, but according to Heinle (1972) their ecology is not well known. *Pseudodiaptomus coronatus* is an important fall and winter form, in salinities between 15 and 20‰, reaching 1000 m^{-3} in the Delaware. It is similarly common in the Chesapeake system. Epibenthic Malacostraca, the amphipod *Gammarus fasciatus* and the mysid *Neomysis americana*, are very important in Delaware Bay plankton hauls at night when they enter the plankton. Cronin et al. (1962) found they were 54% and 6% by volume in night and day samples respectively of the Delaware plankton although the time of sampling is not well specified.

(3) **Fresh water:** copepods — *Cyclops viridis*; cladocerans — *Bosmina longirostris*. Cronin et al. group *Gammarus fasciatus* with the fresh-water forms. The seasonal maximum for fresh-water forms is spring, the season of large fresh-water flow. They can be abundant to salinities as high as 5‰.

Heinle (1972) states the remarkable extent to which the plankton of the Chesapeake is dominated by a few species of copepods:

"Of the 73 species [of copepods] listed from Chesapeake Bay about 5 are abundant. The calanoid *Acartia tonsa* (Dana) is found throughout the year at all salinities but is most abundant from April to October. Population densities are commonly 100 000 to 200 000 individuals per m^3 and sometimes exceed 1 000 000 per m^3 during that period. During winter and early spring *A. tonsa* is exceeded by the calanoids *A. clausi* at higher salinities and *Eurytemora affinis* (Poppe) at lower salinities. *Acartia clausi* occurs to salinities as low as 5‰ (Baltimore Harbor) and *E. affinis* to salinities as high as 12‰ (mid-Bay). Densities of both of these species are commonly in excess of 200 000 per m^3, and *E. affinis* at times exceeds 3 000 000 per m^3.

The relatively small *Oithona brevicornis* (Giesbrecht) is found at higher salinities >10‰ all year but appears to breed during the summer months when it achieves greatest densities. This species is epibenthic in habit and has not been adequately sampled. Adults and immature copepodites of this species at times exceed 100 000 per m^3.

The harpacticoid *S. canadensis* has pelagic nauplii which at times exceed 1 500 000 per m^3. ... Two other calanoids are common, but not abundant, *Pseudodiaptomus coronatus* (Williams) and *Pseudocalanus* sp. Alleged juveniles of a cyclopoid *Hemicyclops americana* (C.B. Wilson) are common during summer and fall at mid-Bay."

The annual density cycles of the dominant estuarine copepods alternate. The only data presented in terms of absolute density estimates are those of Heinle (1969a) and Van Engel and Joseph (1968). The latter data are from 0.5-mm mesh nets, too coarse for good estimates of copepods. Fig. 5.14 from Heinle (1969a) shows the seasonal cycles of *A. tonsa*, *A. clausi*, *E. affinis* and *S. canadensis* determined from sampling with 0.074-mm nets. *Acartia tonsa* is continuously present in the Patuxent, and the continuous presence of nauplii implies continuous reproduction. Conditions do not appear to be continuously suitable for growth, however, since copepodites are absent during January and February, and even longer upstream. At the peaks in mid-summer nauplii exceed 10^5 m^{-3}, and adults run to 10^4 m^{-3} or more. *Acartia clausi* intrudes from downstream in late winter and spring. Its peak is not in any sense complementary to a dearth of *A. tonsa*, whose population is building up through the time *A. clausi* is present. A number of authors have referred to a congeneric replacement, but that does not really occur. If there is a seasonal complementarity, it is between *A. tonsa* and *E. affinis* (Fig. 5.14A–C, H–J), whose spatial and seasonal peaks of abundance occur where and when *A. tonsa* is absent. *Scottolana canadensis* (Fig. 5.14F, G) is an upstream, summer form (April to November) with density peaks in May and August–September. It, like *Pseudodiaptomus coronatus* (see Cronin et al., 1962), spends its adult phase out of the water column.

It is clear that the peak seasonal count for zooplankton (see Fig. 5.15 from Herman et al., 1968) occurs when all the dominants are present at once, that is in March and April. This peak is probably cropped back in April–May by ctenophores (Fig. 5.15), primarily *Mnemiopsis leidyi*, as in Narragansett Bay. Such cropping was suggested by Nelson (1925) for the New Jersey estuaries and by Cronin et al. (1962) for the Delaware Bay, but sampling was not frequent enough to show the closeness of the relationship. Burrell and Van Engel (1976) used an 0.50-mm mesh (the paper says 0.75 mm, but see Van Engel and Joseph, 1968) to estimate zooplankton numbers and compare them to the volume density of *M. leidyi* in York River (see Fig. 5.10) and its Pamunkey River tributary. In general they found an inverse relationship both along the river and temporally. However, the 0.5-mm nets are too coarse for useful estimation of estuarine plankton, particularly in summer.

Maximal densities of *M. leidyi* occurred in June (33 ml m^{-3}) in 1966, timing comparable to that in the Patuxent in 1964 (Herman et al., 1968, see Fig. 5.15). *Mnemiopsis leidyi* in York River had the same range and mean of size in all seasons with a trend toward shorter lengths upstream (see Table 5.2). Length thus varies directly with salinity. Burrell and Van Engel (1976) suggest this may be caused by the smaller average size of the prey available upstream, but there are no direct data. Copepods were the preferred food of ctenophores at all stations: 2101 of 3769 food items. Other food items included the entire range of zooplankton up to sizes

TABLE 5.2

Mean length of *Mnemiopsis leidyi* in two seasons at two positions in York River, tributary of Chesapeake Bay (data of Burrell and Van Engel, 1976)

	June 1966	Dec. 1966
York River mouth	62 mm	78 mm
River Mile 25	20 mm	16 mm

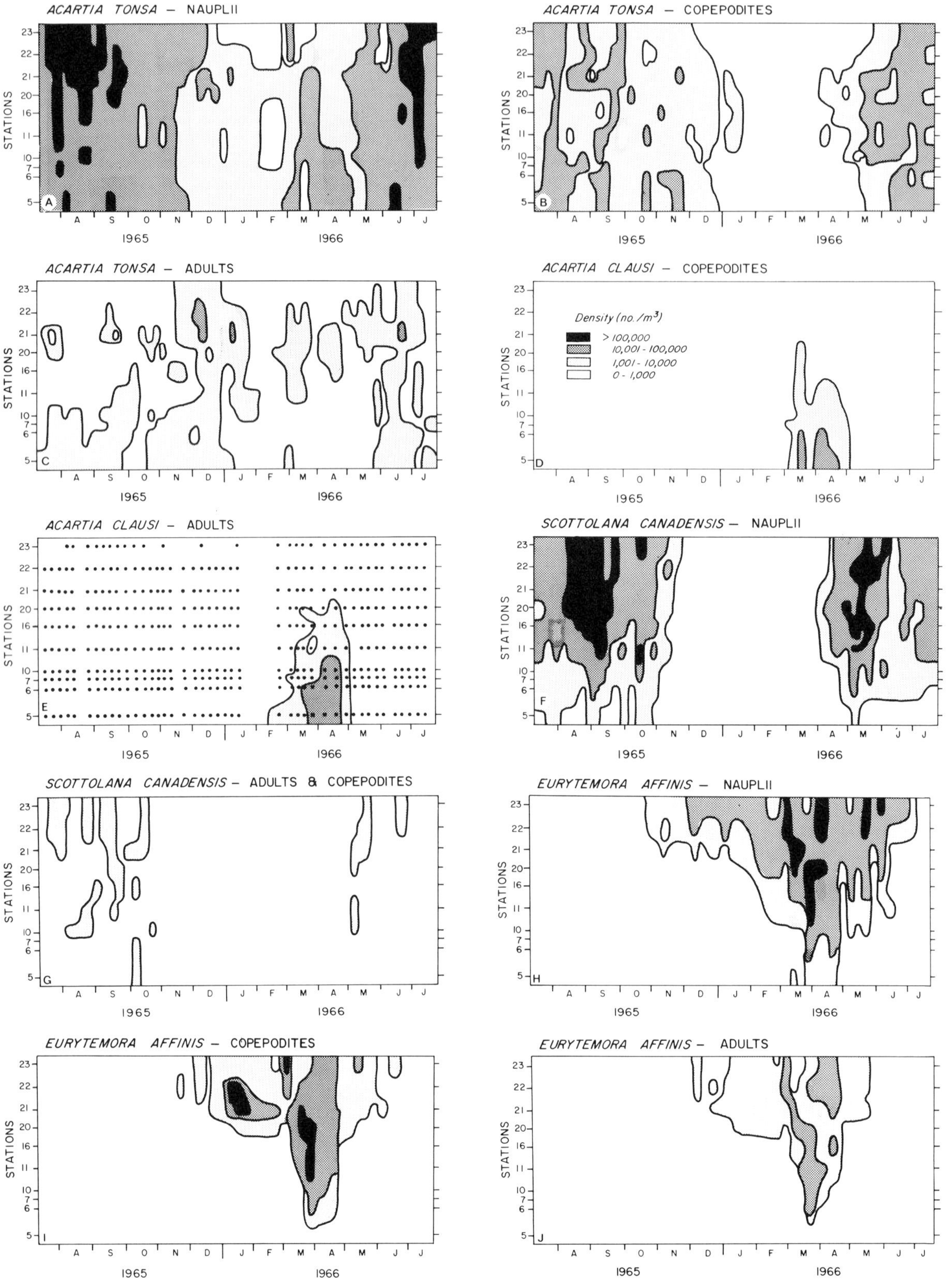
ACARTIA TONSA — NAUPLII
ACARTIA TONSA — COPEPODITES
ACARTIA TONSA — ADULTS
ACARTIA CLAUSI — COPEPODITES
Density (no./m³)
> 100,000
10,001 - 100,000
1,001 - 10,000
0 - 1,000
ACARTIA CLAUSI — ADULTS
SCOTTOLANA CANADENSIS — NAUPLII
SCOTTOLANA CANADENSIS — ADULTS & COPEPODITES
EURYTEMORA AFFINIS — NAUPLII
EURYTEMORA AFFINIS — COPEPODITES
EURYTEMORA AFFINIS — ADULTS
STATIONS
23
22
21
20
16
11
10
7
6
5
A S O N D J F M A M J J
1965
1966
A
B
C
D
E
F
G
H
I
J

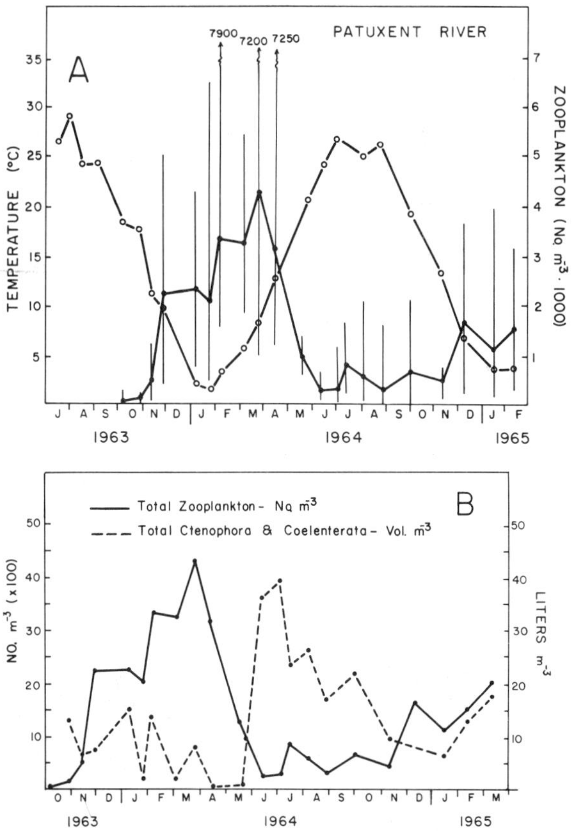

Fig. 5.15. A. Seasonal cycle of mean zooplankton density (●) compared to mean temperature (○) for the Patuxent River (Maryland, U.S.A.). Vertical lines are the range of observed densities. B. Mean zooplankton density compared to displacement volume of ctenophores for the Patuxent River. (Redrawn from Herman et al., 1968.)

of 5 to 6 mm. Heinle (1974) has shown that the stocks and replacement rates of *Acartia tonsa* alone are sufficient in the Patuxent to support the feeding rates expected for *M. leidyi*, given the densities estimated by Herman et al. (1968) and the respiration rates determined for the ctenophore by

Fig. 5.14. Seasonal and spatial variation in the numerical density of various species of zooplankton from July 1965 to July 1966 in the Patuxent River. The dates of observation and the key for shading are the same for all, but are shown only in D and E. Station 5 is 18.7 km and station 23 is 50.9 km from the mouth of the estuary. (Data after Heinle, 1969a.) A–C. *Acartia tonsa*: nauplii, copepodites, and adults. D,E. *Acartia clausi*: copepodites and adults. F,G. *Scottolana canadensis*: nauplii, and copepodites plus adults. H–J. *Eurytemora affinis*: nauplii, copepodites, and adults.

Miller and Williams (1972) using specimens from Pamlico Sound, North Carolina. However, Heinle included no terms for ctenophore growth and turnover, so the trophic budget is not yet clear. It is fair to conclude that the *A. tonsa* population must be the principal support of the summer bloom of *M. leidyi*.

Control of the *M. leidyi* population is partly attributable to predation by *Beröe ovata*. Burrell and Van Engel (1976) show this ctenophore to be present in mid-summer up to river mile 25 (40 km) in York River. It was present at the mouth from July through December. Burrell and Van Engel show almost complete exclusion of *M. leidyi* from the places and times that *B. ovata* is present. The scyphozoan *Chrysaora quinquecirrha* is another important ctenophore predator. The pelagic medusae of this form are present from early July through September or even mid-November. They have been well studied by Cargo and Schultz (1967), who show that all life cycle stages of *C. quinquecirrha* will eat *M. leidyi*. A report they cite of the coincidence of decline of *M. leidyi* with the summer increase of *C. quinquecirrha* is not yet confirmed by quantitative data.

Bosch and Taylor (1970, 1973) have evaluated the distribution and timing of the population cycle of *Podon polyphemoides* (Cladocera) in Chesapeake Bay. The annual cycle begins with the appearance of parthenogenetic females in March at the heads of the tributary rivers. This occurs when temperature exceeds about 4°C. The bloom proceeds downstream, reaching an overall peak in mid-bay by April or May (Figs. 5.16 and 5.17). The very highest densities (10^4 to $3 \times 10^4\,m^{-3}$ and occasionally higher) were shown to occur between 12°C and 25°C. Higher summer temperatures, apparently, reduce the population in some years (1968 but not 1964 or 1965) to nil. This is then followed by a fall resurgence. Males and sexual females are present at the peak of both pulses, producing resting eggs. The dormancy of resting eggs is not yet well studied for *P. polyphemoides*.

Very convincing sets of data were gathered by Bosch and Taylor (1970) on several occasions in fall which show that *P. polyphemoides* is an inverse vertical migrator (Fig. 5.18). The bulk of the fall population spends the night at the top of the halocline, which is, more importantly, the oxycline as well. They spend the daylight period at about

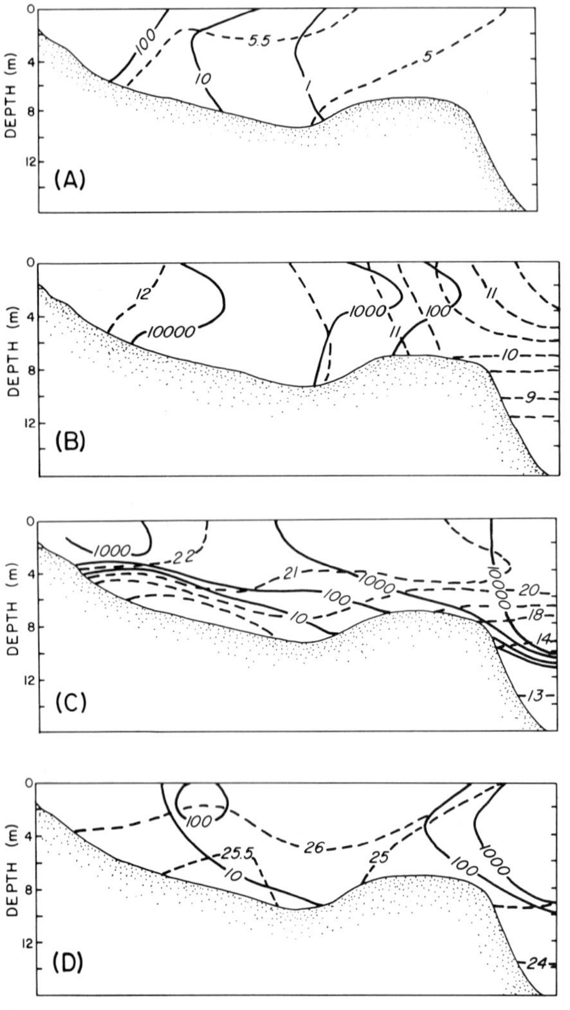

Fig. 5.16. A time series of distributions of *Podon polyphemoides* along the axis of the Severn River (Maryland, U.S.A.) in 1965 compared to the temperature distributions. A. March 15, 1965. B. April 19, 1965. C. May 24, 1965. D. August 2, 1965. The solid contours are numbers m^{-3}. The dashed contours are temperature in degrees centigrade. (Redrawn from Bosch and Taylor, 1970.)

4 m. Bosch and Taylor suggest that the top of the halocline (17‰) corresponds to the level of minimal net transport, so that the migration results in position maintenance. It could certainly slow removal downstream, but a return landward would require a deeper migration. Perhaps that occurs in spring when oxygen depletion in the salt wedge is much less severe. There are no data from spring.

Productivity studies of the dominant Chesapeake copepod species will be reviewed in a separate section below.

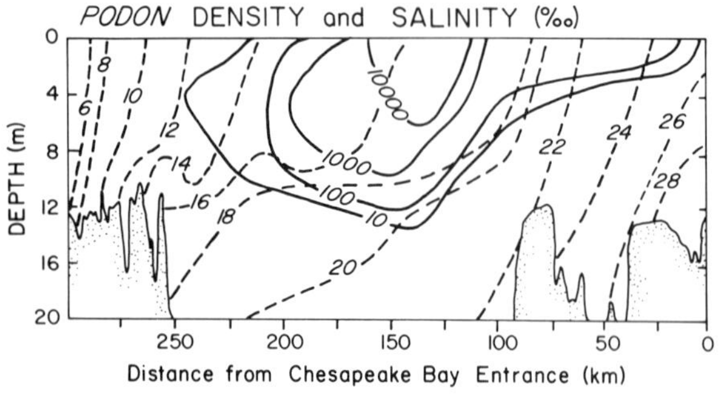

Fig. 5.17. Daytime vertical distribution of *Podon polyphemoides* along the main axis of Chesapeake Bay compared to the salinity distribution. Solid contours are numbers m^{-3}. Dashed contours are salinity (‰). (Redrawn from Bosch and Taylor, 1970.)

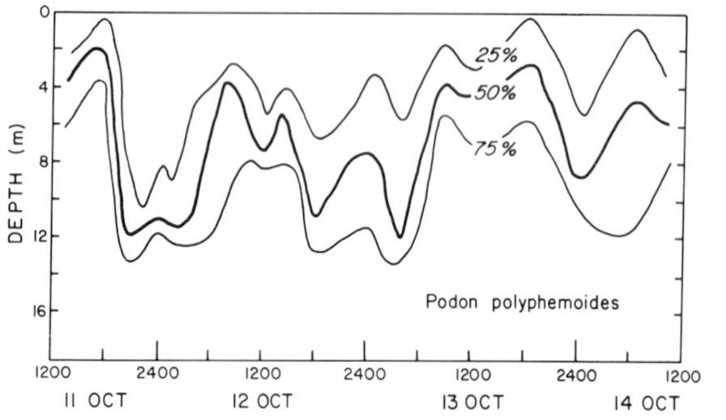

Fig. 5.18. Median and quartiles of the depth distribution of *Podon polyphemoides* at a station 170 km upstream from the mouth of Chesapeake Bay. (From Bosch and Taylor, 1970.)

Yaquina Bay, Oregon (U.S.A.)

Yaquina Bay, Oregon (Fig. 5.19) is one of a series of small estuaries along the coast of the United States south of the Columbia River. The rivers entering these bays drain small sections of the coastal mountain range, in the case of Yaquina Bay an area of only 627 km^2. The head of tide water in Yaquina Bay is about 45 km from the mouth. Weather in the region is monsoonal with heavy winter rains, intermittent rain in spring and fall, and summer drought. This causes a pronounced annual cycle of river flow, which in turn causes an annual cycle of salinity distribution (Fig. 5.20). The monsoonal weather pattern also drives an alternation of flow in the nearby ocean. There is upwelling in the summer with flow alongshore from the north. In winter the flow is from the south and onshore. The upwelling balances the greater solar heating of the summer period, so that there is only a very damped annual cycle of temperature in the

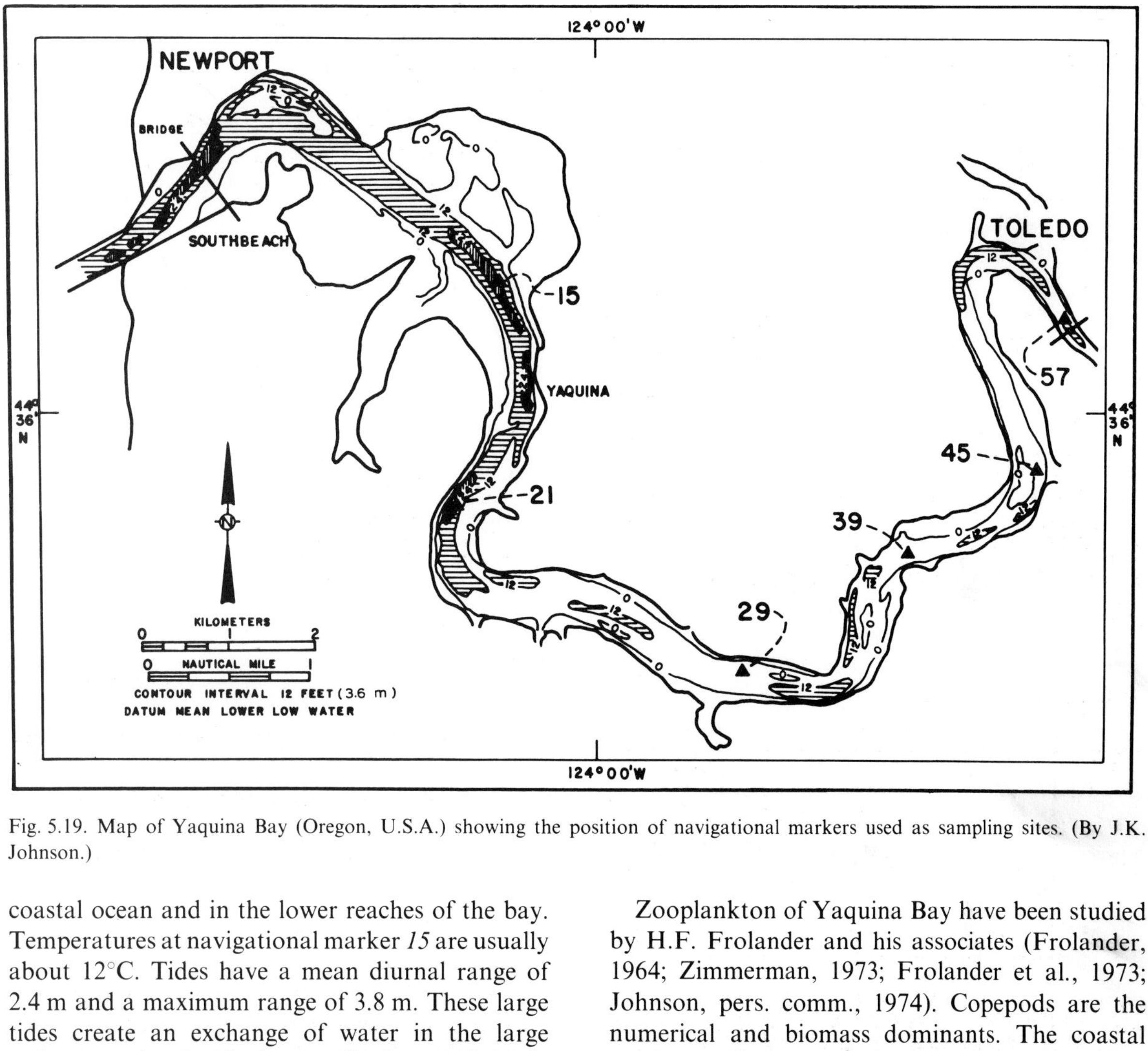

Fig. 5.19. Map of Yaquina Bay (Oregon, U.S.A.) showing the position of navigational markers used as sampling sites. (By J.K. Johnson.)

coastal ocean and in the lower reaches of the bay. Temperatures at navigational marker *15* are usually about 12°C. Tides have a mean diurnal range of 2.4 m and a maximum range of 3.8 m. These large tides create an exchange of water in the large embayment just inside the mouth of approximately 70% per tide (Frolander et al., 1973). Above navigational marker *21* (Fig. 5.19) the exchange depends upon fresh-water flow. This varies, on the average, from 17 $m^3 s^{-1}$ in January to 1 $m^3 s^{-1}$ in August (Reed, 1978). In summer the exchange ratios are low in the upper bay, and the water warms to temperatures around 23°C (Fig. 5.21). The large volumes of water moving on each tide produce high velocities and thorough vertical mixing. That explains the small differences between surface and bottom temperatures and salinities. There is more difference in winter when the flux of fresh water from a single interval of strong rain can sweep out over the surface of the bay. Sometimes almost all of the salt water can be pushed from the bay by an extremely heavy flood.

Zooplankton of Yaquina Bay have been studied by H.F. Frolander and his associates (Frolander, 1964; Zimmerman, 1973; Frolander et al., 1973; Johnson, pers. comm., 1974). Copepods are the numerical and biomass dominants. The coastal marine component of the fauna goes through a seasonal cycle of species composition parallel to that of the nearshore ocean. This cycle is driven by the monsoonal flow patterns (Peterson and Miller, 1977). Summer species are mostly of northern origin: *Acartia clausi* s.l.[1], *A. longiremis*, *Calanus*

[1]The form referred to as *Acartia clausi sensu lato* has been demonstrated by the failure of cross-breeding experiments (Carrillo et al., 1974) to be different from the form called by that name on the American east coast. Bradford (1976) describes several species of *Acartiura*, the subgenus to which all these forms belong, as living along the west coast: a form like *A. hudsonica*, to which she attributes most east coast records; and a form she designates *Acartia* sp. that is found in various locations from Puget Sound southward. Some extremely careful systematic work will be required to make the correct assignments.

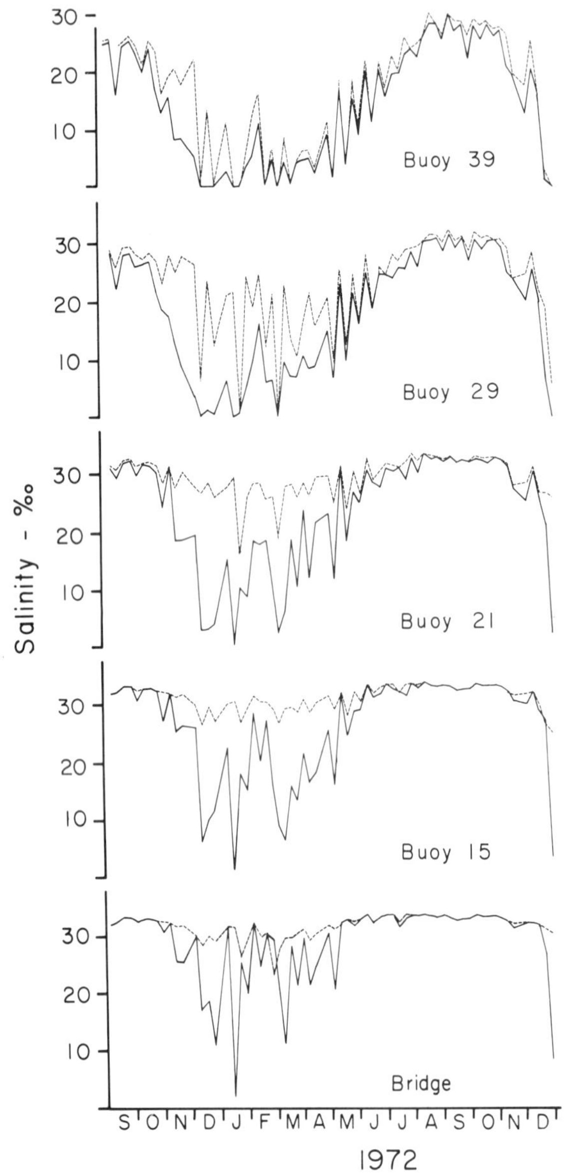

Fig. 5.20. Seasonal cycle of salinity at the surface and near the bottom at a series of stations in Yaquina Bay (Oregon, U.S.A.). Dashed line is for the bottom. Solid line is for the surface. Station positions are given in Fig. 5.19. Original data.

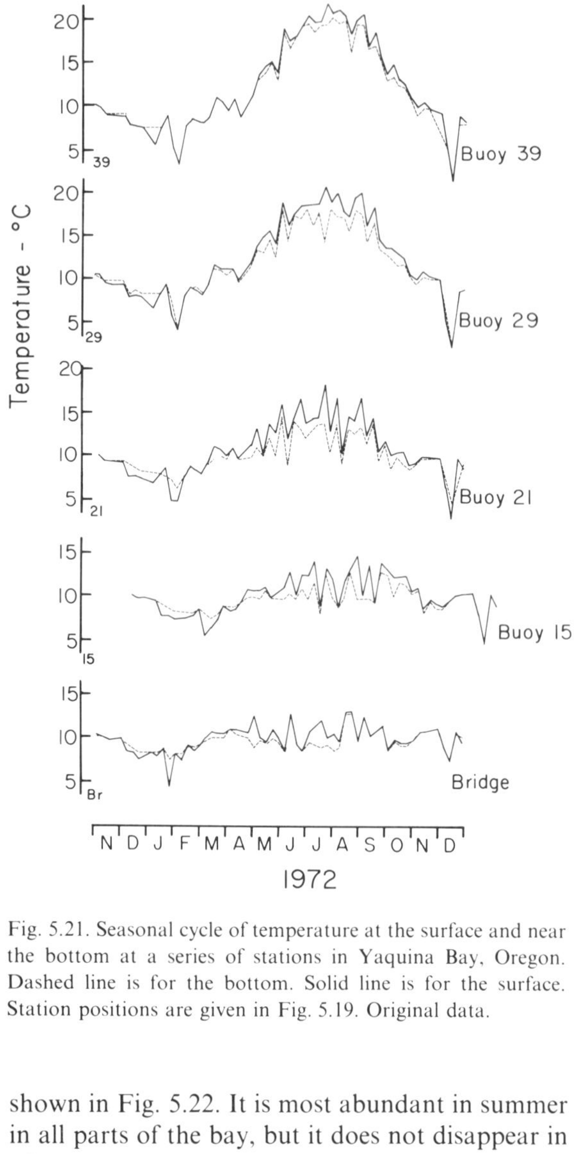

Fig. 5.21. Seasonal cycle of temperature at the surface and near the bottom at a series of stations in Yaquina Bay, Oregon. Dashed line is for the bottom. Solid line is for the surface. Station positions are given in Fig. 5.19. Original data.

marshallae, *Centropages abdominalis*, and *Oithona similis*. Winter species are of southern origin: *Clausocalanus* spp., *Corycaeus anglicus*, *Ctenocalanus vanus*, *Paracalanus parvus*. The summer species are much more abundant than the winter ones, and they do not wholly disappear during winter. The overall cycle up and down the bay for one of the summer forms, *Acartia clausi* s.l., is shown in Fig. 5.22. It is most abundant in summer in all parts of the bay, but it does not disappear in winter.

The plankton upstream from the large embayment is dominated by three species: *Acartia californiensis*, *A. clausi* s.l., and *Eurytemora americana*. Zimmerman (1973) and C.B. Miller and J. Flynn (pers. comm.) have shown that the *Acartia clausi* s.l. population goes through parallel cycles throughout the bay, increasing in spring to a mid-summer high that persists into November, then decreasing to a mid-winter low (Fig. 5.22). The mid-summer drop in abundance shown in Fig. 5.22 for upstream stations is only an artifact. The temperatures rise in spring and the individuals get progressively smaller (see below, Fig. 5.32). This proceeds to such an

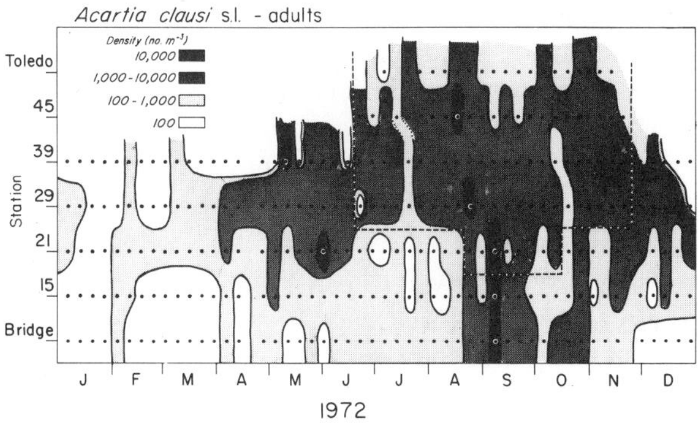

Fig. 5.22. Seasonal cycle of abundance of *Acartia clausi* s.l. at a series of stations in Yaquina Bay, Oregon. Stations are indicated by dots. Stations above the dashed line were sampled with a 120-μm net. Other stations were sampled with a 239-μm net. Only adult densities are shown because younger stages were undersampled. Original.

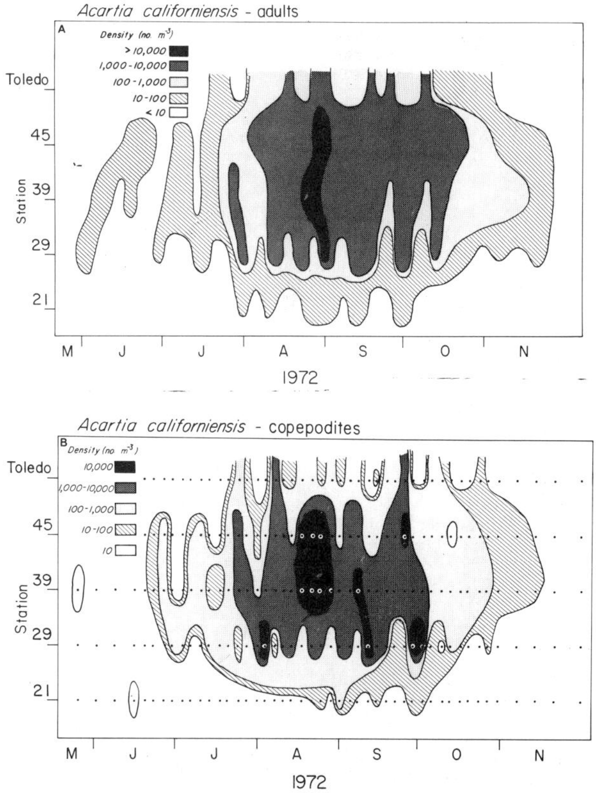

Fig. 5.23. Seasonal cycle of abundance of *Acartia californiensis* at a series of stations in Yaquina Bay, Oregon. Stations are indicated by dots in part B. All sampling was with a 120-μm net. Abundance was nil during months not shown. Original data.

extent that even the adult females slip through the 0.239-mm mesh samplers used in most of the studies. J.K. Johnson (pers. comm.) has shown, using very fine mesh nets, that the upstream population of *A. clausi* s.l. in 1972 increased to a smoothly rounded summer peak in abundance. The size shift demonstrated in Fig. 5.32 only occurs in the upstream stations, with the size found at navigational marker *21* depending in mid-summer upon the stand of the tide at sampling. This marker, then, is at the effective null point in the circulation. Drogue experiments (C.B. Miller, unpublished, 1974) show that the water at marker *21* at high tide does not leave the bay at low tide. Thus, zooplankton populations above that point are largely retained and can develop special characteristics.

Acartia californiensis (Fig. 5.23) develops a population in summer at stations above the null point and extending to Toledo. This species is apparently totally absent from the plankton from late November until June. Johnson (1979) has shown that the winter period is spent as resting eggs in the sediment. The detailed biology of this seasonal cycle will be discussed in several sections below. It is clear that the development of the summer bloom depends upon the low exchange rates and upon the consequent warming and salinity rise that apply to the upper bay during the drought period of summer.

Eurytemora americana is not the only member of this genus that occurs in Yaquina Bay. *Eurytemora affinis* is also found in summer above Toledo, but no very extensive sampling has been done so far upstream. The seasonal cycle of *E. americana* is shown in Fig. 5.24. It reaches maximum abundance in spring right at the zone of the bay that is the summer null point. Flushing rates must be substantial in this reach of the bay in that season, and the mechanisms by which the population maintains itself have not been studied. The situation is comparable to the development of a population of *Eurytemora herdmani* just downstream of the apparent null point in the St. Lawrence (see p. 106 above). It has been observed in the laboratory (C.B. Miller, unpublished) that these animals spend a good deal of time against the sides and bottom of containers. It may be that they are really epibenthic forms whose population is mostly on the bottom. Attempts at sampling very close to the bottom with nets on a sled and with pumps have been unsuccessful. Too much sand contaminated the samples. Perhaps work with a coring sampler like that of Landry (1976) will provide an answer.

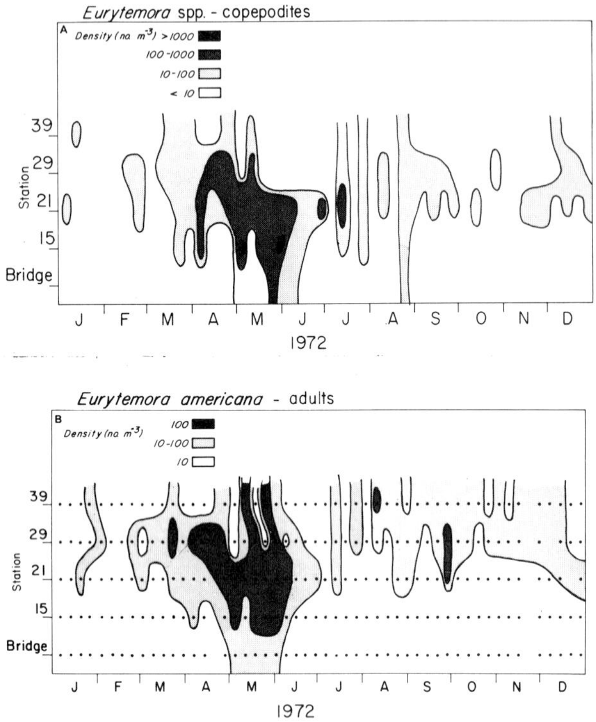

Fig. 5.24. Seasonal cycle of abundance of *Eurytemora* spp. copepodites (A) and *Eurytemora americana* adults (B) at a series of stations in Yaquina Bay, Oregon. Stations are indicated by dots in part B. All sampling was with a 239-μm net. Original data.

The situation in Yaquina Bay in summer is a good illustration of the role of null points in the distribution of plankton along tidal estuaries. Vertical mixing is very thorough in summer, and fresh-water flow is very small; the flow in the bay must be very close to the piston-like model used by Ketchum (1951) to describe the dispersion of suspended materials in estuaries. Water that is thrust out of the bay during the ebb is so greatly diluted with the water outside before the flood that endemic populations do not develop. Water that is too far upstream to reach the mouth on the ebb develops different physical characteristics and provides a habitat for endemic populations.

The role of zooplankton in the pelagic ecology of Yaquina Bay is not yet very thoroughly studied. They are phytoplankton grazers, of course, but their role in determining the species composition and abundance of the phytoplankton needs more work. Deason (1975) compared the populations of phytoplankton that developed in large suspended plastic bags filled with bay water including and not including zooplankton. The experiment was run in the lower bay in spring. The animals included in the bag with zooplankton were mostly copepodites of *Oithona similis* and *Pseudocalanus* sp., and barnacle nauplii. The phytoplankton species composition was different in the two bags: that without zooplankton developed a large population of *Thalassiosira fluviatilis*; that with zooplankton developed a biomass only half as great consisting of equal parts of *T. fluviatilis*, *Thalassiosira decipiens*, and *Chaetoceros debilis*. The zooplankton bag had amounts of ammonia and urea not present in the other, but both bags became nutrient limited in the period of incubation, the zooplankton bag several days later than the other. The experiment did not deal with the basic dominant zooplankton species, nor with a plant population endemic to the bay. Work comparable to that being done in Narragansett Bay is needed to describe the interaction of phytoplankton and zooplankton.

Predators on zooplankton are not well studied in Yaquina Bay. It is clear that the end of the summer–fall peaks of abundance of *A. clausi* s.l. and *A. californiensis* are not connected to ctenophore predation. Virtually no observations of ctenophores have been made. There are important fish predators, and it is probable that they limit the densities reached in summer population peaks. The role of these fish will be discussed below. The fall decline of *A. californiensis* is certainly caused by cooling, which converts their reproduction to resting eggs. The decline of *A. clausi* s.l. in the upper bay is probably attributable mostly to the autumn increase in flushing rates. The overall seasonal decline of this form must also be affected by the lowered primary production in the nearshore ocean, which limits the food supply, and, therefore, growth and reproduction.

Australian estuaries

The zooplankton of Australian estuaries make a curious contrast to those of other regions. As elsewhere, the dominant forms are copepods, and the marine forms are similar in generic or even specific composition to those of other lower-temperate latitude regions: *Oithona*, *Tortanus*, *Paracalanus parvus*, and *Acartia clausi* are all important in the downstream reaches. However, there are no representatives of the genera that provide the bulk of euryhaline species in other regions. Instead of *Eurytemora* spp. and *Acartia tonsa* (or close relatives like *A. californiensis*) there is a genus of the family Centropagidae, *Gladioferens*, and a parapontellid-like form, *Sulcanus conflictus*. *Sulcanus conflictus* is very like *Acartia* in outward appearance. Perhaps the whole situation can be regarded as a case of convergent evolution. It is exactly parallel to the distinctness of the Australian marsupial mammals, reptiles, flowering plants, and so on. It suggests that the dispersive powers of the truly estuarine plankton fauna are limited. The genus *Gladioferens* is also found in New Zealand, but that is its farthest extension.

The centropagid fauna of Australia is complex, including genera both estuarine (*Gladioferens*, *Isias*, *Gippslandia*) and fresh-water (*Calamoecia*, *Hemiboeckella*, *Boeckella*). Some species of *Calamoecia* are specially adapted to and only found in athalassic saline lakes (*C. salina*, *C. clitellata*) (Bayly, 1961). A complete key to these forms and references to their descriptions can be found in Bayly and Arnott (1969). The genus *Gladioferens* is reviewed by Bayly (1963). *Isias* is represented by a single species, *I. uncipes*. This genus is also found in Eurasian coastal and downstream estuarine waters. Bayly (1964) sees it as part of a radiation from the genus *Centropages* that produced much of the Australian copepod fauna:

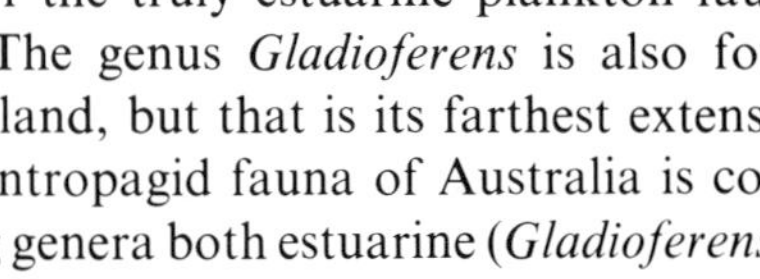

Gippslandia estuarina is another offshoot of *Centropages*. It is somewhat primitive compared to *Gladioferens* or to *Centropages* itself in the reduced complexity of spination and segmentation of the swimming legs. It could be a derivative of any of the marine centropagid genera of the region. Bayly feels that this radiation is similar to another array of centropagid genera termed phylogenetic by Burckhardt (1913):

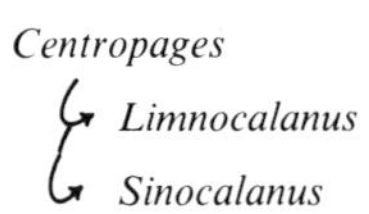

Gurney (1931), it should be noted, doubted Burkhardt's conclusion. Issues of this sort cannot be settled definitively in the absence of a fossil record.

Gladioferens is constituted of five species which differ in geographic distribution, in habitat (some are epibenthic, some are open-water forms), and in salinity tolerance. They are listed in Table 5.3, which was developed from discussion in Bayly (1963).

There have been a number of studies of the longitudinal distribution of the planktonic forms in Australian estuaries, and a number of studies of the seasonal cycles of species composition and abundance. However, none of them are very thoroughly reported in the literature. Most of the papers give only very sketchy diagrams or are otherwise incomplete. The best studies are that by Bayly (1965) of the plankton of Brisbane River estuary in southeast Queensland, and that by Hodgkin and Rippingale (1971) and Rippingale and Hodgkin (1974, 1975, 1977) of the Swan River in Western Australia. There are a number of other, less extensive studies that the reader directly concerned should consult: Lake Macquarie, central coast of New South Wales (Kott, 1955); the Werribee River, Victoria (Arnott and Hussainy, 1972); four rivers draining into Port Phillip Bay, Victoria, including the Werribee (Neale and Bayly, 1974); Hawksbury River, New South Wales (Morris, cited in Rochford, 1951; see comments by Bayly, 1965). Three studies are in thesis form and not generally available: Gippsland Lakes, east coast of Victoria (Arnott, 1968); Swan River, Western Australia (Bhuiyan, 1966); and mysids in Brisbane River (Hodge, 1963).

The seasonal study of Brisbane River reported by Bayly (1965) covered seventeen months from January through May. Sampling was not quantitative. The Brisbane River is colder (15°C) in winter (July) and warmer (27° to 29°C) in summer (November through March). Stations near the limit of salt water are warmer by 2° to 4°C throughout the year than stations near the mouth. There are monsoonal rains in eastern Australia, which begin abruptly in spring or early summer, and cause a very sudden rise in river flow rates with a corresponding drop in salinity throughout the estuary. The rains are usually brief, so that salinity increases gradually through summer, fall, and winter. The annual range at the uppermost sampling station

TABLE 5.3

Ecological characterization of the species of *Gladioferens*, an Australian genus of centropagid copepods (from discussion in Bayly, 1963)

Species	Range	Habitat	Salinity
G. symmetricus	Queensland, New South Wales, Victoria	epibenthic, on aquatic plants	"euryhaline"
G. pectinatus	Queensland, New South Wales, Victoria, Tasmania, New Zealand	open-water plankton	0.4–30‰
G. inermis	South Australia: Blanche Harbor only to date	uncertain	probably euryhaline
G. spinosus	coastal strip in eastern Australia, New Zealand	littoral — among algae	closed lakes very near to the sea, strictly fresh water
G. imparipes	Western Australia only, Swan, Murray, Moore rivers	daytime — among vegetation night — planktonic	euryhaline: from less than 5‰ to 33‰

(number 10, ca. 70 km from the mouth) was about 1 to 6‰, and at the most seaward station (number 1, ca. 6 km from the mouth) was about 26 to 36‰. There were short-term variations with tidal flux as well. The basic results on species distribution and seasonal cycles were presented by Bayly in a figure which has been revised here to make Fig. 5.25. The coastal marine component of the fauna was *Oithona* sp., *Acartia* sp. (usually described as *A. clausi* from Australian waters), three species of *Pseudodiaptomus* given numbers, an undescribed species of *Labidocera*, and *Isias uncipes*. Most of these forms only penetrated the river to salinities in the low 20's, with the exact position corresponding to the seasonal positions of those isohalines. There were only two euryhaline forms: *Gladioferens pectinatus* and *Sulcanus conflictus*. The first of these extended to station 1 (near mouth) in low numbers, but was an important form throughout the upper reaches of the estuary. *Sulcanus conflictus* was present at stations 5 (33 km) through 9 (59 km) only during the summer season just prior to the monsoonal rains which occurred in the year of sampling mostly in January. Bayly states that it was present mostly at salinities less than 3 or 4‰.

The results of Arnott and Hussainy (1972) for the Werribee give the same general spatial succession of copepod species. They report some additional forms upstream, however. These were present at salinities less than 20‰ in very low numbers: *Boeckella triarticulata*, *Mesocyclops leuckarti*, *Gladioferens symmetricus*, *Gladioferens spinosus*.

Bayly (1965) also conducted a quantitative study of vertical distribution in summer and winter using a Clarke–Bumpus sampler (Clarke and Bumpus, 1940). The winter study was conducted near the mouth, while the summer one was conducted in the middle reach. This choice gave comparable salinity ranges (12 to 28‰ in winter before monsoonal rains; and 9 to 20‰ in summer during rains). Summer temperatures were 28.5° to 31°C. Winter temperatures were all between 17° and 18°C. *Gladioferens pectinatus* was the dominant species in winter, being about $2 \times 10^4\ m^{-3}$. Adults avoided the surface, whereas copepodites were evenly distributed. Nauplii were not studied. Sampling was in daytime only. In summer the overall abundance was about an order of magnitude less, and there was no vertical migration evident in a comparison of day and night samples. *Oithona* sp. was important in summer samples, too, although the mesh was too coarse to sample them quantitatively.

Bayly (1965) observed that, as the temperature went up in the Brisbane River, the maximum salinity at which *Gladioferens pectinatus* is the dominant form goes down. This is not necessarily a fact about its biology, since dominance might depend upon the relative abundance of other things. On the other hand, it could be true that the upper range of salinity tolerance could be raised by lower temperatures as Bayly suggests. Some experimentation is called for. Brand and Bayly (1971) have studied osmoregulation in *G. pectinatus*, but not as a function of temperature — a strange omission in

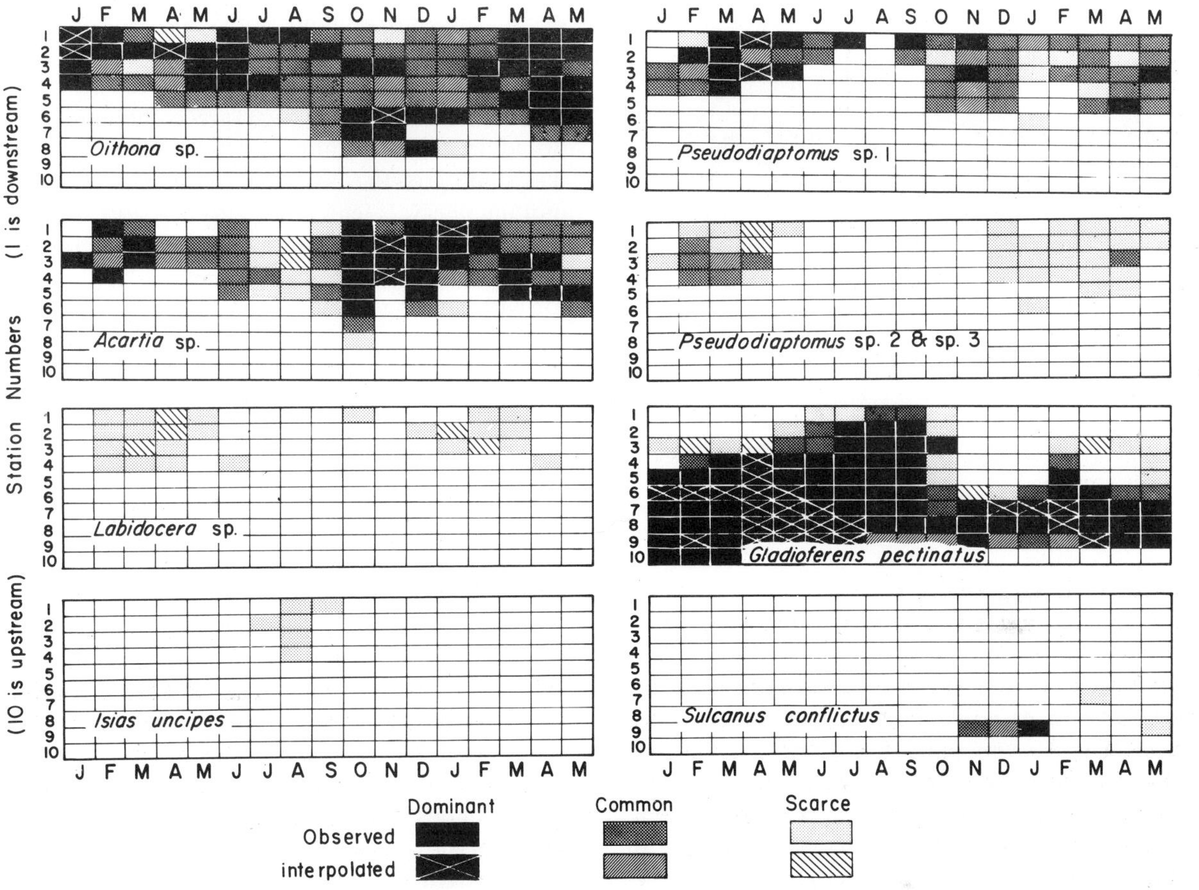

Fig. 5.25. Relative abundance of zooplankton species at a series of stations in Brisbane River, southeast Queensland, Australia. (Drawn from data in Bayly, 1965.)

light of Bayly's (1965) observation.

None of the papers of Rippingale and Hodgkin actually presents the raw data of the study; Rippingale's (1972) dissertation states: "Details of the plankton distribution through the seasonal changes will not be given here" (Rippingale and Hodgkin, 1975). This practice has made it hard to reconstruct the basic information from the study. The general seasonal and spatial pattern is represented in Figs. 5.26 and 5.27. *Gladioferens imparipes* is dominant from May through October, the season of maximum runoff in Western Australia. During that period the section of the estuary studied is apparently almost fresh. As the dry season progresses, *G. imparipes* is abundant only farther and farther upstream, since higher salinities are advancing slowly in that direction. It is essentially replaced by *Sulcanus conflictus* during the mid-summer season. That species stays in the zone with salinities less than 15‰. *Acartia clausi* is apparently important in winter, downstream in the salt wedge, although the data presented are not quantitative. An attempt

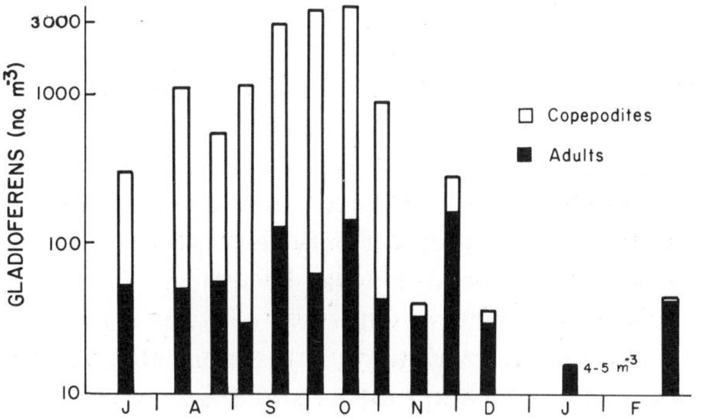

Fig. 5.26. Numerical density of *Gladioferens imparipes* in night surface plankton in lower Swan River, Western Australia, 1968. Exact position not given. (Redrawn from Rippingale and Hodgkin, 1975.)

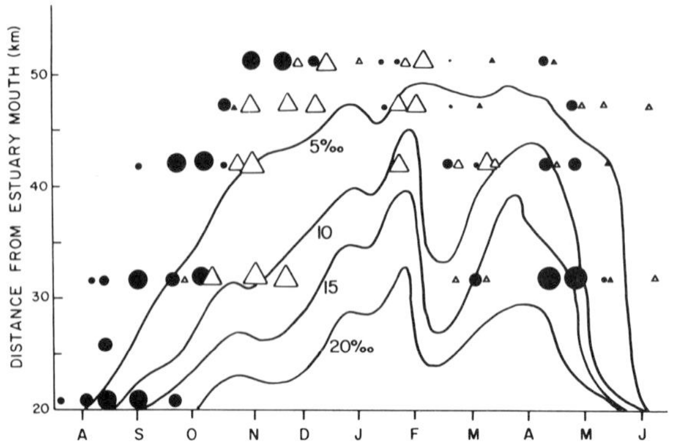

Fig. 5.27. Relative abundance of *Sulcanus conflictus* (triangles) and *Gladioferens imparipes* nauplii (solid circles) by season and position in the Swan River, Western Australia, and compared to salinity variation. Relative abundance is shown by symbol size, but no units were given in the original paper. (Redrawn from Rippingale and Hodgkin, 1975.)

has been made (Hodgkin and Rippingale, 1971; Rippingale and Hodgkin, 1975) to show that the changes in distribution of *G. imparipes* should be attributed in part to predation by *Sulcanus conflictus* and *Acartia clausi*. The demonstration has two parts: (1) *G. imparipes* will survive and increase in the laboratory under salinity conditions where each of the other two species becomes dominant in the field; (2) both of the other species will eat the younger nauplii of *G. imparipes* when they are offered to them in culture. The suggested pair of relationships is certainly possible, if not demonstrated.

SPECIAL BIOLOGY OF ESTUARINE ZOOPLANKTON: COPEPODS

The zooplankton of estuaries everywhere is dominated by copepods. They are the most numerous organisms and the greatest contributors to zooplankton biomass. Moreover, the copepod fauna is made up everywhere of very few species. In the downstream fauna there may be a variety of forms that are also found in the nearby coastal ocean, but in the zone of the estuary that is more affected by reduction in salinity, the calanoid genera *Acartia* and *Eurytemora* are the principal contributors. In Australia *Acartia* and *Eurytemora* are replaced by various centropagid forms (see above). Therefore, we will consider the biology of these organisms in detail. There are a number of typical features that will provide an organizing principle for the discussion: the species are characterized by rapid turnover times; resting eggs are commonly produced in times of most extreme environmental hardship; growth has some special features — molting intervals tend to be constant, and increase in bulk is exponential throughout development to the adult; and physiologic mechanisms are present for withstanding lowered salinity.

The basic biology of copepods

Copepods are one of the so-called entomostracan groups of the Crustacea. The later life stages, termed copepodites, have a torpedo-shaped forebody called the prosome and a tail section which inserts into it termed the urosome. The sizes of adult copepods vary from about 0.6 mm total length up to almost 1 cm. Typical estuarine forms are about 1 to 2 mm. Beneath the prosome are attached a variety of limbs. The most anterior ones, the first antennae, serve to bear sensory organs and act as stabilizers of body position. They are typically as long as or longer than the rest of the body. The biramous second antennae are very flexible and serve for slow swimming or propulsion of water currents during feeding. There are four sets of limbs involved in feeding: the mandibles, two pairs of maxillae, and the maxillipeds. The mandibles have a hardened blade, usually with siliceous teeth (Sullivan et al., 1975), that fits between the medial labrum and the two labial palps, which close the mouth in front and behind respectively. The first maxillae usually bear very heavy spines which rotate backward to clean the filtering setae of the second maxillae and forward to push food between the labial palps. The second maxillae are quite variable between forms, depending upon the exact mode of feeding. The maxillipeds are the longest mouth parts, and often have stout setae that are probably used in predatory feeding.

There are primitively five pairs of biramous swimming legs on the posterior half of the prosome. These are flat, paddle-like in form, and each branch rotates so that the narrow aspect meets the water on forward strokes. The operation of these legs is exactly like that of oars, hence the name copepod from the Greek for "oar foot". The fifth pair of legs is often modified for various roles in copulation, that of the male often having a grasping claw on one side and various tools for spermatophore manipulation on the other.

The urosome is basically rod-like, being a sequence of two to five segments plus two terminal paddles called furcae. The first segment in one large group, the gymnoplea, and the second in another, the podoplea, bears the genital opening. This is particularly conspicuous in females, where the chitin in the area may be variously sculptured, the opening covered with operculae, and so on. The female has seminal receptacles internal to this opening, into which sperm are extruded from the attached spermatophore of the male. The eggs are then fertilized during laying by the female. In some forms the eggs may be carried for a time in a sack formed from the secretion of glands around the genital opening. The caudal furcae bear fans of very long, stiff setae that have many setules. The whole forms a steering vane for the extremely rapid movements that copepods make when escaping predators.

The earlier life cycle stages are termed nauplii. They are lozenge-shaped, and bear only three pairs of limbs: uniramous first antennae, biramous second antennae, and the biramous mandibles. The life cycle consists uniformly of a fixed number of stages produced by a succession of molts: from the egg hatches the first of six naupliar stages. The sixth naupliar stage undergoes a substantial metamorphosis into the first of six copepodite stages. These all resemble the adult (which is termed the sixth copepodite) in form, but have fewer swimming legs and somewhat reduced segmentation. More limbs and more divisions into segments are added at successive molts. There are eleven molts between twelve swimming stages. There are never, so far as is known, any intercalary stages, nor is there ever molting after adulthood is reached.

There are only a few good general references on the biology of copepods. Marshall and Orr's (1955) book is a classic, and deals with all aspects of the life of a common coastal (not estuarine) form, *Calanus finmarchicus*. The anatomy and histology of another, more complex species, *Labidocera amphitrites*, has been examined in a brilliant study by Park (1965). Most of the rest of the literature is scattered, deals with small issues, and has never been summarized. There is an extensive systematic literature. However, for the typically estuarine genera there has been, as yet, no orderly comparison of forms covering the entire ranges in which they are found. Those interested in *Acartia* need to see Steuer's (1923, 1929) notes toward a monograph. There are several more comprehensive surveys underway by Bradford (1976) and by Crisafi (Crisafi and Crescenti, 1972, plus proposed further work). A thorough review of the biology of *Acartia* is being undertaken by Elizabeth Trinast, but is several years from completion.

Egg production in *Acartia* and *Eurytemora*

At least some forms in both of these genera lay two kinds of eggs: ordinary rapidly hatching eggs and overwintering or resting eggs. Resting eggs were first definitely demonstrated by Zillioux and Gonzalez (1972) for *Acartia tonsa* on the American east coast. They have also been found in a very similar form on the American west coast (Uye and Fleminger, 1976). Other *Acartia* species have been shown to produce them as well: *Acartia californiensis* (Uye and Fleminger, 1976, their sp. I; Johnson, 1979), *Acartia clausi* (Kasahara et al., 1974, 1976; Johnson, 1979), and *Acartia erythraea* (Kasahara et al., 1974, 1975). Only one report of resting eggs has been made for *Eurytemora*: *E. affinis* was hatched from bay mud by Johnson (1979). More observations will surely be forthcoming. Resting eggs have been reported from a number of other coastal copepod genera, all of which are heterarthrandrid forms: *Centropages abdominalis* (Kasahara et al., 1974, 1975), *C. yamadai* (Kasahara et al., 1974, 1975), *C. hamatus* (Pertzova, 1974), *Calanopia thompsoni*, and *Tortanus forcipatus* (both by Kasahara et al., 1974, 1975), *Labidocera aestiva* (Grice and Gibson, 1975), *Epilabidocera longipedata* (Johnson, 1979), and *Pontella meadi* (Grice and Gibson, 1977). Sazhina (1968) has demonstrated production of eggs with the typical appearance of resting eggs by *Pontella mediterranea* and *Centropages ponticus* from the Black Sea. She was unable to make them hatch. This typical appearance includes for most forms a much thicker outer pellicle and often spination (see Kasahara et al., 1974). Tranter and Abraham (1971) have suggested on the basis of seasonal abundance cycles in a complex of nine *Acartia* species found in a coastal lagoon system near Cochin, Kerala, India, that at least some of them must produce resting eggs, since they disappear totally from the plankton of the lagoons and cannot be found in the ocean outside. It is likely, therefore, that some tropical forms employ this mechanism as well, although no strong data have been produced.

Some of Johnson's (1979) experimental results for *Acartia californiensis* will be used to typify the conditions of production and hatching for resting eggs. Johnson collected animals from Yaquina Bay, Oregon, and had them lay eggs at a variety of temperatures at 25‰ salinity. These were then transferred from each of the temperatures at laying into a variety of temperatures and salinities for study of their hatching rates. Temperature was a dominant factor. It was found that dormancy can be developed either by the egg itself (transfer from warm laying conditions to cold hatching conditions), or can be installed in the egg by the female (eggs laid in the cold). The two forms of resting egg, which Johnson terms *quiescent* and *dormant* respectively, have different hatching properties. These are represented in Fig. 5.28. In that figure 17°S means "spawned at 17°C," and 17°H means "transferred to 17°C for hatching." In parts A and B it is shown that eggs spawned and hatched at 17°C hatch in about one day, whereas eggs spawned at 13°C or less hatch slowly when warmed. Eggs spawned at 17° or 21°C are shown in part C to still hatch at 13°C, whereas those spawned at 13°C or less do not hatch to any appreciable extent.

When the eggs laid in the cold are given warm conditions, they proceed to hatch, but at a different rate than those laid in warm conditions. There is, thus, some evidence for "stratification", a tendency for dormant eggs to hatch faster if a cold period has passed first. In part D a cessation of hatching is shown for eggs laid in the warm after a brief period in the cold. The partial hatch that is observed is attributable to the fact that the eggs laid earliest before transfer have had a longer period to develop in the warm conditions of laying. The cessation of hatching is quiescence. Hatching resumes upon warming. However, in the long term (120 days) there was no hatching from quiescent eggs upon warming. There was substantial hatching from laboratory-produced dormant eggs.

Johnson also showed that the development of dormant eggs would not begin in response to warming alone: salinities of 5‰ or less would retard it. For eggs laid and maintained in warm water, but held at low salinities, he found that development

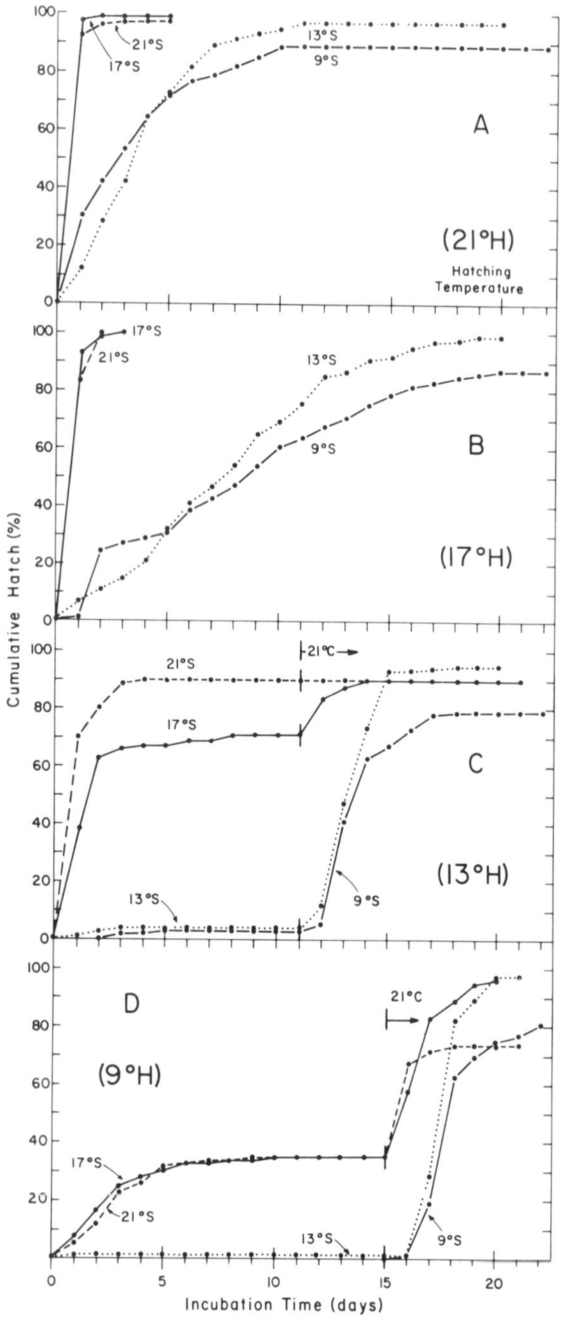

Fig. 5.28. Hatching rates (percentage hatched from 50 eggs) plotted against time for eggs of *Acartia californiensis* under various temperatures at and before spawning (°S) and during incubation prior to hatching (°H). The vertical lines in C and D indicate times of temperature changes. Temperatures (both °S and °H) are degrees centigrade. (Data of J.K. Johnson.)

would proceed, but hatching would not occur until salinity was raised. The duration of this "holding" condition was limited. After about twelve days the nauplii began to die in the shell membranes.

Landry (1975) has observed that eggs of *Acartia clausi* s.l. from Jakle's Lagoon, Washington, have similar post-development "holding" of hatching if they are kept in darkness. Hatching proceeds rapidly after illumination. This is true for *A. clausi* s.l. from Yaquina Bay as well (J.K. Johnson and P. Tester, unpublished). Johnson found that the holding phase in darkness can be maintained for only about twelve days, after which the unhatched nauplii die rapidly. He also observed some hatching (up to 20%) of the holding nauplii in the dark after about eight days. Similar results are reported for *A. clausi* s.l. from the San Diego area, California, by Uye and Fleminger (1976). No such holding in darkness occurs in *A. californiensis* (unpublished data of Johnson and Tester).

All of the typically estuarine copepods belong to genera in which the males have feeding mouth parts, feed, and have life spans that are a substantial fraction of that of the female. The females of these species require repeated mating for continuous production of fertile, non-resting eggs. Overall fecundity and effects of mating in *Acartia tonsa* have been very carefully studied in the laboratory by Wilson and Parrish (1971) and Parrish and Wilson (1978). They followed egg production throughout the life span of individual females from culture stocks maintained at 18°C under various situations of nutrition and mating. After the final molt, a female took one to several days for egg laying to begin, and several more for full production to become established. This was usually achieved by day 5. Since males were never provided until some egg production was observed, the first eggs were always infertile. Typical rates of production from day 5 onward were between 40 and 50 eggs per day. Rates often fell off gradually after two or three weeks. The total adult life span was of the order of 30 to 36 days, with one extraordinary instance of 65 days, eggs being produced on day 64. Absence of males after the initial fertilization resulted in production of only infertile eggs after a period typically of 14 days. Remating at that point could enable immediate resumption of fertility in the eggs. Females that did not have males continuously present more frequently died early, and produced smaller totals of fertile and infertile eggs than those that had males continuously present. Those with males continuously present tended, however, to lay a somewhat larger proportion of infertile eggs throughout adult life than those without them laid over their first 10 productive days. Completely unmated females usually died after 10 to 16 days, having produced a few hundred to seven hundred infertile eggs. Egg production was shown to require the continuous availability of food, production dropping immediately when food was removed, reaching zero eggs per day after three days.

TABLE 5.4

Fecundity of *Acartia clausi* s.l. taken from Jakle's Lagoon and transferred to various temperatures in the laboratory (data from Landry, 1976)

Field temperature (°C)	Animal size, prosome length (mm)	Experimental temperature (°C)	Eggs per day, approximate mean from graphs
c. 7	0.825	10	28
		15	30
		20	53
c. 20	0.600	10	8
		15	16
		20	16

Other fecundity studies show egg production rates of the same general order for *Acartia*: Johnson (1974) found means of 22 to 39 eggs per day for females of *A. californiensis* from Yaquina Bay captured in the field and maintained with males. The longest-lived females survived about one month in the laboratory after transfer from the field. Landry (1976) found that egg production rates for females of *A. clausi* collected from Jakle's Lagoon (San Juan Island, Washington) were a function of both female size and temperature. Size of adults varied throughout the year with temperature (see below), and that influenced fecundity. The general result is summarized in Table 5.4. Overall longevity in the egg-laying phase was not determined. There are similar responses of adult size to temperature in most copepods that have been studied, and fecundity apparently varies proportionally with size.

Seasonal development of populations by estuarine copepods

The overall seasonal cycle for copepod populations that overwinter as resting eggs can be very dramatic. This will be illustrated by data from Johnson (1974) for *Acartia californiensis* in upper Yaquina Bay, Oregon. This species is restricted to the region of the bay above the apparent null point in the circulation. It is only rarely found below that location. Abundance estimates from a series of twice-weekly samples taken in 1972 are shown in Fig. 5.29. There is a tremendous population increase each summer like that of 1972. Population density intermittently reaches levels of 10^4 m^{-3} or more. Johnson sampled twice a week during daytime, ignoring the effects of the tide. He took account of tides later by plotting as a time series the station for each sampling date which had the highest overall abundance, the population "center". The result is a clear sequence of generations such that a peak density of copepodites is followed by a peak of adult density for a total of six or seven discrete cycles. The overall result of numerous discrete generations has been confirmed by multiplying each density estimate by the volume of the estuary it represents and then summing to obtain an estimate of the total population in the river. The decline of the population in the autumn precedes the decrease in salinity resulting from the start of the seasonal rains. It is reasonable to suppose, therefore, that the decline is not caused by an increase of losses to flushing of the estuary. The decline does correspond in time to cooling. The early October temperatures were below 15°C, which, according to Johnson's experimental data, should cause production of dormant eggs.

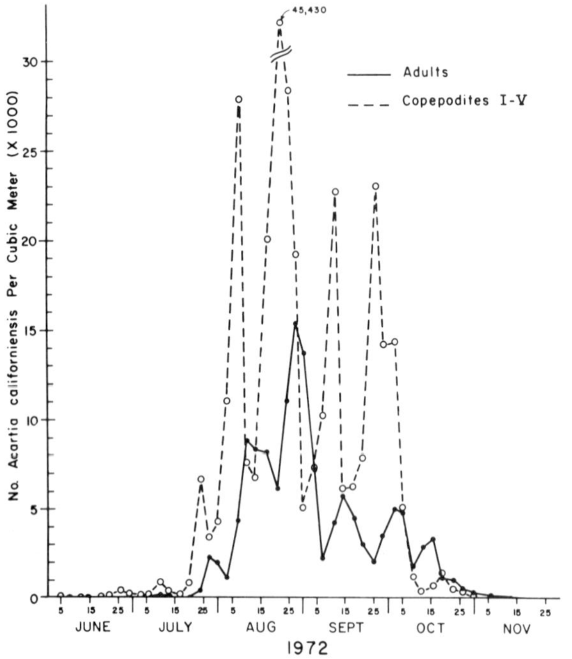

Fig. 5.29. The population cycle of *Acartia californiensis* in Yaquina Bay, Oregon, in 1972. Each datum is from the sample at the station with highest abundance on each sampling date, the population center which moves with the tide. (Data from Johnson, 1974.)

Collections of fish in Yaquina Bay show that the herring (*Clupea harengus pallasi*) and northern anchovy (*Engraulis mordax*), both of which are abundant in the upper bay in summer, along with several other fish, are important predators on the older life stages of *A. californiensis* (and *A. clausi* s.l., as well), particularly on females. This suggests that the generations are kept discrete by rapid cropping of the females. They are not able to lay eggs over a substantial fraction of the development period, so the generations do not overlap greatly. Some overlap has developed by the last adult peak of the 1972 data.

Reed (1978) constructed a numerical model of flow in the entire Yaquina Estuary. It takes account of flow into the estuary from above and of tidal oscillation from below, and includes terms for material diffusion and dispersion. It is made to resemble the bay by adjusting dispersion terms to keep salinity distribution like that of the bay. The progressive change in distribution of a hypothetical *A. californiensis* population is represented in Fig. 5.30. There is some spread of the population away from its original peak (installed from sampling data), but over three days there is very little loss below the null point at marker *21* (Fig. 5.19).

Similarly rapid population development has been demonstrated in Jakle's Lagoon by Landry (1976). This result will be reviewed as an example of a secondary productivity estimation method for an estuarine situation. It is sufficient here to note that Landry also found preferential predation by fish on older stages to be important in maintaining discrete generations.

Growth patterns of estuarine copepods

Miller et al. (1977) have suggested that the capacity for almost explosive increase typical of

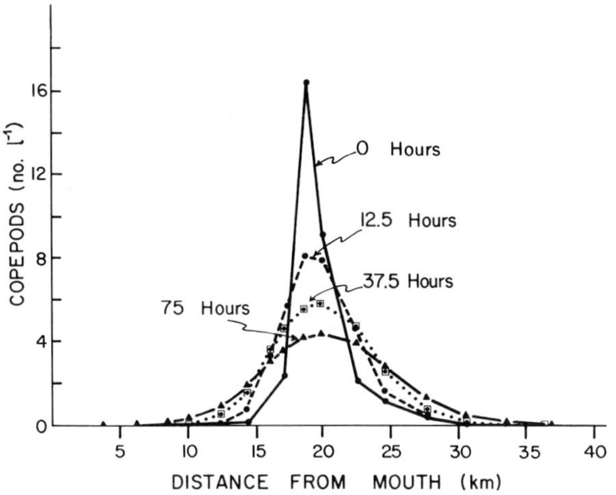

Fig. 5.30. Demonstration of dispersion of a hypothetical population of *Acartia californiensis* in a numerical model of flow in Yaquina Bay, Oregon. Distribution at 0 hours from a longitudinal survey of the Bay by J.K. Johnson (1974). (Result from Reed, 1978.)

estuarine copepod populations is correlated with peculiar growth patterns of the typical species. In particular, they show that under saturating nutrition (all the food the animal will eat is always available) *A. clausi* s.l., *A. tonsa*, and *A. californiensis* (termed *A. tonsa* in the paper), and *Eurytemora affinis* (Katona's, 1971, data) will add biomass as dry weight at an almost constant exponential rate throughout development to adulthood. Development takes place with almost constant time intervals between molts, termed isochronal development. There is some slowing of weight increase in the last naupliar stage. This is presumably because the animals have higher metabolic demands during the metamorphosis to copepodite form. While the data are not, as yet, very clear, it seems that coastal and oceanic copepods of most other genera, particularly amphascandrid forms, spend progressively more time in later life stages. This is particularly true of late copepodite stages relative to early ones.

Several arguments were advanced by Miller et al. (1977) for the selective advantage of the development pattern of estuarine copepods. Compared to successively longer times in each stage, this pattern maximizes the fraction of the life cycle that is spent in the younger, smaller stages, and minimizes the fraction of the development period spent in large pre-reproductive stages. This could cut down on losses to size-selective predators preferring large prey, as indeed estuarine fish have been shown to do. The shortening of the later life stages, on the other hand, may just be part of speeding the development to the maximum extent possible. This permits maximal population growth in short, possibly unreliable, growing seasons.

Deevey (1960b) has shown that a wide variety of marine copepods grow to larger sizes at low than at high temperatures within their range. There are several ways to express the mechanism by which this occurs. The simplest is to note that molting rate must have a higher coefficient of response to increased temperature than does growth. Since the number of molts is fixed, faster molting caused by higher temperatures results in smaller terminal sizes. Less growth can occur in each inter-molt interval. The consequence of this for egg production has been illustrated from Landry's data above. Smaller individuals lay many fewer eggs per day than large individuals. However, this is generally more than compensated by the greater rapidity of population turnover with short generations. Rates of increase are much higher at higher temperatures.

The difference in size that occurs at the extremes of the temperature range is great. Landry's (1976) data for *Acartia clausi*, for example, show

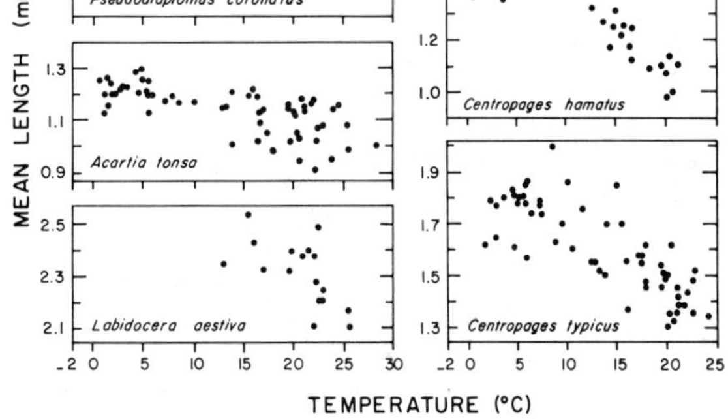

Fig. 5.31. Comparisons of size of adult copepods with temperature at the time of collection for a variety of species found in Delaware Bay. (Redrawn from Deevey, 1960b.)

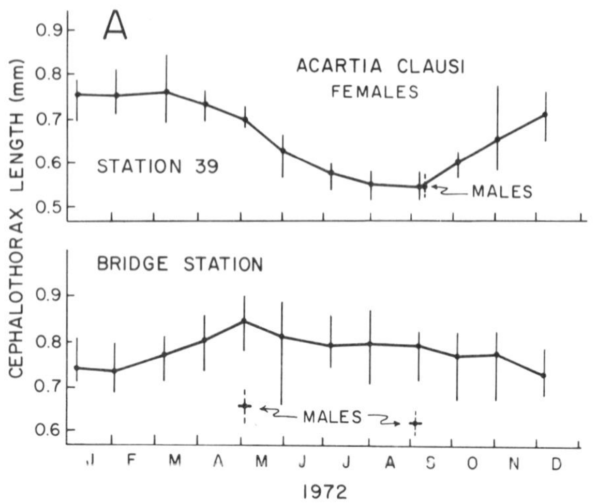

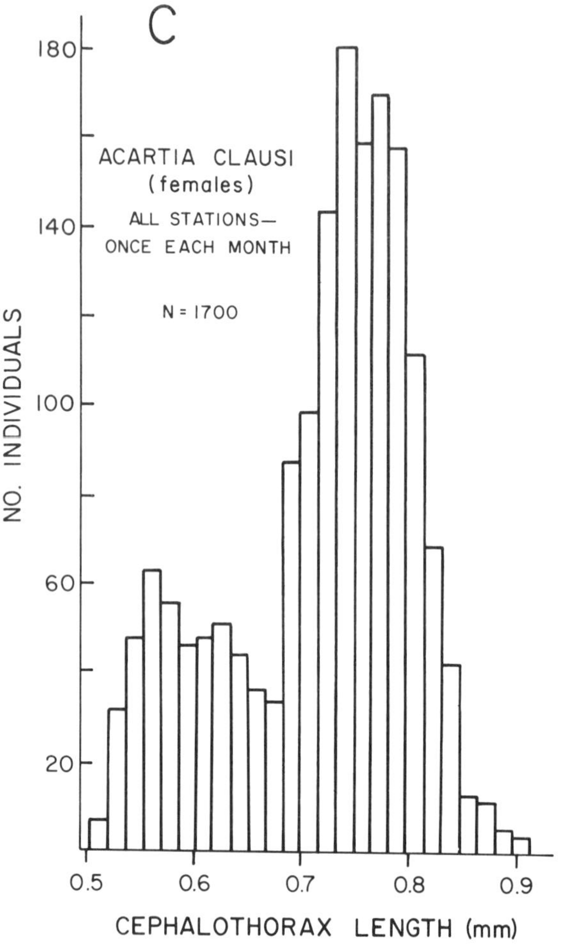

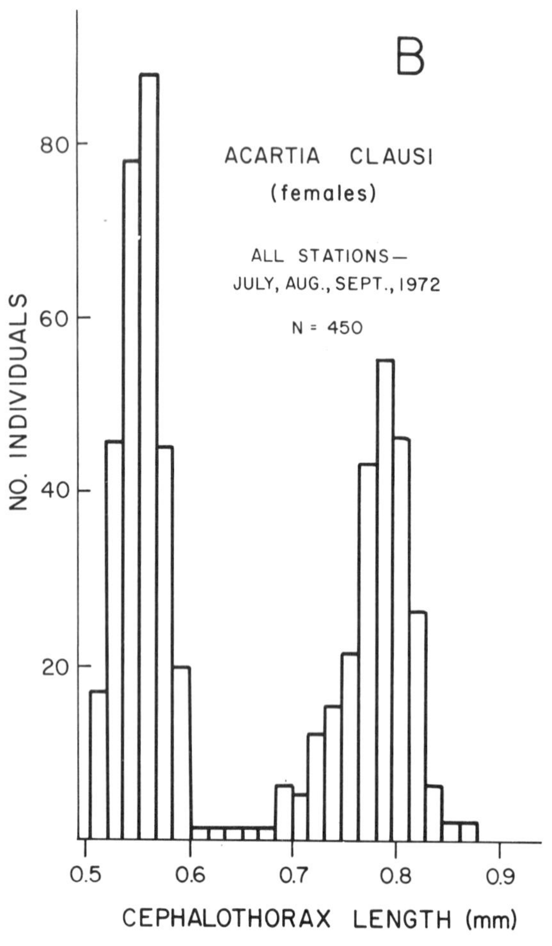

Fig. 5.32. A. Seasonal cycle of size of *Acartia clausi* s.l. at an upstream station (Buoy "39") and at a downstream station ("Bridge") in Yaquina Bay, Oregon in 1972. B. Frequency distribution of sizes over the entire bay during July, August, and September. C. Frequency distribution of sizes over the entire bay over the entire year (1972). (Data of Joan Flynn.)

a summer–winter change in mean prosome length from 0.580 to 0.830 mm. Because bulk is related to length by a roughly cubic proportion, this length change corresponds to a change in dry weight from 2.6 to more than 6 μg (Landry's weight to length comparison). Thus, an increase in length by a factor of 1.43 corresponds to an increase in weight by a factor of 2.3. A variety of comparisons from Deevey's (1960b) study of copepods from Delaware Bay is given in Fig. 5.31.

In Yaquina Bay, Oregon, it has been observed (J. Flynn, unpublished data) that the longitudinal temperature gradient which develops in the bay during summer produces two size classes of *Acartia clausi* s.l. (Fig. 5.32). The temperature tolerance of the species exceeds the range of the gradient. In the lower bay at temperatures typically 12°C, the prosome lengths remain about 0.79 mm throughout the year. In the upper bay, summer temperatures rise to 23°C, and mean lengths become progressively shorter during spring. In summer the upper bay population is distinct from that of the lower bay in size. Mean lengths are 0.55 mm. The transition occurs in a short reach of the bay around its null point. The lack of intermediate sizes in mid-summer is probably explained by the lack of a substantial volume of habitat at intermediate temperatures.

In Narragansett Bay, Durbin and Durbin (1978) have shown that *A. clausi* s.l. changes abruptly in spring from large to small adults. They attribute that to the rapidity of the spring warming. There simply is not a long enough duration of intermediate temperatures for intermediate-size adults to develop. In Jakle's Lagoon the temperature cycle is smoothly sinusoidal, and the transition in size appears to be smooth with definite intermediate sizes present in spring and fall (Landry, 1976). However, Landry's data are presented as a graph of mean sizes with confidence limits for the mean (calculated from standard errors). None of the published data give actual distributions of lengths of individuals through time, only means. Fig. 5.32C shows a combined frequency distribution from Yaquina Bay for stations covering its whole length at each month of the year. There are, in fact, middle-sized individuals in substantial numbers. These animals were collected in spring and fall.

Survival in brackish water

One of the principal features of estuarine habitats is longitudinal and temporal variation in salinity. The planktonic forms of the middle reaches live at salinities that vary from a few parts per thousand to those of the ocean downstream. This must be accomplished either by active osmoregulation or by toleration of low internal osmolality. Surprisingly, few good studies have been done of salinity tolerance and fewer of tolerance mechanisms in estuarine zooplankters. This is at least partly attributable to the difficulty of physiological research on such small forms.

Lance (1963) did a comparative study of the three species of *Acartia* which succeed each other seasonally in Southampton Water (England). These were *A. tonsa* collected at 19.5°C in September, *A. discaudata* at 15.3° in November, and *A. bifilosa* at 8.1°C in March. Captured animals were transferred from the net to filtered water of 36‰ at the field temperature, but field salinity was not determined. Adult females surviving the unknown salinity shock for one day were divided into groups of 50 and subjected to reduction of salinity by 3.6‰ each half-hour to 14.4‰, then by 1.8‰ each half-hour to the final test value. In some cases even slower reduction was used. Groups at each test salinity were maintained at a variety of temperatures. In general there was not substantially reduced survival relative to controls for any group down to 18‰. All three species had increased mortality rates below 9‰, and only *A. tonsa* survived at all well at 5‰, where 50% mortality increase relative to controls required three days. Each species survived more extreme salinity reduction at the experimental temperature closest to that of field collection than at either higher or lower temperatures. Slower acclimation to reduced salinity gave somewhat, but not dramatically, better tolerance in *A. tonsa*.

Lance (1964) has also shown that females of all three *Acartia* species have greater tolerance to dilution than males or copepodite stages. Comparable mortality patterns occur for females at salinities 3 to 4‰ lower than for males and young. However, Lance notes that molting was not normal in the laboratory, and suggests treating the results with caution. A single experiment with *A. clausi* at 15.2°C is reported by Lance in this 1964 paper. Survival was reduced at 20‰, and there was almost no survival after 20 h at 11‰.

The character of euryhalinity in *Acartia tonsa* was investigated by Lance (1965) using a miniaturized freezing point determination method to evaluate the relative osmolality of the medium and of tissue fluids at a wide range of salinities. Body fluid was withdrawn by micropipette from the region of the heart. Because of the small body size, it was not possible to obtain this fluid without some contamination by cellular debris. The result is shown in Fig. 5.33. *Acartia tonsa* apparently does not osmoregulate to any great extent. Reduced internal osmolality is tolerated, or osmoregulation occurs

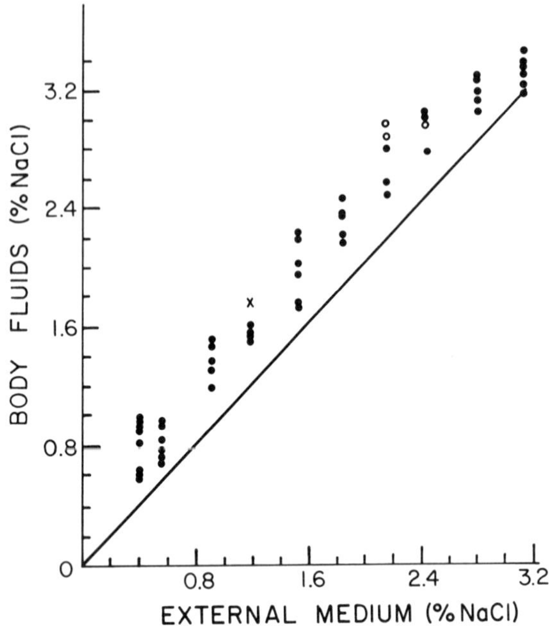

Fig. 5.33. Freezing point of medium and of body fluid compared for *Acartia tonsa* at a variety of salinities. Seawater was the medium; % NaCl is only a calibration comparison. A salinity of 36‰ is approximately equivalent to 3.18% NaCl. Filled circles are 1 datum, open circles represent 2 data, the *X* represents 3 data. (Redrawn from Lance, 1965.)

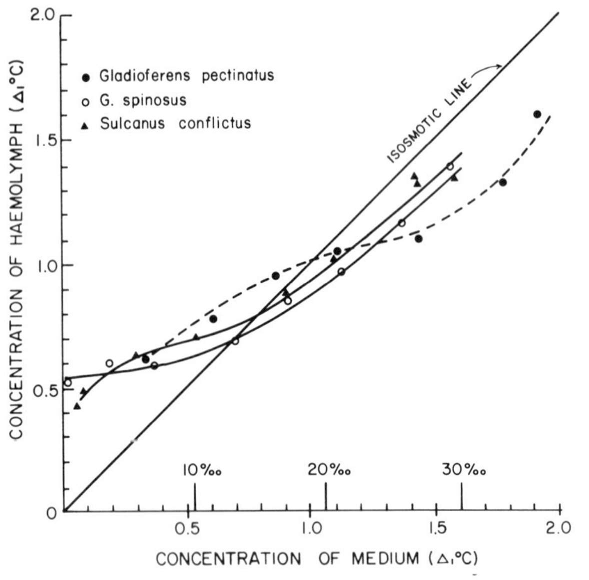

Fig. 5.34. Freezing point depression of body fluid and medium compared for a variety of estuarine copepods from Australia. (Data from Brand and Bayly, 1971.)

only at a cellular level. The modest excess of internal over external osmolality may represent a degree of regulation, or result from enhanced breakdown into small molecules of cellular debris at lower salinities. Lance (1965) employed manometric determinations of oxygen uptake by *A. tonsa* at 20°C to show the effect of reduced salinity on metabolism. At dilutions equal to or less than 30‰ there was progressive increase in respiration compared to that at 36.4‰. At the lower end of the tolerance range, 10‰, uptake was doubled. Either tolerance of body fluid dilution (which may be accomplished by internal osmoregulation) requires energy, or tissue hydration increases metabolism as a direct effect. Lance (1962) has shown that reduced salinity tends to suppress swimming activity, so the respiration change is not caused by a change in activity.

Brand and Bayly (1971) also have used the miniature freezing point method to study osmoregulation in copepods. They studied a number of Australian forms. The results are shown in Fig. 5.34. A greater degree of osmoregulation was found in most of the species than Lance (1965) found in *Acartia tonsa*. Isosmotic points were in the middle of the salinity tolerance ranges. Internal osmolality was below that of the habitat in the upper range and was above it in the lower range. The osmoregulation pattern found by Brand and Bayly for *Sulcanus conflictus* has been experimentally compared to its salinity tolerance by Rippingale and Hodgkin (1977). *Sulcanus conflictus* was captured in environments of less than 2‰ salinity, so the study was restricted to concentrations above those of the usual range. There were increasing mortality rates as salinity was raised above 25‰. However, mortality rates at increased salinities were lower among well-fed than among unfed experimental groups. Hyposmotic regulation, therefore, can be presumed to be metabolically costly. King (1976) used manometric measurements to determine respiration of *Gladioferens spinosus* at 4, 13, and 23‰. The oxygen uptake rates at these salinities were 0.47, 0.29, and 0.25 μl O_2 copepod $^{-1}$h $^{-1}$, respectively. Thus, hyperosmotic regulation in that species requires energy, just as Lance (1965) found it did in *Acartia tonsa*.

It is not likely that estuarine copepods will become accessible to detailed physiologic analysis because of their small size. There are, however, a number of aspects of their osmoregulation that can be studied. These include ionic balance (Battaglia and Bryan, 1964), effects on growth and molting, changes in feeding behavior under osmotic stress, and the rates of acclimation to salinity change. These are ecologically important topics, so there is reason to study them.

Jeffries (1962), Lance (1964), and others have suggested that tolerance of reduced salinity plays a role in the seasonal exchanges of dominance in the field between species of *Acartia* in Atlantic estuaries, possibly by changing their relative success in competitive interaction. Fig. 5.35 is Jeffries's comparison of observed temperatures and salinities during the period of transition in dominance. *Acartia tonsa* is the dominant above the solid line, *A. clausi* below the dashed one. It is, of course, possible that only one, or even neither, of these factors is important. Temperature and salinity simply change in seasonal cycles with the same period. Therefore, a plot of one against the other will automatically give an apparent regression. It is very likely that at least temperature plays a role.

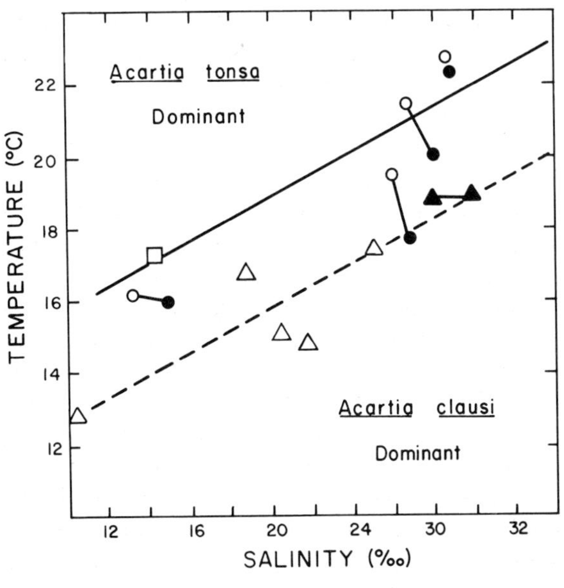

Fig. 5.35. Comparison of observed temperatures and salinities during the periods of transition in dominance from *Acartia clausi* to *Acartia tonsa* or the reverse. Circles are from Charleston and Green Hill ponds (open are surface averages; filled are bottom averages). Open triangles are from Raritan Bay. Filled triangles are the range of surface salinities at a station 16 km inside Narragansett Bay. Square is for York River. Solid and dashed lines are for circles and triangles, respectively. (Redrawn from Jeffries, 1962.)

For example, in the St. Lawrence (Bousfield et al., 1975) suitable habitat for both species is present throughout the summer, and *A. clausi* remains abundant downstream. It is only farther south, where downstream temperatures also rise in summer, that *A. clausi* nearly disappears in that season (p. 109). There is as yet no clear demonstration, as Jeffries (1967) admits, that competition or preferentially interspecific predation by older stages on younger ones are factors in the changeover.

Feeding biology

Literature on the feeding biology of estuarine copepods has been thoroughly reviewed by Odhner (1977). The following depends somewhat upon that more extensive paper.

Acartia and *Eurytemora* are setular filter feeders. Particles are strained from the water by copepodites using a spray of setae, densely covered with setules, on the second maxilla (Fig. 5.36). In the genus *Acartia* these setae reach well forward below the anterior prosome in the form of a basket. Water enters this basket as the animal swims slowly forward. Conover (1956) reports *Acartia* making a seining movement with the basket from front to back. The setae are intermittently folded together

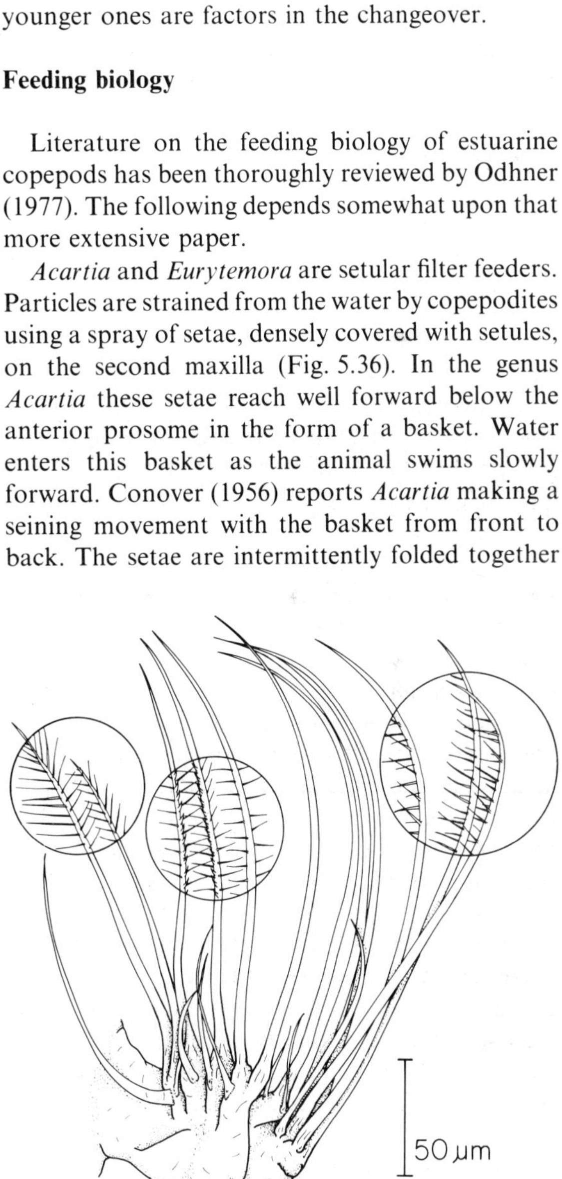

Fig. 5.36. The second maxilla of *Acartia clausi* s.l. viewed from the medial side. Setulation is shown only in the circled portions. Original.

and drawn through a comb of stiff spines on the first maxilla. In *Eurytemora* the setae of the second maxillae form a V-shaped barrier located vertically below the body. A current of water generated by the second antennae loops around and passes through this filtering barrier from behind. Captured food is combed out by the first maxillae in the same way. It is passed into the mouth by swinging of the comb anteriorly through the medial gap between the two labial palps which close the mouth from behind. The second maxillae and the maxillipeds can also be used to capture small animal prey in quick grabbing movements. All of the copepodite stages feed in essentially the same way.

An analysis based on a mechanical model for the filtering process has been suggested by Nival and Nival (1973, 1976) as a means of evaluating the size spectrum of food that copepods can capture. Measures are made of the intersetule spacing along the setae of the second maxillae and a cumulative frequency distribution is constructed for the hole sizes implied. This is couched in an unnecessarily complex index. The suggestion is that no particles smaller than the smallest hole should be retained, and that filtration efficiency should increase for larger and larger particles as the fraction of filter area occupied by holes smaller than those particles increases. Nival and Nival's (1976) graph for *Acartia clausi* shows no holes smaller than 4 μm in adults and few smaller than 3 μm in any copepodite stage. However, it has been shown using electron microscopy (M. Friedman, unpublished data) that copepods with filters this coarse can consume bacteria smaller than 1 μm in diameter. This throws the whole efficiency determination and experimental confirmation of it (Nival and Nival, 1976) into doubt. The whole matter of determining the smallest sizes that can be filtered requires more work. It is already clear from recent studies by R. Strickler using extremely high-speed movies that the whole filtering process needs redescription. It may not involve "screening", as described above, at all.

Nauplii feed with the biramous mandibles. These bear a number of spines and setae, the spacing of which is quite wide in most copepods, including *Acartia*: 15 to 20 μm (Gauld, 1959). Filtration occurs on an outward, then posterior sweep. The spines form a basket beneath the body at the end of the sweep, and food is drawn, possibly by gut expansion, into the mouth. There are in most species siliceous projections on the most proximal parts of the naupliar mandibles which serve as teeth (B.K. Sullivan, pers. comm., 1978). There are few studies on the actual range of particle sizes that can be eaten compared to adults. Present data are contradictory as to whether particles less than 15 μm can be eaten. Petipa et al. (1973) stated that *Acartia clausi* nauplii eat mostly particles from 15 to 50 μm, but did not say that is a lower limit. Hargrave and Geen (1970), on the other hand, found that only particles less than 5 μm were important for nauplii of *A. tonsa*. The latter authors did not describe the details of this particular observation. Experimental work with nauplii is difficult since each one eats so little. The matter may not, therefore, be quickly resolved.

Recent experimental work has made it clear that the capacity of copepods to select among available foods and to evaluate individual captured food particles is considerable. The best published experimental studies are those of Richman et al. (1977) for *Eurytemora affinis* and *Acartia tonsa*. They used the electronic particle-counter techniques that have been rapidly advancing studies of feeding rates and selectivity. Groups of animals were allowed to graze on natural suspensions of particles from Chesapeake Bay. Grazing rates were determined for a wide range of particle-size categories as a function of the difference in particle concentrations between grazed and ungrazed suspensions after a period of grazing. This is not an ideal formulation because it does not take account of possible differential particle growth (see Frost, 1972 for better formulae). However the conclusions of Richman et al. will surely stand. Over a long series of experiments it was found that both species of copepod would graze on particles of a specific size and ignore those larger or smaller, depending upon the occasion. Sometimes (Fig. 5.37) feeding occurred on large and on small particles while middle-sized ones were not consumed at all. It is difficult to see how this could happen in the filtering process itself. It is almost certain that food is captured and evaluated on a particle-by-particle basis. Unwanted particles must be rejected after capture. It has been shown (Ong, 1969; Friedman and Strickler, 1975) that the mandibles and setae of the maxillae have nerve endings similar to those demonstrated in insects to be olfactory. It is probable that food sorting is on the

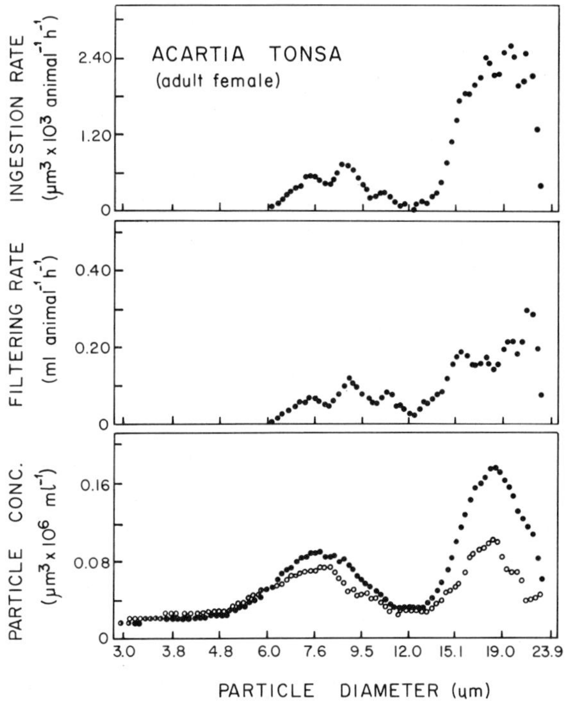

Fig. 5.37. Filtering rate determinations for a range of particle sizes for *Acartia tonsa* feeding on a natural array of particle sizes from Chesapeake Bay. This example was chosen to illustrate the capacity to feed on large and small particles while avoiding the middle-sized ones between. (Redrawn from Richman et al., 1977.)

basis of chemical cues. Experimental work outlining the extent of sophistication in copepod feeding behavior is being very actively pursued. It is already clear that feeding is much more complex than sieving water and swallowing the resulting porridge.

The spectrum of plant foods known to be eaten, to at least some extent, by *Acartia* and *Eurytemora* species includes practically all of the kinds that grow in their habitats. Evaluation of the nutritive value of these foods is a difficult matter. The best criterion of this value is that a food can be used as a sole source of nutrition for culture of the animals. However, many foods may be useful in mixtures that would not sustain growth and reproduction alone. This introduces great uncertainty into the evaluation. *Acartia* spp. have been raised on mixtures including diatoms (e.g. *Chaetoceros*: Heinle, 1969a; *Thalassiosira*: Carrillo et al., 1974), dinoflagellates (e.g. *Exuviella* and *Prorocentrum*: Sazhina, 1960), chlorophytes (e.g. *Chlamydomonas*: Heinle, 1969a), and sundry microflagellates (e.g. *Isochrysis*: Heinle, 1969a; Corkett, 1970; *Rhodomonas*: Zillioux and Wilson, 1966; Carrillo et al., 1974). *Eurytemora affinis* can be cultured on food mixtures identical to those used for *Acartia* (Katona, 1970; Vander Hart, 1976).

The potential role of detritus as a food for estuarine copepods has been investigated by several workers. Heinle and Flemer (1975) suggested that algal food concentrations in the Patuxent River were not great enough from March to May to support the *Eurytemora affinis* population and its production, and that detritus from the tributary marshes could be a supplementary food supply. This led them (Heinle et al., 1977) to assess the value of several sorts of detritus as food. They used the rate of egg production as a measure of food adequacy, since *E. affinis* females produce only one or at most two egg sacs after food is withdrawn. It was found that autoclaved detritus was worthless, but finely divided plant material (primarily wild rice plants collected from the marshes and broken into pieces between 30 and 50 μm) would support egg production when bacteria and protozoa were allowed to grow on it first. All useful diets were shown, using the presence of adenosine triphosphate (ATP) as a criterion, to contain substantial quantities of living matter. Some detritus-based protozoan cultures and some batches of protozoans + detritus were as good for sustaining egg production as mixed planktonic algae. A number of combinations of bacteria (particularly a form cultured from a body smear of *E. affinis*) and detritus were inferior, but would support some egg production. Heinle et al. were unable to show that nauplii could grow eating detritus or microbiota derived from it. It was not clear whether this would be generally true or not.

Roman (1977) has shown that *Acartia tonsa* will ingest detritus made from *Fucus vesiculosus* by drying, grinding, and aging it for various periods in unfiltered sea water, but they would not grow or reproduce with only that material as food. No evaluation of the bacteria or protozoa associated with the *Fucus* detritus is mentioned.

Carnivory is an important source of food for later stages of *Acartia*. Gut content studies usually show some crustacean parts (Marshall and Orr, 1949; Hodgkin and Rippingale, 1971; Petipa et al., 1973). Heinle (1970) has shown cannibalism by older stages on eggs and nauplii to be important in laboratory culture. Anraku and Omori (1963) have shown that *A. tonsa* will eat *Artemia* nauplii in the laboratory. Presumably this crustacean component of the diet is nutritious.

There is no evidence that dissolved organic matter is of any importance as food for copepods.

SECONDARY PRODUCTION IN ESTUARIES

Jakle's Lagoon (Wash., U.S.A.)

A number of attempts have been made to determine the rates of production of biomass by populations of copepods dominant in the estuarine plankton. One of the most sophisticated of these was the study by Michael Landry (1976, 1978) of the population of *Acartia clausi* in Jakle's Lagoon. The approach was that due to Ricker (1946). The growth rates and standing stocks of each life-cycle stage are determined at frequent intervals over the seasonal cycle, and production over each time unit is determined from:

$$P = \sum_{i=1}^{\text{adult}} G_i \cdot \bar{B}_i$$

where P is the daily production of the population, i signifies the life-cycle stage, G_i is the growth rate (weight per unit biomass per day), and $\bar{B}_i$ is the mean biomass of the life cycle stage during the day in the habitat. Growth is evaluated either directly from changes in the field or experimentally, and biomass is determined from field sampling data. These estimates can be cumulated to give production of the population during some suitable period, such as a year or a growing season.

Jakle's Lagoon is a small, almost closed impoundment adjacent to Griffin Bay in Puget Sound, Washington. It is roughly circular, about 200 m across, with sides sloping smoothly to a central basin of 125 m diameter that is 3.5 m deep. The lagoon was initially isolated from the bay, but in the middle of the first year of the study a dam at the connecting creek was broken. After that time there was some exchange, especially at night when tides were highest and older life-cycle stages were near the surface. This introduces some uncertainty into the calculations. However, the study area was more confined and stable than that in any other study to date in a marine environment. Landry showed that the population of *A. clausi* was very strongly concentrated near the bottom during the day. That made sampling with a simple coring tube, which was closed at the sediment surface by a rotating flap, a suitable method. The tube was dropped from a boat, and the enclosed water drained out through a mesh net. There were several horizontal surveys of the population, and a general ratio was worked out between the overall density in the lagoon and that at a central station where a time series was collected. This series ran from March of 1973 to September of 1974. Sampling was twice-weekly in 1973, but only once-weekly in 1974. The sequence of numerical abundance estimates for each life-cycle stage is shown in Fig. 5.38. Meaningful cohorts are distinguishable in the 1973 sampling data, but not in the 1974 data. This is a direct result of the change of sampling interval. It is not possible faithfully to represent the growth of cohorts of roughly two-week duration with only two samples per generation period. However, Landry has been able to provide a reasonable interpretation of the

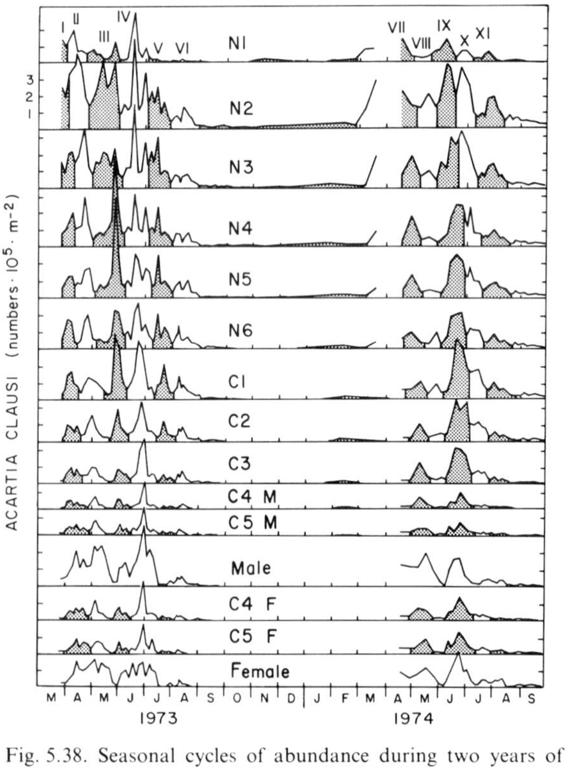

Fig. 5.38. Seasonal cycles of abundance during two years of individual developmental stages of *Acartia clausi* s.l. in Jakle's Lagoon, Washington. (Redrawn from Landry, 1976.)

cohorts of 1974 by extrapolation of the cohort timing of 1973.

Growth rates were directly determined from (1) the time required for shift of the abundance mode of a cohort from one stage to the next, and (2) measurements of the change in length between stages. The length measures were converted to weight by application of a general length: dry weight relationship. Growth was nearly exponential, with varying rates for each cohort. There was a good linear relationship between the logarithm of estimated weight and time. The data of Fig. 5.38 were also used to obtain an approximate estimate of the mortality of each stage within a cohort. This was needed in order to permit interpolation of stage-specific biomass estimates on days without sampling. First, the area under the curve of numerical abundance against time for a stage in a given cohort was determined (A_i). This was divided by the duration (D_i) of the stage, producing an estimate "somewhere between the number of copepods which initially entered the stage and the number which successfully molted to the next developmental stage" (Landry, 1976). This was taken to be "the number of copepods which survived to the midpoint of the stage duration". Second, it is assumed that there is simple exponential mortality during the developmental time of the stage. Thus the mortality rate (M_i) for the stage is given by:

$$\frac{A_{i+1}/D_{i+1}}{A_i/D_i} = \frac{A_{i+1} \cdot D_i}{A_i \cdot D_{i+1}} = S_i$$

$$= \exp[-M_i D_i']$$

where S_i is the fractional survival and D_i' is the duration from the mid-point of the ith to the mid-point of the $(i+1)$th stage. These mortalities were applied to estimation of the dry-weight biomass from abundance and estimated dry-weight data as follows:

$$B_i = \int_0^{x\,\text{days}} N_i W_i \exp(G_i - M_i) t \cdot \mathrm{d}t$$

$$= N_i W_i [\exp(G_i - M_i) - 1]/(G_i - M_i)$$

N_i and W_i are the numerical density and estimated dry-weight of the ith stage. Thus the biomass for a stage for a given day is corrected from the day (days = 0) for which there was a sample estimate by the best mortality estimate (M_i) and the best growth rate estimate (G_i) over the interval to the next sample (days = x).

Ricker's (1946) equation was then applied, and the resulting daily production estimates and the cumulative production estimates are shown in Fig. 5.39, converted to terms of organic carbon using the measured relationship: body carbon = 37% of dry weight. The annual production was 5.0 g C m^{-2} yr^{-1} from April 1973 through March 1974. The cumulative total for the growing season of 1974 was not complete by the time work stopped, but the rate of accumulation was comparable. The growing season rate was 14 mg C m^{-2} day^{-1}. Landry calculated that for the deeper part of the lagoon, which contained most of the population, the rates were much higher — 69.9 mg C m^{-2} day^{-1} during April through July. *Acartia clausi* was the overwhelming dominant species in Jakle's Lagoon, so these numbers essentially are the overall secondary production rates. No primary production estimates were made for Jakle's Lagoon with which to compare the secondary production values.

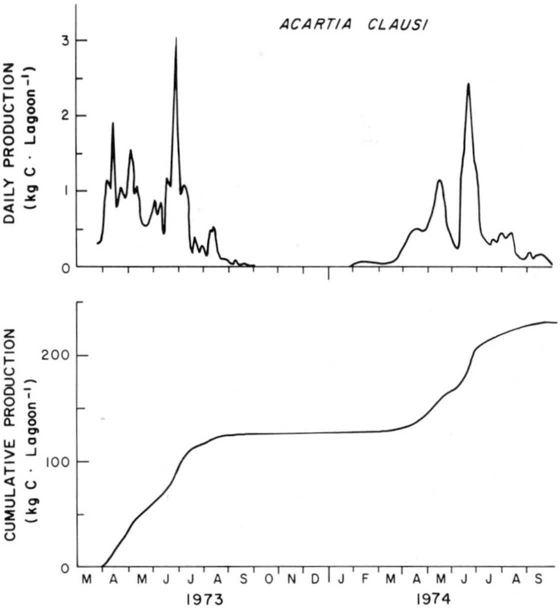

Fig. 5.39. Seasonal cycle of secondary production and cumulative secondary production during two growing seasons in Jakle's Lagoon, Washington. Production is only that of *Acartia clausi*, the overwhelmingly dominant form. (Redrawn from Landry, 1976.)

Chesapeake Bay

Donald Heinle's studies of *Acartia tonsa* and *Eurytemora affinis* in the Patuxent River subestuary of the Chesapeake system provide another secondary production estimate. The most careful calculations are those in Heinle's 1969 paper, which should be consulted rather than any of his earlier publications. The seasonal cycle of stage composition and standing stock of each species was calculated from monthly means of weekly samples. These means were taken to represent the entire month in steady state; that is, constant standing stock and ratios of life-cycle stage abundances were assumed. Stages were combined into only three categories: nauplii, copepodites, and adults. Growth was studied in the laboratory as a function of temperature over the range from 12° to 31°C. Approximate death rates were obtained from the ratios of stage abundances for nauplii, copepodites, and adults: $d_N = (\ln N - \ln C)/t_N$, and $d_C = (\ln c - \ln A)\ t_C$, where N, C, and A are the densities of nauplii, copepodites, and adults respectively, and t_N and t_C are the temperature-dependent durations of the naupliar and larval copepodite phases determined by culture work. This approximation is very rough, since the number of adults is not affected by the survivorship to adulthood alone, but also by the duration of adulthood. Adults of different ages are indistinguishable. The steady-state assumption is also very improbable given the results of Johnson (1974) and Landry (1976) (Figs. 5.29 and 5.38). They show that there can be radical oscillations in the ratios of abundance of stages several times within a month.

Total densities of nauplii and copepodites were next apportioned by Heinle among the successive stages making up the naupliar or copepodite phases by assuming they had the negative exponential survivorship pattern predicted by the phase-specific death rates. Exponential growth rates which varied with temperature were then applied to these densities according to the mean temperature for the month. The experimental data showed this simplification to be quite exact (see discussion of growth features above). These stage-specific production rates were taken to prevail throughout the month, and were summed over the year to give an annual production.

Most of the annual production was concentrated in the months from May through September, the period in which biomass was highest. Estimates were made per cubic meter and then multiplied by the depth to obtain an estimate on an areal basis. The use of surface to near-bottom oblique hauls is sufficient justification for this. The results are compared for several subsections of the estuary and sundry periods in Table 5.5. The totals over the year by area for both species (plus a more approximate estimate for *Scottolana canadensis*) are shown in Table 5.6.

Heinle (1969a) compares his value to the estimates of Stross and Stottlemeyer (1965) of primary production in the Patuxent. The results are given in Table 5.6. Something less than 10% of the primary productivity seems to appear as secondary production in the dominant holoplankters. That does not provide an estimate of the trophic efficiency of the phytoplankton-based food web in the estuary. Much of the phytoplankton is consumed by benthic herbivores (clams, annelids, etc.), and the overall efficiency must be much higher.

Heinle's overall rates of production for the year only differ from Landry's value for central Jakle's Lagoon by a factor of about 2. Given all of the assumptions, this may be taken any way the reader

TABLE 5.5

Production of *Acartia tonsa* and *Eurytemora affinis* in three areas of the Patuxent River, a subestuary of Chesapeake Bay; units are grams carbon per cubic meter for the entire period (from Heinle, 1969a)

	Zone: by quarters from mouth (I) to fresh water (IV)		
	I	II	III
Acartia tonsa			
July–Sept. 1964	1.03	0.42	0.60
July–Sept. 1965	1.11	0.52	1.16
July–Sept. 1966	0.69	0.71	1.12
July, 1965–July, 1966	1.88	1.30	2.62
Eurytemora affinis			
July, 1965–July, 1966	–	0.87	1.14
Scottolana canadensis			
July, 1965–July, 1966	–	0.61	0.57
Total			
July, 1965–July, 1966	1.89	2.77	4.33

TABLE 5.6

Average depth, copepod production by volume and area, and a comparison of secondary and primary production for the Patuxent River, a subestuary of Chesapeake Bay (from Heinle, 1969a)

	Zone: by quarters from the mouth (I) to fresh water (IV)		
	I	II	III
Depth (m)	3.8	2.4	2.7
Copepod production ($g\ C\ m^{-3}\ yr^{-1}$)	1.9	2.8	3.5–4.3
($g\ C\ m^{-2}\ yr^{-1}$)	7.2	6.9	9.5–11.6
Primary production ($g\ C\ m^{-2}\ yr^{-1}$)	271	204	162
Copepod production / Primary production	2.6%	3.4%	5.8–7.2%

chooses. For the time being, our best estimates of secondary production in temperate estuaries suggest that it is of the order of 5 to 10 $g\ m^{-2}\ yr^{-1}$, mostly concentrated in about one-third of the year. No tropical studies are available for comparison.

POPULATION MAINTENANCE AND FLUSHING RATES

In order for endemic zooplankton populations to persist in an estuary, they must have rates of population increase at least equal to the rates of loss due to tidal flushing and net river flow (Rogers, 1940). A means has been available since Ketchum's paper of 1951 to make the physical estimates necessary for a reasonably precise comparison of these rates, but few attempts have been made. The reason is the difficulty of estimating the mortality *in the field* attributable to factors other than flushing. It is possible to make sufficiently good estimates of reproduction and development rates, but the overall biological budget for a population depends critically on the mortality estimates. We are still left with the most general comparison suggested by Ketchum in his 1954 paper: estuaries with high flushing rates should have no endemic populations, merely plankton resembling that of the ocean offshore. Estuaries with slower flushing rates should be the ones with endemic populations. Despite the difficulties in making the desired population budget estimates, the few attempts that have been made are among the most instructive studies in estuarine planktology.

Ketchum (1951) suggested a method for approximately estimating the rate at which a conservative material would be dispersed downstream and lost from an estuary. The estuary is surveyed, and then conceptually divided into a series of segments. The most upstream segment is given a length such that its change in volume on a single tidal cycle is entirely supplied by the flow of the river. Succeeding segments are assigned lengths such that the high-tide volume of any given one is equal to the low-tide volume of the one below. Segments get larger downstream. By assuming that on full flood tide there is complete mixing in all segments, this conceptual division permits calculation of an exchange ratio, $r_n = P_n/(P_n + V_n)$, where P_n is the change in volume of the nth segment between high and low tide, and V_n is the segment volume at low tide. After m tidal cycles the remaining quantity, Q_n, of a conservative constituent introduced to segment n at time zero will be reduced by the fraction $(1-r_n)^m$, an exponential decay. If the constituent is also supplied from the next segment upstream, as in the case of fresh water, the rate of change will be less, depending upon the gradient.

For a zooplankton population of initial size P_0 that is confined by tolerance to the conditions of a given segment, or range of segments, the population remaining after m tidal cycles will be $P_m = P_0(1-r_n)^m$. The population's net biological increase rate must be equal to this loss rate if the population is to persist. If we represent the rate of biological increase by an exponential function, $e^{mk'}$, the net population change per tidal cycle is $P_m = P_0(1-r_n)^m e^{mk}$. At equilibrium $k = -\ln(1-r_n)$.

Ketchum (1954) demonstrated this form of comparison and applied it to several estuaries in the

United States and Canada. Fig. 5.40 shows his calculation of the rates of population increase necessary to permit development of endemic populations in those locations. Moriches Bay is a polluted inlet on Long Island Sound. It develops dense endemic phytoplankton populations (Ryther, 1954). Raritan Bay definitely has endemic zooplankton populations (Jeffries, 1962, 1967). Alberni Inlet on Vancouver Island has very sharp vertical stratification above 10 m depth. The lower levels are replenished seasonally by upwelled water that enters over the sill at the mouth (Bell, 1976). The salinities and temperatures of this water are suitable for fully marine plankton, and those are the forms typically found in British Columbian inlets. Presumably, the exchange ratio shown in Fig. 5.39 applies to the upper layer only. Typical estuarine populations do not develop in this level, but that may be attributable to the very low salinity as much as to the high exchange rates. Ketchum states that Barnstable Harbor has no endemic forms. He cites work by Ayers indicating that successful clam settlement is very infrequent in that bay because too many larvae are carried away by the large exchange.

A fairly exact comparison of flushing rates and population growth rates was carried out by Barlow (1955) in a tiny (3 km × 0.3 km × 1–2 m) estuary on Cape Cod, Massachusetts, called Great Pond. A small stream enters at the head. The principal seasonal event in the planktology of the pond is development of a spring bloom of *Acartia tonsa*, the greatest numbers of which are restricted to the upstream half. Unfortunately, no details of the physical flushing time calculation are given. The result is compared to a determination of the rate of population increase of *A. tonsa* enclosed in glass carboys and incubated in the pond for two to three days. The result is shown in Fig. 5.41. Basically Barlow (1955) found good agreement between the relative observed increase rates and those calculated to be required. The observed increase rates correctly predicted the position in the pond of the highest densities.

A more sophisticated comparison of this general type is done in a mathematical model of pelagic ecosystem processes in the Narragansett Bay developed by Kremer and Nixon (1978). The overall model depended on a simulation of flow and mixing in the bay developed by Hess and White (1974). This was a finite-element, horizontally two-dimensional representation of the shape of the bay with a finite-difference representation of the Navier–Stokes momentum equations and of conservation of mass requiring a 4-min time step. The 324 elements of the simulation were collapsed by

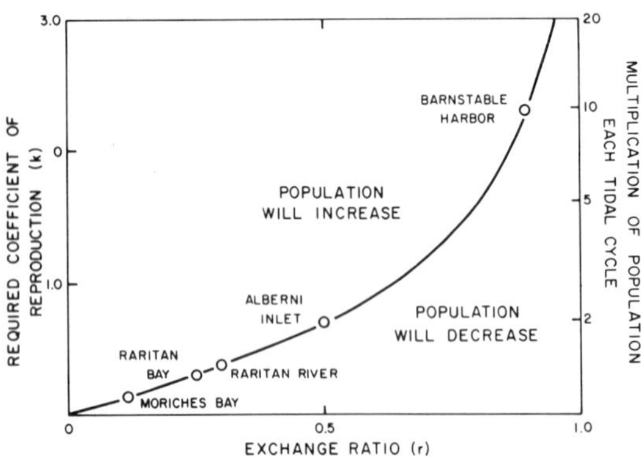

Fig. 5.40. Comparison of the rates of population increase necessary to maintain an endemic population of zooplankton with the exchange ratio of the estuary. Positions of five estuaries are shown. (Redrawn from Ketchum, 1954.)

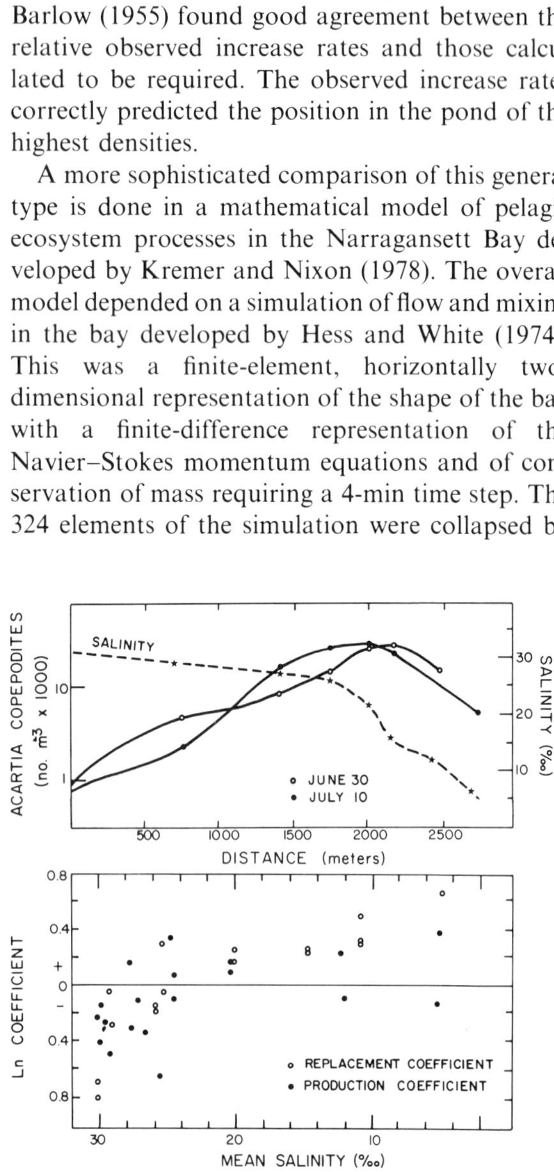

Fig. 5.41. Required replacement rate compared with experimentally observed replacement rate for *Acartia tonsa* copepodites in Great Pond, Massachusetts. Distributions of *A. tonsa* and salinity are also shown. Production coefficients were observed rates of copepodite production in experimental containers. Replacement coefficients are the production rates required to maintain a population despite losses to exchange. Exchange ratios increased upstream. (Redrawn from Barlow, 1955.)

Kremer and Nixon into eight segments, and the overall exchange ratios per complete tidal cycle for neighboring pairs were determined. These exchange ratios for a given cycle were found to be linearly related to the tidal amplitude for that cycle in all of the segments. Therefore, tidal mixing was represented in the overall model by simple fractional exchange ratios derived from the regression of ratios on amplitude. The exchange between the lowermost sections and Rhode Island Sound was evaluated in a variety of ways. For the West Passage segment it was found to be of the order of 0.3 to 0.4 per cycle, and for the East Passage segment 0.35 to 0.45 per cycle. A detailed simulation was then prepared for the phytoplankton and zooplankton biology, using parameters for the latter thought to be appropriate as a function of temperature for a combination of *Acartia clausi* and *A. tonsa*. Biologic mortality was represented in the model by incorporation of field data on predator density and approximately measured feeding rates.

The results of model runs using extreme low and high values of the exchange coefficients with Rhode Island Sound are compared with each other and with field data in Fig. 5.42. The lower exchange rates give the more accurate simulation in the lower reaches. However, the effect of this difference is surprisingly small in other segments. Because the seasonal cycle of flushing rates is not very dramatic in Narragansett Bay, the comparison is less instructive than it would be elsewhere. The seasonal

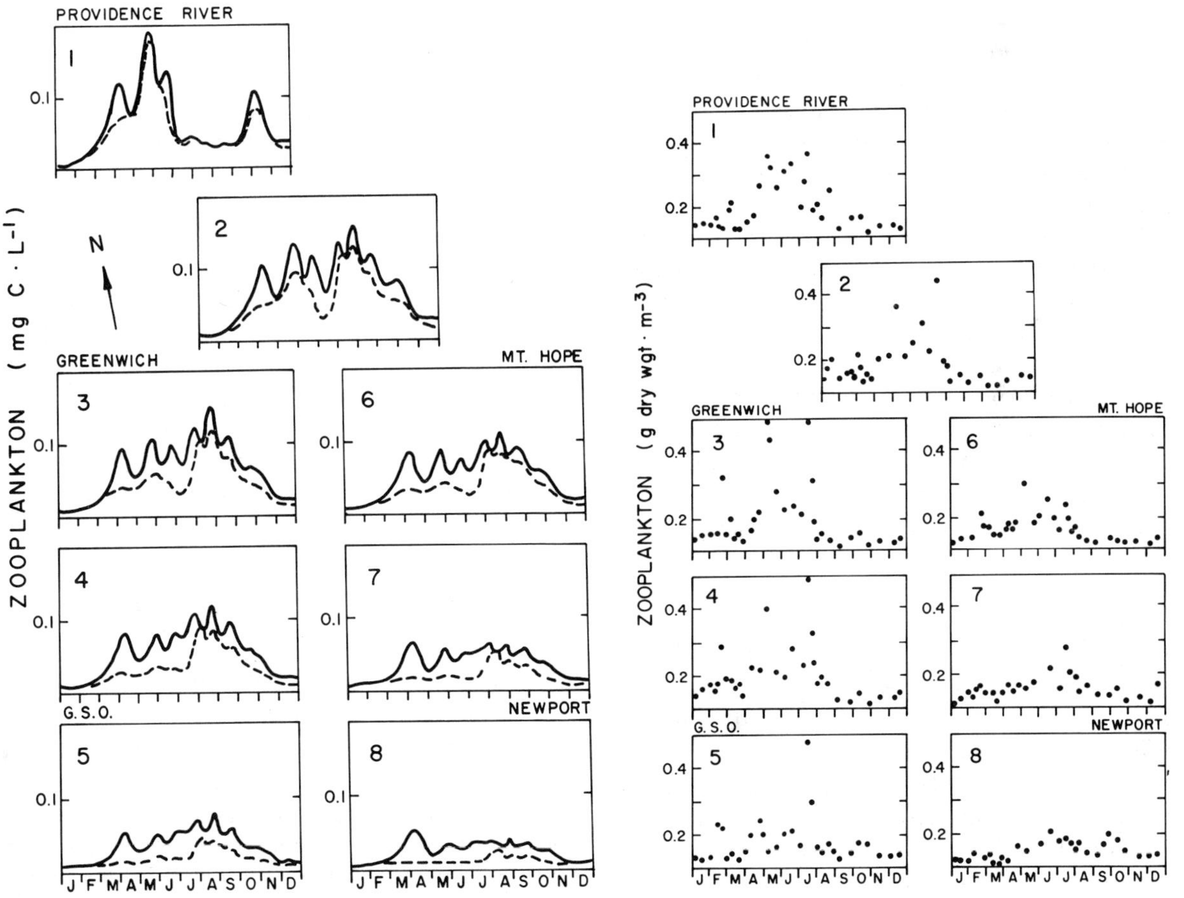

Fig. 5.42. Left: Simulation of abundance cycle for *Acartia* spp. in Narragansett Bay under high (dashed lines) and low (solid lines) exchange rates with Rhode Island Sound. Exchange with the Sound is from the boxes labelled 5 and 8. Each numbered box represents the situation in a segment of the estuary. Box 1 is upstream. Right: Actual density data for zooplankton captured with a 150-μm net at thirteen stations averaged for the eight segments. Data are for August 1972–August 1973. (Redrawn from Kremer and Nixon, 1978.)

cycle is too strongly dominated by changes in biologic processes. In a bay with a greater seasonal variation of fresh-water flow, that factor might be found to be the principal cause of population cycles. A preliminary numerical model of flow developed by Reed (1978) for Yaquina Bay, Oregon, shows that the fresh-water cycle is more important in that bay, but no estimates of field mortality rates are available to allow complete comparison.

Trinast (1975) has attempted to show that an eddy at a bend in Upper Newport Bay (southeast of Los Angeles, California) is sufficiently permanent that a large standing stock of *Acartia californiensis* develops in it. There were higher standing stocks in the eddy than there were downstream from it. Unfortunately, Trinast did no sampling in the extensive estuarine area upstream of the eddy. Better spatial coverage and a longer time series of comparisons are required to establish that local eddies play a major role in population dynamics. Given the physical data provided by Trinast, it is hard to believe the average residence time of plankters in the eddy could approach ten days, a probable generation time at the 22° to 25°C temperatures.

Jacobs (1968) has studied the tidal movement of the population maximum of *Acartia tonsa* in the tidal channels around Sapelo Island, Georgia. He found the maximum was near the ocean entrances to the sounds at low tide and high among the *Spartina* marshes at high tide. Jacobs has tried to explain why the maximum is not progressively dispersed by tidal mixing. Since *A. tonsa* remain preferentially in lower levels, Jacobs suggests that they are carried landward on the flood (which he says occurs mostly by flow along the bottom), but are not equally moved seaward on the ebb (which he says occurs mostly by flow along the surface). This mechanism should, indeed, keep the animals together and upstream, but no quantitative flow or population budgets are offered in support of it. Moreover, it does not explain either (1) why the vertically integrated volume densities (no. m^{-3}) drop to very low values in the marsh channels on the ebb, or (2) why the horizontal maximum on the ebb is at the river mouth. On the other hand, it is very likely that the shear between shallow and deep currents in estuaries is used for position maintenance in some fashion. Most suggested mechanisms, unlike Jacob's, involve directed vertical migration.

In the lower reaches of estuaries that are not thoroughly mixed in the vertical direction there is opposite net flow in the surface layer (seaward) and the bottom layer (landward). This implies that a population of zooplankton could maintain itself at a given longitudinal position in the estuary by migrating vertically between the two layers. This mechanism has been suggested many times, but there are not many data to show that it actually happens. Most of the better information does not pertain to holoplankton, but to barnacle (Bousfield, 1955) and oyster larvae (Carriker, 1951; Wood and Hargis, 1971). Carriker and Bousfield both found that there was a change in vertical distribution with stage of development: younger larvae were high in the water, late larvae were near the bottom. Wood and Hargis (1971) found from a time series of samples that larvae of *Crassostrea virginica* in the James River were only present in the water column during flood tide. They suggested that the larvae rest on or very close to the bottom, swimming upward in response to either velocity increases, or, more likely, increases in salinity.

The data on vertical distribution of holoplanktonic animals in estuaries are shamefully scanty. Landry (1976) has shown that older stages of *Acartia clausi* migrate upward at night in Jakle's Lagoon, while younger stages nestle right near the bottom continuously. The lagoon has no strong flow regime, but in other places the change in behavior with development could promote retention of a population within a zone of an estuary. Bosch and Taylor (1970, 1973) demonstrated an inverse diel vertical migration (up in the day, down at night) of *Podon polyphemoides* (Fig. 5.18) in Chesapeake Bay. That migration could affect its horizontal distribution, although the night-time density mode was not deep enough for actual upstream transport. Grindley (1972) has studied the migratory behavior of the older stages of *Pseudodiaptomus hessi*, a dominant form in South African estuaries, e.g. the Swartkops. He presents no quantitative data, but states that there is a distinct diel migration of older stages with arrival in surface layers after dusk and downward migration before dawn. The older stages actually burrow in the flocculent layer of the sediment during daytime.

Several experimental studies (Lance, 1962; Harder, 1968) have shown that estuarine copepods of many species stop upward migration toward downward directed artificial light sources (phototaxis in directed beams) at salinity discontinuities. In water of reduced salinity phototaxis may be reversed, or swimming ceases and the animals sink until lower layers of higher salinity are reached. This may play a role in the field. Crandell and Horacek (1973), for example, have shown that *Acartia clausi* s.l. and *Eurytemora affinis* remain below the dramatic halocline of Big Lagoon in California, a closed habitat.

The whole matter of population budgets in estuaries is open for study. The potential of numerical modelling to give adequate, inexpensive estimates of net flow and dispersion based upon fitting model parameters to conservative properties (salinity) is good. Field mortality estimates can be obtained in a variety of ways, including study of predation processes, evaluation of population growth in suitable enclosures, and short-term estimation of age-structure change. Progress in the next decade should be dramatic.

ACKNOWLEDGEMENTS

Many people have helped with this chapter, and I thank them. J. Kenneth Johnson gave many hours of discussions, the use of his library, permission to quote from his unpublished work, and drafted many of the figures. M. Joan Flynn counted and measured the animals in the studies of Yaquina Bay that are reported here. Waldo W. Wakefield provided many materials and helpful discussion about the Chesapeake Bay and its plankton. Joan Neumann typed the manuscript. Peter H. Wiebe gave a helpful critical reading. All of the work on Yaquina Bay zooplankton was supported by the U.S. Department of Commerce, NOAA, Office of Sea Grant through the Sea Grant College Program at Oregon State University.

REFERENCES

Allen, J.D., Kinsey, T.G. and James, M.C. 1976. Abundances and production of copepods in the Rhode River subestuary of Chesapeake Bay. *Chesapeake Sci.*, 17: 88–92.

Anraku, M. and Omori, M., 1963. Preliminary survey of the relationship between the feeding habit and the structure of the mouth-parts of marine copepods. *Limnol. Oceanogr.*, 8: 116–126.

Arnott, G.H., 1968. *An Ecological Study of the Zooplankton of the Gippsland Lakes with Special Reference to the Calanoida.* Thesis, Monash University, Clayton, Vic.

Arnott, G.H. and Hussainy, S.U., 1972. Brackish-water plankton and their environment in the Werribee River, Victoria. *Aust. J. Mar. Freshwater Res.*, 23: 85–97.

Barlow, J.P., 1955. Physical and biological processes determining the distribution of zooplankton in a tidal estuary. *Biol. Bull.*, 109: 211–225.

Battaglia, B. and Bryan, G.W., 1964. Some aspects of ionic and osmotic regulation in *Tisbe* (Copepoda, Harpacticoida) in relation to polymorphism and geographical distribution. *J. Mar. Biol. Assoc. U.K.*, 44: 1–15.

Bayly, I.A.E., 1961. A revision of the inland water genus *Calamoecia* (Copepoda: Calanoida). *Aust. J. Mar. Freshwater Res.*, 12: 54–91.

Bayly, I.A.E., 1963. A revision of the coastal water genus *Gladioferens* (Copepoda: Calanoida). *Aust. J. Mar. Freshwater Res.*, 14: 194–217.

Bayly, I.A.E., 1964. A new species of *Isias* (Copepoda: Calanoida) from the Brisbane River estuary and a comparison of the Australasian centropagid genera. *Aust. J. Mar. Freshwater Res.*, 15: 239–247.

Bayly, I.A.E., 1965. Ecological studies on the planktonic copepoda of the Brisbane River with special reference to *Gladioferens pectinatus* (Brady) (Calanoida). *Aust. J. Mar. Freshwater Res.*, 16: 315–350.

Bayly, I.A.E. and Arnott, G.H., 1969. A new centropagid genus (Copepoda: Calanoida) from Australian estuarine waters. *Aust. J. Mar. Freshwater Res.*, 20: 189–198.

Bell, W.H., 1976. *The Exchange of Deep Water in Alberni Inlet.* Pac. Mar. Sci. Rep. 76–22. Inst. Ocean Sci., Patricia Bay, Victoria, B.C., 19 pp.

Bhuiyan, A.L., 1966. *The Biology of Certain Planktonic Organisms of the Swan River Estuary with Special Reference to* Sulcanus conflictus *Nicholls (Crustacea: Copepoda).* Thesis, University of Western Australia, Nedlands, W.A.

Bigelow, H.B. and Sears, M., 1939. Studies of the waters of the continental shelf, Cape Cod to Chesapeake Bay. III. A volumetric study of the zooplankton. *Mem. Mus. Comp. Zool. Harvard*, 54: 183–378.

Bishop, J.W., 1967. Feeding rates of the ctenophore, *Mnemiopsis leidyi*. *Chesapeake Sci.*, 8: 259–261.

Bongers, L.H., Polgar, T.T., Lippson, A.J., Krainak, G.M., Moran, L.R., Holland, A.F. and Richkus, W.A., 1975. *The Impact of the Morgantown Power Plant on the Potomac Estuary: An Interpretative Summary of the 1972–73 Investigations.* Md. Dep. Natl Res. Power Plant Siting Program, PPSP-MP-15, Martin Marietta Corp., Baltimore, Md., 133 pp.

Bosch, H.F. and Taylor, W.R., 1970. *Ecology of* Podon polyphemoides *(Crustacea, Branchipoda) in the Chesapeake Bay.* Chesapeake Bay Inst. Ref. 70–5: 77 pp.

Bosch, H.F. and Taylor, W.R., 1973. Distribution of the cladoceran *Podon polyphemoides* in the Chesapeake Bay. *Mar. Biol.*, 19: 161–171.

Bousfield, E.L., 1955. Ecological control of the occurrence of barnacles in the Miramichi estuary. *Nat. Mus. Can., Bull.*, No. 137: 69 pp.

Bousfield, E.L., Filteau, G., O'Neill, M. and Gentes, P., 1975.

Population dynamics of zooplankton in the middle St Lawrence estuary. In: L.E. Cronin (Editor), *Estuarine Research, 1.* Academic Press, New York, N.Y., pp. 325–351.

Bowman, T.E., 1961. The copepod genus *Acartia* in Chesapeake Bay. *Chesapeake Sci.*, 2: 206–207.

Bradford, J.M., 1976. Partial revision of the *Acartia* subgenus *Acartiura* (Copepoda: Calanoida: Acartiidae). *N.Z. J. Mar. Freshwater Res.*, 10: 159–202.

Brand, G.W. and Bayly, I.A.E., 1971. A comparative study of osmotic regulation in four species of calanoid copepod. *Comp. Biochem. Physiol.*, 38B: 361–371.

Burckhardt, G., 1913. Wissenschaftliche Ergebnisse einer Reise um die Erde von M. Pernod und C.C. Schröter. III. Zooplancton aus ost- und süd-asiatischen Binnengewässern. *Zool. Jahrb. Syst.*, 34: 341–472.

Burrell, V.G., 1972. *Distribution and Abundances of Calonoid Copepods in the York River Estuary, Virginia, 1968 and 1969.* Thesis, College of William and Mary, Williamsburg, Va., 263 pp.

Burrell, B.G., Jr. and Van Engel, W.A., 1976. Predation by and distribution of a ctenophore, *Mnemiopsis leidyi* A. Agassiz, in the York River estuary. *Estuarine Coastal Mar. Sci.*, 4: 235–242.

Cargo, D.G. and Schultz, L.P., 1966. Notes on the biology of the sea nettle, *Chrysaora quinquecirrha*, in Chesapeake Bay. *Chesapeake Sci.*, 7: 95–100.

Cargo, D.G. and Schultz, L.P., 1967. Further observations on the biology of the sea nettle and jellyfishes in Chesapeake Bay. *Chesapeake Sci.*, 8: 209–220.

Carriker, M.R., 1951. Ecological observations on the distribution of oyster larvae in New Jersey estuaries. *Ecol. Monogr.*, 21: 19–38.

Carrillo, B.-G., E., Miller, C.B. and Wiebe, P.H., 1974. Failure of interbreeding between Atlantic and Pacific populations of the marine calanoid copepod *Acartia clausi* Giesbrecht. *Limnol. Oceanogr.*, 19: 452–458.

Clarke, G.L. and Bumpus, D.F., 1940. The plankton sampler — an instrument for quantitative plankton investigations. *Limnol. Soc. Am., Spec. Publ.*, 5: 1–8 (revised 1950).

Cones, H.N., Jr. and Haven, D.S., 1969. Distribution of *Chrysaora quinquecirrha* in the York River. *Chesapeake Sci.*, 10: 75–84.

Conover, R.J., 1956. Oceanography of Long Island Sound, 1952–54. VI. The biology of *Acartia clausi* and *A. tonsa*. *Bull. Bingham Oceanogr. Coll.*, 15: 156–233.

Corkett, C.J., 1970. Techniques for breeding and rearing marine calanoid copepods. *Helgold. Wiss. Meeresunters.*, 20: 318–328.

Crandell, G.F. and Horacek, B., 1973. Stratification in Big Lagoon, California, and its effect on copepod populations. *Crustaceana*, 25: 261–266.

Crisafi, P. and Crescenti, M., 1972. Comportamento, morphologia, sviluppo, misure, confronti e revisione di otto specie della femiglia Acartiidae, incontrate in acque marine inquinate soprattotto da scarichi industriali. *Boll. Pesca Piscic. Idrobiol.*, 27: 221–254.

Cronin, L.E., Daiber, J.C. and Hulburt, E.M., 1962. Quantitative seasonal aspects of zooplankton in the Delaware River estuary. *Chesapeake Sci.*, 3: 63–93.

Deason, E.E., 1975. *An In Situ Experiment on the Effects of Zooplankton Grazing and Nutrient Regeneration on the Phytoplankton of Yaquina Bay, Oregon.* Thesis, Oregon State University, Corvallis, Ore., 69 pp.

Deevey, G.B., 1960a. The zooplankton of the surface waters of the Delaware Bay region. *Bull. Bingham Oceanogr. Coll.*, 17(2): 5–53.

Deevey, G.B., 1960b. Relative effects of temperature and food on seasonal variations in length of marine copepods in some eastern American and western European waters. *Bull. Bingham Oceanogr. Coll.*, 17: 54–85.

Durbin, E.G. and Durbin, A.G., 1978. Length and weight relationships of *Acartia clausi* from Narragansett Bay, R.I. *Limnol. Oceanogr.*, 23: 958–969.

Faber, D.J., 1959. *Studies on the Biology of the Zooplankton of Narragansett Bay.* Thesis, University of Rhode Island, Kingston, R.I., 131 pp.

Forns, J.M., 1972. Biological investigations. In: W.D. Clarke and H.D. Palmer (Editors), *Chester River Study, II.* Westinghouse Electric Corp., pp. 61–73.

Friedman, M.M. and Strickler, J.R., 1975. Chemoreceptors and feeding in calanoid copepods (Arthropoda: Crustacea). *Proc. Natl. Sci. Acad. U.S.A.*, 72: 4185–4188.

Frolander, H.F., 1955. *The Biology of the Zooplankton of the Narragansett Bay Area.* Thesis, Brown University, Providence, R.I., 94 pp.

Frolander, H.F., 1964. Biological and chemical features of tidal estuaries. *J. Water. Pollut. Control. Fed.*, 36: 1037–1048.

Frolander, H.F., Miller, C.B., Flynn, M.J., Myers, S.C. and Zimmerman, S.T., 1973. Seasonal cycles of abundance in zooplankton populations of Yaquina Bay, Oregon. *Mar. Biol.*, 21: 277–288.

Frost, B.W., 1972. Effects of size and concentration of food particles on the feeding behavior of the marine planktonic copepod *Calanus pacificus*. *Limnol. Oceanogr.*, 17: 805–815.

Gauld, D.T., 1959. Swimming and feeding in crustacean larvae: the nauplius larva. *Proc. Zool. Soc. Lond.*, 132: 31–50.

Grice, G.D. and Gibson, V.R., 1975. Occurrence, viability and significance of resting eggs of the calanoid copepod *Labidocera aestiva*. *Mar. Biol.*, 31(4): 335–337.

Grice, G.D. and Gibson, V.R., 1977. Resting eggs in *Pontella meadi* (Copepoda: Calanoida). *J. Fish. Res. Board Can.*, 34: 410–412.

Grindley, J.R., 1972. The vertical migration behavior of estuarine plankton. *Zool. Afr.*, 7: 13–20.

Gurney, R., 1931. *British Freshwater Copepoda, I.* Ray Society, London, 238 pp.

Harder, W., 1968. Reactions of plankton organisms to water stratification. *Limnol. Oceanogr.*, 13: 156–168.

Hargrave, B.T. and Geen, G.H., 1970. Effects of copepod grazing on two natural phytoplankton populations. *J. Fish. Res. Board Can.*, 27: 1395–1403.

Heinle, D.R., 1966. Production of a calanoid copepod, *Acartia tonsa*, in the Patuxent River estuary. *Chesapeake Sci.*, 7(2): 59–74.

Heinle, D.R., 1969a. *Effects of Temperature on the Population Dynamics of Estuarine Copepods.* Thesis, University of Maryland, College Park, Md., 132 pp.

Heinle, D.R., 1969b. Temperature and zooplankton. *Chesapeake Sci.*, 10: 186–209.

Heinle, D.R., 1970. Population dynamics of exploited cultures of calanoid copepods. *Helgol. Wiss. Meeresunters.*, 20: 360–372.

Heinle, D.R., 1972. Free-living copepoda of the Chesapeake Bay. *Chesapeake Sci.*, 13(*Suppl.*): S117–S119.

Heinle, D.R., 1974. An alternate grazing hypothesis for the Patuxent estuary. *Chesapeake Sci.*, 15: 146–150.

Heinle, D.R. and Flemer, D.A., 1975. Carbon requirements of a population of the estuarine copepod *Eurytemora affinis*. *Mar. Biol.*, 31: 235–247.

Heinle, D.R., Harris, R.P., Ustach, J.F. and Flemer, D.A., 1977. Detritus as food for estuarine copepods. *Mar. Biol.*, 40: 341–353.

Herman, S.S. and Mihursky, J.A., 1964. Infestation of the copepod, *Acartia tonsa* with the stalked ciliate, *Zoothamnium*. *Science*, 146: 543–544.

Herman, S.S., Mihursky, J.A. and McErlean, A.J., 1968. Zooplankton and environmental characteristics of the Patuxent River estuary, 1963–65. *Chesapeake Sci.*, 9(2): 67–82.

Hess, K.W. and White, F., 1974. *A Numerical Tidal Model of Narragansett Bay*. Sea Grant Mar. Tech. Rep. 20, University of Rhode Island, Providence, R.I.

Hodge, D., 1963. The distribution and ecology of the mysids in the Brisbane River. *Pap. Dep. Zool. Univ. Qld.*, 2: 91–102.

Hodgkin, E.P. and Rippingale, R.J., 1971. Interspecies conflict in estuarine copepods. *Limnol. Oceanogr.*, 16: 573–576.

Hulsizer, E.E., 1976. Zooplankton of lower Narragansett Bay. *Chesapeake Sci.*, 17: 260–270.

Jacobs, J., 1968. Animal behavior and water movement as co-determinants of plankton distribution in a tidal system. *Sarsia*, 34: 355–370.

Jeffries, H.P., 1962. Succession of two *Acartia* species in estuaries. *Limnol. Oceanogr.*, 7: 354–364.

Jeffries, H.P., 1967. Saturation of estuarine zooplankton by congeneric associates. In: G.H. Lauff (Editor), *Estuaries. Am. Assoc. Adv. Publ.*, 83: 500–508.

Johnson, J.K., 1974. *The Dynamics of an Isolated Population of* Acartia tonsa *Dana (Copepoda) in Yaquina Bay, Oregon*. Thesis, Oregon State University, Corvallis, Ore., 97 pp.

Johnson, J.K., 1979. Effects of temperature and salinity on the production and hatching of dormant eggs of *Acartia californiensis* (Copepoda) in an Oregon estuary. *Fish. Bull.*, 77: 567–584.

Kasahara, S., Uye, S. and Onbe, T., 1974. Calanoid copepod eggs in sea-bottom muds. *Mar. Biol.*, 26: 167–171.

Kasahara, S., Uye, S. and Onbe, T., 1975. Calanoid copepod eggs in sea-bottom muds. II. Seasonal cycles of abundance in the populations of several species of copepods and their eggs in the Inland Sea of Japan. *Mar. Biol.*, 31: 25–29.

Katona, S.K., 1970. Growth characteristics of the copepods *Eurytemora affinis* and *E. herdmani* in laboratory cultures. *Helgol. Wiss. Meeresunters.*, 20: 373–384.

Katona, S.K., 1971. *Ecological Studies on some Planktonic Marine Copepods*. Thesis, Harvard University, Cambridge, Mass., 145 pp.

Ketchum, B.H., 1951. The exchanges of fresh and salt water in tidal estuaries. *J. Mar. Res.*, 10: 19–38.

Ketchum, B.H., 1954. Relation between circulation and planktonic populations in estuaries. *Ecology*, 35: 191–200.

King, I.M., 1976. Respiration in three species of the genus *Gladioferens* (Copepoda: Calanoida). *Aust. J. Mar. Freshwater Res.*, 27: 529–532.

Kott, P., 1955. The zooplankton of Lake Macquarie, 1953–1954. *Aust. J. Mar. Freshwater Res.*, 6: 429–442.

Kremer, J.N.K. and Nixon, S.W., 1978. A coastal marine ecosystem, simulation and analysis. *Ecol. Stud.*, 24: 217.

Kremer, P., 1976. Population dynamics and ecological energetics of a pulsed zooplankton predator, the ctenophore *Mnemiopsis leidyi*. In: *Proc. 3rd Int. Estuarine Res. Conf., Galveston, Texas, 1*. Academic Press, New York, N.Y., pp. 197–215.

Kremer, P. and Nixon, S., 1976. Distribution and abundance of the ctenophore, *Mnemiopsis leidyi* in Narragansett Bay. *Estuarine Coastal Mar. Sci.*, 4: 627–639.

Lance, J., 1962. Effects of water of reduced salinity on the vertical migration of zooplankton. *J. Mar. Biol. Assoc. U.K.*, 42: 131–154.

Lance, J., 1963. The salinity tolerance of some estuarine planktonic copepods. *Limnol. Oceanogr.*, 8: 440–449.

Lance, J., 1964. The salinity tolerance of some estuarine planktonic crustacea. *Biol. Bull.*, 127: 108–118.

Lance, J., 1965. Respiration and osmotic behavior of the copepod *Acartia tonsa* in diluted sea water. *Comp. Biochem. Physiol.*, 14: 5–165.

Landry, M.R., 1975. Dark inhibition of egg hatching of the marine copepod *Acartia clausi* Giesbr. *J. Exp. Mar. Biol. Ecol.*, 20: 43–47.

Landry, M.R., 1976. *Population Dynamics of the Planktonic Marine Copepod* Acartia clausi *Giesbrecht in a Small Temperate Lagoon*. Thesis, University of Washington, Seattle, Wash., 200 pp.

Landry, M.R., 1978. Population dynamics and production of a planktonic marine copepod, *Acartia clausi*, in a small temperate lagoon on San Juan Island, Washington. *Int. Rev. Ges. Hydrobiol.*, 63: 77–119.

McHugh, J.L., 1967. Estuarine nekton. In: G.H. Lauff (Editor), *Estuaries. Am. Assoc. Adv. Sci., Publ.*, 83: 581–620.

Marshall, S.M. and Orr, A.P., 1949. On the biology of small copepods in Loch Striven. *J. Mar. Biol. Assoc. U.K.*, 28: 45–122.

Marshall, S.M. and Orr, A.P., 1955. *The Biology of a Marine Copepod*. Oliver and Boyd, London, 188 pp.

Martin, J.H., 1965. Phytoplankton–zooplankton relationships in Narragansett Bay. *Limnol. Oceanogr.*, 10: 185–191.

Martin, J.H., 1968. Phytoplankton–zooplankton relationships in Narragansett Bay. III. Seasonal changes in zooplankton excretion rates in relation to phytoplankton abundance. *Limnol. Oceanogr.*, 13: 63–71.

Martin, J.H., 1970. Phytoplankton–zooplankton relationships in Narragansett Bay. IV. The seasonal importance of grazing. *Limnol. Oceanogr.*, 15: 413–418.

Miller, C.B., Johnson, J.K. and Heinle, D.R., 1977. Growth rates in the marine copepod genus *Acartia Limnol. Oceanogr.*, 22: 326–335.

Miller, R.J. and Williams, R.B., 1972. Energy requirements and food supplies of ctenophores and jellyfish in the Patuxent River estuary. *Chesapeake Sci.*, 13: 328–331.

Morris, M.C., 1950. *A Study of the Macroplankton of the Hawkesbury River Estuary*. Part A. Thesis, University of Sydney, Sydney, N.S.W.

Neale, I.M. and Bayly, I.A.E., 1974. Studies on the ecology of the zooplankton of four estuaries in Victoria. *Aust. J. Mar. Freshwater Res.*, 25: 337–350.

Nelson, T.C., 1925. On the occurrence and food habits of ctenophores in New Jersey inland waters. *Biol. Bull.*, 48: 92–111.

Nicholls, A.G., 1945. A new calanoid copepod from Australia. *Annu. Mag. Nat. Hist.*, 12: 501–514.

Nival, P. and Nival, S., 1973. Efficacité de filtration des copépodes planctoniques. *Ann. Inst. Océanogr., Paris*, 49: 135–144.

Nival, P. and Nival, S., 1976. Particle retention efficiencies of an herbivorous copepod, *Acartia clausi* (adult and copepodite stages): effects on grazing. *Limnol. Oceanogr.*, 21: 24–38.

Odhner, M.X., 1977. *Life Histories and Nutrition of Copepods at Calvert Cliffs.* Report for Baltimore Gas and Electric Co., Acad. Nat. Sci. Philadelphia, Pa., 48 pp. (unpublished).

Ong, J.E., 1969. The fine structure of the mandibular sensory receptors in the brackish water calanoid copepod *Gladioferens pectinatus* (Brady). *Z. Zellforsch.*, 97: 178–195.

Oviatt, C.A. and Kremer, P., 1977. Predation on the ctenophore, *Mnemiopsis leidyi* by butterfish, *Peprilus triacanthus* in Narragansett Bay, Rhode Island. *Chesapeake Sci.*, 18: 236–240.

Park, T.S., 1965. The biology of a calanoid copepod, *Epilabidocera amphitrites* McMurrich. *La Cellule*, 66: 129–251 (10 plates).

Parrish, K.K. and Wilson, D.F., 1978. Fecundity studies on *Acartia tonsa* (Copepoda: Calanoida) in standardized culture. *Mar. Biol.*, 46: 65–81.

Pertzova, N.M., 1974. Life cycle and ecology of a thermophilic copepod *Centropages hamatus* in the White Sea. *Zool. Zh.*, 53: 1013–1022.

Peterson, W.T. and Miller, C.B., 1977. Seasonal cycle of zooplankton abundance and species composition along the central Oregon coast. *Fish. Bull.*, 75: 717–724.

Petipa, T.S., Pavlova, E.V. and Mironov, G.N., 1973. The food web structure, utilization and transport of energy by trophic levels in the planktonic communities. In: J.H. Steele (Editor), *Marine Food Chains.* University of California Press, Berkeley, Calif., pp. 142–167.

Pratt, D.M., 1965. The winter–spring diatom flowering in Narragansett Bay. *Limnol. Oceanogr.*, 10: 173–184.

Reed, J.C., 1978. *An Application of a Mathematical Hydraulics Model to Yaquina Bay, Oregon.* Thesis, Oregon State University, Corvallis, Ore., 235 pp.

Richman, S., Heinle, D.R. and Huff, R., 1977. Grazing by adult estuarine calanoid copepods of the Chesapeake Bay. *Mar. Biol.*, 42: 69–84.

Ricker, W.E., 1946. Production and utilization of fish populations. *Ecol. Monogr.*, 16: 373–391.

Rippingale, R.J., 1972. *Biology of the Copepod* Gladioferens imparipes *Thomson.* Thesis, University of Western Australia, Nedlands, W.A.

Rippingale, R.J. and Hodgkin, E.P., 1974. Population growth of a copepod *Gladioferens imparipes* Thomson. *Aust. J. Mar. Freshwater Res.*, 25: 351–360.

Rippingale, R.J. and Hodgkin, E.P., 1975. Predation effects on the distribution of a copepod. *Aust. J. Mar. Freshwater Res.*, 26: 81–91.

Rippingale, R.J. and Hodgkin, E.P., 1977. Food availability and salinity tolerance in a brackish water copepod. *Aust. J. Mar. Freshwater Res.*, 28: 1–7.

Rochford, D.J., 1951. Studies in Australian estuarine hydrology. I. Introduction and comparative features. *Aust. J. Mar. Freshwater Res.*, 2: 1–116.

Rogers, H., 1940. Occurrence and retention of plankton within an estuary. *J. Fish. Res. Board Can.*, 5: 164–171.

Roman, M.R., 1977. Feeding of the copepod *Acartia tonsa* on the diatom *Nitzschia closterium* and brown algae (*Fucus vesiculosus*) detritus. *Mar. Biol.*, 42: 149–155.

Ryther, J.H., 1954. The ecology of phytoplankton blooms in Moriches Bay and Great South Bay, Long Island, New York. *Biol. Bull.*, 106: 198–209.

Sage, L.E., Summerfield, J.M. and Olson, M.M., 1976. Zooplankton of the Potomac estuary. In: W.T. Mason and K.C. Flynn (Editors), *The Potomac Estuary, Biological Resources, Trends and Options.* Interstate Commission on the Potomac River Basin Tech. Rep. 76–2: 81–87.

Sazhina, L.I., 1960. Development of the Black Sea copepoda. I. Naupliar stages of *Acartia clausi* G., *Centropages kroyeri, Oithona minuta* Kritsz. *Tr. Sevastopol. Biol. Sta.*, 13: 49–67 (in Russian).

Sazhina, L.I., 1968. Hibernating eggs of marine Calanoida. *Zool. Zh.*, 47: 1554–1556.

Smayda, T.J., 1973. The growth of *Skeletonema costatum* during a winter–spring bloom in Narragansett Bay, R.I. *Nor. J. Bot.*, 20: 219–247.

Steuer, A., 1923. Bausteine zu einer Monographie der Copepodengattung *Acartia. Arb. Zool. Inst. Univ. Innsbruck*, 1: 89–148.

Steuer, A., 1929. Die Arten der Copepodengattung *Acartia* in der Mediterranen Provinz. *Sitzungsber Öst. Akad. Wiss. Mathem. Naturwiss. Kl.*, 138: 497–516.

Stross, R.G. and Stottlemeyer, J.R., 1965. Primary production in the Patuxent River. *Chesapeake Sci.*, 6: 125–140.

Stroup, E.C. and Lynn, R.J., 1963. *Atlas of Salinity and Temperature Distributions in Chesapeake Bay 1952–1961 and Seasonal Averages 1949–1961.* Chesapeake Bay Inst. Ref., 63–1: 410 pp.

Sullivan, B.K., Miller, C.B., Peterson, W.T. and Soeldner, A.H., 1975. A scanning electron microscope study of mandibular morphology of boreal copepods. *Mar. Biol.*, 30: 175–182.

Tranter, D.J. and Abraham, S., 1971. Coexistence of species of Acartiidae (Copepoda) in the Cochin Backwater, a monsoonal estuarine lagoon. *Mar. Biol.*, 11: 222–241.

Trinast, E.M., 1975. Tidal currents and *Acartia* distribution in Newport Bay, California. *Estuarine Coastal Mar. Sci.*, 3: 165–176.

Uye, S. and Fleminger, A., 1976. Effects of various environmental factors on egg development of several species of *Acartia* in Southern California. *Mar. Biol.*, 38: 253–262.

Vander Hart, A.L., 1976. *Factors Influencing Sex Ratio in the Estuarine Copepod Genus* Eurytemora. Thesis, Oregon State University, Corvallis, Ore., 77 pp.

Van Engel, W.A. and Joseph, E.B., 1968. *Characteristics of Coastal and Estuarine Fish Nursery Grounds as Natural Communities.* Final Report to U.S. Fish and Wildlife Service, Virginia Institute of Marine Science, 43 pp.

Williams, L.W., 1906. A List of the Rhode Island Copepoda,

Phyllopoda, and Ostracoda with new species of Copepoda. *37th Annu. Rep. R. I. Comm. Inland Fish., Spec. Pap.*, 30: 69–79.

Wilson, C.B., 1932. The copepod crustaceans of Chesapeake Bay. *Proc. U.S. Natl. Mus.*, 80: 1–54.

Wilson, D.F. and Parrish, K.K., 1971. Remating in a planktonic marine calanoid copepod. *Mar. Biol.*, 9: 202–204.

Wood, L. and Hargis, W.J., Jr., 1971. Transport of bivalve larvae in a tidal estuary. In: D.J. Crisp (Editor), *Fourth European Mar. Biol. Symp.* Cambridge University Press, Cambridge, pp. 29–44.

Zenkevitch, L., 1963. *Biology of the Seas of the U.S.S.R.* George Allen and Unwin, London, 955 pp.

Zillioux, E.J. and Gonzalez, J.G., 1972. Egg dormancy in a neritic calanoid copepod and its implications to overwintering in boreal waters. In: *5th European Mar. Biol. Symp.* Piccini Editore, Padova, pp. 217–230.

Zillioux, E.J. and Wilson, D.F., 1966. Culture of a planktonic calanoid copepod through multiple generations. *Science*, 151: 996–998.

Zimmerman, S.T., 1973. *Succession of Zooplankton Populations in Two Dissimilar Marine Embayments on the Oregon Coast.* Thesis, Oregon State University, Corvallis, Ore., 212 pp.

Chapter 6

ESTUARINE BENTHOS

W.J. WOLFF

INTRODUCTION

The benthos comprises those organisms which live on or in the bottom of any body of water. Phytobenthos is the collective name for all plants among the benthos, such as the diatoms, macroalgae and seagrasses, and zoobenthos for all animals occupying this habitat, though those found in hard substrates, such as rocks, wood and shells, are very different from those of soft sediments such as sand and mud. Both types of substrate are occupied by species which live upon the surface of the bottom, the epifauna, and, especially in soft sediments, there are the infauna, the species living within the bottom. For reasons of methodology the (zoo)benthos is divided into macro-, meio-, and microbenthos. Meiobenthos are those forms passing through a 1- or 2-mm sieve but larger than about 100 μm; macro- and microbenthos are respectively larger and smaller than meiobenthos.

In estuaries, all of these categories may be encountered. This review rests heavily on the author's experience with the soft-bottom macrofauna in a number of estuaries of The Netherlands. This probably results in an overrepresentation of European examples; but many of these examples may serve to explain the patterns and processes among estuarine benthos outside Europe. The meio- and microfauna are underrepresented, partly because of my experience mainly with larger animals, but partly also because of the general lack of information about many aspects of the biology of micro- and meiobenthos in estuaries.

THE ENVIRONMENT OF ESTUARINE BENTHOS

Many factors in estuaries show a continuous and irregular pattern of change, so that the estuarine environment is very unstable as well as unpredictable, particularly in temperate latitudes subject to variable weather conditions. Despite these characteristics, however, estuaries are hospitable to those benthic species which are able to cope with the harsh environment.

The water characteristics within estuaries change continuously under the influence of tidal forces, land runoff and winds. This results in tidal, river- and wind-driven currents which vary in speed from nearly zero to over 10 knots (18.5 km h^{-1}) and, hence, in large-scale movements of the estuarine water types. Due to the heterogeneous nature of estuarine waters, the relatively stationary benthic animals on the bottom have to endure a wide range of environmental changes when the circulation carries different kinds of water over their site or burrow (Stone and Reish, 1965).

The major variable in most estuarine environments is salinity. Usually there is a continuous gradient between the fresh river water and the sea, which moves up and down the estuary under the influence of the tides, as well as river runoff and weather conditions. Estuarine animals thus live under brackish conditions, implying not only a salinity intermediate between fresh and sea water but also a fluctuation of salinity. The fluctuation may be small in some parts of the estuary, or it can amount to the full range between sea and fresh water (about 30‰). In hot, arid climates, where evaporation exceeds precipitation and runoff, estuaries may have salinities greater than that of adjacent coastal sea water. Another extreme is the tidal fresh-water range of rivers, where ebb and flow as well as estuarine animals occur in completely fresh water.

Temperature is also subject to change in estuarine water, where it is influenced by seasonal as well as diurnal changes in air temperature and in the temperature of the adjacent sea and river. Vugts and Zimmerman (1975), moreover, report a spring–neap tide cycle in estuarine water temperatures caused by the interaction of the day–night rhythm of insolation and the spring–neap tide pattern of emergence and submergence of tidal flats. Temperatures in shallow estuaries change relatively fast compared to the open sea, and the extremes are separated by a larger amplitude than in the sea.

Oxygen conditions in estuarine waters also vary widely. Some estuaries are nearly always fully saturated, while others show temporary or constant oxygen shortages, sometimes to the point of completely anoxic conditions with the formation of hydrogen sulfide. Low-oxygen conditions may be restricted to the bottom-water layers because of temperature and/or salinity stratifications.

The sediment load of estuarine waters varies considerably, both in time and from place to place, under the influence of current and wave conditions. Some estuaries have relatively clear water, while others may be very turbid temporarily or continuously. A common feature of many river estuaries is a turbidity cloud or mud section. This section is usually found in the transition zone between fresh and brackish waters, and is characterized by an extremely high sediment content and low transparency. Because of currents and waves, sediments in estuaries are frequently unstable where mud and sand predominate. Bare bedrock, peat, pebbles, empty shells or shell reefs provide a more stable bottom. Both active sedimentation and erosion are continuous and often alternate at the same place, so that benthic animals often are confronted with changes of the sediment in or upon which they live.

At increasing depth within the estuarine sediments, the oxygen usually decreases, and the deeper sediment layers are characterized by low redox potentials and high amounts of hydrogen sulfide. In this sulfide system no macrobenthic animals occur without contact with the overlying water, but meio- and microfaunal species live there in fairly large numbers (Fenchel and Riedl, 1970). In many estuaries the preponderance of fine particles in the sediments and the high contents of organic matter make this sulfide system an important part of the estuarine habitat.

Estuaries also possess some positive advantages for benthic animals. Compared to the open coast, estuaries are relatively sheltered against wind waves and ocean swell. Most estuaries are also rich in food provided by river input, input from salt marshes or mangroves, input from the coastal sea, high primary production *in situ* or, most often, a combination of these factors (Postma, 1954, 1961, 1967; Schelske and Odum, 1961; Teal, 1962; De Jonge and Postma, 1974; Odum and Heald, 1975; Wolff, 1977). Because of the shallow depth of most estuaries, suspended food particles are readily available for benthic animals through sinking as well as through downward transport by turbulent water movements (Wolff et al., 1976a).

This description of the estuarine environment applies primarily to the classical form of the estuary: the tidal mouth of a river. However, many benthic animals considered to be estuarine species are not restricted to estuaries in a geomorphological sense. The same species may also occur in many other coastal areas characterized by environmental instability and an ample supply of food (Remane, 1958, 1971; Muus, 1967). Therefore our discussion also includes the benthic fauna of other water bodies with these estuarine conditions.

ADAPTATIONS OF ESTUARINE BENTHOS

Morphology

Several marine benthic organisms occur in a reduced size in brackish-water populations. This is true for various molluscs, such as *Mytilus edulis* and *Mya arenaria* (Fig. 6.1), a number of echinoderms, and some coelenterates. This does not occur in many other species, such as most crustaceans and polychaetes (Remane, 1958, 1971; Muus, 1967), and Braber and Borghouts (1977) have also failed to find a size reduction with decreasing salinity for the sea anemones *Tealia felina* and *Diadumene cincta* in the Westerschelde Estuary, The Netherlands.

Brackish-water populations may also show re-

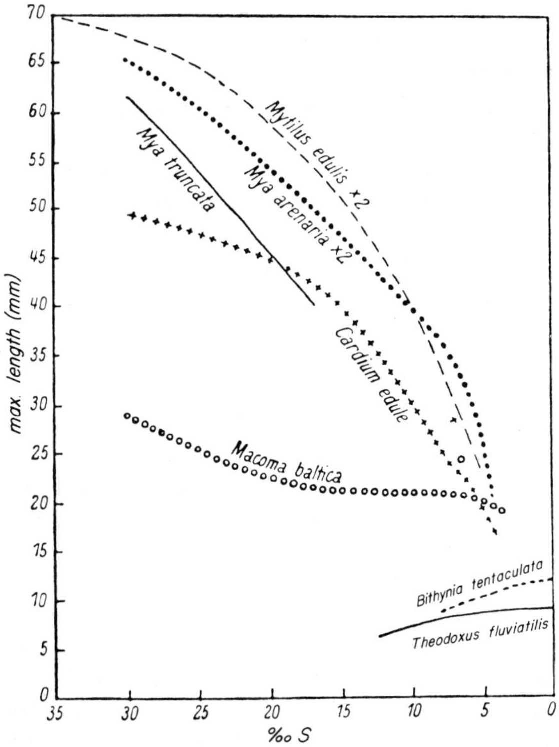

Fig. 6.1. Reduction of maximal length of some molluscs with diminishing salinity in a transect from North Sea to Baltic. (After Remane, 1971.)

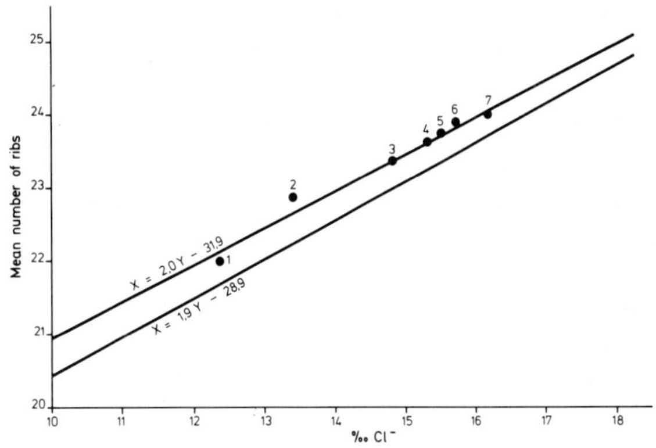

Fig. 6.2. Relationship between mean rib number of the bivalve *Cerastoderma edule* and salinity. The upper line represents data from the Westerschelde Estuary, The Netherlands, the lower line data from many estuarine waters in The Netherlands. (After Koulman and Wolff, 1977.)

ductions or changes of their calcareous parts. This has been observed in molluscs, ostracods and bryozoans (Remane, 1971). The cockle *Cerastoderma edule* (Fig. 6.2) has a reduced number of ribs on the shell with decreasing salinity (Eisma, 1965; Russell, 1972; Boyden, 1973; Eisma et al., 1976; Koulman and Wolff, 1977). Muus (1967) and Rasmussen (1973) draw attention to the high degree of morphological variation of brackish-water species.

Whether any of these changes may be considered to be an adaptation to the estuarine environment is uncertain. Perhaps they should be viewed as morphological responses to physiological stresses. This belief is supported by the high percentage of abnormalities in low salinities (Muus, 1967).

Physiology

Physiological adaptations to the estuarine environment are widespread among benthic animals (Ch. 3). The major adaptations are made in the regulation of the osmotic and ionic composition of the body fluids, and in the maintenance of life under conditions of reduced oxygen. The adaptations observed are discussed in Chapter 3 by Vernberg; hence, no further comments will be made here. A special comment may be appropriate, however, concerning adaptations to conditions of reduced oxygen. Macrobenthic species adapt to low oxygen tensions by a general reduction of activity and metabolism, by developing blood pigments with a very high oxygen affinity and by anaerobic respiration with repayment of an oxygen debt or with production and excretion of respiratory end-products (Brafield, 1972). Obligate anaerobiosis has been claimed to occur among the meio- and the microfauna. Fenchel and Riedl (1970) postulated the existence of an anaerobic sulfide system inhabited by meio- and microbenthic species. Boaden and Platt (1971) coined the term "thiobios" for the entire biota of this biotope. Boaden (1975, 1977) specifies the list of thiobiotic meiofauna species. He postulates even an anoxic origin for the Gnathostomulida, Platyhelminthes and Aschelminthes and assumes that the early evolution of Metazoa has passed through a thiobiotic phase in anoxic marine sediments. These hypotheses are rejected by Reise and Ax (1979) who demonstrate that all meiofauna in the tidal flats of the German Wadden Sea remain in close proximity to oxygenated layers or pockets in the sediment. Apparently, life in anoxic environments is possible when contact with oxygen is possible, at least at regular intervals.

Behavior

Estuarine benthic animals make many behavioral adaptations to their environment. Burying behavior is widespread, although it is not limited to estuaries. Salinity within sediments is usually more stable and, on the average, higher than that in the water flowing over them (Reid, 1930, 1932; Smith, 1956; Sanders et al., 1965; Wolff, 1973). Sanders et al. (1965) demonstrated that infauna species survived in interstitial water but died in the ebb water flowing over the sediments in the locality where they were collected. It is probably for this reason that marine infauna species occur further up the estuary than do marine epifaunal species (Sanders et al., 1965; Carriker, 1967). On the other hand, burying is also a common method to escape predator pressure and desiccation, as is demonstrated in high and constant salinities by many burrowing bivalves and worms which use the water flowing over them as their source of food and oxygen. Burying behavior may be well timed to the tidal cycle of emergence and submergence of tidal flats. Polychaete worms from the Dutch Wadden Sea were found to rise in their burrows even before the incoming tide covered them, and they retreated soon after emergence of the flats on the ebb (Vader, 1964). However, Johnson (1967) was not able to confirm this behavior for other species living in tidal flats in California, U.S.A., perhaps for methodological reasons.

Bivalves are able to survive periods of low salinity by closing their valves tightly. Prosobranch snails do the same by closing their operculum.

Migration of estuarine benthic animals seems to be fairly common. In many cases the species rely more or less on passive transport by tidal currents. Newell (1962) described passive transport of the small mud snail *Hydrobia ulvae* while creeping at the underside of the surface film of the water of the incoming tide. Vader (1964) and Wolff (1973), however, were not able to confirm this behavior. Beukema (1973) found that juvenile specimens of the bivalve *Macoma balthica* leave their nurseries at high-level mudflats during their first winter, and migrate with the tidal currents until they establish a secondary settlement at the lower flats and in subtidal habitats. A similar behavior seems to occur in the lugworm *Arenicola marina* (Wolff and De Wolf, 1977). The shore crab *Carcinus maenas* spends its first year on the tidal flats and migrates next spring to the tidal channels (Klein Breteler, 1976). The shrimp *Crangon crangon* shows a similar behavior during its first summer. The worm *Nereis virens*, however, seems to settle in subtidal areas and migrates later to intertidal areas (Bass and Brafield, 1972).

Migrations into and out of estuaries also occur among benthic invertebrates. The crabs *Carcinus maenas* and *Eriocheir sinensis* walk on the bottom, respectively to the sea and to fresh water (Peters and Panning, 1933; Broekhuyzen, 1936; Kamps, 1937). The swimming crab *Macropipus holsatus* (Venema and Creutzberg, 1973) is transported by the tidal currents while migrating (Fig. 6.3), as are the shrimps *Crangon crangon* (Boddeke, 1976) and *Penaeus duorarum* (Hughes, 1972). These migrations are usually related to reproduction or to a nursery function of estuaries. Species capable of swimming actively move up in the water column when the tide runs in the right direction and settle again and bury at the turn of the tide. These animals must be able to discriminate between ebb and flood water, and this has indeed been demonstrated (Hughes, 1972; Venema and Creutzberg, 1973; Creutzberg, 1974). Stimuli and directional clues for migration have been identified as chemical features of fresh water (*Anguilla anguilla*), a gradient of salinity (*Macropipus holsatus*, *Penaeus duorarum*) and empty stomachs (*Pleuronectes platessa*). By this mechanism animals may be carried many kilometres per day by the water currents.

Bousfield (1955) describes comparable behavior for barnacle larvae. The young stages swim actively to the upper water layers and are transported seaward by the estuarine residual current; the older larval stages would tend to keep near the bottom and thus be transported back by the estuarine residual bottom current. Wood and Hargis (1971) describe similar transport for oyster larvae (*Crassostrea virginica*). De Wolf (1973), who studied the occurrence of barnacle larvae in the completely mixed Dutch Wadden Sea, considered Bousfield's model to be unnecessarily complex. He argued that barnacle larvae are not able to overcome the turbulent water movements, and he described the distribution of barnacle larvae with a purely physical model based on variations in current speed and turbulence, inequality of ebb and flow currents, and the sinking speed of the larvae.

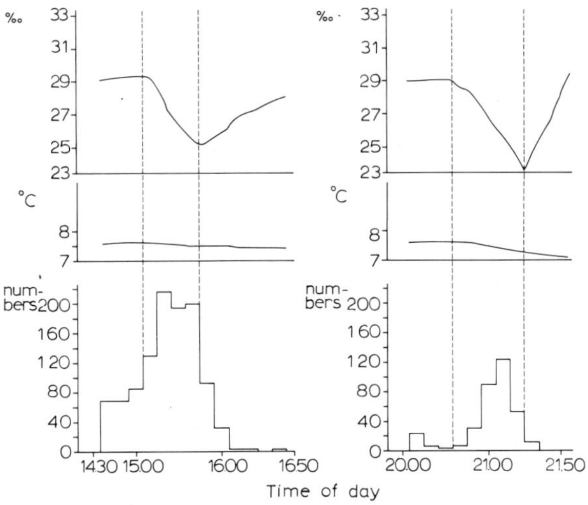

Fig. 6.3. Number of swimming *Macropipus holsatus* in an experimental tidal current in relation to salinity (‰) and temperature (°C). A drop of salinity leads to an increase of the number of swimming crabs. (After Venema and Creutzberg, 1973.)

De Wolf (1973, 1974) indicated that the same mechanisms may be operating in other estuarine benthic species with planktonic larvae.

The settling behavior of the larvae of estuarine animals is insufficiently known, since most cases studied are of marine species.

Ecology

In many seas a large number of benthic organisms have a planktonic larval stage. In shallow seas these larvae, known as meroplankton, form an important proportion of the plankton. It might be expected that estuarine benthic animals would be dominated by non-pelagic development types, as Muus (1967) found for the macrofauna of mesohaline lagoons in Denmark. Data by Wolff (1973), however, do not confirm this idea (Table 6.1). Even in the low-salinity areas of estuaries in The Netherlands half of all the species present have a pelagic phase in their development.

Organisms provided with a large amount of food in an unstable and unpredictable environment may choose different strategies to exploit this food. One strategy is clearly opportunistic and aims at exploitation only during favorable periods, which would be achieved by migration. Species usually migrate into the estuaries during spring or summer and withdraw in autumn, at least in temperate latitudes. This kind of behavior occurs mainly among crustaceans. Another opportunistic method of exploitation is found among polychaetes and molluscs, which colonize an area rapidly, usually by pelagic larvae, during favorable periods, reproduce quickly, and thus are able to build up large populations. When the circumstances become less favorable, these species disappear equally rapidly by mortality. Species showing this kind of behavior are the polychaetes *Capitella capitata*, *Polydora ligni* and *Streblospio benedicti* and the molluscs *Hydrobia ulvae*, *Mytilus edulis* and *Cerastoderma edule* (Grassle and Grassle, 1974; Wolff et al., 1977).

The other strategy available to benthic animals in estuaries is to develop mechanisms that protect them from the unpredictable effects of the environment. Osmoregulatory capacity is a good example of this type of development. Moreover, species may inhabit the most stable and hospitable parts of the estuarine environment — below low-tide level, for instance, and/or buried in the sediment. The isopods of the genus *Cyathura* are a very good example of this type of strategy. It must be stressed that many estuarine species show aspects of both

TABLE 6.1

Types of larval development of soft-bottom macrobenthic animals in the Delta area, The Netherlands (Wolff, 1973)

	Marine	Estuarine			River
		high salinity	low salinity	fresh	
Pelagic larvae	56%	51%	50%	3%	0%
Non-pelagic larvae	30%	34%	44%	94%	100%
Development unknown	14%	15%	6%	3%	0%
Number of species	157	144	34	34	31

strategies in their life-cycles. The shrimp *Crangon crangon*, for instance, shows osmoregulation as well as migration.

Another aspect related to opportunistic life histories is the occurrence of wide niches. Wolff (1973) found in the brackish parts of the estuaries of the Delta area,[1] The Netherlands, that 35% of the species might be considered to be omnivores, whereas this percentage was only 6 to 16% in adjacent fresh-water or sea-water areas. Another example is the polychaete *Nereis diversicolor*; in the high-salinity areas of European estuaries this species is nearly restricted to the intertidal mudflats, but in brackish water it can also be found subtidally and in all kinds of sediments.

Also connected with opportunistic life histories are inefficient ways of feeding. Estuarine bivalves, such as *Mytilus edulis* and *Mya arenaria*, have relatively slow filtration rates (Winter, 1969). In their estuarine habitats this is compensated by the abundance of their food, but this feature apparently restricts their distribution offshore (Wolff, 1973).

DISTRIBUTION PATTERNS OF ESTUARINE BENTHOS

The small-scale distribution of benthic organisms in estuaries is related to such factors as depth, current speed, and sediment characteristics. On the scale of the estuary the patterns are related to the gradual change from the marine to a fresh-water environment. On a global scale, zoogeographical patterns dominate the distributions of estuarine benthic organisms. On all scales, interrelationships with other species play their roles.

Small-scale distribution patterns

Even in habitats which seem horizontally homogeneous, such as some intertidal flats, the horizontal distribution of benthic organisms varies. Aggregated patterns of distribution are common, occurring in many kinds of invertebrates, including Foraminifera, Polychaeta, Oligochaeta, Gastropoda, Bivalvia, Amphipoda, Isopoda, Nematoda, Copepoda, and Ostracoda (Heip, 1975). Less common are randomly distributed benthic organisms. A Poisson-type distribution has been assumed for some species among the polychaetes and bivalves (Holme, 1953; Wolff and De Wolf, 1977). Evenly distributed patterns seem to be unusual. Holme (1950) described such a distribution, apparently caused by intraspecific competition, for the bivalve *Tellina tenuis* in England. The distribution pattern found often is dependent on sample size, since larger samples usually result in a higher frequency of aggregated patterns (Rosenberg, 1974).

In heterogeneous environments, variations in numbers of benthic animals are often correlated with variations of environmental factors. In this respect, the complex of water movements and sediment composition is most important. Many species normally are found only in a limited range of sediments. In most cases, however, it is not clear whether the occurrence of the species is dependent on the type of sediment, or on the type of water movements which also influence the composition of the sediment. Only when the animals show a preference for a certain type of sediment in experiments designed to observe choice can it be concluded that it is the sediment itself which governs the distribution. Thus Gray (1966a, b) concluded that the archiannelid *Protodriloides symbioticus* had a preference for sandy bottoms. Gray and Johnson (1970) found for *Turbanella hyalina* (Aschelminthes: Gastrotricha) that this selection was based on a preference for certain bacteria occurring on sand grains. Wilson (1937) found that settling larvae of the polychaete *Notomastus latericeus* prefer mud and sand to clean shell gravel.

Although the experimental data are scanty, many field observations show that several benthic species have a distinct preference for a certain grade of sediment (Sanders, 1958; Vader, 1965; McNulty et al., 1962; Maurer, 1969; Johnson, 1971; Wolff, 1973). Thus haustoriid amphipods (e.g. *Haustorius*, *Bathyporeia*) and polychaetes of the genus *Ophelia* are characteristic inhabitants of sandy bottoms. The polychaete *Polydora ligni* and the bivalve *Scrobicularia plana* are characteristic for mud bottoms. The polychaete *Polydora ciliata* as well as several pholad bivalves (*Pholas*, *Barnea*, *Zirfaea*) are able to bore into limestone, and the bivalves can also bore into peat. The snails *Rissoa membranacea* and *Lacuna vincta* are attached to seagrass leaves (*Zostera marina*). Longbottom (1970) found that

[1] The estuarine area of the rivers Rhine, Meuse and Scheldt.

the biomass of the lugworm *Arenicola marina* is positively correlated with the amount of organic matter in the sediment. Newell (1965) states a similar relationship for the bivalve *Macoma balthica* and the mudsnail *Hydrobia ulvae*.

Several species live exclusively in salt marshes. In European salt marshes the gastropods *Ovatella myosotis* and *Assiminea grayana* and various amphipods of the genus *Orchestia*, as well as the isopods *Sphaeroma rugicauda* and *Paragnathia formica*, are characteristic. In salt marshes along the Atlantic and Gulf Coast of North America the snail *Littorina irrorata*, the bivalve *Modiolus demissus* and the crab *Uca pugnax* are abundant.

Vertical distribution patterns of benthic organisms are related to vertical environmental gradients. Most benthic species live at the surface or in the better oxygenated upper layers of the sediment, and decrease in numbers with increasing sediment depth and decreasing oxygen supply. Other species, representatives of the meio- and microfauna (Fig. 6.4), are restricted to deeper sediment layers (Fenchel and Jansson, 1966). Some species of polychaetes (*Harmothoe*, *Lepidonotus*) live in the burrows of other organisms, such as the lugworm *Arenicola marina*.

Distribution patterns on the scale of the estuary

On the scale of the whole estuary there are a number of environmental gradients: exposure to wave action; the rate of dilution of sea water by river water; turbidity; the oxygen saturation in some cases; and pollution in some estuaries. Benthic animals respond to most of these gradients in their distribution patterns. Of these variables salinity is the most important. Turbidity caused by suspended particles may affect suspension-feeders negatively (Rhoads and Young, 1970). At very high turbidities such species do not occur (Howell and Shelton, 1970). Oxygen shortages also have mainly negative effects. Anoxic areas are not inhabited by macrobenthic species (Rosenberg, 1972, 1973a, b), although meio- and microbenthos may be found. Areas influenced by organic pollution are often dominated by short-lived, opportunistic species such as *Capitella capitata* and *Polydora ligni* (Reish, 1960; Wass, 1967; Henriksson, 1969; Rosenberg, 1973a; Pearson, 1975; Pearson and Rosenberg, 1976).

Characteristic estuarine species are limited in number. These species are restricted to estuarine brackish waters and hence do not occur in the sea

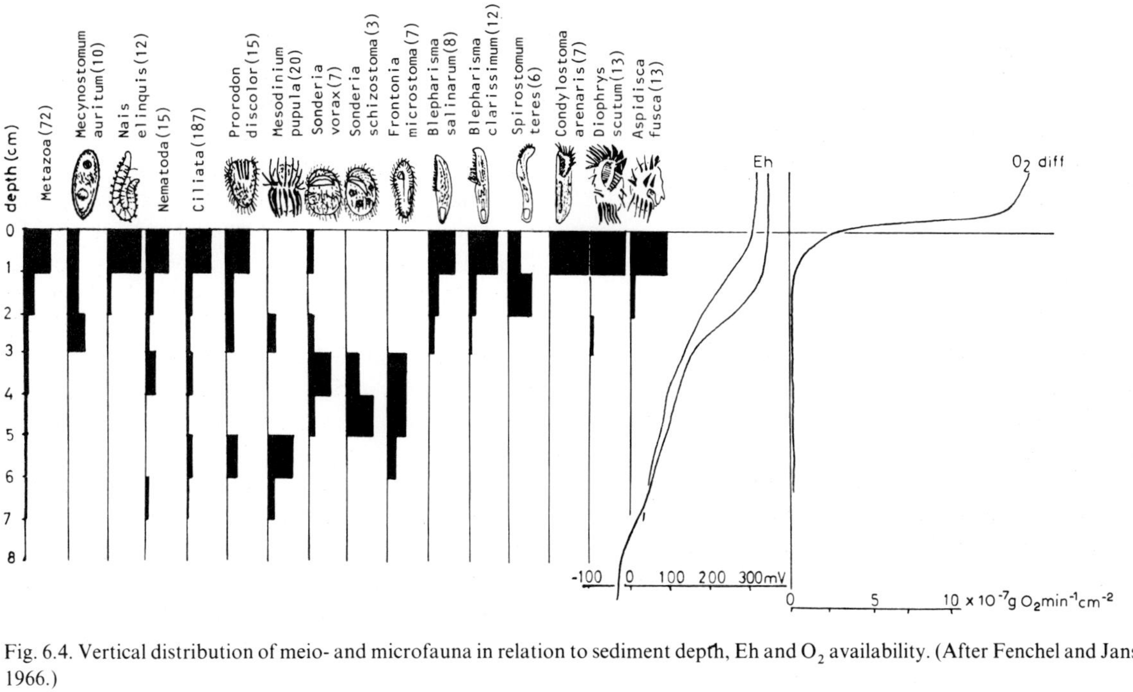

Fig. 6.4. Vertical distribution of meio- and microfauna in relation to sediment depth, Eh and O_2 availability. (After Fenchel and Jansson, 1966.)

or in fresh water. In the Delta area of The Netherlands, Wolff (1973) was able to assign 20 species out of 229 macrobenthos species to this category. Characteristic representatives are the isopod *Cyathura carinata*, the snail *Assiminea grayana* and the polychaetes *Boccardia ligerica* (= *Polydora redekei*) (Fig. 6.5A) and *Streblospio shrubsolii*.

Salinity is an important factor governing the distribution of benthic animals in estuaries. Along the seaward edge of most estuaries a typical beach fauna including talitrid and haustoriid amphipods may occur when salinities are sufficiently high (Fig. 6.5B). This fauna does not occur in the sheltered shores along the estuary where salt-marsh and quiet-water species predominate. Another group of species lives in a zone of varying width along the open coast as well as in the seaward parts of estuaries, where they reach their highest densities. Such species may be considered to be high-salinity estuarine species. European examples are *Mytilus edulis*, *Macoma balthica* (Fig. 6.5C), *Cerastoderma edule*, *Arenicola marina* and *Heteromastus filiformis*. Many other marine species which belong to the shelf or coastal fauna are able to penetrate to some extent in waters of lower salinity (Fig. 6.5D) (Remane, 1934, 1958, 1971).

There seems to be no category of species which lives in fresh water as well as in estuaries, although some fresh-water species are able to penetrate into water of higher salinity, as do many Rotifera and Oligochaeta (Remane, 1958, 1971; Pfannkuche, 1974, 1977). Such species cannot be considered as

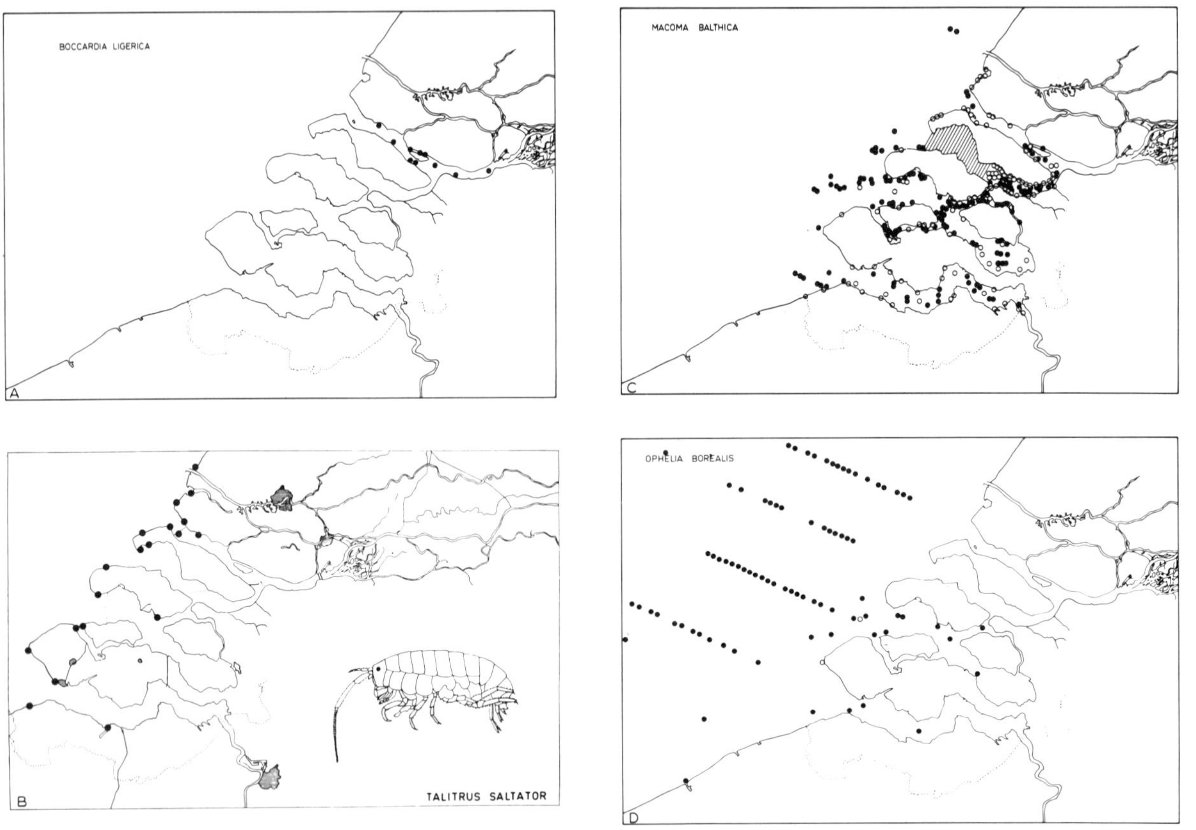

Fig. 6.5. Distributions of some benthic organisms in the estuarine area of the southwest Netherlands. A. The polychaete worm *Boccardia ligerica* was found only in the brackish part of the Rhine–Meuse Estuary. (After Wolff, 1973.) B. The amphipod *Talitrus saltator* was found only at sandy beaches exposed to the North Sea waves. (After Den Hartog, 1964.) C. The bivalve *Macoma balthica* was found only in the brackish and high salinity parts of the estuaries and in a narrow band along the North Sea coast. Dots are subtidal finds, circles intertidal ones. The hatching denotes an area investigated extensively, where *Macoma* occurred everywhere. (After Wolff, 1973.) D. The polychaete *Ophelia borealis* was found only offshore and in the high salinity parts of the estuaries. Dots are subtidal finds, circles intertidal. (After Wolff, 1973.)

estuarine even though they are able to bear some of the stresses which the estuarine environment imposes on them. Usually they do not have special adaptations to this environment, and to them the estuary is a border region of their main area of distribution. The estuarine fresh-water tidal areas seem to be even poorer in characteristic species. The species listed by Wolff (1973) are characteristic only on a local scale.

Distribution patterns on a world scale

To compare estuarine benthic faunas on a world scale, it is useful to distinguish between: (1) a group of fouling species occurring in sea water and brackish water; (2) a group of benthic species living in coastal waters and in estuaries; and (3) a group of benthic species living nearly exclusively in estuaries and other brackish waters.

The first group of species, which includes the polychaete *Mercierella enigmatica*, the barnacles *Balanus improvisus* and *B. amphitrite*, and the entoproct *Barentsia gracilis*, for example, have a worldwide distribution. No doubt these species have been spread by ships and possibly by transplants of shellfish. Their origins mostly are obscure, although several species thrive in tropical waters.

The coastal group of species also is distributed widely. A number of common European species also occurs along the Atlantic and Pacific coasts of North America (Filice, 1958; Tenore, 1972; Wass, 1972; Wolff, 1973).

The third group of exclusively estuarine species are very different. Several brackish-water faunas may be distinguished in various parts of the world. In those parts of the world which have been investigated in some detail, the composition of the brackish-water fauna is more or less homogeneous over the entire area. However, the boundaries of these areas are not well known and for that reason these areas cannot be equated with the zoogeographical provinces described earlier. In the estuaries of Europe and northern Africa, the same 20 or 30 brackish-water macrobenthic species are encountered nearly everywhere (Remane, 1971; Elkaim, 1972; Wolff, 1973). Well-known representatives are the isopod *Cyathura carinata*, the polychaetes *Nereis diversicolor* and *Streblospio shrubsolii* and the amphipod *Gammarus zaddachi*.

In the Black Sea and the Caspian Sea, with their adjacent waters, a very different brackish-water fauna is found (Zenkevitch, 1959, 1963). This fauna includes many endemic genera of crustaceans and molluscs. Only a few species also occur in the other European estuaries, for example *Cerastoderma glaucum* (Russell, 1971).

In the northeastern parts of Europe and in Siberia an arctic brackish-water fauna is found (Segerstråle, 1949; Zenkevitch, 1963; Remane, 1971). Characteristic species are *Mysis relicta*, *Mesidothea entomon*, *Pontoporeia affinis*, *Portlandia arctica*, and *P. aestuariorum*.

Another fauna found in the brackish parts of the estuaries and lagoons along the Atlantic and Gulf coasts of North America has as characteristic forms the polychaete *Streblospio benedicti*, the isopod *Cyathura polita* and the crabs *Callinectes sapidus* and *Rithropanopeus harrisii*.

The estuaries along the Pacific Coast of North America have their own fauna, such as *Nereis limnicola* and *Neomysis awatschensis* (Filice, 1958; Haertel and Osterberg, 1967).

From the work of Day (1964, 1974) it appears that the South African estuarine brackish-water fauna is also unique with many endemic crustaceans.

Perhaps the richest brackish-water fauna of the world is found in Southeast Asia; unfortunately, this fauna is not well known.

Some brackish-water species occur in several of these faunal regions. In several cases, however, it is known from which area these species originated and when they were introduced to other areas. The mud crab *Rhithropanopeus harrisii* is originally an inhabitant of the Atlantic Coast of North America. It is known to have been introduced on the Pacific Coast of North America (Filice, 1958). Further, Buitendijk and Holthuis (1949) were able to synonymize the species *Heteropanope tridentatus*, found endemically in the brackish Zuiderzee (now the IJsselmeer), The Netherlands, with *R. harrisii*, and postulated that this species had been introduced from America. This crab has now spread all over Europe, even as far as the Baltic and the Black Sea, after a population explosion in The Netherlands (Wolff, 1954). Two other cases of introduced species are the American blue crab *Callinectes sapidus* which now is common in Mediterranean brackish-water areas, and the Chinese mitten crab *Eriocheir sinensis* which now spawns in several European estuaries (Peters and Panning, 1933; Kamps, 1937).

Some species are known to have been introduced to other parts of the world. For other species, the possibility of past introductions is only suspected. The European *Crassostrea angulata*, for instance, seems to be closely related to the Japanese *C. gigas*. Its early area of distribution in European waters was so small that there is some doubt as to its origin, although it has been stated that fossil specimens occur over a larger area (Korringa, 1976).

The claim that estuarine species are cosmopolitan (Carriker, 1967; De Lattin, 1967; Maurer and Watling, 1973) is certainly not justified for all species. It is true to some extent for the first two groups mentioned in this section, as well as for those species of the third group that have been distributed by man over large parts of the world.

FOOD RELATIONSHIPS OF BENTHIC ANIMALS

All kinds of organic matter in the estuarine environment — planktonic and benthic organisms, detritus, bacteria and dissolved organic matter — are potential sources of food. A special case is that of benthic animals containing photosynthesizing zooxanthellae, of which the flatworm *Procerodes ulvae* of European estuaries is a famous example. It has a tidal rhythm of migration in and out of the sediment, apparently enabling it to take maximum advantage of solar radiation during low tide.

Many estuarine benthic species employ several feeding techniques. The polychaete *Nereis diversicolor* may act as a deposit feeder, as a suspension feeder, as a scavenger and as a predator (Goerke, 1971). Gerlach et al. (1976) demonstrated suspension feeding in the hermit crab *Pagurus bernhardus*, a species also classified as a predator and a deposit feeder. Wolff (1973) found that in the low-salinity reaches of the estuaries of the Dutch Delta area, 35% of the species present could be classified as omnivores, compared with only 6 to 16% in marine and fresh waters.

Suspension feeders

Species feeding on organic particles suspended in the water are termed suspension feeders; their potential food sources are phytoplankton, zooplankton, bacteria, and detritus. Zooplankton is relatively unimportant when compared to phytoplankton on a weight basis, but some suspension feeders such as *Ostrea edulis* may consume considerable amounts of zooplankton (Orton, 1925). The role of detritus as a food source is still unclear, as will be discussed in the following section.

Examples of estuarine suspension feeders are the bryozoans *Conopeum seurati* and *Bowerbankia gracilis* (Jebram, 1974), molluscs such as *Crepidula fornicata, Crassostrea virginica, Ostrea edulis, Mytilus edulis* and *Cerastoderma edule*, barnacles like *Balanus improvisus*, and spionid polychaetes, e.g. *Polydora ligni* (Hempel, 1957). The amphipod *Haustorius arenarius* is a suspension feeder which takes its food from the interstitial water (Dennell, 1933).

Deposit feeders

Deposit feeders conveniently may be split up in two categories: selective and non-selective. Non-selective deposit feeders ingest the sediment as a whole, use the digestible particles, and pass the remainder of the sediment through their guts. The lugworms *Arenicola marina* and *Abarenicola claparedii* are well-known examples. Tubificid as well as naidid (genus *Paranais*) oligochaetes also are non-selective deposit feeders (Pfannkuche, 1977). Food sources for non-selective deposit feeders are benthic microalgae, meio- and microfauna, bacteria and possibly detritus.

Selective deposit feeders separate their food particles from the sediment. This is achieved in very different ways. Spionid and cirratulid polychaetes gather food from the surface layer of the sediment by ciliated tentacles. Amphipods of the genus *Bathyporeia* take the sand grains one by one and rasp the epiflora from the grains. The bivalve *Macoma balthica* sucks the benthic diatom flora from the surface layer of the sediment by the current through its inhalant siphon. Many nematodes, oligochaetes, and turbellarians eat diatoms and bacteria one by one (Bilio, 1967; Pfannkuche, 1977).

That benthic microalgae are an important source of food is beyond doubt. Beukema et al. (1977), for instance, found indications that the availability of benthic diatoms limits the growth of the bivalve *Macoma balthica* (Fig. 6.6). The nutritional value of detritus and bacteria is less clear. Newell (1965) has shown that the prosobranch *Hydrobia ulvae*

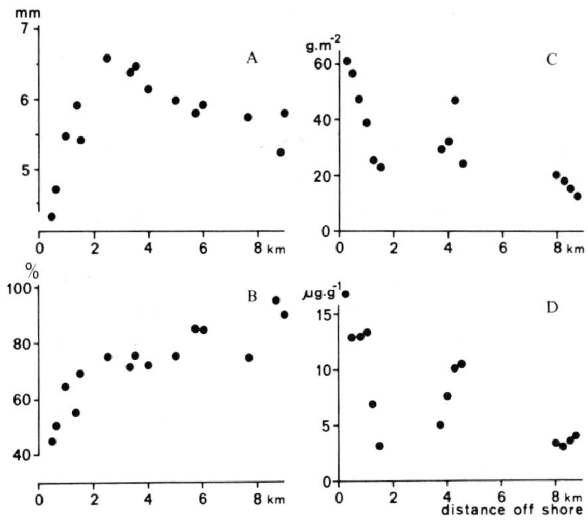

Fig. 6.6. Measures of growth rate, time of immersion, and food supply of the bivalve *Macoma balthica* at various distances from the shore at the Balgzand tidal flats, The Netherlands. A. Annual growth rate as eight-year averages of the length increase of the shell. B. Immersion time as percentages of total time. C. Primary production in g C m^{-2} in April–June 1974. D. Concentration of functional chlorophyll *a* in the top cm of the sediment during the same period (μg chlorophyll *a* per g of sediment). *Macoma* shows an increasing growth rate with increasing food supply until immersion times become too short. (After Beukema et al., 1977.)

and the bivalve *Macoma balthica* were unable to alter or to utilize ingested detritus particles but were capable of digesting decomposer organisms adsorbed on the particles. The relative importance of this food source is not evaluated, however. The results of Tunnicliffe and Risk (1977) from the Bay of Fundy in North America are more conclusive. They found a significant positive correlation between the densities of *Macoma balthica* and of bacteria in the sediment, but not between the numbers of *Macoma* and total carbon. Moreover, bacteria constituted on average one third of the total amount of carbon in their samples, and sometimes nearly reached 100%. Odum and Heald (1975) assumed that a group of characteristic detritus feeders in American estuaries also digest only the bacteria from the detritus particles, but they stated that in addition there may be intestinal microbes which further reduce the detritus particle and supply the host organism with nutrition in the form of excreted organic substances or of the micro-organism itself.

Predators and scavengers

Several benthic species may be classified as predators or scavengers. In many cases the boundary between these categories is rather vague. Examples of this group of species are the echinoderms *Asterias rubens*, the snails *Urosalpinx cinereus*, *Buccinum undatum* and *Nassarius reticulatus*, the polychaete *Nereis virens* and the crabs *Callinectes sapidus* and *Carcinus maenas*.

Estuarine predators generally have a wide choice of prey species. Reise (1977a, b) used exclusion experiments to show that benthic predators such as crabs (*Carcinus maenas*), shrimps (*Crangon crangon*) and gobiid fishes have a profound influence on the structure and biomass of intertidal estuarine benthic communities.

Grazers

A number of species may be classified as grazers, although the difference between grazers and selective deposit feeders may be rather arbitrary. Grazing species eat mostly benthic microalgae (*Littorina*), although some species may eat macroalgae (*Strongylocentrotus*) or seagrasses (*Idotea chelipes*). However, vascular plants (*Zostera*, *Thalassia*, etc.) are only rarely consumed alive by benthic organisms (Wood et al., 1969).

Dissolved organic matter

Although some marine benthic organisms are able to utilize dissolved organic matter for metabolic purposes (Southward and Southward, 1972; Sepers, 1977), this food source seems less important for estuarine animals. In the first place, uptake of dissolved compounds is largely inhibited at lower salinities (Stephens, 1964, 1967); and, in the second place, all estuarine species investigated so far need additional food sources to meet their nutrient requirements.

Trophic interactions

An obvious trophic interaction is the influence of predators on other feeding categories. Reise (1977a, b) showed that benthic predators are able to exclude certain benthic mollusc species from the intertidal benthos of the German Wadden Sea, while other species were reduced considerably in number. Rhoads and Young (1970) suggested that the frequent resuspension of sediments reworked by deposit feeders was limiting for most suspension feeders. They termed this incompatibility of deposit and suspension feeders "trophic group amensalism". This hypothesis was confirmed when an oil spillage killed such a deposit-feeding community. After this event the bottom became firm and suspension feeders colonized the area (Rhoads, 1974).

BENTHIC BIOMASS

The biomass of estuarine benthic animals varies widely from place to place, but for a given estuary it remains fairly constant from year to year (Beukema et al., 1978). Table 6.2 gives some examples of biomasses recorded in investigations of large areas. It appears that the highest values are found among suspension-feeding species, especially those living in dense aggregations. In more local investigations much higher values have been recorded. Dare (1976) gave a series of biomass values for mussel (*Mytilus edulis*) beds in Great Britain with maxima ranging from 400 to 1420 g (dry flesh weight) m^{-2}. Walne (1972) recorded 200 g (dry flesh weight) m^{-2} for *Ostrea edulis* and *Mercenaria mercenaria* in England. Bahr (1976) found 970 g (ash-free dry weight) m^{-2} for *Crassostrea virginica* in Georgia, U.S.A. Dame et al. (1977) recorded 165 g m^{-2} for the same species in South Carolina, U.S.A. Such large biomass values are dependent on a much larger area of estuarine habitat for their food supply, however.

TABLE 6.2

Biomass values for some estuarine macrobenthic animals averaged over large areas; all values expressed as g (ash-free dry weight) m^{-2}

Area investigated	Reference	Species	Feeding type	Biomass
Long Island Sound, U.S.A., subtidal, about 5000 km^2	Sanders (1956)	*Nephtys incisa*	deposit feeder	4.3
		Cistenoides gouldii	deposit feeder	0.9
		Pandora gouldiana	suspension feeder	3.1
		Yoldia limatula	deposit feeder	1.4
Conway Bay, Wales, intertidal, 19 ha	Hughes (1970)	*Scrobicularia plana*	deposit feeder	3.8–40.8
Petpeswick Inlet, Canada, intertidal, about 5 km^2	Burke and Mann (1974)	*Littorina saxatilis*	grazer	0.1–0.8
		Mytilus edulis	suspension feeder	2.7–15.2
		Melampus lineatus	grazer	0.8
		Nassarius obsoletus	grazer	0.9
		Lacuna vincta	grazer	0.0
Tidal flats of the Wadden Sea, The Netherlands, 1300 km^2	Beukema (1976)	*Mytilus edulis*	suspension feeder	6.2
		Arenicola marina	deposit feeder	5.0
		Mya arenaria	suspension feeder	4.6
		Cerastoderma edule	suspension feeder	4.3
		Macoma balthica	deposit feeder	2.2
		Nereis diversicolor	omnivore	1.4
		Lanice conchilega	suspension feeder	0.7
		Scoloplos armiger	deposit feeder	0.4
		Nephtys hombergii	predator?	0.3
		Heteromastus filiformis	deposit feeder	0.3
		Hydrobia ulvae	grazer	0.2
		Carcinus maenas	scavenger/predator	0.2
		Crangon crangon	scavenger/predator	0.1
Grevelingen Estuary, The Netherlands, 140 km^2	Wolff and De Wolf (1977)	*Cerastoderma edule*	suspension feeder	18.7
		Arenicola marina	deposit feeder	1.8
		Macoma balthica	deposit feeder	1.0
		Hydrobia ulvae	grazer	2.2
		Littorina littorea	grazer	0.0

Most species in temperate latitudes show a pronounced seasonal fluctuation of biomass, primarily due to weight changes, with maximum values in spring or summer and minimum values in winter (Fig. 6.7) (Beukema, 1974). The same seasonal trend of biomass values is observed for the total fauna (Fig. 6.8) (Beukema, 1976; Wolff and De Wolf, 1977).

Table 6.3 presents data on the total biomass of all or the most important species in a number of estuaries. Again suspension-feeding species contribute markedly to these values.

Estuarine benthic biomasses appear to reach very high values when compared to other aquatic habitats. This may even result in much higher phytoplankton consumption by the benthic fauna than by the zooplankton (Rosenberg et al., 1977).

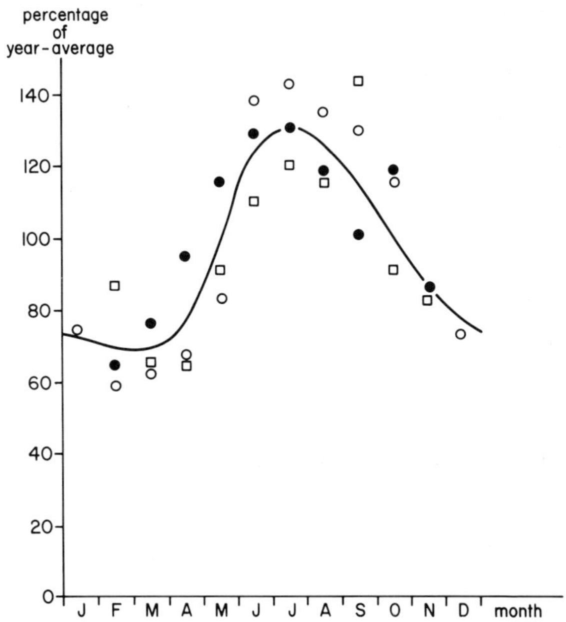

Fig. 6.8. Relative amounts of biomass at three stations at the Balgzand tidal flats, The Netherlands, expressed as mean percentages of year-average of biomass for each month at each station. (After Beukema, 1974.)

PRODUCTION OF BENTHIC ANIMALS

The secondary production of benthic animals varies widely from place to place, but is generally high compared to other ecosystems, possibly because of a high level of food, easy availability of food and a preponderance of opportunistic species characterized by high growth rates and a rapid turnover. Table 6.4 enumerates some production values for estuarine benthic animal species from investigations of large areas. As with the biomass, the highest values are found among suspension feeders. Usually animals grow in spring and summer and lose weight during autumn and winter.

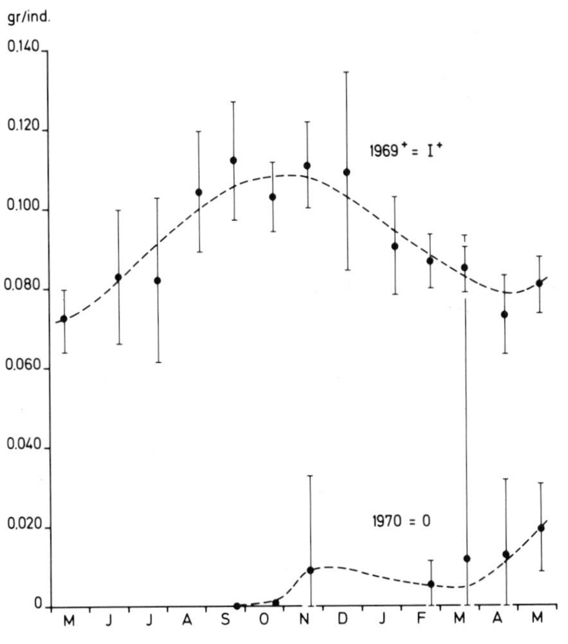

Fig. 6.7. Average weight in grams of class 0 and I (and older-class) individuals of the prosobranch snail *Littorina littorea* in the Grevelingen Estuary, The Netherlands, between May, 1970 and May, 1971. Mean values are shown with their 95% confidence intervals. The curves are fifth-degree polynomials calculated using the mean values, with the reciprocal of the confidence interval as a weighing factor. (After Wolff and De Wolf, 1977.)

Table 6.5 presents production values for all or the most important species of estuarine benthic assemblages, and shows that dense aggregations of suspension-feeding animals have the highest values. Thus, Dare (1976) records annual production values ranging from 210 to 3850 g (dry flesh weight) m^{-2} for beds of mussels (*Mytilus edulis*). Bahr (1976) found about 200 g m^{-2} yr^{-1} for oyster reefs (*Crassostrea virginica*).

The ecological efficiency (net production/food ingested) of benthic animals has been determined a few times. Tenore et al. (1973) recorded 10.01% for *Mytilus edulis*, 18.38% for *Crassostrea virginica* and 23.69% for *Mercenaria mercenaria*. Tenore and Gopalan (1974) found an ecological efficiency of

TABLE 6.3

Biomass values for some estuarine macrobenthic species assemblages

Area	Depth	Biomass ($g\ m^{-2}$)	Reference
Wadden Sea, Denmark	intertidal	174–497 (wet)	Smidt (1951)
Ringkøbing Fjord, Denmark	intertidal	267–450 (wet)	Smidt (1951)
Long Island Sound, U.S.A.	subtidal	54.6 (dry)	Sanders (1956)
Barnstable Harbor, U.S.A.	intertidal	38–40 (dry)	Sanders et al. (1962)
North Carolina Estuary, U.S.A.		8.1 (dry tissue)	Williams and Thomas (1967)
Puget Sound, U.S.A.	12–195 m	8–9 (ash-free)	Lie (1968)
Firemore Bay, Scotland	beach	1.3 (dry)	McIntyre and Eleftheriou (1968)
Firemore Bay, Scotland	subtidal	3.7 (dry)	McIntyre and Eleftheriou (1968)
Biscayne Bay, U.S.A.	intertidal	30 (dry)	Moore et al. (1968)
Narragansett Bay, U.S.A.	intertidal musselbed	1852 (ash-free)	Nixon et al. (1971)
Lower Mystic River, U.S.A.	15 m	1.0–50 (wet)	Rowe et al. (1972)
Balgzand, The Netherlands	intertidal	9.6–25.1 (ash-free)	Beukema (1974)
Kiel Bay, Germany	>15 m	26.3 (ash-free)	Arntz and Brunswig (1975)
Orwell Estuary, England	intertidal	46.0 (ash-free)	Kay and Knights (1975)
Stour Estuary, England	intertidal	28.1–38.7 (ash-free)	Kay and Knights (1975)
Colne Estuary, England	intertidal	7.8 (ash-free)	Kay and Knights (1975)
Dengie Flats, England	intertidal	18.5 (ash-free)	Kay and Knights (1975)
Crouch Estuary, England	intertidal	8.7 (ash-free)	Kay and Knights (1975)
Roach Estuary, England	intertidal	8.7 (ash-free)	Kay and Knights (1975)
Foulness Flats, England	intertidal	19.6 (ash-free)	Kay and Knights (1975)
Thames Estuary, England	intertidal	18.8 (ash-free)	Kay and Knights (1975)
Medway Estuary, England	intertidal	24.1 (ash-free)	Kay and Knights (1975)
Swale Estuary, England	intertidal	32.8 (ash-free)	Kay and Knights (1975)
Tamar Estuary, England	intertidal	13 (ash-free)	Warwick and Price (1975)
Southampton Water, England	intertidal	90–190 (ash-free)	Hibbert (1976)
Wadden Sea, The Netherlands	intertidal	27 (ash-free)	Beukema (1976)
Grevelingen Estuary, The Netherlands	whole estuary	20.8 (ash-free)	Wolff and De Wolf (1977)
Byfjord, Sweden	subtidal	4.5 (dry)	Rosenberg et al. (1977)
Delaware River, U.S.A.	intertidal fresh water	7.0 (dry)	Crumb (1977)
San Francisco Bay, U.S.A.	intertidal	10.1–30.5 (ash-free)	Nichols (1977)

18.0% for the polychaete *Nereis virens* grown on tissue of the clam *Mercenaria mercenaria* and one of 5.7% for the same species grown on biodeposits of oysters (*Crassostrea virginica*). The food chain efficiencies (net production/available food) were found to be lower in these species, suggesting underexploitation of available food. The values for the three bivalves range from 8.22 to 15.12%.

Benthic production in estuaries is quite high when compared to other aquatic habitats because of the relative abundance of food in estuaries, combined with the shallow depths of most estuaries. In such situations food becomes readily available to the bottom-living animals through sinking through shallow water and vertical transport by turbulent water movements caused by waves and tidal currents (Hargrave, 1973; Wolff et al., 1976a; Wolff, 1977). Another possible cause is a predominance of relatively opportunistic species. Muus (1967) pointed out that in Danish brackish lagoons the majority of the species have one or more generations per year, compared to the slower reproducing fauna in more marine environments. This higher turnover of biomass would result in a higher production.

THE PLACE OF ZOOBENTHOS IN THE ESTUARINE FOOD WEB

Few attempts have been made to quantify the place of zoobenthos in the estuarine food web. A major reason is our lack of knowledge of many ecological interrelationships in the estuary, particularly the trophic role of detritus.

In the classical study by Teal (1962) on energy

TABLE 6.4

Production values for some estuarine macrobenthic species found in investigations of large areas; all values are expressed as g (ash-free dry weight) m^{-2} yr^{-1}

Area investigated	Reference	Species	Feeding type	Production
Long Island Sound, U.S.A., subtidal, about 1000 km^2	Sanders (1956)	*Nephtys incisa*	deposit feeder	9.3
		Cistenoides gouldii	deposit feeder	1.7
		Pandora gouldiana	suspension feeder	6.1
		Yoldia limatula	deposit feeder	3.2
Conway Bay, Wales, intertidal, 19 ha	Hughes (1970)	*Scrobicularia plana*	deposit feeder	3–13
Petpeswick Inlet, Canada, intertidal, about 5 km^2	Burke and Mann (1974)	*Mya arenaria*	suspension feeder	11.6
		Macoma balthica	deposit feeder	1.9
		Littorina saxatilis	grazer	3.3
Southampton Water, England, intertidal, 60 ha	Hibbert (1976)	*Cerastoderma edule*	suspension feeder	29–71
		Mercenaria mercenaria	suspension feeder	4–14
Grevelingen Estuary, The Netherlands, intertidal and subtidal, 140 km^2	Wolff and De Wolf (1977)	*Cerastoderma edule*	suspension feeder	10–120
		Macoma balthica	deposit feeder	0.3–7.9
		Arenicola marina	deposit feeder	3.3–6.3
		Hydrobia ulvae	grazer	7.2–12.8
		Littorina littorea	grazer	6.1

TABLE 6.5

Production values for some estuarine macrobenthic assemblages; all values are expressed as g (ash-free dry weight) m^{-2} yr^{-1}, unless otherwise stated

Area	Depth	Production	Reference
Long Island Sound, U.S.A.	subtidal	21.4 (infauna)	Sanders (1956)
Long Island Sound, U.S.A.	subtidal	5.8 (epifauna)	Richards and Riley (1967)
Firemore Bay, Scotland	whole bay	2.5 (g carbon)	McIntyre and Eleftheriou (1968)
Kiel Bight, Germany	>15 m	17.9	Arntz and Brunswig (1975)
Tamar Estuary, England	intertidal	13.3	Warwick and Price (1975)
Southampton Water, England	intertidal	152–225 (?)	Hibbert (1976)
Grevelingen Estuary, The Netherlands	whole estuary	50.3–57.4	Wolff and De Wolf (1977)

flow in an estuarine marsh, about 5.6% of the net production of the salt marsh was consumed by benthic primary consumers. The salt marsh production consists of about 80% by *Spartina alterniflora* and about 20% by benthic microalgae; the ratio in the consumed material was not known. Teal (1962) assumed that about 45% of the primary production was exported from the salt marsh to the adjacent aquatic ecosystems. It has been stated that this export is responsible for the abundance of benthic animals in the tidal channels between the marshes. This belief is supported by the excellent work of Odum and Heald (1972, 1975) on the export and consumption of mangrove detritus in Florida. They found that a whole series of estuarine animals feed on decaying leaves of mangroves. Recent work by Haines (1977) on the $^{12}C/^{13}C$ ratio of detritus in the estuaries of Georgia, U.S.A., has questioned the detritus export postulated by Teal (1962). It appears from Haines' data that the estuarine detritus she studied cannot be derived from the salt marshes. It has also been found for other salt marshes that these do not export organic matter (Moll, 1977; Woodwell et al., 1977; Wolff et al., 1979).

Postma and De Jonge have postulated that there is a considerable import of organic material into the Dutch part of the Wadden Sea from the adjacent North Sea, and that this import trebled in the period from 1950 to 1970 due to the eutrophication of the North Sea (Postma, 1954, 1961; De Jonge and Postma, 1974). At present this import of organic material (detritus and marine phytoplankton?) is about 2.5 to 3 times larger than the primary production within the estuary. This is supported by Cadée and Hegeman (1977), who found an increase of organic carbon on the high tidal flats of the same area which could be only partly explained by primary production of microphytobenthos on the flats. Nevertheless there are no indications that the benthic consumers increased by the same factor during the increase recorded by De Jonge and Postma (1974). This observation is in agreement with the conclusion of Beukema et al. (1977) that the production of the bivalve *Macoma balthica* probably is limited by the availability of benthic microalgae and not by other organic material which is also available.

Wolff (1977) published a survey of food sources for the benthic macrofauna of the salt-water lake formed by damming the Grevelingen Estuary, The Netherlands. This survey is presented in a revised form in Table 6.6. If all sources in this budget deliver high-quality food, the ecological efficiency of the Grevelingen benthos would be between at least 6.8 and 9.6%. If detritus is not available as food these figures would be 16.2 to 19.5%. These values are obtained by neglecting any consumption by zooplankton, meiobenthos or mullets (Wolff and De Wolf, 1977; Wolff, 1977). Since these efficiencies seem high compared to the values found experimentally by Tenore et al. (1973), and Tenore and Gopalan (1974), it is tentatively assumed that at least part of the benthos is able to utilize detritus.

Van Es (1977) gave a carbon budget for a brackish embayment of the Ems Estuary, The Netherlands (Table 6.7). Detritus appears to be the chief source of organic carbon in this estuary. Since the zoobenthos production in this estuary is not very high (M.A. van Arkel, pers. comm., 1978), it seems that here detritus is not very important as a food source. Burke and Mann (1974) computed a transfer efficiency of 4.7% for the food chain from macrophytes to molluscs in a small Canadian estuary, but the number of pathways for food that were not considered in their study is too large to make their figure meaningful.

All available evidence suggests that land- or salt-

TABLE 6.6

Benthic food sources in the Grevelingen Estuary, The Netherlands; revised after Wolff (1977); expressed as g (organic carbon) m^{-2} yr^{-1}

Food source	Amount
Net import of detritus from salt marshes	0
Production of detritus by seagrass beds	5–30
Net import of detritus from North Sea	155–225
Detritus from land runoff	2
Primary production, benthic microalgae	25–57
Primary production, phytoplankton	130
Total	317–444

TABLE 6.7

Carbon sources in the Dollard embayment in the Ems Estuary, The Netherlands, expressed as 10^6 kg (organic carbon) yr^{-1}; after Van Es (1977)

Food sources	Amount
Net import of detritus from salt marshes	0.5
Net import of detritus from North Sea and River Ems	max. 37.1
Import of detritus from River Westerwoldse Aa	33.0
Primary production, benthic microalgae	9.3
Primary production, phytoplankton	0.7
Maximum	80.6

marsh-derived detritus has a less important role as a food source for estuarine benthos than was assumed earlier. Instead, primary production *in situ* and the import of organic matter from the sea seem to be the main food sources.

Benthic animals are consumed by benthic predators (starfish, crabs), fishes, birds and by man. Klein Breteler (1976) estimated the consumption by the crab *Carcinus maenas* at minimally 1.5 g (ash-free dry weight) m^{-2} yr^{-1}, constituting about 5% of the annual production of the benthic macrofauna. Van der Schoot (1974) found that starfish *Asterias rubens* in the salt-water lake formed by damming the Grevelingen Estuary maximally consumed some 0.3 to 1.8 g (ash-free dry weight) m^{-2} yr^{-1} of mussels *Mytilus edulis* and other molluscs. Anger et al. (1977) calculated that in summer the same species took 0.0235 g (ash-free dry weight) m^{-2} day^{-1} in Kiel Bay in the western Baltic, a value resulting in a predation of possibly 2 g (ash-free dry weight) m^{-2} yr^{-1} or more, when the lower temperatures during the rest of the year are taken into consideration.

The consumption of benthic animals on the Balgzand tidal flats in the Dutch Wadden Sea by plaice (*Pleuronectes platessa*) varied between 1.7 and 5.0 g (ash-free dry weight) m^{-2} yr^{-1} in four consecutive years (Kuipers, 1977). Gobiid fishes (*Pomatoschistus minutus* and *P. microps*) may consume some 1.4 g m^{-2} yr^{-1} in the same area (Van Beek, 1976).

Nilsson (1969) recorded that diving ducks in the coastal zone (less than 3 m depth) of the western Baltic, Sweden, consumed about 3% of the benthic biomass that had been present in November during the following winter months. In the same period another 30 to 70% of the biomass disappeared due to other unknown causes. Hibbert (1976) estimated that, from an intertidal area in Southampton Water, England, birds take about 37% of the bivalve production, mainly *Cerastoderma edule*. Hulscher (1975) and Swennen (1976) have estimated the consumption by carnivorous birds in the Dutch Wadden Sea. Their data show that birds consume about 3.3 to 4.3 g (ash-free dry weight) m^{-2} yr^{-1} of benthic animals. This is an average value for the whole Dutch Wadden Sea (2600 km^2). Wolff et al. (1976b) made a similar calculation for the Grevelingen Estuary. They estimated that birds ate about 3.4 g (ash-free dry weight) m^{-2} yr^{-1} of benthic animals, constituting about 6% of the total production of the macrobenthos in this area. Birds cannot be neglected as predators of estuarine benthos.

Finally man fishes for a number of benthic organisms in estuaries — for instance, clams, mussels, oysters, cockles, and other shellfish, some crabs and shrimps, as well as some worms for bait.

CHARACTERISTICS OF THE ESTUARINE BENTHIC ECOSYSTEM

Remane's species–salinity relationship

In 1934 Remane drew attention to what then was called the "paradox of brackish water". He observed that the largest number of species occurs at intermediate values of most ecological factors, but that with salinity the number of species shows two peaks: one in fresh water and another in marine salinities (Fig. 6.9). Remane's species–salinity re-

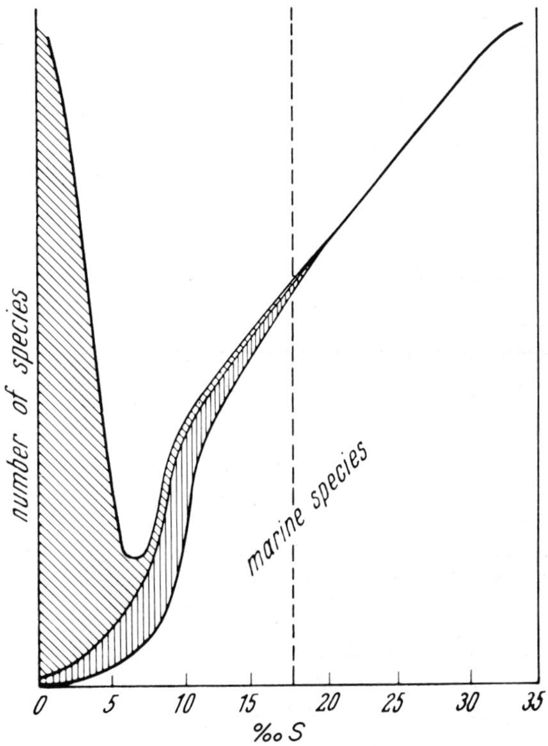

Fig. 6.9. Number of species in relation to salinity. Obliquely hatched: fresh-water species. Vertically hatched: brackish-water species. White: marine species. (After Remane, 1971.)

lationship was established for Baltic waters. Since then several authors have confirmed and improved on Remane's concept (Brattström, 1941; Remane, 1941, 1958, 1971; Dahl, 1956), but the relationship remains essentially unchanged. The same relationship has been found for brackish lagoons in Denmark (Remane, 1934; Muus, 1967), the Tay and the Tees estuaries in Great Britain (Alexander et al., 1935), the estuaries and brackish waters of the Delta area in The Netherlands (Den Hartog, 1961; Heerebout, 1970; Wolff, 1973), and for estuaries in Virginia (Boesch et al., 1976). Indications for it have been presented for Delaware Bay, U.S.A. (Maurer and Watling, 1973), for the Cananeia Estuary, Brazil (Tommasi, 1970), and for Morrumbene Estuary, Moçambique (Day, 1974). Dörjes and Howard (1975), on the other hand, recorded a continuous decrease of the number of species when going from high salinities to tidal fresh water in an estuary in Georgia, U.S.A.

A different situation has been described by Bačescu (1954), who, in a study of mysids and cumaceans along the shores of the Black Sea, found two peaks: one at the full salinity of the Black Sea (about 20‰) and another at about 2‰ (Fig. 6.10). The latter peak is brought about by representatives of the Ponto-Caspian relict fauna. The question may be raised whether the species minimum may be correlated with low salinity values, or with salinity

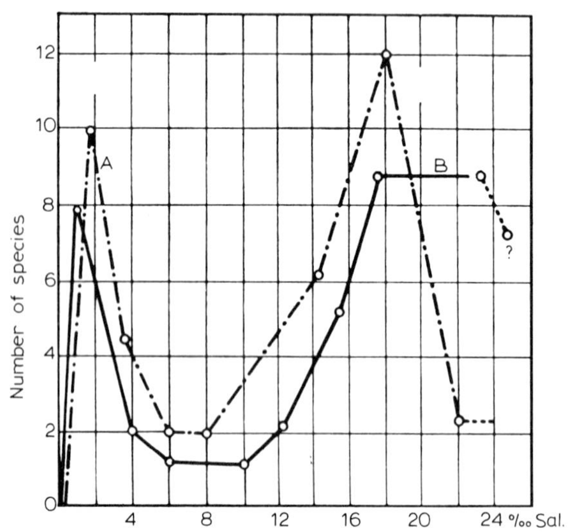

Fig. 6.10. Number of species among Mysidacea (A) and Cumacea (B) in the Black Sea in relation to salinity. (After Remane, 1971.)

fluctuations. Remane (1958, 1971) believed the first explanation to be true, since the area of the Baltic with the largest salinity fluctuations contained many more species than did areas with lower salinities. He quoted similar data for the Black Sea as well. In tidal estuaries, however, where the largest salinity fluctuations usually occur in the middle of the estuarine salinity gradient, the species minimum contains fewer species than are found in the more stable Baltic salinity gradient. The main reason for this seems to be that marine as well as fresh-water species do not penetrate as far into the very unstable brackish water of most estuaries as they do into the stable brackish Baltic Sea (Johansen, 1918; Alexander et al., 1935; Den Hartog, 1960, 1964; Wolff, 1973). Nevertheless, brackish waters, including those with a limited salinity fluctuation such as the central Baltic Sea, are poorer in species than are either marine waters or fresh water.

Diversity of estuarine benthos

Sanders (1968) established that the diversity of estuarine benthos, as measured by means of his rarefaction technique, is less than that of marine benthos (Fig. 6.11). Boesch (1972) observed great diversity on the outer shelf and in the high salinity section of Chesapeake Bay, lower values on the shallow shelf and in the middle parts of estuaries, and the lowest in the low-salinity parts of estuaries. Similarly, Wolff (1973) found in the estuaries of the Dutch Delta area that diversity, measured by various techniques, was large in high salinities in the North Sea and in the estuaries, lower in the surf zone along the North Sea and in stable brackish water and fresh water, and lowest in strongly fluctuating brackish water (Fig. 6.12). Maurer and Watling (1973) found that the diversity of the epifauna of oyster beds in Delaware Bay, U.S.A., decreased with decreasing salinity.

Beukema (1976) measured diversity of intertidal benthos in the Dutch Wadden Sea by number of species per area, and found, as might be expected, a strong correlation between species diversity and biomass. The highest diversities were observed in the middle part of the intertidal zone, and low values coincided with extreme environmental conditions.

Diversity of estuarine benthos, however, is not

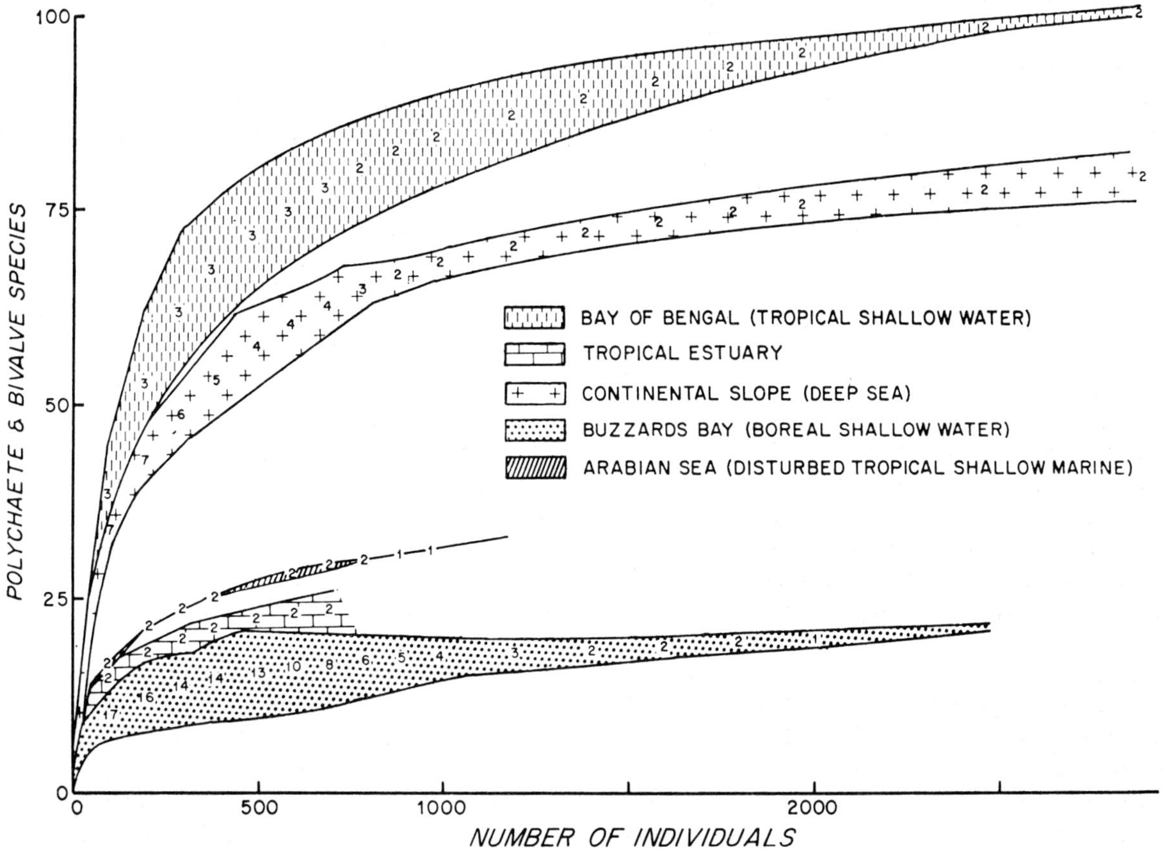

Fig. 6.11. Number of benthic polychaete and bivalve species in relation to the number of individuals in various marine environments. (After Sanders, 1968.) Shown are the envelopes for all curves from the areas considered, the number of curves being indicated in the centre of the envelopes.

exclusively determined by abiotic factors. Young et al. (1976), as well as Reise (1977a, b), have demonstrated by means of exclusion experiments that the absence of predators results in a higher number of species.

Possible causes for the low number of species in brackish waters

As the species minimum apparently occurs in all brackish waters, pollution and turbidity may be ruled out as causes (Wolff, 1973). An alternative explanation has been proposed by Chislenko (1964), who stated that marine organisms have difficulties in passing into brackish and fresh water because there they must adapt to life in relatively small areas and, consequently, in relatively small numbers. This would explain the relative paucity of the brackish-water fauna. Although this difficulty indeed may exist, three objections can be made to this theory. Firstly, brackish seas like the Baltic, the Black Sea, and the Caspian Sea, are also poor in species, despite their large size. Secondly, the same difficulty might exist for the fresh-water fauna, but nevertheless the number of fresh-water species is large. Finally, Chislenko's hypothesis does not explain why brackish waters have not been colonized by fresh-water species.

Khlebovich (1968, 1969) demonstrated that at a salinity of about 5 to 8‰ the relative ionic composition of nearly all brackish waters changes considerably. This salinity range therefore would constitute an important boundary between marine and fresh-water faunas, and Khlebovich (1969) ascribed the poverty of species of brackish waters to the presence of this boundary. However, since a number of species are able to live on either side of this boundary, it is not clear why many other

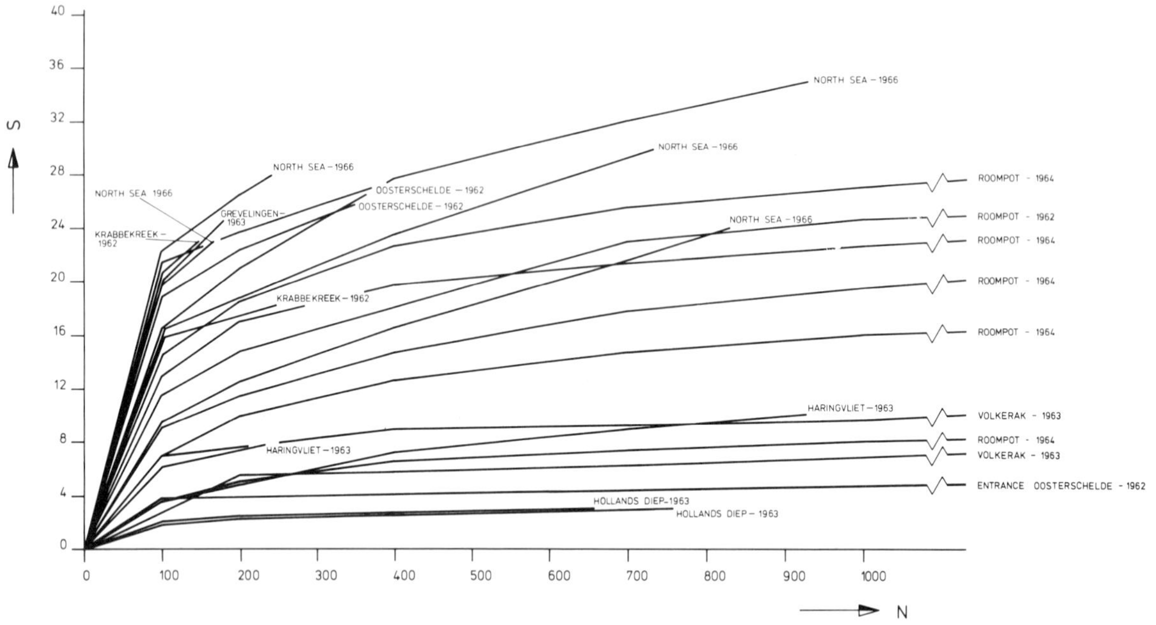

Fig. 6.12. Number of benthic macrofauna species (S) in relation to number of individuals (N) for various estuaries in the southwest Netherlands and the adjacent North Sea. Oosterschelde, Grevelingen and Krabbekreek are sheltered high-salinity areas, Roompot and Entrance Oosterschelde are high-salinity areas exposed to North Sea waves, Haringvliet and Volkerak are brackish areas with fluctuating salinities, Hollands Diep is a fresh-water tidal area with irregular intrusions of brackish water. (After Wolff, 1973.)

species may not do the same. In addition, this boundary does not explain why the number of species is already reduced between 10 and 20‰ salinity.

Sanders (1968) developed a theory based on the stability or predictability of the environment and on its geological history to explain differences in species diversity of benthos. This stability–time theory combines suggestions of several earlier authors who each considered only one aspect (Hedgpeth, 1953; Dahl, 1956, 1959; Klopfer, 1959; Fischer, 1960; Dunbar, 1960; Connell and Orias, 1964; Simpson, 1964; Muus, 1967). The main elements of this theory as applied to brackish waters and estuaries are:

(1) In brackish waters and estuaries the environmental factors usually show an unstable and unpredictable behavior.

(2) Speciation in such an environment is less probable than in more stable environments (Slobodkin and Sanders, 1969).

(3) Extinction in such an environment is more probable than in more stable environments (Slobodkin and Sanders, 1969).

(4) Brackish waters and estuaries are geologically ephemeral phenomena, thus increasing the likelihood of extinction of brackish-water populations.

(5) For these reasons in brackish waters and estuaries the number of species adapted to the environment is lower than in either the marine or the fresh-water environment.

(6) The unstable and unpredictable nature of brackish estuaries also excludes most marine and fresh-water species.

Sanders' theory thus provides an explanation for the low number of species in all kinds of brackish waters, even the very stable ones such as the Baltic and the Black Sea. Gessner and Schramm (1971) indeed ascribed the poverty of benthic algae in the Baltic area to the geological history of this sea. Maurer and Watling (1973) used Sanders' theory to explain the pattern of diversity of the epifauna of oyster beds in Delaware Bay, U.S.A. Wolff (1971, 1973) concluded that Sanders' theory is in agreement with the geological history of northwest European brackish waters. The geological history of other brackish-water faunas also seems to be in agreement with Sanders' theory.

Species packing in estuarine brackish water

Estuarine brackish waters are comparable to the sea or fresh water in many respects, such as sedimentation, water movement, aeration and food supply. However, species packing in estuarine brackish waters may be different from that in the sea or in fresh water since the number of species is much lower in brackish waters. Estuaries may differ either by larger niches, by less overlap of niches or by the occurrence of empty places ("empty niches") in the environment.

No instances are known of reduced overlap of niches in estuarine benthos, although Heip and Engels (1977) found insignificant niche overlap for some brackish-water copepods. On the other hand there are several indications for larger niches. Wolff (1973) found that 35% of all species in estuarine brackish waters could be considered as omnivores, compared with 6 to 16% in marine and fresh waters (see Table 6.2). Many species which are intertidal in the North Sea are able to live down to the greatest depths in the Baltic (Remane, 1971). In the Delta area of The Netherlands the polychaete *Nereis diversicolor*, which at high salinities lives only intertidally, is able to expand to subtidal habitats in low salinities (Fig. 6.13). These habitat changes may be considered as extensions of the realized niche (in the sense of Hutchinson, 1960) in the absence of competitors.

The occurrence of unexploited places in the environment ("empty niches") is more difficult to prove. It might even be stated that an empty niche is theoretically impossible. There have been a few observations with regard to this problem. Until recently European brackish waters did not harbor any species of crabs. Due to human transport the species *Rhithropanopeus harrisii*, *Callinectes sapidus* and *Eriocheir sinensis* were able to settle there. This might be seen as an indication that the estuarine environment previously had one or more empty niches. Another indication is the high number of alien species in brackish-water faunas. For tidal and non-tidal brackish waters in The Netherlands, Wolff (1973) found that at least 20% and 28%, respectively, of the species did not belong to the European fauna. These figures may be compared to those for the marine area (2%), for the fresh-water tidal area (8%) and for rivers (5%) of the same estuaries. Filice (1958) also found a high percentage (15%) of introduced species in the brackish sections of San Francisco Bay, U.S.A. Because the total number of brackish-water species in the world is rather low, these data suggest that the brackish-water environment might be "undersaturated" with respect to its number of species.

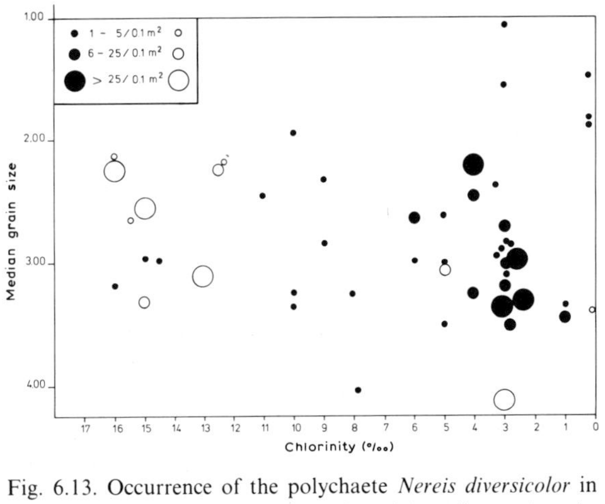

Fig. 6.13. Occurrence of the polychaete *Nereis diversicolor* in relation to tidal level, median grain size of the sediment (φ units $= -\log_2$ of the ratio between the diameter of the grains in mm and a standard grain diameter of 1 mm) and chlorinity (‰). Dots denote subtidal finds, circles intertidal ones. (After Wolff, 1973.)

Competition in estuarine brackish water

Competition has been defined as "any interaction between two or more species populations which adversely affects their growth and survival" (Odum, 1971). This broad definition also includes predator–prey relationships. In fact, most known examples of competition in estuaries and brackish waters are of this kind, obviously because it is most straightforward in its effects. There is the often-cited case of spat of the bivalve *Macoma balthica* being swallowed by the deposit-feeding amphipod *Pontoporeia affinis* in the Baltic (Segerstråle, 1962, 1969). This leads to the extinction of entire year classes of *Macoma*. Larvae of benthic animals seem to be destroyed in large numbers by adult benthic suspension feeders (Mileikovsky, 1971). Wolff (1973) demonstrated in a simple experiment a predator–prey type of competition between *Nereis virens* and *N. diversicolor*. Similarly, Reise (1977a, b) demonstrated in exclusion experiments that several bivalves of the coastal sea were driven to extinction by benthic predators in the estuarine Wadden Sea, Germany. More subtle is the competition between the amphipod *Gammarus duebeni* and several other estuarine species of *Gammarus*. Kinne (1959) showed experimentally that the reproductive rate of *G. duebeni* is much lower than in other species. Furthermore, females of *G. duebeni* may be prevented from taking part in reproduction because males of other species may enter into precopula with *G. duebeni* females, and possibly even mate with them. Thus, many estuarine distribution patterns are ascribed to competition, often of the predator–prey type (e.g. Segerstråle, 1969), but experimental proof is lacking in most cases. Nevertheless, the conclusion that many euryhaline brackish-water species are absent from marine habitats because of superior marine competitors seems logical.

It has not yet been quantified that competition is less severe in estuarine environments than in marine or fresh-water habitats, but it certainly exists. Beukema (1976), however, states that the biomass of intertidal estuarine benthos in the Dutch Wadden Sea is restricted not by biological factors, but by the physical and chemical instability of the environment.

Resistance to stress

It has been maintained for some time that estuaries are vulnerable ecosystems (see for instance W.E. Odum, 1970; E.P. Odum, 1971). This belief has been contested by Copeland (1970), Boesch (1972, 1974) and Jernelöv and Rosenberg (1976), who state that estuarine systems have a high tolerance to stress. Their studies of stress tolerance in estuaries showed that the species already living under stress, such as that of reduced and changing salinities, were more tolerant to pollution than were species in a less fluctuating environment. They attribute this to the low degree of specialization and the high genetic diversity of estuarine species (Grassle and Grassle, 1974). Boesch et al. (1976) describe how drastic salinity reduction in the York and James estuaries, Virginia, U.S.A., had great effects on the zoobenthos of the normally high-salinity parts of the estuary, where many species were eliminated and others greatly reduced. The zoobenthos in the low-salinity parts of the same estuaries, however, were hardly affected by the salinity reductions.

Succession of estuarine benthos

The development of estuarine benthic assemblages after some drastic change in the environment has been studied for several cases (Reish, 1961, 1962, 1963; Dean and Haskin, 1964; Rosenberg, 1972, 1973a, b, 1976; Grassle and Grassle, 1974; Wolff, 1974; Wolff et al., 1977). In a completely abiotic situation the first colonizers are usually opportunistic species such as *Capitella capitata* and *Polydora ligni*. These species are rapidly followed by a series of other organisms (Fig. 6.14) and a

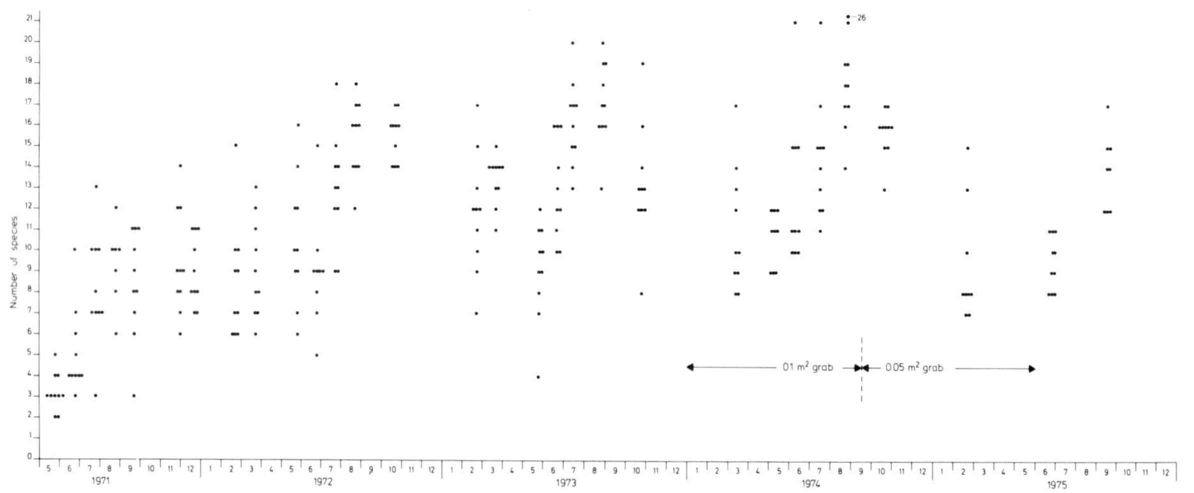

Fig. 6.14. Increase of the number of species per sample after an uninhabited sediment in the saline Lake Grevelingen became available for colonization in May 1971. Each dot represents one sample. (After Wolff et al., 1977.)

sudden increase of biomass usually occurs (Fig. 6.15). After 3 to 8 years the newly colonized areas cannot be distinguished from other areas.

It is still an open question whether there is really any succession in estuaries — a sequence of species made possible because one species paves the way for the next. Benthic succession might merely be a reflection of habitat changes and of differences of abilities for dispersal and colonization (Reish, 1963; Drury and Nisbet, 1973; Wolff et al., 1977). Indeed, after pollution abatement in a fjord-like estuary in Sweden, the order in which the benthic species appeared was the reverse of that in which they disappeared from a similar estuary in Scotland after the introduction of pollution. For instance, *Capitella capitata* appeared first in Sweden and disappeared last in Scotland (Rosenberg, 1973a, b; Pearson, 1975; Pearson and Rosenberg, 1976).

BIOLOGICAL SUBDIVISIONS OF ESTUARIES

Den Hartog (1964) stressed the instability of salinity as the main characteristic of brackish water. Rochford (1951), Dahl (1956) and Caspers (1967) maintained that the short-term fluctuations of hydrographic factors caused by the tides form a very distinct character of estuaries as opposed to other brackish waters, and proposed to reserve the term "estuarine" for brackish water that is significantly influenced by tides.

The seaward limit of estuaries is not indicated by the disappearance of estuarine species — they have already vanished earlier — but by the appearance of marine species. Probably some of these species do not inhabit the estuary because of temperature conditions.

Den Hartog (1961, 1968, 1971) considered the riverward limit of the estuarine environment — that is, the boundary between the tidal fresh water and the non-tidal river. He concluded that this limit is not accompanied by any biological change, a view confirmed by Wolff (1973) for soft-bottom benthos. Thus, it may be concluded that neither the seaward nor the landward limit of estuaries is defined biologically.

Several subdivisions of brackish water that are based on the patterns of distribution of flora and fauna along the salinity gradient from sea to fresh water have been proposed. Redeke's (1922) proposal, elaborated by Välikangas (1926), was widely used until about 1958 (Table 6.8). In that year another classification ("the Venice system") was suggested at an international symposium on the classification of brackish waters (Table 6.8). This system has been criticized as well as refined (Den Hartog, 1961; Kinne, 1971). Den Hartog (1971) distinguished no less than seven sub-areas in the area between the North Sea and the non-tidal rivers of the Delta area of The Netherlands.

Wolff (1973) questioned the basic assumptions for these subdivisions. These are (1) that species are

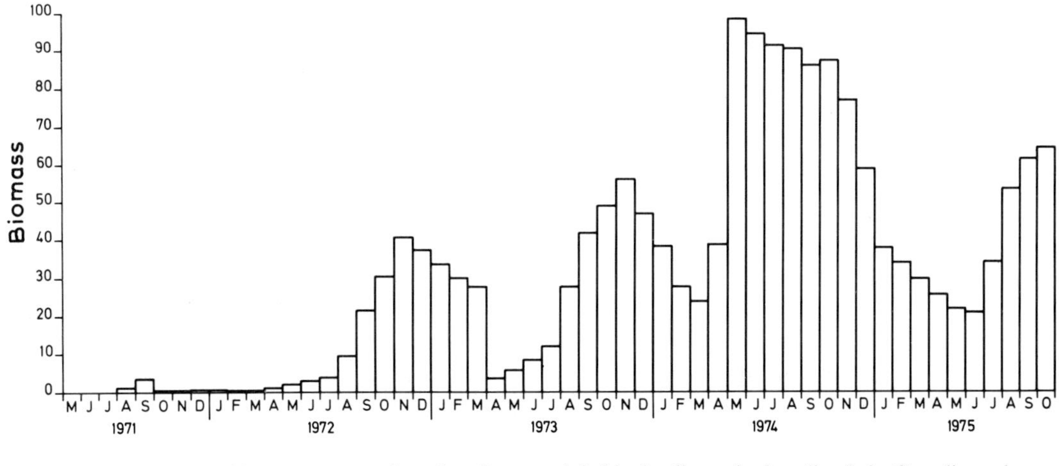

Fig. 6.15. Increase of the biomass at a sampling plot after an uninhabited sediment in the saline Lake Grevelingen became available for colonization in May 1971. Expressed as g (ash-free dry weight) m^{-2}. (After Wolff et al., 1977.)

TABLE 6.8

Brackish-water classifications

Redeke–Välikangas (‰ Cl⁻)	Water type	Venice system	
		‰ salinity	equivalent ‰ Cl⁻
<0.1	fresh water	<0.5	<0.3
0.1–1.0	oligohaline	0.5–5	0.3–3
1.0–10.0	mesohaline	5–18	3–10
10.0–17.0	polyhaline	18–30	10–17
	euhaline	>30	>17 (adjacent euhaline sea)
>17.0	sea water	30–40	17–22
	hyperhaline	>40	>22

The system by Redeke (1922) and Välikangas (1926) has been defined in ‰ Cl^-, the Venice system (Symposium on the Classification of Brackish Waters, Venice, 1958. *Arch. Oceanogr. Limnol.*, XI) in ‰ salinity. For comparison the approximate Cl^- ‰ values are given as well as for the Venice system (see also Table 4.2, p. 68).

restricted to certain salinity intervals; and (2) that groups of species have their limits at about the same salinity, an assumption that is especially difficult to justify. All the boundaries proposed by Den Hartog (1971) for the estuaries of the Delta area coincided with irregularities in the hydrographic pattern, such as shallow sills and confluences of two estuaries. Wolff (1973) therefore concluded that, along the estuarine salinity gradient, the benthic species disappear one by one and that any valid classification must depend on the hydrographical structure of an estuary. Boesch (1977) arrived at a similar conclusion by a mathematical classification technique applied to benthos samples from the James River, Virginia, U.S.A. (Fig. 6.16). He found that different species appeared at different locations and salinities in the estuary. Groups of species showing the same type of distribution could not be identified. Thus it might be better to avoid the terms of the Venice system and to use neutral terms, such as upper, middle and lower reaches for estuaries.

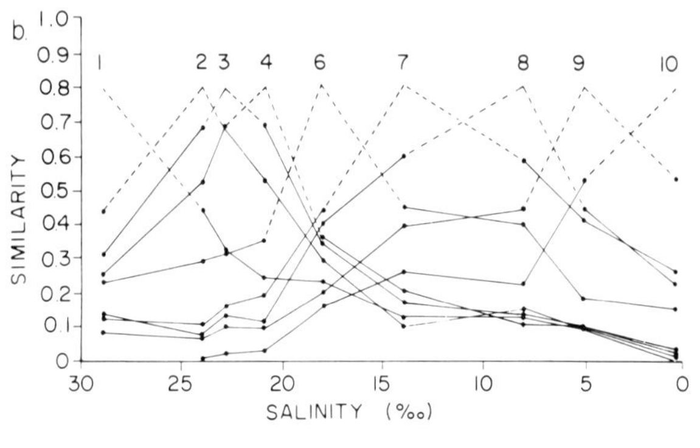

Fig. 6.16. Similarity of benthic macrofauna samples taken along the salinity gradient of Chesapeake Bay and York River, U.S.A. The hypothetical intra-site similarity is set at 0.8. (After Boesch, 1977.)

EXPLOITATION OF ESTUARINE BENTHOS

Human exploitation of estuarine benthos occurs either as fisheries or as cultivation of selected organisms. The main estuarine harvest consists of molluscs and crustaceans, which are captured or grown for human consumption. Polychaete worms are fished or dug as bait for sport fishermen. Various other organisms are fished on a minor scale for shell collectors and the souvenir industry.

The majority of fisheries for benthic organisms, as well as nearly all culture of benthic organisms, occur in estuaries. Benthic animals, especially benthic filter feeders, are relatively large animals occupying a position low in the food chain, which ensures maximum profit from primary food sources. Estuaries also are generally rich in food. In shallow waters this is readily available to benthic organisms through the mixing effects of the estuarine currents, which supports the rapid growth of large numbers of individuals in relatively small areas. An estuarine shellfish bed or aquaculture is not dependent solely upon the area where the animals occur, but instead draws upon food supplied in the water from a much larger area, brought into the estuary by the currents. Finally, estuaries in general are easier to work in than the open sea because of their sheltered nature. Boats can be smaller and less expensive, and permanent structures necessary for culture can be fairly simple.

Reliable figures for the production by estuarine fisheries are difficult to obtain for several reasons. Many authors do not indicate whether their figures relate to a restricted part of the estuary, or to the whole estuary. When the yield of a fishery for mussels is presented as the yield of the area of the mussel beds, an unrealistically high figure results since this mussel bed can exist only in a much larger estuarine area. Also, a number of fisheries may occur within the estuary itself as well as just outside the tidal inlets, and fishermen do not discriminate between their catches in both areas. Some data on estuarine fisheries include real fisheries as well as aquaculture, which is comparable to including sheep and cattle in the list of hares and partridges shot in a meadow. Walne (1972) records the yield of shellfish in some United Kingdom estuaries with figures varying between 32 and 7839 g (wet weight) m^{-2} yr^{-1}, which illustrates some of these problems.

Important estuarine fisheries in Europe are concentrated on the shrimp *Crangon crangon*, the cockle *Cerastoderma edule*, and the mussel *Mytilus edulis*. The latter species is also important as a cultured organism. Locally and on a smaller scale other species are or have been fished. This applies to the quahog *Mercenaria mercenaria*, the soft-shelled clam *Mya arenaria*, the periwinkle *Littorina littorea*, the whelk *Buccinum undatum*, and the blue crab *Callinectes sapidus*. *Mercenaria* and *Callinectes* are American imports. In a few places oysters are harvested as well. Bait worms gathered are *Nereis virens* and *Arenicola marina*, the latter even by hydraulic suction dredges. Postuma and Rauck (1978) present data for the Wadden Sea, a large estuarine area in The Netherlands, Germany and Denmark. Fisheries in the Dutch part yielded annually 1500 to 16 000 tons of shrimps (*Crangon crangon*), at least 8400 tons of cockles (*Cerastoderma edule*) and small amounts of other shellfish in the period 1960 to 1973. In the German part (4400 km^2) the annual yields in the same period were 22 000 to 41 000 tons of shrimps, 1000 to 6000 tons of cockles and as much as 5600 tons of soft-shelled clams (*Mya arenaria*). The benthic fishery yield of this area, excluding mussel culture, is thus about 3 to 12 g (wet weight) m^{-2} yr^{-1}.

American estuarine fisheries concentrate on blue crabs *Callinectes sapidus*, oysters *Crassostrea virginica*, quahog *Mercenaria mercenaria*, soft-shelled clams *Mya arenaria* and shrimp (*Penaeus aztecus*, *P. setiferus*, *P. duorarum*). Locally and on a smaller scale several other species are fished for as well. Bait worms collected are *Nereis virens* and *Glycera dibranchiata*. St. Amant (1973) gives results for the estuarine-dependent fisheries of the Atlantic and Gulf coasts of the U.S.A. and arrives at an average yield of about 4 g (wet weight) m^{-2} yr^{-1} for the period 1965 to 1970. Subtraction of cultured shellfish, and crustaceans caught outside the estuaries, results in a yield of about 2 g (wet weight) m^{-2} yr^{-1}. These figures are based on the total surface area of the estuaries considered.

Crustaceans are not farmed in estuaries, although storage basins sometimes are called farms. The existing shrimp cultures are highly industrial and are not dependent on estuaries once they have started.

Estuarine shellfish culture is an ancient practice. Along the Adriatic the tradition goes back to Roman times, and the Japanese are believed to have started in the seventeenth century. Along the Atlantic coasts of Europe shellfish culture began in France in 1858 with the work of Coste (see Korringa, 1976), who laid the base for the present culture of the European flat oyster (*Ostrea edulis*). Before that time the natural oyster beds were open to a free fishery which nearly everywhere seriously depleted the stocks. Some stocks became extinct, in the Firth of Forth and the Wadden Sea (Korringa, 1976). Coste developed a method to collect the oyster larvae and to grow the young oysters to adult size. One of his main recommendations was the closing of certain areas to the fisheries and leasing these areas to oyster growers, advice that resulted in great success. In the Oosterschelde Estuary, The Netherlands, the free fisheries harvested some 500 000 flat oysters annually, but the culture started in 1870 was able to produce 30 million per year. Similar results were obtained in France, and later the Portuguese oyster *Crassostrea angulata* was cultivated also.

North American and Australian oyster culture was introduced by European emigrants who had learned the methods in their native countries. The species cultured in North America are *Crassostrea virginica*, mainly along the Atlantic and Gulf coasts, and *Ostrea lurida* and *Crassostrea gigas* along the Pacific Coast. In Australia the rock oyster *Crassostrea commercialis* is cultivated, and in India and South Africa respectively *C. gryphoides* and *C. margaritacea* are grown. *C. gigas*, the Japanese oyster, was originally grown only in Japan, but nowadays this fast-growing species is also cultured in North America and Europe. The harvest of *C. virginica* in Long Island Sound, U.S.A., has reached maxima of from 500 to 1000 million oysters per year. The total U.S.A. production of this species was reported to reach more than 80 000 tons of oyster meat in 1890. The production of the Japanese *C. gigas* in Japan is estimated at 40 000 tons of oyster meat (Korringa, 1976).

Mussels are cultured as well, especially in Europe. The *rias* of Galicia, Spain, produce annually more than 100 000 tons of edible mussels (Korringa, 1971), and the Wadden Sea (The Netherlands and Germany) produces up to 80 000 tons annually (Postuma and Rauck, 1978). Mussel cultures are also found in other European countries, as well as in the Philippines (Korringa, 1976).

The evolution of shellfish culture development has been summarized by Korringa (1970, 1976):

(1) Switching over from a free fishery to a system with private grounds, marked delimitation and adequate surveillance.

(2) Promoting spat production by offering surfaces for attachment in suitable places, if spat production is the limiting factor.

(3) Transferring of young shellfish from the site where they settled to places where growth is fast.

(4) Control of predators, parasites and diseases whenever necessary and feasible.

(5) Fattening on special plots for one season prior to marketing.

(6) A system of storage basins and adequate sanitary control to guarantee a gradual and uninterrupted supply to the market.

For several species of shellfish the knowledge and technology are available for the complete biological cycle, but this knowledge is not used, mainly for economic reasons. Instead all present aquaculture of shellfish is a kind of semiculture, relying on nature for important parts of the life cycle of the organism in question. In the most developed cultures, such as those of *Ostrea edulis*, only the larvae are not controlled; once the shellfish grower has caught them, all following life stages are subject to culture measures. In less developed cultures the young individuals are fished either on special plots or in a free fishery once they have settled and grown to a certain size. This is the case in the European mussel culture, for example, which is based on spat at least half a year old. The most "primitive" cultures take only shellfish of marketable size, which are transferred to special plots to obtain a special flavor or color (Korringa, 1976). Adults are also transferred from polluted areas to clean water to enable the meat to be purified from faecal bacteria.

REFERENCES

Alexander, W.B., Southgate, B.A. and Bassindale, R., 1935. Survey of the River Tees. II. The estuary: chemical and biological. *D.S.I.R. Water Pollut. Res., Tech. Pap.*, No. 5: 1–171.

Anger, K., Rogal, U., Schriever, G. and Valentin, C., 1977. In-situ investigations on the echinoderm *Asterias rubens* as a predator of soft-bottom communities in the western Baltic Sea. *Helgol. Wiss. Meeresunters.*, 29: 439–459.

Arntz, W.E. and Brunswig, D., 1975. An approach to estimating the production of macrobenthos and demersal fish in a western Baltic *Abra alba* community. *Merentutkimuslaitoksen Julk. Havsforskingsinst. Skr.*, 235: 195–205.

Băcescu, M., 1954. Mysidacea (Fauna Republicii Popularc Romane). *Acad. Rep. Rom. Crustacea*, 4(3): 126 pp.

Bahr, L.M., 1976. Energetic aspects of the intertidal oyster reef community at Sapelo Island, Georgia (USA). *Ecology*, 57: 121–131.

Bass, N.R. and Brafield, A.E., 1972. The life-cycle of the polychaete *Nereis virens*. *J. Mar. biol. Assoc. U.K.*, 52: 701–726.

Beukema, J.J., 1973. Migration and secondary spatfall of *Macoma balthica* (L.) in the western part of the Wadden Sea. *Neth. J. Zool.*, 23: 356–357.

Beukema, J.J., 1974. Seasonal changes in the biomass of the macrobenthos of a tidal flat in the Dutch Wadden Sea, *Neth. J. Sea Res.*, 8: 94–107.

Beukema, J.J., 1976. Biomass and species richness of the macrobenthic animals living on the tidal flats of the Dutch Wadden Sea. *Neth. J. Sea Res.*, 10: 236–261.

Beukema, J.J., Cadée, G.C. and Jansen, J.J.M., 1977. Variability of growth rate of *Macoma balthica* (L.) in the Wadden Sea in relation to availability of food. In: B.F. Keegan, P. O'Ceidigh and P.J.S. Boaden (Editors), *Biology of Benthic Organisms*. Pergamon Press, New York, N.Y., pp. 69–77.

Beukema, J.J., Bruin, W. and Jansen, J.J.M., 1978. Biomass and species richness of the macrobenthic animals living on the tidal flats of the Dutch Wadden Sea: long-term changes during a period with mild winters. *Neth. J. Sea Res.*, 12: 58–77.

Bilio, M., 1967. Nahrungsbeziehungen der Turbellarien in Küsten-Salzwiesen. *Helgol. Wiss. Meeresunters.*, 15: 602–621.

Boaden, P.J.S., 1975. Anaerobiosis, meiofauna, and early metazoan evolution. *Zool. Scr.*, 4: 21–24.

Boaden, P.J.S., 1977. Thiobiotic facts and fancies (Aspects of the distribution and evolution of anaerobic meiofauna). *Mikrofauna Meeresboden* 61: 45–63.

Boaden, P.J.S. and Platt, H.M., 1971. Daily migration patterns in an intertidal meiobenthic community. *Thalassia Jugosl.*, 7: 1–12.

Boaden, P.J.S., 1974. Three new thiobiotic Gastrotricha. *Cah. Biol. Mar.*, 15: 367–378.

Boddeke, R., 1976. The seasonal migration of the brown shrimp *Crangon crangon*. *Neth. J. Sea. Res.*, 10: 103–130.

Boesch, D.F., 1972. Species diversity of marine macrobenthos in the Virginia area. *Chesapeake Sci.*, 13: 206–211.

Boesch, D.F., 1974. Diversity, stability and response to human disturbance. In: A.J. Cavé (Editor), *Proc. First Int. Congr. Ecology, The Hague, The Netherlands*. Centre for Agricultural Publishing and Documentation, Wageningen, pp. 109–114.

Boesch, D.F., 1977. A new look at the zonation of benthos along the estuarine gradient. In: B.C. Coull (Editor), *Ecology of Marine Benthos*. Belle W. Baruch Library in Marine Science, University of South Carolina Press, Columbia, S.C., pp. 245–266.

Boesch, D.F., Diaz, R.J. and Virnstein, R.W., 1976. Effects of tropical storm Agnes on soft-bottom macrobenthic communities of the James and York estuaries and the Lower Chesapeake Bay. *Chesapeake Sci.*, 17: 246–259.

Bousfield, E.L., 1955. Ecological control of the occurrence of barnacles (Crustacea, Cirripedia) in the Miramichee Estuary. *Bull. Nat. Mus. Can.*, 137: 1–69.

Boyden, C.R., 1973. Observations on the shell morphology of the two species of cockle *Cerastoderma edule* and *C. glaucum*. *Zool. J. Linn. Soc.*, 52: 269–292.

Braber, L. and Borghouts, C.H., 1977. Distribution and ecology of Anthozoa in the estuarine region of the rivers Rhine, Meuse and Scheldt. *Hydrobiologia*, 52: 15–21.

Brafield, A.E., 1972. Some adaptations to low environmental oxygen in marine borders. In: R.B. Clark and R.J. Woulton (Editors), *Essays in Hydrobiology*. University of Exeter, pp. 94–106.

Brattström, H., 1941. Studien über die Echinodermen des Gebietes zwischen Skagerrak und Ostsee, besonders des Öresundes, mit einer Übersicht über die physische Geografie. *Undersökn. Öresund.*, 27: 1–329.

Broekhuyzen, G.J., 1936. On development, growth and distribution of *Carcinus maenas* (L.). *Arch. Néerl. Zool.*, 2: 257–399.

Buitendijk, A.M. and Holthuis, L.B., 1949. Note on the Zuiderzee Crab, *Rhithropanopeus harrisii* (Gould) subspecies *tridentatus* (Maitland). *Zool. Meded. Leiden*, 30: 95–106.

Burke, M.V. and Mann, K.H., 1974. Productivity and production: biomass ratios of bivalve and gastropod populations in an eastern Canadian estuary. *J. Fish. Res. Board Can.*, 31: 167–177.

Cadée, G.C. and Hegeman, J., 1977. Distribution of primary production of the benthic microflora and accumulation of organic matter on a tidal flat area, Balgzand, Dutch Wadden Sea. *Neth. J. Sea Res.*, 11: 24–41.

Carriker, M.R., 1967. Ecology of estuarine benthic invertebrates: a perspective. In: G.H. Lauff (Editor), *Estuaries. Am. Assoc. Adv. Sci. Publ.*, 83: 442–487.

Caspers, H., 1967. Estuaries: analysis of definitions and biological considerations. In: G.H. Lauff (Editor), *Estuaries. Am. Assoc. Adv. Sci. Publ.*, 83: 6–8.

Chislenko, L.L., 1964. Relationship between the size of a body of brackish water and some characteristics of its fauna. *Okeanologiya*, 4: 528–534 (in Russian).

Connell, J.H. and Orias, E., 1964. The ecological regulation of species diversity. *Am. Nat.*, 98: 399–414.

Copeland, B.J., 1970. Estuarine classification and response to disturbances. *Trans. Am. Fish. Soc.*, 99: 826–835.

Creutzberg, F., 1974. 8. Orientation in space: animals. 8.1. Invertebrates. In: O. Kinne (Editor), *Marine Ecology*. Wiley, London, pp. 555–655.

Crumb, S.E., 1977. Macrobenthos of the tidal Delaware River between Trenton and Burlington, New Jersey. *Chesapeake Sci.*, 18: 253–265.

Dahl, E., 1956. Ecological salinity boundaries in poikilohaline waters. *Oikos*, 7: 1–21.

Dahl, E., 1959. Intertidal ecology in the terms of poikilohalinity. *Arch. Oceanogr. Limnol.*, 11(Suppl.): 227–237.

Dame, R., Vernberg, F., Bonnell, R. and Kitchens, W., 1977. The North Inlet marsh-estuarine ecosystem: a conceptual approach. *Helgol. Wiss. Meeresunters.*, 30: 343–356.

Dare, P.J., 1976. Settlement, growth and production of the mussel *Mytilus edulis* L., in Morecambe Bay, England. *Fish. Invest.*, (2)28(1).

Day, J.H., 1964. The origin and distribution of estuarine animals in South Africa. *Monogr. Biol.*, 14: 159–173.

Day, J.H., 1974. The ecology of Morrumbene Estuary, Moçambique. *Trans. R. Soc. S. Afr.*, 41: 43–97.

Dean, D. and Haskin, H.H., 1964. Benthic repopulation of the Raritan River Estuary following pollution abatement. *Limnol. Oceanogr.*, 9: 551–563.

De Jonge, V.N. and Postma, H., 1974. Phosphorus compounds in the Dutch Wadden Sea. *Neth. J. Sea Res.*, 8: 139–153.

De Lattin, G., 1967. *Grundriss der Zoogeographie*. Gustav Fischer, Jena, 602 pp.

Den Hartog, C., 1960. Comments on the Venice-system for the classification of brackish waters. *Int. Rev. Ges. Hydrobiol.*, 45: 481–485.

Den Hartog, C., 1961. Die faunistische Gliederung im südwestniederländischen Delta-gebiet. *Int. Rev. Ges. Hydrobiol.*, 46: 407–418.

Den Hartog, C., 1964. Typologie des Brackwassers. *Helgol. Wiss. Meeresunters.*, 10: 377–399.

Den Hartog, C., 1968. Een karakteristiek van het zoetwatergetijdengebied en de plaats van de Biesbosch in dit geheel. *Meded. Hydrobiol. Ver.*, 2: 173–182.

Den Hartog, C., 1971. The border environment between the sea and the fresh water, with special reference to the estuary. *Vie Milieu*, 22: 739–751.

Dennell, R., 1933. The habits and feeding mechanism of the amphipod *Haustorius arenarius*. *J. Linn. Soc. Lond., Zool.*, 38: 363–368.

De Wolf, P., 1973. Ecological observations on the mechanisms of dispersal of barnacle larvae during planktonic life and settling. *Neth. J. Sea Res.*, 6: 1–129.

De Wolf, P., 1974. On the retention of marine larvae in estuaries. *Thalassia Jugosl.*, 10: 415–424.

Dörjes, J. and Howard, J.D., 1975. Estuaries of the Georgia Coast, USA: N. Fluvial–marine transition indicators in an estuarine environment, Ogeechee River–Ossabaw Sound. *Senckenbergiana Marit.*, 7: 137–179.

Drury, W.H. and Nisbet, I.C.T., 1973. Succession. *J. Arnold Arbor.*, 54: 331–368.

Dunbar, M.J., 1960. The evolution of stability in marine environments. Natural selection at the level of the ecosystem. *Am. Nat.*, 94: 129–136.

Eisma, D., 1965. Shell-characteristics of *Cardium edule* L. as indicators of salinity. *Neth. J. Sea Res.*, 2: 492–540.

Eisma, D., Mook, W.G. and Das, H.A., 1976. Shell characteristics, isotopic composition and trace-element contents of some euryhaline molluscs as indicators of salinity. *Palaeogeogr., Palaeoclimatol., Palaeoecol.*, 19: 39–62.

Elkaim, B., 1972. Contribution à l'étude écologique d'un estuaire Atlantique Marocain. L'Estuaire du Bou Regreg. *Soc. Sci. Nat. Phys. Maroc*, 52: 131–339.

Fenchel, T. and Jansson, B.O., 1966. On the vertical distribution of the microfauna in the sediments of a brackish-water beach. *Ophelia*, 3: 161–177.

Fenchel, T.M. and Riedl, R.J., 1970. The sulfide system: a new biotic community underneath the oxidized layer of marine sand bottom. *Mar. Biol.*, 7: 225–268.

Filice, F.P., 1958. Invertebrates from the estuarine portion of San Francisco Bay and some factors influencing their distribution. *Wasmann J. Biol.*, 16: 159–211.

Fišcher, A.G., 1960. Latitudinal variations in organic diversity. *Evolution*, 14: 64–81.

Gerlach, S.A., Ekstrøm, D.K. and Eckardt, P.B., 1976. Filter feeding in the Hermit Crab, *Pagurus bernhardus*. *Oecologia*, 24: 257–264.

Gessner, F. and Schramm, W., 1971. 4. Salinity, 4.2 Plants. In: O. Kinne (Editor), *Marine Ecology 1(2)*. Wiley, London, pp. 705–820.

Goerke, H., 1971. Die Ernährungsweise der *Nereis*-Arten (Polychaeta, Nereidae) der deutschen Küsten. *Veröff. Inst. Meeresforsch. Bremerhaven*, 13: 1–50.

Grassle, J.F. and Grassle, J.P., 1974. Opportunistic life histories and genetic systems in marine benthic polychaetes. *J. Mar. Res.*, 32: 253–284.

Gray, J.S., 1966a. Factors controlling the localizations of populations of *Protodrilus symbioticus* Giard. *J. Anim. Ecol.*, 35: 435–442.

Gray, J.S., 1966b. Selection of sands by *Protodrilus symbioticus* Giard. *Veröff. Inst. Meeresforsch. Bremerhaven*, Sonderb. 2: 105–116.

Gray, J.S. and Johnson, R.M., 1970. The bacteria of a sandy beach as an ecological factor affecting the interstitial gastrotrich *Turbanella hyalina* Schultze. *J. Exp. Mar. Biol. Ecol.*, 4: 119–133.

Haertel, L. and Osterberg, C., 1967. Ecology of zooplankton, benthos and fishes in the Columbia River Estuary. *Ecology*, 48: 459–472.

Haines, E.B., 1977. The origins of detritus in Georgia salt marsh estuaries. *Oikos*, 29: 254–260.

Hargrave, B.T., 1973. Coupling carbon flow through some pelagic and benthic communities. *J. Fish. Res. Board Can.*, 30: 1317–1326.

Hedgpeth, J.W., 1953. An introduction to the zoogeography of the northwestern Gulf of Mexico with reference to the invertebrate fauna. *Publ. Inst. Mar. Sci. Texas*, 3: 107–224.

Heerebout, G.R., 1970. A classification system for isolated brackish inland waters, based on median chlorinity and chlorinity fluctuation. *Neth. J. Sea Res.*, 4: 494–503.

Heip, C., 1975. On the significance of aggregation in some benthic marine invertebrates. In: H. Barnes (Editor), *Proc. 9th European Marine Biology Symposium*. Aberdeen University Press, Aberdeen, pp. 527–538.

Heip, C. and Engels, P., 1977. Spatial segregation in copepod species from a brackish water habitat. *J. Exper. Mar. Biol. Ecol.*, 26: 77–96.

Hempel, C., 1957. Über den Röhrenbau und die Nahrungsaufnahme einiger Spioniden (*Polychaeta sedentaria*) der deutschen Küsten. *Helgol. Wiss. Meeresunters.*, 6: 100–135.

Henriksson, R., 1969. Influence of pollution on the bottom fauna of the Sound (Öresund). *Oikos*, 20: 507–523.

Hibbert, C.J., 1976. Biomass and production of a bivalve community on an intertidal mud-flat. *J. Exp. Mar. Biol. Ecol.*, 25: 249–261.

Holme, N.A., 1950. Population-dispersion in *Tellina tenuis* Da Costa. *J. Mar. Biol. Assoc. U.K.*, 29: 267–280.

Holme, N.A., 1953. The biomass of the bottom fauna in the English Channel off Plymouth. *J. Mar. Biol. Assoc. U.K.*, 32: 1–49.

Howell, B.R. and Shelton, R.G.J., 1970. The effect of China clay on the bottom fauna of St. Austell and Mevagissey bays. *J. Mar. Biol. Assoc. U.K.*, 50: 593–607.

Hughes, D.A., 1972. Response to salinity change as a tidal transport mechanism of pink shrimp, *Penaeus duoranum*. *Biol. Bull.*, 142: 271–280.

Hughes, R.N., 1970. An energy budget for a tidal-flat population of the bivalve *Scrobicularia plana* (Da Costa). *J. Anim. Ecol.*, 39: 357–381.

Hulscher, J.B., 1975. Het wad, een overvloedig of schaars gedekte tafel voor vogels? In: C. Swennen, P.A.W.J. de Wilde and J. Haeck (Editors), *Symposium Waddenonderzoek*. Arnhem, pp. 57–82.

Hutchinson, G.E., 1960. Concluding remarks. *Cold Spring Harbor Symp. Quant. Biol.* 22: 415–427.

Jebram, D., 1974. Effects of different foods on *Conopeum seurati* (Canu) (Bryozoa Cheilostomata) and *Bowerbankia gracilis* Leidy (Bryozoa Ctenostomata). "Bryozoa 1974". *Doc. Lab. Géol. Fac. Sci. Lyon*, 3: 97–108.

Jernelöv, A. and Rosenberg, R., 1976. Stress tolerance of ecosystems. *Environ. Conserv.*, 3: 43–46.

Johannessen, O.H., 1973. Population structure and individual growth of *Venerupis pullastra* (Montagu) (Lamellibranchia). *Sarsia*, 52: 97–116.

Johansen, A.C., 1918. *Randersfjords Naturhistorie*. Copenhagen, pp. 1–520.

Johnson, R.G., 1967. The vertical distribution of the infauna of a sand flat. *Ecology*, 48: 571–578.

Johnson, R.G., 1971. Animal–sediment relations in shallow water benthic communities. *Mar. Geol.*, 11: 93–104.
Kamps, L.F., 1937. *De chineesche wolhandkrab in Nederland.* Thesis, University of Groningen, Groningen, 112 pp.
Kay, D.G. and Knights, R.D., 1975. The macro-invertebrate fauna of the intertidal soft sediments of South East England. *J. Mar. Biol. Assoc. U.K.*, 55: 811–832.
Khlebovich, V.V., 1968. Some peculiar features of the hydrochemical regime and the fauna of mesohaline waters. *Mar. Biol.*, 2: 47–49.
Khlebovich, V.V., 1969. Aspects of animal evolution related to critical salinity and internal state. *Mar. Biol.*, 2: 338–345.
Kinne, O., 1959. Ecological data on the amphipod *Gammarus duebeni.* A monograph. *Veröff. Inst. Meeresforsch. Bremerhaven*, 6: 177–202.
Kinne, O., 1971. 4. Salinity. 4.3 Animals. 4.3I Invertebrates. In: O. Kinne (Editor), *Marine Ecology, 1(2)*, Wiley, London, pp. 683–1244.
Klein Breteler, W.C.M., 1976. Settlement, growth and production of the Shore Crab, *Carcinus maenas*, on tidal flats in the Dutch Wadden Sea. *Neth. J. Sea Res.*, 10: 354–376.
Klopfer, P.H., 1959. Environmental determinants of faunal diversity. *Am. Nat.*, 93: 337–342.
Korringa, P., 1970. The basic principles of shellfish farming on the continental coast of Europe. In: *Symp. Mollusca, Mar. Biol. Assoc. India, 1968*, part III, pp. 818–823.
Korringa, P., 1971. *Mariculture, Yearbook of Science and Technology.* McGraw-Hill, New York, N.Y., pp. 13–23.
Korringa, P., 1976. *Farming Marine Organisms Low in the Food Chain.* Developments in Aquaculture and Fisheries Science, Elsevier, Amsterdam, 264 pp.
Koulman, J.G. and Wolff, W.J., 1977. The mollusca of the estuarine region of the rivers Rhine, Meuse, and Scheldt in relation to the hydrography of the area. V. The Cardiidae. *Basteria*, 41: 21–32.
Kuipers, B.R., 1977. On the ecology of juvenile plaice on a tidal flat in the Wadden Sea. *Neth. J. Sea Res.*, 11: 56–91.
Lie, V., 1968. A quantitative study of benthic infauna in Puget Sound. *Fiskeridir. Skr., Ser. Havunders.*, 14: 229–556.
Longbottom, M.R., 1970. The distribution of *Arenicola marina* (L.), with particular reference to the effects of particle size and organic matter in sediments. *J. Exp. Mar. Biol. Ecol.*, 5: 138–157.
McIntyre, A.D. and Eleftheriou, A., 1968. The bottom fauna of a flat fish nursery ground. *J. Mar. Biol. Assoc. U.K.*, 48: 113–142.
McNulty, J.K., Work, R.C. and Moore, H.B., 1962. Some relationships between the infauna of the level bottom and the sediments of southern Florida. *Bull. Mar. Sci. Gulf Caribbean*, 12: 322–332.
Maurer, D., 1969. Pelecypod–sediment association in Tomales Bay, California. *Veliger*, 11: 243–249.
Maurer, D. and Watling, L., 1973. Studies on the oyster community in Delaware: the effects of the estuarine environment on the associated fauna. *Int. Rev. Ges. Hydrobiol.*, 58: 161–201.
Mileikovsky, S.A., 1971. Types of larval development in marine bottom invertebrates, their distribution and ecological significance: a re-evaluation. *Mar. Biol.*, 10: 193–213.
Milne, H. and Dunnet, G.M., 1972. Standing crop, productivity and trophic relations of the fauna of the Ythan estuary. In: R.S.K. Barnes and J. Green (Editors). *The Estuarine Environment.* Applied Science Publ., London, pp. 56–196.
Moll, R.A., 1977. Phytoplankton in a temperate zone salt marsh: Net production and exchange with coastal waters. *Mar. Biol.*, 42: 109–118.
Moore, H.B., Davies, L.T., Fraser, T.H., Gore, R.H. and Lopez, N.R., 1968. Some biomass figures from a tidal flat in Biscayne Bay, Florida. *Bull. Sci. Gulf Caribb.*, 18: 261–279.
Muus, B.J., 1967. The fauna of Danish estuaries and lagoons. Distribution and ecology of dominating species in the shallow reaches of the mesohaline zone. *Medd. Dan. Fisk. Havundersøg.*, (*N. S.*), 5(1): 1–316.
Newell, R., 1962. Behavioural aspects of the ecology of *Peringia* (=*Hydrobia*) *ulvae* (Pennant) (Gastropoda, Prosobranchia). *Proc. Zool. Soc. Lond.*, 138: 49–75.
Newell, R., 1964. Some factors controlling the upstream distribution of *Hydrobia ulvae* (Pennant), (Gastropoda, Prosobranchia). *Proc. Zool. Soc. London*, 142: 85–106.
Newell, R., 1965. The role of detritus in the nutrition of two marine deposit-feeders, the prosobranch *Hydrobia ulvae* and the bivalve *Macoma balthica. Proc. Zool. Soc. Lond.*, 144(1): 25–46.
Nichols, F.H., 1977. Infaunal biomass and production on a mudflat, San Francisco Bay, California. In: B.C. Coull (Editor), *Ecology of Marine Benthos*, Belle W. Baruch Library in Marine Science. University of South Carolina Press, Columbia, S.C., pp. 339–357.
Nilsson, L., 1969. Food consumption of diving ducks wintering at the coast of South Sweden in relation of food resources. *Oikos*, 20: 128–135.
Nixon, S.W., Oviatt, C.A., Rogers, C. and Taylor, K., 1971. Mass and metabolism of a musselbed. *Oecologia* (*Berl.*), 8: 21–30.
Odum, E.P., 1971. *Fundamentals of Ecology.* Saunders, Philadelphia, Penn., 573 pp.
Odum, W.E., 1970. Insidious alteration of the estuarine environment. *Trans. Am. Fish. Soc.*, 99: 836–847.
Odum, W.E. and Heald, E.J., 1972. Trophic analysis of an estuarine mangrove community. *Bull. Mar. Sci.*, 22: 671–738.
Odum, W.E. and Heald, E.J., 1975. Mangrove forests and aquatic productivity. In: A.D. Hasler (Editor), *Coupling of Land and Water Systems.* Springer, Berlin, pp. 129–136.
Orton, J.H., 1925. The conditions for calcareous metabolism in oysters and other marine animals. *Nature*, 116: 13.
Pearson, T.H., 1975. The benthic ecology of Loch Linnhe and Loch Eil, a sea loch system on the west coast of Scotland. IV. Changes in the benthic fauna attributable to organic enrichment. *J. Exp. Mar. Biol. Ecol.*, 20: 1–41.
Pearson, T.H. and Rosenberg, R., 1976. A comparative study of the effects on the marine environment of wastes from cellulose industries in Scotland and Sweden. *Ambio*, 5: 77–79.
Peters, N. and Panning, A., 1933. Die chinesische Wollhandkrabbe (*Eriocheir sinensis* H. Milne Edwards) in Deutschland. *Zool. Anz.*, 104: I–VIII, 1–180.
Pfannkuche, O., 1974. Zur Systematik und Ökologie naidomor-

pher Brackwasser-Oligochaeten. *Mitt. Hamburg Zool. Mus. Inst.*, 71: 115–134.

Pfannkuche, O., 1977. *Ökologische und systematische Untersuchungen an naidomorphen Oligochaeten brackiger und limnischer Biotope*. Thesis, University of Hamburg, Hamburg, 138 pp.

Postma, H., 1954. Hydrography of the Dutch Wadden Sea. *Arch. Néerl. Zool.*, 10: 405–511.

Postma, H., 1961. Transport and accumulation of suspended matter in the Dutch Wadden Sea. *Neth. J. Sea Res.*, 1: 148–190.

Postma, H., 1967. Sediment transport and sedimentation in the estuarine environment. In: G.H. Lauff (Editor), *Estuaries. Am. Assoc. Adv. Sci. Publ.*, 83: 158–179.

Postuma, K.H. and Rauck, G., 1978. The fishery in the Wadden Sea. In: N. Dankers, W.J. Wolff and J.J. Zijlstra (Editors), *Fishes and Fisheries of the Wadden Sea*. Balkema, Rotterdam, pp. 139–157.

Rasmussen, E., 1973. Systematics and ecology of the Isefjord marine fauna (Denmark), with a survey of the eelgrass (*Zostera*) vegetation and its communities. *Ophelia*, 11: 1–495.

Redeke, H.C., 1922. Zur Biologie der niederländischen Brackwassertypen. *Bijdr. Dierkd. Amsterdam*, 22: 239–335.

Reid, D.M., 1930–1932. Salinity interchange between salt water in sand and overflowing fresh water at low tide. *J. Mar. Biol. Assoc. U.K.*, 16: 609–613; 18: 299–306.

Reise, K., 1977a. Predator exclusion experiments in an intertidal mud flat. *Helgol. Wiss. Meeresunters.*, 30: 263–271.

Reise, K., 1977b. Predation pressure and community structure of an intertidal soft-bottom fauna. In: B.F. Keegan, P.O. Ceidigh and P.J.S. Boaden (Editors), *Biology of Benthic Organisms*. Pergamon Press, Oxford, pp. 513–519.

Reise, K. and Ax, P., 1979. A meiofaunal "thiobios" limited to the anaerobic sulfide system of marine sand does not exist. *Mar. Biol.*, 54: 225–237.

Reish, D.J., 1960. The use of marine invertebrates as indicators of water quality. In: E.A. Pearson (Editor), *Waste Disposal in the Marine Environment. Proc. First Int. Conf. New York*, pp. 92–103.

Reish, D.J., 1961. A study of benthic fauna in a recently constructed boat harbor in southern California. *Ecology*, 42: 84–91.

Reish, D.J., 1962. A study of succession in recently constructed marine harbors in southern California. In: *Coastal and Shallow Water Research Conference*. Long Beach State College, Long Beach, Calif., pp. 570–572.

Reish, D.J., 1963. Further studies on the benthic fauna in a recently constructed boat harbor in southern California. *Bull. S. Calif. Acad. Sci.*, 62: 23–32.

Remane, A., 1934. Die Brackwasserfauna. *Verh. Dtsch. Zool. Ges.*, 36: 34–74.

Remane, A., 1941. Einführung in die zoologische Ökologie der Nord- und Ostsee. In: *Tierwelt der Nord- und Ostsee*, 1a: 1–238.

Remane, A., 1958. Ökologie des Brackwassers. In: A. Remane and C. Schlieper (Editors), *Die Biologie des Brackwassers*. E. Schweizerbart'sche Verlagsbuchhandlung, Stuttgart. pp. 1–216.

Remane, A., 1971. Ecology of brackish water. In: A. Remane and C. Schlieper (Editors), *Biology of Brackish Water*. E. Schweizerbart'sche Verlagsbuchhandlung, Stuttgart, pp. 1–210.

Rhoads, D.C., 1974. Organism–sediment relations on the muddy sea floor. *Oceanogr. Mar. Biol. Annu. Rev.*, 12: 263–300.

Rhoads, D.C. and Young, D.K., 1970. The influence of deposit-feeding organisms on sediment stability and community trophic structure. *J. Mar. Res.*, 28: 150–178.

Richards, S.W. and Riley, G.A., 1967. The benthic epifauna of Long Island Sound. *Bull. Bingham Oceanogr. Coll.*, 19: 89–135.

Rochford, D.J., 1951. Hydrobiology of the estuarine environment. *Proc. 2nd Meet. Indo-Pacific Fish. Counc.*, 3: 157–168.

Rosenberg, R., 1972. Benthic faunal recovery in a Swedish fjord following the closure of a sulphite pulp mill. *Oikos*, 23: 92–108.

Rosenberg, R., 1973a. Succession in benthic macrofauna in a Swedish fjord subsequent to the closure of a sulphite pulp mill. *Oikos*, 24: 244–258.

Rosenberg, R., 1973b. Recovery of the littoral fauna in Saltkällefjord subsequent to discontinued operations of a sulphite pulp mill. *Thalassia Jugosl.*, 7: 747–757.

Rosenberg, R., 1974. Spatial dispersion of an estuarine benthic faunal community. *J. Exp. Mar. Biol. Ecol.*, 15: 69–80.

Rosenberg, R., 1976. Benthic faunal dynamics during succession following pollution abatement in a Swedish estuary. *Oikos*, 27: 414–427.

Rosenberg, R., Olsson, I. and Ölundh, E., 1977. Energy flow model of an oxygen-deficient estuary on the Swedish west coast. *Mar. Biol.*, 42: 99–107.

Rowe, G.T., Polloni, P.T. and Rowe, J.I., 1972. Benthic community parameters in Lower Mystic River. *Int. Rev. Ges. Hydrobiol.*, 57: 573–584.

Russell, P.J.C., 1971. A reappraisal of the geographical distributions of the cockles, *Cardium edule* L. and *C. glaucum* Bruguière. *J. Conchol.*, 27: 225–234.

Russell, P.J.C., 1972. A significance in the number of ribs on the shells of two closely related Cardium species. *J. Conchol.*, 27: 401–409.

Sanders, H.L., 1956. Oceanography of Long Island Sound, 1952–1954. X. Biology of marine bottom communities. *Bull. Bingham Oceanogr. Coll.*, 15: 345–414.

Sanders, H.L., 1958. Benthic studies in Buzzards Bay. I. Animal–sediment relationships. *Limnol. Oceanogr.*, 3: 245–258.

Sanders, H.L., 1968. Marine benthic diversity: a comparative study. *Am. Nat.*, 102: 243–282.

Sanders, H.L., Mangelsdorf, P.C. and Hampson, G.R., 1965. Salinity and faunal distribution in the Pocasset River, Massachusetts. *Limnol. Oceanogr.* 10 (Suppl.): R 216–R 229.

Sanders, H.L., Goudsmit, E.M., Mills, E.L. and Hampson, G.E., 1962. A study of the intertidal fauna of Barnstable Harbor, Mass. *Limnol. Oceanogr.*, 7: 63–79.

Schelske, C.L. and Odum, E.P., 1961. Mechanisms maintaining high productivity in Georgia estuaries. *Proc. Gulf Caribb. Fish. Inst.*, 14: 75–80.

Segerstråle, S., 1949. The brackish-water fauna of Finland. *Oikos*, 1: 127–141.

Segerstråle, S.G., 1962. Investigations on Baltic populations of the bivalve *Macoma balthica* (L.). *Soc. Sci. Fenn.*, 24(7): 1–26.

Segerstråle, S.G., 1969. The competition factor and the fauna of the Baltic Sea. *Limnologica*, 7: 99–111.

Sepers, A.B.J., 1977. The utilization of dissolved organic compounds in aquatic environments. *Hydrobiologia*, 52: 39–54.

Simpson, G.G., 1964. Species density of North American recent mammals. *Syst. Zool.*, 13: 57–73.

Slobodkin, L.B. and Sanders, H.L., 1969. On the contribution of environmental predictability to species diversity. *Brookhaven Symp. Biol.*, 22: 82–93.

Smidt, E.L.B., 1951. Animal production in the Danish Waddensea. *Medd. Komm. Dan. Fisk.-Havundersøg.*, (*Fisk.*) II(6): 1–151.

Smith, R.I., 1956. The ecology of the Tamar estuary. VII Observations on the interstitial salinity of intertidal muds in the estuarine habitat of *Nereis diversicolor*. *J. Mar. Biol. Assoc. U.K.*, 38: 81–104.

Southward, A.J. and Southward, E.C., 1972. Observations on the role of dissolved organic compounds in the nutrition of benthic invertebrates. III Uptake in relation to organic content of the habitat. *Sarsia*, 50: 29–46.

St. Amant, L.S., 1973. Shellfish and crustacean productivity in marshes and estuaries. In: H. Chabuck (Editor), *Proc. Coastal Marsh and Estuary Management Symp.* Louisiana State University, Baton Rouge, La., pp. 151–161.

Stephens, G.C., 1964. Uptake of organic material by aquatic invertebrates. III. Uptake of glycine by brackish-water annelids. *Biol. Bull.*, 126: 160–162.

Stephens, G.C., 1967. Dissolved organic material as nutritional source for marine and estuarine invertebrates. In: G.H. Lauff (Editor), *Estuaries. Am. Assoc. Adv. Sci., Publ.*, 83: 367–373.

Stone, A.N. and Reish, D.J., 1965. The effect of fresh-water runoff on a population of estuarine Polychaetous Annelids. *Bull. S. Calif. Acad. Sci.*, 64: 111–119.

Swennen, C., 1976. Wadden Seas are rare, hospitable and productive. In: M. Smart (Editor), *Proc. Int. Conf. Conserv. Wetlands Waterfowl, Heiligenhafen, 1974*, pp. 184–198.

Teal, J.M., 1962. Energy flow in a salt marsh ecosystem of Georgia. *Ecology*, 43: 614–624.

Tenore, K.R., 1972. Macrobenthos of Pamlico River estuary, North Carolina. *Ecol. Monogr.*, 42: 51–69.

Tenore, K.R. and Gopalan, U.K., 1974. Feeding efficiencies of the polychaete *Nereis virens* cultured on hard clam tissue and oyster detritus. *J. Fish Res. Board Can.*, 31: 1675–1678.

Tenore, K.R., Goldman, J.C. and Clarner, J.P., 1973. The food chain dynamics of the oyster, clam, and mussel in an aquaculture food chain. *J. Exp. Mar. Biol. Ecol.*, 12: 157–165.

Tommasi, L.R., 1970. Observaçoes sôbre a fauna bêntica do complexo estuarinolagunar de Cananéia (Sp.). *Bol. Inst. Oceanogr. S. Paulo*, 19: 43–56.

Tunnicliffe, V. and Risk, M.J., 1977. Relationships between the bivalve *Macoma balthica* and bacteria in intertidal sediments: Minas Basin, Bay of Fundy. *J. Mar. Res.*, 35: 499–507.

Vader, W.J.M., 1964. A preliminary investigation into the reactions of the infauna of the tidal flats to tidal fluctuations in water level. *Neth. J. Sea Res.*, 2: 185–222.

Vader, W.J.M., 1965. Intertidal distribution of Haustoriid amphipods in the Netherlands. *Bot. Gothob.*, 3: 233–246.

Välikangas, J., 1926. Planktologische Untersuchungen im Hafengebiet von Helsingfors. *J. Acta Zool. Fenn.*, 1: 1–277.

Van Beek, F.A., 1976. Aantallen, groei, produktie en voedselopname van de zandgrondel (*P. minutus*) en de wadgrondel (*P. microps*) op het Balgzand. *Intern NIOZ Rapp.*, 1976–1979.

Van der Schoot, E., 1974. *Onderzoek naar het aandeel van* Asterias rubens *in de koolstofkringloop van het Grevelingenbekken.* Delta Instituut voor Hydrobiologisch Onderzoek Yerseke, Doctoraalverslag D5-1974.

Van Es, F.B., 1977. A preliminary carbon budget for a part of the Ems estuary: the Dollard. *Helgol. Wiss. Meeresunters.*, 30: 283–294.

Venema, S.C. and Creutzberg, F., 1973. Seasonal migration of the swimming crab (*Macropipus holsatus*) in an estuarine area controlled by tidal streams. *Neth. J. Sea Res.*, 7: 97–102.

Vugts, H.F. and Zimmerman, J.T.F., 1975. Interaction between the daily heat balance and the tidal cycle. *Nature*, 225: 113–117.

Walne, P.R., 1972. The importance of estuaries to commercial fisheries. In: R.S.K. Barnes and J. Green (Editors), *The Estuarine Environment.* Applied Science Publishers, London, pp. 107–118.

Warwick, R.M. and Price, R., 1975. Macrofauna production in an estuarine mudflat. *J. Mar. Biol. Assoc. U.K.*, 55: 1–18.

Wass, M.L., 1967. Biological and physiological basis of indicator organisms and communities II. Indicators of pollution. In: Th.A. Olson and F.J. Burgess (Editors), *Pollution and Marine Ecology.* Wiley and Sons, New York, N.Y., pp. 271–283.

Wass, M.L., 1972. *A Check List of the Biota of Lower Chesapeake Bay with Inclusions from the Upper Bay and the Virginian Sea.* Special Scientific Report, No. 65. Virginia Institute of Marine Science, 290 pp.

Williams, R.B. and Thomas, L.K., 1967. The standing crop of benthic animals in a North Carolina estuarine area. *J. Elisha Mitchell Sci. Soc.*, 83: 135–139.

Wilson, D.P., 1937. The influence of the substratum on the metamorphosis of *Notomastus* larvae. *J. Mar. Biol. Assoc. U.K.*, 22: 227–243.

Winter, J.E., 1969. Über den Einfluss der Nahrungskonzentration und anderer Faktoren auf Filtrierleistung und Nahrungsausnutzung der Muscheln *Arctica islandica* und *Modiolus modiolus*. *Mar. Biol.*, 4: 87–135.

Wolff, T, 1954. Tre østamerikanske Krabber fundet i Danmark. *Flora, Fauna, Silkeborg*, 60: 19–34.

Wolff, W.J., 1971. Origin and history of the brackish water fauna of N.W. Europe. In: B. Battaglia (Editor), *Proc. 5th European Marine Biology Symposium 1971, Padova*, pp. 11–18.

Wolff, W.J., 1973. The estuary as a habitat. An analysis of data on the soft-bottom macrofauna of the estuarine area of the rivers Rhine, Meuse, and Scheldt. *Zoöl. Verh. Leiden*, 126: 1–242.

Wolff, W.J., 1974. Benthic diversity in the Rhine–Meuse estuary. *Hydrobiol. Bull., Amsterdam*, 8: 242–252.

Wolff, W.J., 1977. A benthic food budget for the Grevelingen Estuary, the Netherlands, and a consideration of the mechanisms causing high benthic secondary production in estuaries. In: B.C. Coull (Editor), *Ecology of Marine Benthos*. Belle W. Baruch Library in Marine Science, University of South Carolina Press, Columbia, S.C., pp. 267–280.

Wolff, W.J. and De Wolf, L., 1977. Biomass and production of zoobenthos in the Grevelingen Estuary, the Netherlands. *Estuarine Coastal Mar. Sci.*, 5: 1–24.

Wolff, W.J., Vegter, F., Mulder, H.G. and Meijs, T., 1976a. The production of benthic animals in relation to the phytoplankton production. Observations on the saline Lake Grevelingen, the Netherlands. In: G. Persoone and W. Jaspers (Editors), *Proc. 10th European Symposium on Marine Biology, 2, Population Dynamics*. Universa Press, Wetteren, pp. 653–672.

Wolff, W.J., van Haperen, A.M.M., Sandee, A.J.J., Baptist, H.J.M. and Saeijs, H.L.F., 1976b. The trophic role of birds in the Grevelingen estuary, the Netherlands, as compared to their role in the saline Lake Grevelingen. In: G. Persoone and W. Jaspers (Editors), *Proc. 10th European Symposium on Marine Biology, 2, Population Dynamics*. Universa Press, Wetteren, pp. 673–689.

Wolff, W.J., Sandee, A.J.J. and De Wolf, L., 1977. The development of a benthic ecosystem. *Hydrobiologia*, 52: 107–115.

Wolff, W.J., Van Eeden, M.J. and Lammens, E., 1979. Primary production and import of particulate organic matter on a salt marsh in the Netherlands. *Neth. J. Sea Res.*, 13: 242–255.

Wood, E.J., Ferguson, F., Odum, W.E. and Zieman, J.C., 1969. Influence of sea grass on the productivity of coastal lagoons. *Mem. Simp. Int. Lagunas Costeras UNAM–UNESCO, 1967, México, D.F.*, pp. 495–502.

Wood, L. and Hargis, W.J., 1971. Transport of bivalve larvae in a tidal estuary. In: D.J. Crisp (Editor), *Fourth European Marine Biology Symposium*. Cambridge University Press, Cambridge, pp. 29–44.

Woodwell, G.M., Whitney, D.E., Hall, C.A.S. and Houghton, R.A., 1977. The Flax pond ecosystem study: Exchanges of carbon in water between a salt marsh and Long Island Sound. *Limnol. Oceanogr.*, 22: 833–838.

Young, D.K., Buzas, M.A. and Young, M.W., 1976. Species densities of macrobenthos associated with seagrass: a field experimental study of predation. *J. Mar. Res.*, 34: 577–592.

Zenkevitch, L.A., 1959. The classification of brackish-water basins as exemplified by the seas of the USSR. *Arch. Oceanogr. Limnol.*, II (Suppl.): 53–62.

Zenkevitch, L.A., 1963. *Biology of the Seas of the USSR*. Allen and Unwin, London, 955 pp.

Chapter 7

ESTUARINE FISHES

RICHARD L. HAEDRICH

INTRODUCTION

The study of estuaries has received increasingly greater attention during the past fifteen years. This interest is reflected in the new journals, such as *Estuarine and Coastal Marine Science*[1], devoted particularly to the publication of estuarine work, and in the many symposium volumes (i.e. Lauff, 1967; Cronin, 1975; Wiley, 1976) and books (i.e. Remane and Schlieper, 1971; Dyer, 1973; Officer, 1976) on the subject. A part of such interest stems from the fact that estuaries have considerable importance in human affairs. Estuarine systems are used as ports and harbors, as receivers of domestic and industrial wastes, as sites for aquaculture, for commercial fishing, and for a myriad of recreational purposes, including particularly boating and fishing.

The rather recent appreciation (Smith et al., 1966) that five of the six most important commercial fishery species in the United States are in some way dependent on estuaries underscores the theme. The nekton, an often-ignored component in ecosystem studies, have thus received considerable attention in estuaries. Particularly important reviews have been provided by McHugh (1967, 1976) concerning general principles and the fisheries, and by deSylva (1975) concerning food and feeding relationships. The impact of pollutants (Saville, 1966; Livingston, 1975) and the problems associated with power plant siting (Hoss et al., 1974; O'Connor and McErlean, 1975) are recurrent themes. Saila (1975) examined the dynamics and limits of estuarine fish production from a theoretical standpoint.

Because most of these general articles are commonly available, this review will not attempt to summarize them in detail. The concern instead will be with a general overview, mainly from an ecological/evolutionary perspective, of the present state of knowledge concerning estuarine fishes. Emphasis will be as much on what is not known as upon what is. I shall deal with the nature of estuarine fish faunas, their variation in space and time, the life histories of estuarine fishes, and the relationship between estuarine fishes and man.

THE NATURE OF ESTUARINE FISH FAUNAS

Students of estuaries emphasize three points. Estuaries are regions where marine and fresh waters meet, estuaries are regions where environmental stress is pronounced, and estuaries are regions of exceptionally high productivity. These three conditions have a decided effect on the kinds and numbers of fishes that can live in estuaries.

The strong salinity gradient and the abrupt changes that can occur in temperature, oxygen concentration, turbidity, and a host of related factors place considerable physiological demands on an organism. Adaptation to such a variable environment is of course possible (Vernberg and Vernberg, 1976; Vernberg, Ch. 3), but the successful organism must be broadly tolerant in its requirements. Despite the heterogeneity of the estuarine environment, its variability and the requisite broad tolerance of estuarine organisms in effect sharply limit the number of niches available. For this reason, the diversity of estuarine faunas, those assemblages of animals that customarily make their

[1] Renamed *Estuarine, Coastal and Shelf Science* in January, 1981 (Vol. 12).

home there, will tend to be relatively low. Thus the number of kinds of fishes that are truly estuarine is not great. The squids, important members of marine nekton communities, do not occur in estuaries at all, probably because their systems of osmotic balance cannot cope with the estuarine extremes.

The high productivity of the estuary is a favorable condition. As a general rule, high primary productivity tends to promote increased productivity and high standing stocks at all levels in the food web. The fishes, with a wide range of trophic capabilities and as the largest and most motile of the estuarine animals, are in a good position to exploit this production. Thus, the abundance and biomass of fishes in estuaries will be large. This expectation, combined with that of only a modest diversity, implies that estuarine fish faunas should on the whole be dominated by a relatively few forms. We will see that this is so, but first we should examine the question of the sources from whence the estuarine fish faunas are derived.

Types of distributions

The estuary is characterized as a mixing zone on the basis of its physical properties, and this generality extends most certainly to the fauna as well. Faunal zonation along estuarine gradients has been well described for benthic invertebrates (e.g. Boesch, 1977; cf. Ch. 6, pp. 156–157 and pp. 167–168). Attempts to correlate the distributions of animals with physical factors have emphasized salinity, with, in general, the description of schemes wherein each faunal assemblage has its own particular range of tolerances. Arrangement of these assemblages along the salinity gradient produces the zonation observed. From the data of an excellent comparative study of the ecology of South African estuaries, Day (1951) suggested that the steepest faunal gradients were to be found at about 25, 15, 5, and 0.1‰ salinity. This system has undergone a number of revisions, including that of an international symposium on estuarine classifications (see Table 6.8, p. 174), but still, within rough limits, the pattern seems to hold (Carriker, 1967).

For fishes, because of their mobility, determination of patterns and correlation with environmental parameters is more difficult than for the sessile invertebrates. For the estuarine fish fauna, Day (1951) recognized five components, based for the most part on salinity tolerance. These were fresh-water fishes found near the head of the estuary, stenohaline marine forms from the seaward end of the estuary, euryhaline marine forms occurring over wide areas, the truly estuarine fishes found only in the estuary, and migratory forms that either pass through the estuary or enter it only occasionally. In an extensive and important review, McHugh (1967) modified this classification to place it on a more ecological basis. His six categories were:

(1) Fresh-water fishes that occasionally enter brackish waters.

(2) Truly estuarine species which spend their entire lives in the estuary.

(3) Anadromous and catadromous species.

(4) Marine species which pay regular seasonal visits to the estuary, usually as adults.

(5) Marine species which use the estuary primarily as a nursery ground, usually spawning and spending much of their adult life at sea but often returning seasonally to the estuary.

(6) Adventitious visitors which appear irregularly and have no apparent estuarine requirements.

With little change, these classifications continue to be used and seem to be applicable to most estuarine situations. They are important concepts both for practical and for more theoretical concerns. Originally derived from the subjective impressions of experienced field scientists, some indication of their objective reality can be found in a study of Hoff and Ibara (1977). Computer-run factor analysis of fish communities in a small New England estuary showed that, within any one season, the patterns of salinity distribution best explained the distribution of fish assemblages.

The greatest problem with the classification is not in the definitions themselves, but in having sufficient information to assign individual species to a category. Knowledge of the natural history of fishes is relatively complete for only a very few species, and primarily for those important to commercial interests. Even so apparently simple a matter as describing in detail the distribution of any species as a function of developmental stage and season has been done for only a handful of species. Good examples of such work are to be found in the papers by Talbot (1955) on the sparid *Rhabdosargus globiceps* in South Africa, by Hartley (1940) on the flounder *Pleuronectes flesus* in southwest England,

and by Pearcy (1962) on the winter flounder (*Pseudopleuronectes americanus*) in southern New England. More of this type of work is needed, as others have also observed (e.g. Kendall, 1975).

Particularly vexing is the distinction between "estuarine" and "inshore". Conceptually not difficult, it becomes so in application. The circumstances of the inshore situation — shallow water, lowered salinities, varying temperatures, increased turbidity, higher production — are just those found at points within the estuary. Fishes found in and reported from estuaries may in fact be essentially inshore forms which, because of the similar conditions, occur through the estuary as well. Emphasis on estuarine studies, the relative ease of sampling in estuaries as opposed to the open coast, and the concentrating effect of the enclosed waters of the estuary, may all lead to a bias in perception. Are there obligate estuarine stages in fishes? Are there significant differences in growth or mortality in and out of estuaries? What particular and measurable benefits does the estuary provide? These are questions which cannot be answered with full assurance in any quantitative way.

The inshore/estuarine dilemma could be resolved by accepting McHugh's (1967) interesting notion of an extended estuarine zone for fishes. He argues that the conventional definition of the estuary depends too much on a topographic and essentially terrestrial point of view, and that more properly hydrographic criteria should be used. In suggesting the extended estuarine zone, he uses salinity as the defining characteristic with waters of less than 33.5‰ bounding offshore estuarine zones. These areas have considerable extent (Fig. 7.1), for wherever precipitation exceeds evaporation, as in the Equatorial Pacific, vast shallow lenses of lower-salinity water overlie those of higher salinity. The largest of these offshore extended estuarine zones encompasses almost the entire Pacific north of 40°N. The argument is aided in considering the distribution of the Pacific salmons, which in their migrations range across this region, and by reference to an estuarine analogy for the area proposed with respect to the physical oceanography (Tully and Barber, 1960). Nonetheless, it seems essentially a semantic solution, with its focus on a single property too simplistic. Choice of the 33.5‰ isohaline may be almost as arbitrary as the topographic definition, and there would appear to be no physiological basis for its selection. McHugh's concept is certainly arguable, and it does not seem to have been applied by others. But it remains an interesting concept for discussing the distribution of estuarine fishes.

Numbers of species

There are some 20,000 species of fishes (Cohen, 1970) in approximately 410 families (Greenwood et

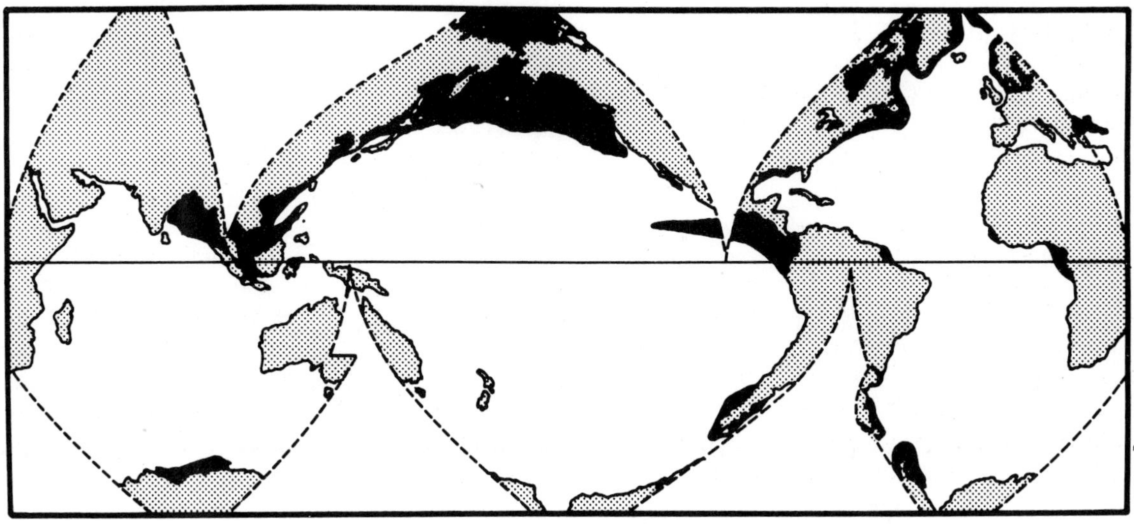

Fig. 7.1. Offshore estuarine zones of the world, bounded by the 33.5‰ isohaline, as proposed by McHugh (1967). These shallow lenses (dark areas) overlie water of higher salinity.

al., 1966). About 45% of the species are shallow-water marine and 41% fresh water, with a mere 0.6% (115 species) recognized as diadromous. Among this last small group, members of which can occur from salt into fresh water, are found most of the estuarine fishes.

Comparing the results from faunal surveys made around the world, in both large and small estuarine regions, the total number of species reported from estuaries is never large. Hartley (1940) recorded about 40 species from the estuary of the Tamar and Lynher in southwestern England. The Avon–Heathcote Estuary in New Zealand contained about 28 (Webb, 1972). There were 149 species recorded from the large tropical Rokan Estuary in Sumatra, but of these only 80 occurred regularly and only 50 to 60 could be considered common (Hardenberg, 1931). Talbot (1955) found 25 species in the Klein River Estuary in South Africa. Bechtel and Copeland (1970) recorded 55 species from Galveston Bay, Texas, and Livingston (1975) found 73 species in the Econfina–Fenholloway system in Florida. Dahlberg and Odum (1970) recorded 70 species from a Georgia marsh–estuary system, and Richards and Castagna (1970) took the same number in Chesapeake Bay. In New England, Pearcy and Richards (1962) collected 51 species in the Mystic River, Connecticut; Oviatt and Nixon (1973) took 99 species in the whole of Narragansett Bay; and Haedrich and Haedrich (1974) took 23 species in the small Mystic River Estuary, Massachusetts. The large Miramichi system in New Brunswick had in all 78 species (McKenzie, 1959).

Horn and Allen (1976), in a survey of fish populations in a number of California bays and estuaries, recorded species numbers ranging from 22 in the very small Los Penasquitos Lagoon to 121 for the entire San Francisco Bay. The average number of species for thirteen areas was 66. For the California data, they show that a significant correlation exists between the size of an embayment and the number of species present. This observation for fish faunas is in keeping with the general rule relating species and area originally put forward by Jaccard (1902).

Regression analysis of the California data showed that the size of the opening of an embayment to the sea was the best predictor of the number of species in the embayment (Horn and Allen, 1976). The implication was that a wider mouth provided greater access to the presumed source fauna, that of the California continental shelf. The shelf fauna, with 439 species (Miller and Lea, 1972), is far richer than that found in any of the embayments. The size and mouth width versus species number relationship has not been investigated elsewhere, so a full understanding of its implications must await further work in other areas. For all but the most unusual situations, however, the rule that the local marine fauna has many more species than an estuary would appear to be a safe generalization. In New England, for example, the number of estuarine species is only about 10% of the number of marine ones (Haedrich and Hall, 1976). The data on which this rule is based come for the most part from surveys made over quite broad areas. Real documentation — faunal transects from the inner estuary out onto the shelf — has not been done in detail for the fishes of any region. Hoese (1973) found inshore/offshore species differences on the Georgia coast were there related to the rather different environmental settings, and relative numbers of species could not be properly compared.

Characteristic families

Because of the problem of definition discussed above, and because of the tendency for fish faunal surveys to be reported as checklists rather than even semi-quantitatively, a fully comprehensive list of estuarine families and species is difficult to compile. Nonetheless, by assimilation and comparison of the faunal lists reported in such useful papers as those cited above for species numbers, a moderately accurate family list can be drawn up. In compiling the list, the criteria for inclusion include the ubiquitousness of a family in estuaries, the abundance of individual species of the family in estuaries, and the number of species in the family that occur within any estuary.

Temperate areas

The best data exist for temperate systems in North America, South Africa, and New Zealand. Important estuarine fish families are the Anguillidae (fresh-water eels), Clupeidae (herrings), Engraulidae (anchovies), Ariidae (salt-water catfishes), Cyprinodontidae (killifishes), Gadidae (cods), Gasterosteidae (sticklebacks), Serranidae (basses), Sciaenidae (croakers), Sparidae (sea-breams), and Pleuronectidae (flounders). Of these, only the Cyprinodontidae belong to a group whose major evolution has been in fresh water. The rest are essentially marine. For a number, diadromy, moving from fresh to salt water as a part of the reproductive behaviour, is the essential reason for their charactization as estuarine. All *Anguilla* eels are catadromous, and migrate from fresh or brackish water to central ocean gyres to spawn. The more common habit is to be anadromous. Members of the Clupeidae (*Alosa*, shads) and Serranidae (*Morone* spp.) live as adults in salt water but move into estuaries and up the rivers on their spawning migrations.

Tropics and subtropics

It is more difficult to assemble a list of characteristic families in tropical and subtropical estuaries. The lists of species and families taken in these areas tend to be long. Hardenberg (1931) records more than 150 species in the commercial landings from the Rokan Estuary in Sumatra. Pillay (1967a, b) lists 128 species in 45 families from West African estuaries and 301 species in 76 families from estuaries along the coast of India. Relative abundance is not given for all groups, and the inshore/estuarine distinction is not made. Important families appear to be Clupeidae, Engraulidae, various families of tropical catfishes, Chanidae (milkfish), Synodontidae (lizardfish), Belonidae (silver gars), Mugilidae (mullets), Polynemidae (threadfins), Sciaenidae, Gobiidae (gobies), Cichlidae (*Tilapia*-like fishes), and various flounder families, especially the Soleidae and Cynoglossidae. As with the temperate forms, these are essentially marine families, with only the Cichlidae having basically a fresh-water heritage. Diadromy appears less pronounced, but life-history information is very inadequate. The shad-like clupeids *Ethmalosa* and *Hilsa* can be anadromous. Other groups, particularly Chanidae, Sciaenidae, Mugilidae, and Soleidae, move offshore to spawn and might be considered catadromous. Although these fishes do make excursions into fresh water, residence there is probably not critical in their development.

Shrimps, particularly the Penaeidae, are important members of semi-nektonic assemblages in tropical and subtropical estuarine situations. As adults, they live in the brackish to fresh sections, but migrate offshore to salt water for spawning.

High latitudes

Boreal estuaries contain much simpler faunas. Important families are few and include the Salmonidae (salmon and trout), Osmeridae (smelt and capelin), Gasterosteidae, Ammodytidae (sand lance), and Cottidae (sculpins) and their relatives. The Salmonidae, Osmeridae, and Gasterosteidae tend to be anadromous. The other families are marine, and probably have no real estuarine dependence. In the cold-water estuaries of Tierra del Fuego, Tasmania, and New Zealand's South Island, the salmonids are replaced by their Southern Hemisphere analogues, the Galaxioidei. Community composition in these regions does not seem to have been reported, and by now may have been markedly changed through the introduction of salmonids.

Some generalizations

A number of points emerge. One is that, even on a worldwide basis, the number of families that make important contributions to estuarine fish faunas is very small, something like 6% of the total number of fish families described. This parallels the low diversity of species, and emphasizes the fact that only relatively few kinds of fishes can live in estuaries. The affinities of almost all the important estuarine families are marine. Thus, it appears true that estuarine fish faunas are derived from the sea, a generalization that was stated some time ago (Annandale, 1922; Gunter, 1961). Nonetheless, the generalization might require some modification when the faunas of the really large tropical systems — the estuaries of the Amazon, Congo, Ganges and the rivers of Southeast Asia — are properly surveyed. These all have extremely rich fresh-water fish faunas upstream, comparable in diversity to the adjacent coastal marine faunas. With such a pool of potential immigrants, the balance of sources might shift. The work has not been done, however. Systematic investigation of the fish faunas in the large tropical estuaries would seem a potentially exciting scientific undertaking.

Another finding supported by all studies is that the degree of dominance by a few species is high in estuarine fish faunas. It is in fact this dominance, combined with the high productivity, that makes the assemblages attractive to the fisheries. In Narragansett Bay, the 10 most abundant species, of

99 recorded, accounted for 90% of the individuals (Oviatt and Nixon, 1973). In the Avon–Heathcote Estuary, New Zealand, 9 species of 28 comprised 90% of all individuals (Webb, 1972). Allen and Horn (1975), through an examination of a number of studies of fish populations in small estuaries in the United States, found it to be true that for such faunas a small number of species (≤ 5) comprises a large percentage (≥ 75) of all individuals. These few abundant species were usually ones low in the trophic structure, being either planktivorous, omnivorous, or low-level carnivores.

In summation, then, we can say that the nature of estuarine fish faunas is to be generally of low species diversity with a tendency towards dominance of the total number of individuals by only a very few species. The number of important estuarine families is very small, and their basic affinities appear to be marine. This apparently simple picture of the estuarine fish fauna becomes more complex when the basic mobility of fishes is taken into account. Fishes can position themselves along estuarine gradients, and, according to their placement, faunal sub-components can be defined. These range from the stenohaline marine forms that occur in the outer reaches of the estuary, perhaps only as adventitious visitors, to essentially fresh-water species that have very slight salinity tolerance and occur in the estuary only as strays. In between are euryhaline forms including a few permanent residents, and the anadromous and catadromous species that pass through the estuary on their spawning migrations. Thus, despite the simplicity, estuarine faunas have the potential for considerable variation. In the next section we will examine the sorts of patterns that arise.

VARIATION IN ESTUARINE FAUNAS

The composition of estuarine fish faunas can be compared along many dimensions. Most basic is the phylogenetic, or systematic, variation, of interest at the level both of species and of higher categories. Diurnal tidal and day/night cycles can produce short-term variations. The seasons themselves, particularly in temperate and boreal regions, may impose a seasonal variation in composition. No two estuaries are exactly the same, and the individual conditions of each are likely to be expressed in the fauna present. The degree of stress is important here, thus comparative studies of polluted systems are of interest. Finally, there is the variation introduced by changes over time, either in the course of the natural physical evolution of the estuary or as it is altered through man's activities.

Zoogeography and phylogenetic differences

The species associated with estuaries are surprisingly widespread. The mullet (*Mugil cephalus*), a quite characteristic estuarine species, occurs around the world in tropical and subtropical regions (Briggs, 1960). The Arctic char (*Salvelinus alpinus*) and the stickleback (*Gasterosteus aculeatus*) are circumglobal in cold northern waters (Dunbar, 1970), and the diadromous *Galaxias maculatus* has a similar widespread pattern in the Southern Hemisphere (McDowall, 1970). In temperate waters along the east coast of North America, the winter flounder (*Pseudopleuronectes americanus*), the menhaden (*Brevoortia tyrannus*), the alewife (*Alosa pseudoharengus*), the silversides (*Menidia menidia*), and the marsh minnow (*Fundulus heteroclitus*) are found from the Maritime Provinces of Canada to the Gulf of Mexico. The bass *Dicentrarchus labrax* occurs from Denmark to Egypt. The breams *Rhabdosargus globiceps* and *Lithognathus lithognathus* seem widespread in Asia and Africa. The list could be continued, but the uncertain systematic status of most species would render it of limited use. The evidence of faunal lists strongly suggests that certain species are encountered again and again, but the genetic relationship of widespread forms under the same name is not known.

Systematic differences at the higher categories are a function of rather broad areas. In boreal estuaries, for example, the greatest numbers and biomass of fishes are to be found in the Salmonidae. Clupeidae and flounders in the Pleuronectidae are dominant in more temperate regions, and a large number of families, including particularly Engraulidae (anchovies), Synodontidae (lizardfish), Mugilidae (mullets), Sciaenidae (croakers), Gobiidae (gobies), and flounders in the Soleidae are important in the tropics and subtropics. Transitions from one condition to the other are not abrupt. There seems to be, rather, a gradual change in familial composition with latitude. From high to low latitudes families drop out and are replaced, and new ones are added. Families of minor importance become dominant, while dominant ones decline. The changes appear clinal, but detailed data on the rate of change have not been assembled.

With change in latitude, there is a shift in the relative phylogenetic position of the dominant estuarine families. Relatively primitive groups tend to be more important toward the poles, while the advanced perciform families are more important in the tropics. This pattern is in keeping with the general distribution of the world's fish fauna. The picture is somewhat beclouded by the importance in all estuaries of the primitive clupeiform fishes (Division I of Greenwood et al., 1966). Clupeidae are dominant in high latitudes and Engraulidae in low ones.

In proceeding towards the Equator, the number of fish families found in estuaries increases. The same is true for the number of species, but in both cases the increase is only a modest one. Table 7.1 shows such data for a series of Atlantic estuaries. In preparing this table, strictly fresh-water groups, for example the Cyprinidae, have been excluded. Of particular interest is the ratio of species to families (S/F). This ratio is, in a sense, an index of the niches available to fishes in an area, for it is a measure of the level of diversification achieved. Despite the wide range in location, type, size, and phylogenetic composition of the estuaries in Table 7.1, the ratio S/F is for the most part quite similar, near 1.8. The rather low value suggests only a moderate degree of diversification is possible in estuaries, and hence niche widths should be broad. The observation is in keeping with the expectation that estuarine species must be tolerant of a wide range of conditions. The higher value, 2.8, for West Africa could reflect an increased tropical diversity, with narrower niches, but some care must be exercised here, for the numbers used (Pillay, 1967a) are a composite from several systems. The S/F ratio for fishes on a coral patch reef in the Caribbean is 2.8 (Smith and Tyler, 1972), and our unpublished data for the mesopelagic fish fauna of the northern Sargasso Sea shows for that assemblage an S/F ratio of 5.5.

Daily and seasonal differences

Short-term changes in the estuarine environment, particularly tidal and day/night variation, can be expected to influence the community structure of motile fishes. The day/night change is of course ubiquitous, although the balance shifts with latitude and season.

A diurnal tidal cycle occurs in almost all systems. The ebb and flow of the tide carries fishes to different parts of the estuary. In a Florida salt marsh, the tide brought *Fundulus* and *Menidia* in to feed over the covered grasses, and at the same time more predatory fish had access to the channels. In the Mystic River, Massachusetts, *Osmerus* moved up and down with the tides, but other species were more sedentary (Haedrich and Haedrich, 1974). On the other hand, Livingston (1975) found that time of day rather than tidal cycle was more important in determining the short-term variation in fish community structure. More species and a greater number of individuals were consistently taken at

TABLE 7.1

Latitude, numbers of families and numbers of species of fishes for several Atlantic estuaries (S/F is the ratio of species to families)

Area	Latitude	Families	Species	S/F	Source
Lake Melville–Hamilton River, Labrador	53°N	14	25	1.8	Backus (1957)
Miramichi River, New Brunswick	47°N	38	69	1.8	McKenzie (1959)
Mystic River, Massachusetts	42°N	13	23	1.8	Haedrich and Haedrich (1974)
Slocum River, Massachusetts	41°N	36	54	1.5	Hoff and Ibara (1977)
Mystic River, Connecticut	41°N	33	51	1.5	Pearcy and Richards (1962)
Chesapeake Bay	38°N	39	70	1.8	Richards and Castagna (1970)
Georgia salt marsh estuary	32°N	37	70	1.9	Dahlberg and Odum (1970)
Florida salt marsh estuary	30°N	29	55	1.9	Subrahmanyam and Drake (1975)
Puerto Rico mangrove swamp	18°N	35	62	1.8	Austin (1971)
West African estuaries	~10°N	45	128	2.8	Pillay (1967a)
Klein River, South Africa	34°S	16	25	1.6	Talbot (1955)

night. Movements of *Acipenser brevirostrum* in a Maine estuary did not show any consistent pattern when tracked for periods of from 5 to 28 h during early summer (McCleave et al., 1977).

The inclusion of a migratory category in the classification of the types of fishes found in estuaries recognizes the seasonal character of the fauna. The migratory component can form a considerable fraction of the species present, for example in the Chesapeake Bay where 59 (84%) of the 70 species recorded were considered migrants (Richards and Castagna, 1970). In Maine, all of the 15 pelagic fish species taken in the small Sheepscot River–Back River Estuary departed in the cold winter months (Recksiek and McCleave, 1973). Most studies of seasonal variation have been done in temperate areas along the east coast of the United States, where the seasons are a dominant feature of the environment. Seasonal variation in tropical estuaries has received little attention and cannot be quantified. Nonetheless, it does occur, as shown, for example, in the seasonal inshore migration of *Chanos chanos* juveniles in India (Pillay, 1967b). On the other hand, Longhurst (1957) found little variation over the year in the euryhaline fishes of the Sierra Leone River estuary. The boreal pattern seems to be one of almost total absence during the winter (Andrews and Lear, 1956; Grainger, 1953), but data directly addressing the question are lacking. There is a seasonal component in inshore Arctic fish faunas, primarily as a function of species' abilities to withstand sub-zero temperatures (Smith and Paulson, 1977).

The seasonal variation in the composition of estuarine fish faunas reflects varying life histories, and, in particular, the role the estuary assumes for any of the different species. Some studies stress reproductive habits. In the Mystic River, Massachusetts, for example, dominant species succeed one another as a function of the time of spawning. *Pseudopleuronectes americanus* are dominant as metamorphosing juveniles in the summer, and as adults moving inshore to spawn in the fall; the anadromous *Osmerus mordax* are dominant as they pass through on the winter spawning migration; and likewise the anadromous *Alosa pseudoharengus* are dominant in the spring (Haedrich and Haedrich, 1974). Migrations associated with feeding are more frequently mentioned. Such, for example, are the spring influx of larval fishes into the Columbia River estuary (Misitano, 1977), the temperature-influenced movements of juvenile demersal fish into the Wadden Sea (Creutzberg and Fonds, 1971), and the movements of both young and adults, particularly of piscivorous species, in Plymouth Sound, England (Hartley, 1940). No one aspect can be singled out, and it is in fact a complex interaction of various ecological factors that determine the orderly temporal succession of dominant forms generally observed (Jeffries and Johnson, 1974; Livingstone et al., 1976b).

The annual cycle in the fish fauna that seems to be so much a part of the ecology of temperate estuaries is an important mechanism for resource sharing. In this manner, more kinds and numbers of fishes can use the estuarine environment than would be the case if the situation were static. It can be imagined, but not as yet proven, that this sharing mechanism would result in the fish-carrying capacity of the estuary being more fully realized throughout the year. Where the pattern of primary production is cyclical, with its attendant cycles of secondary and tertiary production, the changing pattern in the fish fauna should reflect it. This appears to be the case (Dahlberg and Odum, 1970), but the relationship is not a simple one. There may be several peaks in fish abundance and/or diversity throughout the year. These can be associated with spawning migrations that are a part of the anadromous habit, the appearance of young fishes of many species to feed on burgeoning populations of zooplankton and benthos (usually a spring phenomenon), and visits by predatory fishes such as *Pomatomus* (bluefish) and Carangidae (jacks) to feed on smaller fishes (usually a fall phenomenon). Jeffries and Johnson (1974) show a pattern of seasonal replacement for common species in Narragansett Bay, with the overall picture being one of a consecutive arrangement of bimodal abundances (Fig. 7.2). Such a scheme might keep the fish niches filled. A comprehensive body of work on the subject has been carried out by Livingston (1975) and his co-workers (1976b) in the Apalachicola Estuary in northwest Florida, and by Webb (1972, 1973a–d) for the Avon–Heathcote Estuary in New Zealand. Comparable studies in other regions have not been done, but should be.

Hartley (1940) thought that the seasonal cycle of species and abundances produced a situation ranging from virtually no competition to much com-

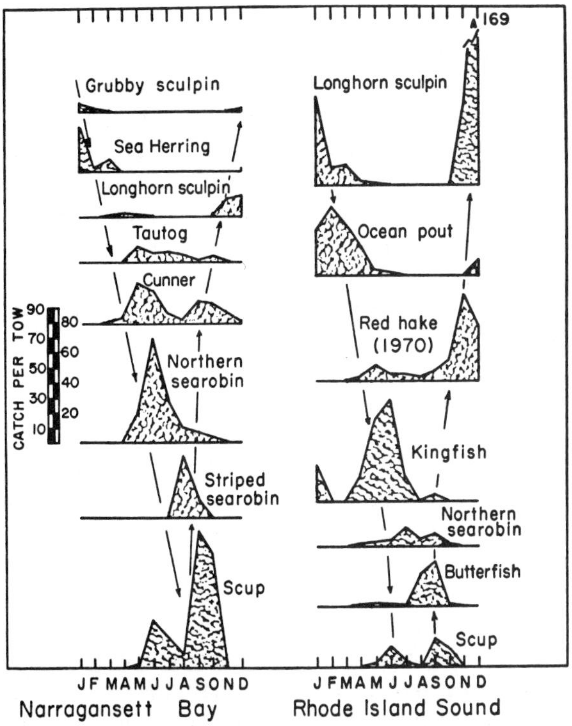

Fig. 7.2. Abundance of fish species taken in bottom trawls in Rhode Island waters during 1969. The seasonal pattern is one of continuous replacement, with bimodal distributions in a number of species (from Jeffries and Johnson, 1974). The fishes are: butterfish, *Poronotus triacanthus*; cunner, *Tautogolabrus adspersus*; red hake, *Urophycis chuss*; sea herring, *Clupea harengus*; kingfish, *Menticirrhus saxatilis*; ocean pout *Macrozoarces americanus*; grubby sculpin, *Myoxocephalus aeneus*; longhorn sculpin, *M. octodecemspinosus*; scup, *Stenotomus chrysops*; northern searobin, *Prionotus carolinus*; striped searobin, *P. evolans*; tautog, *Tautoga onitus*.

petition for the food resource. He had only data on stomach contents, however, and little information on the relative abundance of prey species. In a nice series of studies on a marine community of fishes in Passamaquoddy Bay, New Brunswick, Tyler (1971a, 1972) showed first that seasonal components were an important part of the demersal assemblage, and second that, when principal prey was considered, there was relatively little overlap in the diets of the species present. The generality for estuarine systems of Tyler's findings remains to be confirmed.

Variation over the long term

A limitation of almost all annual studies is that they are conducted over only one year. McErlean et al. (1973) showed that the cyclical pattern in abundance and diversity of fishes in the Patuxent River estuary, Maryland, repeated itself over the course of their four-year study. Other studies have shown, however, that while overall abundance and diversity may be predictable as a function of season, the actual species composition may not. Creutzberg and Fonds (1971) reported that the dominant goby in the Wadden Sea, *Pomatoschistus microps*, was suddenly replaced one year by another goby, designated as *Pomatoschistus minutus/lozanoi*. In Apalachicola Bay, *Anchoa mitchilli* was far more abundant in the first year of a three-year study than it was in the following two (Livingston et al., 1976b). Similar rather abrupt changes in community composition were found in a six-year study of the fishes of shore-zone communities in Long Island Sound (Hillman et al., 1977). The object of this long study was to examine the influence of local environmental changes, particularly those associated with the operation of a nuclear power plant, on fish communities. Even so, no cause for the changes observed could be identified, and the investigators suggested that they were only manifestations of natural, but as yet unexplained, long-term fluctuations. Variations in the catch of *Pseudopleuronectes americanus* in Narragansett Bay, where that flounder is the overwhelmingly dominant form in bottom fish assemblages, appeared correlated with a very subtle change in the annual temperature regime (Jeffries and Johnson, 1974). As temperature increased, catches declined, but the abundance of the second most abundant species, another flounder *Scophthalmus aquosus*, remained unchanged. The faunal changes that take place over periods of years are only beginning to be examined in estuaries, an observation that surely pertains to all ecosystems.

Variations in the composition of local estuarine faunas over the long term of historical and geological time of course occur, but have been little studied and are difficult to actually document. The influence of temperature is a common area of focus (e.g. Taylor et al., 1957). The remains of fishes, many of them possibly estuarine, from Indian middens in southern California showed that a warm-water fauna, unlike that presently in the area, probably existed there for considerable periods of time as little as 300 years ago (Hubbs, 1948; Follett, 1976).

Over geological time, any particular estuary must be considered an ephemeral feature. The processes that produce them, in which coastal submergence plays an important role, have been active over recent time, and most of the estuarine systems of today are probably no older than several thousand years (Gorsline, 1967; Russell, 1967). This is a short period with respect to the time scale over which speciation takes place. For this reason, the incidence of endemic fish species in estuaries should be low, and it is.

On the basis of their physical characteristics, estuaries in the United States fall into four major groups (Folger, 1972a; see also Ch. 2). But within this relatively simple classification, each of the 45 systems studied in a coast-to-coast survey had its own particular characteristics with respect to the texture and composition of the bottom sediments, and these were a function of the individual geologic, bathymetric, and hydrologic settings (Folger, 1972b). It is to be expected that each estuary might thus harbor its own somewhat unique fish fauna, but, unfortunately, the needed biological companion to Folger's geological work has not been assembled. The fish faunas of three nearby but physically different estuaries in Florida were unlike one another (Subrahmanyam and Drake, 1975). Horn and Allen (1976) compared the fish faunas in thirteen bays and estuaries along the California coast and found that the greatest similarity between any two areas was that 62% of the species present in them were shared. In Tomales–Bodega Bay, one of the larger systems, 18% of the species occurred only there, but the proportion was much less in the other systems and five areas had no unique species at all. A similar study in New England showed in general a high degree of similarity between the faunas of Massachusetts bays and estuaries, although the individual characteristics of each area did seem to have an influence on the composition and, particularly, the patterns of dominance (Haedrich, 1975).

One factor in the individuality of estuarine fish faunas, if such is found to be the rule, could be the pattern of seasonal occurrence discussed above. Such seasonality could be effectively damped by increasing the environmental stress in an estuary, and thereby eliminating the seasonal species. The few remaining species, presumably those most resistant to stress, would soon dominate the fauna at all seasons, with the result that any individuality on the part of the estuarine fish fauna would be lost. An expectation would be that polluted and thus stressed estuaries along a coastline should have a high degree of faunal similarity in terms of the fishes present. As a working hypothesis, this rule seems to hold for small systems in Massachusetts where the winter flounder (*Pseudopleuronectes americanus*) is the dominant species (Haedrich, 1975), but its generality remains to be tested.

Summing up

Variation in estuarine fish faunas is present on many scales. Geologically, estuaries are ephemeral features and their faunas come and go. The lifetime of an estuary is short when considered in evolutionary time, and thus estuaries have not been for fishes important centers of speciation. On a worldwide basis, the composition of estuarine fish faunas reflects the general patterns found in the fish fauna as a whole, but the number of families and species found in estuaries is quite low. Dominant estuarine species tend to be quite widespread and are encountered again and again in faunal lists, a reflection of the broad tolerances and range of adaptations necessary in estuarine species and of the resulting low diversity of the estuarine fish fauna. Because fish are motile, they can and do follow the daily tidal and other cycles in estuaries. Since the responses of individual species vary, this introduces a short-term variation in community structure. The composition of estuarine faunas changes over the year as seasonal migrants move in and out. Both reproductive and feeding cycles are important in these movements. Each estuary has its own set of unique physical characteristics, with the potential to produce its own particular balance of species. The role of seasonal migrants could be large here. Environmental stress, as for example through increased pollution, would tend to dampen the seasonal change and differences between estuaries by favoring only the most widely tolerant, and therefore dominant and characteristic, estuarine species. At the heart of understanding variation in estuarine fish faunas is an appreciation in detail of how fishes use estuaries. The next section will look at life histories to see some of the ways in which they do.

LIFE HISTORIES OF ESTUARINE FISHES

Work on the ecology of estuarine fishes has focussed on the commercial species. There is, thus, a good body of information concerning species such as the striped bass (*Morone saxatilis*; e.g. Talbot, 1966), the spotted sea trout (*Cynoscion nebulosus*; Tabb, 1966), the menhaden (*Brevoortia*; Reintjes and Pacheco, 1966; Gunter, 1967), the white stumpnose (*Rhabdosargus globiceps*; Talbot, 1955), the winter flounder (*Pseudopleuronectes americanus*; Pearcy, 1962), the mullet (*Mugil*; El-Maghraby et al., 1973, 1974; Moore, 1974, 1976), and any of the various species of salmon (Hart, 1973). None of these spend their entire lives in the estuary, although some such as the mullets may not stray far from it. *Morone* and the salmons are anadromous and run up into fresh water to spawn. The distance may be only a short one, as it is for the bass, or involve movements of more than 3200 km as it does for the chum salmon (*Oncorhynchus keta*) in the Yukon River in northern Canada. *Brevoortia*, *Rhabdosargus*, and *Mugil* spawn in the sea, and *Pseudopleuronectes* spawns in shallow brackish water. Even in such a small sample of species — and the literature on natural history does not exist to expand it substantially — there is a diversity of approaches in the way that fishes use estuaries.

Anadromy

The anadromous habit, moving from salt into fresh water to spawn, figures largely in the reproductive use of estuaries by fishes. Diverse species are involved. The Acipenseridae (sturgeons), Clupeidae, Salmonidae, Osmeridae (smelts), Serranidae (basses), and Petromyzontidae (sea lampreys) all have important anadromous forms. For the adults, the estuary is best thought of as a resting place where the fish pause to acclimate themselves to the fresh water before continuing further on their spawning migration. For the young of anadromous species, however, the estuary assumes a greater importance. In most cases, as in the shads, smelts, and bass (Fig. 7.3), the young move rather soon after hatching into estuarine waters, where they feed and grow. The salmons show a range of behaviors, from the pink salmon (*Oncorhynchus gorbuscha*), which moves at once towards the sea upon hatching to the coho (*O. kisutch*) and Atlantic salmon (*Salmo salar*) where the young spend a year or more in fresh water before migrating to the estuary (Hart, 1973). Even in the latter case, feeding in the estuary before going out to sea may be an important aspect for the young fish.

The nature of the eggs of anadromous species follows no one pattern. They are large, few in number, and buried in nests (redds) by the salmons. Bass lay a large number of smaller semi-buoyant eggs into the water, where they float with the currents. Smelts deposit great mats of sticky eggs on the bottom in running water.

Anadromy as an important feature in the lives of estuarine fishes declines with latitude. In boreal estuaries where the fauna is dominated by Salmonidae and Osmeridae, anadromy is the rule, but in tropical regions only a few Clupeidae, such as *Ethmalosa dorsalis* and *Hilsa ilisha*, are thought at times to move into fresh water for spawning, but may not need to. Of the 27 species of native fresh-water fish in New Zealand, 63% move at some stage into salt water (McDowall, 1976). The trend could be related to the relative richness and development of the local fresh-water faunas. Predation pressure in the highly diverse tropical fresh-water faunas could easily be sufficient to discourage anadromy. Furthermore, the demands of a diverse fauna for spawning areas would place a premium on space, probably to the exclusion of species, such as anadromous ones, from other regions. At high latitudes, the fresh-water fish fauna is typically extremely depauperate. Potential predators on eggs are few and space for spawning abounds.

Temperate regions show intermediate conditions, dependent in part on the particular local situation. That the anadromous spawner niche is not always filled is shown by the success of certain transplants of species, for example of *Morone saxatilis* and *Alosa sapidissima* from the east to the west coast of the United States. Attempted introductions of new anadromous species into waters where anadromy is well-developed, for example of Atlantic salmon into British Columbia rivers, have not succeeded (Hart, 1973). In *Morone saxatilis*, anadromy decreases in southern populations in favor of a wholly riverine existence (Dudley et al., 1977). This appears to be temperature-related, for the river waters, particularly below large impoundments, remain colder year-round than the local shallow sea.

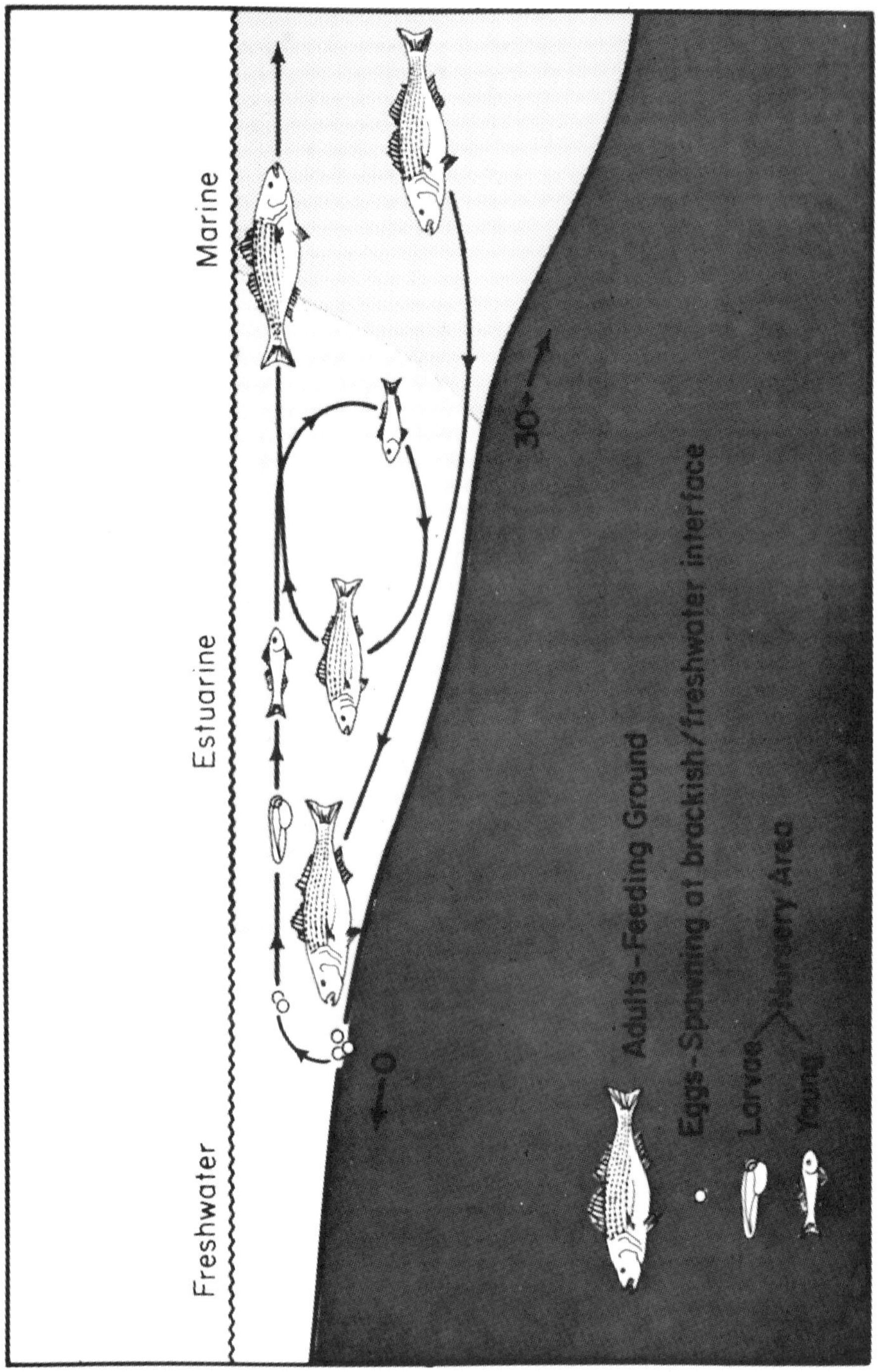

Fig. 7.3. The life history of the striped bass, *Morone saxatilis*, an important estuarine species. The pattern is similar in many anadromous forms (from Haedrich and Hall, 1976).

The pattern of anadromy seems to illustrate the principle of competitive exclusion, but a rigorous examination of the situation has not been undertaken. It may be that the interactions of faunas are more important in establishing estuarine distributional patterns than strictly physiological tolerances of individual species. Some insight is provided by the very few observations that wherever in estuarine situations the fresh-water fauna is absent, marine species intrude into the fresh water. This is true both in coastal ponds of Hawaii (Brock, 1977) and small estuaries of Newfoundland (Scott and Crossman, 1964).

Life history of the smelt

As an example of an anadromous estuarine species, we consider the North American smelt, *Osmerus mordax*. There is little recent work concerning this species. Most studies of its biology, distribution, and systematics were done before 1940, and many of them before 1900. A general summary and bibliographic listing appears in Bigelow and Schroeder (1963).

The smelt ranges from Labrador to Virginia, but is most abundant in the Canadian Maritime Provinces (Liem and Scott, 1966). It is a fish of the estuaries and shallow water, rarely occurring deeper than 3 to 5 fathoms (5.4–9 m). Numerous local varieties have been described; land-locked populations are common. In marine waters, smelt seem to be a natural constituent of the community dominated by the winter flounder. In the Mystic River, the smelt was the second-ranking fish both in numbers (32% of all specimens taken) and in biomass (Haedrich and Haedrich, 1974).

Smelt spawn in fresh water, from immediately above the tide in small coastal brooks to quite far up the tributaries of large rivers like the Miramichi, New Brunswick. Where obstructions occur, they have been reported to spawn in the brackish water of the intertidal zone. Salt water kills the eggs. Maturing fish, generally two years old, begin to move into fresh water in February–March in southern New England. The spawning run lasts only a few weeks, with optimum egg production occurring in temperatures from 10 to 14°C. The eggs are laid usually on a pebbly bottom where there is some current, often in water no more than a few centimeters deep. Spawning occurs at night. Many die after spawning; the remainder drop back at once to the estuary. The number of eggs produced by a female depends on the size of the fish; numbers in the range from 7000 to 60 000 are common. The eggs (1.3 mm diameter) are adhesive, and cling to the bottom in thick carpets with as many as $2.6 \times 10^6 m^{-2}$ reported. The eggs hatch in about two weeks, dependent on the temperature, and the larvae 5 to 6 mm in length then rise to the surface and drift downstream. They grow very rapidly, reaching 18 mm within a month and 40 mm in about three months. New Hampshire smelt are about 90 mm long at one year, and about 145 mm when they reach maturity in their second year. Maximum size is 350 mm for a five-year-old fish.

Very little is known about the movements of smelt in the sea. It appears that many remain quite near the place where they were spawned, drifting up and down the estuaries with the tide. Exceptions are known, however; a fish tagged in the Miramichi Estuary was recaptured 160 km away (Liem and Scott, 1966). Similar-sized fish form large schools, and, where water temperature remain cool enough, these can be seen year-round in estuaries and small harbors. The fish hold position off the bottom, and feed mainly on decapods and mysids (*Crangon septemspinosum* is an important food) and small fishes. Benthic food items are found in their stomachs only rarely, and they do not feed during the spawning season. In lakes, smelt can be very cannibalistic.

Little or nothing is known of their physiology. Because they are anadromous, and seem to move rather rapidly from salt to fresh water, their ability to osmoregulate must be well-developed. Although they occur in very cold situations, and in fact are commonly angled for through the ice, there are no reports of their ever having been killed by winter chilling.

Because of their brook-spawning habit, smelt have lent themselves well to management efforts. Plantings of both eggs and fry have been quite successful in establishing populations where none were present. Large hatchery and stocking programs of this sort were carried out around the turn of the century. For years, the Palmer Hatchery in Massachusetts was an important source of eggs and fry, with a peak production of about 110 million in 1919. For some reason, however, the smelt program there was abandoned after 1920. Good smelt fishing can be reported within as little as three years after the beginning of a stocking program.

Smelt, in addition to being a desirable food fish for humans, are probably preyed upon heavily by larger fish in salt water. They also figure in the diet of cormorants (*Phalacrocorax*) and mergansers (*Mergus*). There are few data on the extent of such predation, however. In fresh water, they are known to be an important forage fish for salmonids, and are often used in this role in salmon stocking/ management plans. There are no studies as to what specifically limits their production, although the extent of good spawning areas would seem to be quite critical.

Catadromy

Catadromy, moving from fresh to salt water to spawn, is a rare phenomenon. The best known example of the catadromous habit is found in the fresh-water eels (*Anguilla* spp.), which move from fresh and brackish habitats to spawn far at sea in the central ocean gyres. The Mugilidae, although not really fresh-water fish, do move out of brackish areas to spawn some little distance in the sea. In high latitudes of the Southern Hemisphere, *Galaxias maculatus* is a catadromous species (McDowall, 1970), and the flounder *Rhombosolea retiaria* appears to be catadromous in New Zealand estuaries (McDowall, 1976). Two species of *Anguilla* are found in New Zealand, so that 4 of the 27 species of native fresh-water fishes that occur on those large islands are catadromous. This is a substantial fraction, and it may be that catadromy is really a feature of cold-temperate waters in the Southern Hemisphere. In the Northern Hemisphere, only the eels and mullets are catadromous. A comparison might prove interesting, but the data are lacking.

Estuarine spawners

Other than the resident species — and these of course are in any case rather few — there are not many fishes that actually spawn in estuaries. Gobies, *Menidia*, and *Fundulus* spp. are fishes that do. These are small species with relatively short lifespans. The common goby (*Pomatoschistus microps*) lives on an average only from 12 to 20 months in British waters. The adults move to the saltier stretches of the estuary to spawn, and after spawning die; newly demersal young are more prevalent in lower salinities (Miller, 1975). Chesney and Iglesias (1979) found that the Ría de Arosa Estuary in Spain, in contrast to most systems, contained a strong resident component. This group was dominated by small gobies, some of which were annual species. Gobies were among the few fish found spawning in the Avon–Heathcote Estuary (Webb, 1973a). The white goby (*Glossogobius giurus*) spawns year-round in both brackish and fresh water in the Philippines (Marques, 1968). On the United States west coast, *Clupea harengus pallasi*, *Cottus asper*, *Leptocottus armatus*, and *Lepidogobius lepidus* spawn in some estuaries (Pearcy and Myers, 1974; Misitano, 1977). *Galaxias maculatus* migrates from fresh water to lay its eggs on grassy estuarine river flats (McDowall, 1976).

A feature common to all estuarine spawners is the production of non-buoyant, sticky eggs. *Pseudopleuronectes americanus* and *Microgadus* (tomcod) both lay large masses of demersal eggs during fall and winter. That this must be a particular adaptation to estuarine life — for non-buoyant adhesive eggs will be kept in the estuary instead of being swept out with the currents — is suggested by the fact that the offshore relatives of both these species lay pelagic eggs. Large schools of *Fundulus* crowd into small tidal creeks to spawn in spring. The males become very brightly colored and ardently court the females. The large eggs, spawned a few at a time, are deposited in shallow water and stick to algal mats on the bottom. *Menidia* spawns in early summer over sandy bottoms and near the base of the *Spartina* grasses. The abundant eggs sink and adhere to the grass with long sticky filaments. In all these species, incubation is quite rapid, of the order of weeks, usually dependent on temperature. Differences in time of spawning by these species, all of which can be taken in the same estuary, are one means of avoiding competition and sharing the estuarine resources (Haedrich and Haedrich, 1974; Subrahmanyam and Drake, 1975).

Life history of the winter flounder

As an example of an estuarine spawning fish, we consider the winter flounder (*Pseudopleuronectes americanus*). This is a well-studied fish. It occurs from Labrador to Georgia (Liem and Scott, 1966), but is most abundant in the middle portions of its range. It is rarely found in water deeper than 20 fathoms (37 m). It is the dominant form in the inshore fish community of muddy to muddy sand bottoms from Connecticut to Newfoundland (Pearcy and Richards, 1962; Levings, 1973; Jeffries and Johnson, 1974; Haedrich, 1975; Van Guelpen and Davis, 1979). In the Mystic River Estuary (Boston), winter flounder comprised 52% of all fish specimens and 89% of total fish biomass taken in a seasonal trawl survey (Haedrich and Haedrich, 1974).

Bigelow and Schroeder (1953) have summarized well the life history of the winter flounder, and subsequent studies have served only to refine details in their account. Winter flounder move into shallow water (2–6 m) to spawn on sandy bottoms from January to May. Optimum spawning occurs at temperatures near 3 or 4°C. Spawning occurs at night. The sex ratio in the adult population off southern New England is about 7 females: 3 males (Saila, 1961a; Briggs, 1965; Howe and Coates, 1975). On the average, a female produces as many as 500 000 eggs per spawning. The eggs are about 0.8 mm in diameter. They are not buoyant, in contrast to the situation in most flounders, but sink to the bottom where they adhere in clusters. Hatching occurs in about two to three weeks at 4°C in New England; Canadian studies suggest that hatching there takes nearly 26 days at 4.4°C and 18 days at 15.5°C (Scott, 1929). Rate of development appears governed by temperature, with metamorphosis from the larval stage occurring from $2\frac{1}{2}$ to $3\frac{1}{2}$ months after hatching. The larvae are only semi-planktonic. Growth is rapid, the fish growing from about 3 mm at hatching to between 100 and 150 mm within a year. Young flounders from the Weweantic River, Massachusetts, when fed on clam siphons had a mean conversion efficiency from food to biomass of 21.1% (range 13.9–36.8%) and a growth rate between 12 and 16°C of 6.1 g per month (Frame, 1973b). A 12-year-old fish, which is about the maximum age, from south of Cape Cod is about 45 cm long (Howe and Coates, 1975).

Winter flounders spend, in general, the first two years of their lives in estuaries or shallow bays (Saila, 1961b). They feed on a variety of invertebrates, both planktonic and benthic. The intertidal zone appears to be an important foraging area (Tyler, 1971b; Wells et al., 1973). In the Weweantic River, Frame (1974) found a dependence on copepods and amphipods up to about the midpoint of their first year, when, with a shift to a more benthic existence, their diet tended more to polychaetes and bivalves. Pelagic feeding did not cease, however, and amphipods continued to be an important food in spring, while molluscs (*Laevicardium*, *Retusa*) and polychaetes were important in the summer. The same pattern has been reported in Connecticut (Pearcy, 1962). Clam siphons are an important food for larger fish wherever food habits have been studied.

The winter flounder is considered a typical estuarine fish, and physiological studies have focussed on its adaptability in this variable regime. There is a seasonal change in the serum chemistry of adults (Umminger and Mahoney, 1972; Frame, 1973a; Fletcher, 1977) that depresses the freezing point. This protects the fish in the condition of low temperature and generally low salinity during the spawning season. High temperatures combined with high salinities can be fatal (Bigelow and Schroeder, 1953; Frame, 1973b).

Perlmutter (1947) claimed that winter flounder did not move very far from their place of spawning, and were hence quite susceptible to local management. Recent studies in Massachusetts waters (Howe and Coates, 1975) do not support this view. In particular, fish tagged in upper Buzzards Bay had moved on the average from 10 to 32 km at the time of recapture. Most returns were from the Nantucket Shoals area. Movement back inshore as spawning season approaches does not appear to be place-specific (McCracken, 1963; Howe and Coates, 1975), although recent work suggests that in some areas distinct groups comprising adjacent estuarine spawning units can be recognized (Pierce and Howe, 1977).

There are no explicit studies as to what limits flounder production. Since fecundity is high, greatest mortality must occur at the larval and juvenile stages. Since the fish are quite adaptable to the physical environment, food alone may be limiting. The major predator on winter flounder may be man. Winter flounder do not seem to be important in the diets of any fishes, but ospreys (*Pandion*) fish for them in shallow water.

Fishes that spawn in the sea

The great majority of estuarine fish neither spawn in estuaries nor pass through on spawning migrations. These are representatives of the essentially marine groups that make up the bulk of estuarine populations, especially among those which occur seasonally. Important examples are *Brevoortia* spp., and other Clupeidae, Engraulidae, Sciaenidae, *Rhabdosargus globiceps* and other Sparidae, Gadidae, Chanidae, Synodontidae, Belonidae, *Pomatomus saltatrix*, and most flounders (Pleuronectidae, Bothidae, Soleidae, Cynoglossidae). The common pattern is offshore spawning, often as far away as the edge of the continental shelf, with the production of great numbers of small pelagic eggs. Upon hatching, the larvae move inshore, and many of them to take up residence in estuaries. Details of the early life history for most species are poorly known. Analysis of data from plankton hauls made in the Middle Atlantic Bight extending from Cape Cod to Cape Hatteras, U.S.A., identified particular larval assemblages (Kendall, 1975). These assemblages were not cohesive, for, even though the larvae occurred together, the adults often occupied quite different habitats. Some were later estuarine, some demersal on the shelf, and some pelagic offshore. Stratification, and thus separation, of certain groups was found in the water column, with the implication that larval feeding habits could be somewhat different. How larvae from offshore find and move into estuaries is not known. Within the estuarine influence, larvae can remain within the estuary by vertically migrating to take advantage of the varying current flow (Graham, 1972). By the time most offshore larvae reach the estuary, however, they have metamorphosed into juveniles. For this reason, larval surveys may not reveal the true degree to which young fishes are using an estuarine region (Pearcy and Myers, 1974).

In general, knowledge of the life histories of estuarine species that spawn in the sea is quite limited. Because so many species are involved, it is difficult to pick any one species as typical of this type. The example we consider is the striped mullet, *Mugil cephalus*.

Life history of mullets

Mugil cephalus occurs around the world in subtropical waters (Briggs, 1960), and aspects of its life history seem quite similar wherever studied. There is good information on this species available from Texas waters (Moore, 1974, 1976), and from Egypt for *M. cephalus* and the quite similar *M. capito* (El-Zarka et al., 1970; El-Maghraby et al., 1973, 1974; Hashem et al., 1973; Helmy et al., 1974; Rafail and Hamid, 1974).

Mugil cephalus is one of the larger species of mullets, reaching a meter in length although most specimens are in the 30 to 50 cm range. The usual habitat of the adult is the shallow brackish bays and estuaries of the subtropics. In the western Atlantic, it has been recorded from Cape Cod to southern Brazil, but its greatest abundance is achieved in the Gulf of Mexico (Gunter, 1967). Towards the equator, in true tropical waters, it is replaced by the warmer-adapted *M. curema* (Moore, 1976). In south Texas, it is found year-round and is widespread in all estuarine habitats. The mullet is widely tolerant of a great range of conditions, including salinity to 80‰ (Hedgpeth, 1967). The adults feed in shallow waters on small invertebrates and diatoms strained from the sediments, on planktonic organisms, and on plant material including floating algae and detritus (Odum, 1970a); Odum (1968) reports mullet grazing in a dinoflagellate bloom. Growth rate is very fast, and the fish reach maturity at an age of from one to two years. Ripe fish migrate from the estuaries and lagoons in the late summer and fall in order to spawn in the sea. Spawning is thought to take place rather near the coast, but Arnold and Thompson (1958) found *M. cephalus* spawning at sea 80 km off the Mississippi River mouth. The eggs are pelagic and float near the surface. Rather soon after hatching, in winter and into early spring, the young mullets return in great quantities to brackish water where they will feed and grow. How the little fishes find the narrow passes into the lagoons — and these openings comprise, for example, less than 2% of the Texas coastline — is not known. Some response to the current flow must be involved, as shown by the fact that sufficient mullet fry are obtained by hand in the artificial drainage canal from Lake Maryut, Egypt, to stock the entire lake, which is otherwise not connected to the sea (Rafail and Hamid, 1974).

Young fishes in estuaries

By far the great preponderance of fishes in estuarine communities, regardless of the species, are juveniles. This point is made again and again, so much so that estuaries are often referred to as nursery grounds for fishes (Gunter, 1961). This is certainly true for temperate regions, although whether shallow bays are significantly different from true estuarine areas has not been unequivocally established. The Wadden Sea in Europe can certainly be called a nursery area (Creutzberg and Fonds, 1971).

Gunter (1967) pointed out the great importance of the estuarine nursery areas in the maintenance of the commercial offshore stocks of fishes. McHugh (1967) thought that the U.S. Atlantic coast and Gulf of Mexico estuaries constituted extremely important nursery grounds, but that Pacific coast estuaries did not. Studies such as those of Haertel and Osterberg (1967), Pearcy and Myers (1974), and Misitano (1977), however, have shown that in fact large numbers of juvenile fishes do move into estuaries in the Pacific northwest. On the west coast, the young of anadromous forms coming from fresh water play a more dominant role in the young fish community than is the case along the Atlantic and Gulf coasts.

The extent to which boreal and tropical estuaries are nursery grounds for fishes is not so clear. The question does not seem to have been addressed in really northern areas, but Sutcliffe (1972, 1973) has shown that an extremely interesting relationship exists between river discharge from the St. Lawrence and the subsequent catch of commercial species that spend their early lives in that large estuary. The argument is that greater discharge increases the nutrient flux, and the resulting increase in food improves fish survival during the critical first few months of their lives. The improvement in survival should make itself felt when the fish first enter the fishery several years later. The correlation seems to hold (Fig. 7.4), with the inference that this fairly northern area does serve an important nursery function. Juvenile fishes did dominate the fauna of a mangrove swamp in Puerto Rico (Austin, 1971), but data from other tropical areas have not been reported.

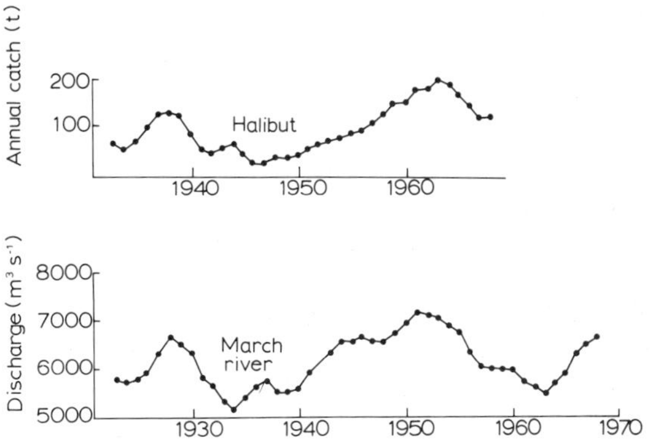

Fig. 7.4. Annual Atlantic halibut (*Hippoglossus hippoglossus*) catch of Quebec and the March discharge of the St. Lawrence River ten years earlier. It takes about ten years for halibut to enter the fishery (after Sutcliffe, 1973).

Trophic considerations

Clearly, the high productivity of estuarine regions is a basic reason for their use by fishes as nursery areas. The emphasis is thus on feeding and growth, and it is the superiority of estuaries in respect to these two activities that really accounts for their importance to fishes. An excellent summary of fish feeding and trophic relationships in estuaries is provided by deSylva (1975). His comparative study recognized fifteen possible food webs, six based on phytoplankton and nine beginning with detritus (Table 7.2). Depending both on season and the state of their development, fishes shift from one sort of feeding habit to another (Austin and Austin, 1971), thus complicating the overall picture and making detailed quantification extremely difficult. The situation will vary also in regard to the individual conditions of each estuary. Diener et al. (1974) found no well-defined food chains and much overlap in Clear Lake, Texas, with dependence on crustaceans being particularly marked. In a Louisiana estuary where fish weirs had created a special habitat dominated by submerged vegetation, Weaver and Holloway (1974) found two distinct communities, with quite different feeding relationships. An open-water group dominated by *Brevoortia*, *Anchoa*, and *Micropogon* fed on detritus, zooplankton, and benthos, and the fauna near the weirs, dominated by *Poecilia* and *Lucania* (Cyprinodontidae), fed on the small crustaceans living on the plants. Haertel and Osterberg (1967) found that young fish were at their most abundant, and the diversity was much higher, at that point in the estuary where the major food species, the copepod *Eurytemora*, reached its greatest densities.

Although some specialized feeding habits can be found in estuarine fishes, for example adult

TABLE 7.2

Classification of nektonic food webs in estuaries (After deSylva, 1975)

Phytoplankton-generated food webs
Phytoplankton→zooplankton→planktivorous pelagic and benthopelagic fish
Phytoplankton→zooplankton→planktivorous fish→large fish predators
Phytoplankton→phytoplanktonic fishes such as menhaden (*Brevoortia*): summer
Phytoplankton→zooplankton→menhaden: winter
Phytoplankton→zooplankton→large carnivores (*Manta*)
Phytoplankton (dinoflagellates)→mullet (Mugilidae): alteration of usual feeding habits
Detritus-generated food webs
Detritus→benthos epifauna)→benthophagous fishes
Detritus→benthos (infauna)→benthophagous fishes
Detritus→benthos→benthophagous fishes→large fish predators (sharks)
Detritus→small benthos→larger invertebrates and small benthic fishes→large fishes
Detritus→large detritivorous fishes (mullet): "telescoping" of food chain
Detritus→benthos→large predators
Detritus→micronekton→intermediate predators such as snappers (Lutjanidae) and croaker (Sciaenidae)
Detritus→zooplankton→small fishes and invertebrates
Detritus→zooplankton→small fishes and invertebrates→larger fishes

Pomatomus which depend largely on *Brevoortia*, the general rule seems to be that estuarine fishes are rather generalized in their habits (Longhurst, 1957). In part this may stem from the fact that, because of the high productivity, food is never a limiting factor and competition for it is not an issue. Certainly fishes at the lower trophic levels, for example *Fundulus* and *Mugil*, are quite omnivorous, even eating detritus. A detailed study of the feeding of an estuarine fish community, with emphasis on principal prey and energetics, has not been done completely, but Odum and Heald (1972, 1975) have made a beginning in southern Florida.

The larger fish that enter estuaries most often as seasonal migrants tend to feed on the small fish which have profited directly from the estuarine production. These, and the subsequent growth stages of the small fish, eventually depart from the estuary. In this way, feeding by fish is an important pathway for the export from estuaries of the nutrients that might otherwise be trapped there.

Besides the feeding opportunities, other reasons have been suggested that might explain the great abundance of juvenile fishes in estuaries. One is that estuarine areas tend to be more protected than open coasts, and this relatively quiet environment has its benefits (Day, 1951). The other is that predators may be somewhat less abundant, or at least only seasonally present, in the estuary, thus reducing pressure from that quarter (Gunter, 1961). The estuary offers opportunities for concealment from predators, particularly in the generally turbid water, that the open coast might not (Livingston, 1975). On the other hand, as Austin (1971) noted, juvenile fishes may be more tolerant of salinity variation than are adults, and the reason could thus be simple exclusion.

Feeding, some shelter, and the reduction of predation would apply as benefits to large fishes as well as small ones, although the relative degree of importance would vary. Boreal estuaries where fresh water continues to run in winter offer a refuge from the super-cooled conditions of the sea to large *Salvelinus alpinus*.

In summary, then, it can be said that estuarine fishes display a wide range in aspects of their life histories. As a site for spawning, the estuary is important for only a very few species. For those, the production of large numbers of adherent demersal eggs is the rule, with staggered spawning seasons and areas reducing competition. The estuary is most important as a feeding or nursery area for the young of a great many species, regardless of where they spawn. Detritus- and phytoplankton-based food webs are both important to fishes in estuaries, and a wide range of feeding types is present. Generalized feeding seems to be the rule, but the data available on this point are insufficient for a conclusive statement. Nonetheless, specialized adaptations related to feeding are not apparent in estuarine fishes.

ESTUARINE FISHES AND MAN

The affairs of estuarine fishes are closely intertwined with the affairs of men. The value of the estuarine-dependent fisheries, which in the United States in 1970 exceeded 350 million dollars (McHugh, 1976), is basic to this relationship. The fisheries, whether for strictly estuarine species or for species which use the estuary mainly as a nursery, depend on the high productive potential of the estuaries. The desire to maintain this potential, and the accompanying necessity to keep estuarine areas from becoming over-stressed, underlies the burgeoning concern and literature dealing with the effects of pollution in near-shore environments (e.g. Livingston, 1976; Vernberg et al., 1977). When interests other than productivity are paramount, for example the construction of electric power plants or the damming of rivers for irrigation purposes, estuarine fishes often are the losers (Clark, 1967).

The often-cited estuarine dependence of many species is in fact an intuitive observation, and has yet to be unequivocally demonstrated. Nonetheless, there can be no doubt that estuarine conditions exert considerable influence on the abundance of certain fish stocks. Sutcliffe (1972, 1973) showed that commercial landings in the large semi-estuarine Gulf of St. Lawrence tracked variations in the annual river flow. Increased flow was thought to result in a greater availability of nutrients, both from the river and from entrainment and upward mixing from the salt water, and the increased primary production would be reflected at all levels, including the fishes. A direct relationship between photosynthesis and fish production, conceptually satisfying at that time, has only recently been shown to hold, and then only for a few situations of pond culture (McConnell et al., 1977). Stevens (1977) found a direct correlation between river flow and year-class strength for *Morone saxatilis* in California's Sacramento River, but could not relate this to changes in productivity. He suggested that a complex of factors beyond Sutcliffe's simple model must be responsible for the observed relation. But when the High Dam at Aswan stopped the Nile flood, with its load of nutrients, the annual Egyptian near-shore catch of *Sardinella* dropped from 20 000 t to around 600 t (Aleem, 1972; Ben-Tuvia, 1973). At the same time, the increased catch from the newly-formed Lake Nasser almost exactly matched the decline in marine landings (Ryder and Henderson, 1975).

Estuaries are bottlenecks through which diadromous species must pass on their spawning migrations. For these species, and to a lesser extent those that customarily spawn in estuaries, estuarine degradation can have serious consequences. Dredging, filling, rechanneling, and the construction of dams all have an obvious detrimental impact. The St. Lucia Estuary in South Africa, once with an abundant and diverse biota, was rendered almost sterile when siltation resulting from man's activities in the drainage system closed the opening to the sea (Day et al., 1955). Less obvious are what have been termed the insidious effects of pollution (Odum, 1970b). Pollutants can be trapped in the estuary in the same way nutrients are, and sublethal, chronic exposure of the biota to pollutants can be debilitating to all estuarine populations, both resident and seasonal. A manifestation of such exposure is thought to be a high incidence of fin-rot and tumors in fishes that encounter domestic sewage wastes (Mearns and Sherwood, 1974; Mahoney et al., 1973). Free chlorine in concentrations less than 0.20 ppm can result in abnormalities during early growth in fishes, but the time when exposure occurs can be critical. Eggs of several species of Chesapeake Bay fishes were very sensitive when first spawned, but the sensitivity declined with age (Morgan and Prince, 1977). After 48 h, eggs were not harmed by as much as 0.40 ppm Cl_2. Larvae of the same species showed just the opposite response, becoming increasingly sensitive to chlorine the older they were. Heavy metals were not assimilated to any great extent by the three dominant fishes during the summer in a North Carolina estuary, but the activities of the three were important in the cycling there of those materials (Cross et al., 1975).

From field observations it may be very difficult to separate pollution-induced changes in fish populations from naturally occurring and little understood long-term variations (McHugh, 1976; Hillman et al., 1977). Extrapolations from laboratory studies may be misleading because of the confined area and because the experiments cannot simulate the heterogeneous estuarine environment. Synergisms can be quite important in determining fish distribution (Copeland and Bechtel, 1974), and fish, being motile, can influence their surroundings behaviorally by moving (Livingston et al., 1976a).

Experiments on the effects of pollutants not uncommonly give mixed results (Nimmo and Bahner, 1976). Benzene in low concentrations alters the metabolic rate of larval fishes, causing death in some individuals but apparently beneficial effects in others. Eldridge et al. (1977) point out that stress, as by pollutants at low levels, can exercise the adaptive physiological functions otherwise perhaps latent in young estuarine fishes, thereby producing healthier fry. Domestic sewage in estuaries could have a beneficial effect in increasing nutrients and thereby fish stocks, a relationship suggested by Mansueti (1961) for *Morone saxatilis* in Chesapeake Bay. But too much nutrient can produce eutrophication, with the result that large areas of the bottom water become anoxic. This is what apparently happened in the New York Bight off the Hudson River in the summer of 1976 (Officer and Ryther, 1977).

Power plants influence estuarine fishes both mechanically, through entrainment in the cooling water, and physiologically by heat shock in the plant and the discharge of warm effluent (O'Connor and McErlean, 1975). In addition, the toxic materials that are on occasion passed through the system to discourage the growth of fouling organisms are discharged as well. Passage through a power plant can cause mortalities near 90% in eggs and larvae of fishes (Marcy, 1971), and, where effects are apparently sublethal, disorientation renders young fishes more susceptible to predation (Goodyear, 1972; Sylvester, 1972). The trapping of larger fishes on intake screens has resulted in enormous kills (O'Connor and McErlean, 1975).

Heat cannot be considered a deleterious influence on adult fishes near power plants, for they seem to avoid effluent plumes during warm weather, but congregate there during cold spells (Gallaway and Strawn, 1974). Although the thermal shock associated with entrainment can be fatal to larval fishes, the rate of mortality varies with species, duration of exposure, and other conditions (Hoss et al., 1974). The magnitude of temperature elevation is important. Schubel (1974) found that from 6 to 10°C excess temperature had no influence on the hatching success of the eggs of *Alosa* and *Morone* species. An important point made by Hoss et al. (1974), and subsequently by others (Merriman and Thorpe, 1976), is that no *a priori* set of criteria can be used in power plant siting or indeed in any other environmental management scheme (Rickert and Hines, 1978). Each must be considered individually, with appropriate field studies, if environmental impact is to be minimized.

Hickling (1970) provides a general review of fish aquaculture in estuaries. The practice is widespread, particularly in the warmer waters of the Indo-Pacific. Important species include those in the Mugilidae, *Chanos chanos*, *Anguilla* spp., and, in brackish lagoons, *Tilapia mossambica*. In northern Europe, the rearing of flatfish in Scottish sea lochs has been the subject of considerable research for many years (Saila, 1975), and the artificial culture of salmonids is widely used to maintain stocks in the Northern Hemisphere.

The use of estuaries for the culture of fish would seem to be more a matter of convenience than of necessity. Many of the species used in estuarine aquaculture are diadromous, and the bottleneck effect of the estuary makes capture of ripe adults or of migrating young stages a simple matter. Portions of estuaries are more easily diked off or blocked than other regions might be. Markets, particularly in the Orient, are nearby. Nonetheless, most estuarine aquaculture for fish is essentially pond culture, or even in brackish lakes far from the sea, as for example is successfully done in Lake Qârûn (Birket Qârûn) in Egypt. An estuarine or near-estuarine location is more necessary where management schemes are involved. A salmon or smelt project takes advantage of the species' ability to home to a particular estuary, providing an opportunity for management not available elsewhere. The fish are spawned artificially and are released to grow in the sea, from whence they return to be harvested. Other sorts of management, for example pulsing river flow to inject nutrients at appropriate times in accord with Sutcliffe's (1972, 1973) hypothesis, are possible. Most water control measures have as their goal, however, the reduction of pulses in flow, and projects of this sort seem unlikely.

Man's activities have considerable impact on the lives of estuarine fishes. Agricultural and engineering projects alter the shape and nature of estuaries through marsh destruction and siltation. The reopening of closed estuaries improves the water quality. Domestic and industrial discharges add pollutional stress to an already stressed condition, but such discharges can also enhance production. Interest in maintaining estuarine fisheries is a force to keep estuaries clean and productive. Heavy fishing pressure has a significant effect on abundance and community structure. Aquaculture enhances the production of one species, but probably almost always at the expense of another. In all these areas, scientific input is necessary. Quantification has only begun, and much information is still not available. Ultimately, however, management involves values, and must be agreed upon in the political, rather than the scientific, arena.

ACKNOWLEDGEMENTS

My thanks to R. John Gibson, Pamela T. Polloni, and Richard H. Backus for useful critiques of the manuscript. Support was provided by the NOAA Office of Sea Grant, Department of Commerce, grant number 04-7-158-44104.

REFERENCES

Aleem, A.A., 1972. Effect of river outflow management on marine life. *Mar. Biol.*, 15: 200–298.

Allen, L.G. and Horn, M.H., 1975. Abundance, diversity and seasonality of fishes in Colorado Lagoon, Alamitos Bay, California. *Estuarine Coastal Mar. Sci.*, 3: 371–380.

Andrews, C.W. and Lear, E., 1956. The biology of Arctic char (*Salvelinus alpinus* L.) in northern Labrador. *J. Fish. Res. Board Can.*, 13 (6): 843–860.

Annandale, N., 1922. The marine element in the fauna of the Ganges. *Bijdr. Dierkd.*, 22: 143–154.

Arnold, E.L., Jr. and Thompson, J.R., 1958. Offshore spawning of the striped mullet, *Mugil cephalus*, in the Gulf of Mexico. *Copeia*, 1958 (2): 130–132.

Austin, H.M., 1971. A survey of the ichthyofauna of the mangroves of western Puerto Rico during December 1967–August 1968. *Caribb. J. Sci.*, 11(1)2): 27–39.

Austin, H.M. and Austin, S., 1971. The feeding habits of some juvenile marine fishes from the mangroves in western Puerto Rico. *Caribb. J. Sci.*, 11(3/4): 171–178.

Backus, R.H., 1957. The fishes of Labrador. *Bull. Am. Mus. Nat. Hist.*, 133(4): 273–338.

Bechtel, T.J. and Copeland, B.J., 1970. Fish species diversity indices as indicators of pollution in Galveston Bay, Texas. *Contrib. Mar. Sci.*, 15: 103–132.

Ben-Tuvia, A., 1973. Man-made changes in the eastern Mediterranean and their effect on the fishery resources. *Mar. Biol.*, 19: 197–203.

Bigelow, H.B. and Schroeder, W.C., 1953. Fishes of the Gulf of Maine. *U.S. Fish Wildl. Serv., Fish. Bull.*, 74: 1–577.

Bigelow, H.B. and Schroeder, W.C., 1963. Family Osmeridae. Fishes of the Western North Atlantic. *Sears Found. Mar. Res. Mem. 1, part 3*: 553–598.

Boesch, D.F., 1977. A new look at the zonation of benthos along the estuarine gradient. In: B.C. Coull (Editor), *Ecology of Marine Benthos*. University of South Carolina Press, Columbia, S.C., pp. 245–266.

Briggs, J.C., 1960. Fishes of worldwide (Circumtropical) distribution. *Copeia*, 1960(3): 171–180.

Briggs, P.T., 1965. The sport fisheries for winter flounder in several bays of Long Island. *N. Y. Fish. Game J.*, 12(1): 48–70.

Brock, R.H., 1977. Occurrence and variety of fishes in mixohaline ponds of the Kona, Hawaii, Coast. *Copeia*, 1977(1): 134–138.

Carriker, M.R., 1967. Ecology of estuarine benthic invertebrates: a perspective. In: G.H. Lauff (Editor), *Estuaries. Am. Assoc. Adv. Sci., Publ.*, No. 83: 442–487.

Chenoweth, S.B., 1973. Fish larvae of the estuaries and coast of central Maine. *Fish. Bull., U.S.* 71(1): 105–113.

Chesney, Jr., E.J. and Iglesias, J., 1979. Influence of mussel rafts on the distribution, abundance, and diversity of demersal fishes in the inner Ria de Arosa. Northwest Spain. *Estuarine Coastal Mar. Sci.*, 8: 227–239.

Clark, J., 1967. Fish and man. Conflict in the Atlantic estuaries. *Am. Littoral Soc., Spec. Publ.*, No. 5: 1–78.

Cohen, D.M., 1970. How many recent fishes are there? *Proc. Calif. Acad. Sci., 4th Ser.*, 38(17): 341–346.

Copeland, B.J. and Bechtel, T.J., 1974. Some environmental limits of six Gulf Coast estuarine organisms. *Contrib. Mar. Sci.*, 18: 169–204.

Creutzberg, F. and Fonds, M., 1971. The seasonal variation in the distribution of some demersal fish species in the Dutch Wadden Sea. *Thalassia Jugosl.*, 7(1): 13–23.

Cronin, L.E. (Editor), 1975. *Estuarine Research, I. Chemistry, Biology, and the Estuarine System.* Academic Press, New York, N.Y., 738 pp.

Cross, F.A., Willis, J.N., Hardy, L.H., Jones, N.Y. and Lewis, J.M., 1975. Role of juvenile fish in cycling of Mn, Fe, Cu, and Zn in a coastal plain estuary. In: L.E. Cronin (Editor), *Estuarine Research, I. Chemistry Biology, and the Estuarine System.* Academic Press, New York, N.Y., pp. 45–63.

Dahlberg, M.D. and Odum, E.P., 1970. Annual cycles of species occurrence, abundance, and diversity in Georgia estuarine fish populations. *Am. Midl. Nat.*, 83(2): 382–392.

Day, J.H., 1951. The ecology of South African estuaries. Part I. A review of estuarine conditions in general. *Trans. R. Soc. S. Afr.*, 33: 53–91.

Day, J.H., 1964. The origin and distribution of estuarine animals in South Africa. *Monogr. Biol.*, 14: 159–173.

Day, J.H., Millard, N.A.H. and Harrison, A.D., 1952. The ecology of South African estuaries. Part III. Knysa: a clear open estuary. *Trans. R. Soc. S. Afr.*, 33: 367–413.

Day, J.H., Millard, N.A.H. and Broekehuysen, G.J., 1955. The ecology of South African estuaries. Part IV. The St. Lucia System. *Trans. R. Soc. S. Afr.*, 34: 129–156.

deSylva, D.P., 1975. Nektonic food webs in estuaries. In: L.E. Cronin (Editor), *Estuarine Research, I. Chemistry, Biology, and the Estuarine System.* Academic Press, New York, N.Y., pp. 420–447.

Diener, R.A., Inglis, A. and Adams, G.B., 1974. Stomach contents of fishes from Clear Lake and tributary waters, a Texas estuarine area. *Contrib. Mar. Sci.*, 18: 7–17.

Dudley, R.G., Mullis, A.W. and Terrell, J.W., 1977. Movements of adult striped bass (*Morone saxatilis*) in the Savannah River, Georgia. *Trans. Am. Fish. Soc.*, 106(4): 314–322.

Dunbar, M.J., 1970. On the fishery potential of the sea waters of the Canadian north. *Arctic*, 23(3): 150–174.

Dyer, K.R., 1973. *Estuaries: A Physical Introduction.* Wiley, London, 140 pp.

Eldridge, M.B., Echeverria, T. and Whipple, J.A., 1977. Energetics of Pacific herring (*Clupea harengus pallasi*) embryos and larvae exposed to low concentrations of benzene, a monoaromatic component of crude oil. *Trans. Am. Fish. Soc.*, 106(5): 452–461.

El-Maghraby, A.M., Hashem, M.T. and El-Sedfy, H.M., 1973. Some biological characters of *Mugil capito* (Cuv) in Lake Borollus. *Bull. Inst. Oceangr. Fish. A.R.E.*, 3: 58–82.

El-Maghraby, A.M., Hashem, M.T. and El-Sedfy, H.M., 1974. Sexual maturity, spawning migration, and fecundity of *Mugil capito* (Cuv.) in Lake Borollus. *Bull. Inst. Oceanogr. Fish. A.R.E.*, 4:33–56.

El-Zarka, S., El-Maghraby, A.M. and Abdel-Hamid, K., 1970. Studies on the distribution, growth, and abundance of migrating fry and juveniles of mullet in a brackish coastal lake (Edku) in the United Arab Republic. *Stud. Rev. G.F.C.M.*, No. 46: 1–19.

Ezzat, A. and El-Serafy, S., 1977. The migration of *Anguilla anguilla* L. in the Mex Canal, Alexandria, Egypt. *J. Fish. Biol.*, 11: 249–256.

Fletcher, G.L., 1977. Circannual cycles of blood plasma freezing point and Na^+ and Cl^- concentrations in Newfoundland winter flounder (*Pseudopleuronectes americanus*). *Can. J. Zool.*, 53: 197–206.

Folger, D.W., 1972a. Texture and organic carbon content of bottom sediments in some estuaries of the United States. *Geol. Soc. Am. Mem.*, 133: 391–408.

Folger, D.W., 1972b. Characteristics of estuarine sediments of the United States. *U.S. Geol. Surv. Prof. Pap.*, 742: 1–94.

Follett, W.I., 1976. Fish remains from an archeological site at Rancho Carillo on the Silver Strand, San Diego County, California. *Bull. S. Calif. Acad. Sci.*, 75(2): 131–137.

Frame, D.W., 1973a. Biology of young winter flounder, *Pseudopleuronectes americanus* (Walbaum): metabolism under simulated estuarine conditions. *Trans. Am. Fish. Sci.*, 102(2): 423–430.

Frame, D.W., 1973b. Conversation efficiency and survival of young winter flounder (*Pseudopleuronectes americanus*) under experimental conditions. *Trans. Am Fish. Soc.*, 102(3): 614–617.

Frame, D.W., 1974. Feeding habits of young winter flounder (*Pseudopleuronectes americanus*): prey availability and diversity. *Trans. Am. Fish. Soc.*, 103(2): 261–269.

Gallaway, B.J. and Strawn, K., 1974. Seasonal abundance and distribution of marine fishes at a hot-water discharge in Galveston Bay, Texas. *Contrib. Mar. Sci.*, 18: 71–137.

Goodyear, C.P., 1972. A simple technique for detecting effects of toxicants or other stresses on a predator prey interaction. *Trans. Am. Fish. Soc.*, 101: 367–370.

Gorsline, D.S., 1967. Contrasts in coastal sediments on the Gulf and Pacific coasts. In: G.H.Lauff (Editor), *Estuaries. Am. Assoc. Adv. Sci., Publ.*, 83: 219–225.

Graham, J.J., 1972. Retention of larval herring within the Sheepscot estuary of Maine. *Fish. Bull., U.S.*, 70: 299–305.

Grainger, E.H., 1953. On the age, growth, migration, reproductive potential and feeding habits of the Arctic char, *Salvelinus alpinus*, of Frobisher Bay, Baffin Island. *J. Fish. Res. Board., Can.*, 10(6): 326–370.

Greenwood, P.H., Rosen, D.E., Weitzman, S.H. and Myers, G.S., 1966. Phyletic studies of teleostean fishes with a provisional classification of living forms. *Bull. Am. Mus. Nat. Hist.*, 131(4): 339–456.

Gunter, G., 1956. A revised list of euryhalin fishes of North and Middle America. *Am. Midl. Nat.*, 56(2): 345–354.

Gunter, G., 1961. Some relations of estuarine organisms to salinity. *Limnol. Oceanogr.*, 6(2): 182–190.

Gunter, G., 1967. Some relationships of estuaries to the fisheries of the Gulf of Mexico. In: G.H. Lauff (Editor), *Estuaries. Am. Assoc. Adv. Sci., Publ.* 83: 621–638.

Haedrich, R.L., 1975. Diversity and overlap as measures of environmental quality. *Water Res.*, 9: 945–952.

Haedrich, R.L. and Haedrich, S.O., 1974. A seasonal survey of the fishes in the Mystic River, a polluted estuary in downtown Boston, Massachusetts. *Estuarine Coastal Mar. Sci.*, 2: 59–73.

Haedrich, R.L. and Hall, C.A.S., 1976. Fishes and estuaries. *Oceanus*, 19 (5): 55–63.

Haertel, L. and Osterberg, C., 1967. Ecology of zooplankton, benthos, and fish in the Columbia River estuary. *Ecology*, 48(3): 459–472.

Hardenberg, J.D.F., 1931. The fish fauna of the Rokan mouth. *Treubia*, 13: 81–168.

Hart, J.L., 1973. Pacific fishes of Canada. *Fish. Res. Board Can., Bull.*, 180: 1–739.

Hartley, P.H.T., 1940. The Saltash tuck-net fishery and the ecology of some estuarine fishes. *J. Mar. Biol. Assoc. U.K.*, 24(1): 1–68.

Hashem, M.T., El-Maghraby, A.M. and El-Sedfy, H.M., 1973. The grey mullet fishery of Lake Borollus. *Bull. Inst. Oceanogr. Fish. A.R.E.*, 3: 29–54.

Hedgpeth, J.W., 1967. Ecological aspects of the Laguna Madre, a hypersaline estuary. In: G.H. Lauff (Editor), *Estuaries. Am. Assoc. Adv. Sci., Publ.* 83: 408–419.

Helmy, A.M., Badawi, H.K. and El-Bishry, A., 1974. Seasonal variations in the protein composition of blood serum of *Anguilla vulgaris* and *Mugil cephalus*. *Bull. Inst. Oceanogr. Fish A.R.E.*, 4: 367–381.

Hickling, C.F., 1970. Estuarine fish farming. *Adv. Mar. Biol.*, 8: 119–213.

Hillman, R.E., Davis, N.W., and Wennemer, J., 1977. Abundance, diversity and stability of shore-zone fish communities in an area of Long Island Sound affected by the thermal discharge of a nuclear power station. *Estuarine Coastal Mar. Sci.*, 5(3): 355–382.

Hoese, H.D., 1973. A trawl study of nearshore fishes and invertebrates of the Georgia coast. *Contrib. Mar. Sci.*, 17: 63–98.

Hoff, J.G. and Ibara, R.M., 1977. Factors affecting the seasonal abundance, composition and diversity of fishes in a southeastern New England estuary. *Estuarine Coastal Mar. Sci.*, 5(5): 665–678.

Horn, M.H. and Allen, L.G., 1976. Numbers of species and faunal resemblance of marine fishes in California bays and estuaries. *Bull. S. Calif. Acad. Sci.*, 75(2): 159–169.

Hoss, D.E., Hettler Jr., W.F. and Costen, L.C., 1974. Effects of thermal shock on larval estuarine fish — ecological implications with respect to entrainment in power plant cooling systems. In: J.H.S. Blaxter (Editor), *The Early Life History of Fish*. Springer-Verlag, Berlin, pp. 357–371.

Howe, A.B. and Coates, P.G., 1975. Winter flounder movements, growth, and mortality off Massachusetts. *Trans. Am. Fish. Soc.*, 103(1): 13–29.

Hubbs, C.L., 1948. Changes in the fish fauna of western North America correlated with changes in ocean temperature. *J. Mar. Res.*, 7(3): 459–482.

Jaccard, P., 1902. Lois de distribution florale dans la zone alpine. *Bull. Soc. Vaud. Sci. Nat.*, 38: 69–130.

Jeffries, H.P. and Johnson, W.C., 1974. Seasonal distributions of bottom fishes in the Narragansett Bay area: seven-year variations in the abundance of winter flounder (*Pseudopleuronectes americanus*). *J. Fish. Res. Board Can.*, 31: 1057–1066.

Kendall Jr., A.W., 1975. Patterns of larval fish distributions in the Middle Atlantic Bight. In: *Effects of Energy-Related Activities on the Atlantic Continental Shelf.* Brookhaven National Laboratory, Upton, N.Y., pp. 126–145.

Lauff, G.H. (Editor), 1967. *Estuaries. Am. Assoc. Adv. Sci., Publ.*, 83: 723 pp.

Levings, C.D., 1973. Dominance and movements of winter flounder (*Pseudopleuronectes americanus*) at St. Margaret's Bay, Nova Scotia. *Nat. Can.*, 100: 337–345.

Liem, A.H. and Scott, W.B., 1966. Fishes of the Atlantic Coast of Canada. *Fish. Res. Board Can., Bull.*, 155: 1–485.

Livingston, R.J., 1975. Impact of kraft pulp-mill effluents on estuarine and coastal fishes in Apalachee Bay, Florida, USA. *Mar. Biol.*, 32: 19–48.

Livingston, R.J.., 1976. Dynamics of organochlorine pesticides in estuaries: effects on estuarine biota. In: M. Wiley (Editor), *Estuarine Processes, I. Uses, Stresses, and Adaptation to the Estuary*. Academic Press, New York, N.Y., pp. 507–522.

Livingston, R.J., Cripe, C.R., Laughlin, A.R. and Lewis, III, F.G. 1976a. Avoidance responses of estuarine organisms to storm water runoff and pulp mill effluents. In: M. Wiley (Editor), *Estuarine Processes, I. Uses, Stresses, and Adaptation to the Estuary*. Academic Press, New York, N.Y., pp. 313–332.

Livingston, R.J., Kobylinski, G.J., Lewis, III, F.G. and Sheridan, P.F., 1976b. Long-term fluctuations of epibenthic fish and invertebrate populations in Apalachicola Bay, Florida. *Fish. Bull. U.S.*, 74(2): 311–321.

Longhurst, A.R., 1957. The food of the commercial demersal fish of a West African estuary. *J. Anim. Ecol.*, 26(2): 369–387.

Lowe-McConnell, R.H., 1975. *Fish Communities in Tropical Freshwaters*. Longman, London, 337 pp.

MacArthur, R.H. and Wilson, E.O., 1967. *The Theory of Island Biogeography*. Princeton University Press, Princeton, N.J., 203 pp.

McCleave, J.D. and Fried, S.M., 1975. Nighttime catches of fishes in a tidal cove in Montsweag Bay near Wiscasset, Maine. *Trans. Am. Fish. Soc.*, 104(1): 30–34.

McCleave, J.D., Fried, S.M. and Towt, A.K., 1977. Daily movements of shortnose sturgeon, *Acipenser brevirostrum*, in a Maine estuary. *Copeia*, 1977(1): 149–157.

McConnell, W.J., Lewis, S. and Olsen, J.E. 1977. Gross photosynthesis as an estimator of potential fish production. *Trans. Am. Fish. Soc.*, 106(5): 417–423.

McCracken, F.D., 1963. Seasonal movements of the winter flounder, *Pseudopleuronectes americanus* (Walbaum), on the Atlantic Coast. *J. Fish Res. Board Can.*, 20(2): 551–585.

McDowall, R.M., 1970. The galaxiid fishes of New Zealand. *Bull. Mus. Comp. Zool. Harvard*, 139(7): 341–432.

McDowall, R.M., 1972. The taxonomy of estuarine and brackish-lake *Retropinna* from New Zealand (Galaxioidei: Retropinnidae). *J. R. Soc. N.Z.*, 2(4): 501–531.

McDowall, R.M., 1976. The role of estuaries in the life cycles of fishes in New Zealand. *Proc. N.Z. Ecol. Soc.*, 23: 27–32.

McErlean, A.J., O'Connor, S.F., Mihursky, J.A. and Gibson, C.I., 1973. Abundance, diversity, and seasonal patterns of estuarine fish populations. *Estuarine Coastal Mar. Sci.*, 1: 19–36.

McHugh, J.L., 1966. Management of estuarine fisheries. *Trans. Am. Fish. Soc., Suppl.*, 95(4): 133–154.

McHugh, J.L., 1967. Estuarine nekton. In: G.H. Lauff (Editor), *Estuaries. Am. Assoc. Adv. Sci., Publ.*, 83: 581–620.

McHugh, J.L., 1976. Estuarine fisheries: are they doomed? In: M. Wiley (Editor), *Estuarine Processes, I. Uses, Stresses, and Adaptation to the Estuary*, Academic Press, New York, pp. 15–27.

McKenzie, R.A., 1959. Marine and freshwater fishes of the Miramichi River and estuary, New Brunswick. *J. Fish. Res. Board Can.*, 16(6): 807–829.

Mahoney, J.B., Midlige, F.H. and Deuel, D.G., 1973. A fin rot disease of marine and euryhaline fishes in the New York Bight. *Trans. Am. Fish. Soc.*, 102(3): 596–605.

Mansueti, R.J., 1961. Effects of civilization on striped bass and other estuarine biota in Chesapeake Bay and tributaries. *Proc. Gulf Caribb. Fish. Inst., 14th Annu. Sess.* pp. 110–136.

Marcy Jr., B.C., 1971. Survival of young fish in the discharge canal of a nuclear power plant. *J. Fish Res. Board Can.*, 28: 1057–1060.

Marques, J.R.S., 1968. Age and size at sexual maturity of white goby (*Glossogobius giurus*), a common species of fish of Laguna de Bay, with notes on its food habits. *Phillip. J. Fish.*, 8(1): 71–97.

Mearns, A.J. and Sherwood, M., 1974. Environmental aspects of fin erosion and tumors in Southern California Dover sole. *Trans. Am. Fish. Soc.*, 103(4): 799–810.

Merriman, D. and Thorpe, L.M., 1976. The Connecticut River ecological study, the impact of a nuclear power plant. *Monogr. Ser., Am. Fish Soc.*, 1: 1–252.

Millard, N.A.H. and Scott, K.M.F., 1955. The ecology of South African estuaries. Part VI. Milnerton estuary and the Deep River, Cape, *Trans. R. Soc. S. Afr.*, 34: 279–324.

Miller, D.J. and Lea, R.N., 1972. Guide to the coastal marine fishes of California. *Calif. Dep. Fish Game, Fish. Bull.*, 157: 1–235.

Miller, P.J., 1975. Age-structure and life-span in the common goby, *Pomatoschistus microps*. *J. Zool.* (*London*), 177(3): 425–448.

Misitano, D.A., 1977. Species composition and relative abundance of larval and post-larval fishes in the Columbia River estuary, 1973. *Fish. Bull., U.S.*, 75(1): 218–221.

Moore, R.H., 1974. General ecology, distribution, and relative abundance of *Mugil cephalus* and *Mugil curema* on the south Texas coast. *Contrib. Mar. Sci*, 18: 241–255.

Moore, R.H., 1976. Seasonal patterns in the respiratory metabolism of the mullets. *Mugil cephalus* and *Mugil curema*. *Contrib. Mar. Sci.*, 20: 133–146.

Morgan, R.P. and Prince, R.D., 1977. Chlorine toxicity to eggs and larvae of five Chesapeake Bay fishes. *Trans. Am. Fish. Soc.*, 106(4): 380–385.

Nimmo, D.R. and Bahner, L.H., 1976. Metals, pesticides, and PCBs: toxicities to shrimp singly and in combination. In: M. Wiley (Editor), *Estuarine Processes, I. Uses, Stresses, and Adaptation to the Estuary*. Academic Press, New York, N.Y., pp. 523–532.

O'Connor, S.G. and McErlean, A.J., 1975. The effects of power plants on productivity of the nekton. In: L.E. Cronin (Editor), *Estuarine Research, I. Chemistry, Biology and the Estuarine System*. Academic Press, New York, N.Y., pp. 494–517.

Odum, W.E., 1968. Mullet grazing on a dinoflagellate bloom. *Chesapeake Sci.*, 9: 202–204.

Odum, W.E., 1970a. Utilization of the direct grazing and plant detritus food chains by the striped mullet *Mugil cephalus*. In: J.H. Steele (Editor), *Marine Food Chains*. University of California Press, Berkeley, Calif., pp. 222–240.

Odum, W.E., 1970b. Insidious alteration of the estuarine environment. *Trans. Am. Fish. Soc.*, 99: 836–847.

Odum, W.E. and Heald, E.J., 1972. Trophic analyses of an estuarine mangrove community. *Bull. Mar. Sci.*, 22(3): 671–738.

Odum, W.E. and Heald, E.J., 1975. The detritus-based food web of an estuarine mangrove community. In: L.E. Cronin (Editor), *Estuarine Research, I. Chemistry, Biology, and the Estuarine System*. Academic Press, New York, N.Y., pp. 265–286.

Officer, C.B., 1976. *Physical Oceanography of Estuaries and Associated Coastal Waters*. Wiley, New York, N.Y., 465 pp.

Officer, C.B. and Ryther, J.H., 1977. Secondary sewage treatment versus ocean outfalls: an assessment. *Science*, 197: 1056–1060.

Oviatt, C.A. and Nixon, S.W.., 1973. The demersal fish of Narragansett Bay: an analysis of community structure, distribution, and abundance. *Estuarine Coastal Mar. Sci.*, 1: 361–378.

Pearcy, W.G., 1962. Ecology of an estuarine population of winter flounder (*Pseudopleuronectes americanus*) (Walbaum). Parts I–IV. *Bull. Bingham Oceanogr. Coll.*, 18(1): 1–78.

Pearcy, W.G. and Myers, S.S., 1974. Larval fishes of Yaquina Bay, Oregon: a nursery ground for marine fishes? *Fish. Bull., U.S.*, 72(1): 201–213.

Pearcy, W.G. and Richards, S.W., 1962. Distribution and ecology of fishes of the Mystic River estuary, Connecticut. *Ecology*, 43: 248–259.

Perlmutter, A., 1947. The blackback flounder and its fishery in New England and New York. *Bull. Bingham Oceanogr. Coll.*, 11(2): 1–92.

Peters, D.S. and Kjelson, M.A., 1975. Consumption and utilization of food by various postlarval and juvenile fishes of North Carolina estuaries. In: L.E. Cronin (Editor), *Estuarine Research, I, Chemistry, Biology, and the Estuarine System*. Academic Press, New York, N.Y., pp. 448–472.

Pierce, D.E. and Howe, A.B., 1977. A further study on winter flounder group identification off Massachusetts. *Trans. Am. Fish. Soc.*, 106(2): 131–139.

Pillay, T.V.R., 1967a. Estuarine fisheries of West Africa. In: G.H. Lauff (Editor), *Estuaries. Am. Assoc. Adv. Sci., Publ.*, 83: 639–646.

Pillay, T.V.R., 1967b. Estuarine fisheries of the Indian Ocean Coastal zone. In: G.H. Lauff (Editor), *Estuaries. Am. Assoc. Adv. Sci., Publ.*, 83: 647–657.

Rafail, S.Z. and Hamid, E.M., 1974. The abundance of mullet fry at the sides of Mex-Canal. *Bull. Inst. Oceanogr. Fish. A.R.E.*, 4: 99–129.

Recksiek, C.W. and McCleave, J.D., 1973. Distribution of pelagic fishes in the Sheepscot River-Back River Estuary, Wiscasset, Maine. *Trans. Am. Fish. Soc.*, 102(3): 541–551.

Reintjes, J.W. and Pacheco, A.L., 1966. The relation of menhaden to estuaries. *Trans. Am. Fish. Soc., Suppl.* 95(4): 50–58.

Remane, A. and Schlieper, C., 1971. *Biology of Brackish-Water*. Wiley–Interscience, New York, N.Y., 372 pp.

Richards, C.E. and Castagna, M., 1970. Marine fishes of Virginia's eastern shore (inlet and marsh, seaside waters). *Chesapeake Sci.*, 11(4): 235–248.

Rickert, D.A. and Hines, W.G., 1978. River quality assessment: implications of a prototype project. *Science*, 200: 1113–1118.

Russell, R.J., 1967. Origins of estuaries. In: G.H. Lauff (Editor), *Estuaries. Am. Assoc. Adv. Sci., Publ.*, No. 83: 93–99.

Ryder, R.A. and Henderson, H.F., 1975. Estimates of potential fish yield for the Nasser Reservoir, Arab Republic of Egypt. *J. Fish. Res. Board Can.*, 32: 2137–2151.

Saila, S.B., 1961a. A study of winter flounder movements. *Limnol. Oceanogr.*, 6(3): 292–298.

Saila, S.B., 1961b. The contribution of estuaries to the offshore winter flounder fishery in Rhode Island. *Proc. Gulf Caribb. Fish. Inst., 14th Annu. Sess.*, pp. 95–109.

Saila, S.B., 1975. Some aspects of fish production and cropping in estuarine systems. In: L.E. Cronin (Editor), *Estuarine Research, I. Chemistry Biology, and the Estuarine System*. Academic Press. New York, N.Y., pp. 473–493.

Saville, T., 1966. A study of estuarine pollution problems on a small unpolluted estuary and a small polluted estuary in Florida. *Eng. Progr. Univ. Fla.*, 125: 1–202.

Schubel, J.R., 1974. Effects of exposure to time-excess temperature histories typically experienced at power plants on the hatching success of fish eggs. *Estuarine Coastal Mar. Sci.*, 2(2): 105–116.

Scott, K.M.F., Harrison, A.D. and Macnae, W., 1952. The ecology of South African estuaries. Part II. The Klein River estuary, Hermanus, Cape. *Trans. R. Soc. S. Afr.*, 33: 283–331.

Scott, W.B. and Crossman, E.J., 1964. *Fishes Occurring in the Fresh Waters of Insular Newfoundland*. Canada Dep. Fish., Ottawa, Ont., 124 pp.

Scott, W.C.M., 1929. A note on the effect of temperature and salinity on the hatching of the eggs of the winter flounder (*Pseudopleuronectes americanus*) (Walbaum). *Contrib. Can. Biol. Fish.*, 4(1): 137–141.

Smith, C.L. and Tyler, J.C., 1972. Space resource sharing in a coral reef fish community. *Los Angeles County Mus. Sci. Bull.*, 14: 125–170.

Smith, R.F., Swartz, A.H. and Massman, W.H. (Editors), 1966. A symposium on estuarine fisheries. *Trans. Am. Fish. Soc., Suppl.*, 95(4): 1–154.

Smith, R.L. and Paulson, A.C., 1977. Osmoregulatory seasonality and freezing avoidance in some fishes from a subarctic eelgrass community. *Copeia*, 1977(2): 362–369.

Stevens, D.E., 1977. Striped bass (*Morone saxatilis*) year-class strength in relation to river flow in the Sacramento–San Joaquin Estuary, California. *Trans. Am. Fish. Soc.*, 106(1): 34–42.

Subrahmanyam, C.B. and Drake, S.H., 1975. Studies of the animal communities in two north Florida salt marshes. Part 1, Fish communities. *Bull. Mar. Sci.*, 25(4): 445–465.

Sutcliffe Jr., W.H., 1972. Some relations of land drainage, nutrients, particulate material, and fish catch in two eastern Canadian bays. *J. Fish. Res. Board Can.*, 29: 357–362.

Sutcliffe Jr., W.H., 1973. Correlations between seasonal river discharge and local landings of American lobster (*Homarus americanus*) and Atlantic halibut (*Hippoglossus hippoglossus*) in the Gulf of St. Lawrence. *J. Fish. Res. Board Can.*, 30: 856–859.

Sylvester, J.R., 1972. Effect of thermal stress on predator avoidance in sockeye salmon. *J. Fish. Res. Board Can.*, 29: 601–603.

Tabb, D.C., 1966. The estuary as a habitat for spotted sea trout. *Trans. Am. Fish. Soc., Suppl.*, 95(4): 59–67.

Talbot, F.H., 1955. Notes on the biology of the white stumpnose *Rhabdosargus globiceps* (Cuvier), and on the fish fauna of the Klein River estuary. *Trans. R. Soc. S. Afr.*, 34: 387–407.

Talbot, G.B., 1966. Estuarine environmental requirements and limiting factors for striped bass. *Trans. Am. Fish. Soc., Suppl.* 95(4): 37–49.

Taylor, C.C., Bigelow, H.B. and Graham, H.W., 1957. Climatic trends and the distribution of marine animals in New England. *U.S. Fish Wildl. Serv., Fish. Bull.*, 115: 293–345.

Tully, J.P. and Barber, F.G., 1960. An estuarine analogy in the sub-Arctic Pacific Ocean. *J. Fish Res. Board Can.*, 17: 91–112.

Tyler, A.V., 1971a. Periodic and resident components in communities of Atlantic fishes. *J. Fish. Res. Board Can.*, 28: 935–946.

Tyler, A.V., 1971b. Surges of the winter flounder, *Pseudopleuronectes americanus*, into the intertidal zone. *J. Fish. Res. Board Can.*, 28: 1727–1732.

Tyler, A.V., 1972. Food resource division among northern, marine, demersal fishes. *J. Fish. Res. Board Can.*, 29(7): 997–1003.

Umminger, B.L. and Mahoney, J.B., 1972. Seasonal changes in the serum chemistry of the winter flounder, *Pseudopleuronectes americanus*. *Trans. Am. Fish. Soc.*, 101(4): 746–748.

Vaas, K.F., 1968. De visfauna van het estuariumgebied van Rijn en Maas. *Biol. Jaarb.*, 36: 115–128.

Van Guelpen, L. and Davis, C.C., 1979. Seasonal movements of the winter flounder, *Pseudopleuronectes americanus*, in two contrasting inshore locations in Newfoundland. *Trans. Am. Fish. Soc.*, 108: 26–37.

Vernberg, F.J., Calabrese, A., Thurberg, F.P. and Vernberg, W.B. (Editors), 1977. *Physiological Responses of Marine Biota to Pollutants.* Academic Press, New York, N.Y., 467 pp.

Vernberg, W.B. and Vernberg, F.J., 1976. Physiological adaptations of estuarine animals. *Oceanus*, 19(5): 48–54.

Voyer, R.A. and Morrison, G.E., 1971. Factors affecting the respiration rates of winter flounder (*Pseudopleuronectes americanus*). *J. Fish. Res. Board Can.*, 28(12): 1907–1911.

Weaver, J.E. and Holloway, L.F., 1974. Community structure of fishes and microcrustaceans in ponds of a Louisiana tidal marsh influenced by weirs. *Contrib. Mar. Sci.*, 18: 57–69.

Webb, B.F., 1972. Fish populations of the Avon–Heathcote Estuary. 1. General ecology, distribution and length frequency. *N.Z. J. Mar. Freshwater Res.*, 6(4): 570–601.

Webb, B.F., 1973a. Fish populations of the Avon–Heathcote Estuary. 2. Breeding and gonad maturity. *N.Z. J. Mar. Freshwater Res.*, 7(1/2): 45–66.

Webb, B.F., 1973b. Fish populations of the Avon–Heathcote Estuary. 3. Gut contents. *N.Z. J. Mar. Freshwater Res.*, 7(3): 223–234.

Webb, B.F., 1973c. Fish populations of the Avon–Heathcote Estuary. 4. Parasites *N.Z. J. Mar. Freshwater Res.*, 7(4): 301–306.

Webb, B.F., 1973d. Fish populations of the Avon–Heathcote Estuary. 5. Records of less common species. *N.Z. J. Mar. Freshwater Res.*, 7(4): 307–322.

Wells, B., Steele, D.H. and Tyler, A.V., 1973. Intertidal feeding of winter flounder (*Pseudopleuronectes americanus*) in the Bay of Fundy. *J. Fish. Res. Board Can.*, 30: 1374–1378.

Wiley, M. (Editor), 1976. *Estuarine Processes. I. Uses, Stresses, and Adaptation to the Estuary. Vol. II. Circulation, Sediments and Transfer of Material in the Estuary.* Academic Press, New York, N.Y., 424 pp.

Wootton, R.J., 1977. *The Biology of the Sticklebacks.* Academic Press, London, 387 pp.

Chapter 8

ENCLOSED SEAS — INTRODUCTION

BOSTWICK H. KETCHUM

The major oceans of the world, the Atlantic, Indian and Pacific, are ringed by adjacent seas with characteristics intermediate between those of estuaries and of the open ocean. Eight enclosed seas and two gulfs have been selected for discussion in this volume, namely the Mediterranean, the Red, the Black, the Baltic, the Bering, the Okhotsk, the Japan and the East China Seas, the Gulf of St. Lawrence and the Gulf of California. These were selected by the editor; the choice was based upon both subjective and objective criteria.

Nineteen adjacent seas, ranging from large mediterranean seas to marginal seas, straits and gulfs are recognized by classical oceanographers and their characteristics were listed by Sverdrup et al. (1942) quoting data after Kossinna (1921). Some of these adjacent seas are not included here because their ecosystems are more similar to coastal waters than they are to estuaries and they can be more properly considered in the volume in this series on the ecosystems of the continental shelves. For example, the Andaman Sea, the North Sea, the Irish Sea, the English Channel, Bass Strait and the American mediterranean, consisting of the Caribbean and the Gulf of Mexico, have not been included in this volume since conditions in these adjacent seas are essentially coastal in character. Not enough is known about the ecosystems of other enclosed seas to warrant inclusion; examples of these are the Arctic Ocean, the Asiatic mediterranean including the South China and adjacent seas, Hudson Bay and the Persian Gulf.

The surface area, enclosed volume and average depth of each of the enclosed seas and gulfs discussed are given in Table 8.1. The total area of about ten million square kilometers represents about 3% of the area of all oceans. Because of the comparatively shallow average depth of the en-

TABLE 8.1

Area, volume, mean and maximum depth of enclosed seas

Chapter	Sea	Area (10^6 km^2)	Volume (10^6 km^3)	Mean depth (m)	Maximum depth (m)
9, 10	Mediterranean	2.523	3.708	1470	4600
11	Black Sea	0.420	0.537	1280	2212
12	Red Sea	0.450	0.251	558	2920
13	Baltic Sea	0.370	0.021	57	300
14	Bering Sea	2.292	3.750	1636	4150
15	Okhotsk Sea	1.58	1.23	778	3374
	Japan Sea	1.01	1.36	1350	3695
	East China Sea	1.24	0.28	226	2717
16	Gulf of St. Lawrence	0.226	0.035	152	>500
17	Gulf of California	0.162	0.132	813	>2500
Total		10.311	10.772		

closed seas, the volume is only about 0.8% of the sea water volume on earth. The largest of these enclosed seas is the European Mediterranean with a surface area of about 2.5 million square kilometers, closely followed by the Bering Sea, which has the greatest average depth (1636 m) and enclosed volume (3.75 million km^3). The Baltic Sea is the shallowest (57 m), and the Gulf of California is the smallest with a surface area of 162 000 km^2.

Each of the enclosed seas described in this volume shares characteristics with some or all of the others and each has some characteristic estuarine-type processes representative of a class of other enclosed seas or of large estuaries. Each sea, however, is also unique in the relationships between the physical system and the distribution and productivity of the biota. It is hoped that the discussions will point the way to more general application of basic principles to other areas, and especially to those enclosed seas which are not yet adequately studied and described.

GENERAL CHARACTERISTICS

Geological history and structure

The enclosed seas with shallow sills in the passages connecting the seas with the open ocean have been isolated from the ocean by the lowered sea level associated with glacial periods. Milliman and Emery (1968) concluded that sea level was about 130 m below present about 15 000 years ago. The Black Sea, the Red Sea and the Japan Sea all have shallow sills which were presumably dry land during glacial periods. The continental shelf in the Bering Sea once connected the continents of Asia and North America and formed a migration route for the population of the North American continent. The Strait of Sicily in the Mediterranean was also a land bridge between Europe and Africa in the distant past.

The Black Sea and the Japan Sea were probably great inland lakes at times of low sea level. The Euxine Sea, the ancient name for the Black Sea, was a shallow, brackish lake in the Early Quaternary. During the latest glacial epoch the Neo-Euxine was a freshwater lake until the connection with the Mediterranean was established about 7000 years ago. In the Far East most of the area occupied by the present-day Okhotsk, Japan and East China Seas were vast dry lands embracing shallow inland basins or gulfs during the Late Pliocene and most of the Pleistocene. During this era the Japan Sea probably communicated with the Pacific only through a narrow channel; the Okhotsk and East China Seas may have been large estuaries emptying into the Pacific through their deeper and wider channels. The Bering Sea was also characterized by repeated episodes of subaerial exposure alternating with massive glacial encroachment during the Pleistocene. The Baltic Sea and the Gulf of St. Lawrence were covered by ice during the most recent glacial epoch. Most of these enclosed seas have been inundated or connected with the ocean as a result of the rising sea level during the present interglacial period.

Two of the areas considered, the Red Sea and the Gulf of California, have been formed by the spreading of the sea floor as the continents or parts of continents drift apart. In the Red Sea, several periods of sea-floor spreading are postulated; the most recent one probably began about 3 million years ago. The present rate of spreading along the central portions of the Red Sea is estimated to be about 1 cm yr^{-1}. The present spreading episode began in the Gulf of California about 4 million years ago, and the tip of Baja California has been separating from the mainland of Mexico at an average rate of 6 cm yr^{-1}.

Geography and climate

All of the enclosed seas discussed lie within the Northern Hemisphere. The only adjacent seas in the Southern Hemisphere among those listed by Sverdrup et al. (1942) are the Bass Strait between the southeastern tip of Australia and Tasmania, which has wide openings at both ends to the ocean, and the southern part of the Asiatic mediterranean (e.g. the Java Sea and Banda Sea). The Southern Hemisphere thus contains few enclosed seas, even though it is often called the "Water Hemisphere" because 81% of the total area is covered by ocean and only 19% by land; this contrasts to the Northern Hemisphere where 61% is covered by ocean and 39% by land (Kossinna, 1921).

These enclosed seas include examples ranging from the Tropics to the Arctic. The southern entrance to the Red Sea through the Strait of Bab al Mandab is at 13°N, about 10° south of the Tropic of Cancer. The northern end of the Bering Strait, connecting the Bering Sea with the Arctic Ocean, lies on the Arctic circle (66°32′N), and the northern end of the Gulf of Bothnia in the Baltic Sea is only a few kilometers south of the Arctic circle. This wide geographical range is associated with large differences in the climate of the various seas.

The characteristics of the surrounding land further modify the climatic conditions. For example, the Red Sea is bordered by high plateaus and mountains and the average monthly winds are generally either north or south and are dominated by the monsoon wind system. The entire area has a hot, dry climate so that the Red Sea is very warm with high salinities as a result of intense evaporation. The Gulf of California is separated from the Pacific Ocean by high mountains along Baja California so that the climate is continental rather than oceanic in character. The Mediterranean Sea lies between mountainous areas to the north and deserts to the south, with a wide range of climatic variations. The circulation patterns in the Mediterranean are so dependent on climate and weather that it has been suggested that it could be used as an indicator of climatic change. Emiliani (1955) for example concluded from a study of the oxygen isotope ratios of foraminifera skeletons in a Mediterranean core that the minimum temperatures during glacial periods ranged from 8°C to 13°C while the interglacial temperatures were 27°C to 33°C, a little warmer than present-day August mean temperature.

The Black Sea and the Baltic both receive large amounts of fresh water in river drainage which exerts a major control on environmental conditions in these two seas. All of the northern seas, the Okhotsk, the Bering, the Baltic and the Gulf of St. Lawrence, are subject to intense winter cooling and to various degrees of ice cover, both of which phenomena modify the character of the circulation and the environmental characteristics of these seas.

Salinity

Variations of salinity are a general characteristic of estuaries, and also of these enclosed seas. The vertical change of salinity may be as abrupt in an enclosed sea as it is in some smaller estuaries, but in general the horizontal gradients from the fresh water in the rivers to the coastal sea-water conditions are more gradual in the enclosed seas because of the greater distance involved.

Both the Black Sea and the Baltic Sea are permanently stratified as a result of the large amounts of fresh-water inflow and the development of a permanent halocline. Even the deep water in the Baltic is brackish (<20‰), whereas the brackish layer in the Black Sea is limited to the upper 150 to 200 m. In the Black Sea, the deep water is permanently anoxic, a condition which has been developing over the last 3000 years or so. Anoxic conditions also occur in the Baltic Sea, but occasional flushing of the deep waters about every twenty years results in an intermittent, rather than a permanent anoxic condition.

In three of these enclosed seas — the Mediterranean, the Red Sea and the Gulf of California — evaporation exceeds precipitation on the surface and the addition of fresh water from rivers. The water in these three seas has a higher salinity than that of the adjacent ocean water. The Red Sea has the highest salinity of any large body of water with a direct connection to the ocean; the salinities exceed 40‰ in comparison to the oceanic mean salinity of 35‰. The average salinity of the Mediterranean water is 38.45‰— significantly higher than the average ocean salinity but lower than that of the Red Sea. In the northern part of the Gulf of California, surface salinities may reach 38.5‰ in the summer due to the excess evaporation there.

Biological aspects

It is impossible to generalize about the biological characteristics of these enclosed seas. The biota vary from tropical to boreal as would be expected from the wide geographical range of the seas. The plankton and fishes are generally similar to those found in the offshore coastal waters which participate in the circulation within each sea. The Sea of Okhotsk, the Baltic Sea and the Gulf of St. Lawrence are important migration routes for the anadromous salmon, and the latter two are also routes for the catadromous eel. The unique biological aspects of each of the seas such as the anoxic zones of the Black Sea, the deep brine areas of the Red Sea and the cold-adapted abyssal fauna of the Okhotsk and Japan Seas are presented in detail in the appropriate chapters.

Historical perspectives

Some of these seas have been known and used since ancient times; others have only recently been discovered and explored by modern civilizations. The Mediterranean was the center of the development of western civilizations and the Mediterranean Sea, the Black Sea and the Red Sea have been used by man since the beginning of recorded history. Similarly, the Okhotsk Sea, the Japan Sea and East China Sea have been the source of important fishery products for eastern civilizations for several thousand years. The Vikings were famed sea explorers making transatlantic crossings at least as far as Greenland where a Danish colony was established over a thousand years ago. Certainly, the Baltic Sea was well known to these seafaring people; the knowledge is contained in sagas and legends, but not in recorded history.

The other enclosed seas were no doubt known by primitive people, but have been more recently discovered and explored by modern civilizations. The Gulf of St. Lawrence was found early in the exploration of the North American continent. Cabot Strait, one of the main entrances to the Gulf, was named for John Cabot, an English navigator who, with his son, Sebastian, discovered Labrador in 1497 and plotted much of the outer coastline of this part of Canada. Jacques Cartier of France in 1535 explored the Gulf and the St. Lawrence River as far inland as Quebec and Montreal. Hernán Cortés of Spain discovered Baja California, which he thought to be an island, in 1524. De Ulloa, serving with Cortés, circumnavigated the Gulf in 1539 and showed that Baja California is a peninsula. The Bering Sea was explored and charted by Vitus Bering, a Danish explorer commissioned by the Russian Empress Catherine for the expedition. In 1728, Bering sailed into the Arctic Ocean through the Strait which now bears his name and determined that Asia and America were separate continents. Bering died of scurvy in 1741 during the second expedition, which charted much of the Arctic coast of Asia, Alaska and the Aleutian Island arc. Later, Captain James Cook of England confirmed these observations and sailed through the Bering Strait into the Arctic Ocean in 1778.

Oceanographic exploration of these enclosed seas has an even more recent history. The globe-encircling expeditions of the last half of the nineteenth century, especially the *Challenger* expedition of 1872 to 1876, were the precursors of modern oceanography. The Mediterranean and the Red Seas were studied in some, but not all, of these expeditions; for example, only two of these world-wide expeditions visited the Red Sea, the *Vitiaz* (1886–1889) and the *Valdivia* (1898–1899). Two major marine stations were established in the nineteenth century on the shores of these seas; the Naples Zoological Station in the Mediterranean (1874) and the Sevastopol Biological Station in the Black Sea (1872). Virtually all of our knowledge of the environment and ecosystems of these seas has thus been developed in the current century, much of it in the last few decades. It is the purpose of this volume to summarize the present state of our knowledge of these seas, and to show where further studies are needed.

SPECIFIC CHARACTERISTICS OF ENCLOSED SEAS

The Mediterranean Sea (Chapters 9 and 10)

The Mediterranean Sea connects with the Atlantic Ocean through the Strait of Gibraltar. This narrow (14 km), shallow (330 m) strait isolates the deeper waters of the Mediterranean from direct exchange with the Atlantic. Evaporation from the surface of the Mediterranean Sea exceeds fresh-water additions from rainfall and river outflow so that the water level is maintained by a net inflow of surface Atlantic Ocean water. The dense deeper Mediterranean water flows through the Strait of Gibraltar into the Atlantic Ocean as a subsurface current. Because of excess evaporation within the Mediterranean, the salinity is increased so that the deep water in the eastern part of the Mediterranean has a high salinity (40‰) compared to the inflowing surface water in the Strait of Gibraltar (34.8‰). The flows through the Strait of Gibraltar are the reverse of those characteristic of estuaries where the surface flow is generally seaward and the deeper flow landward. This has been termed an inverse estuarine-type circulation, and it is also found in other estuaries where evaporation exceeds precipitation and runoff (see p. 3).

One important consequence of this inverted circulation is that the Mediterranean Sea is generally impoverished in nutrients since the inflowing surface water from the Atlantic Ocean is already low in nutrients. In the Western Mediterranean, the total phosphorus content of the surface layers is generally less than 0.2 μg-at. l^{-1}, and it rarely exceeds 0.5 μg-at. l^{-1} in the deep waters throughout the Mediterranean. Similarly, the nitrate concentration in the deeper waters of the central Mediterranean range from 4 to 5 μg-at. l^{-1} and the surface concentrations are generally of the order of 0.1 μg-at. l^{-1} (Miller et al., with McGill, 1970). The surface concentrations of phosphate and nitrate do not differ remarkably from those of the open Atlantic, but the concentrations at 1000 to 2000 m depth are only about one fifth of the concentration found at these depths in the Atlantic.

The Mediterranean Sea is divided into two major basins, the Western and Eastern Mediterranean, and it is surrounded by adjacent seas. The Alboran Sea lies between the Strait of Gibraltar and the Western Mediterranean, and the Tyrrhenian and Adriatic Seas are also adjacent to the Western Basin. The Aegean Sea lies between the Eastern Mediterranean and the Black Sea. These adjacent seas, except for the Black Sea, are considered in Chapter 9 as integral parts of the Mediterranean Sea.

The biota of the Mediterranean consists primarily of Atlanto-Mediterranean species (62%) derived from the adjacent Atlantic biogeographic provinces beyond the Strait of Gibraltar. Many Mediterranean species are endemic (20%) while others are cosmopolitan, or circumtropical (13%) or Indo-Pacific (5%). These proportions differ for different major taxonomic groups, and also for different parts of the Mediterranean Sea, but the pattern remains essentially the same.

The deep-sea fauna is particularly impoverished in the Mediterranean, both in species and biomass. There is no direct connection with the deep Atlantic waters through the Strait of Gibraltar and the temperature of the deep waters in the Mediterranean is considerably higher (12–14°C) than at comparable depths in the Atlantic Ocean.

Species diversity, biomass and yield of fisheries all decrease from west to east. The southeastern corner of the Mediterranean Sea, the Levant Basin, is the most impoverished area. The number of species in different taxa in the Levant Basin range from about one quarter to one half of the total number of species known for the Mediterranean as a whole. The benthic biomass throughout the Mediterranean is considerably less than that found at comparable depths in the eastern Atlantic, with minimum values being found in the Levant Basin, along the Libyan coast and in the southern part of the Aegean Sea. The yield of fisheries is also comparatively low except for the extensive coastal lagoons in the eastern Mediterranean which are highly productive. The fish yields of four Nile Delta lagoons can exceed 10 000 kg km^{-2}. This is nearly ten times greater than the average for the shelf areas of the Mediterranean.

Two major engineering developments, the Suez Canal and the Aswan Dam, have had major impacts on the biology of the Eastern Mediterranean. The Suez Canal connects the species-rich tropical Red Sea with the impoverished Levant Basin. A large number of species have immigrated so that now, among the fish, the immigrants constitute about 12% of the total number of species in the Levant Basin. A variable proportion of Red Sea immigrants has been identified among other taxonomic groups. Some of the immigrants have extended their range to the southern part of the Aegean Sea and to the central Mediterranean, but the low winter temperatures probably prevent the spread of Red Sea immigrants to the western and northern Mediterranean. The timing, development and spread of Red Sea immigrants in the Eastern Mediterranean and the interactions of these immigrants with resident populations are still poorly understood.

The Nile floods, prior to the development of the Aswan Dam, annually carried millions of tons of sediment, with the associated plant nutrients, into the Eastern Mediterranean. This enrichment produced enormous blooms of phytoplankton and attracted large concentrations of sardines to the area off the delta of the Nile. Following the closing of the Aswan High Dam, phytoplankton blooms were reduced to about a tenth of the previous values and the sardine fishery dropped to a fraction of its former volume. There is no clear evidence that other fisheries were adversely affected. It is postulated that fish production in Nasser Lake may partially compensate for this loss of the sardine fisheries in the Eastern Mediterranean. Much remains to be learned about the long-term effects of the diversion of the Nile River by the Aswan Dam.

The Black Sea (Chapter 11)

The entrance to the Black Sea, the Bosporus, is long (31 km), very narrow (0.73–3.6 km) and shallow (36–92 m). The Bosporus connects the Black Sea with the Sea of Marmara which in turn is connected with the Aegean Sea through the Dardanelles. Thus, the Black Sea, through this complex of straits and intervening seas, is an adjacent sea to the eastern Mediterranean. Sverdrup et al. (1942) include the Black Sea with the Mediterranean in their listing of adjacent seas.

The Black Sea differs greatly from the Mediterranean, however, because precipitation and runoff exceed evaporation from the sea surface. Although the eastern Mediterranean is more saline than the adjacent Atlantic Ocean with which it is connected, the Black Sea is greatly freshened by this excess river runoff and precipitation. The surface salinities of the waters of the Black Sea (about 17.5‰) are about half of those of the surface Atlantic waters and only about 44% of the salinity of the eastern Mediterranean. In a typical estuarine-type circulation freshened water (<18‰) flows seaward in the surface layers of the Bosporus and more saline deep water (>34‰) flows into the Black Sea.

The freshened surface layer produces a very intense density gradient at 150 to 200 m which isolates the deep water from the surface layers. Organic matter brought in by the rivers or produced in the surface layers sinks and decomposes below the density gradient. As a result the deep waters of the Black Sea are relatively stagnant; they are anoxic and there has been an accumulation of hydrogen sulphide and of organic and inorganic forms of phosphorus. Since nitrates are reduced in the oxidation of organic matter, nitrogen compounds do not accumulate in the deep water to the same extent as the phosphorus moiety of the organic molecules.

The energy of the bacterial oxidation of hydrogen sulphide is used to produce microbial organic matter by chemosynthesis. This chemosynthetic process is a significant part of the productivity of the Black Sea. Because of this complex sulphur cycle and the importance of bacterial activity in the Black Sea, it has also been termed the "microbial sea".

The Red Sea (Chapter 12)

The Red Sea is connected through the Gulf of Aden with the adjacent Indian Ocean by the narrow (24 km), shallow (137 m) Strait of Bab al Mandab. Like the Mediterranean, evaporation exceeds precipitation and runoff in the Red Sea. It is surrounded by deserts and has a very hot, very dry climate which produces a large evaporation from the sea surface, estimated at as much as 2150 mm yr^{-1}. As a consequence of these climatic phenomena, the Red Sea is the warmest and saltiest body of water on earth with an open connection to the ocean.

Geologically, a rift valley occupies the central axis of the Red Sea and the sea floor is slowly spreading as the African continent and the Arabian Peninsula drift apart. Associated with this central rift valley are deep basins which are occupied by very hot and very salty water. Temperatures as high as 61.3°C and salinities as great as 257‰ have been found. These brines are not evaporated sea water but rather they contain salts which have been dissolved from rocks within the earth's mantle. The evidence to date suggests that these hot brines are devoid of life; even the bacteria do not live there.

The Baltic Sea (Chapter 13)

The Baltic Sea is connected with the North Sea through the Kattegat and the Skagerrak. The Baltic Sea is the shallowest sea considered, with an average depth of 57 m. There are two major entrances to the Baltic, both of which are narrow and shallow. The sill depth of the Öresund is 7 to 8 m, and that of the Belt Sea is 17 to 18 m. Precipitation and runoff greatly exceed evaporation so that there is a net outflow of freshened surface water and an inflow of deeper, more saline water in a characteristic estuarine circulation. The salinity of the surface layers of water vary from practically fresh in the northern end of the Gulf of Bothnia to about 20‰ in the outflowing waters near the Kattegat; even the deep waters in the Baltic are brackish (<20‰).

Several comparatively deep basins (up to 300 m) are separated from adjacent basins by shallow sills. The deeper waters of these basins are isolated from the active circulation and tend to become anoxic with accumulation of hydrogen sulphide. However, they are periodically flushed so there is no long-term buildup of hydrogen sulphide as is found in the Black Sea. The geological records show alternate reduced and oxidized layers in the sediments with an estimated cyclical time period of about twenty years. During the anoxic periods, the benthic populations are obviously exposed to unfavorable conditions from which they recover slowly during the period when the deep waters are renewed.

The Bering Sea (Chapter 14)

The Bering Sea lies at the northern apex of the Pacific Ocean from which it is separated by the chain of Aleutian Islands. Numerous shallow passages through these islands connect the surface waters of the Bering Sea with the North Pacific and there are two deeper passages between the end of the island chain and the Asiatic mainland. The Sea connects with the Arctic Ocean to the north through the narrow (85 km) and shallow (45 m) Bering Strait.

The Bering Sea is characterized by extensive shallow continental shelves (<200 m) in the eastern part which occupy nearly half of the area of the bottom. This is separated from the western basin by a continental slope which runs in a northwest–southeast direction. The abyssal basin to the west is deep with most of the bottom lying at a depth of 3800 to 3900 m.

The Sea ranges from subarctic to arctic in climate and is covered by pack ice throughout six to eight months of the year. This pack ice does not provide a uniform coverage. Little ice occurs over the deep southwest portion of the sea, and it is more abundant over the shallower continental shelf area.

There are extensive fishing operations in the Bering Sea, mostly for species living on or close to the bottom, by American, Soviet and Japanese fishing fleets. Sea birds and sea mammals are also abundant in the Bering Sea. Heavy exploitation of these stocks has led to depletion of the stocks and international agreements will be necessary for the preservation and sustained utilization of these living resources.

The Okhotsk, Japan and East China Seas (Chapter 15)

These three interconnected seas are separated from the Pacific Ocean by the Kuril Island chain in the north and by the Ryukyu Island chain in the south. The Sea of Japan, in the central position, is almost completely isolated from the Pacific by the Japanese Islands. The circulation is complex, with exchanges between the Sea of Japan and the two adjacent seas. Thirteen discrete water masses are identified, some of which originate in the Pacific Ocean and others are formed within the enclosed seas by winter cooling and convection.

The Japan Sea has a large deep central basin which contains water that is largely isolated from the circulation with the adjacent seas. The basin is periodically flushed by dense waters formed by winter cooling and convection so that it does not develop the anoxic conditions associated with stagnation. In contrast, the East China Sea, the adjacent Yellow Sea and the Gulf of Bo Hai are comparatively shallow (<200 m).

The three seas combined extend from 25°N to 65°N so that the biota is extremely diverse. It ranges from Indo–West Pacific species characteristic of the East China Sea to the north-temperate and boreal species found in the Okhotsk Sea to the north. The Okhotsk Sea may have been the center for the development and evolution of the cold-adapted secondary deep-sea benthic fauna of the North Pacific Ocean.

The Gulf of St. Lawrence (Chapter 16)

The Gulf of St. Lawrence receives large amounts of fresh water from the St. Lawrence River, and it is also fed along its northern coast by a number of intermediate and small rivers. The drainage basin is about nine times greater than the surface area of the Gulf proper. The circulation takes place in three characteristic layers. The freshened surface layer is underlain by an intermediate layer of very cold water which extends to a depth of approximately 150 m. Below this a warmer, more saline body of water extends to the bottom and occupies the deeper channels in the Gulf. The intermediate water is continuous with the superficial waters of the Labrador Shelf and is similar in its temperature and salinity characteristics. The deepest layer is connected with the deep water in the channels of the continental shelf through Cabot Strait.

Geologically, the Gulf of St. Lawrence is a drowned river valley or lowland dating back to the Paleozoic. The major channels were probably formed in the Late Cretaceous with the deep Laurentian Channel extending through Cabot Strait and across the continental shelf some 250 km to the continental slope. There are two main entrances to the Gulf connecting it with the Atlantic Ocean. The larger of these, Cabot Strait on the southeast, is 104 km wide with a maximum depth of 480 m. The smaller strait of Belle Isle, at the northeast corner, is 16 km wide with a sill depth of 60 m.

Vigorous mixing occurs near the mouth of the St.

Lawrence River where the river bottom deepens sharply from 40 m to a depth of more than 350 m at the head of the Laurentian Channel. This vigorous mixing enriches the surface layers with nutrients, a process that is important in supporting the high productivity of the Gulf of St. Lawrence. The major fishing grounds in the Gulf occupy the comparatively shallow southeastern quadrant. Here local upwellings around the islands enrich the surface water and extend the period of high primary production to provide a rich source of food for the fishes. Even though the Gulf is ice-covered for two months of the year, the primary productivity is high in comparison to that of other coastal waters. Many of the commercially important species leave the Gulf during the period of ice cover and are abundant in summer and early autumn only. Others appear to remain in the Gulf throughout the year.

Strong correlations have been demonstrated between the total river flow and the later abundance of commercially available fish and shellfish. The effect appears to be on larval and early development. Presumably the relationship is mediated by the effect of river flow on the nutrient distribution and on the subsequent primary and secondary production of the system. The effect is widespread and it extends well beyond the mouth of the Gulf of St. Lawrence into the coastal waters nearby.

Gulf of California (Chapter 17)

The Gulf of California lies between the peninsula of Baja California and the mainland of Mexico. It is long (1000 km) and narrow (150 km) with a wide (200 km) and deep (>1000 m) opening to the Pacific Ocean. There are six types of water in the entrance, only one of which, the warm saline surface water, originates in the Gulf. The remaining five water masses, two at the surface and three subsurface layers, originate in the Pacific Ocean. The salinity range of these six water masses is very slight (34.3 to 34.9‰); the temperature decreases gradually with depth from about 25°C at the surface to about 4°C at 1200 m.

The northern third of the Gulf is comparatively shallow (<100 m), whereas the remainder of the Gulf contains a series of basins separated from each other by transverse ridges and gradually deepening to the southward. Both the wind fields and the currents show marked seasonal variations. The winds blow generally from the northwest from November to May and the midwinter surface currents are southerly. Conversely, the winds are from the southeast at other times of the year and the mid-summer surface currents are directed northerly. Because of the mountain ranges in Baja California and on the mainland of Mexico, the offshore winds generally blow parallel to the long axis of the Gulf.

The Colorado River, entering the extreme northern end of the Gulf, is the main source of fresh water. The flow of the Colorado has been progressively decreased as a result of the construction of massive dams and the diversion of water for irrigation and for domestic water supplies for the large population in the southwestern part of the United States. Direct rainfall on the Gulf is variable, both geographically and from year to year and ranges from average annual amounts of less than 100 mm yr^{-1} to about 1000 mm yr^{-1}. In contrast, evaporation is large, ranging from 1300 mm yr^{-1} in the southern part of the Gulf to as much as 2500 mm yr^{-1} in the arid northern part of the Gulf.

The primary productivity of the Gulf is exceptionally high being comparable to that found in upwelling areas off the coast of Baja California or of North Africa. It is a subtropical area and most of the fish species in the Gulf are tropical or subtropical species of the tropical American or Panamic faunal provinces. Ninety-two of the fish species (17% of the total) are endemic to the Gulf and most of the species are shore-fishes. The northern part of the Gulf supports an important commercial fishery and the area adjacent to the Colorado River mouth, where an abundance of juveniles is found, has been declared a sanctuary by the Mexican government.

IMPACT OF HUMAN ACTIVITIES

While people have occupied the shores of these seas for millennia and have used them for transportation, human-induced changes in the seas were negligible prior to the industrial revolution of the eighteenth and nineteenth centuries. So long as ships were propelled by sail and oars the pollution impact was negligible. Similarly, primitive fishing methods probably have had little impact on the biota. Within the last century, and principally during the last few decades, mankind's impact on these seas has dramatically increased.

Living resources

From time immemorial people have depended on the living resources of the sea as a part of their food supply. Marine fishery products are an important source of animal protein, and in the Far East they constitute more than half of the annual supply of animal protein for the population. Seven nations border the Baltic Sea, and fish is an important item in the diet of these people. The Bering Sea is the site of extensive fishing operations by American, Soviet and Japanese fleets, and heavy exploitation has led to a depletion of these stocks. The Gulf of St. Lawrence is an exceptionally productive area for fisheries. More than 25 species have been landed commercially at one time or another. The productivity of these fisheries is closely linked with the dynamics of the circulation. The Black Sea also produces an abundant fisheries catch, with a significant part of it (about 20%) coming from a relatively small area by the Kerch Straits in the northeastern corner of the sea.

The tropical seas are comparatively impoverished. In the Mediterranean and Red Seas fishing from shore or from small boats predominates. In the Gulf of California the commercial fishery is concentrated in the northern end of the Gulf.

There is evidence that the living resources in most of these enclosed seas are being over-exploited, with a consequent danger of the depletion of the resource. International agreements will be necessary, in most cases, to reverse this trend. The present draft of the Law of the Sea Conference provides for an economic zone extending 200 nautical miles (370 km) from the coast. The coastal state would have the responsibility to manage and preserve the natural living resources within this economic zone. In many of the enclosed seas the economic zone would place virtually all of the water within the jurisdiction of one or another of the adjacent countries. Unfortunately, migrating fishes pay no attention to artificial national boundaries and this complicates independent management.

The first international commission for the protection of living resources was the North Pacific Fur Seal Commission which was established in 1911. The fur seal herd on the Aleutian Islands adjacent to the Bering Sea had been seriously reduced by over-exploitation from a population of about 5 million to only 125 000. Enforced conservation measures based on scientific study have permitted recovery of the population so that it is no longer endangered (Fye et al., 1968). More recently, the Baltic Fishing Convention entered into force in 1974. In the Far Eastern seas the regulation of fishing has already been in practice on some commercial items within and between nationalities. In other cases, the need for international agreements to protect and preserve the living resources is recognized. The recently imposed 200 mile territorial limits of the United States include the major fishing grounds of the eastern Bering Sea, and an intensive effort is being made to reduce the total catch to a sustainable yield basis.

Modification of river flow

In both the Gulf of California and the Gulf of St. Lawrence artificial impoundments of water have greatly modified the natural fresh-water supply and consequently have had an impact upon the general circulation and the biota. In the Gulf of California, impoundment of the waters of the Colorado River within the territory of the United States has drastically decreased the river discharge into the Gulf. The river water is diverted for irrigation purposes and is consequently removed from the normal hydrologic cycle of the Gulf system. This has changed the ecological conditions from a typical estuarine system to one in which evaporation exceeds precipitation and fresh-water drainage.

In the Gulf of St. Lawrence large dams and reservoirs have been constructed for hydroelectric power and extensive canal systems have been built for transportation in the St. Lawrence Seaway. Since the water is not diverted for other purposes, this has little impact on the annual average discharge but there is a major impact on the natural seasonal variation. This has been reduced to about half of the original value by an increase in the midwinter discharge and a decrease in the spring discharge. The impact of this change and of further planned modifications on the ecosystem are not yet fully evaluated.

Marine pollution

The impact of marine pollution is most obvious in the enclosed seas which are major routes of transportation for the industrial world. With the opening of the Suez Canal in 1869, the Red Sea and the Mediterranean became a major shipping artery connecting European and North American countries with the oil-rich nations of the Near East. Although the Canal was blocked temporarily in recent years, it is now being dredged to greater depths in an effort to regain some of the shipping which has been routed around the southern tip of Africa. The Mediterranean has for some time been subjected to serious oil pollution which now may be expected to increase again as a result of the widening and deepening of the Suez Canal.

The Gulf of St. Lawrence and the St. Lawrence River provide a major seaway for access to the interior of the North American continent. Although the seaway is closed to transportation by ice cover for several months during the year, it is subject to the pollution inherent in modern-day transportation.

Oil pollution of the sea has been of international concern for many years. In 1954 an international conference convened in London and prepared a draft of the Convention for the Prevention of Pollution of the Sea by Oil. This was opened for signature in 1954 and, supervised by the Intergovernmental Maritime Consultative Organization (IMCO), an independent agency of the United Nations, it entered into force in 1958 (Barros and Johnston, 1974; Ketchum, 1975). The convention has been amended several times, and the latest convention extends the control to include other substances carried by ships (IMCO, 1974).

Another pollution source subject to international control is the dumping of waste materials at sea. The first regional international action to take effect on this subject was the Oslo Convention in 1972, and more recently the London Dumping Convention came into force in 1975 (Waldichuk, 1977).

The United Nations Environment Programme adopted a Regional Seas Programme in 1975 (Anonymous, 1978). In addition to the Mediterranean, for which an international convention is already in force (UNEP, 1978), seven other regional seas are identified as posing critical pollution problems. Much of the pollution originates on land within territorial boundaries of the adjacent countries and is, consequently, not subject to international control. Regional programs are proposed to help concerned governments develop an action plan so that the governments can gradually assume full responsibility for the control of pollution in their region. A thorough understanding of the ecology of each environment will be necessary to assure success of these programs.

REFERENCES

Anonymous, 1978. *Regional Seas*. Brochure issued by Regional Seas Programme Activity Centre, United Nations Environment Programme, Geneva.

Barros, J. and Johnston, D.M., 1974. *The International Law of Pollution*. The Free Press, Macmillan, New York, N.Y., 476 pp.

Emiliani, C., 1955. Pleistocene temperature variations in the Mediterranean. *Quaternaria*, 3: 87–98.

Fye, P.M., Maxwell, A.E., Emery, K.O. and Ketchum, B.H., 1968. Ocean Science and Marine Resources. In: E.A. Gullion (Editor), *Uses of the Sea*. The American Assembly, pp. 17–68.

IMCO, 1974. International Conference on Marine Pollution 1973. *International Convention for the Prevention of Pollution from Ships, 1973*. Intergovernmental Maritime Consultative Organization, London, England.

Ketchum, B.H., 1975. Marine pollution and international law. In: J.M. Sharpley and A.M. Kaplan (Editors), *Proceedings of the Third International Biodegradation Symposium*. Applied Science Publishers, London, pp. 11–19.

Kossinna, E., 1921. Die Tiefen des Weltmeeres. *Inst. Meereskd., Berlin, Veröff. N.F., A. Geogr., Naturwiss. Reihe* 9: 70 pp.

Miller, A.R., Tchernia, P. and Charnock, H. with McGill, D., 1970. *Mediterranean Sea Atlas. Temperature, Salinity, Oxygen Profiles and Data from Cruises of RV ATLANTIS and RV CHAIN*. Woods Hole Oceanographic Atlas Series, 3, Woods Hole Oceanographic Institution, Woods Hole, Mass., 190 pp.

Milliman, J.D. and Emery, K.O., 1968. Sea level during the last 35,000 years. *Science*, 162: 1121–1123.

Sverdrup, H.U., Johnson, M.W. and Fleming, R.H., 1942. *The Oceans, Their Physics, Chemistry, and General Biology*. Prentice-Hall, New York, N.Y., 1087 pp.

UNEP, 1978. Mediterranean Pollution Treaties Enter into Force. *The Siren*, 1. United Nations Environment Programme, Regional Seas Programme, Geneva 10.

Waldichuk, M., 1977. *Global Marine Pollution: An Overview*. Intergovernmental Oceanographic Commission, Techn. Ser. 18, UNESCO, 1977, Paris, 96 pp.

Chapter 9

THE MEDITERRANEAN SEA, A. PHYSICAL ASPECTS

ARTHUR R. MILLER

INTRODUCTION

The Mediterranean Sea (Fig. 9.1) is a complex environment consisting of interacting bodies of water sensitive to climatic and meteorological events. As an ecological entity it changes progressively from west to east — from Atlantic character to a wholly indigenous Mediterranean character, with high salinity and narrow ranges of temperature. In the north and south, differences of environmental condition are governed by seasonal change. In the north, winter cooling promotes massive seasonal overturn. This results in the production and storage of dense water which replenishes the deep water that becomes entrained in the flow out of the Mediterranean (Miller, 1976a). Because of the circulation patterns, the Mediterranean Sea is more subject to change than are other bodies of water of comparable size.

Within the Mediterranean Sea the Strait of Sicily is the major channel separating the Eastern from the Western Mediterranean. This and other channels play a significant role in the development of the oceanographic characteristics of the individual

Fig. 9.1. The Mediterranean Sea.

seas. Flow through each channel is constricted laterally by the land and vertically by topographical features where the bathymetry rises to partially shut off one basin from another. Circulation through these channels is further complicated by the requirements for return flow in order to maintain conservation of mass. This two-way circulation effectively reduces free water movements between adjoining basins, thereby promoting residence times of sufficient duration to develop water characteristics peculiar to each basin. The long-term stability or ecological balance with respect to the environment is consequently dependent upon the exchanges through these passages. The state of sea level, in part, determines these exchanges. There is an active circulation, and the Mediterranean is far from becoming stagnant.

In the distant past the Strait of Sicily must have been a land bridge between Europe and Africa. East of the Strait on the island of Malta the remains of mastodons and other prehistoric creatures have been identified. There is evidence of significant sea-level changes throughout the Mediterranean within historic times. The ancient port city of Carthage is now far removed from harbor facilities. The harbor of Split in Yugoslavia demonstrates high water encroachments. Since the Emperor Diocletian (245–313 AD) vacated the apartments of his palace at Split, they have been put to continuous use by various inhabitants, with the exception of the ground level. The basement has accumulated all the debris from the upper floors through the years. This debris or garbage has preserved Roman stone masonry and artifacts. The floor itself was too close to the sea to be inhabited. It is well known that the city of Venice has been sinking into the sea. In the last two thousand years it has been estimated that sea level, particularly in the northwest Mediterranean, has risen about 0.5 m. Eustatic changes account for significant variations in this average value (Pirazzoli, 1976). This rise or fall can affect the ecological environment by changing the amounts of circulation through connecting passages and, hence, the quality of the water in various parts of the Mediterranean and the adjacent seas. The Mediterranean is a series of interacting parts and adjacent seas. In addition to the major eastern and western basins of the Mediterranean proper, the adjacent smaller seas include the Ligurian, Tyrrhenian, Adriatic, and Aegean Seas, interconnected by various straits and passages.

The size of the Mediterranean Sea, from west to east from Gibraltar to Syria, is about 4000 km. At its greatest breadth, from the coast of France to that of Algeria, the distance is 900 km. The area of the Mediterranean, including all of its adjacent seas except the Black Sea, is 2 523 000 km^2 (Miller and Stanley, 1965), and its volume is 3 708 000 km^3, giving a mean depth of 1470 m.

Menard and Smith (1966) give figures which differ from these by only a small percentage (area $= 2510 \times 10^6$ km^2; volume $= 3771 \times 10^6$ km^3). The average depth is 1487 m according to Kossinna (1921) or 1502 m after Menard and Smith. The latter authors tabulate the areas within various depth boundaries, and calculating the respective volumes based upon the mean depth of the range gives:

depth (km)	0–0.2	0.2–1	1–2	2–3	3–4	4–5	Total
area (10^6 km^2)	0.513	0.564	0.437	0.766	0.224	0.006	2.510
volume (10^6 km^3)	0.051	0.338	0.656	1.915	0.784	0.027	3.771

About 30% of the surface area and 50% of the total water volume is contained between the 2 and 3 km depth contours. In contrast, the area shallower than 200 m constitutes more than 20% of the total area of the Mediterranean Sea but contains less than 1.5% of the total volume.

The Strait of Gibraltar connects the Mediterranean Sea with the major ocean systems. It feeds the Mediterranean and its lesser seas and channels with a surface flow of Atlantic sea water. In turn, deep Mediterranean water spills over the sill of Gibraltar into the deep Atlantic and penetrates the North Atlantic circulation system at mid-depths. This mutual exchange through the Gibraltar Strait represents nearly an equality of flow into and out of the Mediterranean. However, precipitation and runoff contributions to water volume are exceeded by evaporation by about 1 m per year. Since Gibraltar is the only access for renewal of sea water lost through evaporation, the increased salinity of the Mediterranean water is due to evaporation of the lower-salinity water introduced from the Atlantic Ocean.

There is no cold, deep sea water in the Mediterranean. Deep water temperatures vary between

12.5°C and 13.5°C in the west and between 13.5°C and 15°C in the east, and deep salinity throughout varies between 38.4‰ and 39‰. Surface values are more or less transitory, varying with the season or with surface transport. The present average salt content of 38.45‰ represents the cumulative effect of evaporation upon resident Atlantic water for a period of about eighty years. This is the estimated turnover time for Mediterranean water.

Thus, within a little more than modern human life expectancy, the Mediterranean Sea can flush itself out and be renewed. Over a long period the ecological balance must contain elements of uncertainty since the quality of Mediterranean water depends upon the Atlantic Ocean and upon climatological variables. The eighty-year turnover time, however, may be long enough to act as a buffer in the system. If sea level changes, the volumes of flow would also change. The character of Mediterranean water is determined, in part, by the amount of Atlantic water fed into the system, so that any significant change in volume would have an effect upon resident Mediterranean sea water. If one could screen out seasonal effects, it might be possible to use the Mediterranean as an indicator or predictor of climatic change.

THE CIRCULATION

The surface current system of the Mediterranean is illustrated in the chart produced by Lacombe and Tchernia (1971) shown in Fig. 9.2. The surface currents show a migration of Atlantic water toward the east with numerous spin-off eddies along the path. There is no surface return system. Wüst's diagrams in Fig. 9.3 (Wüst, 1959) and Fig. 9.4 (Wüst, 1960) show that the return is subsurface by way of the Levantine intermediate water and bottom flow.

The basic nature of the Mediterranean circulation system thus contains components of strong vertical convections which determine the distribution of salinity and produce vertical recycling of nutrients and other dissolved substances. Since the Mediterranean Sea is characteristically active in three dimensions, the forces moving the water about need to be examined in some detail. Such an

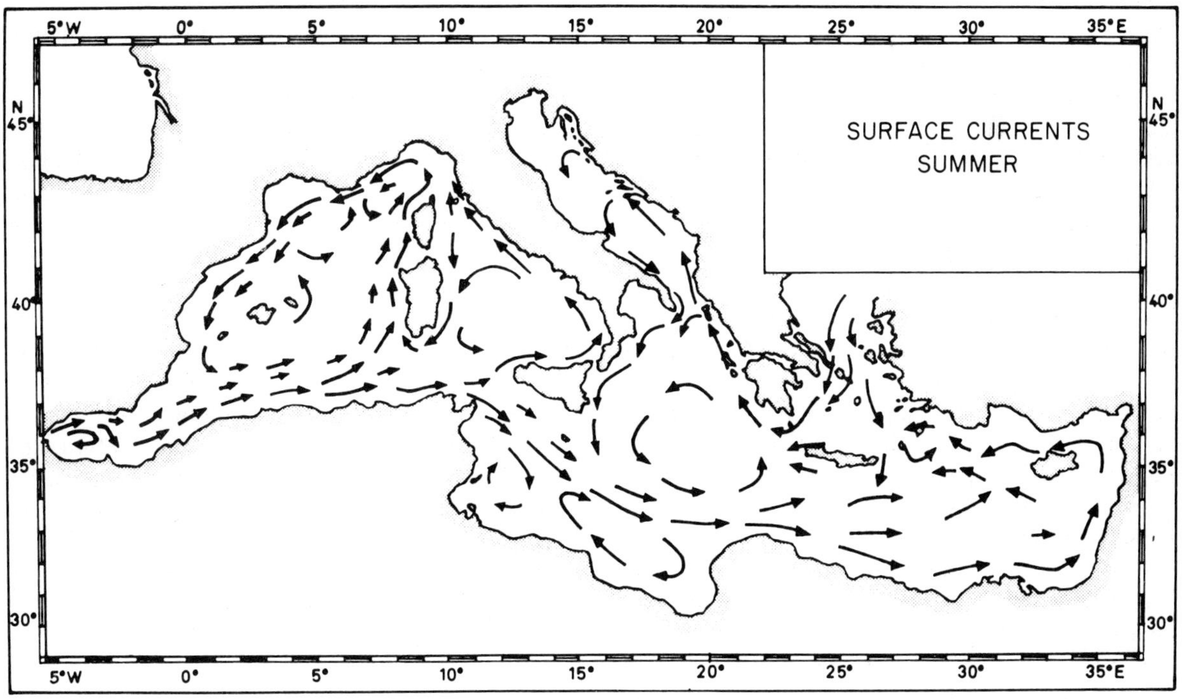

Fig. 9.2. Surface currents in the Mediterranean Sea in summer (after Lacombe and Tchernia, 1971).

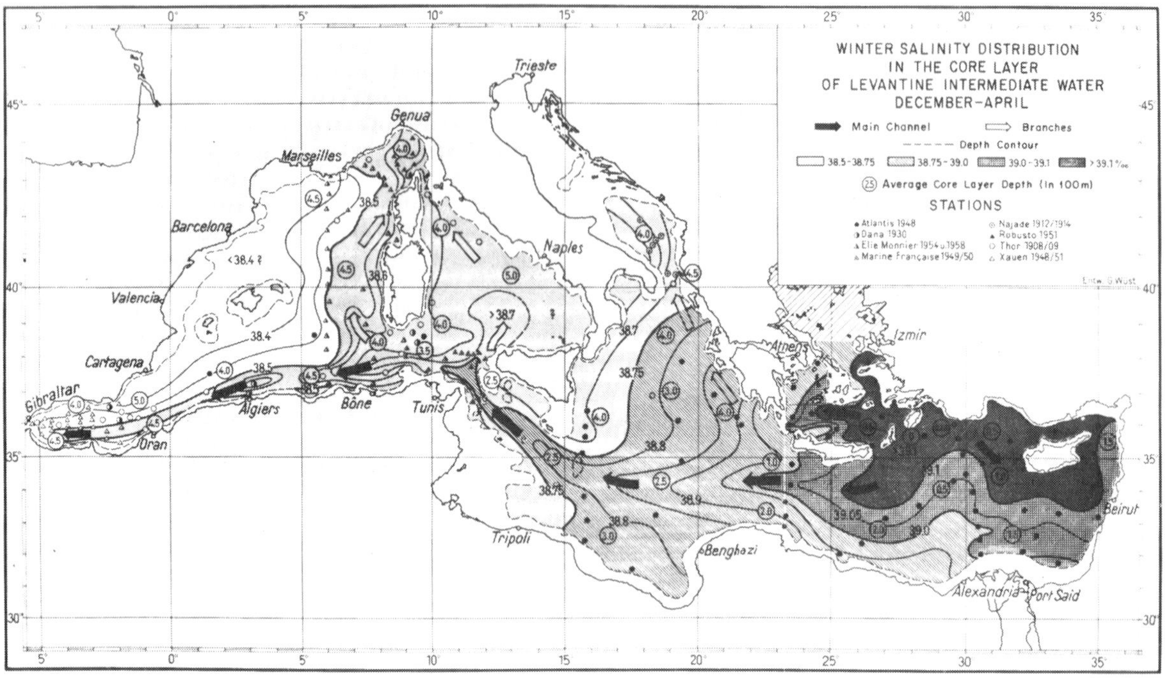

Fig. 9.3. Deep circulation of the Mediterranean Sea within the core layer (*Kernschicht*) of intermediate water (after Wüst, 1959, 1960).

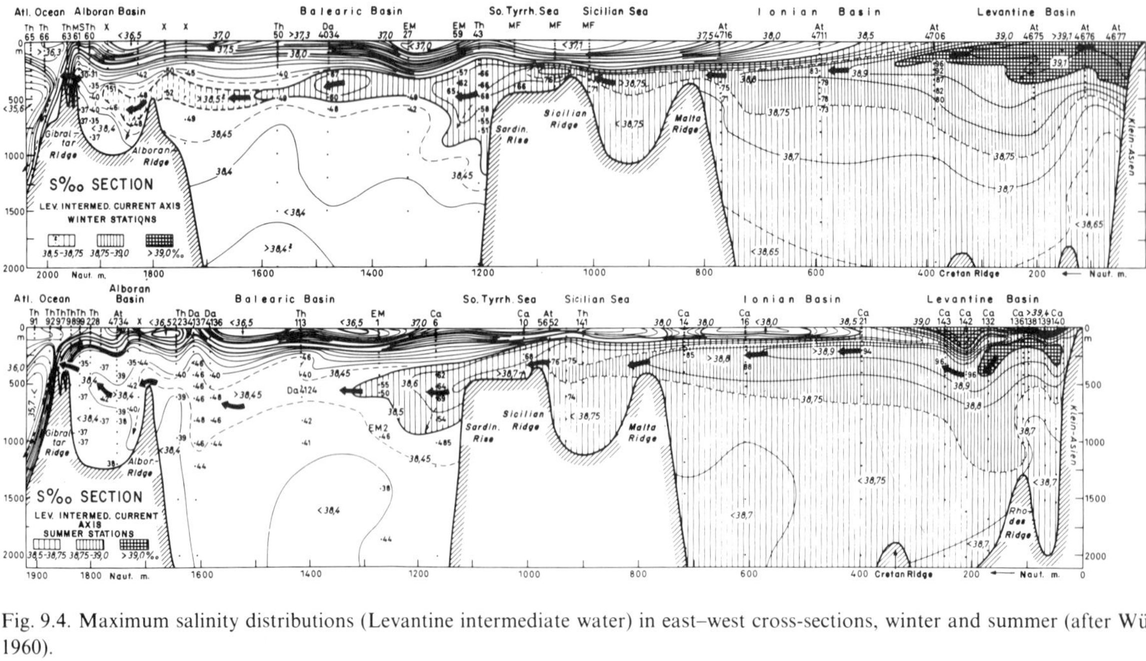

Fig. 9.4. Maximum salinity distributions (Levantine intermediate water) in east–west cross-sections, winter and summer (after Wüst, 1960).

examination may help to determine how delicate or stable the ecological balance is within the Mediterranean environment. For instance, if sea levels were to become significantly different from the present, water transports through connecting straits with their changed cross-sectional areas would be modified. Climatic influence, corollary to substantial sea-level change, would act to modify the water characteristics and change the character of the circulation system. The sensitivity of the water quality of the Mediterranean Sea to sea-level changes has been demonstrated recently in an indirect but positive recognition that there are two modes of Mediterranean-type water flowing into the Atlantic. Continuous recordings of temperature and conductivity with depth using *in situ* measuring devices have clearly defined two deep high salinity maxima that can be identified as Mediterranean water. Zenk (1975) has ascribed these maxima to interfacial mixing at high and low tidal periods of the Atlantic Ocean where the incoming Atlantic and the outgoing Mediterranean waters meet at the Strait of Gibraltar.

Sea level and tides

Tides in the Mediterranean are weak. Consequently, they are dominated by the Atlantic tide at Gibraltar. The rise and fall of tide here is sufficiently strong to produce a different salinity, and hence density, stratum to mix with the relatively homogeneous Mediterranean water at and above the Gibraltar sill at each opposing phase of tide. In this case, the ordinary sense of ebb and flow does not apply, for the Atlantic water is continually pumped into the Mediterranean. The result of this bimodal interfacial mixing is the production of two separate densities of Mediterranean outflow in the Bay of Cadiz. The profile of water types in Fig. 9.5 shows that within the Mediterranean side of Gibraltar the Atlantic water enters and persists for some distance as pure Atlantic water. Its lower surface appears to be the only contact with Mediterranean water, which quickly loses its own pure quality as it pours down the Atlantic slope. The Mediterranean water levels off after reaching an appropriate density stratum. There are two recognizable strata at depths of about 800 and 1200 m where the Mediterranean quality is strongest.

Sea level in the Mediterranean Sea is significantly lower than in the Atlantic Ocean, a fact that is made obvious by the continual inflow of Atlantic water in spite of the large ranges of tide at the Gibraltar entrance. Lisitzin (1974), on the basis of dynamic computations from observed data, has determined that Mediterranean sea level decreases from west to east with markedly significant height differences, comparing the Eastern Mediterranean with the Atlantic Ocean west of Gibraltar.

The constriction at the Strait of Gibraltar and differences in specific volumes of water on either side of the Strait establish the flow patterns typified with incoming currents at the surface and outgoing currents at the bottom. Defant (1961) gives the sill depth as 333 m and the boundary between incoming and outgoing currents as 100 m in winter and 200 m in summer, or, on the average, 150 m, roughly one-half sill depth. Based on a current boundary of 142 m, Defant (1961) estimated the slope of the sea surface to slant downwards in the Mediterranean at 0.6 cm per 100 km. Lisitzin (1974), remarking that this slope appears to be too slight, revises the slope in the transitional region of Atlantic and Mediterranean water to 5 cm per 100 km with a continuous decrease in mean sea level proceeding eastward. Lisitzin (1974) estimates that at 5°E longitude the decrease is 25 cm, with a further decrease between 5° and 20°E longitude of 15 cm. From this point there is a slow increase of 10 cm in the Levantine portion of the Mediterranean (see Fig. 9.6). These estimates were compared with the precise-leveling network and were found to agree substantially in part, and to conflict in other respects. Lisitzin (1974) admits that the problem of oceanographic leveling is still unsolved.

Defant (1961) has summarized the general knowledge about Mediterranean tides which, despite their weak amplitudes, have been recognized and studied by many investigators, even by ancient philosophers. Tides in the Mediterranean are complicated by forced oscillations from the Atlantic tide entering the Strait of Gibraltar and by independent tides, characteristic of closed basins. These latter tides co-oscillate, similarly to the entering Atlantic tide, with adjoining lesser seas such as the Adriatic and Aegean Seas.

Studies made early in this century have been used to explain much of the observed tidal phenomena in the Mediterranean (Wegemann, 1913; Sterneck, 1915a; Defant, 1916). The natural period of oscil-

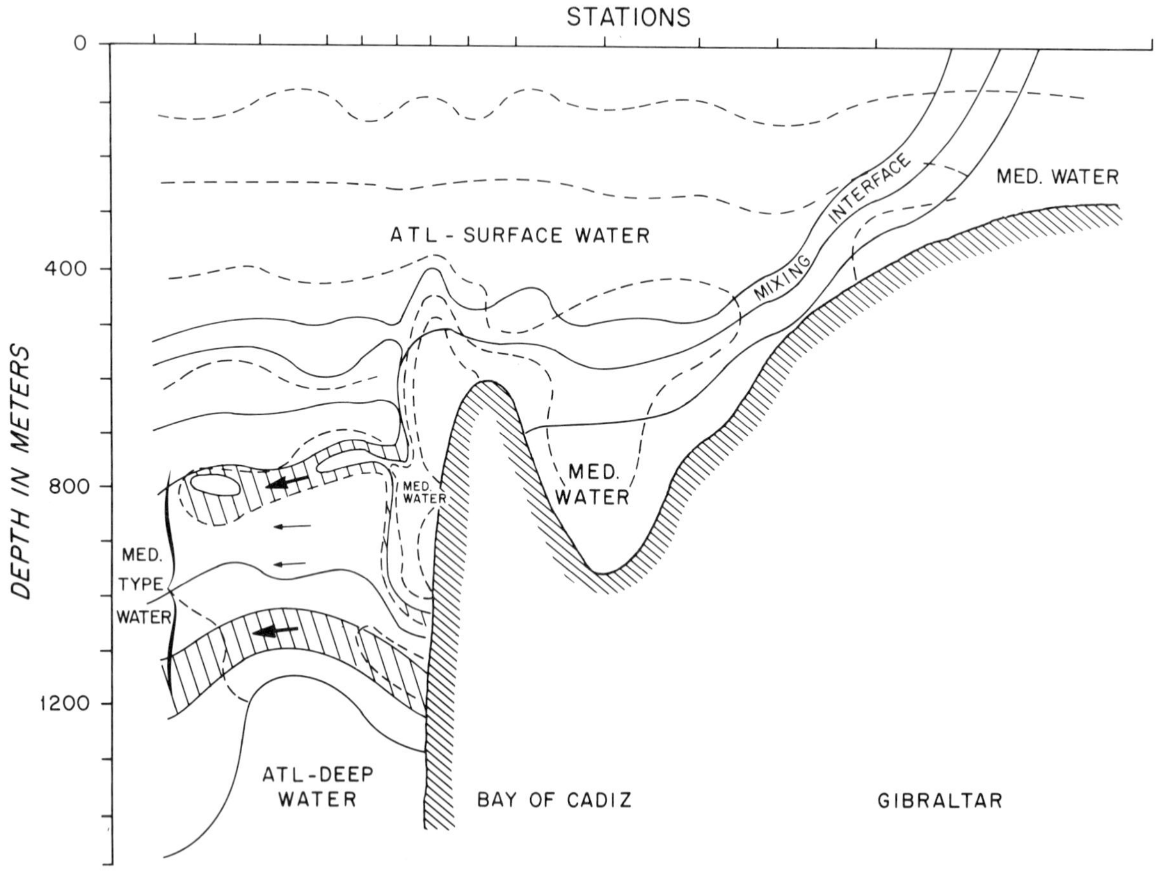

Fig. 9.5. Water-type distributions in the Atlantic, the Bay of Cadiz, and the Strait of Gibraltar. (Each area bounded by dashed or full lines represents a uniform water type.) (Unpublished data obtained during the International Geophysical Year, 1958.)

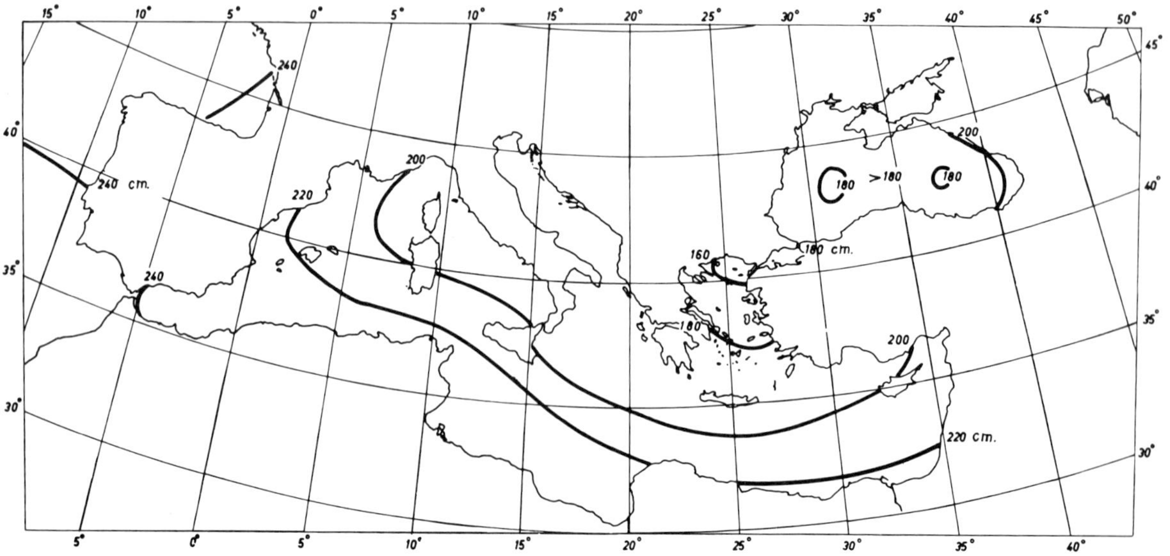

Fig. 9.6. Mean sea level (in cm) in the Mediterranean and Black Seas (after Lisitzin, 1974).

lation for the western basin of the Mediterranean is 5.96 h and for the eastern basin, 8.54 h. The length of the Mediterranean can be assumed to have three nodal cross-sections — the western line at the east end of the Alboran Sea, the middle line at the strait between Sicily and Tunisia, and the eastern line between Crete and Libya.

Figs. 9.7 and 9.8 from Defant (1961) show the major tidal constituents along with the longitudinal distribution of amplitudes and phases for the semi-diurnal and diurnal tides. The diurnal tides comprise both the co-oscillatory Atlantic tide and the independent tide. The tide in the eastern basin has been determined to be completely independent from that in the west by reason of the conservation of water mass throughout a single tidal period (Sterneck, 1915b). At time of syzygies (periods when astronomical forces are in conjunction or opposition) the loss of water to the Aegean and Adriatic Seas during a tidal period is balanced by gain through the Straits of Sicily and Messina except for an amount less than 5 km^3.

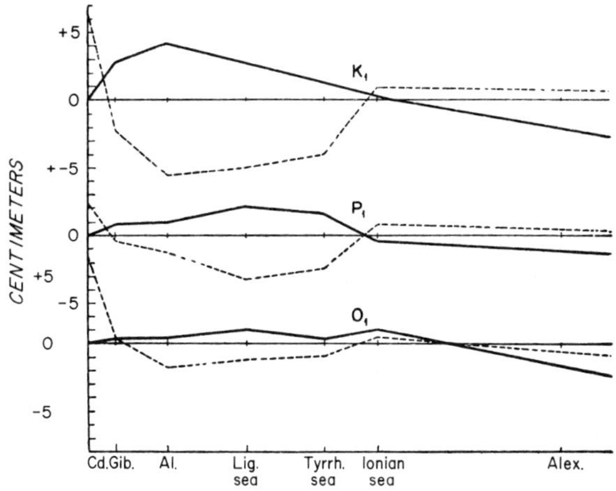

Fig. 9.8. Longitudinal distribution of amplitudes of the diurnal partial tides in the Mediterranean. Legend: -------, co-oscillation with Atlantic Ocean, ———, independent tide (Defant, 1961). (Amplitudes in centimeters referenced to mean sea level.) The constituents of the diurnal tides are: K_1 = lunar–solar; P_1 = principal solar: O_1 = principal lunar.

Currents through straits

In the Strait of Messina the currents are tidal. The extremely narrow passage at Messina produces currents that dramatically illustrate the characteristic interchanges in Mediterranean passages where heavy water of one kind exchanges with lighter water of another sort. The basic current consists of southward-flowing lighter water (4.3 cm s^{-1}) from the Tyrrhenian Sea, to a depth of 30 m, and northward-flowing (9.3 cm s^{-1}) water from the Ionian Sea from that depth to the bottom at a mean sill depth of about 80 m. Part of the inequality of flow, north and south, is due to the return of Tyrrhenian water to the northward flow after it has already passed southward through the Strait. Essentially, the density differences are not very great and are due, for the most part, to salinity differences of one part in a thousand (37‰ vs. 38‰). Consequently, within the Messina channel there is a strong potential for turbulent mixing, particularly when strong tidal currents add further turbulence to the basic currents.

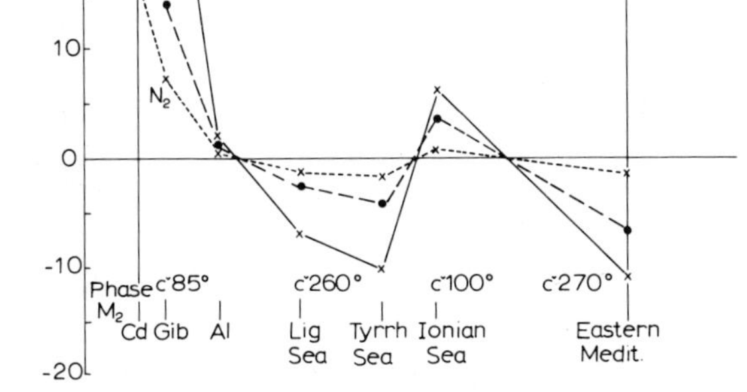

Fig. 9.7. Longitudinal distributions of amplitudes and phase lags of semidiurnal partial tides in the Mediterranean (from Defant, 1961). M_2 = principal constituent; S_2 = principal solar constituent; N_2 = principal lunar–solar constituent.

Atlantic waters in the Mediterranean

The circulatory processes driven by these tidal conditions and other oscillations and the drop in sea level to the east seem to require a corollary — extensive penetration of Atlantic sea water into the Mediterranean. However, Atlantic water anywhere in the Mediterranean is not readily identified, even near its passage through the Strait of Gibraltar. Nowhere in the Mediterranean can one observe Atlantic-type sea water of about 36‰. Such low salinities can be produced in some local situation where they represent dilution from runoff or estuarine outflow. In order to express continuity of Atlantic water into the Mediterranean, it is convenient, though arbitrary, to designate 38‰ as a dividing salinity value in which water fresher than 38‰ is assumed to contain water of recent Atlantic origin.

Surface values at the western end of the Gibraltar sill are approximately 36.25‰ to 36.45‰, the greater value representing the interfacial contact with outflowing Mediterranean water. The Alboran Sea and its semipermanent anticyclonic circulation provides a mixing area in which the incoming Atlantic water is immediately blended with resident Mediterranean water to produce 37‰ water which, as it flows eastward, is lighter than resident water and therefore subject to modification through atmospheric processes. The designation of water with salinities between 37‰ and 38‰ as recent Atlantic water permits one to trace the course of this water throughout the Western Mediterranean without difficulty. While in other oceans and seas the natural tracing parameter is temperature, in the Mediterranean, because of its narrow temperature ranges, it is more instructive to rely on salinity differences.

The Western Mediterranean

The Western Mediterranean deep water is typically homogeneous, having a range of potential temperature (Θ) of approximately 12.7° to 13.0°C and a salinity range of 38.40 to 38.46‰. Also, the Western Mediterranean Sea has no permanent thermocline as is generally found in other comparable areas. The potential temperature is derived by correcting for the pressure effect, and is particularly useful in the Mediterranean because of the narrow temperature range. For example, deep *in situ* temperatures in the Mediterranean are generally warmer than those above because of adiabatic warming; potential temperature, however, decreases gradually to the bottom (Miller et al., 1970). Also, conversion to potential temperature helps to demonstrate the origins of water producing the observed mixtures (Mamayev, 1975). Without this precaution the temperature distribution in the Mediterranean sets up pitfalls and snares for the unwary observer, particularly if compared with those elsewhere.

Vertical movement is particularly strong in the Western Mediterranean, especially during the cold winter season when deep convective sinking and turnover take place (MEDOC Group, 1970). The barrier layer which inhibits this turnover is the Levantine intermediate water which originates in the Eastern Mediterranean. This is warm, high-salinity water which, on passing over the Sicilian sill, sinks to depths of 400 m or greater. The further it is removed from the sill in time and distance, the more the Levantine intermediate water is diluted with west Mediterranean deep water and with surface water of recent Atlantic origin.

With a potential density of approximately 29.10 σ_Θ, there is very little density difference in most of the water column to inhibit vertical circulation. When winter storms lower surface temperatures to 12°C or so, deep convection can take place. This phenomenon occurs in the Algero-Provençal Basin of the Western Mediterranean, where the convective process has been observed and traced to depths of 2000 m.

The Tyrrhenian Sea is a major basin in the Western Mediterranean. It is separated from the western Algero-Provençal Basin by Corsica and Sardinia and apparently has no measurable winter turnover except for the slow erosion of the intruding water masses. The Tyrrhenian Sea is the first recipient of Levantine intermediate water after it passes across the Sicilian sill. If the isohaline of 38.5‰ is used as a lower limit for identification, Levantine water can be observed as deep as 1300 m in the Tyrrhenian Sea, rising a few hundred meters in the northern part (Miller, 1971a). The presence of so much Levantine water in the Tyrrhenian Sea probably affects the deep bottom water, which is 0.2°C warmer than bottom water in the western basin. On the other hand, lack of winter turnover may also serve as an explanation for the 0.2° difference.

Between Sardinia and Tunis, Levantine intermediate water (LIW) appears to have two modes, indistinguishable as far as temperature and salinity are concerned but separated geographically as demonstrated in north–south profiles (see Fig. 9.9).

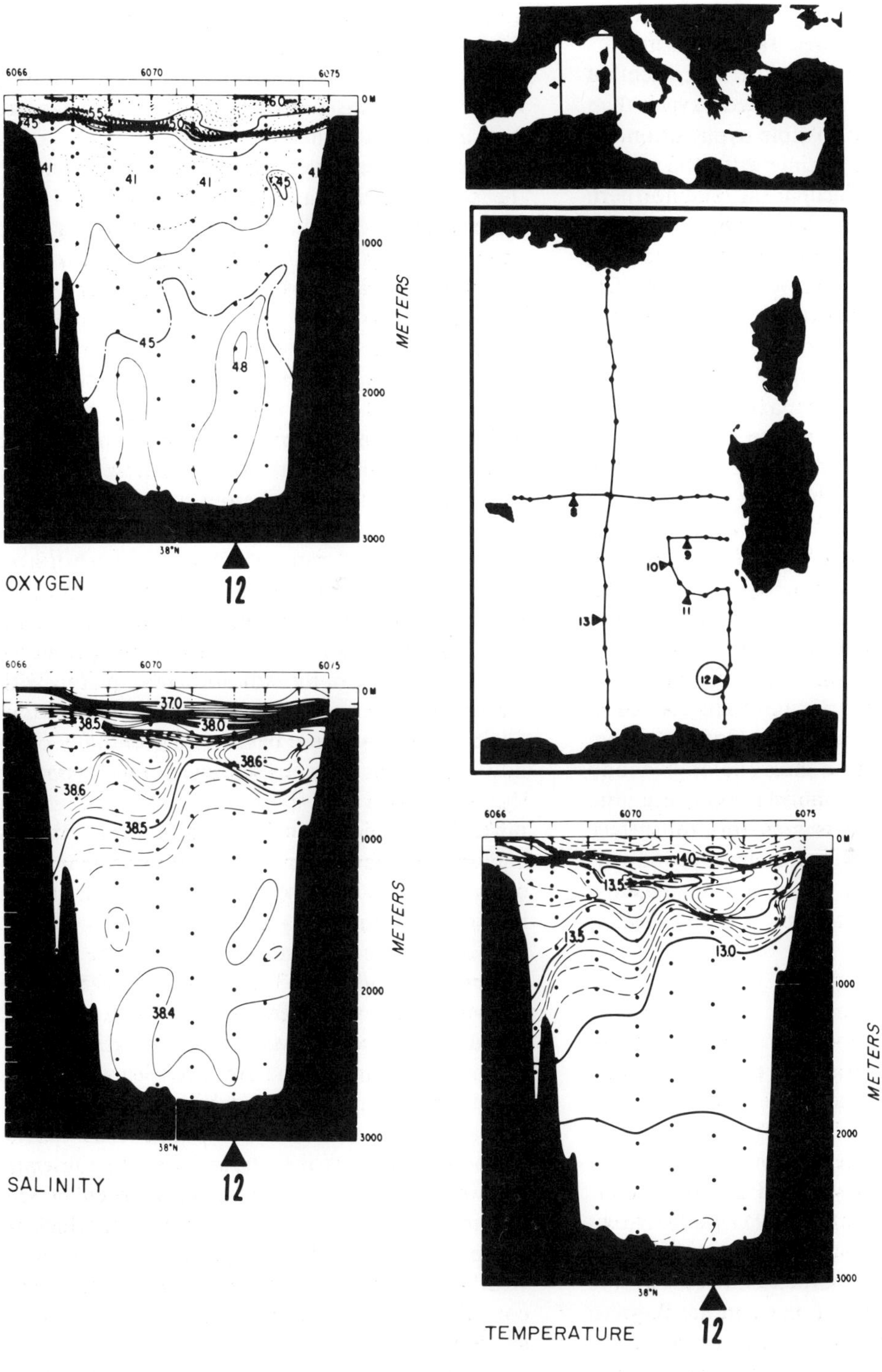

Fig. 9.9. Distribution of temperature (°C), salinity (‰), and oxygen ml l^{-1} in the waters between Sardinia and Tunis in winter (from Miller et al., 1970). South is to the right in each profile.

In a temperature vs. salinity diagram, which can be used to identify water masses, the two segments appear to be alike, but the northern segment of LIW has a lower content of dissolved oxygen than the southern segment. A possible explanation for these differences may lie in time differences since the water masses left the Sicilian sill. The northern segment of LIW travels the greater distance, around the Tyrrhenian Sea, and the southern segment, off Africa, takes a more direct route. The oxygen differences in the two segments are slight; however they occur within the oxygen-minimum layer, which is well-defined in the Western Mediterranean and coincides with the maximum salinity of Levantine intermediate water. The Eastern Mediterranean does not have a distinguishable oxygen minimum layer (Miller et al., 1970).

The temperature-salinity characteristics of Levantine intermediate water appear to change very little during its traverse through the Tyrrhenian Sea (Miller, 1970). On the other hand, in the Algero–Provençal Basin LIW lies like a thick wedge hugging the west coasts of Sardinia and Corsica. Its lower boundary rises from 1300 m in the south to 800 m off northern Corsica. From Corsica, LIW continues to hug the Italian and French Riviera coasts, diminishing in size and intensity. The vestigial effects of this subsurface warm saline water mass continue around the Spanish mainland coast, skirting the Balearic Islands and traceable into the Alboran Sea at depths of 400 to 500 m.

The Eastern Mediterranean

The Eastern Mediterranean consists of two major basins, the Levant and the Ionian Sea. One connects with the Aegean Sea in the north and the other with the Adriatic Sea. Although a number of straits join the Aegean Sea to the Eastern Mediterranean, only one strait, the Strait of Otranto, allows Adriatic water to flow into the Ionian Sea. The water columns in the Eastern Mediterranean are warmer and more saline than those in the Western Mediterranean. The total water column below the seasonal thermocline is comparatively homogeneous (13.3–13.7°Θ, and 38.70–38.75‰). Levantine intermediate water is identified as a layer of greater salinity, 38.8–39‰, extending from the entrance to the Sicilian passage at depths of 400 m and breaking out at the surface in the Levant near the island of Rhodes.

From the Sicilian passage in the west, Atlantic water moves eastward at the surface near the African coast. It is identified as a water mass having less than 38‰ salinity. Atlantic water sinks beneath the surface after passing Libya. East of Libya, Atlantic water continues to pervade the eastern waters as a thin stratum at 100 m depth. This layer is not to be confused with the Nile effluent.

The Adriatic Sea

The Adriatic is a shallow sea, except for its southern portion, and is subject to the vagaries of weather and other external influences. The Adriatic water budget is maintained by inflow from the Ionian Sea and by runoff from streams and rivers. The fresh-water input is a determining factor affecting the density of water which leaves the Adriatic. The deep lower basin of the Adriatic collects dense water which is formed during the winter. This dense water is the product of cooling in the shallow northern reaches of the Adriatic and of mixing with high-salinity Ionian Sea water. This dense water fills the lower Adriatic basin, and eventually overflows across the Otranto sill. Because of its density the Adriatic water spills down into the deeper portions of the Ionian Sea in sporadic outbursts and spreads along the bottom in a slow counterclockwise circulation. According to Pollak (1951), this irregular outflow is the source of deep water in the Eastern Mediterranean.

Zoré-Armanda (1963) and Buljan (1953) have shown that the winter environment in the Adriatic fluctuates from year to year. This fluctuation relates to the sensitivity of the Adriatic Sea to various climatic and meteorological conditions. Zoré-Armanda (1972) has demonstrated that over a period of fifty years or so the salinity of the Adriatic has varied with the amount of icebergs and sea ice around Greenland. This is not an unreasonable comparison when one considers that European weather, influencing the Adriatic, is affected by conditions in the North Atlantic. Zoré-Armanda suggests that the lower basin acts as a collector of dense water produced by the mixing of Ionian Sea water with waters from the shallow reaches of both the upper and middle Adriatic. The dense water is characterized by temperatures of 13°C and salinities of 38.6‰. The quantity formed each season is dependent upon the prevailing environment. If Ionian water is weakly advected into the Adriatic, large amounts of dense water are produced in the southern Adriatic. Conversely, with strong northern penetrations of Ionian Sea water, smaller amounts of dense water are formed. On such occasions intermediate flows may return to the Ionian Sea and contribute to the upper deep structure of the Eastern Mediterranean.

Fresher Adriatic water also enters the Ionian Sea at the surface. This water leaves the Adriatic at the western end of the strait and tends to concentrate off the southern Italian coasts. The three identifiable separate flows out of the Adriatic make the general circulation problem of the central Mediterranean more complex. When there are persistent or semipersistent shifts in the climatology of the region, the water is vulnerable to change. Pollak (1951) was unable to show any appreciable effect of Aegean Sea water upon deep water of the Eastern Mediterranean. The scientific literature has offered only speculations or inferences (Nielsen, 1912; Wüst, 1961) that Aegean water may influence the character of the deep Eastern Mediterranean. In 1962, however, the Research Vessel *Atlantis* made a section south of the Strait of Scarpanto and observed Aegean Sea water in the Eastern Mediterranean at depths of 1300 m (*Atlantis* Station 6286). This water had the characteristics of the deep Aegean water (14.3°C, 38.9‰) found on the northern side of the Strait. Adjoining and nearby stations supported this identification of an overflow. Apparently sporadic contributions of Aegean Sea water and Adriatic Sea water make up the upper and lower sources of Eastern Mediterranean deep water.

The Aegean Sea

The Aegean Sea differs considerably from its parent sea, the Eastern Mediterranean. It is characterized by strong deep seasonal overturn (Miller, 1972). Surface salinity is usually greater than 39‰, but winter stirring and convection reduce this salinity to 38.9‰, a characteristic salinity even at great depths. Connecting straits allow fresher (38.7‰) Eastern Mediterranean water to enter the Aegean. This infusion spreads at mid-depths and separates the upper high-salinity water from the deeper high-salinity water. Density differences in the water column are so slight that massive turnovers during the winter season should be common. For instance, north of Scarpanto, Bruce and Charnock (1965) observed a sink consisting of a large, relatively cold eddy. The deep-water oxygen content of the Aegean Sea is greater (>5 ml l^{-1}) than in any other part of the Mediterranean (Miller, 1972), indicating rapid and deep convective phenomena.

In contrast to the high oxygen content of the Aegean Sea, oxygen is extremely low in the Sea of Marmara, which connects the Black Sea and the Aegean. The Black Sea is brackish and almost devoid of oxygen in its lower depths (cf. Ch. 11). However, the Sea of Marmara, with the exception of a thin layer of fresh water flowing into the Aegean, has salinities of 38.5‰. When there is insufficient flow of Black Sea water through the Straits of Bosporus and Dardanelles, Aegean Sea water can penetrate the Sea of Marmara. The extremely low density of the surface water acts as an insulating barrier, preventing any aerated water from convecting downwards, thereby isolating the deep water in the Sea of Marmara. Thus the Sea of Marmara contains basic Mediterranean water properties (14.5°C, 38.5‰) with a unique environment of less than 2 ml l^{-1} of oxygen.

NUTRIENT PROPERTIES

A fundamental characteristic of Mediterranean water is its impoverished nutrient concentration. None of the deep nutrient-rich Atlantic waters take part in the Mediterranean circulation. Since only the upper 150 m or so of Atlantic water provide replenishment for the Mediterranean Sea, any increase of concentration of nutrients in excess of amounts contained within the surface layers of the neighboring Atlantic Ocean are probably due to local phenomena.

Nutrients enter the Mediterranean in the runoff of rivers, together with their land-derived salts, sediments, and pollutant materials. Even this potential source is characteristically Mediterranean in time and space. Runoff from rivers is concentrated on the northern or European coasts (see Fig. 9.10). There is no comparable southern contribution, with the exception of the Nile effluent. The total budget losses of water due to evaporation have been estimated at 1.44 m per year. Approximately

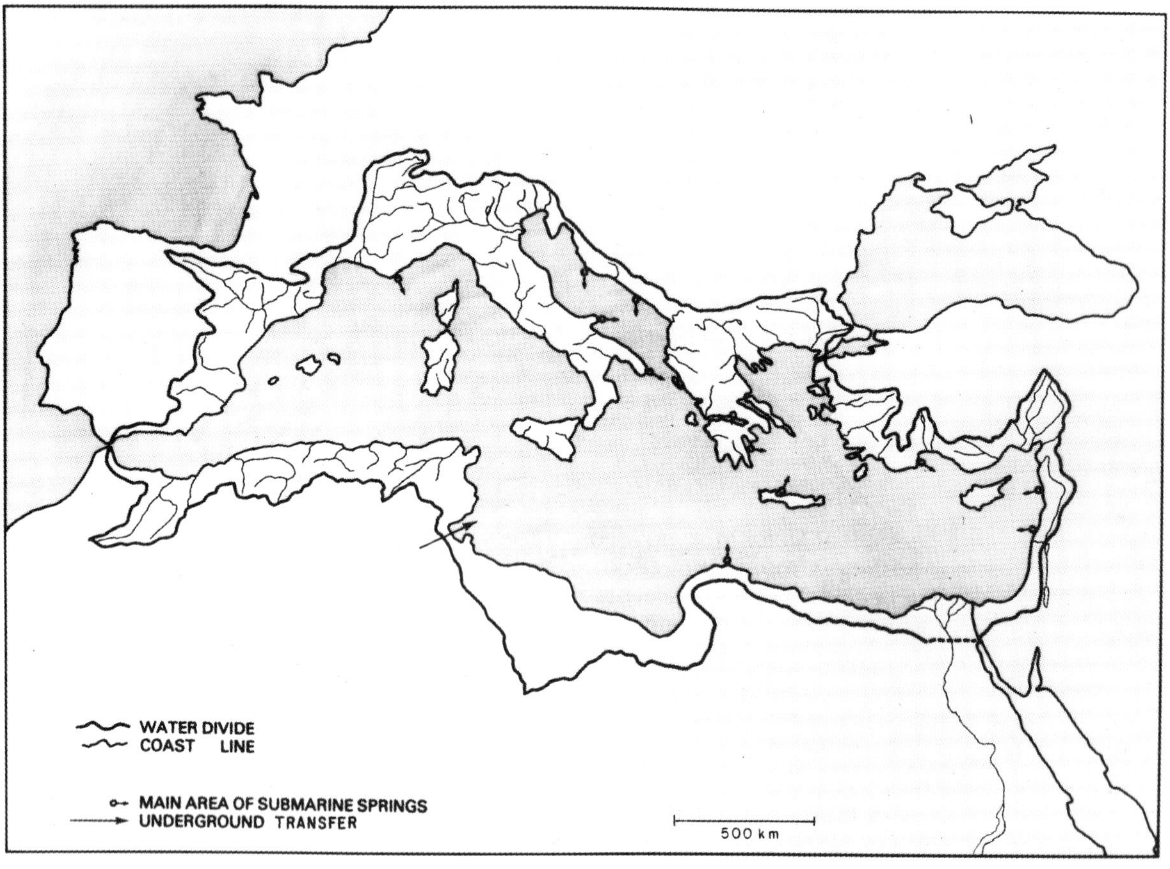

Fig. 9.10. The watershed of the Mediterranean Basin (Ambroggi, 1977). Drainage areas of the Black Sea and the Nile River are not included.

25% of this loss is balanced by contributions of runoff and precipitation. Rainfall is substantial only during the fall and winter seasons; consequently, contributions of nutrients from land-derived sources can be assumed to occur locally and seasonally.

The north–south profile in Fig. 9.11 is typical of the phosphate content throughout the Mediterranean. Phosphate values vary from 0.1 to 0.5 μg-at. l^{-1} with very few definitive patterns other than the occurrence of higher values in deep water. The Eastern Mediterranean has a smaller range of phosphate content than the Western. The isolated Sea of Marmara (not shown in Fig. 9.11) is an exception, with values greater than 1.0 μg-at l^{-1}. The Mediterranean is remarkably deficient in other nutrients and lacks any distinguishing features. The lack of definition is interpreted by some investigators as symptomatic of a naturally efficient recycling of nutrients. In spite of the relative nutrient poverty, the turnover rate of deep and surface waters that is so characteristically strong in the Mediterranean seas allows for efficient use of all the nutrients in the water column.

EXTERNAL INFLUENCES

The geography of the entire Mediterranean area, including the land masses surrounding its seas (Fig. 9.12), suggests that there is a strong geomorphic influence upon the movements of the water masses and the character of water properties. The two major basins are divided into two basic water regimes, west and east, and are less distinctly separated north and south. In the north the sea areas are surrounded by mountains from the Sierra Nevada ranges and the Pyrenees in the northwest to

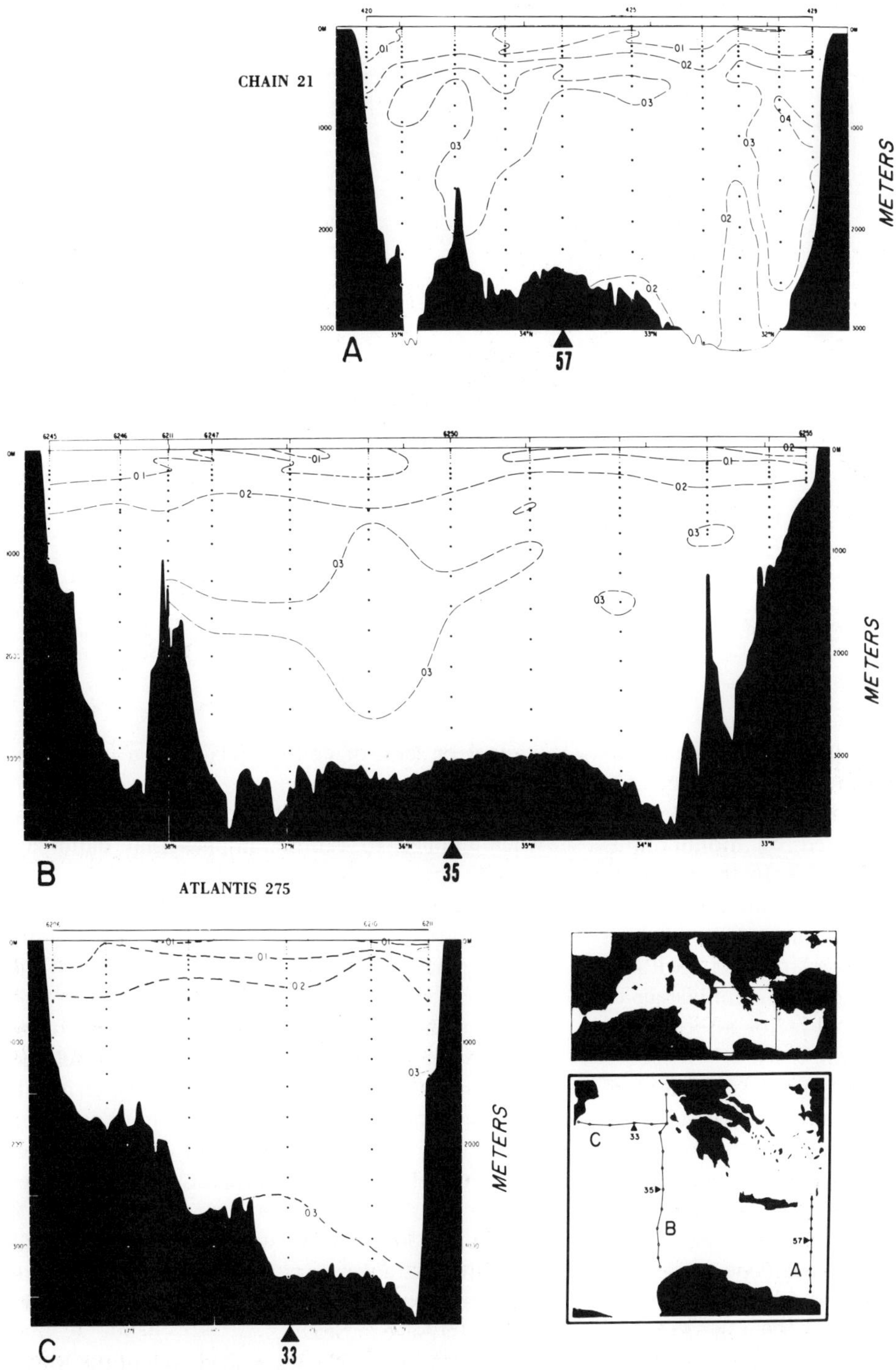

Fig. 9.11. Inorganic phosphate distributions (μg-at. l^{-1}) in the Eastern and Central Mediterranean from the *Mediterranean Sea Atlas* (Miller et al., 1970). Section A: R.V. *Chain*, 21 October, November 1961; sections B, C: R.V. *Atlantis*, February, March 1962.

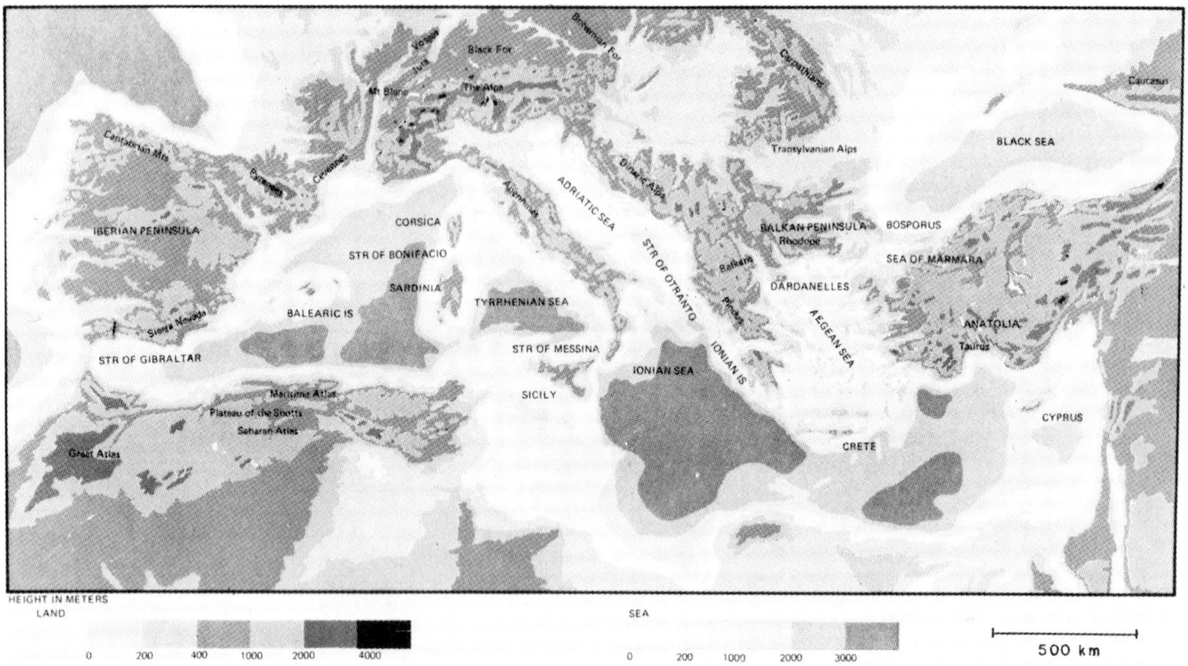

Fig. 9.12. Physiographic map of the Mediterranean; its seas and its land masses (Henry, 1977).

the Balkan and Anatolian Mountains in the northeast. While there are large alluvial plains associated with the Ebro, Rhône, and Po Rivers, the Mediterranean is essentially bounded by mountainous coasts sloping steeply to the sea. In the southern regions the African Atlas Mountains are barriers along the Western Mediterranean, but the remainder of the southern coasts are deserts with no high barriers to affect air flow. This imbalance of geomorphology is an external factor affecting the ecology of the Mediterranean.

Henry (1977) has related the development of Mediterranean peoples to the ecological constraints imposed upon them by the natural environment. The Mediterranean is a transitional area climatically, with a temperate, damp climate in the north and an extreme arid climate in the south. Orographic rains on the European coasts contrast markedly with minimal annual rainfalls in the southern regions of less than 100 mm. Fig. 9.13 (from Henry, 1977) shows the duration of dry seasons for the Mediterranean, and Fig. 9.14 gives the distribution and amounts of rainfall. The disparity in the amounts and occurrences of precipitation is influenced by the topography of the land and the climatic conditions of the area. Water control or lack of control has been a constraint upon human development.

The term orographic, denoting mountain effects, when applied to rainfall implies that damp air forced upwards by terrain leads to fallout of moisture when the air reaches heights conducive to condensation and coalescence. This phenomenon, in the Mediterranean context, implies that the source of moisture for orographic rain is the Mediterranean Sea surface. The orographic rain, linked with Mediterranean evaporation, is a closed system contributing nothing to the long-term salt budget, because the fallout and runoff are soon returned to the sea. Nevertheless, orographic precipitation tends to redistribute the water properties within the Mediterranean Sea. Seasonal runoff adding new fresh water with acquired local temperature characteristics affects the northwestern Mediterranean coasts, the northern Adriatic and the north Aegean Sea.

Conventionally, the watershed area of the Mediterranean Basin is limited by the mountains that come down to the coast and comprises about 1.8 million km^2, which is somewhat less than the sea area. However, the Black Sea drainage which enters the Mediterranean by way of the Bosporus has a

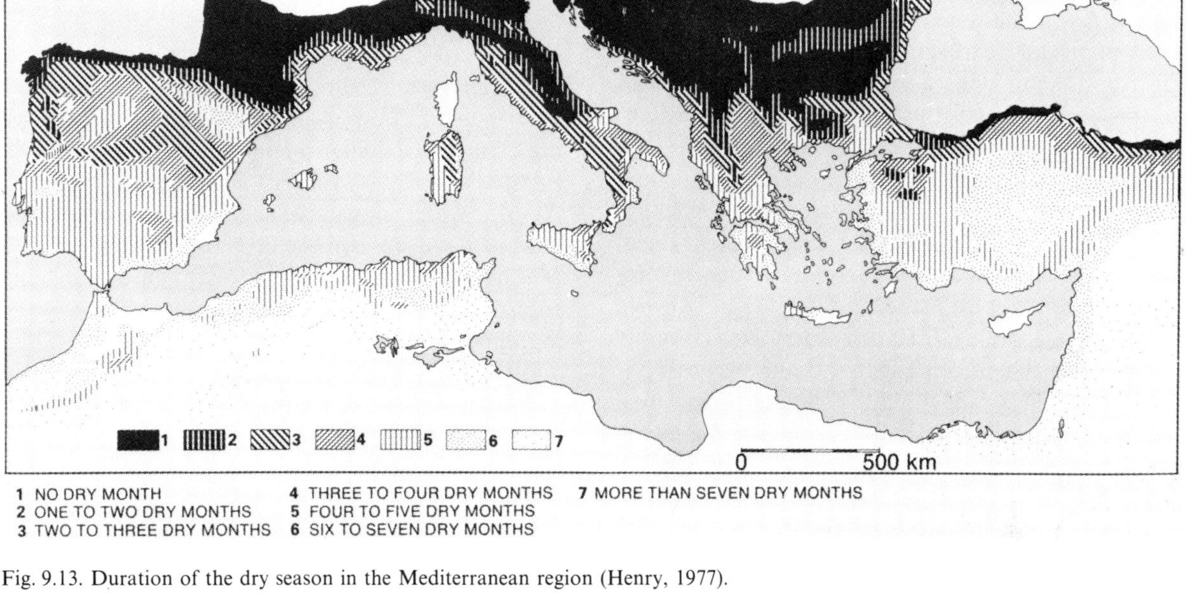

Fig. 9.13. Duration of the dry season in the Mediterranean region (Henry, 1977).

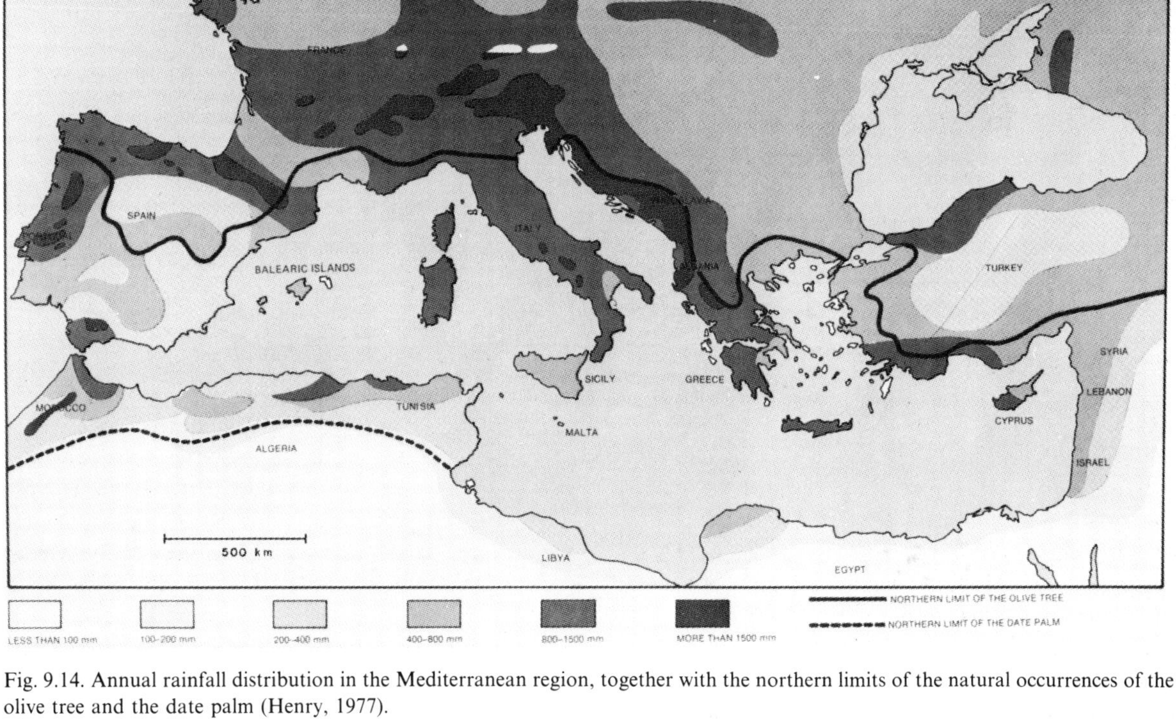

Fig. 9.14. Annual rainfall distribution in the Mediterranean region, together with the northern limits of the natural occurrences of the olive tree and the date palm (Henry, 1977).

watershed which extends as far north as 60°N, and the Nile watershed reaches as far south as the Equator. Throughout most of the southern coast runoff is minimal or nonexistent.

Climate and weather are important factors in the distribution and consistency of Mediterranean water properties. Fig. 9.15 shows that annual surface winds, a strong influence in determining rates of evaporation and the transport of heat and moisture to the atmosphere, are generally from the north and west. Mean winds mask local storm occurrences but help to determine general trends. In contrast to the orographic rains associated with local storms, mean winds coming from the northwest connote indirect influence from North Atlantic weather.

With winds approaching the Mediterranean Sea surface from the north and west, the surface air in contact with the sea surface would ordinarily be dry, for it has had to descend from the mountains (Bunker, 1972). The combination of dry winds and sunny days, which occurs as often as 250 times a year, produces a strong evaporating influence over the entire surface of the Mediterranean that counteracts the effects of precipitation and runoff.

There has been little emphasis here on the quantification of the various elements and factors that affect the condition of Mediterranean water. Rather, attempts have been made to illustrate that there are various external influences on the Mediterranean environment which are variable and subject to change. The Mediterranean Sea characteristics result from processes associated with time and from a series of progressive modifying changes which act upon Atlantic water that is isolated from its origin. The conditioning of Atlantic water is as stable as the external influences permit. Each stage in the processing of Atlantic to Mediterranean water is dependent upon, and responsive to, the sum of local external effects working upon its surface.

COASTAL CONDITIONS AND POLLUTION

The present climate guarantees that for a long time to come the Mediterranean Sea will continue to be an evaporating area. As long as this is so, Atlantic water will continue to flow in from the west, replenishing the moisture lost by excess evaporation. This means that almost any substance introduced into the surface environment of the Mediterranean Sea, unless it is volatile or miscible

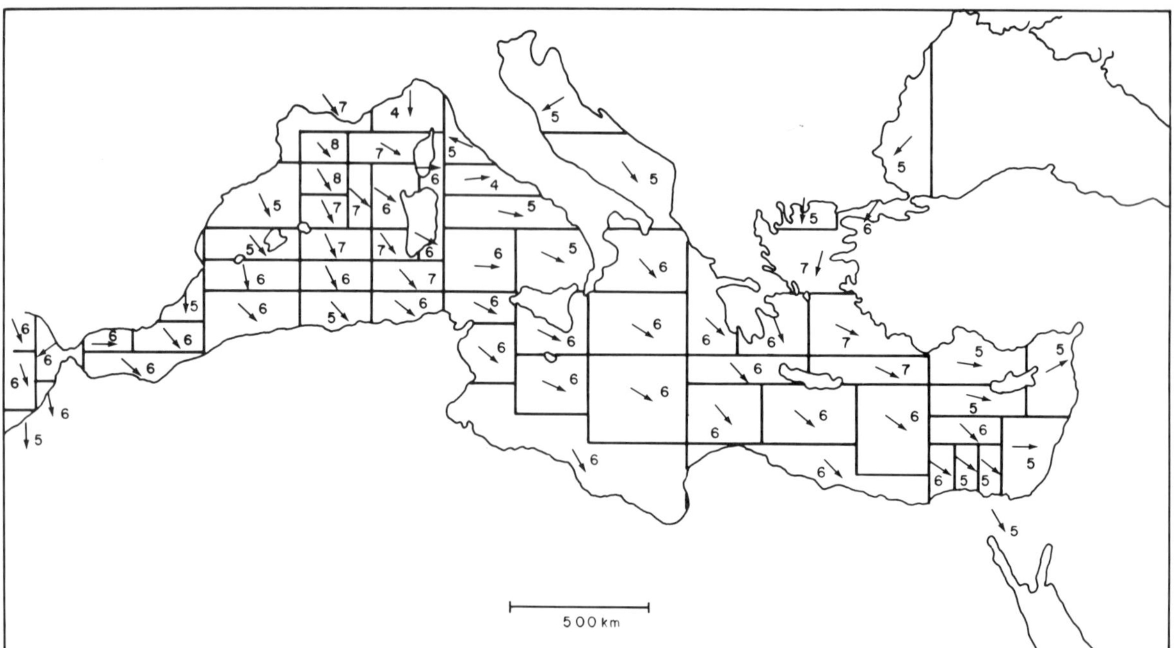

Fig. 9.15. Annual observed mean surface winds ($m\ s^{-1}$) for the Mediterranean (1941–1972) determined from surface marine observations. (Data supplied by U.S.A. National Oceanographic Data Center.)

within the layer of water that leaves the Mediterranean, will remain within its boundaries.

The introductions of substances foreign to Mediterranean waters are essentially, though not exclusively, from land-based sources. Airborne particles can reach the sea by way of such atmospheric transport as dust storms or fallout from precipitation. Tritium, for instance, is a tracer for radioactive contamination which marks the events of nuclear bomb test activity (Roether and Weiss, 1975). The Aegean Sea contains three times the tritium concentration observed in other parts of the Mediterranean (W. Weiss, 1978, pers. comm.). This excessive tritium must come from the Black Sea runoff which receives its river discharge from a broad drainage system. Otherwise, the Mediterranean drainage areas remain very close to the coasts and extend only as far north as the Swiss Alps.

Atmospheric processes could introduce foreign substance by way of dust storms which occur from time to time in both the Western and Eastern Mediterranean seas. The fallout of "blood rains" consists of airborne particles from the African deserts (Miller, 1971b). Compared to the waterborne discharges from land, atmospheric pollution does not seem to be as significant a factor affecting Mediterranean waters. Waste loads of domestic sewage, industrial discharges, and agricultural runoff are probably the major contributors polluting the Mediterranean. The uneven distribution of runoff and precipitation along the northern coasts of the Mediterranean Sea, combined with the northern concentrations of population and industrial activity, contribute a waste load of pollutants to Mediterranean waters that are confined first to the northern coasts and then are spread and recirculated through the natural processes of advection and convection. The estimated annual pollution loads are shown in Table 9.1 (from Helmer, 1977).

A major effort is now in force to combat and alleviate the pollution of the Mediterranean Sea. Protocols and treaties have been agreed to by the Mediterranean nations to assess and monitor the extent of pollution in the Mediterranean Sea (Thacher, 1977). Laws have been passed or are under consideration to alleviate the polluting activities of industry, urban centers, and ship traffic. This effort forms the Mediterranean Action Plan, coordinated by the United Nations Environment Programme.

In addition to land-based sources of pollution, which primarily affect the coastal regions, the demands for energy, reflected in the heavy transport of oil and shipping-lane traffic, make the entire Mediterranean Sea vulnerable to chronic oil pollution. Fig. 9.16 demonstrates the complete vulnerability created by the traffic of oil where paths of transport crisscross the Mediterranean. In the north–south direction there are tanker routings from crude oil sources in the south to refinery and user destinations in the north. East to west, the tanker routes extend from Middle East sources to destinations in northern Europe and America. Oil pollution is derived from tank washings, ballasting operations, bilge discharges, discarded lubricants, refineries, oil explorations, and accidents. As tanker traffic increases, so does the vulnerability of the Mediterranean to toxicity from hydrocarbons.

SOME PROBLEMS

The Suez Canal, an engineering achievement which has affected the environment, has been in a state of change ever since it was first opened more than a century ago. As a connecting link between eastern and western oceans it has been a potential warm water route for the migration of species between the Red Sea and the Mediterranean. However, migrations were at first inhibited by the extreme salt content of the intervening Bitter Lakes where dissolving salt beds set up an ecological barrier. As the salt beds dissolved, the saline content of the Bitter Lakes was gradually lowered to a few parts per thousand higher than Red Sea water. The remaining inhibiting factor is the high salt concentration produced by local evaporation (Miller and Munns, 1974). The areal ratio of the Bitter Lakes to the Canal complex is sufficient to maintain a salinity varying seasonally from 43‰ to 46‰.

The demand for shorter routings of tanker traffic has made it economically feasible to dredge and widen the Suez Canal again in order to accommodate the huge tankers carrying Middle East oil. This shorter travel route will increase the traffic through the Mediterranean and the danger of collisions and oil spills. It will also lower the

TABLE 9.1

Estimated pollution loads (t yr^{-1}) from land-based sources (Helmer, 1977)

	Pollutant:	Pollution loads originating in the coastal zone: domestic	industrial	agricultural	sub-total	Loads carried by rivers into the Mediterranean: pollution	background	sub-total (range)	Total Mediterranean loads: pollution	total (incl. background) (range)
1. Volume:										
total discharge	$\times 10^9$	2	6	–*	(8)	–	420	420 (400–500)	–	430 (400–500)
2. Organic matter:										
BOD	$\times 10^3$	500	900	100	1500	1000	(800)	1800 (1200–2300)	2500	3300 (2700–3800)
COD	$\times 10^3$	1100	2400	1600	5100	2700	800	3500 (2300–4700)	7800	8600 (7400–9800)
3. Nutrients:										
phosphorus	$\times 10^3$	22	5	30	57	260	40	300 (200–400)	320	360 (260–460)
nitrogen	$\times 10^3$	110	25	65	200	600	200	800 (600–1000)	800	1000 (800–1200)
4. Specific organics:										
detergents	$\times 10^3$	18	–	–	18	42	0	42 (9–75)	60	60 (30–90)
phenols	$\times 10^3$	–	11	–	11	1	0	1 (0.5–1.8)	12	12 (6–18)
mineral oil	$\times 10^3$	–	120	–	120	(–)	0	(–)	(120)	(–)
5. Metals:										
mercury		0.8	7	–	8	90	30	120 (40–200)	100	130 (50–200)
lead		200	1400	–	1600	2200	1000	3200 (2700–3800)	3800	4800 (4300–5400)
chromium		250	950	–	1200	1200	400	1600 (500–2700)	2400	2800 (1700–3900)
zinc		1900	5000	–	6900	14 000	4000	18 000 (14000–22000)	21 000	25 000 (21000–29000)
6. Suspended matter:										
TSS	$\times 10^6$	0.6	2.8	50	53	–	300	300 (100–500)	–	350 (100–600)
7. Pesticides:										
organochlorines		–	–	–*	–	90	0	90 (50–200)	90	90 (50–200)
8. Radioactivity:										
tritium	Ci yr^{-1}	–	400	–	400	2100	(–)	2100 (1600–3100)	2500	(–)
other radio-nuclides	Ci yr^{-1}	–	25	(–)	25	15	(–)	15 (10–25)	40	(–)

Sources are coastal (defined as a 20 km strip inland) and riverine, inclusive of small coastal rivers. Total discharge approximates the annual fresh-water inflow into the Mediterranean, exclusive of the Black Sea drainage and the Nile River.

Legend: – contributions from this source negligible; (–) insufficient data base for estimate; –* included in river assessment.

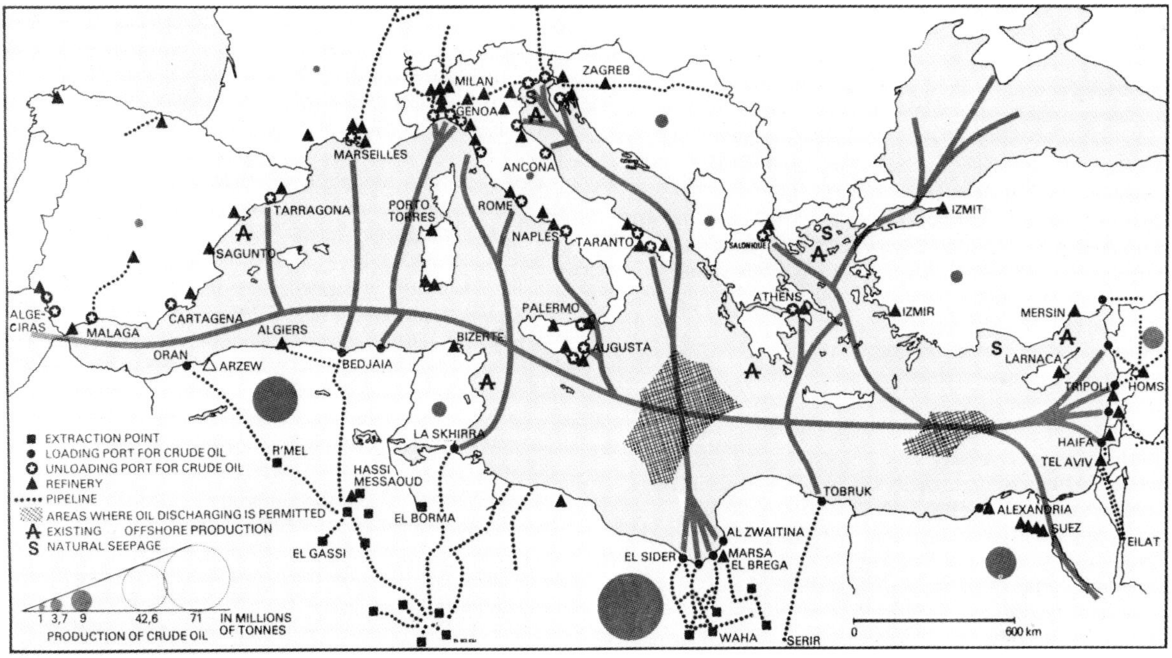

Fig. 9.16. Production areas and transport paths of oil in the Mediterranean region (Le Lourd, 1977).

barriers to species migrations, for the dredging of the Canal will increase the water volume and lessen the salinities of the Bitter Lake by altering the area:volume ratio.

In January of 1965 the Danish biologist Gunnar Thorson observed that the fish markets of Beirut, Lebanon, were displaying large baskets of living pearl oysters for sale (Christensen, 1972). The evidence that this Indo-Pacific species had now become so abundant as to be an important fishery surprised and concerned him. He concluded that the faunal composition of the Eastern Mediterranean would be changed radically by the lowered salinity that would result from the deepening of the Suez Canal.

Biologists and marine chemists should find the Mediterranean environment a particular challenge with respect to the geographic distribution of species and properties. Sources and origins can almost be declared as definable within certain points and areas. Perhaps the least definable point source might be oil pollution since its origins may occur anywhere there is traffic. However, it has been clearly shown that the Mediterranean has an active vertical circulation. If some way were found to fertilize the entering Atlantic water by inducing upwelling in the Bay of Cadiz, the chemistry of the water might be sufficiently enriched to support substantially increased fisheries.

Since the Mediterranean circulation system is characterized by winter turnovers in certain regions, it follows that the waters which sink carry with them amounts of dissolved oxygen as determined by the solubilities at winter surface temperatures and salinities. In the Mediterranean these solubilities would not be as great by any means as those in the polar regions, but the oxygen content could be relatively high and therefore be useful as a tracer. For example, the highest oxygen content in the Mediterranean is in the Aegean Sea where the entire water column contains 5 ml l^{-1}, or more, of oxygen. This is no great amount compared to Atlantic Ocean standards, but when these higher values are followed as they diminish by methods such as the "Kernschicht" core layer of Wüst (1959, 1960), it becomes apparent that the oxygen-minimum layer does not develop until the Eastern Mediterranean water passes over the Sicilian sill to the west as Levantine intermediate water. It is possible that the question of the oxygen-minimum layer may have its solution within the Mediterranean Sea.

REFERENCES

Ambroggi, R.P., 1977. Freshwater resources of the Mediterranean basin. *Ambio*, 6(6): 371–373.

Bruce, J.G. and Charnock, H., 1965. Studies of winter sinking of cold water in the Aegean Sea. *Rapp. P.–V. Réun. Comm. Int. Explor. Sci. Mer*, 18(3): 773–778.

Buljan, M., 1953. Fluctuation of salinity in the Adriatic. *Fis. Biol. Exped. "HVAR", Izvjesca-Rep., Split*, 2(3).

Bunker, A.F., 1972. Wintertime interactions of the atmosphere with the Mediterranean Sea. *J. Phys. Oceanogr.*, 2(3): 225–238.

Christensen, A.M., 1972. Gunnar Thorson. *Isr. J. Zool.*, 21(3–4): 135–141.

Defant, A., 1916. Theoretische Überlegungen über Seespiegelschwankungen in Seen und Meeresbuchten. *Ann. Hydr. Mar. Meteorol.*, 44: 29.

Defant, A., 1961. *Physical Oceanography, II.* Pergamon Press, New York, N.Y., 598 pp.

Helmer, R., 1977. Pollutants from land-based sources in the Mediterranean. *Ambio*, 6(6): 312–316.

Henry, P.-M., 1977. The Mediterranean: A threatened microcosm. *Ambio*, 6(6): 300–307.

Kossinna, E., 1921. Die Tiefen des Weltmeeres. *Meerested., Inst. Meereskd., Berlin, Veröff., N.F., A Geogr. Naturwiss. Reihe*, 9: 70.

Lacombe, H. and Tchernia, P., 1971. Caractères hydrologiques et circulation des eaux en Méditerranée. In: D.J. Stanley (Editor), *The Mediterranean Sea: A Natural Sedimentation Laboratory.* Dowden, Hutchinson, and Ross, Stroudsburg, Pa., pp. 25–36.

Le Lourd, P., 1977. Oil pollution in the Mediterranean Sea. *Ambio*, 6(6): 317–320.

Lisitzin, E., 1974. *Sea-Level Changes.* Elsevier Oceanography Series, 8. Elsevier, Amsterdam, 286 pp.

Mamayev, O.I., 1975. *Temperature–Salinity Analysis of World Ocean Waters.* Elsevier Oceanography Series, 11. Elsevier, Amsterdam, 374 pp.

MEDOC Group, 1970. Observation of formation of deep water in the Mediterranean Sea, 1969. *Nature*, 227(5262): 1037–1040.

Menard, H.W. and Smith, S.M., 1966. Hypsometry of ocean basin provinces. *J. Geophys. Res.*, 71: 4305.

Miller, A.R., 1970. La circolazione generale del mare Mediterraneo e le sue relazione con la circolazione del mare Tirreno. *Ann. Ist. Univ. Navale Napoli*, 38: 8 pp.

Miller, A.R., 1971a. Temperature–salinity regime in a north–south section of the Tyrrhenian Sea. *Ann. Ist Univ. Navale Napoli*, Aliverti Anniversary Volume. 10 pp.

Miller, A.R., 1971b. Speculations concerning bottom currents in the Mediterranean Sea. In: D.J. Stanley (Editor), *The Mediterranean Sea: A Natural Sedimentation Laboratory.* Dowden, Hutchinson, and Ross, Stroudsburg, Pa., pp. 37–43.

Miller, A.R., 1972. Deep convection in the Aegean Sea. In: *Processus de formation des eaux océaniques profondes. Coll. Int. C.N.R.S.*, No. 215: 9 pp.

Miller, A.R., 1976a. The significance of the Mediterranean Sea to global climatology. *Rapp. P.-V. Réun. Comm. Int. Explor. Sci. Mer*, 23(5): 53–54.

Miller, A.R., 1976b. Deep variability in the Western Mediterranean Sea. *Rapp. P.-V. Réun. Comm. Int. Explor. Sci. Mer*, 23(5): 35–39.

Miller, A.R. and Munns, R.G., 1974. The Bitter Lake salt barrier. In: *L'Océanographie physique de la Mer Rouge. Act. Coll. CNEXO*, No. 2: 295–310.

Miller, A.R. and Stanley, R.J., 1965. Volumetric T–S diagrams for the Mediterranean Sea. *Rapp. P.–V. Réun. Comm. Int. Explor. Sci. Mer*, 18(3): 755–759.

Miller, A.R., Tchernia, P. and Charnock, H. with McGill, D., 1970. *Mediterranean Sea Atlas. Temperature, Salinity, Oxygen Profiles and Data from Cruises of RV ATLANTIS and RV CHAIN.* Woods Hole Oceanographic Atlas Series, 3. Woods Hole Oceanographic Institution, Woods Hole, Mass., 190 pp.

Nielsen, J.N., 1912. Hydrography of the Mediterranean and adjacent waters. *Rep. Dan. Oceanogr. Exped. Medit.*, 1: 77–192.

Pirazzoli, P.A., 1976. Sea level variations in the Northwest Mediterranean during Roman times. *Science*, 194: 519–521.

Pollak, M.J., 1951. The sources of the deep water of the Eastern Mediterranean Sea. *J. Mar. Res.*, 10(1): 128–152.

Roether, W. and Weiss, W., 1975. On the formation of the outflow through the Strait of Gibraltar. *Geophys. Res. Lett.*, 2(7): 301–304.

Sterneck, R., 1915a. Zur Theorie der Gezeiten des Mittelmeeres. *Sitzungsber. Akad. Wiss. Wien., Abt. IIa*, 122: 561.

Sterneck, R., 1915b. Hydrodynamische Theorie der halbtägigen Gezeiten des Mittelmeeres. *Sitzungsber. Akad. Wiss. Wien., Abt. IIa*, 124: 905.

Thacher, P.S., 1977. The Mediterranean action plan. *Ambio*, 6(6): 308–311.

Wegemann, G., 1913. Die Gezeiten des Mittelmeeres. *Ann. Hydr. Mar. Meteorol.*, 41: 555.

Wüst, G., 1959. Remarks on the circulation of the intermediate and deep water masses in the Mediterranean Sea and the methods of their future exploration. *Ann. Ist. Univ. Navale Napoli*, 28: 1–16.

Wüst, G., 1960. The deep circulation in the core layers of the intermediate and the deep water masses in the Mediterranean Sea. *Dtsch. Hydrogr. Z.*, 13(3): 105–131.

Wüst, G., 1961. On the vertical circulation of the Mediterranean Sea. *J. Geophys. Res.*, 66(10): 3261–3271.

Zenk, W., 1975. On the origin of the intermediate double-maxima in T/S profiles from the North Atlantic. *"Meteor" Forsch.-Ergebn., A*, 16: 35–43.

Zoré-Armanda, M., 1963. Mixing of three water types in the South Adriatic. *Rapp. P.-V. Réun. Comm. Int. Explor. Sci. Mer*, 17(3): 879–995.

Zoré-Armanda, M., 1972. Response of the Mediterranean to the oceanographic–meteorological conditions of the North Atlantic. *Rapp. P.-V. Réun. Comm. Int. Explor. Sci. Mer*, 21(4): 203–205.

Chapter 10

THE MEDITERRANEAN SEA, B. BIOLOGICAL ASPECTS

ADAM BEN-TUVIA

INTRODUCTION

Mediterranean flora and fauna are rich in species and many of them, although not cosmopolitan, are widely distributed throughout the major east Atlantic provinces. Almost every phylum has its representatives described as endemic to the Mediterranean Sea or to one of its subregions. Endemic genera are very few, 1 to 2%, and usually monospecific, often described from one specimen only (Fredj, 1974). Most of the endemic genera, species and subspecies show close affinities to the eastern Atlantic fauna. Several authors (Fredj, 1974) classify them with some justification as "neo-endemic" in order to stress the low level of differentiation of the Mediterranean endemic forms from their Atlantic or Indo-Pacific relatives.

From prehistoric times throughout the splendid civilizations of ancient Egypt, Greece and Rome, fisheries played an important role in the economy of the nations occupying the shores and islands of the Mediterranean. Marine life attracted the attention of both coastal people and scholars. In *Corpus Hippocrateum*, a collection of writings from the fifth century B.C., edible animals are described, among them several fishes and crustaceans from the sea and rivers. Aristotle in the third century B.C. was the first great observer of nature and described hundreds of sea animals. Many unusual biological observations have been described such as the migrations of bluefin tuna, the mammalian character of dolphins, the hermaphroditic nature of *Serranus* and the viviparity of *Mustelus*. Oppian in his *Halieutica* book on fishing, written in the second century A.D., comments on hundreds of Mediterranean fishes and describes fishing methods among which handlining, spearing and fishing with light at night survive to the present day.

At the beginning of the Tertiary the Paleomediterranean was a part of the Tethys (or Mesogean) Sea forming a link between the Indo-Pacific and Atlantic fauna. The Eocene deposits of Monte Bolca near Verona in Italy, studied by generations of fish paleontologists, reveal rich tropical fauna of obvious Indo-Pacific affinities. Families like Berycidae, Holocentridae, Chaetodontidae, Ephippidae, Zanclidae, Acanthuridae and Pomacentridae were well represented (Blot, 1970).

Towards the end of the Tertiary, the tropical fauna progressively disappeared (Menzies, 1973). In the Oligocene and especially in the Miocene, several subtropical and boreal genera of apparent Atlantic affinities appeared. These have been considered as precursors of the present Mediterranean fish fauna in the broad sense of an Atlanto-Mediterranean entity (Arambourg, 1965; Blot, 1970; Cappetta and Ledoux, 1970).

According to Ruggieri (1967), Hsü et al. (1973) and Benson and Ruggieri (1974) the tectonic and climatic changes in the Late Miocene caused a complete separation of the Paleomediterranean from the Atlantic and its partial or total drying up. The obvious consequence was the destruction of the fauna during the saline crisis of the Miocene (Klausewitz, 1973). In Por's (1978) view it is the tropical east Atlantic fauna that shelters relicts of the Tethyian Paleomediterranean. Menzies (1973) views the whole tropical belt fauna with coral reefs as the heritage of Tethys and its Mediterranean branches. In the Pliocene and Pleistocene a permanent connection with the Atlantic was re-established and the Mediterranean Sea was repopulated by Atlantic species. The climatic fluctuations during the Pleistocene (Por, 1975a, b) and the regional, bathygraphic and seasonal differences in temperatures and salinities of the Mediterranean Sea favored adaptive spreading of subtropical, temperate and even some boreal species from the Atlantic into the Mediterranean Sea.

It may be assumed that the process of penetration of Atlantic species still takes place as evident from the occasional appearance of rare species in the western Mediterranean Sea that apparently are brought with the Atlantic Current (Allain et al., 1966; Maurin, 1968; Casanova, 1977). During glacial periods cold-water species could easily have spread throughout the Mediterranean. Sprat (*Sprattus sprattus*), and Norway lobster (*Nephrops norvegicus*) that extend in their distribution to the cold North Atlantic waters, are restricted in the Mediterranean to the northern and western sections of the sea. Spiny or picked dogfish (*Squalus acanthias*) and the starfish (*Marthasterias glacialis*) occur throughout the whole Mediterranean and in the Atlantic from Gibraltar to Norway beyond the Arctic Circle (Luther and Fiedler, 1976). In the Mediterranean they inhabit the bottom depths where temperatures do not exceed 16°C.

Explorations of deep water layers reveal additional representatives of cold-water fauna such as a polychaete *Chaetoparia nilssoni* (Laubier, 1972–73) reported from a depth of 360 to 1110 m off Corsica and previously known only from the coasts of Sweden and the Skagerrak.

The Mediterranean fauna as a whole is richer in species than the adjacent Ibero-Moroccan subregion. The exception is an impoverished deep-sea fauna, both in species and biomass. Maurin (1970), for example, stresses the poor representation of deep-sea fishes on the trawling grounds in depths of 300 to 450 m as compared to similar grounds along the Ibero-Moroccan coast. The comparatively high and constant temperatures (12–14°C) of the Mediterranean deep water masses below 300 m prevent the acclimation of the majority of the Atlantic bathyal species. The bathyal benthonic animals in the Mediterranean Sea are mostly of eurytopic character and their bathymetric distribution extends from shallow waters to a depth of 500 m. Among macrobenthic invertebrates about 20% of the investigated species were found below the depth of 500 m and 1% were found only at depths of more than 1000 m (Fredj, 1974).

BIOGEOGRAPHY

The terms "Mediterranean–Atlantic" (Briggs, 1974) or "Atlanto-Mediterranean" have been used to describe biotic elements common to the two marine regions (Ben-Tuvia, 1966; Ekman, 1953; Peres, 1967; Peres and Picard, 1958, 1964; Tortonese, 1964). Gibraltar divides the Atlanto-Mediterranean Province into the Mediterranean and the Atlantic Lusitanian and Mauritanian Regions (Ekman, 1953; Fredj, 1972; Peres and Picard, 1964). Both the Lusitanian and Mauritanian Regions are part of a wide belt of "warm-temperate waters" in the Atlantic Ocean (Briggs, 1974). The northern boundary of the Lusitanian Region is the entrance to the English Channel; the southern boundary of the Mauritanian Region is Cape Verde. The cold-temperate Boreal Province extends north of the English Channel and the tropical Senegalian Region extends south of Cape Verde.

Faunal characteristics

Fredj (1972, 1974), in his extensive study of 1244 species of Mediterranean macrobenthic invertebrates, calculated that about 66% of the animals have also been recorded from the Lusitanian Region, about 50% from the Mauritanian Region, 47% from the Boreal, 28% Senegalian, 18% Indo-Pacific (including Suez Canal immigrants) and 13% Amphi-Atlantic (found in both the east and west Atlantic). In the category of Indo-Pacific and Amphi-Atlantic, there are many species that occur in both oceans and therefore can be classified as cosmopolitan (or circumtropical). In another summary (Fredj, 1974, p. 14, table 2) the following percentages are given for the same Mediterranean species divided into four broad biogeographic categories: Atlanto-Mediterranean, 62%; endemic, 20%; cosmopolitan, 13%; Indo-Pacific, 5%. Data obtained from other sources do not differ much from the above pattern (Peres and Picard, 1964; Casanova, 1977).

Among 640 species of four benthonic groups analysed by Peres and Picard (1964) the number of endemic forms amounted to 29%. The highest percentage was found among small sessile animals such as Ascidiacea 50.4%, Hydroidea 27.1%, Echinodermata 26.1% and much less among

Decapoda–Reptantia 13.2%. Menzies (1973), mainly on the basis of Echinodermata, estimated the Mediterranean endemism to be 15 to 16%. Tortonese (1963) estimated that the number of endemic fishes amount to 75 out of a total of 540 (13%). For littoral fishes only, the proportion of endemic forms is higher, estimated by Klausewitz (1968) to be about 20%. The endemic species are mainly small and sluggish fishes, usually benthonic, like Gobiidae, Blenniidae, Tripterygiidae, Soleiformes, Anguilliformes and Rajidae. Names of subspecies are in use (Hureau and Monod, 1973) for several commercial fishes like clupeoids, gadoids or sparoids that differ in some morphometric characters from Atlantic populations.

A few species described from the eastern Mediterranean are clearly related to the Indo-Pacific fauna. For example *Alpheus migrans* (Crustacea–Natantia) has been described by Lewinsohn and Holthuis (1978) from the Mediterranean coast of Israel and Sinai. As stated by the authors the new species is most closely related to *Alpheus acutocarinatus* from Thailand and Indonesia and if intermediate forms are discovered it may be reduced to the rank of a subspecies. The same authors assume that the new species belongs to Suez Canal immigrants.

Some endemic species described from the Adriatic and eastern Mediterranean show close affinity to the Ponto-Caspian fauna, as for example *Didogobius bentuvii* (Miller, 1966). In several cases species described as endemic and thought to be restricted in their distribution to a small area were found later in other sections of the Mediterranean Sea and also often in adjacent seas and oceans. Thus, for example, *Blennius* (*Pictiblennius*) *incognitus* and *Blennius* (*Lipophrys*) *dalmatinus* were found recently off the coast of Portugal (Almeida et al., 1980) and *Blennius* (*Pictiblennius*) *zvonimiri* (under the synonym of *B. ponticus*) off the Atlantic Coast of Spain (Ibañez and Motos, 1977).

The Levant Basin

There is a definite gradient in species diversity from west to east. The number of species among all major groups of plants and animals is much lower in the eastern Mediterranean than in the western and central parts of the sea. The southeastern corner, the Levant Basin, is the most impoverished area. The ratio between the number of species from the Levant Basin and the total number of species known from the Mediterranean Sea varies considerably among different taxa. Some examples are given in Table 10.1. The data show that the number of species in the Levant Basin range from 26% to 57% of the total number of species known for the Mediterranean. Schmidt (1976) suggests, for the Hydromedusae at least, that this may partly be related to the limited number of studies carried out in the Levant Basin and that additional species may be found with more collecting effort and taxonomic studies.

The benthonic and littoral populations show a similar change in species diversity and abundance which decrease from west to east and from the north Adriatic to the south. The pelecypod mussel *Mytilus galloprovincialis*, very common in the upper infralittoral zone of the western Mediterranean and the Adriatic Sea, is absent in the Levant. The commercial red "coral" *Corallium rubrum*, found mostly around caves in the circumlittoral zone, has never been collected in the Levant Basin although on one occasion a piece of dead coral was sampled near Megísti (Kastellórizzon) on the south coast of Turkey (Zibrowius, 1979). Gorgonians have not been recorded from the coast of Israel and Egypt. Peres (1967) points out that *Hacellia attenuata* and *Centrostephanus longispinus* among echinoderms and *Rhodosoma verucundum* among ascidians are much more common in the Levant Basin than in the western Mediterranean.

TABLE 10.1

Examples of the impoverished biota of the Levant Basin

Taxa	Numbers of species	
	Mediterranean	Levant Basin
Hydromedusae[a]	115	30 (26%)
Polychaeta		
Errantia[b]	328	159 (48%)
Errantia and Sedentaria	790[c]	303 (38%)[d]
Fishes	565[e]	320 (57%)[f]
Molluscs[g]	1457	702 (48%)

[a]Schmidt (1976); [b]Fredj (1974); [c]Amoureux (1976); [d]Ben-Eliahu (1977, and pers. comm., 1981); [e]Tortonese (1963), Hureau and Monod (1973), Tortonese and Hureau (1979); [f]Ben-Tuvia (1971, 1978), Mouneimne (1977, 1979); [g]Barash (pers. comm., 1981).

Por (1975b) stresses the "unbalanced and depleted animal inventory" of the Levant Basin and follows Peres (1967) in calling the very end of the Mediterranean Sea a cul-de-sac. The impoverished and climatically much warmer and saltier Levant Basin is inhospitable to a great number of Lusitanian species of boreal character that dominate the Atlantic approaches to the Straits of Gibraltar. In terms of abiotic conditions the southeastern corner of the Mediterranean Sea theoretically could accommodate a significant fraction of the tropical Atlantic (Senegalian) fauna, but the cul-de-sac position of the Levant Basin prevents their eastward migration. The strong upwelling along the Saharan-Moroccan Coast cools the coastal waters and inhibits the majority of Senegalian species from reaching the Straits of Gibraltar. Also, the western Mediterranean with its temperate or warm-temperate climate constitutes an additional barrier for tropical species. In addition to this, the tropical eastern Atlantic fauna is considered to be the poorest among the faunas of the tropical seas of the world (Briggs, 1974). With the opening of the Suez Canal, a connection was created between the undersaturated Levant Basin and the species-rich tropical Red Sea (see pp. 246–247).

According to Gilat (1964) *Pennatula phosphorea*, common in the western Mediterranean and Adriatic on muddy bottom, is replaced in the Levant Basin by *Pennatula rubra*. Among echinoderms *Astropecten irregularis pentacanthus* and *Echinocardium cordatum* replace *Astropecten johnstoni* and *Echinocardium mediterraneum*. *Stichopus regalis* has not been reported from the Levant Basin. Gorenshtein-Galil and Lewinsohn (1979) found that the stomatopod *Oratosquilla massavensis* and the gastropod *Cerithium kochi*, both of Red Sea origin, are the dominating organisms on muddy-sand bottom at a depth of 30 to 40 m off Tel-Aviv. It should be pointed out that in deeper water, 50 to 80 m, all the common species were Atlanto-Mediterranean, namely *Parapenaeus longirostris*, *Brissopsis lyrifera*, *Dentalium dentalis*, *Turritella communis*, *Nucula sulcata*, *Sternaspis scutata*, *Sabella pavonina* and *Antedon mediterranea*.

Benthic biomass

It is generally accepted that the Mediterranean benthos is poorer in biomass than the Atlantic coast of Europe although rich in species diversity. Sparck (1931) was probably the first to sample with the same gear (Petersen's grab) several localities of the western Mediterranean and the eastern Atlantic. The calculated biomass expressed as live weight was on the average very low off Naples and Salerno (6 g m^{-2}), three times higher in the Bay of Algiers and about 40 times higher (226 g m^{-2}) in the Atlantic off Faro and Lisbon.

It is difficult to compare directly the quantitative data from different areas without referring to the character of the community (biocoenosis), type of substrate and depth. Standardization of the unit of biomass and of the collection method is essential for best results (Peres, 1967). Peres and Picard (1958) and Peres (1967) for the Mediterranean and Gamulin-Brida (1967) for the Adriatic review studies of the Mediterranean benthos. The available data indicate that the regional differences in the values of biomass are in line with the general pattern of the productivity gradient (Gilat, 1964, 1969; Kiseleva and Chukhchin, 1965; Peres and Picard, 1958; Vatova, 1975; Zavodnik, 1971). The biomass is higher in the western Mediterranean and the north Adriatic than in other sections of the Mediterranean. The Levant Basin, Libyan Coast and Aegean Sea seem to have the lowest biomass. High biomass was recorded near the deltas of rivers like the Rhône, Po and Nile (before the construction of the Aswan Dam). Polychaetes and molluscs usually contribute the greatest share of individuals and biomass of circumlittoral communities in the region of Marseilles (Peres, 1967). Pronounced seasonal fluctuations of ten orders of magnitude were sometimes recorded between the August minimum and December maximum values for the biomass.

Extensive studies of the Adriatic Sea (Gamulin-Brida, 1967; Vatova, 1949, 1975) provide data on the biomass of the main benthonic communities in various sections. The biomass of *Schizaster chiajei* community at the River Po estuary amounted to 283 g m^{-2} decreasing southward along the western coast of Italy. A very high biomass, reaching almost the value of River Po estuary was found also near the coast of Rovinj in northern Yugoslavia. Television cameras and direct divers' observations can be useful in studies of macro-epibenthic communities as carried out on the Gulf of Trieste (Fedra, 1977).

Gilat (1964) compared his data from the circumlittoral shelf of Israel with data from similar grounds in the Adriatic Sea. The biomass of macrobenthonic invertebrates on sand-silt bottom along the coast of Israel was 8 to 16 g m^{-2}, in comparison with 48 to 126 g m^{-2} found in the Adriatic Sea. According to Dinet et al. (1973) and Dinet (1976) the density of bathyal meiobenthos in the north Aegean Sea, calculated as a number of individuals per 10 cm^2, varied between 87 and 427. Nematodes constituted 86.3% of the total number of the meiobenthic animals. The results were similar to those obtained in the western Mediterranean. According to Gilat (1969) the biomass of polychaetes, molluscs and echinoderms is greater in the western Mediterranean than in the Levant Basin but decapods seem to be more abundant in the southeastern Mediterranean due to the influx of the Indo-Pacific species.

FISHERIES RESOURCES

The Mediterranean fisheries are as important today as they were to ancient civilizations in the past. The annual yield for 1977 was estimated to be about 700 000 metric tons, constituting some 1.2% of the world catch. However, the value of the fish on the local markets is much higher than in other world fisheries (Holt, 1974; Levi and Troadec, 1974) constituting 5% of the value of the world catch (Troadec, 1979). In general the catch is composed of a great variety of species but the yield per unit of area is low, about 300 kg km^{-2}, calculated for the whole area of the sea, or 1400 kg km^{-2}, calculated for the area of the continental shelf only (FAO, 1967; Gulland, 1971).

One reason for the low fish catches may be the low primary productivity that apparently cannot support large stocks. As discussed in Chapter 9 (pp. 229–231), the surface Atlantic water entering the Mediterranean through the Strait of Gibraltar is already impoverished in plant nutrients and the concentration in the surface waters decreases to the eastward. In the Alboran Sea close to Gibraltar, Bernard (1967) observed phosphate phosphorus concentrations of 0.03 μg-at l^{-1} and more than 350 000 phytoplankton cells l^{-1}. In other parts of the Mediterranean, not within the direct influence of the Atlantic current, the values were seldom more than half of this amount except for the area off the Nile Delta during the period of floods as will be discussed below (p. 247). The lowest phosphate concentrations were recorded in the Aegean Sea followed by the southern Adriatic Sea, the Ionian Sea and the Levant Basin (FAO, 1967). The primary phytoplankton productivity showed similar gradients being very low in the Levant Basin and off the coast of Israel in comparison to the area south of the Balearic Islands and Sardinia and along the African coast (Furnestin, 1973; Jacques, 1970; O.H. Oren, pers. comm., 1970). Year-to-year fluctuations in the nutrient content and primary and secondary production in the Adriatic Sea were analyzed by Zore-Armanda and Pucher-Petkovic (1976).

Another possible explanation for the low fisheries yield is the generally narrow continental shelves. Except for the northern Adriatic and the east Tunisian coast, there are no extensive banks that normally support large commercial fisheries in other seas. The available data indicate that there is a gradient in the yield if calculated as catch per unit of area or catch per unit of effort, decreasing from west to east and from north to south (Levi and Troadec, 1974; GFCM, 1978). On this basis, the shallower northern Adriatic yields higher catches than the southern areas.

Among commercial stocks regional differences are prominent. Common gadids of the western and northern Mediterranean like poor cod (*Trisopterus minutus capelanus*), whiting (*Merlangius merlangus euxinus*), silvery pout (*Gadiculus argenteus argenteus*), blue ling (*Molva dypterygia macrophthalma*) and shore rockling (*Gaidropsarus mediterraneus*) are absent from the southeastern Mediterranean; poutassou (*Micromesistius poutassou*) is extremely rare (Ben-Tuvia, 1971). Among the commercial flat fishes brill (*Scophthalmus rhombus*), fluke (*Lepidorhombus whiffiagonis*), turbot (*Psetta maxima*) and flounder (*Platichthys flesus*) have never been found in the Levant Basin. In contrast, several demersal fishes of Red Sea origin such as lizardfish (*Saurida undosquamis*), yellowbanded goatfish (*Upeneus moluccensis*) and green-and-gold barracuda (*Sphyraena chrysotaenia*) are now common enough to be of commercial importance.

Not less prominent are the differences among commercial decapods (Massuti, 1971). The European lobster (*Homarus gammarus*) and the common spiny lobster (*Palinurus elephas*) do not occur in the Levant Basin. *Nephrops norvegicus* must be extremely rare since there is only one record from this area (Holthuis and Gottlieb, 1958). On the other hand, the Red Sea immigrants like the shrimps *Penaeus japonicus*, *P. semisulcatus*, *Metapenaeus monoceros*, *M. stebbingi* and the sand crab *Portunus pelagicus* are very common and exploited in local fisheries. The commercial Atlanto-Mediterranean shrimps are limited to triple-grooved shrimp (*Penaeus kerathurus*), mostly in shallow waters, and pink shrimp (*Parapenaeus longirostris*) in deeper water (Holthuis, 1980).

It has been estimated (Ben-Tuvia, 1973) that fishes of Red Sea origin (Erythrean fishes) contribute about 20% in weight of the total trawl fishery landings in the Levant Basin. However, the bulk of the catch is composed of Atlanto-Mediterranean fishes. Hake (*Merluccius merluccius*) is commercially exploited in depths ranging from 20 to 700 m (Ben-Yami and Glaser, 1974; Maurin, 1970). In winter months hake, especially young ones, are often found in shallow waters. Red mullets (*Mullus barbatus* and *M. surmuletus*) are found on trawling grounds of the Levant Basin in various depths from a few meters up to 500 m (Wirszubski, 1953; Gorgy, 1966). Various sparids, centracanthids and elasmobranchs contribute to the demersal catches.

Among non-demersal, schooling fish the most important in the western Mediterranean are European anchovy (*Engraulis encrasicolus*), European pilchard (*Sardina pilchardus*), bluefin tuna (*Thunnus thynnus*) and Atlantic mackerel (*Scomber scombrus*). The first three species occur in the Levant but not in significant commercial quantities. *Scomber scombrus* is not known from the southeastern Mediterranean. The common pelagic species in the Levant Basin are gilt sardinella (*Sardinella aurita*) and chub mackerel (*Scomber japonicus*), both circumtropical and the Atlanto-Mediterranean little tuna *Euthynnus alleteratus*. None of the pelagic Red Sea immigrants (Ben-Tuvia, 1973, 1978) occurs in large commercial quantities.

Cephalopods, caught mostly by trawl, are highly esteemed on the local markets. The yearly catch taken in the Mediterranean Sea reaches about 50 000 t (Worms, 1979) constituting about 7% of the total catch, and 12 to 15% of the demersal fisheries. Italy has the largest share of the cephalopod catch. The main commercial species among cuttlefish, squids and octopuses are *Sepia officinalis*, *Loligo vulgaris*, *Sepiola rondeleti*, *Octopus vulgaris*, *O. macropus*, *Eledone moschata*, *E. cirrosa* and *Todarodes sagittatus*. *Sepia elegans* and *Allotheuthis media* were the most common cephalopods in the northern Adriatic constituting by weight over 50% of the total cephalopod catch (Legac, 1974).

COASTAL LAGOONS

Coastal lagoons are areas of high productivity and many of them, like the "valli" of the Lagoons of Venice (De Angelis, 1960), the Etang de Thau in southeastern France (Amanieu, 1973) and the Bardawil Lagoon in the Sinai (Ben-Tuvia, 1979) yield large commercial catches. Lagoons are also potential sites for aquaculture and as such need protection and management based on applied research (FAO, 1973). The Mediterranean lagoons vary greatly in their size and character. The largest among them, Lake Maryut in Egypt, covers an area of about 1670 km^2 and the total area of the Mediterranean lagoons (without the Sea of Marmara) is estimated at 7000 km^2 (De Angelis, 1960). The yearly production from the lagoons and "valli" of the region of Venice (D'Ancona, 1954) of an area of about 625 km^2 amounted in 1952 to 3810 t or 6100 kg km^{-2}. The four Nile Delta Lagoons, namely Maryut, El Idku, El Burullus and El Manzala with a combined area of 2230 km^2 yielded an annual catch of 26 000 t (El-Saby, 1951) or 11 700 kg km^{-2}.

Salinity is one of the main factors determining the productivity and the composition of the flora and fauna of a lagoon (D'Ancona, 1959). The productivity of the lagoons on the Adriatic coast of Italy increases from the low salinity (oligohaline) lagoons to that found in lagoons where the salinity approaches that of sea water as shown by the data in Table 10.2 (Kiener, 1978). The fish yield of these lagoons ranges from 300 to nearly 20 000 kg km^{-2} in comparison to the values given above of 300 to 1400 kg km^{-2} as calculated for the Mediterranean as a whole or for the area of the continental shelf.

In many lagoons the salinity depends on the maintenance of canals leading to the open sea and periodical dredging is essential. A diversion in the course of a river may strongly affect the salinity of a lagoon as in the case of the two largest lagoons of

TABLE 10.2

The biomass and yield of some North Adriatic coastal lagoons of different salinities (Kiener, 1978)

Type of lagoon	Salinity (‰)	Organic matter ($g\ m^{-2}$)	Biomass ($kg\ ha^{-1}$)	Fish yield ($kg\ ha^{-1}$)
Polyhaline	20–35	141–395	283–789	70–197
Mesohaline	5–20	55–125	109–251	27–63
Oligohaline	0.5–5	6–24	11–47	3–12

the Mediterranean coast of France, the Etang de Berre and the Etang de Vaccarès. The hydrological transitions in these lagoons were followed by pronounced changes in species composition (De Angelis, 1960; Huvé et al., 1973). Several euryhaline species are commonly encountered in most lagoons. Peres (1967) includes among them several molluscs as for example *Cerastoderma glauca* (quoted as *Cardium lamarcki*), *Abra ovata*, *Loripes lacteus* and *Brachidontes marioni*; the brachyuran crab *Carcinus mediterraneus*; isopods *Sphaeroma hookeri* and *Cyathura carinata*; amphipods *Gammarus locusta* and *Microdeutopus gryllotalpa*. In deep lagoons with salinities similar to the adjacent open sea, the marine phanerogam, *Ruppia maritima*, is common. In some shallow lagoons a serpulid polychaete, *Ficopomatus aenigmaticus* (quoted by Peres, 1967, as *Mercierella aenigmatica*), of wide distribution in the Indo-Pacific, forms encrustations on stones, piers and even on the leaves of seagrasses.

The shallow Bardawil Lagoon on the northern (Mediterranean) coast of Sinai with an area of about 650 km^2 can serve as an example of a metahaline and hyperhaline environment (salinities varying between 40 and 120‰) in an area of increasing influence of the Red Sea biota (Por and Ben-Tuvia, 1980). The productivity is high and the annual catch in the years 1974–1978 averaged about 2500 t, or 3800 $kg\ km^{-2}$.

The majority of the common molluscs, like *Cerithium scabridum*, *Brachidontes variabilis*, *Mactra olorina*, *Modiolus glaberrimus*, and *Diala varia* are of Red Sea origin (Barash and Danin, in press). *Pirenella conica*, treated by the same authors and by H. Mienis (pers. comm., 1981) as a synonym of *Pirenella cailliaudi*, is evidently a cosmopolitan species. All those mollusc species were encountered as common food items in stomachs of *Sparus aurata*, caught by commercial gear in the Bardawil Lagoon (Barash and Danin, 1971).

The number of Red Sea fish immigrants is high in the Bardawil Lagoon, constituting about one quarter of the total number of species collected. In terms of biomass the Mediterranean fishes completely dominate both commercial and non-commercial species. The common Bardawil fishes are almost identical with those found in other Mediterranean lagoons, namely gilthead seabream (*Sparus aurata*), various grey-mullets (*Mugil cephalus*, *Liza ramada*, *L. aurata*, *L. saliens*, *Chelon labrosus*), sea basses (*Dicentrarchus labrax*, *D. punctatus*), Dover sole (*Solea solea*) and silverside (*Atherina boyeri*). A needle fish (*Syngnathus abaster*) and a gobi (*Pomatoschistus marmoratus*) are also permanent inhabitants of the Lagoon. As in all other Mediterranean lagoons *Aphanius fasciatus* is very common but no less common is the Erythrean *Aphanius dispar* known only from a few localities of the Levant Basin. The euryhaline character of the genus *Aphanius* and its Paleomediterranean distribution justify viewing it as a Tethyian relict that could survive the Mediterranean desiccation during the saline crisis of the Miocene. The euryhaline mysid shrimp, *Diamysis bahirensis*, can also be classified in the same category of organisms.

The bottom of the lagoon is densely covered with the phanerogam *Ruppia spiralis* with a rich epiphitic overgrowth of *Cladophora* spp. and diatoms, some of them new to science as for example *Cocconeis bardawilensis* and *Mastogloia sirbonensis* (Ehrlich, 1975). The benthonic fauna is rich in species and in biomass represented by a variety of harpacticoid copepods, ostracods and nematodes. The euryhaline mysidacean *Diamysis bahirensis* was found both in bottom samples (Almeida Prado-Por, in press) and in the stomach content of *Sparus aurata* fry. The chironomid *Cricotopus mediterraneus* was an important element in the food web of the lagoon. Kimor (1975) stressed the euryhaline character of the phytoplankton composed predominantly of such dinoflagellates as *Ceratium fusus* and *Ceratium aegyptiacum*. Decapods were represented by Red Sea species only, *Metapenaeus stebbingi*, *Penaeus semisulcatus*, *Portunus pelagicus* and *Charybdis* sp.

IMPACT OF HUMAN ACTIVITIES

The Suez Canal

The opening of the Suez Canal in 1869 provided a route for the migration of Red Sea species of the Indo-Pacific fauna into the impoverished Levant Basin in the eastern Mediterranean. An ancient canal at the time of the pharaohs of Egypt (about 2000 B.C. to 600 B.C.) formed a connection via the Nile River, at least at times of flood. It is not known whether there was any exchange of species in those days but the fresh waters of the Nile would form an effective barrier to the migration of species adapted to the high salinities of the Red Sea or the eastern Mediterranean.

The present-day canal is at sea level and is 162.5 km (about 100 miles) long extending from Port Said at the Mediterranean end to Suez at the Red Sea end. Some of the present-day canal follows the course of the ancient canal including a passage through the Bitter Lakes. With the closing of the ancient canal, these lakes were isolated and dried up forming thick salt deposits which have gradually dissolved over the course of the last hundred years. The high salinity of the Great Bitter Lakes presumably formed an effective barrier for the migration of species during much of this period of time (Miller and Munns, 1974). In recent years there has been growing interest in the immigration or transport of species from the Red Sea to the faunistically impoverished Levant Basin at the eastern end of the Mediterranean.

The main colonizing route of the Erythrean or Red Sea immigrants[1] passes from the Suez Canal along the Asiatic Coast in the north and northwest direction. The largest number of immigrants were found along the coast of Sinai and Israel (Por, 1978); most of the fishes are also present in Lebanon (George et al., 1964; Mouneimne, 1977, 1979). Much less is known about the occurrence of the immigrants along the African coast, west of the Suez Canal. In general the colonizers are confined to the Levant Basin characterized by water masses of elevated temperatures and salinities that resemble the hydrographic conditions of the Gulf of Suez. About 20% of the immigrants have reached the southern outskirts of the Aegean Sea and a few have crossed the Ionian Sea and appeared in southern Italy and Tunisia.

The colonization of the Mediterranean Sea by Red Sea fishes has been reviewed recently by Ben-Tuvia (1978) who listed 36 Red Sea immigrants. Three additional species have been found since. The biogeographic affinities of fish populations in the Levant Basin are compared in Table 10.3 with the fish populations of the entire Mediterranean and the central Mediterranean as characterized by the fish populations along the coast of Italy. The Red Sea immigrants constitute 12% of the population in the Levant Basin and 7% for the Mediterranean as a whole, but only 1% of the population in the central part of the Sea. None of the Red Sea immigrants has reached the western basin.

The percentage of the Red Sea immigrants varies among taxonomic groups. No immigrants are known among 21 cephalopods recorded from the Levant Basin, very few among phytoplanktonic organisms. *Halophila stipulacea* is the only successful colonizer among sea grasses (Lipkin, 1975). It has been reported from the coast of Egypt, Israel, Cyprus, in the Aegean Sea as far as Piraeus and as far west as Malta. The Red Sea stomatopod *Oratosquilla massavensis* is very common along the coast of Egypt, Israel, Lebanon and the southern part of the Anatolian Coast of Turkey but has not been found in the coastal waters, including trawling grounds, of the Gulf of Gabès in southeastern Tunisia (Lewinsohn and Manning, 1980). It is assumed that low winter sea temperatures of the western and northern Mediterranean prevent the Red Sea species from spreading outside the Levant Basin.

Ben-Eliahu (1977, and pers. comm., 1981) recorded 303 species of polychaetes from the Levant

[1] Also called Lessepsian migrants, a name derived from Ferdinand de Lesseps, a French engineer and diplomat who was responsible to a large extent for the development of the Suez Canal.

TABLE 10.3

Affinities of fish populations in the Mediterranean Sea

Area	Reference	Atlanto-Mediterranean	Cosmo-politan	Endemic	Red Sea immigrants
Total	a	71%	9%	13%	7%
Italian Coast	b	74%	11%	15%	<1%
Levant Basin	c	68%	13%	7%	12%

[a]Tortonese (1963), Hureau and Monod (1973), Tortonese and Hureau (1979); [b]Tortonese (1963, 1970, 1975); [c]Ben-Tuvia (1971, 1978), Mouneimne (1977, 1979).

Basin (152 Errantia and 151 Sedentaria); among them 26 species (11%) were of Red Sea origin. According to A. Barash (pers. comm., 1981) 702 molluscan species have been recorded from the Levant Basin; 66 species (9.4%) are Red Sea immigrants.

The precise time of immigration of the Red Sea organisms and the manner of their spreading in the eastern Mediterranean is in general unknown. Records of the sequence were well enough documented in only a few cases to present a clear picture of the pattern of colonization. Steinitz (1967, 1970) was one of the first to attempt to summarize the available records of all Suez Canal immigrants. A comprehensive review and an analysis of the data are given by Por (1978) in his monograph on Lessepsian migration, including a discussion of the speed and the radius of dispersion of the new immigrants.

Time is an important factor in the process of immigration and colonization by the Red Sea organisms. It increases the chance that Red Sea species will pass through the Suez Canal and form a beach-head population along the Mediterranean Coast. It may also influence the genetic changes and the adaptability of the organisms to the new environment (Kosswig, 1974). In some cases immigrants constitute a large, well-established population when discovered. Thus, for example, *Sillago sihama* appeared recently along the coast of Israel and Lebanon and now is being caught occasionally in some numbers by commercial fishermen.

The Red Sea colonizers interact with the native fauna and flora. Some aspects of interaction among fishes were discussed by Ben-Tuvia (1966, 1973, 1978), Ben-Yami and Glaser (1974), and George and Athanassiou (1967). Observations on competitive relationships between Red Sea immigrants and native species among sedentary molluscs have been reported by Felsenburg and Safriel (1974) and Safriel et al. (1980), and the vertical distribution of 21 Red Sea colonizers among the intertidal rocks of Mikhmoret, Israel, are described by Safriel and Lipkin (1975). Potential colonizers might be identified on the basis of genetic factors or population dynamics (Ritte and Safriel, 1977; Lavee and Ritte, 1977; Safriel and Ritte, 1980).

Colonizing immigrants will also introduce parasites in a newly occupied region. Some Red Sea fish immigrants such as *Siganus rivulatus*, *S. luridus*, *Aphanius dispar* and *Pranesus pinguis* continue to host in the Mediterranean Sea their monogenean ectoparasites of Erythrean origin (Paperna, 1972). In the case of *Crenidens crenidens* the Erythrean monogeneans were replaced by newly acquired Mediterranean parasites, mostly those common among the local sparids. Fischtal (1980) investigated the taxonomy and distribution of digenetic trematodes of fishes from the Mediterranean coast of Israel and described one new genus and eleven new species. A number of digenetic trematodes of Erythrean origin continue to infect such Red Sea immigrants as *Saurida undosquamis* and *Siganus rivulatus* but in addition the fish acquired local parasites of Atlanto-Mediterranean or cosmopolitan origin.

The Aswan High Dam

The construction of the Aswan High Dam on the River Nile in the years 1964–1970 (Mancy, 1981) has also modified the ecosystem of the Levant Basin (Halim et al., 1967; Oren, 1969, 1970; Aleem, 1969, 1972; Ben-Tuvia, 1973). The Nile previously discharged annually about 35 thousand million tons of water into the Mediterranean Sea and carried mud and silt estimated at 140 million tons (Halim, 1960). The bulk of the Nile water was discharged during the seasonal floods between August and December with the peak in September–October. During the flood the concentrations of phosphates and silicates in the Nile Delta region increased about tenfold causing spectacular blooms of phytoplankton (Halim, 1960; Halim et al., 1967; Aleem and Dowidar, 1967). The phosphates and nitrates contributed by the Nile during the pre-Aswan years constituted only about 0.5% of the total input to the Mediterranean Sea from all sources (McGill, 1969; Oren, 1970), but for the impoverished Levant Sea, this addition of nutrients was of considerable importance. Phytoplanktonic organisms, mostly diatoms, reached concentrations of 10 million cells l^{-1}, and zooplankton, mostly copepods, reached concentrations of up to 80 thousand individuals m^{-3} (Aleem, 1972). The influence of the Nile flood extended to the coasts of Israel and Lebanon (Oren and Komarovsky, 1967).

The Nile flood also attracted large concentrations of gilt sardinella (*Sardinella aurita*) and to a less extent of shortbody sardinella (*Sardinella maderensis*). The sardine fishery based on the two species exploited the concentrations near the mouths of the two main branches of the river, Rashîd (Rosetta) and Dumyât (Damietta) during the flood period.

After closing the Aswan High Dam the phytoplankton concentrations during the months of the former Nile flood period were reduced to 10% of the previous values (Halim, 1976), and the collapse of the sardine fishery followed. In 1962 the catch amounted to about 18 000 t, in 1966 it dropped to 1200 t and in 1969 only 600 t were taken. The Egyptian Mediterranean sardine fishery has not fully recovered and the annual catch now varies between 1000 and 2000 t. There is no clear evidence that other fisheries were affected (El-Hehyawi, 1974).

The loss of the pelagic fishery catches was partly compensated for by the fresh-water fishery in the Nasser Lake, created by the construction of the Aswan High Dam (Ryder and Henderson, 1975). The potential yield of Lake Nasser was estimated by the same authors on the assumption that the nutrients that formerly fertilized the Nile Delta are now being stored in the lake in the process of its creation and maintenance.

ACKNOWLEDGEMENT

I wish to thank Professor Dov F. Por for his useful comments.

REFERENCES

Aleem, A.A., 1969. Marine resources of the United Arab Republic. *Stud. Rev. Gen. Fish. Coun. Medit.*, No. 43: 1–22.

Aleem, A.A., 1972. Effect of river outflow management on marine life. *Mar. Biol.*, 15(3): 200–208.

Aleem, A.A. and Dowidar, N., 1967. Phytoplankton production in relation to nutrients along the Egyptian Mediterranean Coast. *Stud. Trop. Oceanogr. Miami*, 5: 305–323.

Allain, C., Furnestin, M.L. and Maurin, C., 1966. *Essai de synthèse sur l'océanographie physique et biologique dans la zone d'influence du courant atlantique en Méditerranée du Sud et du Levant.* Comm. Int. Explor. Sci. Mer Méditerranée, Monaco, 38 pp.

Almeida, A.J., Gomes, J.A. and Ré, P., 1980. Trois Blenniidae nouveaux pour la faune du Portugal (Pisces: Perciformes). *Tethys*, 9(3): 235–241.

Almeida Prado-Por, M.S., 1981. Two new taxa of the *Diamysis bahirensis* species group (Mysidacea) from extreme environments of the Levantine Coast. *Isr. J. Zool.*, in press.

Amanieu, M., 1973. Ecologie et exploitation des étangs et lagunes saumâtres du littoral français. *Ann. Soc. R. Zool. Belg.*, 103(1): 79–94.

Amoureux, L., 1976. *Serpula* (*Paraserpula*) *israelitica*, nouvelle espèce de Serpulidae (Annélides Polychètes) et une petite collection annélidienne de la Méditerranée orientale. *Bull. Mus. Nat., 3° Sér., Zool.*, 281(404): 1047–1059.

Arambourg, C., 1965. Considérations nouvelles au sujet de la faune ichthyologique paléoméditerranéenne. *Senckenb. Leth.*, 46a: 13–17.

Barash, A. and Danin, Z., 1971. Mollusca from the stomach of *Sparus aurata*. *Argamon*, 2(34): 307–324.

Barash, A. and Danin, Z., in press. Contribution to the knowledge of Mollusca in the Bardawil Lagoon. *Boll. Malacol., Milano.*

Ben-Eliahu, M.N., 1977. *A Comparison of Species Diversities of Polychaete Cryptofauna From Similar Intertidal Vermitid Reef on the Mediterranean Coast of Israel and in the Gulf of Elat.* Thesis, The Hebrew University of Jerusalem, Jerusalem, 115 pp.

Benson, R.H. and Ruggieri, G., 1974. The end of the Miocene, a time of crisis in the Tethys–Mediterranean history. *Ann. Geol. Surv., Egypt*, 4: 237–250.

Ben-Tuvia, A., 1966. Red Sea fishes recently found in the Mediterranean. *Copeia*, 1966(2): 254–275.

Ben-Tuvia, A., 1971. Revised list of the Mediterranean fishes of Israel. *Isr. J. Zool.*, 20: 1–39.

Ben-Tuvia, A., 1973. Man-made changes in the eastern Mediterranean Sea and their effect on the fishery resources. *Mar. Biol.*, 19(3): 197–203.

Ben-Tuvia, A., 1978. Immigration of fishes through the Suez Canal. *Fish. Bull.*, 76(1): 249–255.

Ben-Tuvia, A., 1979. Studies of the population and fisheries of *Sparus aurata* in the Bardawil Lagoon, eastern Mediterranean. *Invest. Pesq.*, 43(1): 43–67.

Ben-Yami, M. and Glaser, T., 1974. The invasion of *Saurida undosquamis* into the Levant Basin — an example of biological effect of interoceanic canals. *Fish. Bull.*, 72(2): 359–373.

Bernard, I., 1967. Research on phytoplankton and pelagic Protozoa in the Mediterranean Sea from 1953 to 1966. In: H. Barnes (Editor), *Oceanography. Mar. Biol. Annu. Rev.*, 5: 205–229.

Blot, J., 1970 (1973). Étude des faunes ichthyologiques fossiles et interprétation de l'évolution du milieu marin en Méditerranée. In: *Journées Ichthyologie, Rome, C.I.E.S.M.*, pp. 13–20.

Briggs, J.C., 1974. *Marine Zoogeography*. McGraw-Hill, New York, N.Y., 475 pp.

Cappetta, H. and Ledoux, J.C., 1970 (1973). Comparaison de la faune Ichthyologique miocène avec la faune actuelle de Méditerranée. In: *Journées Ichthyologie, Rome, C.I.E.S.M.*, pp. 21–23.

Casanova, J.P., 1977. *La faune pélagique profonde (Zooplancton et Micronecton) de la province Atlanto-Mediterranéenne. Aspects taxonomique, biologique et zoogéographique*. Thesis, Université de Provence (Aix–Marseille), Marseille, 460 pp.

D'Ancona, U., 1954. Fishing and fish culture in brackish-water lagoons. *FAO Fish. Bull. Rome*, 7(4): 1–28.

D'Ancona, U., 1959. The classification of brackish waters with reference to the North Adriatic lagoons. *Arch. Oceanogr. Limnol. (Suppl.)*, 11: 93–109.

De Angelis, R., 1960. Mediterranean brackish water lagoons and their exploitation. *Stud. Rev. Gen. Fish. Counc. Medit.*, No. 12: 1–41 (21 figs.).

Dinet, A., 1976. Etude quantitative du meiobenthos dans le secteur mord de la Mer Egée. *Acta Adriat.*, 18(5): 83–88.

Dinet, A., Laubier, L., Soyer, J. and Vitiello, P., 1973. Résultats biologiques de la campagne Polymede. II. Le méiobenthos abyssal. *Rapp. Comm. Int. Mer Médit.*, 21(9): 701–704.

Ehrlich, A., 1975. Les Diatomées benthiques et epiphytes de la lagune de Bardawil (Sinai septentrional). *Rapp. Comm. Int. Explor. Mer Médit.*, 23(3): 121–123.

Ekman, S., 1953. *Zoogeography of the Sea*. Sedgwick and Jackson, London, 417 pp.

El-Hehyawi, M.L.E., 1974. Changes in salinity and landings of six fish species in the shelf north to the Nile Delta. *Bull. Inst. Oceanogr. Fish., Cairo*, 4: 431–458.

El-Saby, M.K., 1951. The lake Fisheries of Egypt. *Proc. U.N. Sci. Conf. Util. Resour. 1949*, 7: 126–130.

FAO, Marine Biology and Environment Branch, 1967. *Fishery Resources Review of the Mediterranean and the Black Sea*. FAO, Rome, GFCM/9/67/13E, 15 pp.

FAO, 1973. Aquaculture en eau saumâtre dans la région méditerranéenne. *Etud. Rev. Cons. Gén. Pêches Médit.*, 52: 1–129.

Fedra, K., 1977. Structural features of a nortn Adriatic community. In: B.F. Keegan, P.O. Ceidigh and P.J.S. Boaden (Editors), *Biology of Benthic Organisms*. Pergamon Press, London, pp. 233–246.

Felsenburg, T. and Safriel, U., 1974. Colonization of the eastern mediterranean intertidal zone by the Indo-Pacific mussel, *Brachidontes variabilis*. *Isr. J. Zool.*, 23: 212–213.

Fischthal, J.H., 1980. Some digenetic trematodes of marine fishes from Israel's Mediterranean Coast and their zoogeography, especially those from Red Sea immigrant fishes. *Zool. Scr.*, 9: 11–23.

Fredj, G., 1972. Stockage et exploitation des données en écologie marine. A — Un fichier sur ordinateur des invertébrés macrobenthiques. *Mém. Inst. Océanogr., Monaco*, 4: 1–61.

Fredj, G., 1974. Stockage et exploitation des données en écologie marine. C — Considerations biogéographiques sur le peuplement benthique de la Méditerranée. *Mém. Inst. Océanogr., Monaco*, 7: 1–88.

Furnestin, M.L., 1973. Phytoplancton et production primaire dans la secteur sud-occidental de la Méditerranée. *Rev. Trav. Inst. Pêches Marit.*, 37(1): 19–68.

Gamulin-Brida, H., 1967. The benthic fauna of the Adriatic Sea. In: H. Barnes (Editor), *Oceanography Mar. Biol. Annu. Rev.* 5: 535–568.

George, C.J. and Athanassiou, V., 1967. A two-year study of the fishes appearing in the seine fishery of St. George Bay, Lebanon. *Ann. Mus. Civ. Stor. Nat. Giacomo Doria*, 76: 237–294.

George, C.J., Athanassiou, V. and Boulos, I., 1964. The fishes of the coastal waters of Lebanon. *Misc. Pap. Nat. Sci. Am. Univ. Beirut*, No. 4: 1–27.

GFCM/FAO, 1978. Report of the Seventh Session of the Working Party on Resources Appraisal and Fishery Statistics. *FAO Fish. Rep.*, No. 204: 1–141.

Gilat, E., 1964. The macrobenthic invertebrate communities on the Mediterranean continental shelf of Israel. *Bull. Inst. Océanogr., Monaco*, 62(1290): 1–46.

Gilat, E., 1969. Study of an ecosystem in the coastal waters of the Ligurian Sea. III. Macrobenthic communities. *Bull. Inst. Océanogr., Monaco*, 69(1396): 1–76.

Gorenshtein-Galil, B. and Lewinsohn, Ch., 1979. A numerical analysis of zonation and faunal composition of the epibenthic macrofauna off the southern Mediterranean coast of Israel. *Rapp. Comm. Int. Mer Médit.*, 25/26(4): 271–272.

Gorgy, S., 1966. Les pêcheries et le milieu marin dans le secteur méditerranéen de la République Arabe Unie. *Rev. Trav. Inst. (Sci. Tech.) Pêches Marit.*, 30(1): 25–80.

Gulland, J. (Editor), 1971. *The Fish Resources of the Oceans. C. Mediterranean and Black Sea*. Fishing News (Books) Ltd., London, pp. 36–43.

Halim, Y., 1960. Observations on the Nile bloom of phytoplankton in the Mediterranean. *Rapp. Comm. Int. Mer Médit.*, 26(1): 57–67.

Halim, Y., 1976. Marine biological studies in Egyptian Mediterranean waters: a review. *Acta Adriat.*, 18(2): 29–38.

Halim, Y., Guergues S.K. and Saleh, H., 1967. Hydrographic conditions and plankton in the south east Mediterranean during the last normal Nile flood (1964). *Int. Rev. Ges. Hydrobiol.*, 52(3): 401–425.

Holt, S.J., 1974. Mediterranean fisheries. In: N. Ginsburg, S. Holt and W. Murdoch (Editors), *The Mediterranean Marine Environment and the Development of the Region*. The Royal University of Malta Press, Valetta, pp. 154–172.

Holthuis, L.B., 1980. *FAO Species Catalogue, 1. Shrimps and Prawns of the World. An Annotated Catalogue of Species of Interest to Fisheries*. FAO Fish. Synop., No. 125. FAO, Rome, 271 pp.

Holthuis, L.B. and Gottlieb, E., 1958. An annotated list of the Decapod Crustacea of the Mediterranean coast of Israel, with an appendix listing the Decapoda of the eastern Mediterranean. *Bull. Res. Counc. Isr.*, 2B(1–2): 1–126.

Hsü, K.J., Ryan, W.B.F. and Cita, M.B., 1973. Late Miocene desiccation of the Mediterranean. *Nature*, 242: 240–244.

Hureau, J.C. and Monod, Th. (Editors), 1973. *Check-List of the Fishes of the North-eastern Atlantic and of the Mediterranean.* Clofnam, UNESCO, Paris, Vol. 1: 683 pp.; Vol. 2: 331 pp.

Huvé, H., Kiener, A. and Riouall, R., 1973. Modifications de la flore et des populations ichtyologiques des étangs de Berre et de Vaine (Bouches-du-Rhône) en fonction des conditions hydroliques créés par le déversement de la Durance. *Bull. Mus. Hist. Nat. Marseille*, 33: 123–134.

Ibañez, M. and Motos, L., 1977. *Blennius ponticus* (Slastenenko, 1934), nueva especie para la fauna Atlantica. *Vie Milieu, Sér. A*, 27(3): 377–383.

Jacques, G., 1970. Aspects quantitatifs du phytoplancton de Banyuls-Sur-Mer (Golfe du Lion). IV. Biomasse et production, 1965–1969. *Vie Milieu, Sér. B, Océanogr.*, 21 (1B): 37–102.

Kiener, A., 1978. *Ecologie, physiologie et économie des eaux saumâtres. Collection de biologie des milleux marins 1.* Masson, Paris, 220 pp.

Kimor, B., 1975. Euryhaline elements in the plankton of the Bardawil Lagoon *Rapp. Comm. Int. Mer Médit.*, 23(3): 119–120.

Kiseleva, M.I. and Chukhchin, V.D., 1965. Nekotorye dannye o kolichestvennom razvitii makro- i meiobentosa v vostochnoi chasti Sredizemnogo moria. [Some data on the quantitative development of macro- and meiobenthos in the eastern Mediterranean.] In: L.U. Tomin (Editor), *Osnovnye cherty*. Akad. Nauk S.S.S.R., Okeanol. Kom. pp. 192/197.

Klausewitz, W., 1968. Remarks on the zoogeographical situation of the Mediterranean and the Red Sea. *Ann. Mus. Civ. Stor. Nat. Giacomo Doria*, 77: 323–328.

Klausewitz, W., 1973. New aspects of the paleogeography of the Mediterranean ichthyo-fauna. *Ichthyologia*, 5(1): 79–86.

Kosswig, C., 1974. Modificability, a neglected factor for area expansion in marine fish. *Istanbul Univ. Fen. Fak. Mecm., Seri B*, 39(1–2): 1–7.

Laubier, L., 1972–1973. Découverte d'une annélide polychète de l'Atlantique boréal dans l'étage bathyal de Méditerranée occidentale. *Vie Milieu*, 23(2A): 255–261.

Lavee, D. and Ritte, U., 1977. Genetic parameters as indices of preadaptation for successful colonization in mytilids (Mollusca: Pelecypoda). *Isr. J. Zool.*, 26: 259.

Legac, M., 1974. Glavonošci (Cephalopoda) u Kocárskim lovinama sjevernog jadrana. [Cephalopoda in the northern Adriatic trawl catches.] *Acta Adriat.*, 16(11): 201–206.

Levi, D. and Troadec, J.P., 1974. The fish resources of the Mediterranean and the Black Sea. *Stud. Rev. Gen. Fish. Counc. Medit.*, No. 54: 29–52.

Lewinsohn, Ch. and Holthuis, L.B., 1978. On a new species of *Alpheus* (Crustacea, Decapoda, Natantia) from the eastern Mediterranean. *Zool. Meded. Leiden*, 53(7): 75–82.

Lewinsohn, Ch. and Manning, R.B., 1980. Stomatopod Crustacea from the eastern Mediterranean. *Smithson. Contrib. Zool.*, No. 305: 1–22.

Lipkin, Y., 1975. *Halophila stipulacea*, a review of a successful immigration. *Aquat. Bot.*, 1: 203–215.

Luther, W. and Fiedler, K., 1976. *A Field Guide to the Mediterranean Sea Shore.* Collins, London, 272 pp. (translated and edited by P.J. Miller).

McGill, D.A., 1969. A budget of dissolved nutrient salts in the Mediterranean Sea. *Cah. Océanogr.*, 19(6): 543–554.

Mancy, K.H., 1981. The environmental and ecological impacts of the Aswan High Dam. In: H. Shuval (Editor), *Developments in Arid Zone Ecology and Environmental Quality.* Balaban I.S.S., Philadelphia, Pa., pp. 83–99.

Massuti, M., 1971. Evolution de la pêche sur le talus du plateau continental des Iles Baléares entre 1940 et 1969. *Etud. Rev. Cons. Gén. Pêch. Médit.*, No. 49: 1–39.

Maurin, C., 1968. Ecologie ichthyologique des fonds chalutables atlantiques (de la Baie Ibéro-Marocaine à la Mauritanie) et de la Méditerranée occidentale. *Rev. Trav. Inst. Pêches Marit.*, 32(1): 1–147.

Maurin, C., 1970 (1973). Quelques aspects de la faune ichthyologique méditerranéenne. In: *Journées Ichthylogie, Rome, C.I.E.S.M.*, pp. 27–38.

Menzies, R.J., 1973. Biological history of the Mediterranean Sea with reference to the abyssal benthos. *Rapp. Comm. Int. Mer Médit.*, 21(9): 717–723.

Miller, A.R. and Munns, R.G., 1974. The Bitter Lake salt barrier. In: *L'océanographie physique de la Mer Rouge, Symposium de l'Association Internationale des Sciences Physiques de l'Océan (IAPSO).* Publ. CNEXO (Actes Colloq.), No. 2, Paris, pp. 295–310.

Miller, P.J., 1966. A new genus and species of gobiid fish from the eastern Mediterranean. *Ann. Mag. Nat. Hist. Ser*, 13(8): 161–172.

Mouneimne, N., 1977. Liste des poissons de la côte du Liban (Méditerranée orientale). *Cybium, 3e Sér.*, (1): 37–66.

Mouneimne, N., 1979. Poissons nouveaux pour les côtes Libanaises. *Cybium, 3e Sér.*, (6): 105–110.

Oren, O.H., 1969. Oceanographic and biological influence of the Suez Canal, the Nile and the Aswan Dam on the Levant Basin. In: M. Sears (Editor), *Progress in Oceanography, S.* Pergamon Press, New York, N.Y., pp. 161–167.

Oren, O.H., 1970. The Suez Canal and the Aswan High Dam, their effect on the Mediterranean. *Underwater Sci. Technol. J.*, 2: 222–229.

Oren, O.H. and Komarovsky, B., 1961. The influence of the Nile flood on the shore waters of Israel. *Rapp. Comm. Int. Mer Médit.*, 16: 655–659.

Paperna, I., 1972. Parasitological implications of fish migration through interoceanic canals. In *17e Congr. Int. Zool., Monte Carlo, 1972, Thème No. 3: Les conséquences biologiques des canaux interocéans*, 9 pp. (mimeo).

Peres, J.M., 1967. The Mediterranean benthos. In: H. Barnes (Editor), *Oceanography. Mar. Biol. Annu. Rev.*, 5: 449–533.

Peres, J.M. and Picard, J., 1958. Manuel de bionomie benthique de la Mer Méditerranée. *Rec. Trav. Stn. Mar. Endoume–Marseille*, 23(14): 5–122.

Peres, J.M. and Picard, J., 1964. Nouveau manuel de bionomie benthique de la Mer Méditerranée. *Rec. Trav. Stn. Mar. Endoume–Marseille*, 47(31): 3–137.

Por, F.D., 1975a. Pleistocene pulsation and preadaptation of biotas in Mediterranean seas: consequences for Lessepsian migration. *Syst. Zool.*, 24: 72–78.

Por, F.D., 1975b. An outline of the zoogeography of the Levant. *Zool. Scr.*, 4: 5–20.

Por, F.D., 1978. *Lessepsian Migration. The Influx of Red Sea Biota Into the Mediterranean by Way of the Suez Canal.* Ecological Studies, 2. Springer-Verlag, Berlin, 228 pp.

Por, F.D. and Ben-Tuvia, A., 1980. The Bardawil (Sirbonian) Lagoon of North Sinai — A summing up. *Rapp. Comm. Int. Mer. Médit.*, 27(4): 101–107.

Ritte, U. and Safriel, U.N., 1977. The theory of island biogeography as a model for colonization in more complex systems. *Proc. 8th Meeting, Isr. Ecol. Soc., 1977*, pp. 288–294.

Ruggieri, G., 1967. The Miocene and later evolution of the Mediterranean Sea. *Syst. Assoc. Publ.*, 7: 283–290.

Ryder, R.A. and Henderson, H.F., 1975. Estimates of potential fish yield for the Nasser Reservoir, Arab Republic of Egypt. *J. Fish. Res. Board. Can.*, 32(11): 2137–2151.

Safriel, U.N. and Lipkin, Y., 1975. Patterns of colonizations of the Eastern Mediterranean intertidal zone by Red Sea immigrants. *J. Ecol.*, 63(1): 61–63.

Safriel, U.N. and Ritte, U., 1980. Criteria for the identification of potential colonizers. *Biol. J. Linn. Soc., Lond.*, 13(4): 287–297.

Safriel, U.N., Gilboa, A. and Felsenburg, T., 1980. Distribution of rocky intertidal mussels in the Red Sea coasts of Sinai, the Suez Canal and the Mediterranean coast of Israel, with special reference to recent colonizers. *J. Biogeogr.*, 7: 39–62.

Schmidt, H.E., 1976. Comparison between the Hydroida fauna of the eastern Mediterranean and the Red Sea. *Acta Adriat.*, 18(15): 257–266.

Sparck, R., 1931. Some quantitative investigations on the bottom fauna at the West Coast of Italy, in the Bay of Algiers, on the Coast of Portugal. *Rep. Dan. Oceanogr. Exped. 1908–1910, Mediterranean and Adjacent Seas*, 3(7): 1–11.

Steinitz, H., 1967. A tentative list of immigrants via the Suez Canal. *Isr. J. Zool.*, 16(3): 166–169.

Steinitz, H., 1970. A critical list of immigrants via the Suez Canal. In: *The Hebrew University — Smithsonian Institution Joint Program, Biota of the Red Sea and the Eastern Mediterranean, Progress Report.* Department of Zoology, Hebrew University, Jerusalem, pp. 64–74 (mimeo).

Tortonese, E., 1963. Elenco riveduto del leptocardi, ciclostomi, pesci cartilagine e ossei del mare mediterraneo. *Ann. Mus. Civ. Stor. Nat. Giacomo Doria*, 74: 156–185.

Tortonese, E., 1964. The main biogeographical features and problems of the Mediterranean fish fauna. *Copeia*, 1964(1): 98–107.

Tortonese, E., 1970. *Osteichthyes (pesci ossei) Parte prima. Fauna Italica, 10.* Calderini, Bologna, 565 pp.

Tortonese, E., 1973. Facts and perspectives related to the spreading of Red Sea organisms into the eastern Mediterranean. *Ann. Mus. Civ. Storia Nat. Giacomo Doria*, 79: 322–329.

Tortonese, E., 1975. *Osteichthyes (pesci ossei). Parte seconda. Fauna Italica, 11.* Calderini, Bologna, 636 pp.

Tortonese, E. and Hureau, J.C. (Editors), 1979. *Check-List of the Fishes of the North-eastern Atlantic and of the Mediterranean.* Clofnam. Supplément 1978. UNESCO, Cybium, 3e Sér., No. 5, pp. 5(333)–66 (394).

Troadec, J.P., 1979. Resources appraisal and fisheries management in the Mediterranean: the activities of GFCM in this field. *Invest. Pesq.*, 43(1): 261–271.

Vatova, A., 1949. La fauna bentonica dell'Alto e Medio Adriatico. *Nova Thalassia*, 1(3): 3–110.

Vatova, A., 1975. Ricerche qualitative e quantitative sulla fauna bentonica del golfo di Taranto. *Boll. Pesca Piscic. Idrobiol.*, 30: 1–38.

Wirszubski, A., 1953. On the biology and biotope of the red mullet, *Mullus barbatus L. Bull. Sea Fish. Res. Stn. Isr.*, (7): 1–20.

Worms, J., 1979. L'exploitations des Cephalopodes en Méditerranée: évolution et perspectives. *Pêche Marit.*, No. 1212: 169–172.

Zavodnik, D., 1971. Contribution to the dynamics of benthic communities in the region of Rovinj (Northern Adriatic). *Thalassia Jugosl.*, 7: 447–514.

Zibrowius, H., 1979. A propos du corail rouge en Méditerranée orientale. *Rapp. Comm. Int. Explor. Mer Médit.*, 25/26(4): 121–122.

Zoré-Armanda, M. and Pucher-Petkovic, T., 1976. Some dynamic and biological characteristics of the Adriatic and others basins of the eastern Mediterranean Sea. *Acta Adriat.*, 18(1): 17–27.

Chapter 11

THE BLACK SEA

YU.I. SOROKIN

INTRODUCTION

The Black Sea is the largest of the world's meromictic basins — those bodies of water characterized by a permanent halocline. Below 150 to 200 m, it is filled with stagnant anoxic waters which contain hydrogen sulphide. The stagnation is produced by the permanent halocline, which separates lower water layers with a higher density from an upper layer of low density. Vertical mixing and seasonal turnover of the water column is limited to the upper layer down to the halocline, so that denser waters that are isolated from sources of oxygen become stagnant and anoxic.

The higher density of water in the stagnant layer of meromictic basins is usually generated by the solution of components of the rocks of a basin's bed, calcium sulphate for example, producing the denser bottom water, while the upper water layer receives rain and river waters. This mechanism of the formation of meromictic basins is peculiar to some inland lakes, situated either in regions subjected to the karst process, such as Lake Belovod in the vicinity of Moscow, or in mountain lakes such as Lakes Gek Gel or Maral Gel in the Caucasus Mountains (Sorokin, 1970).

An increase of the water density in the hypolimnion of lakes subjected to organic pollution can take place as a result of the intensive processes of mineralization of a mass of organic debris settled down to the bottom. In this case, during warm winters the seasonal mixing is incomplete, and the process which forms a stagnant anoxic layer is initiated (Sorokin, 1975). This modern man-induced development of the features of meromixis in deep lakes subjected to pollution has stimulated interest in the study of natural meromictic basins.

Alternatively, a meromictic basin can be formed where sea water flows into a coastal basin separated from the sea by a sand bar or a narrow or shallow pass. This is the origin of the Black Sea. There are many smaller meromictic estuaries or coastal brackish lakes and fjords, such as Lakes Faro, Nitinat, Mogilnoye, some coastal lakes in Japan and some Norwegian fjords (Richards et al., 1965; Takahashi and Ishimura, 1968; Sorokin and Donato, 1975). The Black Sea is the largest basin of this type, and contains the greatest amount of anoxic water on the planet. The presence of the anoxic layer influences the chemical and biological regimes and indeed the entire ecosystem of this unique basin. The stagnant condition and the formation of a permanent chemocline is caused by the constant inflow of the salty Mediterranean waters through the Bosporus (Kara Deniz Boğazi) into the bottom of the basin; the salinity of the surface waters is diluted by the inflow of fresh water from rivers and of brackish water from the Sea of Azov (Azovskoye More).

The interest of many investigators has been stimulated by the economic importance of the Black Sea. It supports a valuable fishery and contains ports for ocean transport and facilities for recreation, served by large harbors and industrial and resort cities. Oil is piped from inland fields to the harbors of the Black Sea for transportation, and the drilling of its sea bed for oil and gas may occur in the near future. All these industries are expanding, with a concomitant increase in the pollution of the sea; eventually this may destroy the valuable natural resources and may result in a rise in the level of the anoxic waters.

The stagnant waters of the Black Sea have been suggested as a site for the dumping of radioactive wastes, a proposal which lends a sense of urgency to the need for estimations of the rates of vertical mixing and their indirect calculation from the chemical or biological parameters (Vodyanitzki, 1958).

Studies of the nature of this sea were started by P. Pallas and his description of its fish population in the late eighteenth century. Later, in 1860–1880, the fauna of the Black Sea were observed by the Russian zoologists K. Kessler and W. Ulyanin. Studies of the hydrology of the Black Sea were made by Admiral S. Makarov in 1880, and later in 1890, by a special deep-sea expedition organized by N. Andrusov and I. Spindler. Their studies described the main features of the origin, morphometry, circulation, chemistry and zoogeography of the sea up to the beginning of the twentieth century. The existence of the hydrogen sulphide zone was found by N. Andrusov in 1890. At the same time N. Zelinski realized that it is perpetuated by the activity of sulphate-reducing bacteria, and M. Egunov postulated the existence of a layer of sulphide-oxidizing microflora which barred the penetration of hydrogen sulphide to the upper water layers.

Later, extensive studies of hydrology and hydrochemistry were performed by a special Azov–Black Sea expedition of the Academy of Science of the U.S.S.R. in 1927–1930 under the leadership of N. Knipovich. The main features of the geological history of the basin were delineated by Arkhangelski and Strakhov (1938). The regularities of the hydrological regime were described by Danilchenko and Chigirin (1927), Bruevich (1953), Dobrzhanskaya (1960) and Skopintzev (1975). The basic problems of the hydrology of the Black Sea have been analyzed by Antipa (1941), Neumann (1942), Caspers (1957), Bogdanova (1959b), Vodyanitzki (1948), Vladimirtzev (1961) and Filippov (1968).

Fundamental biological studies of the Black Sea have been made by the Sevastopol biological station, founded in 1872 especially for these investigations by the Russian Academy of Science. The biologists who worked there have not limited themselves to the biology of the Black Sea alone; they have made important contributions to marine biology in general including the theory of marine biocoenoses (Zernov, 1913), the problems of marine hyponeuston (Zaitzev, 1970), of marine biochemistry (Vinogradova, 1964), of the physiology of marine phytoplankton and of ecological metabolism (Finenko, 1971; Khailov, 1971), of marine trophology and secondary productivity (Grese, 1968; Petipa, 1975; Zaika et al., 1976), of marine radioecology (Polikarpov, 1966), of mathematical modeling of marine ecosystems (Belyaev et al., 1974), and of hydrokinematics of marine animals (Aleev, 1968). The basic studies of the phytoplankton and its productivity in the sea have been made by Proshkina-Lavrenko (1955), Morozova-Vodyanitzkaya (1957), Sorokin (1964a), Finenko (1967), Pyzik (1968) and Roukhiyainen (1975); of the phytobenthos, zoobenthos by Zernov (1913), Arnol'di (1949), Morozova-Vodyanitzkaya (1959), Kiseleva (1967) and Kalugina-Gutnik (1975); of the microbial populations and their activity by Lebedeva (1957), Kriss (1963), Sorokin (1964a, 1972), Psenin (1966) and Tsiban (1970); of the zooplankton by Nikitin (1945), Galadjiev (1948), Petipa et al. (1963), Grese (1968) and Pavlova (1967); and of the ichthyofauna by Vodyanitzki (1956), Svetovidov (1964), Aleev (1968), Duka (1969) and Dekhnik (1970). The subjects of the biology and the productivity of the Black Sea have been discussed and generalized by Zenkevich (1963). Studies of problems of oceanography, hydrobiology and modern changes in the ecosystem influenced by eutrophication have been conducted by the Institute of Oceanographic Research of the Romanian Academy of Science in Constanţa. Reviews of these studies have been presented by Bacescu (1965), Bodeanu (1969) and Bacescu et al. (1971). A series of studies of the Black Sea were conducted during the expedition of the Woods Hole Oceanographic Institution's (U.S.A.) R.V. *Atlantis II* in 1969. The results of these works and monographical information about this basin are given in the book edited by Degens and Ross (1974).

GEOGRAPHICAL, PHYSICAL AND CHEMICAL CHARACTERISTICS

Morphometry and history of the basin

The Black Sea lies between 40°55′ and 46°32′N, and between 27°27′ and 41°42′E, in the east–west intermontane depression between two alpine fold belts, the Pontic Mountains to the south and the Caucasus Mountains on the northeast. The northwest coast, except for the Crimea (Krym), is relatively low. The shallow Kerch Strait (Kerchenskiy Proliv) connects the northern reaches of the sea with the Sea of Azov, and to the south, Mediterranean waters flow first into the Sea of Marmara and then through the narrow Bosporus Strait to the Black Sea. The steep coastal slopes of 4 to 6° drop to a maximum depth of 2212 m in the basin, with an average depth of 1300 m for the entire sea. Within the total sea area of 420×10^3 km^2, more than 30% has depths greater than 2000 m. Most of the area shallower than 200 m, 27% of the total, is in the northwest. The total volume of the sea is 537×10^3 km^3, 87% of it filled with deep anoxic waters; the water down to 200 m is only 13% of the total. The bottom of the basin can be subdivided into the shelf, basin apron and abyssal (euxine) plain (Fig. 11.1). The sedimentary rocks filling the bed of the sea compose a layer 8 to 14 km thick (Fig. 11.2).

The physical characteristics of the Black Sea took their present shape in the Late Quaternary. In the Early Quaternary the sea existed as a shallow brackish euxine sea lake. In the Middle Quaternary a connection with the Mediterranean was broached, with a resulting increase in the salinity of the basin's water. This connection later filled in, and during the latest glacial period (10 000 to 20 000 years ago) the Black Sea was a fresh-water lake with no anoxic hydrogen sulphide zone. About 7000 years ago the connection with the Mediterranean again opened, and the formation of the present chemical regime of the Black Sea was started. The salinity of the deep waters increased as salt water flowed in from the Sea of Marmara; simultaneously the salinity of the upper layers was diluted by an inflow of fresh water from melting glacial ice. A constant stratification developed and with it the formation of the anoxic hydrogen sulphide zone. The present salinity structure of the Black Sea waters was formed some 3000 years ago when the melting of the glacial ice was complete. Earlier, between 8000 and 3000 years ago, the ancient Black Sea basin was considerably more productive than is the modern Black Sea; bottom sediments formed during this period were enriched with organic matter (Deuser, 1971).

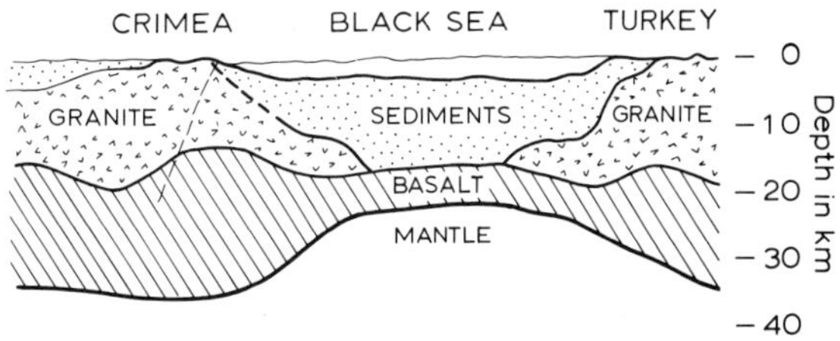

Fig. 11.2. Crustal N–S cross-section of the Black Sea (after Neprochnov and Yel'nikov, 1969).

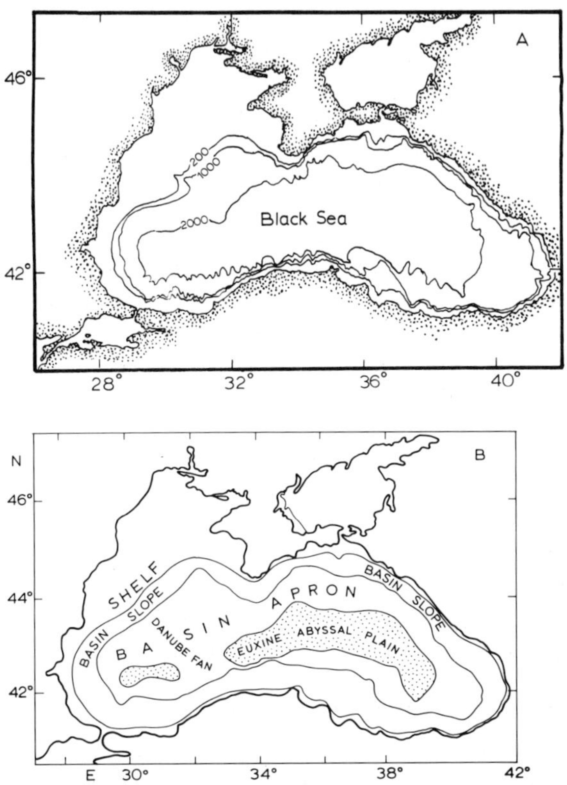

Fig. 11.1. Morphometric characteristics (A) and the main physiographic provinces (B) of the Black Sea.

Features of hydrology

The balance of water and of salt in the Black Sea is near equilibrium (Table 11.1). Most of the river water inflow comes from the River Danube (200 km^3 yr^{-1}); brackish water, about 55 km^3 yr^{-1}, comes through the Kerch Strait by a surface current. Some highly saline Black Sea water is moved by a deep current northward into the Sea of Azov, and surface water from the Black Sea (340 km^3 yr^{-1}), which has a salinity of 18‰, moves to the Marmara through the Bosporus Strait. The deep current through the Bosporus annually transfers about 180 km^3 of saline (34.3‰) water from the Sea of Marmara into the Black Sea.

TABLE 11.1

Elements of annual water balance in the Black Sea (after Soljankin, 1963)

Gains		Losses	
source	km^3	cause	km^3
Rivers	346	evaporation	332
Rains	129	outflow into the Azov Sea	32
Inflow from the Azov Sea	53	outflow into the Sea of Marmara	340
Inflow from the Sea of Marmara	176	–	
Sum	704	Sum	704

The exchange of water through the Bosporus is the most influential factor in the hydrological and chemical regime of the Black Sea. The characteristics of the Bosporus itself dictate the amount and flow of both exchanges of water. This strait lies in the northern 10% of the length of the Turkish straits, which extend about 300 km from the Aegean Sea, through the Dardanelles (Çanakkale Boğazi), the Sea of Marmara and the Bosporus into the Black Sea. This narrow, shallow meandering strait is about 31 km long, with a maximum width of 3.6 km, minimum width 730 m, and an average depth of 65 m (maximum 92 m, minimum 36 m, Fig. 11.3A, B). At the bottom of the strait are

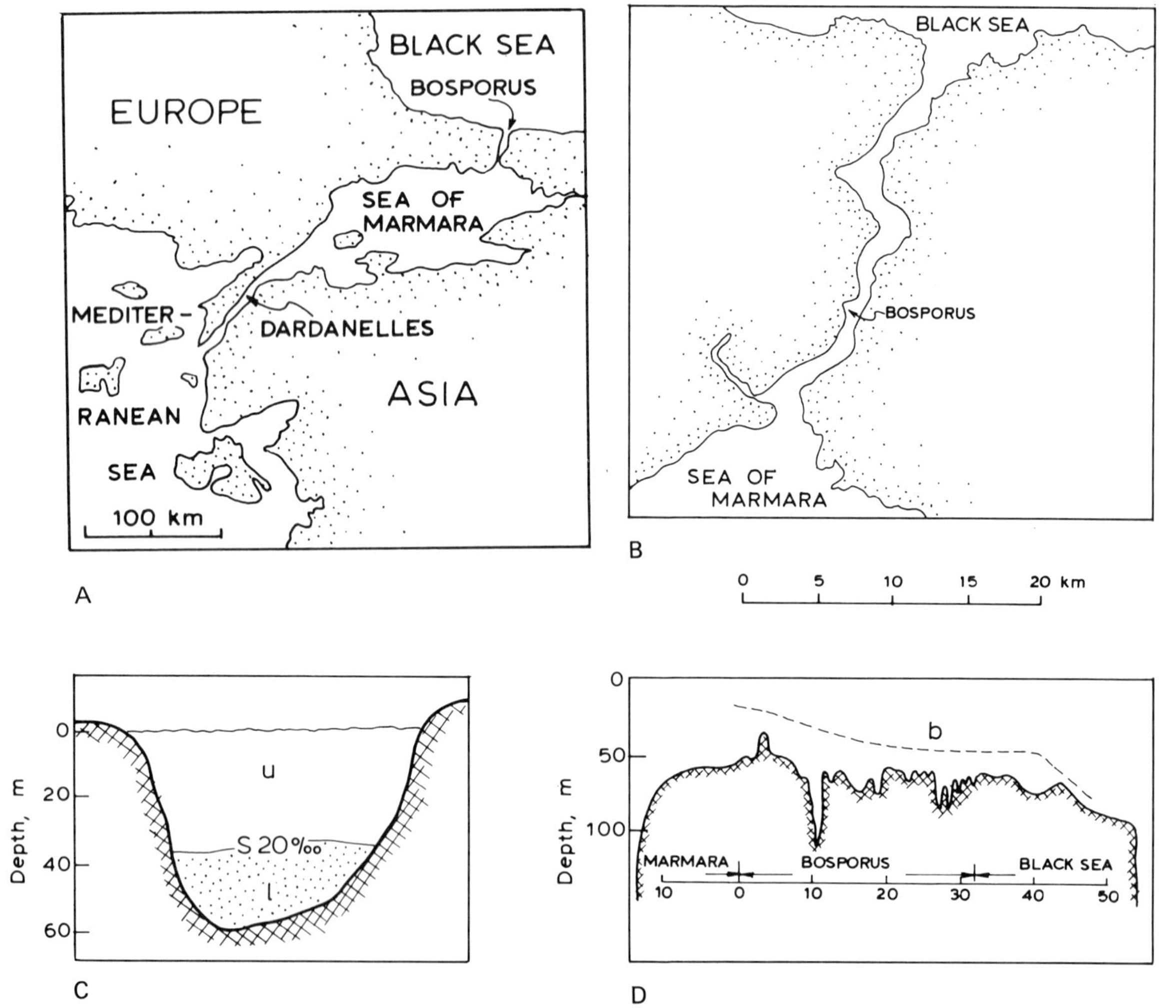

Fig. 11.3. A. The Black Sea straits. B. The Bosporus. C. A section across the Bosporus at its narrowest location (0.8 km) [S = the salinity boundary between the upper southward current from the Black Sea (u) and the northward lower current (l) from the Sea of Marmara]. D. Section along the Bosporus (b = boundary between upper Black Sea water and lower waters from the Sea of Marmara).

elongated depressions of depths up to 110 m. Thresholds at the northern and southern entrances of the Bosporus limit the water exchange to the surface waters above the thresholds (Fig. 11.3D). Subsea canyons opposite the Bosporus can be traced to a depth of 70 m in the Sea of Marmara, and to a depth of 2000 m in the Black Sea. Borings that have been made in the bed of the Bosporus indicate a complex history of cutting and filling, which occurred at alternating periods of changing sea level during the glacial and interglacial epochs of the Quaternary. The level of the Black Sea is now a little higher than that of the Mediterranean, so the water surface in the Bosporus slopes toward the south about 35 cm. This slope causes the outflow of surface water having a salinity of 17 to 18‰ from the Black Sea to the Sea of Marmara. The depth of this surface current is 20 to 40 m (Fig. 11.3C). Its boundary surface has a slope declining from south to north (Fig. 11.3D); beneath it, flowing north, is an opposite current caused by the difference in water density between the Sea of Marmara and the Black Sea. The rate of water exchange through the Bosporus depends upon the differences in sea levels and water densities at both ends of the strait. Under normal wind conditions, the velocity of both currents is about 2 km h^{-1}, and the volume of water flowing both ways is about equal. However, their annual balance is influenced by the seasonal climatic situation and by the size of river inflow to the Black Sea, which changes the sea level. Therefore, the inflow of saline waters into the Black Sea in the deeper Bosporus current decreases markedly from February to August when the level of the Black Sea near the Bosporus is at its highest. The annual outflow of Black Sea water usually exceeds the inflow of saline waters from the Sea of Marmara (Table 11.1). This balance is one of the key factors which determines the formation of the salinity and density profiles of the Black Sea. The Mediterranean waters pass northeast through the Bosporus as a thin layer of dense warm water via the canyons, which are actually the prolongation of the strait into the Black Sea. The waters change direction to west-northwest and appear as a thin layer of water with increased salinity sloping downwards to a depth of more than 500 m, 50 km from the Bosporus entrance (Bogdanova, 1959b, 1961). The halocline can be observed at the boundary of the Black Sea and Mediterranean waters (Fig. 11.4A). The temperature/salinity (T/S) curves have a specific form in this region, reflecting the existence of strict gradient zones at the interface of waters of different origin (Fig. 11.4B).

The water exchange through the Bosporus influences not only the physical and chemical

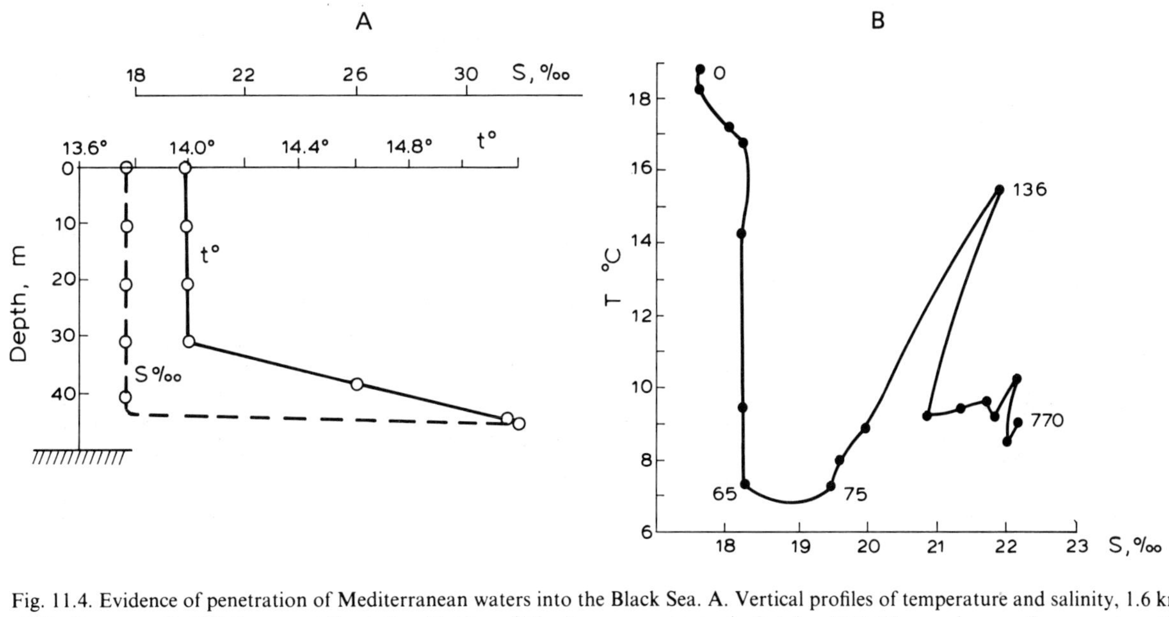

Fig. 11.4. Evidence of penetration of Mediterranean waters into the Black Sea. A. Vertical profiles of temperature and salinity, 1.6 km off the Bosporus. B. T/S diagram at the station 14.5 km off the Bosporus entrance in October 1960. The numbers on the curve show the depth of observations (m) (after Bogdanova, 1961; and Bogdanova and Stepanov, 1976).

characteristics of the Black Sea, but also its plankton population. About twenty species of Mediterranean zooplankton ride in from the Sea of Marmara in the deeper Bosporus current, and have been found in the Black Sea near the Bosporus. Some have also been found in the upper waters far from the Bosporus, up to coast of the Crimea, indicating that some of these species can not only survive but in fact multiply in the Black Sea. Plankton in the Sea of Marmara are influenced in the upper water layer by the transfer of algae and zooplankton from the Black Sea by the upper Bosporus current.

The surface currents of the Black Sea have a circular motion (Fig. 11.5), forming several halistatic eddies, one in the western part of the sea, and three (two cyclonic and one anticyclonic) in its eastern part. In these halistatic areas the isosurfaces of the fields of equal temperature, salinity, oxygen and hydrogen sulphide concentration form a cupola, elevating in the centers toward the surface. The upper boundary of the hydrogen sulphide zone in the centers of the halistatic areas rises up to 110 to 130 m; at the edges it drops down to 150 to 170 m (see Fig. 11.11).

Extensive horizontal currents in the Black Sea cause movement in the deeper water layer which decreases with depth. Nevertheless, the speed of the water is considerable, even at great depths. Instrument observations show that at 700 m depth the current velocity can be as high as 10 to 15 cm s^{-1}, an intensive horizontal motion (Filippov, 1968).

The vertical thermohaline structure is characterized by a steep gradient of salinity and density with depth. This is caused by the constant inflow of fresh water from rivers and rain and of brackish water from the Sea of Azov into the surface layer, together with the simultaneous deeper inflow of salt water from the Bosporus. The increase of salinity is very sharp in the upper 200 m (Fig. 11.6); it amounts to 3‰ in the upper layer and an additional 1‰ in the balance of the water column down to 2000 m. At the surface the salinity is between 17.5 to 18.3‰, that is, half of that in the ocean. The seasonal variations of salinity are within this range and extend down to 150 m.

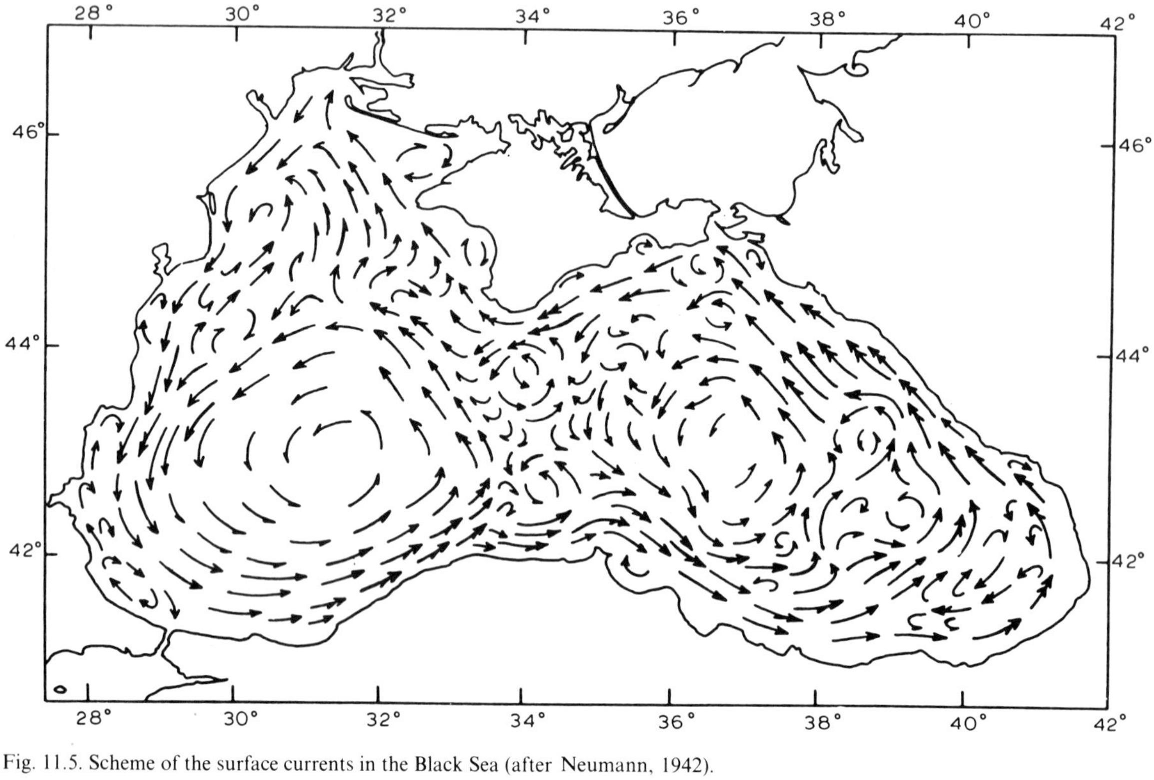

Fig. 11.5. Scheme of the surface currents in the Black Sea (after Neumann, 1942).

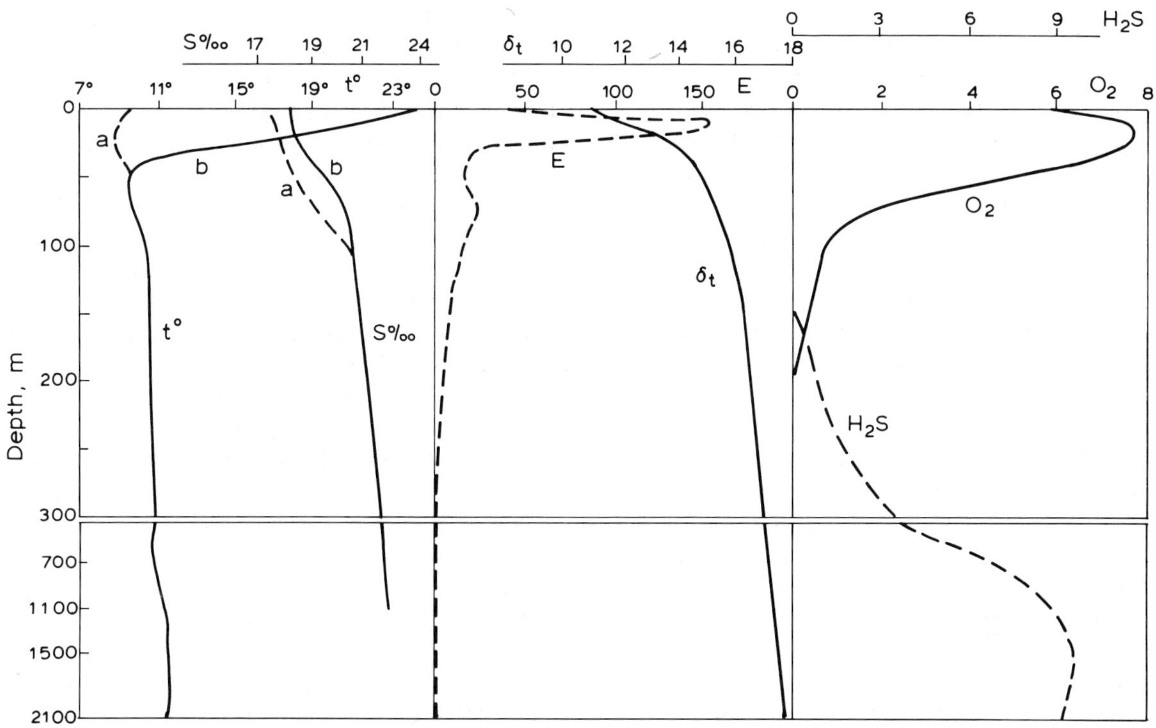

Fig. 11.6. Vertical profiles of the main parameters of stratification: thermohaline structure in winter (*a*) and in summer (*b*) (t°C, salinity S‰), water density (σ_t), coefficient of vertical water stability, $E \times 10^6$ (where $E = 10^{-3}\, d\sigma_t/dz$), oxygen ($O_2$ mg l^{-1}) and H_2S (mg l^{-1}) in the Black Sea.

The vertical distribution of density (σ_t) and of a water mass stability coefficient (E) identify two main boundary zones: the chemocline at depths of 200 to 150 m, where the increase of the density gradient and water stability is dependent upon the increase of salinity alone with no temperature change; and the thermocline where the density gradient increases sharply mostly because of a change in water temperature. The chemocline zone forms the boundary between aerobic and anoxic zones (Fig. 11.6). Below the chemocline the waters of deeper layers have a very low density gradient, and consequently a low vertical stability, which results in vertical mixing of water within this zone. This mixing, especially in the deep water layers below 1000 to 1200 m, causes an even vertical distribution of salinity, water density, hydrogen sulphide, and other parameters. Mixing is stimulated by the flux of geothermal heat from the bottom, which increases the temperature of water in the bottom layers (Fig. 11.6). This increase of temperature in near-bottom waters may also result from the inflow of warm water from the Mediterranean, which then spreads in the deep layers. In summer the temperature minimum in the Black Sea, as well as in some other meromictic basins, lies near the boundary of the euphotic zone at 50 m depth.

The existence of a constant chemocline and density gradient acts as a barrier to seasonal mixing processes, which reach only to depths of 120 to 150 m. This layer defines the boundary of the redox gradient zone in the basin. Although the chemocline perpetuates stagnation conditions that are sufficient to preserve an anoxic zone, it cannot prevent the gradual vertical mixing of the total water mass of the sea. Calculations based on the water exchange through the Bosporus and on ^{14}C age dating suggest that the mean residence time of water in the Black Sea is about 2250 years. Most hydrologists, however, now accept that there is extensive vertical mixing of waters in this basin. Vodyanitzki (1948) calculated that the time of elevation of deep waters to the surface may be an order of magnitude less than this mean residence time, or 100 to 130 years. The main forces that induce vertical water movement and mixing despite a large density gradient he cited are: the influence of the Coriolis force on the wind-driven horizontal currents which intensifies vertical turbulence; the seasonal cooling of surface layers and the warming of bottom layers by geothermal heat; seiches, which are large in the Black Sea; and turbulent diffusion. Vodyanitzki based his conclusions about vertical mixing on biological observations of the abundance of life in pelagic areas of the Black Sea, and, indeed, the rate of vertical circulation is a key factor to the productivity of the Black Sea. Reasonably, the more isolated the deep layers, the less productive should be the basin. The assumption of a low productivity for this sea which dominated early stages of investigation was surely based on an assumption of a low rate of vertical mixing.

Vertical mixing has been studied in more detail by different analytical methods and computations. These include calculations based on the flux of salt (Bogdanova, 1959b) or of other mineral components of water from layer to layer (Skopintzev, 1975), the rates of oxygen consumption and hydrogen sulphide oxidation (Skopintzev, 1953; Dobrzhanskaya, 1967; Aizatullin and Skopintzev, 1974), the vertical component of currents and the coefficients of vertical turbulence (Belyaev et al., 1974), the vertical distributions of hydrogen sulphide and ammonia (Skopintzev, 1953, 1975; Belyaev et al., 1974). All these calculations resulted in similar vertical rates of water exchange of about 10^{-2} to 10^{-4} cm s^{-1} in the deep waters, which means that the time of the movement of the deep waters to the surface should range from a few decades to about 200 years.[1] This more vigorous vertical circulation within the deep waters implies that the large vertical gradients of salinity, hydrogen sulphide, oxygen, nutrients and other chemical components which appear on vertical profiles result from dynamic equilibrium and are supported by active dynamic processes, such as the constant salt inflow from the Bosporus, and the microbiological hydrogen sulphide and ammonia production and oxidation. The position of the gradient layer results from the dynamic interaction of these processes with the hydrodynamical factors of vertical and horizontal water motions.

Hydrochemistry

Ionic composition. The ionic composition of the salts in the Black Sea is similar to that in the ocean (Table 11.2). The only marked difference is in the presence of carbonate ion and consequent increased alkalinity, which are significantly greater in the Black Sea, especially at depths, as compared with the ocean. A high concentration of carbonates is maintained in Black Sea surface waters because of a large inflow of river water containing these compounds. In the deep waters it is even more concentrated because of the production of large quantities of carbon dioxide during anoxic decomposition of organic matter. The absolute content of the major ions in Black Sea water increases down the vertical profile, as does its salinity, because of vertical stratification (Table 11.3). This pattern of vertical distribution is somewhat different than in the ocean. The vertical distribution of potassium and sodium is comparable to the vertical distribution of salinity, but the distribution of calcium, sulphate and carbonate ions varies from the normal ratio to salinity; its decline reflects the biochemical processes of carbon dioxide formation, sulphate reduction and hydrogen sulphide oxidation which occur in the reducing and anoxic zones of the basin.

TABLE 11.2

Major ion content (% of total by weight) of waters of the Black Sea and of the ocean (modified from Skopintzev, 1975)

Ions	Black Sea		Ocean
	0–300 m	1000–2000 m	
Chloride (Cl^-)	55.23	55.25	55.29
Bromide (Br^-)	0.183	0.185	0.190
Sulphate (SO_4^{2-})	7.54	7.41	7.69
Carbonate (CO_3^{2-})	(0.46)	(0.95)	0.207
Sodium (Na^+)	31.36	31.23	30.59
Potassium (K^+)	1.02	1.07	1.10
Calcium (Ca^{2+})	1.31	1.28	1.20
Magnesium (Mg^{2+})	3.34	3.74	3.72

[1] Editor's comment: These times for vertical circulation are not directly comparable to the mean residence time. There can be an active vertical circulation within the deep water without removal or replacement of the water or its constituents if recycling occurs. Additional recent estimates of mean residence time for the water and its constituents in the Black Sea are presented in Degens and Ross (1974).

TABLE 11.3

Average content of major ions ($g\ kg^{-1}$) and total alkalinity (meq. l^{-1}) in the water column of the Black Sea (after Skopintzev, 1975)

Depth (m)	Alkalinity	Cl^-	SO_4^{2-}	Ca^{2+}	Mg^{2+}
0	3.32	10.14	1.4293	0.2504	0.6968
50	3.32	10.31	1.4539	0.2536	0.7134
100	3.33	11.11	1.5621	0.2685	0.7674
200	3.47	11.79	1.6551	0.2792	0.8014
500	3.90	12.17	1.6853	0.2850	0.8302
1000	4.26	12.31	1.6933	0.2890	0.8367
2000	4.35	12.34	1.6817	0.2870	0.8384

Trace elements. Strong stratification and the redox gradient zone also significantly affect the content and the vertical distribution of trace elements. Those which form particulate oxides (iron, manganese) are concentrated above the density discontinuity layer above the depth of zero oxygen (Brewer and Spencer, 1974). The dissolved form of iron is also abundant in the anoxic zone, while the dissolved forms of zinc and copper predominate in the aerobic zone, because of their precipitation as sulphides. The total content of iron and manganese is largest in the anoxic zone, where they accumulate as sulphides or carbonates (Table 11.4), and as ions in their reduced forms (Fig. 11.7).

Redox conditions: oxygen, hydrogen sulphide. The most distinctive feature of the Black Sea as a meromictic basin is the distribution of the redox potential (Eh) and of the factors controlling it: dissolved oxygen and hydrogen sulphide. The water column can be subdivided into three main zones according to their vertical profiles (Fig. 11.8). The

TABLE 11.4

The average content ($\mu g\ l^{-1}$) of some trace elements in waters of the Black Sea (after Skopintzev, 1975)

Element	oxygenated zone	Anoxic zone
Cobalt	0.2–0.5	0.5–3.0
Copper	1–7	0.3–1.8
Mercury	0.3	0.3–0.8
Iron	1–10	5–45
Manganese	<4	250–450
Molybdenum	3.1–3.4	1.3–1.8
Nickel	1–2	1.8–3.0

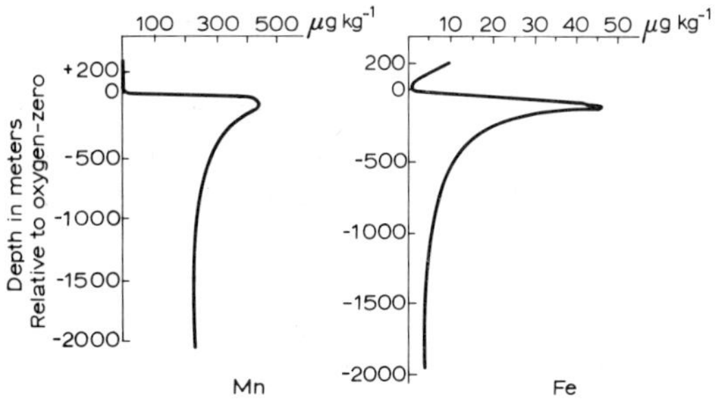

Fig. 11.7. Profiles of dissolved manganese and iron (adapted from Brewer and Spencer, 1974).

oxygen zone extends from the surface down to 100 m, or as much as 200 m depth near the coast. In this zone the redox potential is about the same as in ordinary marine basins: +350 to +430 mV, decreasing slightly down to the boundary of the zone. In summer the oxygen maximum usually can be found in the thermocline layer where the saturation of oxygen sometimes reaches 130 to 140%. Its formation is connected with the photosynthetic activity of phytoplankton (see Fig. 11.19); dissipation of oxygen may also be retarded there by a decrease of turbulence in the thermocline, where the stability of the water mass is high.

In winter the vertical distribution of oxygen is uniform down to the boundary depth of seasonal mixing; its concentration decreases at the end of winter to 70 to 80% saturation. Below this depth the

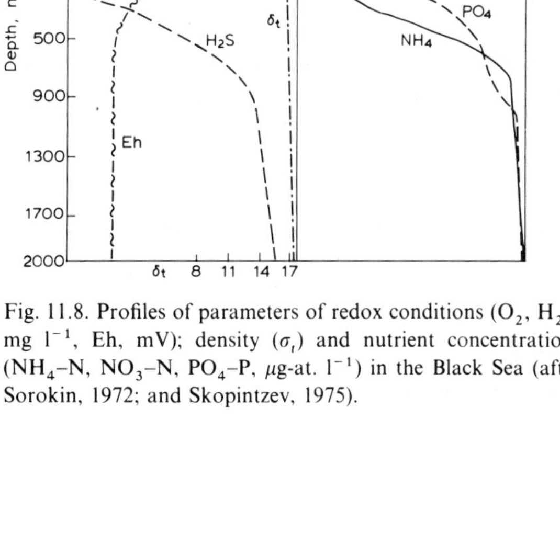

Fig. 11.8. Profiles of parameters of redox conditions (O_2, H_2S, mg l^{-1}, Eh, mV); density (σ_t) and nutrient concentrations (NH_4–N, NO_3–N, PO_4–P, μg-at. l^{-1}) in the Black Sea (after Sorokin, 1972; and Skopintzev, 1975).

concentration of dissolved oxygen decreases, and at 100 m depth it is usually only 5 to 10% of saturation. The decrease of oxygen content coincides well with the vertical distribution of salinity (Bogdanova, 1959a) because of the vertical water exchange which transports the oxygen down under the thermocline and halocline. This oxygen is consumed during the processes of plankton respiration and the oxidation of reduced compounds (mostly hydrogen sulphide) originating in the anoxic zone. The depth of the oxygen zone varies, depending on the region of the sea and the climatic situation. For example, the concentration of oxygen at a depth of 70 m, in the center of the halistatic region (0.8 mg l^{-1}) is found in the coastal regions only at depths of 150 to 200 m.

Below the oxygen zone lies the redox gradient zone, intermediate between the oxygen and anoxic hydrogen sulphide zones. The redox gradient zone is formed at the lower boundary of the chemocline by intensive chemical and biological oxidation of the reduced compounds that have been produced by anoxic microbial decomposition of organic matter in the hydrogen sulphide zone; therefore, high gradients of redox potential exist in this zone, especially in its upper part (Fig. 11.8, 11.9). The absolute values of Eh in the intermediate zone vary from +150 to −50 mV. The depth of the redox gradient zone and the absolute values of Eh are determined by the intensity of turbulence and of processes of oxidation, which vary during the season, as they are influenced by the hydrological situation, and by the geographical location. In the halistatic centers where the isolines rise to the surface, the intermediate zone is narrow (50–70 m), and lies between depths of 130 to 180 m. The use of new methods to estimate dissolved oxygen in the presence of hydrogen sulphide has made it possible to show the coexistence of both oxygen and hydrogen sulphide in the intermediate zone (Fig. 11.9). Their concentration here is usually around 0.1 to 0.3 mg l^{-1} (Sorokin, 1972). A definite decrease of the Eh gradient, which sometimes occurs in the middle of the intermediate zone (Fig. 11.9), reflects the intensive process of biological oxidation of hydrogen sulphide that is taking place here (see Fig. 11.22).

The upper boundary of the anoxic hydrogen sulphide zone corresponds to the layer where Eh drops below −40 to −50 mV, where the concentration of hydrogen sulphide reaches 0.3 mg l^{-1} and oxygen is nearly absent. Below 250 to 300 m, Eh drops to its minimum values (−170 to −200 mV), and this level of redox continues down to the bottom. The level of the redox potential itself is determined in the anoxic zone by the system: $S^{2-} - 2e \rightarrow S^0$. The measured values of the redox compare well with the calculated ones for this system.

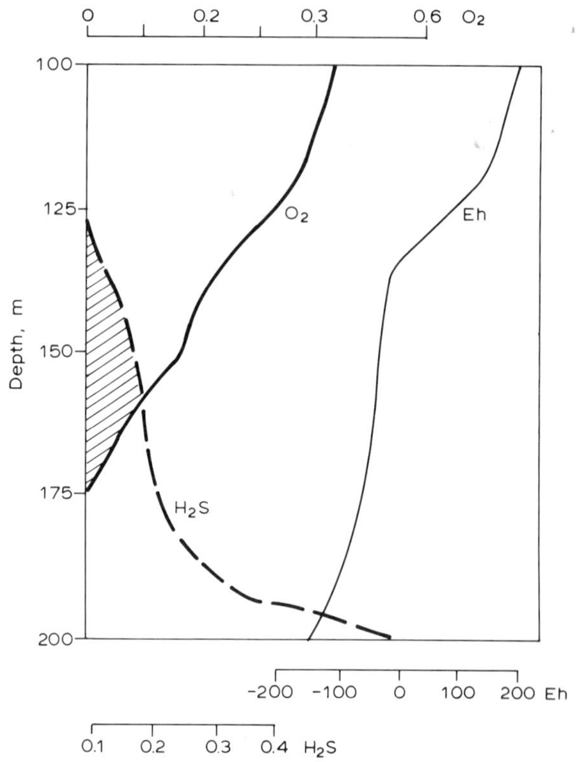

Fig. 11.9. Profiles of hydrogen sulphide concentration (H_2S, mg l^{-1}), oxygen content (O_2, mg l^{-1}), redox potential (Eh, mV) in the west halistatic area (after Sorokin, 1972).

The content of hydrogen sulphide in the anoxic zone increases with depth down to 1500 m, and then stabilizes at the level 9.5 mg l^{-1}. The vertical distribution of hydrogen sulphide coincides well with the distribution of salinity (Skopintzev, 1975). This indicates a common mechanism which regulates the vertical distribution through mixing and turbulence processes (Fig. 11.10). A difference is detectable between the corresponding curves only in the upper part of the hydrogen sulphide zone. It corresponds with the additional formation of hydrogen sulphide via sulphate reduction at the upper boundary of the anoxic zone (see Fig. 11.23). The

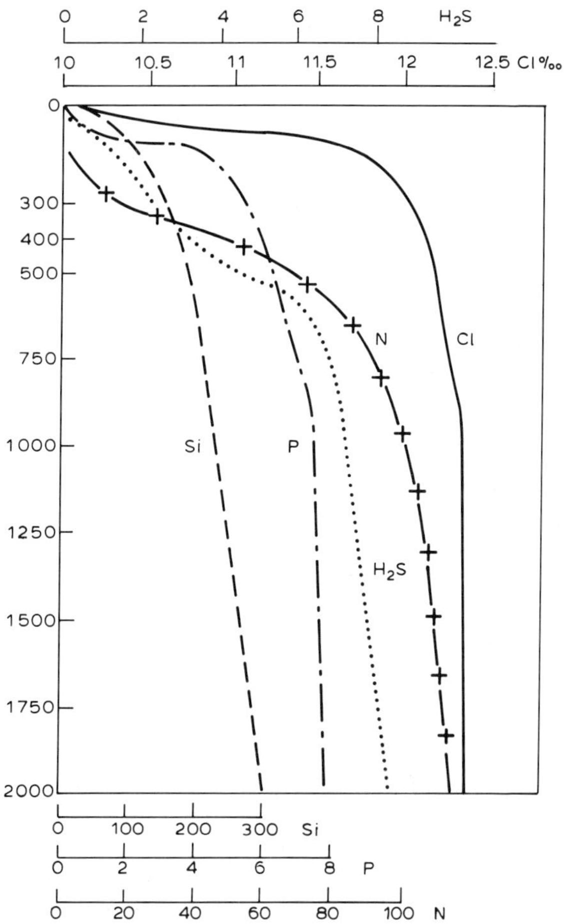

Fig. 11.10. Vertical profiles of salinity (Cl, ‰), H_2S (mg l^{-1}), nutrients: Si, NH_4–N, PO_4–P (μg-at. l^{-1}) in the Black Sea (after Skopintzev, 1975).

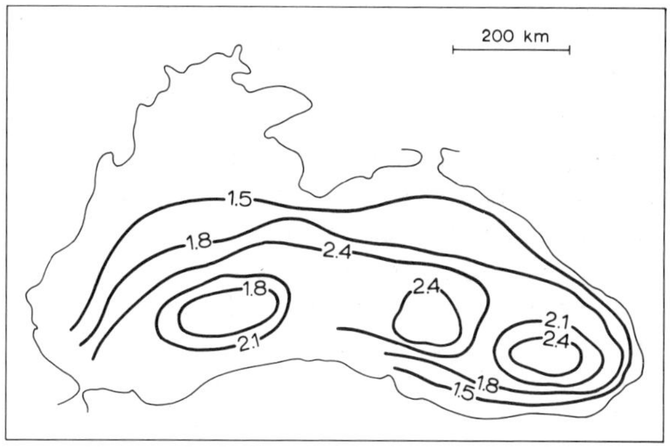

Fig. 11.11. Isolines of hydrogen sulphide concentration (mg l^{-1}) at the depth of 300 m in February 1956 (after Dobrzhanskaya, 1967).

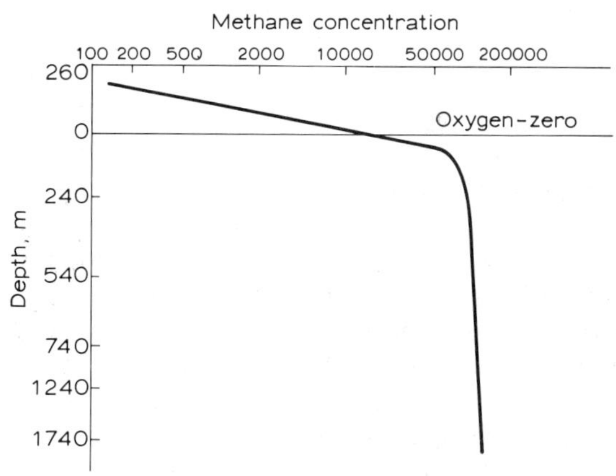

Fig. 11.12. Concentration of dissolved methane (10^{-6} ml l^{-1}) in the water column (after Degens and Ross, 1974).

position of the upper boundary of the anoxic zone varies geographically as well. A "cupola" of sulphide isolines can be observed in the centers of the east and west eddies. As the bed slopes up toward shore, these lines go down (Fig. 11.11) because of processes of downwelling and vertical mixing near the coast.

The main form in which hydrogen sulphide appears in the waters of the anoxic zone is as the hydrosulphide ion SH^- (about 80–90%); the remainder is in the form of the sulphide ion S^{2-}. In addition, some thiosulphate has been found (Bruevich, 1953), formed as an intermediate product of the chemical oxidation of hydrogen sulphide. Among the products of anoxic decomposition and the transformation of organic matter in the anoxic zone are the gases nitrogen, carbon dioxide, methane, ethane and ethylene. The methane was found in concentrations of up to 0.2 ml l^{-1} (Fig. 11.12), the ethane plus ethylene in concentrations of about 0.5 ml l^{-1} (Hunt, 1974). The partial pressure of carbon dioxide in the anoxic zone is about 10 to 17×10^{-4} bar, which is three to four times greater than in the deep oceanic waters. The content of nitrogen gas exceeds 100% level of saturation by 3 to 5%.

Nutrients

The total content, dynamics and vertical distribution of inorganic forms of nitrogen in the oxygen zone of the Black Sea is typical for a mesotrophic inland deep sea, with variations resulting from seasonal development of phytoplankton and the intensity of vertical mixing. But in the anoxic zone their content and distribution deviate from the regime in any ordinary ocean basin. In the Black Sea, ammonium nitrogen accumulates up to concentrations of 1.2 to 1.4 mg l^{-1} (75 to 100 μg-at. l^{-1}), about two orders of magnitude more than in deep oceanic waters, where most of the ammonia is transformed into nitrates during microbial nitrification. Profiles of the ammonium ion, phosphate and silicate distributions in the waters of the Black Sea are given in Fig. 11.10, and in Table 11.5. The accumulation of ammonia in the anoxic zone is the result of anoxic decomposition and ammonification of particulate organic matter settling down from the euphotic zone. The usual oceanic processes of transformation into nitrates in the water column cannot occur in the anoxic zone of the Black Sea, and the ammonium ions that are formed accumulate. Comparison of the curves of the vertical distribution of salinity and of the total inorganic nitrogen, which consists mostly of ammonia (Fig. 11.10), suggests that the ammonium ions are formed mainly at the bottom and are then distributed during vertical mixing (Skopintzev, 1975).

The content of nitrate-nitrogen varies, depending on the season, between zero and 6 μg-at. l^{-1} (average about 1 to 2 μg-at. l^{-1}). Its concentration usually increases in the intermediate redox zone up to 2 to 3 μg-at. l^{-1} as a result of the activity of nitrifying bacteria (Fig. 11.13, Table 11.5). In anoxic conditions below the redox zone nitrates disappear completely, due to the inhibition of nitrification. All traces of nitrates which penetrate into the upper part of the anoxic zone are rapidly

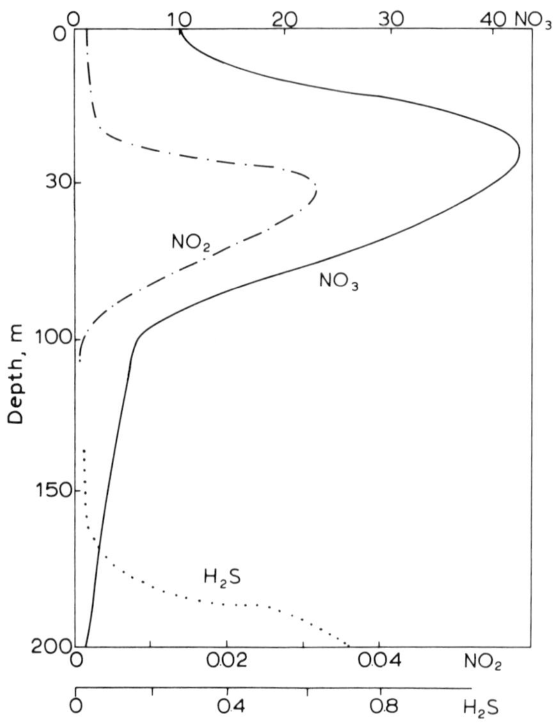

Fig. 11.13. Vertical profiles of NO_3, NO_2 (μg-at. l^{-1}N) and H_2S (mg l^{-1}) in the upper layer of the central part of the Black Sea in October (after Sorokin, 1964a).

TABLE 11.5

Average content of the inorganic nutrients, of total organic carbon and organic nitrogen and phosphorus in water of the Black Sea, vertical profile (N, P, Si as mg-at. l^{-1}; organic carbon as mg C l^{-1}) (based on Bruevich, 1953; Datsko, 1954; Skopintzev, 1975)

Depth (m)	Inorganic nutrients						Organic C, N, P		
	N			$P-PO_4$ (P)	N:P	$Si-SiO_2$	C	N	P
	NH_4	NO_3	NH_4+NO_3 (N)						
1–10	2.8	0.8	3.6	0.4	9	50	3.6	17.0	0.13
25	3.3	0.8	4.1	0.3	14	50	3.6	–	–
50	3.6	0.7	4.3	0.5	9	60	3.5	17.8	0.13
100	5.8	1.8	7.6	1.3	6	80	3.2	16.3	0.50
200	12.8	0.8	13.6	4.0	3	120	3.5	14.2	0.80
300	17.5	0.2	17.7	4.4	4	150	3.0	15.3	0.84
500	58.5	0	58.5	5.6	10	180	2.8	16.0	0.90
1000	81.8	0	81.8	7.0	12	220	2.5	16.0	1.10
2000	96.0	0	96.0	8.4	11	300	2.3	14.8	1.10

denitrified by the thiobacilli, which use the oxygen of nitrates for respiration and reduce the nitrates to nitrites and nitrogen gas. The process of thiodenitrification in the Black Sea has been cited as a cause of the loss of inorganic nitrogen in the upper layers, which results finally in a nitrogen deficit in the euphotic zone (Vodyanitzki, 1948; Sorokin, 1972).

Nitrites usually are formed in the ocean during the microbial nitrification of ammonia and, in part, during the denitrification of nitrates. In the Black Sea the nitrification occurs at the lower boundary of the euphotic zone (50–100 m depth), as is reflected by the vertical distribution of nitrites (Fig. 11.13). The vertical distribution of total inorganic nitrogen reflects the process of the accumulation of ammonia in the anoxic zone. The total content here is more than 80 μg-at. l^{-1}, three to four times greater than that in deep oceanic water (Table 11.5).

The dynamics and vertical distribution of inorganic phosphate in the Black Sea resemble those of the total inorganic nitrogen (Fig. 11.10, Table 11.5). In the euphotic zone the phosphate concentration is low; its level is controlled here by phytoplankton growth and by the intensity of vertical mixing. In the anticyclonic eddy near the city of Batumi where the surface waters sink, the concentration of phosphate decreases to 0.2 μg-at. l^{-1}, while in the cyclonic eddy situated near this anticyclonic eddy the concentration of phosphate can exceed 1 μg-at. l^{-1} (Fig. 11.14). Usually in midwinter this concentration extends to the surface waters; during the summer in the area above the thermocline it drops to concentrations that limit phytoplankton growth (0.2–0.3 μg-at. l^{-1}). The amount of inorganic phosphate at the lower boundary of the euphatic zone is about 1 to 1.3 μg-at. l^{-1}, which is usual for a mesotrophic temperate sea. This distribution of nutrients in the upper water layer of the Black Sea is now changing in some coastal and shallow areas which are subjected to manmade eutrophication. For example, the concentration of phosphates in surface waters along the west coast opposite Constanţa in 1974 to 1975 remained extremely high (3 to 6 μg-at. l^{-1}) even in summer and was not consumed by the intensively blooming phytoplankton (Cociasu and Popa, 1976). In the deeper anoxic layers it increases up to 7 to 8 μg-at. l^{-1}. Thus the content of phosphate phosphorus in the anoxic zone is three to four times greater than in the deep waters of oceans and of other deep inland seas, which is 2 to 3 μg-at. l^{-1} (Skopintzev, 1975). These data indicate that phosphates accumulate in the anoxic zone of the Black Sea, as does inorganic nitrogen, which is typical for stagnant anoxic waters of meromictic basins (Richards et al., 1965).

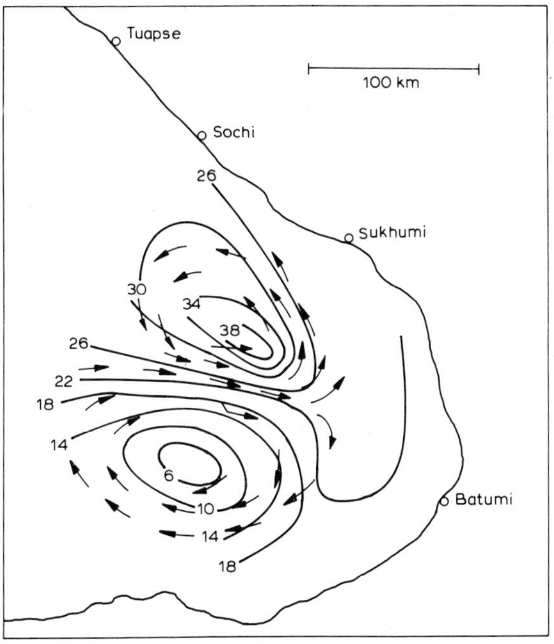

Fig. 11.14. Isolines of phosphate concentrations (μg P l^{-1}) and direction of current at 100 m depth in the southeastern part of the Black Sea (after Bibik, 1964).

The quantity of silicate in the euphotic zone of the Black Sea is sufficient to maintain a normal growth of phytoplankton; it varies within the limits of 0.8 and 1.5 μg-at. l^{-1}. In deep waters it increases up to 7 to 9 μg-at. l^{-1}, which is two to five times greater than in deep oceanic waters.

The general level of the nutrients in the Black Sea waters in the euphotic zone, at its lower boundary, is similar to that in mesotrophic temperate marine basins. The vertical and horizontal motions of the water and the seasonal mixing in the upper layers are intensive enough to provide a flow of nutrients from the deep layers through the discontinuity zone. Therefore, former opinions about the oligotrophy of the pelagic areas are not correct (Vodyanitzki, 1948). The vertical distribution of the inorganic nitrogen, phosphorus and silicon is in good agreement with the salinity. The main source of these nutrients is at the bottom in the processes of organic matter mineralization; they are transferred up by vertical turbulence. The ratio N:P in the vertical profile (Table 11.5) has some specific features in the Black Sea; it decreases at the upper boundary of the anoxic zone because of the loss of inorganic nitrogen via a thiodenitrification process (Sorokin, 1972). The concentrations of nutrients present in the anoxic zone might be used in a future mariculture industry by bringing water from the anoxic zone to rear algae or purple bacteria.

Organic matter

The quantity of dissolved organic matter in the surface layers of the Black Sea varies within the limits 3.2 to 3.7 mg C l^{-1}, and decreases gradually with depth down to 2.2 to 2.6 mg C l^{-1} between 1000 and 2000 m depth (Datsko, 1959; Skopintzev, 1975). Organic forms of nitrogen and phosphorus accumulate in the deep waters, as do their inorganic forms (Table 11.5); the amount of organic nutrients in deep waters is about 15 μg-at. l^{-1} N and 1 μg-at. l^{-1} P. These concentrations of organic carbon, nitrogen and phosphorus exceed by 1.5 to 2 times the amounts in deep oceanic waters. The high organic concentrations in the deep layers of the Black Sea result from the inflow from the rivers of waters enriched with terrestrial organic matter, mostly stable humic compounds.

The composition of dissolved organic matter does not seem to change with depth. The content of total carbohydrates was found to be 1.5 to 2 mg l^{-1}, that of the total amino acids 0.15 to 0.2 mg l^{-1}, of the fatty acids 0.1 to 0.2 mg l^{-1}, and of vitamin B_{12} 0.1 to 0.5 mg m^{-3}.

Suspended organic matter in the upper 200 m is about 10% of the total organic carbon (0.2–0.3 mg C l^{-1}). At the lower boundary of the oxygen zone (110–150 m) in most parts of the Black Sea there exists a layer of a very dense accumulation of particulate organic matter composed mostly of faeces and of decaying phytoplankton (Neuymin, 1973; Fig. 11.15). The total amount of suspended material increases with depth, while the content of the particulate organic matter decreases 1.3 to 1.5 times, as occurs in the deep waters of the oceans.

An approximate balance of the organic matter in the Black Sea (g C m^{-2} yr^{-1}) can be summarized as follows: gains of 100 g of primary pelagic production and 7 g exchange with the adjacent seas and terrestrial input; losses of 80 g by mineralization in the oxygen zone, 10 g by mineralization in the anoxic zone, 4 g by outflow to adjacent seas, 6 g by enrichment of stock of dissolved organic matter, and 7 g by sedimentation (modified from Deuser, 1971).

Bottom sediments

During the last 25 000 years three main sediment layers have been deposited in the Black Sea. The recent (0–3000 years B.P.) layer of microlaminated sediments is enriched with carbonate and is about 30 cm thick in the deep parts of the sea. The second layer of old Black Sea sapropel sediments is enriched with organic matter (the sapropel bed), and was formed between 3000 to 7000 years ago. The third layer is composed of the lutite sediments deposited during the Neoeuxine period (7000–25 000 years B.P.) when the basin was a great fresh-water lake (Strakhov, 1947).

The upper sediments in the central parts of the sea are very fine particles. Microphotographs show that the surface of the bottom is covered with a thin layer of fine black sediment in which the processes of microbial sulphate reduction take place. The slope sediments contain a lot of coarse carbonate material formed by benthic animals, mostly by bivalve mollusks. The distribution of the main types of modern sediments is shown in Fig. 11.16.

The most important components of the sediments are organic matter and sulphur compounds. The content of the organic carbon, as calculated per dry sediment, varies between 1 and 4% (Table 11.6).

The main form assumed by accumulated reduced sulphur is pyrite (FeS_2), which is formed from iron monosulphide hydrotroilite ($FeS{\cdot}H_2O$) reacting with molecular sulphur ($FeS + S^0$). In the deep sea sediments pyrite comprises more than 90% of the total reduced sulphur. The black color of the sediments results from hydrotroilite. It accumulates in the slope sediments, where the rate of sulphate reduction is at its maximum. (Ostroumov et al., 1961).

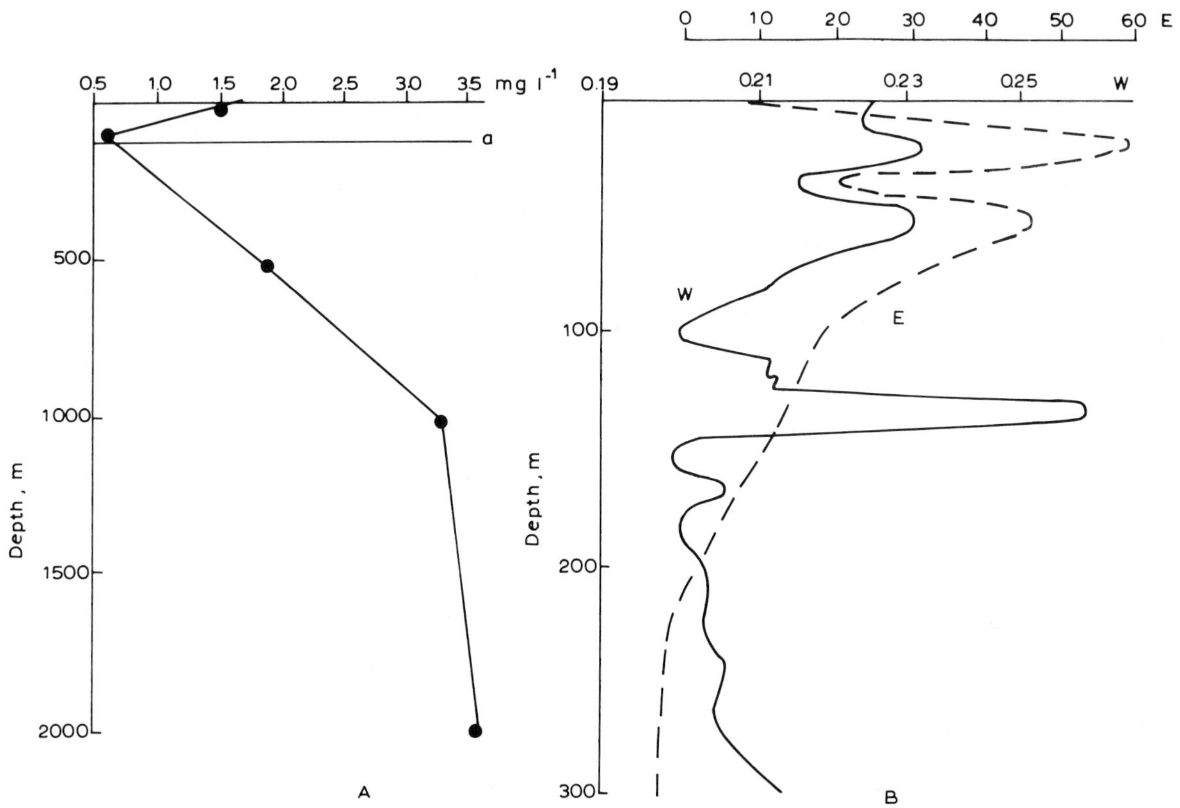

Fig. 11.15. Suspended organic matter: A. In total water column; $a = H_2S/O_2$ interface (after Shimkus and Trimonis, 1974). B. In the upper layer; W=distribution of suspended particles in arbitrary units; $E\ (\times 10^6)$=coefficient of vertical stability of water mass (where $E = 10^{-3}\ d\sigma_1/dz$) (after Neuymin, 1973).

PHYTOPLANKTON AND PRIMARY PRODUCTION

The planktonic algal flora of the Black Sea include more than 200 species of Bacillariophyta, more than 150 species of Pyrrophyta, about 80 species of Chlorophyta, about 30 species each of Chrysophyta and of Cyanophyta, 13 species of Euglenophyta, and 3 species of Xanthophyta. The composition of the phytoplankton has a pronounced similarity to the phytoplankton of other inland temperate seas and fjords, such as the North and Caspian Seas, and Norwegian fjords. Most of the phytoplankton biomass is formed by diatoms which occur in their marine euryhaline neritic forms in the Black Sea. The single oceanic species is *Planktoniella sol*. In the coastal regions some freshwater or halophilic species are present. More than 80% of the diatom species occur in their centric forms, and among them the *Chaetoceros* species are most abundant. The species of diatoms which appear during the spring (February–April) and the autumn phytoplankton maxima are several species of *Chaetoceros* (*C. affinis*, *C. compressus*, *C. curvisetus*, *C. danicus*, *C. rigidus*, *C. socialis*); and several other diatom species including *Cerataulina bergonii*, *Cyclotella caspia*, *Nitzschia seriata*, *Rhizosolenia alata*, *R. calcar-avis*, *R. fragilissima*, *Skeletonema costatum*, *Thalassionema nitzschioides* (Morozova-Vodyanitzkaya, 1950, 1957; Proshkina-Lavrenko, 1955; Bodeanu, 1969; Kovaleva, 1969). The species composition of phytoplankton is regulated largely by temperature and salinity (Pyzik, 1968). The range of euryhalinity of most of the diatom species is very wide. Some of these marine euryhaline forms can flourish within a salinity range of 6 to 18‰ (*Skeletonema costatum*, *Chaetoceros halistaticus*, *C. subtilis*). A large seasonal change of temperature from 0 to 6°C in winter up to

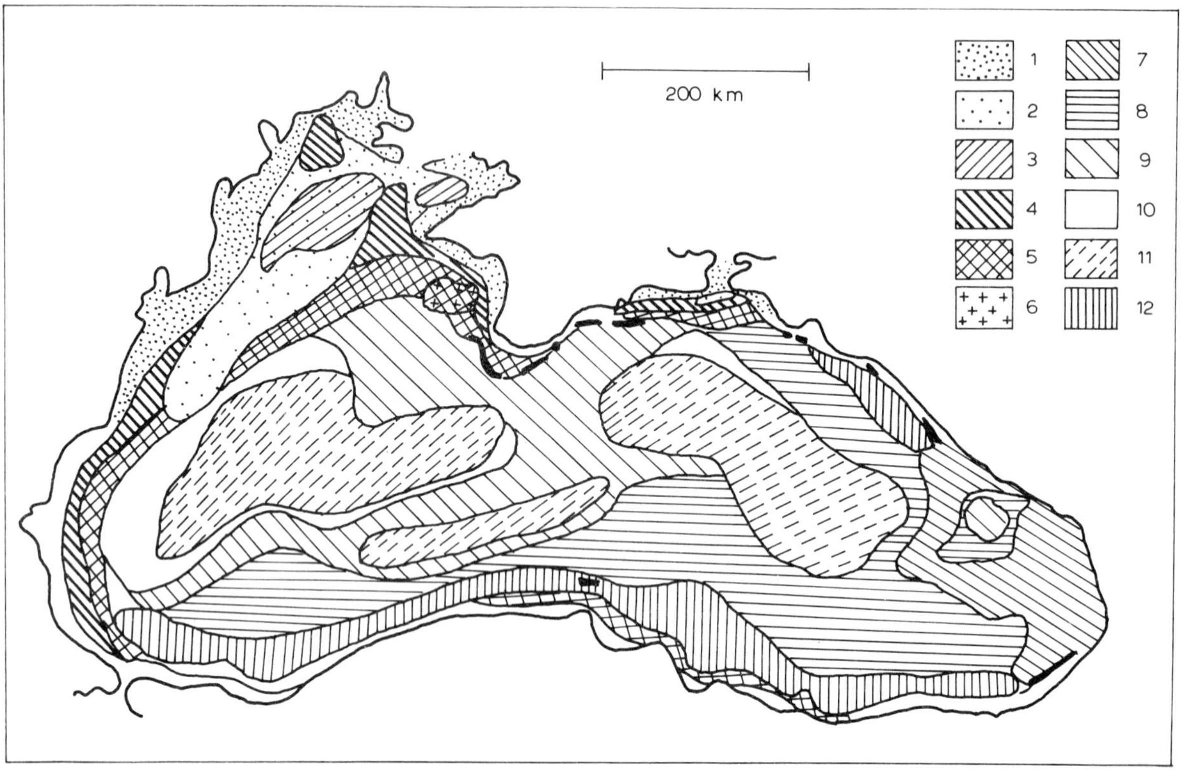

Fig. 11.16. Types of modern Black Sea sediments. *Shallow-water sediments*: *1* = organogenic–clastic coarse-grain sediments; *2* = shell-containing sediments; *3* = sediments covered with alga *Phyllophora*; *4* = organogenic–terrigenous mytilid muds; *5* = iron–manganese nodules area. *Deep-sea sediments*: *7* = terrigenous carbonate-free sediments; *8* = organogenic–terrigenous muds; *9* = finely dispersed muds; *10* = finely dispersed coccolith muds; *11* = coccolith muds enriched with organic matter; *12* = diverse sediments with predominance of organogenic–terrigenous sediments. (From Shimkus and Trimonis, 1974.)

25°C in summer affects the species composition. In winter arctic and subarctic eurythermic forms develop, such as *Chaetoceros danicus*, *C. densus*, *C. subsecundus*, *Nitzschia seriata*, *Skeletonema costatum*. In summer subtropical and even tropical forms of diatoms flourish, such as *Chaetoceros diversus*, *C. peruvianus*, *Coscinodiscus janischii*. Depending on the annual thermal conditions, coccolithophorids develop during late spring, represented mostly by *Coccolithus huxleyi*. During the summer

TABLE 11.6

Content of organic carbon and of reduced sulphur compounds in the surface layer (0–5 cm) of bottom sediments collected at various depths in the Black Sea (modified from Ostroumov et al., 1961)

Depth (m)	$C_{org.}$ (% of dry weight)	Reduced sulphur compounds (mg kg^{-1} of wet sediment)				
		free hydrogen sulphide (H_2S)	molecular sulphur, (S°)	thiosulphate ($S_2O_3^{2-}$)	hydrotroilite (FeS)	pyrite (FeS_2)
150	0.85	0	430	0	140	1420
300	1.51	39	380	6	150	2760
320	4.62	103	210	16	440	2180
470	1.88	3	180	–	270	2430
910	5.35	0	290	–	1330	320
1620	2.96	4	2	3	40	3490
2220	1.36	20	290	15	40	3080

populations of dinoflagellates sometimes flourish, represented by *Exuviaella compressa*, *E. cordata*, *Gymnodinium simplex*, *G. variabile*, *Peridinium trichodium*, *P. globulus*, *Prorocentrum micans*. The small phytoflagellates are an important component of the Black Sea phytoplankton; they flourish abundantly during the late spring or early autumn, especially in the neritic areas, often exceeding 1×10^6 cells l^{-1} (Roukhiyainen, 1975).

The quantity and biomass of phytoplankton in the Black Sea is typical for an inland deep mesotrophic basin. The amount and biomass can change during the seasonal succession from 10×10^6 cells l^{-1} and 2 to 3 g m^{-3} down to 10^4 cells l^{-1}, and biomass 10 to 20 mg m^{-3}. In the western coastal areas of the sea, which are subjected to eutrophication processes, dense blooms of phytoplankton were observed in 1974 to 1975. The main species was *Exuviaella cordata*. The blooms began in the spring, extended into the summer and reached the level of a red tide. The biomass of blooming *Exuviaella* ranged from 20 to 80 g m^{-3} (Bodeanu and Roban, 1975).

The character of the seasonal succession of the phytoplankton community in the Black Sea is also typical for a temperate marine basin. There are usually two main maxima: a larger one in early spring, February to April, and a lesser one in autumn, August and September. The biomass of algae during the spring maximum varies within ranges of 1 to 3 g m^3 in the coastal regions and in the shallow northwest part of the sea. In pelagic areas it is about 0.2 to 0.6 g m^{-3}. During the early summer minimum it drops to 10 to 50 mg m^{-3} (Fig. 11.17). These numbers (Pyzik, 1968) refute former opinions about the oligotrophy of the Black Sea, views which possibly were the result of the preference by planktonologists of taking samples largely during the good weather season, which usually coincides with the natural phytoplankton minima.

Reliable data on the annual primary production of this basin are still lacking because existing sampling has missed the main seasonal maximum of phytoplankton growth, late winter–early spring. Estimations of phytoplankton productivity in the Black Sea have been made using the ^{14}C technique in August and October by Sorokin (1964a), and in September by Finenko (1967). The results are summarized in Fig. 11.18. They show that, during

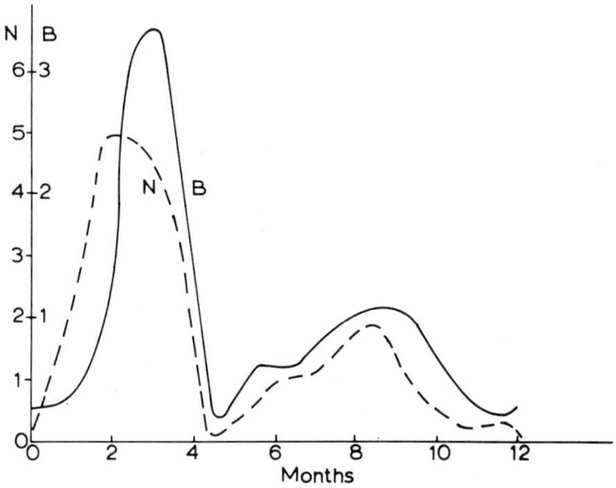

Fig. 11.17. Seasonal dynamics of phytoplankton in the Black Sea: N = number of cells ($\times 10^5$ l^{-1}) opposite the Crimea; B = biomass (g m^{-3} wet weight) in the northwestern part of the Black Sea (after Pyzik, 1968; and Roukhiyainen, 1975).

this period of the second phytoplankton maximum, considerably smaller than the spring one, the most productive areas were the coastal regions and the shallow northwest part of the basin where the vertical mixing processes were most active. Another cause of high productivity in this area, 0.5 to 2.0 g C m^{-2} day^{-1}, is the influence of inflow from rivers enriching the surface layers with nutrients. Although it occupies only 25% of the total area of the basin, more than 60% of the primary production is formed here during August through October. Relatively high production was also recorded in the area between the east and the west cyclonic eddies opposite the Crimea. In the central parts of these eddies, which occupy about half of the total area of the basin, only 20% of the primary

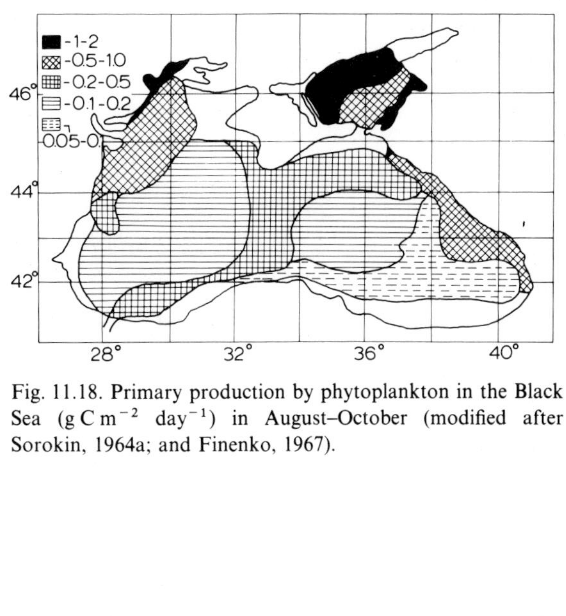

Fig. 11.18. Primary production by phytoplankton in the Black Sea (g C m^{-2} day^{-1}) in August–October (modified after Sorokin, 1964a; and Finenko, 1967).

organic matter was produced. The level of productivity here was as low as 0.1 to 0.2 g C m^{-2} day^{-1}.

Calculations of the annual production of phytoplankton in the Black Sea based upon these measurements gave a value of about 120 g C m^{-2} yr^{-1}. Datsko (1961) calculated a comparable number from the average of phytoplankton biomass (180 mg m^{-3}, wet weight), accepting an annual productivity/biomass ratio (P/B) of 300 yr^{-1}. These results certainly underestimate the primary production of the Black Sea by a factor of two. Direct measurements underestimate it by 1.3 to 1.5 times because of errors in the ^{14}C method itself (Sorokin, 1971). These measurements also lack data on the seasonal phytoplankton maxima and on the local vertical profiles during sampling. Moreover, the estimations by Sorokin (1964a) were made in autumn 1960 when, according to Pyzik (1968), the phytoplankton biomass was two to three times smaller than usual for this time of the year. At present the actual primary production in the Black Sea could be approximately estimated as 200 g C m^{-2} yr^{-1}; the productivity of the coastal and the northwest parts should be evaluated as 250 g C m^{-2} yr^{-1}, while that of the central parts as 150 to 170 g C m^{-2} yr^{-1}.

The depth of the euphotic zone in the open sea is generally 50 to 60 m (Fig. 11.19). The optimum submarine illumination in summer varies from 5 to 10 m in depth, while the maximum rate of photosynthesis occurs 10 to 20 m deeper in the water column (Fig. 11.19B). This is caused by the accumulation of much of the active phytoplankton population in the upper part of the thermocline (Fig. 11.19C), resulting in increased turbidity in this layer.

It is unlikely that microbial photosynthesis occurs in the redox gradient zone of the Black Sea because of insufficient light at the upper boundary of the anoxic zone where the corresponding microflora function. Even when this boundary layer (Eh ≈ -20 mV) is situated as high as 130 m depth, only about 0.005% of the surface light penetrates to it. An attempt by the author to measure microbial photosynthesis in the central part of the western area gave a negative result.

MICROBIAL POPULATIONS

The Black Sea has long been regarded as the "microbial sea" because of the great importance of microbial processes in the formation of its present

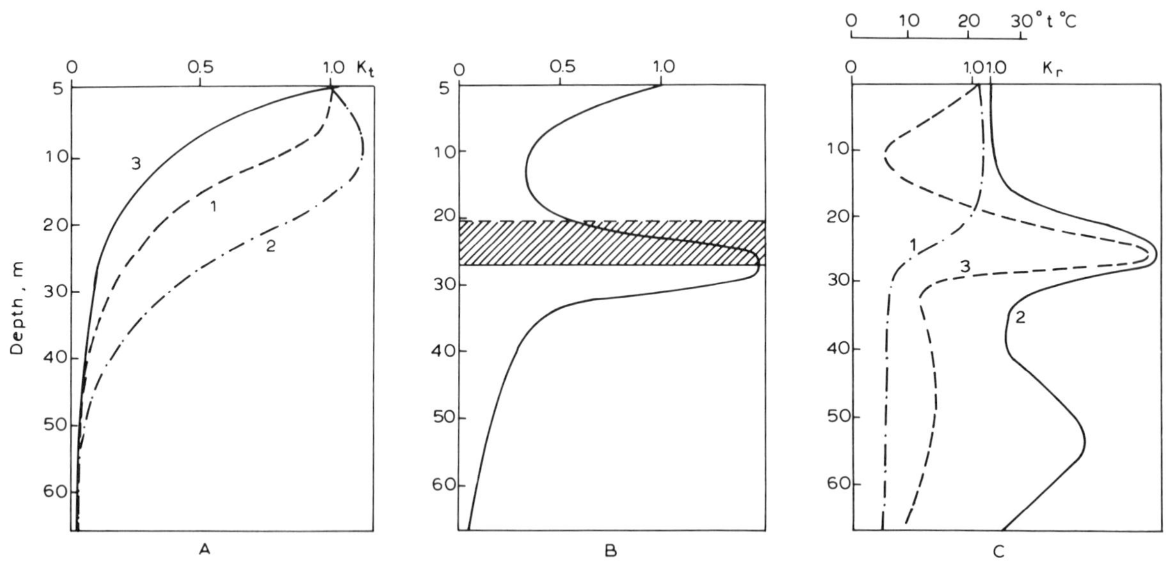

Fig. 11.19. A. Photosynthesis and submarine illumination relative to the values observed at the surface: *1* = phytoplankton from the surface bottled and resuspended at various depths; *2* = phytoplankton from 100 m bottled and resuspended at various depths; *3* = submarine illumination. B. The relative rate of photosynthesis (compared to surface samples) of phytoplankton collected from various depths and resuspended at the same depth. The position of the thermocline is shown by cross hatching. C. *1* = The vertical distribution of temperature (°C); *2* = the relative turbidity of the water; *3* = relative phytoplankton abundance. (After Sorokin, 1964a.)

regime; it supports bacteria in more than 80% of its volume, often where other organisms cannot survive. Extensive studies of different aspects of the composition and functioning of microbial populations have been conducted in this basin for eighty years, and now the Black Sea is probably one of the marine basins most carefully studied in this respect.

Amount, biomass and productivity

The total amount of bacteria and their biomass in the sea have been measured by direct microscopical counts and sizing on stained membrane filters. Abundant data in this field and also on plate counts were reported by Kriss (1963), but, as it was shown later (Bogoyavlenski, 1962; Sorokin, 1964b, 1968), these measurements were unreliable because of his use of the metallic Nansen bottles for microbiological sampling. Such data concerning direct microscopic but not plate counts would be more or less acceptable for the oxygen zone, but they are not credible for the anoxic zone where the surface of the bottles is corroded by the hydrogen sulphide. Particles washed out from the corroding walls contaminate the filters and cause the appearance of artifacts during microscopic counts. Therefore, the data by Kriss (1963) and his assistants, Lebedeva and Markianovich (1958), here are used only for the oxygen zone.

The total number of bacteria in the euphotic zone of the open sea usually varies between 30 and 200×10^3 cells ml^{-1}, and their biomass from 0.5 to 5.0 mg C m^{-3} (Table 11.7). In coastal regions and in the shallow northwest part of the sea it can increase up to 0.5×10^6 cell ml^{-1} (biomass up to 20 mg C m^{-3}); along cross sections from the coast to the open sea it decreases rapidly to a level that is characteristic of the open sea (Fig. 11.20). The

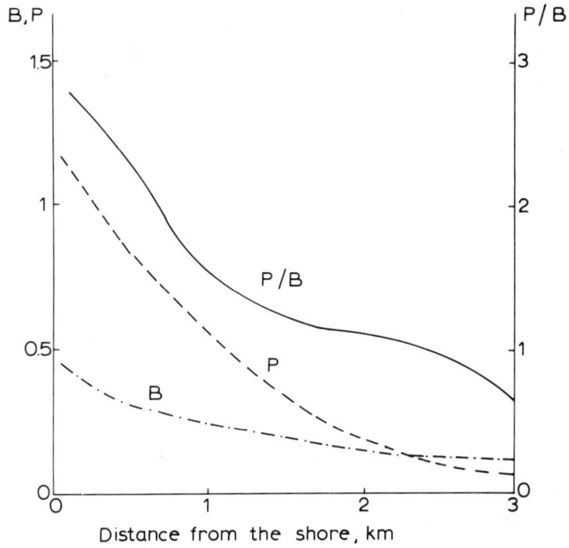

Fig. 11.20. The changes in the density of microbial population and in its production at the section from the shore to the open sea opposite the Crimea in August: B=microbial biomass (g m^{-3}); P=production per day (g m^{-3} day^{-1} wet weight); and P/B ratios.

microbial number and biomass usually decrease during the winter, but sometimes in the center of cyclonic eddies where water rises from the intermediate redox gradient zone, the microbial biomass increases as it is enriched with chain form (filamentous) bacteria which inhabit the redox gradient zone (Lebedeva, 1959; Sorokin, 1964a).

Most cells of planktonic bacteria in the oxygen zone are small rods $0.3 \times 1.0 \times 1.5$ μm, and coccoid forms 0.5 to 1.0 μm diameter. Often they accumulate into aggregates. Occasionally spirillae and filamentous forms have been found, especially in the surface film layer. In samples collected from this layer with a special device, an abundance of microflora was found (Tsiban, 1970). Studies of this layer

TABLE 11.7

Ranges of the total numbers, wet biomass and production of planktonic bacteria in the main biotopes of the Black Sea (original)

Biotope	Cells (10^3 ml^{-1})	Biomass (mg m^{-3})	Production (mg m^{-3} day^{-1})
Inshore areas	300–1500	200–700	100–500
Shallow area of NW part	150–500	50–200	30–100
Area near the Bosporus	100–300	30–150	20–100
Open deep sea, oxygen zone	50–200	20–80	15–40
Same, intermediate redox gradient zone	300–500	100–250	60–100
Same, anoxic zone	20–40	5–10	0.1–0.3

of hyponeuston (Zaitzev, 1970) showed that the total number of bacteria and their wet biomass here reached 2 to 20×10^6 cells ml^{-1}, and 1 to 10 mg l^{-1}, which is one to two orders of magnitude more than that present in the underlying water layers (Table 11.8). A second maximum of microbial biomass within the oxygen zone is situated in the layer of the thermocline (Fig. 11.21), where biologically active layers usually form in stratified basins (Sorokin, 1964a, 1971; Kriss and Lebedeva, 1968).

The basic maximum of the microbial biomass always appears in the redox gradient zone of vertical profiles, at depths of 140 to 180 m in the open sea. It is formed by active populations of chemoautotrophs oxidizing reduced sulphur compounds and methane. The total number of bacteria in this zone, which is also the zone of maximum chemosynthesis (Fig. 11.21, Table 11.9), reaches 0.25 to 0.5×10^6 ml^{-1}, and its biomass 10 to 15 mg C m^{-3}. The bulk of microbial cells are present here in their chain forms, forming long filaments; these are the thiobacilli (Sorokin, 1972).

In the anoxic zone the microbial biomass decreases to about one tenth of the biomass in the intermediate redox gradient zone. In deep waters the same chain-filamentous microbial forms are common; they settle down from the intermediate zone where they developed, or they are adsorbed by settling particles of debris which pass through this zone. Generally, the total microbial quantity and biomass in the upper layer of the Black Sea is similar to that of the surface mesotrophic waters of temperate seas.

The microbial production in the oxygen zone in summer is 5 to 20 mg m^{-3} day^{-1} (wet weight) in the open sea. The coefficient of productivity per unit biomass (P/B) of the total microbial population

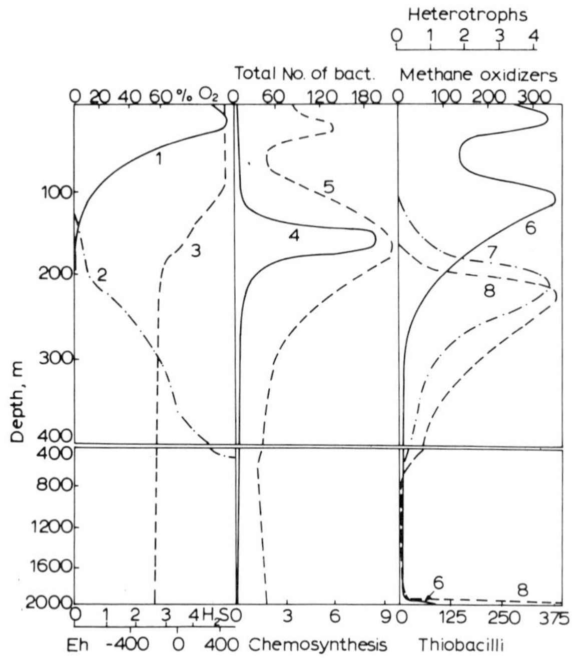

Fig. 11.21. Vertical distribution of microbial populations in the central part of the sea: *1*=O_2 (% of saturation); *2*=H_2S (mg l^{-1}); *3*=Eh (mV); *4*=chemosynthesis rate (mg C m^{-3} day^{-1}); *5*=total number of bacteria ($\times 10^3$ ml^{-1}); *6*=number of plate-counted heterotrophic bacteria ($\times 10^2$ per 100 ml); *7,8*=relative activity of methane-oxidizing bacteria and of thiobacilli (chemosynthesis μg C l^{-1}). (After Sorokin, 1964a, 1972.)

varies within a range of 0.5 to 1.0 day^{-1}. In coastal areas and in the shallow northwest part of the sea, the microbial production increases to 20 to 100 mg m^{-3} day^{-1}, and near the shore to 1000 mg m^{-3} day^{-1} (Table 11.9, Fig. 11.20). Within the water column in the open sea, the permanent maximum of microbial production is in the intermediate redox gradient zone at the hy-

TABLE 11.8

Numbers (per 40 ml) of plate-counted heterotrophic bacteria in surface film and in the water column in the northwestern part of the Black Sea in summer (modified from Tsiban, 1970)

Depth (m)	Positions of stations			
	opposite the mouth of Danube River	opposite Odessa	opposite Dnestr Bay	opposite the mouth of Bug River
Surface film	1330	28 800	44 800	1720
0.5	520	20	92	62
10	420	20	240	130
25	120	130	120	130

TABLE 11.9

Distribution of redox potential (Eh) of chemical and microbiological parameters in the water column of the central part of the Black Sea; microbial biomass and production are given as wet weight (modified from Sorokin, 1964a, 1972; and Skopintzev, 1975)

Depth (m)	Redox characteristics			Microbial populations								
				total number (*N*), biomass (*B*), chemosynthesis production (*Ch*) and total production (*P*)					specific groups of bacteria			
									plate counts (numbers per 10 ml)		relative activity (% of maximum)	
	hydrogen sulphide ($mg\ l^{-1}$)	oxygen ($mg\ l^{-1}$)	Eh (mV)	*N* ($10^3 ml^{-1}$)	*B* ($mg\ m^{-3}$)	*Ch* ($mg\ m^{-3}\ day^{-1}$)	*P* ($mg\ m^{-3}\ day^{-1}$)	*P/B* (day^{-1})	hetero-trophs	sulphate reducers	thiobacilli	methane oxidizers
0	0	5.60	+360	70	14	0	7	0.5	30	0	0	0
50	0	7.8	+380	120	24	0	12	0.5	10	0	0	0
100	0	0.80	+300	50	10	0	3	0.3	20	0	2	0
150	0.05	0.16	+10	170	60	8	12	0.2	5	0	10	0
160	0.07	0.10	−10	490	250	42	45	0.2	25	0	40	10
170	0.15	0.08	−20	380	200	68	70	0.4	15	10	60	50
180	0.30	0.05	−30	260	120	25	30	0.3	10	20	80	100
200	0.60	0	−40	140	65	12	15	0.2	5	35	90	80
250	1.40	0	−100	110	45	5	8	0.2	3	45	100	30
300	2.20	–	−125	40	15	0	3	0.2	1	20	30	10
500	5.10	–	−160	30	12	0	3	0.3	0.5	0	20	5
1000	8.40	–	−170	20	8	0	2	0.3	0.2	0	2	0
1500	9.60	–	−190	20	8	0	2	0.3	0.2	0	0	0
2000	9.50	–	−190	30	9	0	2	0.2	0.5	45	0	0

drogen sulphide–oxygen boundary layer (140–180 m depth) where Eh is around 20 to 40 mV (Table 11.9, Fig. 11.21). About 95% of this production (usually 40–80 mg m^{-3}) results from the chemosynthesis of thiobacilli, and some from the methane-oxidizing bacteria (Sorokin, 1964a, 1972). The depth of the maximum rates of chemosynthesis, as measured with $C^{14}O_2$, coincides well with the maximum rate of hydrogen sulphide oxidation measured with labeled sulphide (Fig. 11.22). The values of P/B coefficients here reach 2.0 to 2.5 per day. This intensive production of microbial biomass in the intermediate zone is a characteristic feature of all meromictic basins (Sorokin, 1970).

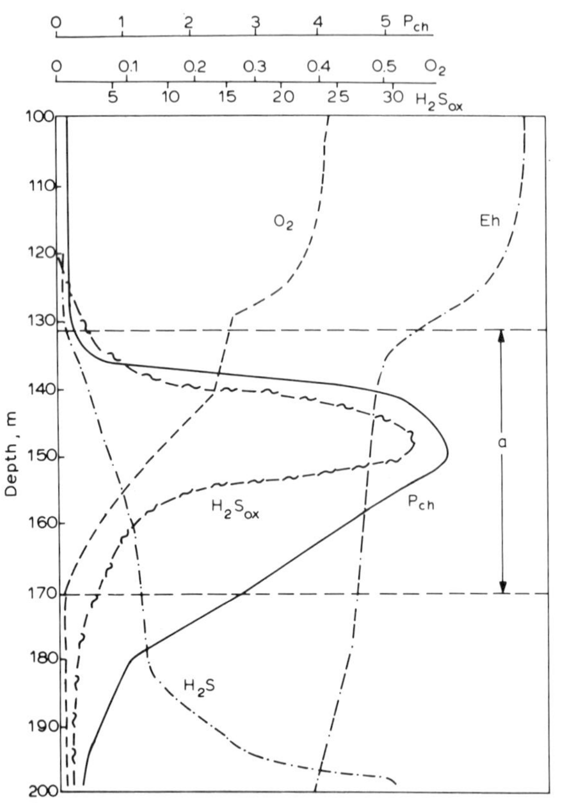

Fig. 11.22. Processes of chemosynthesis and hydrogen sulphide oxidation in the redox gradient zone in the open sea opposite Crimea: O_2 (mg l^{-1}); H_2S (mg l^{-1}); H_2S_{ox} = rate of H_2S oxidation (μg S l^{-1} day^{-1}); P_{ch} = rate of chemosynthesis production (mg C m^{-3} day^{-1}); Eh (mV); a = layer of coexistence of O_2 and H_2S. (After Sorokin, 1972.)

In the anoxic zone the rate of microbial production is negligible because of the lack of hydrogen acceptors and of toxic action by hydrogen sulphide. In this zone it is two orders less than in the redox gradient zone (0.3 to 0.5 mg m^{-3} day^{-1} of wet biomass). The coefficient P/B here is extremely low (0.02), which results from the accumulation in the anoxic zone of starving and inactive microbial cells.

Specific groups of bacteria in water column

The amount of heterotrophic bacteria (counted on broth agar plates) in the surface water layers of the sea varies from 0.3 to 3.0 colonies ml^{-1} in the pelagic regions, and up to 50 to 500 cells ml^{-1} in summer in the shallow coastal areas (Tsiban, 1970; Sorokin, 1972). They are especially abundant in the layers of hyponeuston where they can exceed 10^3 cells ml^{-1} (Table 11.8). The vertical distribution of plate counted heterotrophs is characterized by two maxima, ne at the surface and the second in a layer near the lower boundary of the oxygen zone at 100 to 110 m where there is an accumulation of detritus consisting of the remains of phytoplankton and faeces. In the anoxic zone they decrease sharply (Fig. 11.21).

The vertical distribution of thiobacilli and methane-oxidizing bacteria, which participate in the oxidation of the products of the anaerobic decomposition occurring in the anoxic zone, is given in Fig. 11.21. The maxima of activity and of chemosynthesis production were observed at the lower boundary of the redox zone, 180 to 200 m (Fig. 11.22).

Several hundred cultures of bacteria were isolated from Black Sea water, and their species composition and biochemical features, such as antibiosis and the vitamin B_{12} auxotrophy, have been studied (Lebedeva, 1971). About half of the cultures studied display antibiosis to the microscopic fungi *Candida* or *Fusarium*. Numerous studies on the microbiology of detritus and periphytonic bacteria (Khailov, 1971; Finenko, 1971; Gorbenko, 1977) describe the role of dissolved organic matter in sea water on the metabolic interconnections between planktonic and benthic algae on one side and the microbial populations of periphyton and detritus on the other.

Studies of nitrogen-fixing bacteria in the Black

Sea were made by Psenin (1966), who isolated several genera from water and sediment samples. Among them there were different species of *Azotobacter* (*A. chroococcum*, *A. beijerinckii*, *A. miscellum*, *A. nigricans*), *Clostridium*, *Pseudomonas*, *Vibrio*, *Spirillum*, *Treponema* and *Rhodotorula*. A new species of marine planktonic spirochaete (*T. reponema hyponeustonicum*) was described, which is able to grow in a nitrogen-free medium. The most active nitrogen fixers appeared to be *Spirillum magnum*, *Treponema* sp. and *Azotobacter* sp. (Table 11.10). They were found extensively in the water column, but they usually occurred in the surface film. The nitrogen fixers, and especially *Azotobacter*, were abundant in detritus particles and at the surface of decaying seaweeds.

Mironov (1969, 1971) has studied the microbial decomposition of oil in the waters and sediments of the Black Sea. He isolated 850 strains of bacteria capable of oxidizing oil and its derivatives from different biotopes of the sea; the bacteria from the *Pseudomonas* group predominated. The most widespread species were *P. sinuosa* and *P. desmolyticum*. The second most common group was the genus *Pseudobacterium*, and especially *P. biforme* and *P. cocciformis*. Among the oil-oxidizing species of the genus *Bacterium*, *B. albus* was the most common. The greatest amount of oil-decomposing microflora has been found in the shallow waters near ports. An estimate of 10 to 1000 cells ml^{-1} was made, using the minimum dilution method with a medium containing diesel fuel as the single source of organic matter. In the open sea the quantity decreased by one to two orders of magnitude. An extremely rich oil-oxidizing microflora is present in the surface layer of crude sediments near ports, where counts reach 10^4 to 10^6 g^{-1} of sediment.

A comparison of the total amount of heterotrophic bacteria, estimated by the method of minimum dilution in seawater with peptone, with the quantity of oil-oxidizing bacteria showed that the latter comprise from 20 to 30% of the total microbial population in shallow areas subjected to permanent oil pollution.

TABLE 11.10

Nitrogen-fixing activity of microorganisms from the Black Sea (modified from Psenin, 1966)

Species isolated from the sea water	N_2 fixed (mg)	
	per 100 ml of medium	per 1 g of glucose used
Azotobacter chroococcum	12.8	6.6
Azotobacter miscellum sp. nov.	12.7	6.4
Treponema hyponeustonicum sp. nov.	18.4	9.3
Spirillum magnum sp. nov.	32.2	16.1
Spirillum nana sp. nov.	9.6	4.8
Vibrio nonhydrosulphureus sp. nov.	7.0	5.7
Vibrio frequens	2.9	2.0
Clostridium pasteurianum	0.8	1.0
Pseudomonas radiobacter	0.7	2.3

Microflora of bottom sediments

Data on microbial life in the bottom sediments are given in Table 11.11. The most plentiful and active microbial populations inhabit the sediments in the inshore zone (100–300 m depth), where Eh does not decrease below −150 mV, and where some electron acceptors (free oxygen, nitrate ions) penetrate as a result of turbulence from the upper layers. The total number of bacteria here was 3 to 5×10^9 g^{-1} of wet silt, and their biomass about 0.5 to 1.0 mg g^{-1}. These figures are typical for the upper layer of sediments of lakes or reservoirs (Sorokin, 1962b). The quantity of plate-counted heterotrophic bacteria here was 2 to 10×10^5 g^{-1}, so the ratio (direct count):(plate count) was around 3 to 25×10^3, typical for the microbial count in bottom deposits of lakes (Sorokin, 1958). In deep-sea deposits of the central plain where Eh drops to −200 mV, the total number of bacteria decreases by about an order of magnitude. Their activity as indicated by $^{14}CO_2$ assimilation and by colony counts decreases here by two orders. This indicates that the microflora of deep-sea sediments of the central plain consist mostly of nonactive or starving cells which have settled to the bottom on the surface of faeces or other particles of detritus. Even anoxic microflora here show very little activity. Nitrifying and denitrifying bacteria were found in the surface layer of silts from inshore and slope stations; none appeared in deep-sea sediments. Thio-oxidizing bacteria were present in relatively high concentrations in the slope silts, and were also found in the deep-sea silts deposited by settling particles, but they cannot function in the deep sea because of the lack of any

TABLE 11.11

Microflora in bottom sediments (after Sorokin, 1962a, 1964a)

Depth (m)	Core depth (cm)	Total number of bacteria (10^9 g^{-1})	Plate counts of heterotrophic bacteria (10^3 colonies g^{-1})		Thiobacilli			$^{14}CO_2$ assimilation, (mg O l^{-1} day^{-1})	Nitrifiers (numbers g^{-1}), dilution method[1]	Denitrifiers (numbers g^{-1}), dilution method[1]
					number g^{-1}, dilution method[1]	activity, $^{14}CO_2$ fixation in the presence of thiosulphate, c.p.m. (10^2 g^{-1})				
			aerobes	anaerobes		aerobes ($+O_2$)	anaerobes ($+NO_3^-$)			
180	0–2	4.90	130	28	10^3	88	52	0.69	10^3	10^2
230	0–2	3.76	980	90	10^3	178	79	1.72	10^3	10^2
300	0–2	2.96	450	20	10^2	92	92	0.84	0	0
1600	0–2	1.07	10	25	10^3	17.6	17.6	0.08	0	0
	0–1	0.98	6	0.5	0	2.6	2.6	0.07	0	0
2150	20	0.18	0	0	0	0	0	0	0	0

[1] Determined by serial dilution and inoculation into culture media.

electron acceptors. In fact, their actual activity as indicated by the $^{14}CO_2$ assimilation rate is negligible in the deep-sea sediments. The conclusion by Kriss (1963) of active microbial processes occurring in the deep-sea bottom sediments of the Black Sea has not been proved by direct measurements of the intensity of microbial metabolism. Living microflora are completely absent within the column of sediments of the deep-sea bottom below 5 to 7 cm; bacteria do not play any significant role in the process of the transformation of organic matter of deep-sea deposits which are situated more than 10 cm below the bottom surface.

Microbial formation of hydrogen sulphide

The microbial origin of the hydrogen sulphide in the Black Sea was discovered by N. Zelinski in 1895. He stated that it is produced by the microbial reduction of sulphate by anaerobic bacteria which use the oxygen of sulphate for respiration:

$$Na_2SO_4 + 2CH_2O \rightarrow Na_2S + 2CO_2 + 2H_2O$$

The presence of these sulphate-reducing bacteria in the bottom sediments of the sea was proved by Issachenko in 1926 (Issachenko, 1951). About 99.5% of hydrogen sulphide in this basin is formed by sulphate reduction and only 0.5% by the anoxic decomposition of proteinaceous substances settling down to the anoxic zone. The location and the rate of the process of sulphate reduction have been studied, both with the aid of labeled sulphate $^{35}SO_4^{2-}$, and by an estimation of the quantity of sulphate-reducing bacteria based on a count of colonies in agar medium with lactate and iron sulphate (Sorokin, 1962c, 1964a). The active process of hydrogen sulphide formation occurs in the upper layer of bottom sediments, in the water column near the bottom, and at the upper boundary of the anoxic zone (Fig. 11.23). All these sites within the anoxic zone are places where there is an accumulation of fresh organic material which is necessary for the functioning of sulphate-reducing bacteria.

The source of fresh organic matter in the upper boundary layer of the anoxic zone is detritus — dead phytoplankton, zooplankton faeces and microbial mass produced during chemosynthesis in the redox gradient zone. A maximum of suspended organic matter is retained here by the density

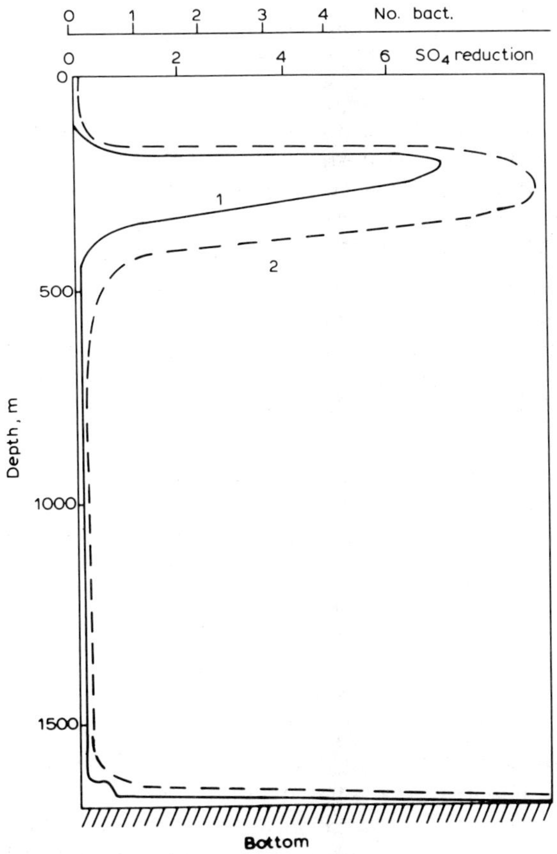

Fig. 11.23. Sulphate reduction in the water column: *1* = rate of SO_4^{2-} reduction (H_2S formed) (mg m^{-3} day^{-1}); *2* = number of sulphate-reducing bacteria (× 10^3 l^{-1}) (modified from Sorokin, 1964a, 1972).

gradient of the chemocline (Datsko, 1959). The rate of sulphate reduction in this layer is about 6 mg H_2S m^{-3} day^{-1} (Sorokin, 1964a). The process of sulphate reduction is indicated by a decrease in this layer (160–220 m) of the ratio SO_4:Cl, showing consumption of sulphate (Skopintzev, 1975).

In the near-bottom water the source of fresh organic matter is organic debris retained by the bottom surface, as if by a screen. The rate of sulphate reduction here is *c.* 10 mg H_2S m^{-3} day^{-1}.

In the bottom deposits it is the surface layer of the continental slope (Table 11.12, Fig. 11.24) where the most active and numerous sulphate-reducing bacteria were found and hydrogen sulphide was formed most intensively. Its rate of formation here is from 1 to 2 mg H_2S l^{-1} day^{-1}. In the central parts of the sea significant rates of hydrogen sulphide production were found in the sediments of the frontal zone between the two halistatic regions (0.3 to 0.5 mg l^{-1} day^{-1}). In the central parts of the halistatic regions the rate of sulphide formation was much less (0.2 to 0.08 mg), and the number of sulphate-reducing bacteria varied from 6000 colonies g^{-1} in the slope sediments down to 1000 in the central parts of the sea. Within the bottom deposits their numbers and activity decreased rapidly, and at 10 to 20 cm below the bottom these bacteria were absent, as was all microbial life (see Tables 11.11, 11.13). Even at 4 to 5 cm below the bottom, the rate of hydrogen

TABLE 11.12

The rate of microbial sulphate reduction in the surface layers of bottom deposits (after Sorokin, 1964a)

Position of stations	Depth (m)	Production of H_2S from SO_4^{2-} during sulphate reduction	
		mg l^{-1} day^{-1}	mg m^{-2} day^{-1}
Coastal areas	180	1.45	72.5
	230	1.89	94.5
	250	0.97	48.5
	300	2.31	115.5
Continental slope	1600	0.89	44.5
	1700	1.29	64.5
Deep sea, boundary area between E and W halistatic regions	2000	0.24	12.0
	2000	0.35	17.8
	2150	0.45	23.0
Deep sea, centers of halistatic regions	2000	0.03	1.3
	2000	0.09	4.2

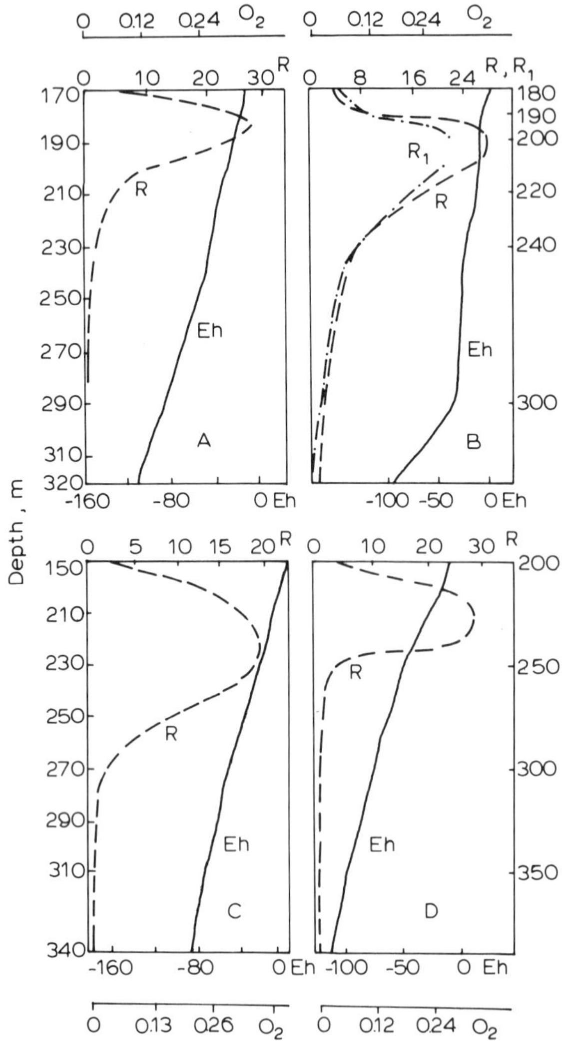

Fig. 11.24. Rates of hydrogen sulphide oxidation (R, mg m^{-3} day^{-1}) in the water column at several stations: A=open sea opposite the Kerch Strait; B=same opposite Feodosiya; C=slope area opposite Anapa; D=open sea at the boundary area between the eastern and western halistatic regions. Eh=redox potential (mV); O_2=oxygen content (mg l^{-1}); R_1=rate of H_2S oxidation calculated by Aizatullin and Skopintzev (after Sorokin, 1972).

sulphide formation and the number of sulphate reducers was greatly decreased (Sorokin, 1962c).

In calculations of hydrogen sulphide production below one square meter of the bottom surface it has been assumed that there is a 5-cm layer of active sulphate reduction (Table 11.12). The total production of hydrogen sulphide varied from 1.35 in the central halistatic parts of the sea to 100 mg m^{-2} day^{-1} in slopes. The cause of these variations lies in differing rates of sedimentation of fresh organic matter. Near shore it is much more intensive than in the central parts. In general the horizontal distribution of the rate of hydrogen sulphide formation in the bottom sediments corresponds well with the distribution of primary production. The total production of hydrogen sulphide in bottom sediments as estimated by Deuser (1970) on the basis of the above experimental data by Sorokin (1964a) is 3.6×10^6 t yr^{-1}, or 12 g m^{-2} yr^{-1}. These numbers are in the same range as those previously calculated by Datsko (1959) from an estimate of the amount of organic matter settling annually (2.4×10^6 t yr^{-1}). The total production of hydrogen sulphide in the water column appears to be equal to that in the bottom sediments (Sorokin, 1964a), so the general production of hydrogen sulphide in the Black Sea can be estimated at 7×10^6 t yr^{-1} or 20 g m^{-2} yr^{-1}.

Sulphide oxidation

Continuous studies for several decades of hydrogen sulphide distribution in the Black Sea have shown the stability of the vertical depth position of the hydrogen sulphide zone. During vertical turbulence this stability can be maintained only when there is a dynamic equilibrium between its formation by sulphate-reducing bacteria and oxidation in the intermediate redox gradient zone (Bogdanova, 1959a; Sorokin, 1972; Skopintzev, 1975). Measurements and calculations of the components of the balance of the sulphur cycle processes could provide a key to evaluations of the rates of vertical mixing, and of the amount of the total organic production in the sea because the level of this production alone limits the rate of hydrogen sulphide formation. It is also important to distinguish between chemical and biological oxidation of hydrogen sulphide *in situ*, and to evaluate their relative significance. These two processes are not equivalent so far as the energetics of the ecosystem are concerned. The energy produced by hydrogen sulphide oxidation, which can be used via microbial chemosynthesis by the ecosystem, is greater than that of the oxidation of an equivalent amount of organic matter. During chemical oxidation this energy is lost as heat; during biological oxidation of hydrogen sulphide by autotrophic bacteria up to 25% of it is utilized for the production of organic matter, and so is included in

TABLE 11.13

The rate of sulphate reduction (*R*), the number (*N*) and relative activity (*A*) of sulphate reducers in the columns of sediments

Depth (m)	Core depth (cm)	Eh (mV)	*R*, H_2S formed (mg l^{-1} day^{-1})	*N* (colonies g^{-1})	*A*, radioactivity of S^{2-} formed in the presence of lactate (c.p.m. g^{-1})
2000	0	−224	0.35	1000	256.10^3
	5	−266	0.01	25	360
	60	−244	0	0	0
2150	0	−213	0.45	6600	593.10^3
	20	−257	0	0	0
	40	−233	0	0	0

the ecosystem energy budget. The microbial biomass formed during chemosynthesis is partially used by filtering mesozooplankton (Datsko, 1954; Lebedeva, 1959), and by protozoa. Recent observations showed that during the daytime protozoans and the filtering copepod *Calanus helgolandicus* form a dense maximum of biomass (up to 3 g m^{-3}) in the upper part of the redox gradient zone between 120 to 140 m. The remainder of the microbial biomass produced is carried by turbulence to the intermediate zone where it supports the process of sulphate reduction (Fig. 11.23), which in turn helps to maintain the stability of the hydrogen sulphide zone (Skopintzev, 1953; Sorokin, 1964a).

Studies of the rate and mechanism of hydrogen sulphide oxidation in the Black Sea were begun by M. Egunov (1895). He proposed the theory of the existence of the "microbial plate" (a layer of a dense population of thiobacilli), which prevents the distribution of hydrogen sulphide to the upper layer. However, studies by microbiologists did not support this theory. Numerous analyses made by B. Issachenko and other investigators failed to reveal an accumulation of thiobacilli in the intermediate zone. These bacteria were rarely found, so that hydrochemists (Bruevich, 1953; Skopintzev, 1953) tended to account for the hydrogen sulphide oxidation in the Black Sea as a purely chemical process. This point of view was supported by experiments on hydrogen sulphide oxidation in samples of natural waters and bottom sediments (Wheatland, 1954; Ivanov, 1957). In these experiments the rate of hydrogen sulphide oxidation was measured by the rate of its disappearance, and this rate was found to be equal in sterile samples, or in the presence of an antiseptic, to the non-sterile samples in the presence of natural microflora.

Subsequent experimental studies of this problem by Sorokin (1967, 1970, 1972) with the aid of ^{14}C and ^{35}S reveal that these conclusions on the development of thiobacilli and on the mechanism of hydrogen sulphide oxidation in natural waters were not correct. Tests using ^{14}C and media with gradients of redox potential along the test tubes have shown that a rich population of thiobacilli exists in the redox gradient zone of the Black Sea. Experiments on the mechanism of hydrogen sulphide oxidation have been made with the aid of labeled $H_2{}^{35}S$ on natural samples, taking all precautions to preserve the redox potential. The sulphate and thiosulphate ions formed were precipitated as the barium salt and the radioactivity was estimated and used as an indication of the products resulting from the oxidation. The addition of antiseptic made it possible to distinguish the chemical from the biological processes of hydrogen sulphide oxidation. The absolute rate of hydrogen sulphide oxidation was calculated, giving values close to the *in situ* rates.

This procedure has provided profiles for the rate of hydrogen sulphide oxidation at several deep-sea stations opposite the Crimea (Sorokin, 1972). The hydrology of these stations was similar to that of the halistatic region. The anoxic layer was relatively shallow (170 m); the intermediate redox gradient zone where both oxygen and hydrogen sulphide are present was 40 to 60 m thick. Within the intermediate zone the redox potential (Eh) changed from 40 to −60 mV and the hydrogen sulphide concentration ranged from 0 to 0.5 mg l^{-1}. The maximum rate of active hydrogen sulphide oxidation (15–25 μg S^{2-} l^{-1} day^{-1}) was observed in

TABLE 11.14

Evaluation of total rates of hydrogen sulphide oxidation in water column of the Black Sea as estimated experimentally by Sorokin (1972) and calculated for the same stations by Aizatullin and Skopintzev (1974)

No. of station	Depth of upper boundary of H_2S zone (m)	Rate of H_2S oxidation (g m^{-2})		
		estimated experimentally		calculated per day
		per day	per year	
1	135	0.53	193	0.26
3	140	0.39	142	0.42
7	150	0.39	144	0.06
5	170	0.67	244	0.52
4	180	0.74	270	0.47
6	190	0.97	354	0.47
2	200	1.10	401	0.66

the layer where Eh values ranged from −10 to −20 mV and hydrogen sulphide concentration was 0.2–0.4 mg l^{-1} (Table 11.14, Figs. 11.22, 11.24). The integrated rates of hydrogen sulphide oxidation in the water column were calculated (Table 11.14) from the vertical profiles; they varied between 0.5 to 1.0 g S m^{-2} day^{-1}.

For the same stations, Aizatullin and Skopintzev (1974) have calculated the possible rates of hydrogen sulphide oxidation on the basis of the vertical distribution of oxygen and hydrogen sulphides, using the second order equation of the rate of chemical reactions, which describes the hydrogen sulphide oxidation process in water. The profiles and the absolute rates of hydrogen sulphide oxidation were comparable to those obtained experimentally. The same range of rates of hydrogen sulphide oxidation can be derived from the rate of vertical mixing (Bogdanova, 1959a), and from the rates of chemosynthesis measured by Sorokin (1964a).

The energy efficiency of chemosynthesis calculated from the above rates of hydrogen sulphide oxidation was found to be 12 to 40%, which is quite reasonable (Table 11.15). Therefore, the annual rate of hydrogen sulphide oxidation evaluated directly (150–200 g m^{-2} yr^{-1}) and that calculated by a different method (80–150 g m^{-2} yr^{-1}) agree reasonably well. However, they exceed by several times the estimated rates of hydrogen sulphide formation in the water column and in bottom sediments. The causes of this apparent disparity in the sulphur cycle could be the absence of annual estimations of the rate of hydrogen sulphide oxidation, which may vary during the year under the influence of changes in the rates of vertical turbulence. The possibility of the participation of ancient organic matter in the formation of additional hydrogen sulphide has been also indicated (Sorokin, 1972).

Studies of interrelationship between products of labeled sulphide oxidation in samples from the intermediate zone indicate that the *in situ* oxidation proceeds in two steps. The first step is the chemical oxidation of hydrogen sulphide to sulphate and thiosulphate. This is a rapid chemical reaction of the first order, and the thiobacilli practically cannot participate in it. When the same process was observed in cultures of these bacteria (Sorokin, 1967), the rate of disappearance of hydrogen sulphide was not influenced by the addition of an antiseptic (chloroform). This explains the con-

TABLE 11.15

The rate of hydrogen sulphide oxidation (K), and the efficiency (E) of the use of the energy of this oxidation for chemosynthesis on vertical profiles at a station opposite the Crimea (after Sorokin, 1972)

Depth (m)	H_2S (mg l^{-1})	O_2 (mg l^{-1})	Eh (mV)	K (mg S l^{-1} day^{-1})	Chemosynthesis production (mg C l^{-1} day^{-1})	E (%)
135	0.01	0.17	0	3.2	0.2	12
145	0.05	0.12	−10	10.3	2.1	36
155	0.15	0.08	−15	25.4	4.4	33
165	0.20	0.06	−20	19.8	4.3	41
180	0.43	0.02	−36	7.0	0.15	39
210	0.86	0	−100	2.3	0.3	25
240	1.36	0	−120	0	0.02	–

troversial results of previous authors. But further oxidation of thiosulphate formed during chemical oxidation proceeds biologically with the participation of thiobacilli. In the absence of an antiseptic, thiosulphate is oxidized (Fig. 11.25). The total rate of hydrogen sulphide oxidation depends more on the presence of hydrogen acceptors than on the presence of active thiobacilli (Fig. 11.26).

ZOOPLANKTON

Microzooplankton

Microzooplankton includes organisms of less than 0.3 to 0.5 mm, mostly the juvenile stages of crustaceans, tunicate larvae, including young *Oikopleura*, and the ciliates and zooflagellates. In the neritic zone it also includes rotifers (*Keratella*, *Trichocerca*) and small representatives of the planktonic stages of larvae of benthic animals like veligers, trochophores, etc. The important role of microzooplankton, and especially of ciliates and zooflagellates, in the productivity and metabolism of marine planktonic

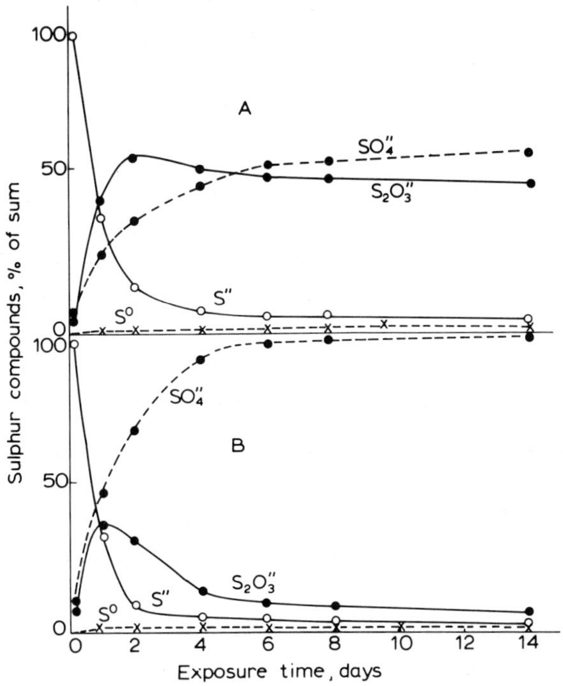

Fig. 11.25. Dynamics of the oxidation of sulphur compounds in a sterile medium (A) and in culture of thiobacilli (B). The initial concentration of H_2S was 15 mg l^{-1}. (After Sorokin, 1970.)

communities has been documented (Beers and Stewart, 1971; Sorokin, 1977). Studies of the microzooplankton in the Black Sea were made by Pavlovskaya (1976) and Zaika et al. (1976). Most of the seasonal observations were made in the Bay of Sevastopol'. The numbers of ciliates varied from 1000 to 5000 l^{-1} and the biomass from 30 to 80 mg m^{-3}. The loricated ciliates (tintinnids) dominated in spring, with species from the genera *Tintinnopsis*, *Tiarina* and *Favella*. The naked ciliates were most abundant in summer, represented mostly by Oligotrichidae. Among them species of *Strombidium* dominated, with species from the genera *Chilodonella*, *Lohmaniella*, *Mesodinium*, *Metacystis*, *Prorodon* and *Uronema* also present. The number of nauplii in this bay in summer was 20 to 30 l^{-1}. The biomass of microzooplankton in summer here was 30 to 100 mg m^{-3}. About 50 to 80% of it was composed of ciliates. These are very approximate numbers because of the usual underestimation of the number of naked ciliates counted in samples after their concentration on the membrane filters.

Recently Mamaeva and Moiseev (1979) have found that in addition to the surface maximum of biomass of protozoa in the layer of thermocline in the Black Sea, a second maximum exists at the upper boundary of the redox gradient zone at depths of 100 to 130 m. A similar phenomenon of a rich fauna of protozoa just above the hydrogen sulphide zone had been described by Sorokin and Donato (1975) for the meromictic Lake Faro (Sicily), formed from a marine lagoon.

Mesozooplankton

In the pelagic food chains of the Black Sea the most important species of zooplankton are the copepods, especially in the deep central part. They include the following mass species: *Acartia clausi*, *A. lattisetosa*, *Calanus helgolandicus*, *Centropages kröyeri*, *Oithona nana*, *O. similis*, *Paracalanus parvus*, *Pontella avirostris* and *Pseudocalanus elongatus*. In the summer, and especially in neritic areas, the epiplankton community contains a large number of Cladocera, which are represented by the following species: *Evadne spinifera*, *E. tergestina*, *Penilia avirostris*, *Podon polyphaemodes*. In coastal waters they compose 50 to 70% of the total zooplankton biomass (Pavlova, 1967). In coastal areas

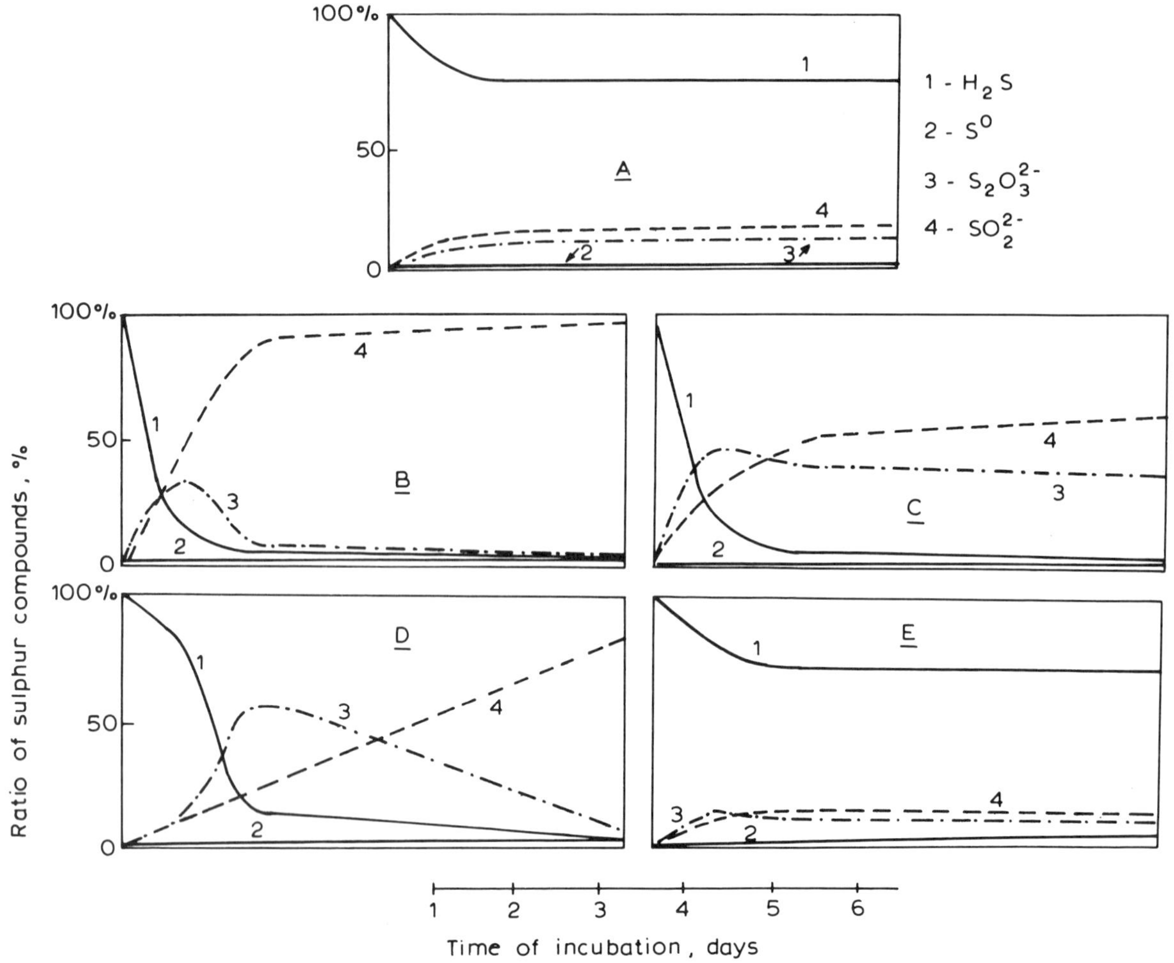

Fig. 11.26. The proportions of sulphur compounds formed during sulphide oxidation in samples taken from the depth 220 m opposite Crimea under the following conditions: A, +antiseptic; B, +air bubbles; C, +antiseptic+air bubbles; D, +$NaNO_3$; E, +$NaNO_3$ +antiseptic. (After Sorokin, 1972.)

a rich population of larvae flourishes, mostly veligers of bivalve mollusks and trochophores. A significant role in zooplankton populations is played by the appendicularian *Oikopleura dioica* and the cystoflagellate *Noctiluca miliaris*. In midsummer the latter sometimes comprises most of the total wet zooplankton biomass. Important components of the larger predatory zooplankton in summertime are the chaetognath *Sagitta setosa*, the ctenophore *Pleurobrachia pileus* and the jellyfishes *Aurelia aurita*, *Rhizostoma plumo* and *Sarsia setosa*.

Average numbers of the main species in the upper 100-m layer are given in Table 11.16. The most plentiful throughout the year is the copepod *Oithona nana*, with a maximum population of

TABLE 11.16

Average numbers per cubic meter of the main species of mesozooplankton in the Black Sea in layer 0 to 100 m (modified from Grese et al., 1973)

Species	Time of year		
	winter	spring	summer
Acartia clausi	310	840	1910
Calanus helgolandicus	160	140	110
Cladocera sp.	0	0	450
Oithona nana	1180	1840	2690
Oithona similis	470	2640	1000
Paracalanus parvus	350	1670	1480
Pseudocalanus elongatus	1770	1500	1160
Sagitta setosa	2	4	42

2500 m^{-3} in August. The number of *Oithona similis* is usually at its peak in May (2500 m^{-3}). The volume of *Calanus helgolandicus* and *Pseudocalanus elongatus* does not vary much during the year, but the age composition of the populations alters, so that in February, for example, the number of *C. helgolandicus* nauplii was 93% of the total, but in summer it was only 44%. Populations of *Paracalanus parvus* and *Acartia clausii* were at a maximum in May and August (1500 m^{-3}). The Cladocera as well as the copepod *Centropages kröyeri* appear in shallow waters in spring; in the open sea they develop in midsummer. Cladocera are represented mostly by *Penilia avirostris*. The average number here is about 400 m^{-3}, while the remaining Cladocera do not exceed 50 m^{-3}. The number of larvae of benthic animals in summer in neritic plankton reaches 1000 to 2000 m^{-3} (Grese et al., 1973). Areas of constantly abundant zooplankton have been described by Nikitin (1945) and by Petipa et al. (1963). Their existence is connected with the acceleration of the vertical turbulence caused by the mesoscale circulation. These areas are situated mostly near the coast, where the circular current is disturbed by ridges, or in the shallow northwest part of the sea, including the coastal areas near the Crimea, the coastal regions of Anatolia (Ereğli–Inebolu), of the Caucasus (Novorossiysk–Tuapse), the pelagic regions at the boundary of the east and the west cyclonic eddies between the Crimea and Anatolia, and in the anticyclonic area in the southeast part of the sea. The number and biomass of zooplankton in these areas can exceed the average by several times. In the coastal waters of Sevastopol' Bay and in the shallow northwest part of the sea the biomass of zooplankton can exceed the average by 5 to 10 times. The number of *Oithona nana* here reaches 15×10^3 m^{-3} in autumn, and the number of *Penilia avirostris* can exceed 1000 m^{-3}.

Most of the pelagic zooplankton population accumulates in the upper 50 m layer, which usually contains 80 to 85% of the total zooplankton biomass (Fig. 11.27). The lower boundary of zooplankton distribution is at the depth of 150 to 170 m. Groups of zooplankton within the 150 to 170 m depth include the relatively thermophilic epiplanktonic group of zooplankton — *Acartia clausii*, *Cladocera* spp., *Oithona minutula*, *Paracalanus parvus* and *Sagitta setosa* (Petipa et al., 1960). This group of mostly non-migrating species inhabits the

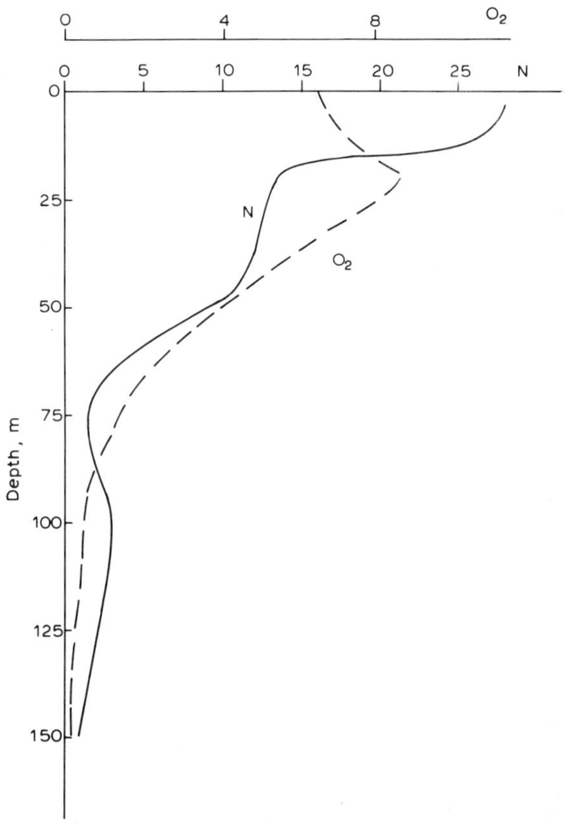

Fig. 11.27. Vertical distribution of zooplankton in the Black Sea in summer. N = total number of zooplankton ($\times 10^3$ m^{-3}); O_2 = oxygen (mg l^{-1}).

upper warm layers of water. A psychrophilic bathyplanktonic group includes *Oithona similis*, *Pleurobrachia pileus*, *Pseudocalanus elongatus*, and, in part, *Calanus helgolandicus*, *Noctiluca miliaris* and *Sagitta euxina*. Presumably these inhabit the lower colder water layers below the thermocline and contain actively migrating plankters such as *C. helgolandicus*, *P. elongatus*, *S. setosa*. The separation of these two groups increases with the temperature gradient. Some of the species accumulate within the thermocline, such as *Paracalanus* or *Noctiluca*. In winter the boundary between these two groups disappears, and the main biomass of zooplankton accumulates in the layer 25 to 50 m. The epiplanktonic organisms only move down to the thermocline, and bathyplanktonic organisms only move up to the same boundary. The exceptions are active species such as *Calanus helgolandicus* or *Sagitta euxina*, which can cross the thermocline during vertical migrations (Petipa et al., 1960). The copepod *Calanus helgolandicus* accumulates within a narrow layer between 130 to 135 m depth at the boundary of the oxygen zone during daytime; its concentration here exceeds 2.5 g m^{-3}. This finding was made using discrete sampling with the aid of 100-l capacity water bottles (Vinogradov and Shushkina, 1980). Using the scientific submarine *Argus* in the open sea near Novorossiysk, an extremely dense layer of mass accumulation of copepods was observed at this depth.

The average wet biomass in the upper 100 m layer is 100 to 300 mg m^{-3}, or around 10 g m^{-2} (Table 11.17). In layers of zooplankton accumulation, usually near the thermocline, it is considerably more — up to 500 to 1000 mg m^{-3}. In summer a significant part of the total zooplankton biomass is composed of *Pleurobrachia pileus*, *Sagitta setosa* and of jellyfishes, which are very active consumers of crustacean zooplankton.

The production of the populations of the separate species has been thoroughly investigated in studies of growth rates at different stages of ontogenesis (Petipa et al., 1970a; Grese et al., 1973). The mass species of copepods have seven to eight generations per year. The values of P/B coefficients obtained are given in Table 11.17, together with average data on the biomass of the main components of zooplankton and their production in summer as calculated with the use of the given P/B coefficients. The largest coefficient was for the cladoceran *Penilia* (0.18 per day), and the minimum for *Calanus* (0.05). An average daily coefficient P/B for the total crustacean zooplankton was 0.11 in summer; in winter it decreased to 0.04. Using these data, annual values were calculated (Table 11.18). They show that the annual production of wet biomass by crustacean zooplankton alone exceeds 200 g m^{-2}, and its total annual production in the sea averages 90 million tons. A significant proportion of the production and biomass of zooplankton in the Black Sea is composed of predatory organisms which are used only partially by fish, and indeed are usually their competitors when feeding upon the crustacean zooplankton. Among them are *Sagitta*, *Noctiluca*, ctenophores and jellyfishes. The production of the predatory chaetognath *Sagitta setosa* alone is evaluated as 30×10^6 t, while fish production is only around 1×10^6 t. The biomass of jellyfishes is evaluated now at more than 200×10^6 t (wet weight). The development of these components of zooplankton, which are mostly the trophic dead ends of the ecosystem, decreases the total efficiency of utilization of zooplankton production by the fish population. It is in fact only one half to one third of that in the Caspian Sea or in the Sea of Azov.

TABLE 11.18

Characteristics of biomass and productivity of total crustacean zooplankton in the Black Sea

Parameter	Annual data for the total Black Sea
Average wet biomass, g m^{-2}	6.10
Total biomass, 10^6 tons	2.56
Annual P/B coefficient	35.4
Average production, g m^{-2}	215.9
Total production, 10^6 tons	90.6

TABLE 11.17

Average wet biomass and production of zooplankton in the Black Sea in summertime (modified from Grese et al., 1973)

Species	Production (mg m^{-3} day^{-1}) (*P*)	Biomass (mg m^{-3}) (*B*)	*P/B*
Acartia clausi	1.86	12.8	0.14
Calanus helgolandicus	1.70	33.2	0.05
Cladocera sp.	3.78	21.0	0.18
Oithona sp.	0.51	5.4	0.09
Paracalanus parvus	0.47	4.3	0.11
Pseudocalanus elongatus	3.04	26.0	0.12
Total crustaceans	11.45	102.6	0.11
Sagitta setosa	5.4	32.0	0.16

ZOOBENTHOS, PHYTOBENTHOS

The species composition of benthic fauna is poor compared to that of the Mediterranean Sea because of the lower salinity of Black Sea water and the presence of hydrogen sulphide in deeper layers. The lower boundary of the distribution of bottom fauna is usually situated at depths of 125 to 135 m, and only near the Bosporus is it at a depth of 170 to 200 m. The area of the bottom inhabited by benthic fauna is about 23% of the total. The species composition of benthic fauna in the Black Sea is a mixture of North Atlantic and of relict Caspian faunas, with Mediterranean fauna present to a lesser degree. Although some Mediterranean species have become abundant in the Black Sea, they occur in smaller sizes and include: *Cardium edule*, *Melinna palmata*, *Nephtys hombergii*, *Nereis succinea*, *Teredo navalis*, and others.

The main bottom biocoenoses were described by S. Zernov in 1913 and later by Bacescu et al. (1971) and Vinogradov (1968). The supralittoral biocoenosis is situated only in the surf zone. It is inhabited mostly by crustaceans: amphipods and isopods (*Euridice*, *Gammarellus*, *Idothea*, *Orchestia*, *Pontogammarus*, *Talorchestia*), mysids (*Gastrosaccus*). They often form a large biomass of up to 120 g m^{-2}. Other main species here are bivalves (*Donacilla cornea*) and polychaetes (*Nerine*, *Ophelia*).

Benthic communities

The biocoenosis of the rocks is situated between the surface and depths of 20 m. The main species here are bivalves *Mytilus galloprovincialis*, gastropods *Nassa*, *Rissoa* and *Trochus*, and different sponges, polychaetes and bryozoans.

The biocoenosis of the sandy and muddy coast is situated down to the depth 15 to 30 m. It is inhabited by a very diverse fauna, including bivalves (*Cardium edule*, *Cardium exiguum*, *Divaricella*, *Syndesmia* and *Venus gallina*), gastropods (*Nassa*, *Rissoa*), polychaetes (*Glyceria*, *Nereis*, *Spio*, *Nerine*), nemertines (*Eunemertes*, *Lineus*), isopods (*Idothea*, *Sphaeroma*) and holothurians (*Synapta*). At the depth of 5 to 9 m a rich development of the sea grass *Zostera marina* often takes place. The area covered with *Zostera* is inhabited by rich benthic fauna, containing abundant populations of Bryozoa (*Lepralia*, *Membranipora*). Along the Romanian coast the sandy shallow bottom biocoenoses are inhabited by the mollusk *Corbula mediterranea*. Since 1971 *Mya arenaria*, a new species of bivalves for the Black Sea, has appeared in these habitats (Gomoiu and Petran, 1973). Another new species of mollusk — the large predatory bivalve *Rapana thomasiana* — appeared in the Black Sea in 1947 to 1950 and since that time has become widely distributed along its coast.

The shell biocoenoses are formed by the accumulation of shells of living and dead mollusks, mostly of bivalves: *Cardium*, *Meretrix*, *Modiolus*, *Mytilus*, *Pecten*, *Tapes*, *Venus*. It is also inhabited by numerous polychaetes, crustaceans (*Balanus*), sponges (*Cliona*) and hydroids.

The biocoenosis of the Mytilus mud is situated in the open sea between depths of 50 to 70 m. It is also inhabited by *Cardium*, *Meretrix*, *Mytilus* and polychaetes. In areas occupied by *Mytilus* — mud in the middle of the northwest shallow part of the sea at depths of 30 to 50 m — enormous masses of the algae *Phyllophora* accumulate, the so-called "Phyllophora field" of Zernov. The benthic fauna here is very poor, and is composed mostly of amphipods and isopods.

The biocoenosis of phaseolina mud is at a deep level and is one of the most specific bottom biotopes in the Black Sea. It begins at about 60 m and spreads down to the lower boundary of bottom fauna, 130 to 160 m. Formed by the bivalve *Modiolus phaseolinus*, the density of the population here is several hundreds per square meter.

A significant part of the total benthic population is composed of microbenthos. It consists of small benthic animals which can pass through 1-mm sieves. The bulk of the microbenthos is composed of nematodes and harpacticoids. Their total number sometimes reach 250×10^3 m^2. The microbenthos is the dominant group of bottom fauna in the microaerophilic conditions of the deeper slope zone between 120 to 200 m depth.

The main stock (85%) of the total benthic biomass of the sea is situated in the shallow northwest part (Table 11.19). The average biomass in this part of the sea is in fact very high (200–400 g m^{-2}). The stock of edible mollusks such as *Mytilus* has been evaluated at around 10 million tons, and the total stock of benthos in the sea at 28.3×10^6 t, of which 27.3×10^6 t is in the northwest. About 27% of this

TABLE 11.19

The average wet biomass of benthos in the west part of the Black Sea (modified from Vinogradov, 1968)

Region	Area ($10^3 km^2$)	Benthos biomass	
		$g\ m^{-2}$	10^6t per total area
NW shallow	57.6	433	25.0
West coast of Crimea	4.1	57	0.2
Bulgarian coast	12.0	74	0.9
Coast near the Bosporus	3.0	386	1.2
Total	76.7	355	27.3

amount is available for consumption by fishes (Vinogradov, 1968).

The benthic algal flora of coastal areas of the Black Sea is also very abundant and is composed of several main species: representatives of Phaeophyta (*Corynophloea*, *Cystroseira*, *Dilophus*, *Ectocarpus*, *Stilophora*), Chlorophyta (*Chaetomorpha*, *Cladophora*, *Enteromorpha*, *Entodadia*, *Ulva*), Rhodophyta (*Ceramium*, *Dasya*, *Gelidium*, *Kylinia*, *Laurencia*, *Melobesia*, *Phyllophora*, *Polysiphonia*); *Zostera marina* and *Z. nana* are also present. About 90% of the total biomass of macrophytes in the Black Sea is formed by *Phyllophora*. The algae form broad fields with a very dense growth of plants which compose specific biocoenoses together with the benthic animals. The biomass of algae in the shallows of the northwest varies between 1.3 to 8 kg m^{-2}. The total stock of algae, mainly *Phyllophora rubens*, was evaluated here at about 10^7 t. This alga is used by industry to produce alginic acids and agar. Of great importance in the food chains of the Black Sea are not only the bottom macrophytes, which are an important source of feeding for zoobenthos, but also the rich periphyton which develop on their surface. The periphyton are consumed intensively by benthic invertebrates and by detritus-feeding fishes such as Mugilidae.

FISHES

The ichthyofauna of the Black Sea includes 108 genera of fishes. Most of them are immigrants from the Mediterranean Sea (57%) and fresh-water species (22%). Some of the species are the relicts of the Pontic basin, including some Gobiidae, Acipenseridae, and some species of herring, *Caspialosa*. In the spring large schools of pelagic fishes migrate through the Bosporus into the Black Sea — *Pelamys*, *Pomatomus*, *Sardinella*, *Scomber*, *Sprattus*, *Trachurus*.

Among the benthos-feeders the dominant species are *Corvina*, *Crenilabrus*, Gobiidae, *Mullus*, *Ophiodon*, *Rhombus*, *Smaris* and *Solea*. The food resources of the benthos are also used by some large predators like Acipenseridae (*Acipenser* spp., *Huso huso*). A large stock of periphyton and detritus in the coastal areas and shallows is used by Mugilidae (*Mugil cephalus*, *M. saliens*), which are among the most valuable commercial fishes in this basin. Pelagic plankton feeders predominate in the commercial fishing industry of the Black Sea: *Caspialosa* spp., *Clupeonella* sp., *Engraulis* spp., *Sardinella* sp., *Sprattus sprattus* and *Syngnathus schmidtii*. Predatory fishes are also abundant, particularly *Acipenser*, *Diplodus*, *Gadus*, *Huso*, *Pomatomus*, *Scomber*, *Trachurus*.

The main fishery area of the Black Sea is located near the Kerch Strait and in the Strait itself. Large schools of *Engraulis*, herring, *Mullus* and *Mugil* gather here and migrate through the strait into the Sea of Azov in spring, and out of it in late autumn. The annual catch reaches 30 to 50×10^3 t. Other rich fishery areas are the estuaries and bays of the northwest part of the sea. The total catch in the Black Sea in 1976 was around 0.24×10^6 t. About 85% of this is composed of the pelagic fishes *Engraulis*, *Sprattus* and *Trachurus*, but among the benthic fishes there are also some very valuable species such as *Acipenser*, *Huso* and *Mugil*. In addition to these, the catch of *Engraulis* in the Kerch Strait amounts to about 60×10^3 t. Important consumers of fish (0.3×10^6 t yr^{-1}) in the Black Sea are the dolphins, of which three species occur: *Delphinus delphis*, *Phocaena phocaena* and *Tursiops tursio*. In 1957 the catch of dolphins amounted to 35×10^3 individuals.

ECOSYSTEM FOOD CHAINS AND ENERGETICS

The structure and functioning of the pelagic ecosystem of the Black Sea has been thoroughly studied by Petipa et al. (1970a). These studies included investigations of the feeding of the main species, made by direct observations as well as experimentally with the aid of ^{14}C (Petipa, 1968; Petipa et al., 1970b; Pavlova and Sorokin, 1970). Among the fine filterers are the appendicularians *Oikopleura*, cladocerans *Penilia*, copepods *Paracalanus*, and larvae such as veligers or nauplii that feed on bacteria and small algae. The crude filterers which feed on larger algae and on bacteria in aggregates and in detritus particles are *Calanus helgolandicus*, *Pseudocalanus elongatus*, and many older copepodites. They can also feed on larger objects such as small animals of 0.2 to 0.5 mm, catching them with second maxillae. Copepods such as *Centropages* and *Pontella* are mixed feeders. They both filter and actively hunt for animals of 0.7 to 1 mm size. The mixed feeders like the copepod *Acartia* cannot really filter, they pick up or catch plant and animal objects 0.02 to 0.5 mm large by the moving of maxillae and maxillipeds. The species of a small copepod *Oithona* are the real predators, and catch nauplii and copepodites.

All of these species of planktonic crustaceans have a wide spectrum of feeding habits, depending on the food sources present. The amount of food consumed by most planktonic crustaceans per day ranges from 50 to 150% of their body weight. Studies of the rate of growth and reproduction of the main species, together with analyses of their trophology and metabolism and a description of their population structure, make estimates of their energy balance possible (Petipa et al., 1970a). An example of the energy balance of the pelagic community in the west halistatic region in early summer is given in Table 11.20. On the basis of this

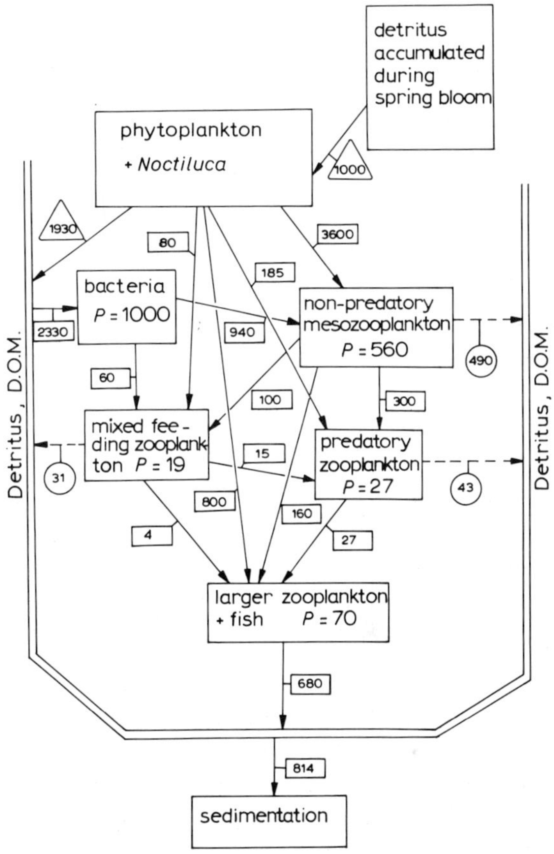

Fig. 11.28. Scheme of energy flow in the pelagic ecosystem in the western part of the Black Sea in June, using the balance data given in Table 11.20: P = production; numbers in triangles = non-grazed production; numbers in squares = food ration of subsequent component of trophic chain; numbers in circles = non-assimilated food.

TABLE 11.20

Elements of the energy balance in the pelagic ecosystem of the western halistatic region of the Black Sea in June (cal. m^{-2} day^{-1}) (original, after Petipa et al., 1970a; Sorokin, 1972)

Components of planktonic community	Biomass	Production	Metabolic loss	Assimilated food	Food consumption	Non-assimilated part of consumption
Phytoplankton + *Noctiluca*	10 650	6660	–	–	–	–
Bacteria	2000	1000	2330	3330	3330	–
Non-predatory zooplankton	5050	560	3490	4050	4540	490
Mixed feeding zooplankton	350	20	190	210	240	30
Predatory crustacean zooplankton	398	27	430	457	500	43
Larger predators (*Sagitta, Pleurobrachia*)	1290	60	320	380	1030	450

1 cal. = 4.1855 J.

TABLE 11.21

Average biomass (B) and annual production (P) of the main ecosystem components in the Black Sea, and elements of the annual energy balance (R=metabolic loss, A=assimilated food, C=food ration, F=non-assimilated part of consumption); all data are given for the whole Basin (original, based on Datsko, 1954; Zenkevitch, 1963; Petipa, 1968; Petipa et al., 1970; Sorokin, 1972; Grese et al., 1973)

Components of ecosystem	Empirical data for calculations of elements of energy balance						Elements of energy balance of the ecosystem (10^{12} kcal yr^{-1})				
	average biomass (B) and annual production (P) (10^6 tons, wet weight)			coefficients for calculations			P	R, $\frac{P(1-K_2)}{K_2}$	A, $(P+R)$	C, A/I	F, $(C-A)$
	B	P	P/B	caloric equivalents of wet biomass[1]	K_2, P/A	I, A/C					
Phytoplankton	6	1500	250	0.5	–	–	750	–	–	–	–
Phytobenthos	20	40	2	0.8	–	–	32	–	–	–	–
Bacteria	0.8	160	200	1.0	0.3	–	160	380	540	540	–
Microzooplankton	0.5	50	100	0.8	0.5	0.6	40	40	80	130	50
Noctiluca	2.5	150	60	0.08	0.3	0.6	12	28	40	67	17
Non-predatory zooplankton	2.3	80	35	1.0	0.35	0.6	80	150	230	380	150
Zoobenthos	28.0	56	2	0.3	0.20	0.6	17	68	85	140	55
Predatory mesozooplankton	0.3	10	33	1.0	0.2	0.7	10	40	50	70	20
Larger predatory zooplankton	1.0	32	32	0.25	0.3	0.7	8	18	26	38	12
Non-predatory fishes	1.0	1.9	2	1.0	0.3	0.7	1.9	4.6	6.6	9.6	3
Predatory fishes	0.3	0.3	0.8	1.0	0.2	0.7	0.3	1.0	1.3	1.8	0.5
Catch	–	0.24	–	–	–	–	0.24	–	–	–	–

[1] 10^6 kcal t^{-1} (wet weight).

table, a scheme of energy flow has been drawn (Fig. 11.28). This scheme shows that among heterotrophs the main role in energy flow is taken by non-predatory mesozooplankton (Crustacea and *Noctiluca*) and by the larger predators. The latter includes *Sagitta*, ctenophores and jellyfishes, which feed mostly on the heterotrophic peridinean *Noctiluca*, which in turn consumes detritus accumulated in the water during the previous spring phytoplankton bloom. It has a large role in the food chain in the Black Sea during the summer period.

Long-term observations on the biomass and production of the main components of the ecosystem in the Black Sea are the basis of calculations of values for the total basin area (Table 11.21). These data and the results of trophological and physiological studies of the main species which form the food chains of the Black Sea can be used to calculate the approximate values of the components of the energy balance for the total ecosystem of the sea. The results of such an attempt are given in Table 11.21, and the scheme of energy flow based on the data given in this table is presented in Fig. 11.29. It shows that the main energy flow passes through phytoplankton, bacteria, non-predatory zooplankton and zoobenthos. A significant place in the functioning of the community is occupied by organisms such as *Noctiluca*, *Sagitta*, *Pleurobrachia* and jellyfishes which have a low food value for higher trophic levels. The ratio between primary production and fish production is about 200, and between fish production and catch it is 12.

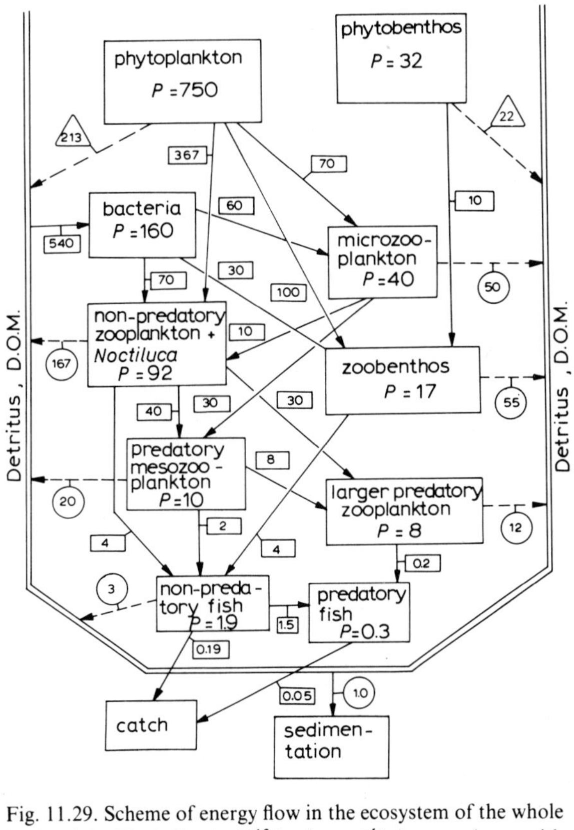

Fig. 11.29. Scheme of energy flow in the ecosystem of the whole basin of the Black Sea ($\times 10^{12}$ kcal year^{-1}), in accordance with data given in Table 11.21; definitions as in Fig. 11.28. (1 kcal = 4.1855 kJ.)

REFERENCES

Aizatullin, T.A. and Skopintzev, B.A., 1974. Studies of the rate of oxidation of hydrogen sulphide in the Black Sea waters. *Okeanologiya*, 14: 403–420 (in Russian).

Aleev, Yu. G., 1968. On the results of ichthyological investigations of the Black Sea. In: *Biological Studies of the Black Sea*. Nauka, Moscow, pp. 156–161 (in Russian).

Antipa, G., 1941. Marea Neagra. Vol. 1. Oceanografia, bionomia si biologia generala a Marii Negre. (The Black Sea. Vol. I. Oceanography, bionomy and general biology of the Black Sea.) *Acad. Rom. Publ. Fond "V. Adamachi"*, 10(55): 313 pp.

Arkhangelski, A.D. and Strakhov, N.M., 1938. *Geological Structure and History of the Evolution of the Black Sea*. Izv. Akad. Nauk S.S.R, Moscow, Leningrad, 310 pp. (in Russian).

Arnol'di, L.V., 1949. Data on the quantitative studies of zoobenthos of the Black Sea. *Tr. Sevastop. Biol. Stn., Akad. Nauk Ukr. S.S.R.*, 7: 127–192 (in Russian).

Bacescu, M., 1965. Bibliographie Roumaine de la Mer Noire. (A comprehensive bibliography of Romanian studies relating to the Black Sea.) Comm. Natl. Repub. Pop. Roum. l'UNESCO, Bucarest, 122 pp.

Bacescu, M., Muller, G.I. and Gomoiu, M.T., 1971. Cercetari de ecologie bentala in Marea Neagra. *Ecol. Mar.* (*Contanța*). 4: 357 pp.

Beers, J.R. and Stewart, G.L., 1971. Micro-zooplankters in the plankton communities of the upper waters in the eastern tropical Pacific. *Deep-Sea Res.*, 18: 861–883.

Belyaev, V., Lenin, A. and Petipa, T.S., 1974. Modeling of the pelagic part of the Black Sea. In: *Mathematical Models of Marine Ecosystems*. Naukova dumka, Kiev (in Russian).

Bibik, V.A., 1964. Properties of water dynamics in the south-eastern Black Sea and the distribution of oceanographic elements. *Tr. Azovo-Chernomorsk. Nauchno-Issled. Inst. Morsk. Rybn. Khoz. Okeanogr.*, 23: 23–33 (in Russian).

Bodeanu, N., 1969. Cercetari asupra fitoplanctonului din zona de mica adincime de la litoralul romanesc al Marii Negre. (Researches on the phytoplankton in the shallow waters along the Rumanian littoral of the Black Sea.) *Ecol. Mar.* (*Constanța*), 3: 65–147.

Bodeanu, N. and Roban, A., 1975. Données concernant la floraison des eaux du littoral Roumain de la Mer Noire avec le péridinien *Exuviaella cordata*. *Cercet. Mar.* (*Constanța*), 8: 63–70.

Bogdanova, A.K., 1959a. On the distribution of oxygen in the Black Sea. *Tr. Sevastop. Biol. Stn., Akad. Nauk Ukr. S.S.R.*, 9: 279–315 (in Russian).

Bogdanova, A.K., 1959b. Water exchange through the Bosporus and its role in the mixing of the Black Sea water. *Tr. Sevastop. Biol. Stn., Akad. Nauk Ukr. S.S.R.*, 12: 401–427 (in Russian).

Bogdanova, A.K., 1961. The distribution of Mediterranean water in the Black Sea. *Okeanologiya*, 1: 983–991 (in Russian) [English transl., 1963, *Deep-Sea Res.*, 10: 665–672].

Bogdanova, A.K. and Stepanov, V.N., 1976. Transformation of waters of the Black and Mediterranean Seas passing through the Bosporus. In: *Studies of Water Exchange Through the Bosporus and Canal de Sicile*. Naukova dumka, Kiev, pp. 94–124 (in Russian).

Bogoyavlenski, A.N., 1962. On the distribution of heterotrophic bacteria in Indian Ocean and Antarctic waters. *Okeanologiya*, 2: 403–420 (in Russian).

Brewer, P.G. and Spencer, D.W., 1974. Distribution of some trace elements in the Black Sea and their flux between dissolved and particulate phases. In: E.T. Degens and D.A. Ross (Editors), *The Black Sea — Geology, Chemistry, and Biology*. American Association of Petroleum Geologists, Tulsa, Okla., pp. 137–143.

Bruevich, V.C., 1953. Chemistry and biological productivity of the Black Sea. *Tr. Inst. Okeanol. Akad. Nauk S.S.S.R.*, 7: 11–51 (in Russian).

Caspers, H., 1957. Black Sea and Sea of Azov. In: J.W. Hedgpeth (Editor), *Treatise on Marine Ecology and Paleoecology. Geol. Soc. Am., Mem*, 67, Vol. I, pp. 801–890.

Cociasu, A. and Popa, L., 1976. Dynamique des phosphates et des silicates des eaux marines du littoral Roumain de la Mer Noire dans la période 1959–1975. *Cercet. Mar.* (*Constanța*), 9: 41–58.

Danilchenko, P.T. and Chigirin, N.I., 1927. On the origin of hydrogen sulphide in the Black Sea. *Tr. Sevastop. Biol. Stn., Akad. Nauk Ukr. S.S.R., Ser. 2*, 11: 141–191 (in Russian).

Datsko, V.G., 1954. Some chemical indicators of the productivity of the Black Sea. *Tr. Vses. Nauchno-Issled. Inst. Morsk. Rybn. Khoz. Okeanogr.*, 28: 188–202 (in Russian).

Datsko, V.G., 1959. *Organic Matter in the Southern Seas of U.S.S.R.* Izv. Akad. Nauk S.S.S.R., Moscow, 268 pp. (in Russian).

Datsko, V.G., 1961. Value of phytoplankton production in the balance of organic matter in marine basins of the Azov, Black and Caspian Seas. In: G.G. Vinberg (Editor), *Primary Production of the Seas and Oceans*. Minsk, pp. 50–58 (in Russian).

Degens, E.T. and Ross, D.A., 1974. *The Black Sea — Geology, Chemistry and Biology*. American Association of Petroleum Geologists, Tulsa, Okla., 635 pp.

Dekhnik, T.V., 1970. *Ichthyoplankton of the Black Sea*. Naukova dumka, Kiev, 234 pp. (in Russian).

Deuser, W.G., 1970. Carbon-13 in Black Sea waters and implications for the origin of hydrogen sulphide. *Science*, 168: 1575–1577.

Deuser, W.G., 1971. Organic carbon balance of the Black Sea. *Deep-Sea Res.*, 18: 995–1004.

Dobrzhanskaya, M.A., 1960. The main features of the hydrochemical regime of the Black Sea. *Tr. Sevastop. Biol. Stn., Akad. Nauk Ukr. S.S.R.*, 13: 325–345 (in Russian).

Dobrzhanskaya, M.A., 1967. Distribution of hydrogen sulphide in the upper layer of the anoxic zone of the Black Sea. In: *Hydrophysical and Hydrochemical Studies in the Black Sea*. Naukova dumka, Kiev, pp. 104–127 (in Russian).

Duka, L.A., 1969. The feeding of *Callynimus* larvae in the Black Sea. In: *Biology of the Sea*. Naukova dumka, Kiev, pp. 208–224 (in Russian).

Filippov, D.M., 1968. *Circulation and Structure of the Waters in the Black Sea*. Nauka, Moscow, 260 pp. (in Russian).

Finenko, Z.Z., 1967. Primary production in southern seas. In: V.A. Vodyanitskii (Editor), *Voprosy biookeanografii*. (*Problems of Biooceanography*.) Naukova dumka, Kiev, pp. 69–74 (in Russian).

Finenko, Z.Z., 1971. *Ecological Physiology of Marine Planktonic Algae*. Naukova dumka, Kiev, 206 pp. (in Russian).

Galadjiev, M.A., 1948. Comparative composition and distribution of zooplankton in the bay and in the open sea in the region of the Crimea. *Tr. Sevastop. Biol. Stn., Akad. Nauk Ukr. S.S.R.*, 6: 173–223 (in Russian).

Gomoiu, M.T. and Petran, A., 1973. Dynamics of settlement of bivalve *Mya arenaria* on the Romanian shore of the Black Sea. *Cercet. Mar.* (*Constanța*), 5–6: 263–289.

Gorbenko, Yu.A., 1977. *Ecology of Marine Periphytonic Microorganisms*. Naukova dumka, Kiev, 250 pp. (in Russian).

Grese, V.N., 1968. Studies of zooplankton in the Black Sea. In: *Biological Studies of the Black Sea*. Nauka, Moscow, pp. 47–52 (in Russian).

Grese, V.N., Fedorina, A.I. and Chmyr, V.D., 1973. Production of the main components of the zooplankton food for the plankton-feeding fishes in the Black Sea. In: *Biology of the Sea*. Naukova dumka, Kiev, pp. 3–22 (in Russian).

Hunt, J.M., 1974. Hydrocarbon geochemistry of the Black Sea. In: E.T. Degens and D.A. Ross (Editors), *The Black Sea — Geology, Chemistry and Biology*. American Association of Petroleum Geologists, Tulsa, Okla., pp. 499–504.

Issachenko, B.L., 1951. *Selected Works*. Akad. Nauk S.S.S.R., Moscow, 388 pp. (in Russian).

Ivanov, M.V., 1957. The role of microorganisms in the formation of sulphur deposits in hydrogen sulphide springs of Sergiev mineral waters. *Mikrobiologiya*, 26: 338–345 (in Russian) [English transl. in *Microbiology*, 26: 342–348].

Kalugina-Gutnik, A.A., 1975. *Phytobenthos of the Black Sea*. Naukova dumka, Kiev, 245 pp. (in Russian).

Khailov, K.M., 1971. *The Ecological Metabolism in the Sea*. Naukova dumka, Kiev, 252 pp. (in Russian).

Kiseleva, M.I., 1967. Comparative characteristics of the analogous bottom biocoenoses in different regions of the Black Sea. In: *Bottom Biocoenosis and Biology of Benthic Organisms of the Black Sea*. Naukova dumka, Kiev, pp. 18–27 (in Russian).

Kovaleva, T.M., 1969. Seasonal variation of phytoplankton in the neritic zone of the Black Sea. In: *Biology in the Sea*.

Naukova dumka, Kiev, pp. 18–31 (in Russian).

Kriss, A.E., 1963. *Marine Microbiology (Deep Sea)*. Oliver and Boyd, Edinburgh and London, 536 pp.

Kriss, A.E. and Lebedeva, M.N., 1968. Microorganisms and biological productivity of the Black Sea. In: *Biological Studies of the Black Sea*. Nauka, Moscow, pp. 10–17 (in Russian).

Lebedeva, M.N., 1957. Abundance and distribution of bacteria in the oxygen zone of the Black Sea. *Dokl. Akad. Nauk S.S.S.R.*, 115: 186–189 (in Russian).

Lebedeva, M.N., 1959. Filamentous bacteria from the depth of the Black Sea as a possible food for filter-feeding zooplankton. *Tr. Sevastop. Biol. Stn., Akad. Nauk Ukr. S.S.R.*, 11: 29–72 (in Russian).

Lebedeva, M.N., 1971. Microbiological studies in the south seas. In: *Problems of Marine Biology*. Naukova dumka, Kiev, pp. 55–61 (in Russian).

Lebedeva, M.N. and Markianovich, Ye.M., 1958. Microorganisms of the Black Sea observed by methods of direct microscopy . *Tr. Sevastop. Biol. Stn., Akad. Nauk Ukr. S.S.R.*., 10: 175–194 (in Russian).

Mamaeva, N.W. and Moiseev, E.V., 1979. The layer enriched with protozoa above the H_2S zone in the Black Sea. *Dokl. Akad. Nauk S.S.S.R.*, 248:506–508.

Mironov, O.G., 1969. Microorganisms growing on hydrocarbons from the Black Sea. *Mikrobiologiya*, 38: 728–730 (in Russian) [English transl. in *Microbiology*, 38: 608–610].

Mironov, O.G., 1971. *On the Oil Oxidizing Microorganisms in the Sea*. Naukova dumka, Kiev, 218 pp. (in Russian).

Morozova-Vodyanitzkaya, N.V., 1950. Abundance and biomass of phytoplankton in the Black Sea. *Dokl. Akad. Nauk S.S.S.R.*, 73: 821–824 (in Russian).

Morozova-Vodyanitzkaya, N.V., 1957. Phytoplankton in the Black Sea and its quantitative development. *Tr. Sevastop. Biol. Stn., Akad. Nauk Ukr. S.S.R.*, 9: 3–57 (in Russian).

Morozova-Vodyanitzkaya, N.V., 1959. Plant associations in the Black Sea. *Tr. Sevastop. Biol. Stn., Akad. Nauk Ukr. S.S.R.*, 11: 3–42 (in Russian).

Neprochnov, Yu.P. and Yel'nikov, I.N., 1969. The results of seismic investigations of the structure of the sedimentary deposits of the Black Sea basin along the Yalta–Sinope profile. *Okeanologiya*, 9: 823–833 (in Russian) [English transl. in *Oceanology*, 9: 667–676].

Neumann, G., 1942. Die absolute Topografie des physikalischen Meeresniveaus und die Oberflächen-Strömungen des Schwarzen Meeres. *Ann. Hydrogr. Berl.*, 70: 265–282.

Neuymin, G.G., 1973. Optical studies in the Black Sea waters. In: *Materials of Symposium on the Studies of the Black and Mediterranean Seas*. Naukova dumka, Kiev, pp. 81–88 (in Russian).

Nikitin, V.N., 1945. Distribution of the biomass of plankton in the Black Sea. *Dokl. Akad. Nauk S.S.S.R.*, 47: 510–512 (in Russian).

Ostroumov, E.A., Volkov, I.I. and Fomina, L.S., 1961. Distribution of sulphur compounds in sediments of the Black Sea. In: N.M. Strakhov (Editor), *Recent Sediments of the Seas and Oceans*. Izd. Akad. Nauk S.S.S.R., Moscow, pp. 597–633 (in Russian). [Also in: *Tr. Inst. Okeanol. Akad. Nauk S.S.S.R.*, 50: 93–129 (in Russian).]

Pavlova, E.V., 1967. Consumption of food and energy transformation by the cladoceran populations in the Black Sea. In: *Structure and Dynamics of Aquatic Communities and Populations*. Naukova dumka, Kiev, pp. 66–86 (in Russian).

Pavlova, E.V. and Sorokin, Yu.I., 1970. Bacterial feeding of the planktonic crustacean, *Penilia* from the Black Sea. In: *Biology of the Sea*. Naukova dumka, Kiev, pp. 166–182 (in Russian).

Pavlovskaya, T.V., 1976. Distribution of microzooplankton in the coastal waters of the Black Sea. In: *Biology of the Sea*. Naukova dumka, Kiev, 36, pp. 75–83 (in Russian).

Petipa, T.S., 1968. Feeding of planktonic crustaceans in the Black Sea. In: *Biological Studies in the Black Sea*. Nauka, Moscow, pp. 67–71 (in Russian).

Petipa, T.S., 1975. Origin and classification of main ecological types of feeding of Calanoida. In: *Biology of the Sea*. Naukova dumka, Kiev, pp. 27–49 (in Russian).

Petipa, T.S., Sajina, L.I. and Delalo, E.P., 1960. Vertical distribution of zooplankton in the Black Sea as dependent upon the hydrological conditions. *Dokl. Akad. Nauk S.S.S.R.*, 133: 964–968 (in Russian).

Petipa, T.S., Sajina, L.I. and Delalo, E.P., 1963. Distribution of zooplankton in the Black Sea in 1951–1956. *Okeanologiya*, 3: 110–120 (in Russian).

Petipa, T.S., Pavlova, E.V. and Mironov, G.N., 1970a. Structure of food chains and energy utilization in the plankton communities of the Black Sea. In: *Biology of the Sea*. Naukova dumka, Kiev, pp. 3–42 (in Russian).

Petipa, T.S., Sorokin, Yu.I. and Lanskaya, L.A., 1970b. Studies of feeding of *Acartia clausii* with the use of ^{14}C. In: *Biology of the Sea*. Naukova dumka, Kiev, pp. 166–181 (in Russian).

Polikarpov, G., 1966. *Radioecology of Aquatic Organisms*. Elsevier, Amsterdam, New York, N.Y., 314 pp.

Proshkina-Lavrenko, A.I., 1955. *Planktonic Diatoms of the Black Sea*. Izd. Akad. Nauk S.S.S.R., Moscow, 243 pp. (in Russian).

Psenin, L.N., 1966. *Biology of Nitrogen Fixing Bacteria*. Naukova dumka, Kiev, 266 pp. (in Russian).

Pyzik, G.K., 1968. Studies of phytoplankton in the Black Sea from 1953–1963. In: *Biological Studies of the Black Sea*. Nauka, Moscow, pp. 30–39 (in Russian).

Richards, F.A., 1965. Anoxic basins and fjords. In: J.P. Riley and G. Skirrow (Editors), *Chemical Oceanography, 1*. Academic Press, London, New York, N.Y., pp. 611–645.

Richards, F.A., Cline, J.D., Broenkow, W.W. and Atkinson, L.P., 1965. Some consequences of the decomposition of organic matter in Lake Nitinat, an anoxic fjord. *Limnol. Oceanogr.*, 10: R185–R201.

Roukhiyainen, M.I., 1975. On the seasonal dynamics of phytoplankton in the Black Sea. In: *Biology of the Sea*. Naukova dumka, Kiev, pp. 3–15 (in Russian).

Shimkus, K.M. and Trimonis, E.S., 1974. Modern sedimentation in Black Sea. In: E.T. Degens and D.A. Ross (Editors), *The Black Sea — Geology, Chemistry, and Biology*. American Association of Petroleum Geologists, Tulsa, Okla., pp. 249–278.

Skopintzev, B.A., 1953. On the distribution of hydrogen sul-

phide in the Black Sea. *Meteorol. Gidrol.*, 7: 37–42 (in Russian).
Skopintzev, B.A., 1975. *Formation of Modern Chemical Composition of Waters in the Black Sea.* Hydrometeoisdat, Leningrad, 335 pp. (in Russian).
Soljankin, E.V., 1963. On the water balance of the Black Sea. *Okeanologiya*, 3: 986–993 (in Russian).
Sorokin, Yu.I., 1958. Microflora and chemical composition of the bottom sediments of Rybinsk reservoir. *Tr. Biol. Stn. "Borok", Akad. Nauk S.S.S.R.*, pp. 89–105 (in Russian).
Sorokin, Yu.I., 1962a. Vertical distribution of bacteria in the Black Sea and the methods for taking samples. *Mikrobiologiya*, 31: 684–693 (in Russian). [English transl. in *Microbiology*, 31: 558–565.]
Sorokin, Yu.I., 1962b. Microflora of Black Sea bottom deposits. *Mikrobiologiya*, 31: 899–903 (in Russian). [English transl. in *Microbiology*, 31: 729–732.]
Sorokin, Yu.I., 1962c. Experimental investigation of bacterial sulfate reduction in the Black Sea using S^{35}. *Mikrobiologiya*, 31: 402–410 (in Russian). [English transl. in *Microbiology*, 31: 329–335.]
Sorokin, Yu.I., 1964a. On the primary production and bacterial activities in the Black Sea. *J. Cons. Int. Explor. Mer*, 29: 41–65.
Sorokin, Yu.I., 1964b. A quantitative study of the microflora in the central Pacific Ocean. *J. Cons. Int. Explor. Mer*, 29: 25–40.
Sorokin, Yu.I., 1967. On the mechanism of chemical and biological oxidation of sulphides at low concentration. *Tr. Biol. Stn. "Borok", Akad. Nauk S.S.S.R.*, 19: 75–83 (in Russian).
Sorokin, Yu.I., 1968. Studies of microbial activity in the Black Sea. In: *Biological Studies of the Black Sea and Its Resources.* Nauka, Moscow, pp. 18–25 (in Russian).
Sorokin, Yu.I., 1970. Interrelations between sulphur and carbon turnover in meromictic lakes. *Arch. Hydrobiol.*, 66: 391–446.
Sorokin, Yu.I., 1971. On the role of bacteria in the productivity of tropical oceanic waters. *Int. Rev. Ges. Hydrobiol.*, 56: 1–48.
Sorokin, Yu.I., 1972. The bacterial population and the processes of hydrogen sulphide oxidation in the Black Sea. *J. Cons. Int. Explor. Mer*, 34: 423–454.
Sorokin, Yu.I., 1975. Sulphide formation and chemical composition of bottom sediments of some Italian lakes. *Hydrobiologia*, 47: 231–240.
Sorokin, Yu.I., 1977. The heterotrophic phase of plankton succession in the Japan Sea. *Mar. Biol.*, 41: 107–117.
Sorokin, Yu.I. and Donato, N., 1975. On the carbon and sulphur metabolism in the meromictic Lake Faro (Sicily). *Hydrobiologia*, 47: 241–252.
Strakhov, N.M., 1947. Toward understanding the mechanisms of marine sedimentation. Station 1. The Black Sea. *Izv. Akad. Nauk S.S.S.R., Ser. Geol.*, 22: 49–90 (in Russian).
Svetovidov, A.M., 1964. *Fish of the Black Sea.* Nauka, Moscow, Leningrad.
Takahashi, M. and Ishimura, S., 1968. Vertical distribution and organic matter production of photosynthetic sulfur bacteria in Japanese lakes. *Limnol. Oceanogr.*, 13: 644–655.
Tsiban, A.V., 1970. *Bacterioplankton and Bacterioneuston in Coastal Waters of the Black Sea.* Naukova dumka, Kiev, 270 pp. (in Russian).
Vinogradov, K.A., 1968. Some results of the studies of benthos in the Black Sea. In: *Biological Studies of the Black Sea.* Nauka, Moscow, pp. 90–96 (in Russian).
Vinogradov, M.E. and Shushkina, F.A., 1980. On the vertical structure of zooplankton community in the Black Sea. In: M. Vinogradov (Editor), *Study of Pelagic Ecosystem of the Black Sea*, Nauka, Moscow, in press.
Vinogradova, Z.A., 1964. Biochemical aspects of comparative studies of plankton of the Black Sea, Caspian Sea and the Sea of Azov. *Okeanologiya*, 4: 232–242 (in Russian).
Vladimirtzev, Yu.A., 1961. On the processes of mixing of waters in the Black Sea. *Okeanologiya*, 1: 976–982 (in Russian).
Vodyanitzki, V.A., 1948. The basic water exchange and the formation of salinity in the Black Sea. *Tr. Sevastop. Biol. Stn., Akad. Nauk Ukr. S.S.R.*, 6: 386–432 (in Russian).
Vodyanitzki, V.A., 1956. On the problem of biological productivity of the Black Sea. *Tr. Sevastop. Biol. Stn., Akad. Nauk Ukr. S.S.R.*, 8: 347–438 (in Russian).
Vodyanitzki, V.A., 1958. On the possibility of atomic waste disposal in the Black Sea. *Priroda*, 2: 46–52 (in Russian).
Wheatland, A.B., 1954. Factors affecting the formation and oxidation of sulphides in a polluted estuary. *J. Hyg.*, 52: 194–210.
Zaika, V.E., Averina, T.Yu., Ostrovskaya, N.A. and Zalkina, A.V., 1976. *Distribution of Marine Microzooplankton.* Naukova dumka, Kiev, 88 pp. (in Russian).
Zaitzev, Yu.P., 1970. *Marine Neustonology.* Naukova dumka, Kiev, 250 pp. (in Russian).
Zenkevitch, L.A., 1963. *Biology of the Seas of the USSR.* Interscience, New York, N.Y., and Allen and Unwin, London, 955 pp.
Zernov, S.A., 1913. On the question of life in the Black Sea. *Zap. Imp. Akad. Nauk, St. Petersb., Ser. 8*, 32: 1–150 (in Russian).

Chapter 12

THE RED SEA

DAVID A. ROSS

INTRODUCTION

The Red Sea is situated between the Arabian Peninsula and the African continent, and is an elongated basin that is being formed by sea-floor spreading. Bottom topography consists of smooth shelves separated by a large, central main trough that itself has a deep and deformed axial valley. It is bounded by topographically high areas that influence the weather and generally restrict atmospheric circulation to winds parallel to the axis of the sea. Climate is arid and hot and evaporation is of the order of 2000 mm yr^{-1}. North of 19°N the winds are generally from a north to northeasterly direction; to the south their direction changes according to the monsoons. From October to May winds are mainly from the southeast, while during June to September they are mainly from the northwest. The combination of climate, low river runoff and rainfall, high evaporation rate and restricted circulation has produced relatively dense, high temperature (up to 30°C) and high salinity (40‰ or more) water.

The Red Sea is about 1800 km long; about 250 km in average width (Fig. 12.1); and about 560 m in average depth, with a maximum depth of 2920 m. Thus the surface area is about 450 000 km^2, and the volume of the Red Sea is about 251 000 km^3 (Muromtsev, 1960). It is connected to the Mediterranean Sea via the Gulf of Suez and the Suez Canal (although relatively little water exchange occurs there), and to the Indian Ocean via the Strait of Bab al Mandab (often spelled Mandeb, but Mandab is more similar to Arabian pronunciation). With the opening of the Suez Canal in 1869 the Red Sea became a main artery for shipping to and from southeastern Asia.

The general circulation pattern of the Red Sea shows very distinct seasonal changes that are influenced primarily by prevailing winds and to a lesser extent by seasonal changes in density and evaporation. In the winter (October–May) surface waters from the Gulf of Aden flow to the north and the subsurface flow is to the south. This pattern reverses in the summer. Upwelling occurs in the northern part of the sea during the summer months, and sinking of dense water occurs in the winter period. The renewal time for the upper 150 m of Red Sea water may be as little as six years, whereas for the entire Red Sea a value of about 200 years may be more appropriate.

The Sinai Peninsula (Fig. 12.1) separates the northern part of the Red Sea into two gulfs: the deep Gulf of Aqaba (also called Elat) to the east and the shallow Gulf of Suez to the west. The Gulf of Aqaba is extremely narrow with steep submarine slopes whereas the Gulf of Suez is wider and has gentle topography. Geologically, the Gulf of Aqaba is part of an active strike-slip fault system (Dead Sea — Jordan Rift), while the Gulf of Suez is tectonically inactive and contains a relatively thick sedimentary sequence. To the south, the Red Sea narrows to its minimum at the Strait of Bab al Mandab where it is only about 25 km wide. The shallowest area, somewhat to the northwest, is 137 m deep (Werner and Lange, 1975). Seismic profiles show this general area to have considerable normal faulting.

Oceanographic observations and measurements have been made in the Red Sea since 1837. Among the early worldwide oceanographic expeditions only two, the *Vitiaz* (1886–1889) and the *Valdivia* (1898–1899), went through the Red Sea. Important studies were made during the two Austrian *Pola* expeditions in the late 1890's which occupied over

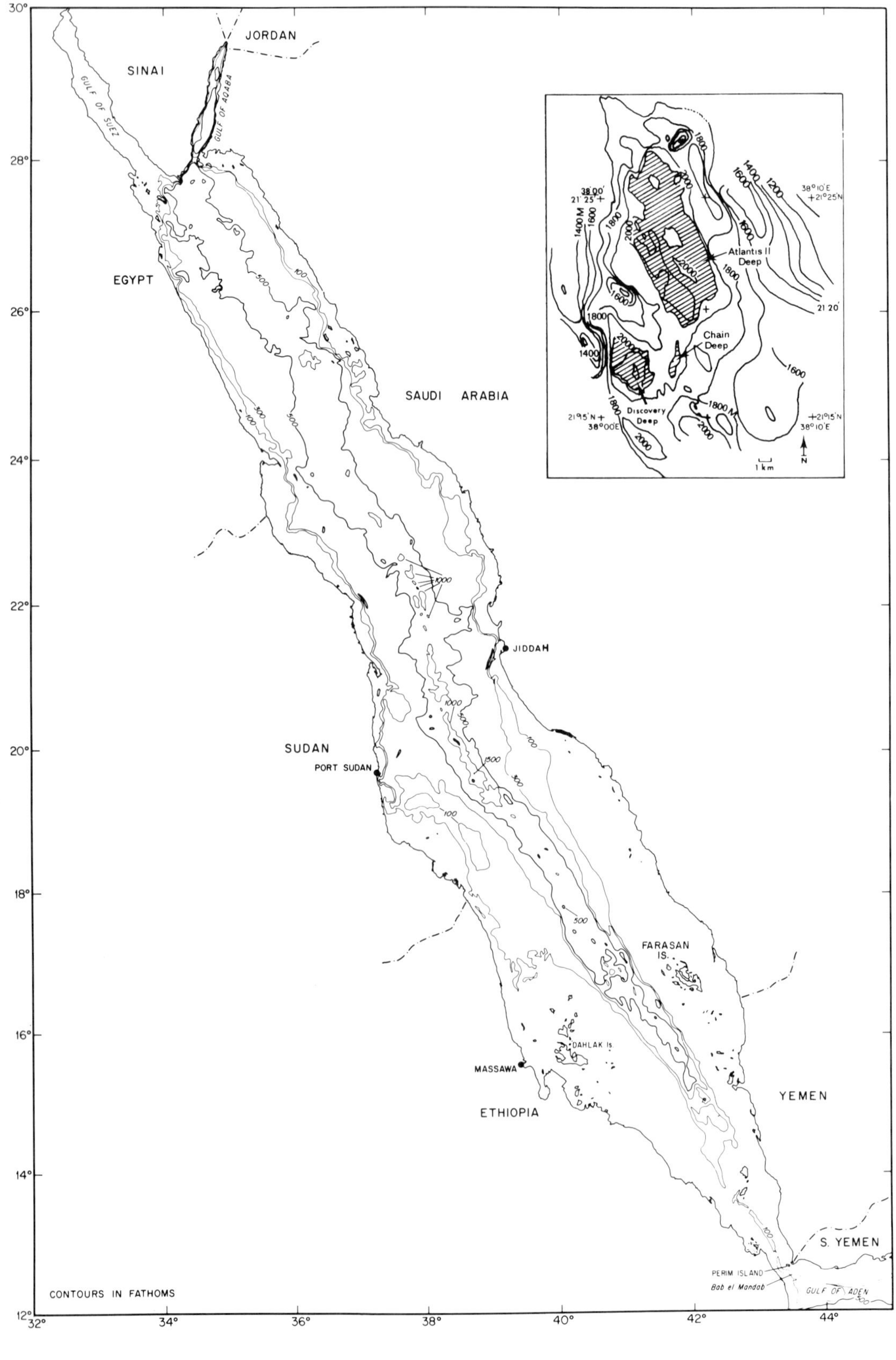

SINAI
JORDAN
GULF OF SUEZ
GULF OF AQABA
EGYPT
SAUDI ARABIA
SUDAN
PORT SUDAN
JIDDAH
FARASAN IS.
DAHLAK Is.
MASSAWA
ETHIOPIA
YEMEN
S. YEMEN
PERIM ISLAND
Bab el Mandab
GULF OF ADEN
CONTOURS IN FATHOMS
Atlantis II Deep
Chain Deep
Discovery Deep
1 km
N

250 stations in the Red Sea. The quality of their measurements, although good for the time, is below the accuracy needed for modern work. In the early 20th century, several expeditions were made to the Red Sea area, mostly as research vessels were in transit to or from the Indian Ocean. These studies were so numerous that by 1910 the Red Sea could be considered to be one of the best known seas. At present, however, the opposite is true. An excellent history of Red Sea exploration is given by Morcos (1970).

GENERAL CHARACTERISTICS

Bathymetry

Probably the first bathymetric chart of the Red Sea was by Luksch (1901a, b) who used about 190 lead line soundings obtained during cruises of the *Pola*. Echosounding in the Red Sea has always been difficult because of steep slopes and narrow valleys in the deeper central portions of the sea and numerous shallow, uncharted reefs in the coastal regions. During the International Indian Ocean Expedition, a physiographic chart of the Indian Ocean that included the Red Sea was made by Heezen and Tharp (1964). Several bathymetric charts subsequently have been prepared, including those of Allen (1966); Belousov (1965); Geologisch-Paläontologisches Institut der Universität Freiburg; and the British Admiralty (Chart C 6359). The recent chart of Laughton (1970) and the Soviet Geological–Geophysical Atlas of the Indian Ocean (Udintsev, 1975) present the most comprehensive bathymetric picture of the Red Sea. Bottom topography consists of a large, main trough between the coastal shelves of Yemen and Saudi Arabia on the east and Ethiopia, Sudan and Egypt on the west. The main trough has its best development in the southern and central portions of the sea where it generally has a more deeply and intensely deformed axial trough (Fig. 12.2). In the northern and extreme southern parts of the Red Sea the axial

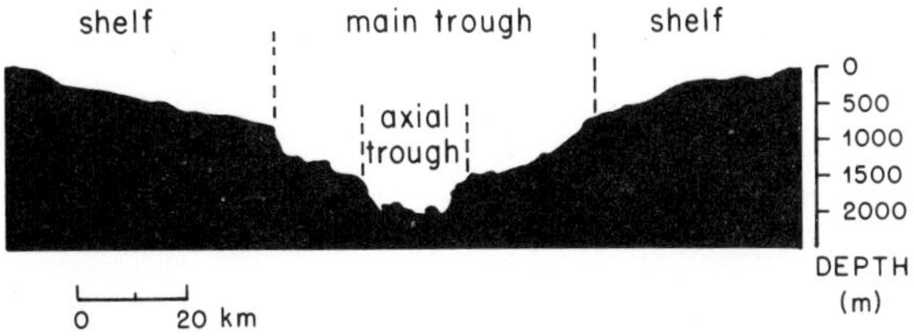

Fig. 12.2. Typical topographic profile across the Red Sea.

trough becomes topographically indistinct. In many places the axial trough and portions of the marginal zone (or collectively the main trough) are displaced in a direction essentially normal to the Red Sea axis by fracture zones. In the north the Red Sea bifurcates around the Sinai Peninsula forming the Gulfs of Aqaba and Suez.

The shelves of the Red Sea are not as well studied as the deeper portions of the basin. Shelf widths are generally about 50 km or less in the northern and central portions but become considerably wider to the south (Fig. 12.1). In many places broad banks or submerged reefs form a shelf complex. There are large island systems, the Farasan Islands off Yemen and the Dahlak Islands off Ethiopia. The shelf often gently slopes down toward the main trough, with a distinct break in topography between the two features occurring at water depths of about 600 m.

Origin of the Red Sea

The Red Sea proper, as discussed in numerous papers, appears to have been formed by the process of sea-floor spreading, with the African continent and the Arabian Peninsula slowly moving apart (McKenzie et al, 1970; Lowell and Genik, 1972; Ross and Schlee, 1973; Girdler and Styles, 1974; and Coleman, 1974). A central crack — the axial trough — is a result of this movement and is a zone of volcanic activity, with material rising from below the earth's crust to fill in the void resulting from the slow horizontal movement that is separating the land masses (Fig. 12.3). Several periods of spreading are hypothesized of which the most recent probably began about three million years ago. The Red Sea is generally considered as a model for the sea-floor spreading mechanism and may represent a basin in the beginning stages of its formation. If the spreading along the central portions of the sea

Fig. 12.1. The general bathymetry of the Red Sea (in fathoms; 1 fathom = 1.828 m), adapted from Laughton (1970). The insert chart shows the location of the hot brine region of the Red Sea (contours in m).

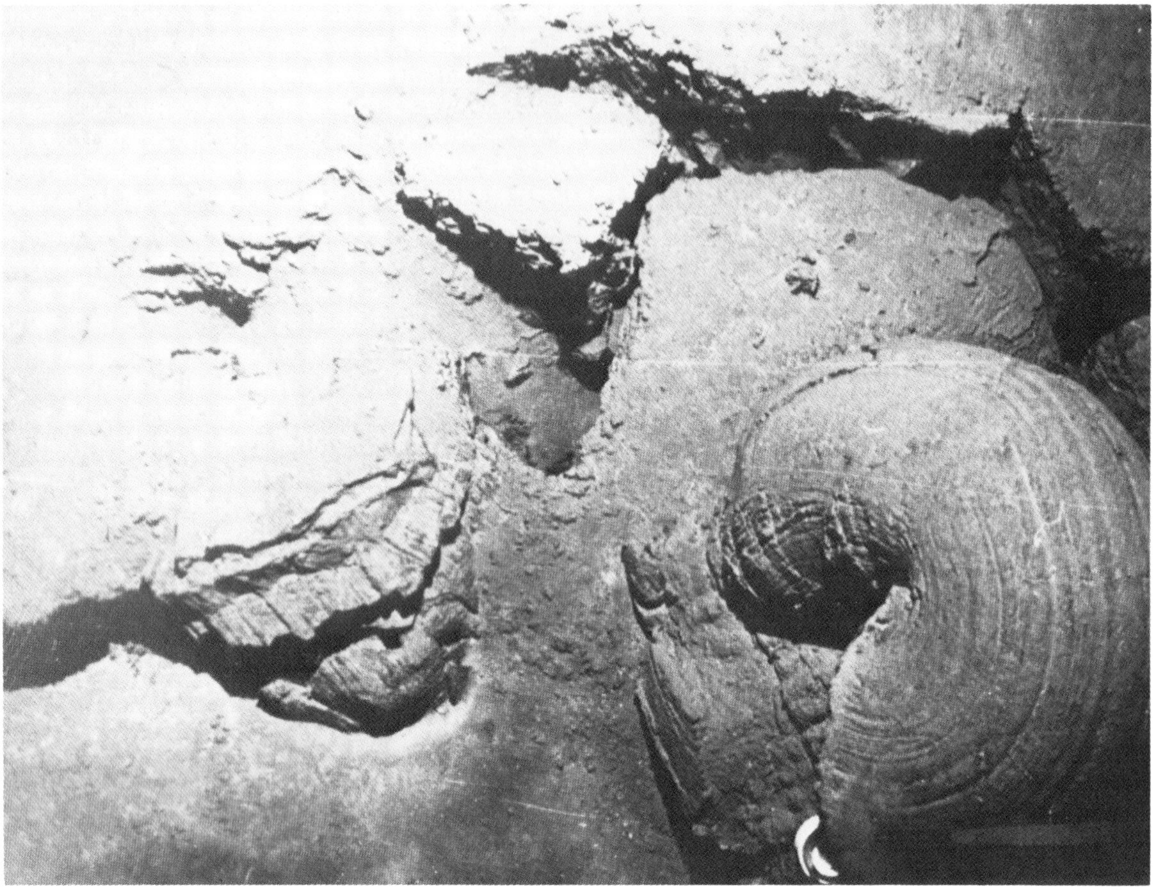

Fig. 12.3. Picture taken in the axial valley of the Red Sea showing basaltic rock forcibly extruded like toothpaste from a tube onto the sea floor. Similar pictures have been taken from other areas where sea-floor spreading is known to occur. (Photograph from Young and Ross, 1974.)

continues at its present rate (about 1 cm yr^{-1}), in about 200 million years the Red Sea will resemble the Atlantic Ocean in width. As spreading occurs, Saudi Arabia is moving eastward and slowly pushing the Persian Gulf under Iran; eventually it will disapear and the two countries will collide.

Biological activity

The Red Sea does not have any major fisheries. A small sardine and shrimp fishery exists in the north part of the sea and some bottom trawling is done in the south. The presence of numerous coral reefs often restricts trawling activities. Data on fisheries and phytoplankton are sparse, although the FAO Atlas of Living Resources of the Sea indicates high amounts of phytoplankton production (>500 mg C m^{-2} day^{-1}) from the southern part of the Red Sea and the Gulf of Aden, probably reflecting the circulation and mixing of the water in this region (described in a later section).

ENVIRONMENTAL SETTING OF THE RED SEA

Weather and wind

Much of the Red Sea is bordered by topographically high areas of plateaus and mountains that influence weather and restrict atmospheric circulation to a direction generally parallel to the axis of the sea (Flohn, 1965).

Climate in the Red Sea area is extremely arid, and rainfall, when it occurs, is usually in small thunder or dust storms. Average rainfall for some localities is shown in Table 12.1. Evaporation is considerably higher than precipitation, and no permanent rivers presently enter the Red Sea. Air temperatures increase considerably southward along the Red Sea, and the coastal areas of the southern Red Sea and Gulf of Aden are among the hottest regions in the world (Table 12.1).

Average monthly winds are generally either north or south and vary only slightly in speed and direction (Koninklijk Nederlands Meteorologisch Instituut, 1949). The wind field is dominated by the monsoon wind system. During the summer period there is usually a broad area of low atmospheric pressure over central portions of Asia that is connected to a North African low pressure trough and a high pressure area over eastern Africa (Fig. 12.4a). This results in a rather steady, southwesterly wind over the Arabian Sea which then changes to north/northwesterly in the southern part of the Red Sea. In the wintertime, however, the pattern is reversed (Fig. 12.4b) because high pressure areas are formed and generated over central Asia and northern Africa, resulting in a low pressure field being formed over eastern Africa (Siedler, 1969).

In the northern portion of the Red Sea, north of 19°N, winds are generally from a north to northwesterly direction whereas south of 19°N they tend to be more influenced by the monsoons of the Arabian Sea and reverse direction twice during the year. During the October–May period (the northeast monsoon of the Arabian Sea), the winds are mainly from the southeast, whereas during June–September (the southwest monsoon of the Arabian Sea) they are mainly from the northwest (Patzert, 1974). In the period from October to December the stronger southeasterly winds (6–9 m s^{-1}) converge near 19°N with the weaker northwesterly winds (2–4 m s^{-1}); the convergence area tends to move southward until by the middle of the year, around June, the winds are entirely from the north/northwest for the entire length of the Red Sea (Patzert, 1974).

Physical and chemical properties of the Red Sea

No rivers of any size enter the Red Sea although some small streams formed by local storms may occasionally flow into it. Precipitation over the area is extremely low and probably does not exceed 50 mm a year (Table 12.1). Estimates of annual evaporation rates range from 1820 mm (Privett, 1959) to 2660 mm (Bogdanova, 1966). Siedler (1968) and Grasshoff (1975) have assumed an average rate of 2000 mm per year. Obviously, the combination of climate, high evaporation rate, low runoff and rainfall, and restricted circulation of the Red Sea to the north and south has produced relatively dense, high temperature and salinity water as compared to that of waters of similar depth outside the Red Sea.

Temperature distribution. Morcos (1970) has summarized much of the early temperature and salinity data from the Red Sea region, and Siedler (1968, 1969) has reported on more recent observations. The observations can be summarized as follows.

TABLE 12.1

General characteristics of Red Sea weather; data from *Handbuch für das Rote Meer* (Anonymous, 1963), as reproduced by Morcos (1970)

	Suez	Abu-el-Kizan	Jeddah	Massawa	Perim
Rainfall					
average, mm yr^{-1}	21	11	63	193	43
average number of days with at least 1 mm rain	5	2	5	21	7
Temperature (°C)					
January: mean maximum	23.5	27.0	32.0	32.0	30.0
mean minimum	6.0	17.0	13.5	19.0	24.0
August: mean maximum	39.5	35.5	42.0	43.0	39.5
mean minimum	20.5	26.5	29.5	26.5	27.5

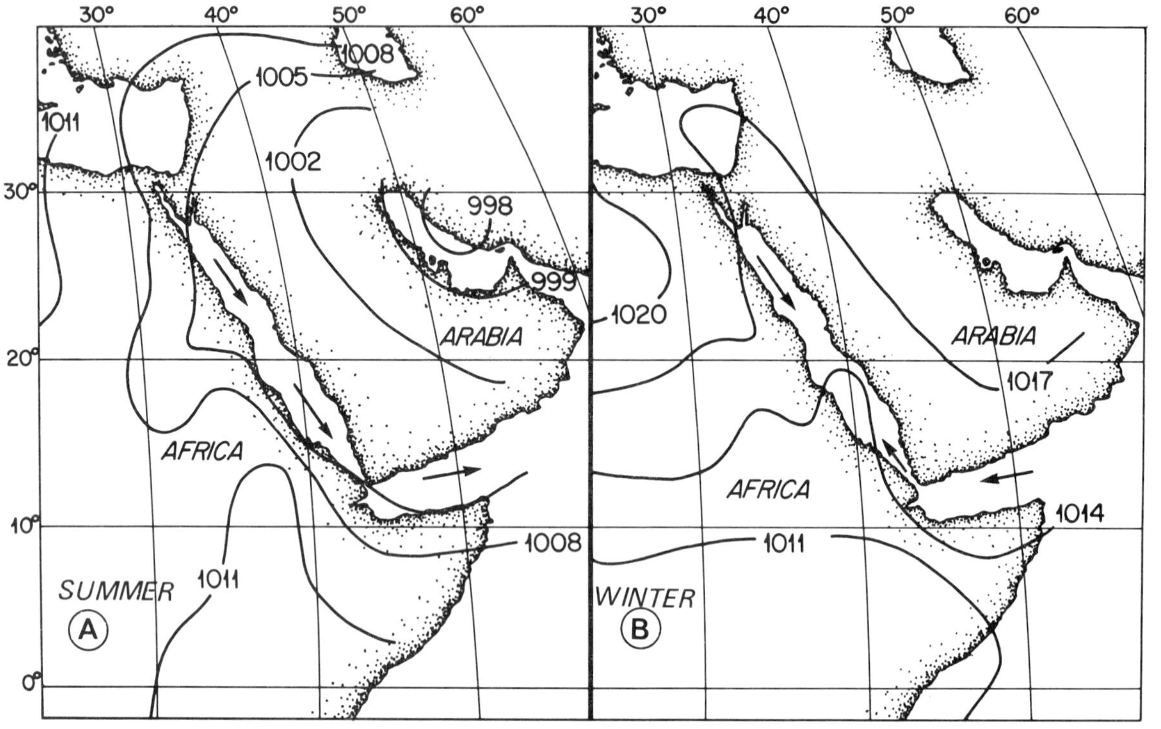

Fig. 12.4. General atmospheric pressure field pattern over the Red Sea and the Gulf of Aden during summer (A) and winter (B). Arrows indicate wind directions, numbers at isobars indicate pressure in millibars (according to Preliminary Climatic Atlas of the World, Weather Information Branch, U.S. Army Air Force, Special Series No. 1, Chicago, 1943). (From Siedler, 1969.)

Sea-water temperature, in general, shows a good correlation with air temperature and increases towards the south until cooled by incoming Gulf of Aden water. This incoming Gulf of Aden water forms a thin surface layer of relatively low salinity and temperature. Surface water temperatures reach 30°C in the southern and central parts of the sea and range between 24 and 26°C in the northern portion (Fig. 12.5). In the winter, surface waters are seldom warmer than 28°C in the central Red Sea nor below 24°C elsewhere (Fig. 12.5B). Maximum surface temperatures generally tend to occur between 19 and 20°N, with another maximum between 16 and 18°N in the summer (Fig. 12.5A). Locally, in lagoons, shallow areas or around islands temperatures can exceed 30°C.

A modest near-surface thermocline is present in both seasons, but generally is more evident in the summer than in winter. Below the thermocline the Red Sea water is essentially isothermal and has a temperature of between 21.5 and 22°C; the major exception is the hot brine pools (discussed later) but they are not really hydrographically related to the overlying normal Red Sea water. Grasshoff (1975) has found that within deeper waters the warmest portions tend to occur at depths of between 500 and 1000 m, except for the winter months when the deep water temperature is usually constant and averages about 21.8°C. This results from the convection of dense, surface water and its movement from north to south. The intermediate and deep waters of the Red Sea (exclusive of the hot brine pools) have higher temperatures for their depth than those of any other portion of the world's oceans. This phenomenon is caused by the restricted exchange with the Indian Ocean, the strong surface heating, and local circulation.

Luksch (1901a, b) made a somewhat controversial observation that temperatures appear to be higher on the Asiatic side of the basin than on the African side. He explained this situation as the result of a cyclonic circulation that brought warmer water from the south, carrying it northward along the Asiatic coast, and moved relatively colder water

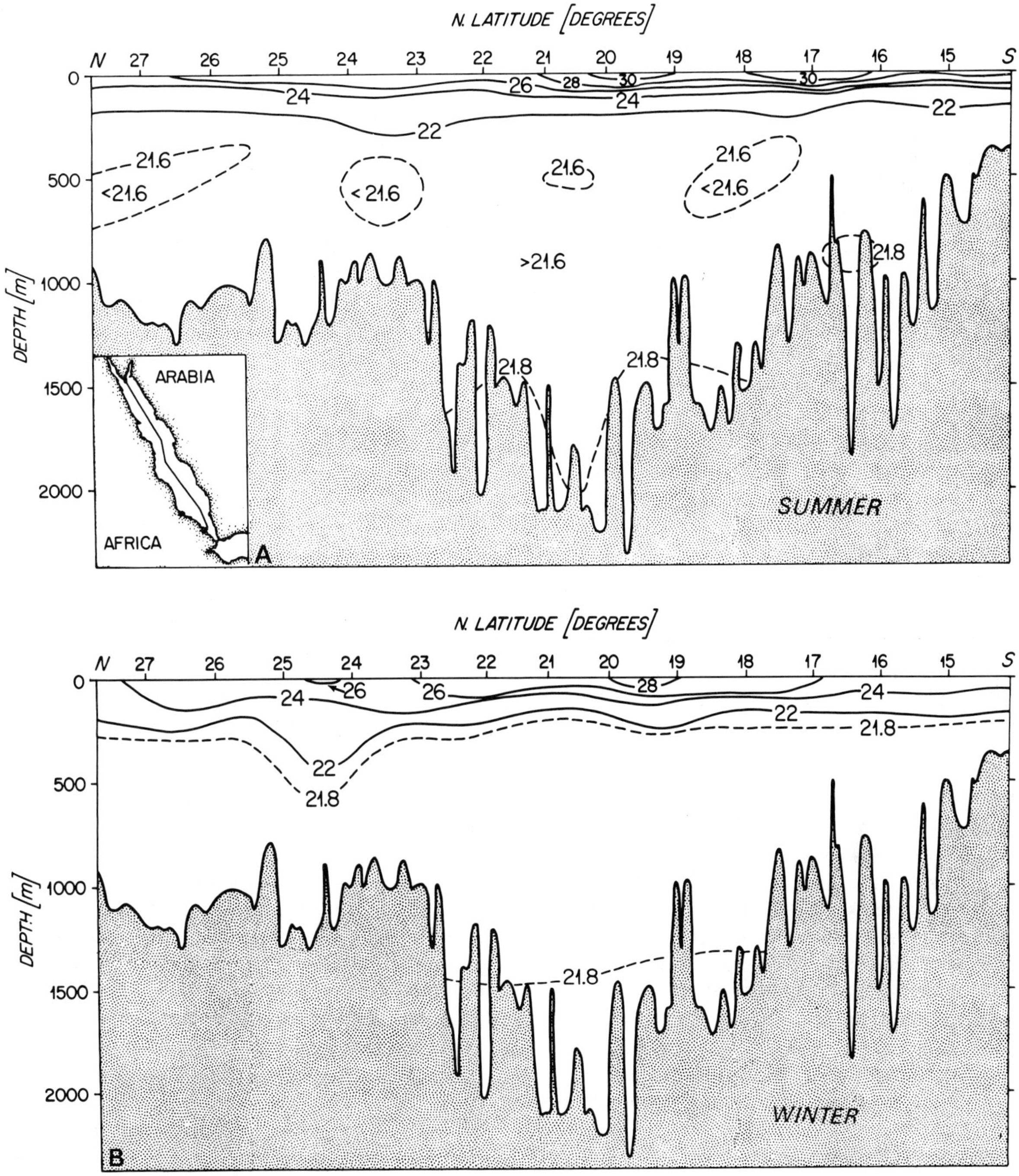

Fig. 12.5. Average summer (A) and winter (B) temperature distributions in the Red Sea in °C (from Siedler, 1968).

from the northern Red Sea to the south along the African coast. Later observations by Vercelli (1927) suggested that this pattern was not necessarily regional but could be due to local climatic and topographic conditions. Data from the Egyptian *Mabahiss* expedition to the northern Red Sea in 1934–35 support Luksch's observations, although Mohamed (1940a, b) explained the temperature patterns as being due to local effects from the northwesterly wind. Siedler (1969) considered that small differences in temperature structure in deep water might not be significant and might simply reflect sampling patterns.

Salinity distribution. The Red Sea has the highest salinity of any major sea or body of water. Its surface salinity often exceeds 40‰; minimum values found in the south and reflecting Gulf of Aden inflow are about 37.5‰ (Fig. 12.6). Generally, the seasonal changes are small, with a modest salinity decrease observable in the winter months (Fig. 12.6) through much of the Red Sea. A fairly strong halocline occurs in the southern and central portions of the Red Sea but is generally absent in the northern areas. Below the surface, the water is generally isohaline with a salinity range of 40.0 to 40.6‰. Vertical sections generally show little salinity structure, although what is present may be more pronounced during the summer (Grasshoff, 1975). The general southward decrease in salinity is only faintly evident in the subsurface waters.

Luksch's (1901a, b) east–west variation in temperature also may apply to salinity patterns, especially in the northern parts of the basin. The previous discussion on this point is applicable for salinity.

Density. The general, near-surface decrease in temperature and increase in salinity cause a corresponding density increase. There is also a progressive increase in surface density from the south (σ_t[1] about 23.00) to the north (σ_t about 28.00) as shown by the work of Neumann and McGill (1961). The deeper waters, because of their isothermal and isohaline characteristics, have a relatively uniform density, but vertical density gradients are stronger in the south (see Figs. 12.5 and 12.6).

Oxygen distribution. The oxygen concentration patterns in the Red Sea form three or four distinct layers. Surface waters are essentially saturated in oxygen, but absolute concentrations are low because of the high temperature and salinity. An oxygen minimum zone occurs at intermediate depths below the thermocline and halocline. This zone appears to be deeper toward the north than the south (Van Riel, 1932), but oxygen content is less in the south. Neumann and McGill (1961) noted a minimum containing less than 0.5 ml l^{-1} of oxygen at a depth of about 400 m, whereas in the northern part of the sea the minimum at about 500 m is about 1.50 to 1.75 ml l^{-1}. This variation may reflect higher biological activity in the south, a point supported by the higher rate of sedimentation (containing biological components) in the south as shown from seismic profiles and drilling results from the *Glomar Challenger* (Whitmarsh et al., 1974). Although the concentrations of oxygen in the minimum zone may be very low, no hydrogen sulfide has been found, nor is there any area with a complete absence of oxygen (Grasshoff, 1975). This is due to the short residence time of Red Sea water and the general circulation pattern (discussed later). Within the deeper waters of the Red Sea the oxygen concentration rises, due to deep circulation, to values of 2 ml l^{-1} or so. Mohamed (1940a, b) noted a possible additional, near-surface zone that coincided with the photosynthetic maximum.

Nutrients. The nutrient distribution in the Red Sea is not completely known, but surface waters of the Red Sea are generally depleted of nutrients. A similar situation exists in the Gulf of Aden, where much of the surface water of the Red Sea originates, so it cannot be a nutrient source. Turbulent mixing in surface layers may provide nutrient enrichment in the Straits of Bab al Mandab region. The distribution of nutrients such as phosphate generally reflects the water circulation. Intermediate waters show a phosphate maximum which results from the oxidation of organic matter; towards the bottom, phosphate concentration decreases. This pattern results from sinking of nutrient-poor surface waters from the northern part of the Red Sea, which then travel southward along the bottom (Grasshoff, 1975). As the water gets near the Straits it moves nearer the surface and increases the nutrient content of the intermediate waters there. Grasshoff (1969) made a study of the phosphate-phosphorus deficit, basing it on the average phosphate concentrations of the inflowing and outflowing waters, and concluded that there was a deficit of 153 000 t per year in the Red Sea. This, he considered, is probably partially counterbalanced by an occasional intrusion of nutrient-rich deeper waters from the Gulf of Aden.

Nitrate is generally present in low levels in the

[1]See p. 31 for definition.

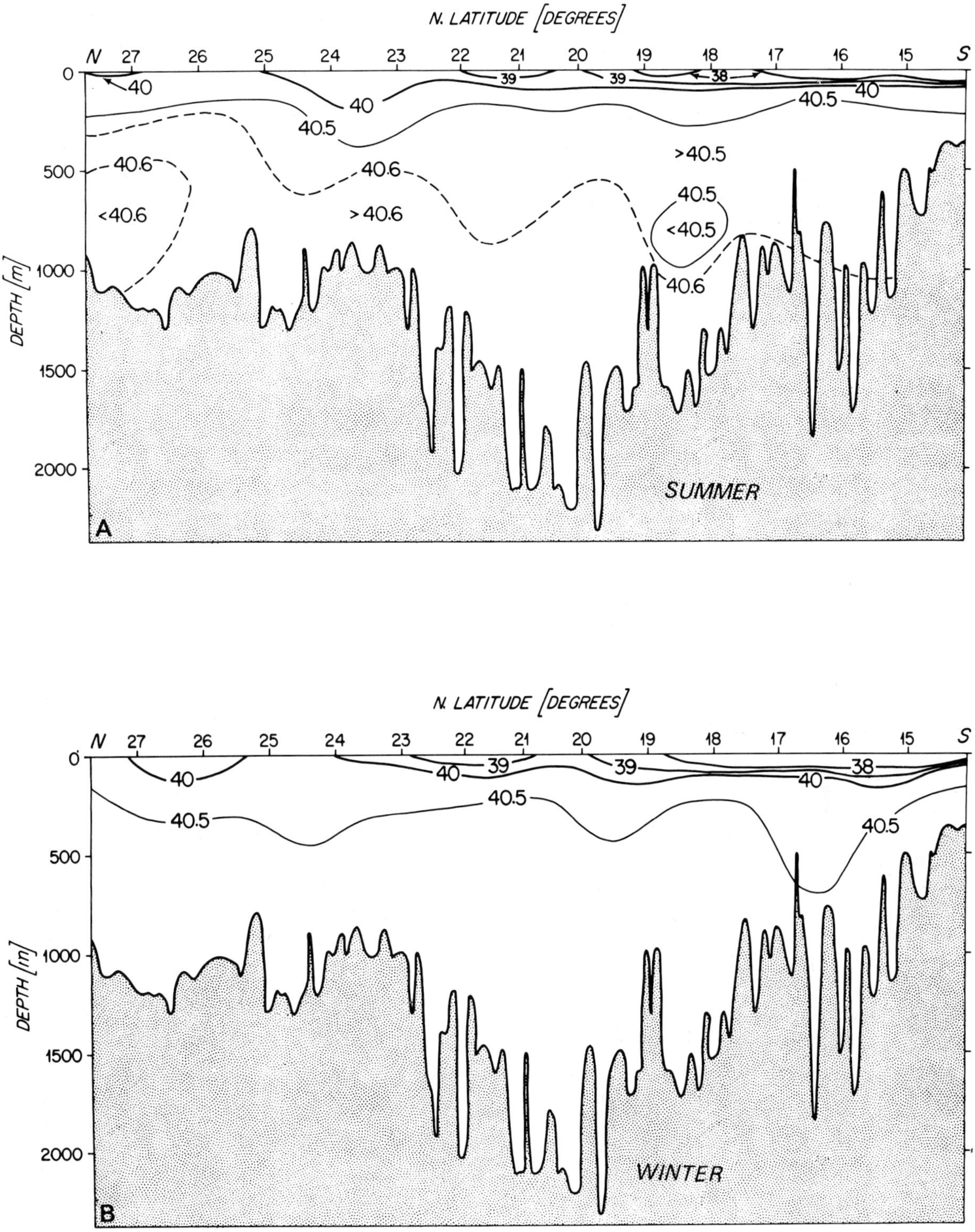

Fig. 12.6. Average summer (A) and winter (B) salinity distributions in the Red Sea. Numbers at isohalines are salinity in parts per thousand. (From Siedler, 1968.)

surface waters of the Red Sea, although turbulent mixing of outgoing Red Sea water can result in nitrate enrichment at the Straits (Grasshoff, 1969). The concentration of reactive silicate in deep water is very low, about one tenth of the amount found at similar depths in the Indian Ocean (Grasshoff, 1969). This characteristic can be useful for tracing the outflowing Red Sea water (see, for example, the *Oceanographic Atlas of the Indian Ocean*: Wyrtki, 1971).

Circulation

The circulation pattern of the Red Sea shows clear seasonal changes (Fig. 12.7). Most of the mixing, as it can be traced by nutrient distribution, appears to occur in the Strait of Bab al Mandab area. The general circulation system in the Red Sea appears to be influenced by four factors (Thompson, 1939a, b): tides, prevailing winds, evaporation, and density differences due to seasonal changes in the evaporation rate. According to Barlow (1934) the most important may be wind direction, which varies with the monsoon period. During June to September the prevailing wind is from the north to northwest. For the remainder of the year in the northern portion the winds come mainly from the north, while to the south the wind direction between October and May is mainly from the south to southeast, and during the transition time winds are essentially weak and variable. The seasonal change in atmospheric patterns results in winds that go in the same direction as the surface water flow in the Strait of Bab al Mandab during the winter but in opposition to the surface water flow in the summertime (Siedler, 1969). Wind speed also tends to be higher in winter than in summer; thus, the inflow from the Gulf of Aden is stronger in winter because of this wind stress.

Apparently there have been only two published reports on current measurements in the Bab al Mandab area. One was made during a fifteen-day period in March, 1924 (Vercelli, 1927), and the second was during November, 1964 (Siedler, 1968). Vercelli (1927) noted a maximum inflow into the Red Sea of about 66 cm s^{-1} at the surface, with a subsurface outflow of 68 cm s^{-1} at a depth of 150 m. Siedler (1968), in a different season, measured a maximum surface inflow of about 80 cm s^{-1} and an outflow of 50 cm s^{-1} at a depth of 120 m.

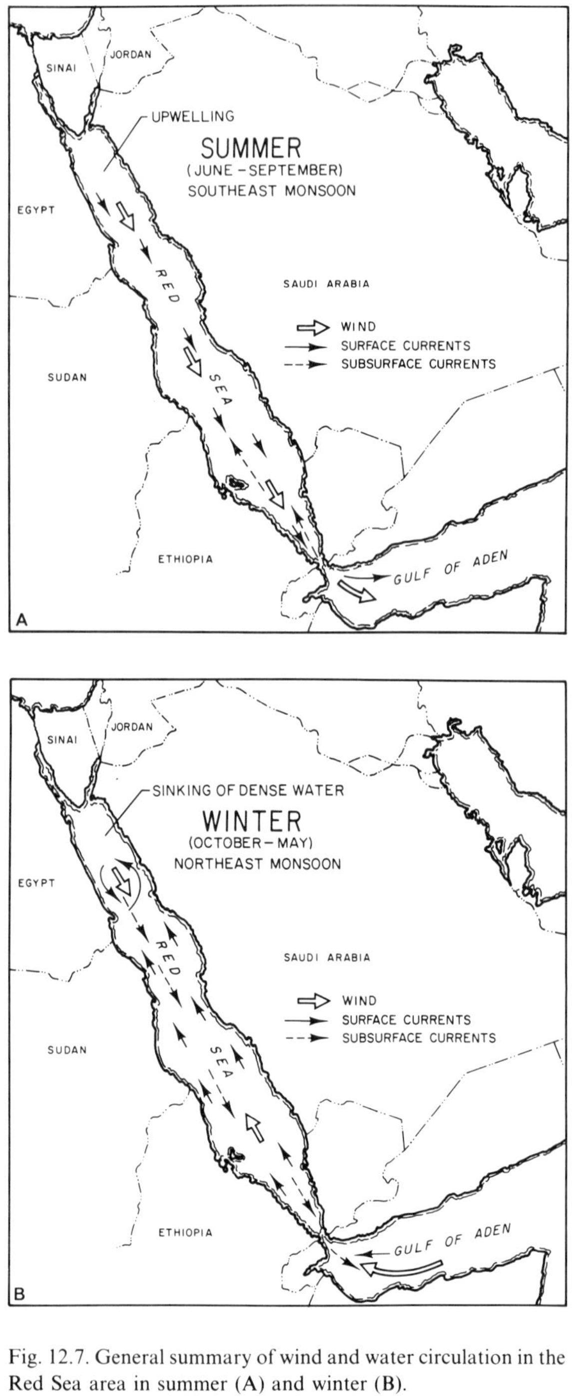

Fig. 12.7. General summary of wind and water circulation in the Red Sea area in summer (A) and winter (B).

During the northeast monsoon period the generally easterly winds of the Gulf of Aden tend to become more south to southeasterly and strengthen as they enter the Strait of Bab al Mandab area. This results in the surface water being driven into the

Red Sea and flowing northward. A surface flow of as much as 15 to 20 cm s^{-1} can occur when mean winds of 6.7 to 9.3 m s^{-1} are blowing in the Strait (Patzert, 1974). Further to the north, between 19 and 25°N, the surface flow is generally north or northwest against the wind.

Water loss from the Red Sea by evaporation is replaced by inflow from the Gulf of Aden and Indian Ocean. According to Neumann and McGill (1961) the inflow from the Gulf of Aden during the summer months tends to occur within intermediate layers because of the surface, wind-driven current flowing out of the Red Sea. Grasshoff (1975) considered that during the winter the inflow occurs mainly in the surface layer down to depths of about 80 m or so.

Near the bottom of the Strait of Bab al Mandab, Red Sea water appears to be moving into the Gulf of Aden. This deep water circulation probably results from evaporation which produces dense, highly saline water (mainly in the north) that eventually sinks to the bottom and moves southward. Neumann and McGill (1961) considered that most of the deep water formed in the northern part of the Red Sea originated in the wintertime when evaporation and cooling are at their maximum.

Patzert (1972, 1974) has written on those seasonal variations in the circulation and structure of the Red Sea due to wind stress. He has noted that during the winter months of October to May the surface waters generally flow north, where they eventually sink and then move southward as a warm, high salinity, subsurface current that eventually flows into the Gulf of Aden over the sill at the Strait of Bab al Mandab. During the summer months of June to September the pattern reverses; the surface waters of the Red Sea flow south, resulting in upwelling in the northern portions of the Red Sea, and in the southern part there is a subsurface inflow of cool, low salinity, Gulf of Aden water. This reversal is closely associated with changes in the monsoon winds as they act on the sea surface and resulting changes in sea level in the southern Red Sea and Gulf of Aden (Fig. 12.7).

A silinity section across the Strait of Bab al Mandab (Fig. 12.8) shows high salinity water overflowing from the Red Sea down into the depths of the Gulf of Aden. This water enters the Indian Ocean as an intermediate water layer. The bottom topography in the Strait of Bab al Mandab area divides the water into two distinct branches (Grasshoff, 1969) that can be traced by their high salinity and low silicate content as far as the coasts of Somalia and Tanzania.

Water budget

An estimate of the water budget is difficult because of seasonal variations in the flow at the Strait of Bab al Mandab. Any measurements made over a period of less than an entire year would be strongly influenced by the prevailing meteorological conditions at the time of the observations. According to Grasshoff (1975), an estimate of water balance would be further complicated by coastal countercurrents, which make any calculation of water exchange on the balance of current measurements in the middle of the Straits somewhat difficult. The best method for calculating water exchange may have to rely on the assumption of constant salinity and constant average sea level. However, this calculation requires data on the evaporation rate but no such measurements have actually been made at sea because of considerable technical difficulties. Some other estimates of water balance have been compiled by Morcos (1970) and are shown in Table 12.2. Two of the estimates based upon current measurements do not provide conservation of volume. Vercelli (1927) estimates a net transport into the Red Sea which is 1.8 times the loss from evaporation; for the data of Siedler (1968) the ratio is 5.7.

Siedler's (1968, 1969) estimate is described in some detail below as it may be one of the better estimates. He has summarized aspects of the Red Sea circulation which are caused principally by evaporation and wind stress, and considered that these could be well approximated by the water exchange through the main channel of the Strait of Bab al Mandab and by the evaporation rate. Within the Red Sea the increase in density caused by salinity increases is not compensated by any corresponding decrease in density due to temperature increases. Thus the density of Red Sea water is higher than that of waters at similar depth in the Arabian Sea. Such a pattern will remain stationary only when there is circulation of Arabian Sea water in through the Strait of Bab al Mandab on the surface, vertical convection within the Red Sea and outflow of Red Sea water near the sill.

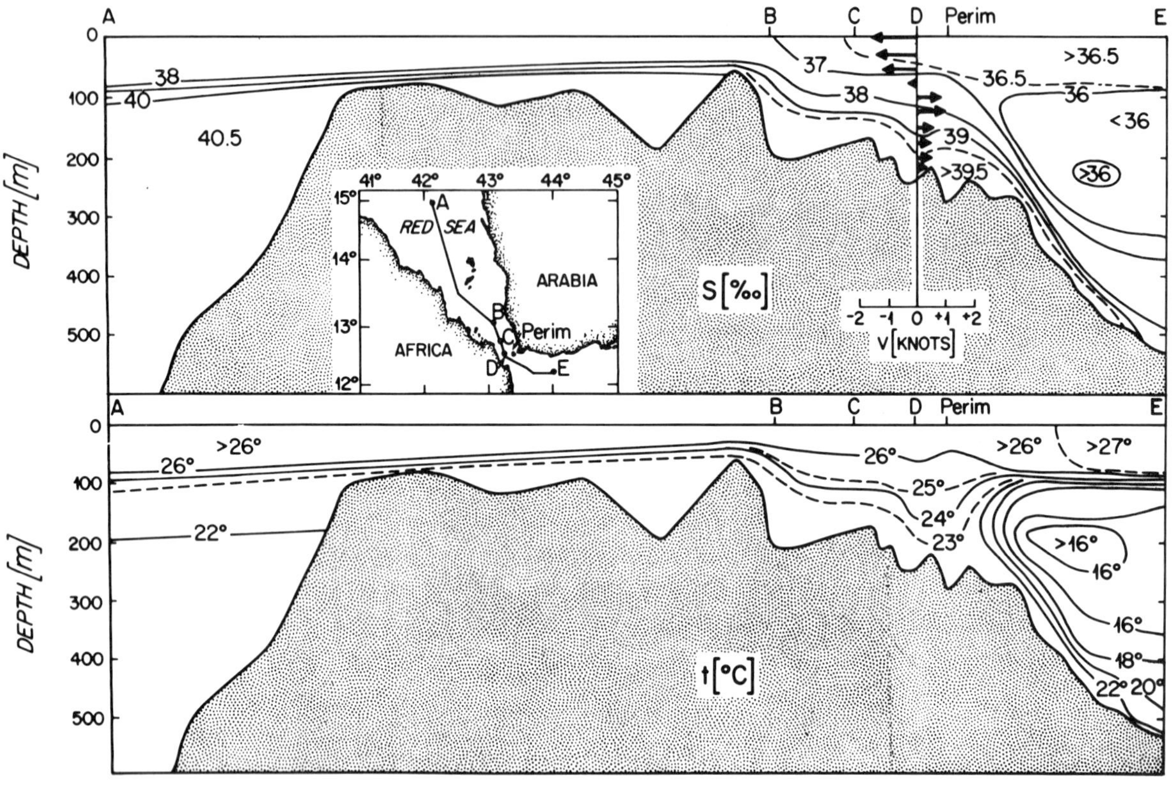

Fig. 12.8. Salinity (upper) and temperature (lower) in the southern Red Sea–Gulf of Aden area. Average direct current measurements made during November 1964 are shown in the upper figure (1 knot = 1852 m h^{-1}). (From Siedler, 1968.)

Siedler (1969) noted that only indirect methods can be used to determine the total water budget of the Red Sea. He assumed that the total volume of the Red Sea has remained constant in light of the fact that no significant change in sea level or salinity has been detected in the Red Sea for about a century or so. He concluded that the net inflow per year through Bab al Mandab should be equal to the total volume of the water evaporated. Privett in 1959 calculated a mean evaporation per year of 1830 mm. Neumann (1952) calculated a value of 2150 mm per year, and thus Siedler assumed an average rate of 2000 mm per year. In doing so, the total volume of water evaporated could be determined by multiplying the evaporation rate times the area of the Red Sea. Using Kossinna's 1921 value for the surface area, $0.44 \times 10^{12} m^2$, gives an average evaporation per year of $0.88 \times 10^{12} m^3$ (0.028×10^6 m^3 s^{-1}). Siedler made estimates of the average volume of inflowing and outflowing water using the continuity equation for constant values of annual averages of sea level, and salt content from the Red Sea; he obtained a volume of outflowing water of 10.5×10^{12} m^3 yr^{-1} (0.33×10^6 m^3 s^{-1}) which agrees well with a value of outflowing water obtained by direct measurements (Siedler, 1968) of 13.14×10^{12} m^3 yr^{-1} (0.43×10^6 m^3 s^{-1}). Using these numbers, Siedler made a rough estimate of the renewal time of Red Sea water, assuming that only the water in the upper 150 m of the Red Sea takes part in the circulation process. Following this assumption and the estimate for the outflow given above, the total renewal time for this layer is six years. If the entire Red Sea were to take part in this mixing, the renewal time would be about twenty years[1]. Since the deep water probably does not take part in this process at the same rate as the surface waters, probably a larger number for the total circulation is appropriate; and Craig's (1969) value of 200 years based on ^{14}C dating of Red Sea waters may be appropriate for the entire Red Sea.

TABLE 12.2

Compilation of various estimates of the water budget of the Red Sea (rearranged from Morcos, 1970); i = ingoing current, u = outgoing undercurrent

A. From the principles of conservation of volume and salt

	Author:	Morcos (1970)	Bogdanova (1966)	Grasshoff (1969)
Salinity	S_i, ‰	36.7	–	36.6
	S_u, ‰	39.7	–	40.5
Volume flow	T_i, 10^6 m^3 s^{-1}	0.358	0.730	0.29
	T_u, 10^6 m^3 s^{-1}	0.330	0.692	0.26
	$T_i - T_u$, 10^6 m^3 s^{-1}	0.028	0.038	0.03
Evaporation	E, 10^6 m^3 s^{-1}	0.0284	0.038	0.028
Water balance	$(T_i - T_u)/E$	1.0	1.0	1.0

B. From direct measurements of currents

	Author:	Krummel (1911)	Vercelli (1927)	Siedler (1968)
Salinity	S_i, ‰	36.9	–	–
	S_u, ‰	39.7	–	–
Current velocity	V_i, cm s^{-1}	43	46.3	–
	depth range, m	0–74	0–50	–
	V_i, cm s^{-1}	–	23.1	–
	depth range, m		50–100	–
	V_u, cm s^{-1}	43	69.4	–
	depth range, m	74–196	100–196	–
Volume flow	T_i, 10^6 m^3 s^{-1}	0.461	0.579	0.59
	T_u, 10^6 m^3 s^{-1}	0.424	0.486	0.43
	$T_i - T_u$, 10^6 m^3 s^{-1}	0.037	0.093	0.16
Evaporation	E, 10^6 m^3 s^{-1}	0.037	0.051	0.028
Water balance	$(T_i - T_u)/E$	1.0	1.8	5.7

HOT BRINE AREA

A unique environment exists in some deep, isolated parts of the axial valley of the Red Sea — within these areas hot, salty water has been found. The first such locality, at about 21°N, was detected as a result of a series of British, U.S., and German expeditions in the mid-1960's (Charnock, 1964; Miller, 1964; Swallow and Crease, 1965; Miller et al., 1966; Krause and Ziegenbein, 1966; Hunt et al., 1967). In 1966 a seven-week U.S. expedition made a detailed survey of the area that resulted in a symposium volume on the subject (Degens and Ross, 1969). Three brine pools (see Fig. 12.1) were found and the largest, the Atlantis II Deep, had a temperature of 56.5°C and a "salinity" of 257‰ in 1966. By 1971 its temperature had increased to 59.2°C and by 1977 to 61.3°C (Ross, 1977). This temperature increase, and a corresponding water volume increase, indicates a dynamic input of hydrothermal waters that appears to be clearly related to sea-floor spreading. The salinity values do not represent a true salinity since the ratio of the major elements is not similar to that of sea water, and some elements, like iron and other metals, are enriched several orders of magnitude; whereas magnesium and sulfate are found in lower quantities than in normal Red Sea water. The sediments

[1] Editor's note: Renewal times for the upper 150 m of the water column can also be estimated from Siedler's (1968) data in Table 12.2. Using the net exchange through the Strait of Bab al Mandab ($T_i - T_u = 5.05 \times 10^{12}$ m^3 yr^{-1}) gives a residence time of 12.5 years; using the evaporation loss, which must be replaced by the circulation (0.88×10^{12} m^3 yr^{-1}) gives a residence time of about 71 years. The mean residence time for the entire volume of the Red Sea based on these two estimates would be 46 and 264 years, respectively. Since the data of Table 12.2 do not provide for continuity of volume, it seems probable that the residence times lie within these ranges, including those quoted in the text from Siedler (1969).

underlying the Atlantis II Deep were found to be also enriched in metals such as iron, copper, zinc, lead, silver and gold and in 1969 were estimated by Bischoff and Manheim (1969) to have a value in excess of $2 400 000 000 (*in situ*, at a depth of about 2000 m).

Both Trüper (1969) and Watson and Waterbury (1969) have made microbiological studies of the hot brine area. Neither of them found any positive evidence of bacteria within the sediments or brines.

Since these early studies a very detailed and comprehensive series of expeditions have been made in the German research vessel *Valdivia* (Bäcker and Schoell, 1972). Their work has defined at least thirteen more deeps containing waters with different high temperatures and salinity values.

Probably the most interesting aspect of the Red Sea hot brine areas is that the hydrothermal processes forming the pools and the underlying heavy metal deposits are, at least in the Atlantis II Deep, an active process. This process results in the deposition of a very valuable mineral deposit. Similar metal deposits on land are usually tens to hundreds of millions of years old, so by studying the Red Sea material we will understand better how such deposits occur. In addition, the presence of the Red Sea deposit along an area of active sea-floor spreading has spurred research and investigations into other similar locations, and similar phenomena have been found, for example, along parts of the spreading areas of the Pacific. It is possible that mining of the Red Sea deposit may begin by the early 1980's.

ACKNOWLEDGEMENTS

I wish to acknowledge Drs. Peter Brewer, Bostwick H. Ketchum and Werner Deuser for reviewing this article. Mrs. Ellen Gately typed the manuscript. Research support was provided by contract with the Office of Naval Research Woods Hole Contribution Number 4092.

REFERENCES

Allan, T.D., 1966. A bathymetric chart of the Red Sea. *Int. Hydrogr. Rev.*, XLIII: 33.

Bäcker, H. and Schoell, M., 1972. New deeps with brines and metalliferous sediments in the Red Sea. *Nature (Phys. Sci.)*, 240: 153–158.

Barlow, E.W., 1934. Currents of the Red Sea and part of the Indian Ocean north of Australia. *Mar. Obs.*, 110: 67–68; 110: 150–154.

Belousov, I.M., 1965. *Oceanological Research Articles, Soviet Geophys. Comm., Acad. Sci. U.S.S.R.*, No. 19, pp. 201–214 (in Russian).

Bischoff, J.L. and Manheim, F.T., 1969. Economic potential of the Red Sea heavy metal deposits. In: E.T. Degens and D.A. Ross (Editors), *Hot Brines and Recent Heavy Metal Deposits in the Red Sea*. Springer-Verlag, New York, N.Y., pp. 535–541.

Bogdanova, A.K., 1966. *Oceanological Research Articles, Soviet Geophys. Comm., Acad. Sci. U.S.S.R.*, No. 15, pp. 45–68 (in Russian).

Charnock, H., 1964. Anomalous bottom water in the Red Sea. *Nature*, 203: 591–592.

Coleman, R.G., 1974. Geologic background of the Red Sea. In: C.A. Burk and C.L. Drake (Editors), *The Geology of Continental Margins*. Springer-Verlag, New York, N.Y., pp. 743–751.

Craig, H., 1969. Geochemistry and origin of the Red Sea brines. In: E.T. Degens and D.A. Ross (Editors), *Hot Brines and Recent Heavy Metal Deposits in the Red Sea*. Springer-Verlag, New York, N.Y., pp. 208–242.

Degens, E.T. and Ross, D.A. (Editors) 1969. *Hot Brines and Recent Heavy Metal Deposits in the Red Sea*. Springer-Verlag, New York, N.Y., 600 pp.

Flohn, H., 1965. Klimaprobleme am Roten Meer. *Erdkunde*, 19(3): 179–191.

Girdler, R.W. and Styles, P., 1974. Two stage Red Sea floor spreading. *Nature*, 247: 7–11.

Grasshoff, K., 1969. Zur Chemie des Roten Meeres und des Inneren Golfs von Aden. *"Meteor" Forschungsergeb., Reihe A*, No. 6: 76 pp.

Grasshoff, K., 1975. The hydrochemistry of landlocked basins and fjords. In: J.P. Riley and G. Skirrow (Editors), *Chemical Oceanography, 2*. Academic Press, New York, N.Y., 2nd ed., pp. 455–597.

Heezen, B.C. and Tharp, M., 1964. *Physiographic Diagram of the Indian Ocean, Red Sea, the South China Sea, the Sulu Sea and the Celebes Sea*. Geol. Soc. Am., Boulder, Colo.

Hunt, J.M., Hays, E.E., Degens, E.T. and Ross, D.A., 1967. Red Sea, detailed survey of hot brine areas. *Science*, 156: 514–516.

Koninklijk Nederlands Meteorologisch Instituut, 1949. *Red Sea and Gulf of Aden Oceanography and Meteorological Data*. Publication 129. K.N.M.I., Den Bilt, 26 pp.

Kossinna, E., 1921. Die Tiefen des Weltmeeres. *Veröff. Inst. Meereskd. Berlin, Reihe A*, 9: 63 pp.

Krause, G. and Ziegenbein, J., 1966. Die Struktur des heissen salzreichen Tiefenwassers im zentralen Roten Meer. *"Meteor" Forschungsergeb., Reihe A*, 1: 53–58.

Krummel, O., 1911. *Handbuch der Ozeanographie, 2*. J. Engelhorn, Stuttgart, 766 pp.

Laughton, A.S., 1970. A new bathymetric chart of the Red Sea, *Philos. Trans. R. Soc. Lond., Ser. A*, 267: 21–22.

Lowell, J.D. and Genik, G.J., 1972. Seafloor spreading and structural evolution of southern Red Sea. *Bull. Am. Assoc. Pet. Geol.*, 54: 247–259.

Luksch, J., 1901a. Expedition S.M. POLA in das Rote Meer, nördliche Hälfte (October 1895–Mai 1896). Wissenschaftliche Ergebnisse, physikalische Untersuchungen. *Denkschr. K. Akad. Wiss. Wien*, 69: 400–485.

Luksch, J., 1901b. Expedition S.M. Schiff POLA in das Rote Meer, südliche Hälfte (Sept. 1897–April 1898). Wissenschaftliche Ergebnisse, physikalische Untersuchungen. *Denkschr. K. Akad. Wiss. Wien*, 69: 337–398.

McKenzie, D.P., Davies, D. and Molnar, P., 1970. Plate tectonics of the Red Sea and East Africa. *Nature*, 226: 243–248.

Miller, A.R., 1964. Highest salinity in the world ocean? *Nature*, 203: 590–591.

Miller, A.R., Densmore, C.D., Degens, E.T., Hathaway, J.C., Manheim, F.T., McFarlin, P.F., Pocklington, R. and Jokela, A., 1966. Hot brines and recent iron deposits in deeps of the Red Sea. *Geochim. Cosmochim. Acta*, 30: 341–359.

Mohamed, A.F., 1940a. Chemical and physical investigations. The distribution of hydrogen-ion concentration in the north-western Indian Ocean and adjacent waters. *John Murray Exped. 1933–1934, Sci. Rep.*, 2: 121–202.

Mohamed, A.F., 1940b. The Egyptian exploration of the Red Sea. *Proc. R. Soc. Lond.*, B, 128: 306–316.

Morcos, S.A., 1970. Physical and chemical oceanography of the Red Sea. *Oceanogr. Mar. Biology.*, 8: 223–235.

Muromtsev, A.M., 1960. A contribution to the hydrology of the Red Sea. *Dokl. Akad. Nauk. S.S.S.R.*, 134: 1443–1446.

Neumann, A.C. and McGill, D.A., 1961. Circulation of the Red Sea in early summer. *Deep-Sea Res.*, 8: 223–235.

Neumann, J., 1952. Evaporation from the Red Sea. *Isr. Explor. J.*, 2: 153–162.

Patzert, W.C., 1972. *Seasonal Variations in Structure and Circulation in the Red Sea.* Hawaii Institute of Geophysics Report, No. HIG-72-13, 58 pp. (unpublished manuscript).

Patzert, W.C., 1974. Wind-induced reversal in Red Sea circulation. *Deep-Sea Res.*, 21: 109–121.

Privett, D.W., 1959. Monthly charts of evaporation from the North Indian Ocean, including the Red Sea and Persian Gulf. *Q. J. R. Meteorol. Soc.*, 85(366): 424–428.

Ross, D.A., 1977. Temperature increases in the Atlantis II Deep hot brine area of the Red Sea. *1977 Am. Geophys. Union Ann. Fall Meeting. EOS*, 58 (12): 1176.

Ross, D.A. and Schlee, J., 1973. Shallow structure and geologic development of the southern Red Sea, *Geol. Soc. Am. Bull.*, 84: 3827–3848.

Siedler, G., 1968. Schichtungs und Bewegungsverhältnisse am Südausgang des Roten Meeres. *"Meteor" Forschungsergeb., Reihe A*, No. 4, 76 pp.

Siedler, G., 1969. General circulation of water masses in the Red Sea. In: E.T. Degens and D.A. Ross (Editors), *Hot Brines and Recent Heavy Metal Deposits in the Red Sea.* Springer-Verlag, New York, N.Y., pp. 131–137.

Swallow, J.C. and Crease, J., 1965. Hot salty water at the bottom of the Red Sea. *Nature*, 205: 165–166.

Thompson, E.F., 1939a. Chemical and physical investigations. The general hydrography of the Red Sea. *John Murray Exped. 1933–34, Sci. Rep.*, 2: 83–103.

Thompson, E.F., 1939b. Chemical and physical investigations. The exchange of water between the Red Sea and the Gulf of Aden over the "Sill." *John Murray Exped. 1933–34, Sci. Rep.*, 2: 105–119.

Trüper, H.G., 1969. Bacterial sulfate reduction in the Red Sea hot brines. In: E.T. Degens and D.A. Ross (Editors), *Hot Brines and Recent Heavy Metal Deposits in the Red Sea.* Springer-Verlag, New York, N.Y., pp. 263–271.

Udintsev, G.B. (Editor), 1975. *Geological–Geophysical Atlas of the Indian Ocean.* Academy of Sciences of the U.S.S.R., Moscow, 151 pp.

Van Riel, P.M., 1932. Einige ozeanographische Beobachtungen im Roten Meer, Golf von Aden und Indischen Ozean. *Amm. Hydrogr. Marit. Meteor.*, 60: 401–407.

Vercelli, P., 1927. Campagna idrografica della R.N. AMMIRAGLIO MAGNAGHI in Mar Ross 1923–24, Ricerche de oceanografia fisica. *Ann. Idrogr.*, 11.

Watson, S.W. and Waterbury, J.B., 1969. The sterile hot brines of the Red Sea. In: E.T. Degens and D.A. Ross (Editors), *Hot Brines and Recent Heavy Metal Deposits in the Red Sea.* Springer-Verlag, New York, N.Y., pp. 272–281.

Werner, F. and Lange, K., 1975. A bathymetric survey of the sill area between the Red Sea and the Gulf of Aden. *Geol. Jahrb., Reihe D*, 13: 125–130.

Whitmarsh, R.B., Ross, D.A., Ali, S.A., Boudreaux, J.E., Coleman, R., Fleisher, R.L., Girdler, R.W., Manheim, F.T., Matter, A., Nigrini, C., Stoffers, P. and Supko, P.R., 1974. Site 225–230. In: R.B. Whitmarsh, O.E. Weser, D.A. Ross et al., *Initial Reports of the Deep Sea Drilling Project, 23.* U.S. Government Printing Office, Washington, D.C., pp. 537–812.

Wyrtki, K., 1971. *Oceanographic Atlas of the International Indian Ocean Expedition.* National Science Foundation, Washington, D.C., 531 pp.

Young, R.A. and Ross, D.A., 1974. Volcanic and sedimentary processes in the Red Sea axial trough. *Deep-Sea Res.*, 21: 289–297.

Chapter 13

THE BALTIC SEA

GUNNAR KULLENBERG

AREA DEFINITION AND GENERAL TOPOGRAPHY

The entire Baltic Sea includes the Gulf of Bothnia from the Åland Sea northward, the Gulf of Finland, and the Baltic proper, consisting of the Eastern and Western Gotland Sea, the Gulf of Riga, the Bornholm and Arkona Seas. For many purposes the so-called transition area, namely the Danish Sounds and the Kattegat, are also included in the Baltic Sea (Fig. 13.1). However, the transition area constitutes a different oceanographic regime and will not here be considered as part of the Baltic.

The Baltic is an intra-continental mediterranean-type sea. The area of this enclosed sea is about 370 000 km^2 and its volume is about 21 000 km^3 (Ehlin et al., 1974). The mean depth is 57 m but roughly 17% of the area is shallower than 10 m.

The Baltic proper is connected to the North Sea via the Skagerrak and a transition area consisting of the Belt Sea, the Öresund and the Kattegat (Wattenberg, 1949). The sill depth of the Öresund is 7 to 8 m, and of the Belt Sea is 17 to 18 m. Thus, connections between the Baltic Sea and the North Sea through the transition area are both shallow and narrow. The area of the cross section over the sill in the Öresund is 0.1 km^2 and in the Belt Sea 0.16 km^2 (Jacobsen, 1925). The transition area is generally very shallow with the mean depth of the Kattegat being 23 m (Svansson, 1975). An extension penetrates from the Norwegian Trench in the Skagerrak down into the Kattegat along the Swedish coast, with a depth of 75 m at Vinga, gradually becoming as shallow as 30 m north of the Öresund. In the Öresund the greatest depth is about 50 m and large areas of the western Kattegat have depths of less than 10 m. In the Belt Sea there are

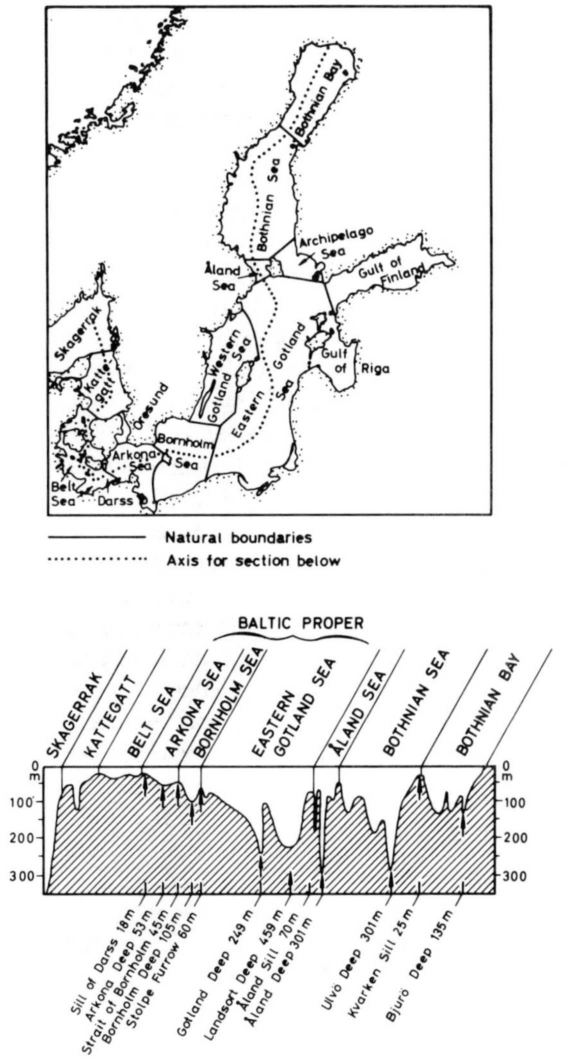

Fig. 13.1. Baltic Sea area (upper) showing subdivisions into various seas and gulfs. A section (lower) showing water depths of the various basins (from Jansson, 1978).

isolated depths down to 80 m in narrow trenches, but generally the area is less than 20 m deep, with a mean depth of 14 m. The area and volume of the transition area are, respectively, 42 000 km^2 and 800 km^3 (Ehlin et al., 1974). The Belt Sea has an area of about 5.5% of the Baltic Sea area.

One of the outstanding characteristics of Baltic topography is its division into a series of basins separated by shallow areas or sills (Table 13.1, and Fig. 13.1). The western part of the Baltic proper is connected through the Stolpe Furrow or Channel, with a maximum depth of 60 m to the central basin, consisting of the Eastern and Western Gotland Seas. The basin constitutes the largest part of the Baltic proper, and it is bordered by several smaller basins. The Gulf of Riga and the Gulf of Finland are connected directly to the central basin. The Gulf of Bothnia, which is divided into three basins, is connected to the central basin by a narrow trench through the Åland Sea, with a sill depth of 70 m. This trench continues into the Bothnian Sea.

The greatest depths in the Baltic Sea occur in the central parts and are separated from the North Sea by several shallow regions and sills inside the Baltic proper, as well as by the transition area. The depths in the Gulf of Bothnia are also considerable relative to the sill depth to the central areas. These topographic features must be borne in mind in any discussion of oceanographic conditions in the Baltic Sea. A lateral exchange of deep and bottom waters can be generated only by water which has come from the North Sea, a great distance, and has passed over several shallow sills. This, in combination with weak vertical mixing, implies long residence times of the order of several years for the bottom waters. In turn this leads to low oxygen concentrations and frequent anoxic conditions, which influence the ecology greatly.

Past alternation of oxic and anoxic conditions in the bottom can be traced in the sediments, which are muddy and contain a considerable fraction of organic material over large areas. The southern areas of the Baltic Sea are dominated by sandy sediments of alluvial origin, while the northern parts consist of bedrock from Precambrian and Paleozoic eras. Land runoff from rivers has supplied mineral deposits as well as organic sediments, which are also derived from production in the sea. In the deeper parts of the basins sediments are mainly postglacial mud lying over varved glacial clay. The sedimentation rate appears to be in the range 10 to 20 cm per 100 years. Sand and gravel are exposed by strong currents and wave motion in shallow areas. The southern and eastern coasts, including the Gulf of Finland, are generally shallow and sandy, whereas the Swedish east coast is rocky

TABLE 13.1

Major Baltic divisions with maximum depths, sill depths at outer boundary, volumes and mean depths

Name	Max. depth (m)	Sill depth (m)	Volume (km^3)	Mean depth (m)
Baltic Sea (total)	459	17	20 900	57
Arkona Sea	55	17	430	23
Bornholm Sea	106	45	1780	46
Eastern Gotland Sea				
Gdańsk/ or Danzig Bay	116	88	1460	57
Gotland Deep	249	60	3570	81
Farö Deep	205	140	1270	–
Northern Central Sea	459	115	2090	72
Landsort Deep	459	138	780	–
Western Gotland Sea	205	100	1640	61
Gulf of Riga		–	410	23
Gulf of Finland		–	1100	37
Åland Sea	300	70	410	75
Archipelago Sea	40	40	170	19
Bothnian Sea	293	90	4300	67
Bothnian Bay	126	25	1490	41

with archipelagoes. This is also the case along the northern Gulf of Finland and the Archipelago Sea. In the Gulf of Bothnia stony and sandy sediments dominate along the coasts.

THE WATER BALANCE

The water balance of the Baltic depends upon fresh-water supply from the rivers, a balance between direct precipitation into the Baltic and evaporation from the surface, and upon exchanges between the Baltic and the North Sea through the transition zone and the Kattegat. For all practical purposes, the groundwater supply is insignificant.

The Baltic is the largest brackish water sea on earth, and can be characterized as a typical mediterranean-type sea with narrow connections to the ocean and a large river runoff giving it a marked positive fresh-water balance. The water supply is 400 to 500 km^3 per year, or about 2% of the total volume.

The most important factor in the Baltic water balance is river inflow, which has been studied by Witting (1918), Hela (1944), Simojoki (1949), Brogmus (1952), Soskin (1963), and Mikulski (1970, 1972). These authors have calculated a long-term average inflow by using series of observations from several rivers. Witting gave 467 $km^3 yr^{-1}$, Brogmus found 471 $km^3 yr^{-1}$ for the period 1900–1940 with the exception of some important rivers for which the period 1921–1930 was used; Soskin gave 473 $km^3 yr^{-1}$ for the period 1898–1944; and Mikulski gave 440 $km^3 yr^{-1}$ for the period 1951–1969. Although the values are remarkably close, they indicate long-term variations in supply, which is only to be expected. The river flows show considerable long-term variability (Fig. 13.2).

The supply is far from evenly distributed over the area. Using the values given by Brogmus (1952), the Gulf of Bothnia receives 32%, the Gulf of Finland 24% and the Baltic proper 49% of the total inflow. By far the largest portion is received by the northern parts more than 1000 km from the Belt Sea. Regional differences in runoff are clearly demonstrated by potentially equivalent water level changes. The supply to the Baltic as a whole implies 110 to 130 cm yr^{-1}. The Gulf of Finland receives the most relative to its area — 400 to 420 cm yr^{-1} — and the Åland Sea the least, with 27 cm yr^{-1}. For the Gulf

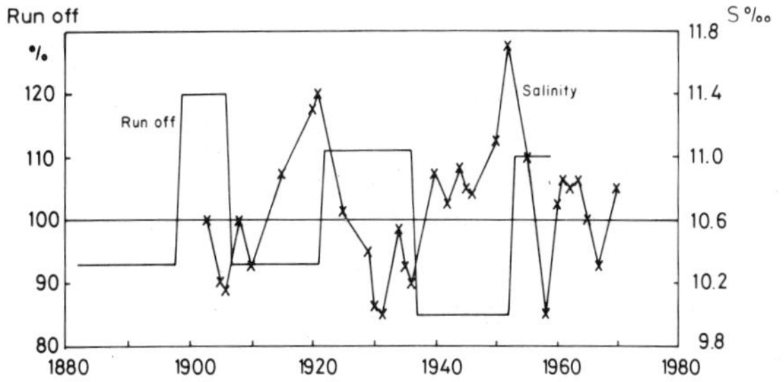

Fig. 13.2. River runoff values (bar graph, as percent of the mean) and mean salinity values (×) for water column 0–90 m in Bornholm Deep (Kaleis, 1976).

of Bothnia and the Baltic proper the values are 170 cm yr^{-1} and 50 cm yr^{-1}, respectively (Brogmus, 1952; Mikulski, 1970).

There are also considerable seasonal variations in runoff. A rather marked main maximum extends over the months of April–June, with a monthly maximum occurring in May, and a minimum occurring in January or February (Brogmus, 1952). There is also a secondary maximum in the fall during October to November. There are considerable time differences between the basins. For example, the Gulf of Riga has a short, rather intense spring runoff and a minimum runoff during the summer as well as in late winter, whereas the spring runoff in the Gulf of Bothnia comes later and is more extended, and the minimum runoff there occurs during the winter months. The latest main maximum occurs in the Bothnian Sea during June. The yearly course of the runoff and the regional time differences are due mainly to meteorological factors (Brogmus, 1952). The spring or early summer maximum is caused by snow-melting, which starts earlier in the south than in the north. The late arrival in the Bothnian Sea may be due to the relatively high altitude of the runoff sources, implying a delayed spring maximum relative to the neighboring regions. The secondary runoff maximum during the fall is probably related to temperature decreases, implying less evaporation over land.

Long-term variations of river runoff are considerable, with an amplitude of 10 to 20% of the long-term mean. These variations have been studied in particular by Soskin (1963), who ascribes them primarily to climatic variations. The runoff is greatest during periods of cyclonic (zonal) circulation with heavy precipitation, and at a minimum during anticyclonic (meridional) circulation with low precipitation. Hela (1966a) studied air pressure fluctuations over Stockholm for 100 years and found a dominating period of 42 years and a secondary period of 19 years. Hallberg (1974) investigated the paleo-redox conditions in the Eastern Gotland basin and found alternating oxic and anoxic conditions, with a periodicity of approximately 20 years. A periodicity of about 20 years also appears to be present in fluctuations of river flow. This fresh-water supply from rivers is one of the essential features of the Baltic area. It reflects the importance of continental influence on the Baltic, the significance of seasonal and long-term climatic variations, and the effects of the considerable north–south extension of the Baltic.

Precipitation and evaporation have been studied in particular by Witting (1918), Simojoki (1949), Brogmus (1952) and Wyrtki (1954a). Brogmus' precipitation values were obtained using data from 161 stations along the Baltic coasts and on islands. The resulting value was 471 mm yr^{-1} or 172 km^3 yr^{-1} over the entire Baltic, excluding the Belt Sea. The maximum occurs during the summer and peaks in August, with considerable regional variations. Generally northern areas receive less than the eastern and southern parts. The yearly maximum occurs in the Gulf of Riga. For comparison, Witting (1918) obtained a total of 565 mm yr^{-1} and Simojoki (1949) 525 mm yr^{-1}, which demonstrates the uncertainties produced by calculations taken at coastal stations. There is a fairly well defined yearly course, with maximum precipitation in August and minimum in March. The greatest uncertainties in determining the fresh-water budget are in estimations of evaporation over the sea. No direct observations from the open Baltic are as yet available. Brogmus (1952) used the bulk aerodynamic method based on meteorological surface data to obtain monthly mean values. Based on these, he found a value of 471 mm yr^{-1} or 172 km^3 for the Baltic, which can be compared with values of 182 and 168 km^3 given by Witting (1918) and Simojoki (1949), respectively. The agreement is surprisingly good, and the results have also been in general corroborated by Palmén and Söderman (1966), who obtained their results from calculations of the divergence of the water vapor flux using radio sonde data. Evaporation also shows regional variations; it is at a minimum in the northern parts of the Gulf of Bothnia and a maximum over the Gotland basin.

The data imply that on a yearly basis evaporation and precipitation over the Baltic are about equal, and that the yearly total fresh-water supply is provided by river runoff. However, on a regional and seasonal basis this does not hold true since the precipitation–evaporation balance is positive in the northeastern areas and negative in the south. Seasonally the balance is generally positive during the summer and negative during the winter.

The exchange between the Baltic and the Kattegat–North Sea must also be considered in order to arrive at the total water balance. This part of the problem has been the subject of many studies (Knudsen, 1899, 1900; Witting, 1918; Jacobsen, 1925; Brogmus, 1952; Wyrtki, 1954a; Soskin, 1963; Svansson, 1972). The water exchange is driven by the fresh-water supply, the estuarine component, and by the meteorological conditions over the North Sea–Baltic Sea area. The yearly fresh-water supply to the Baltic generates an outgoing surface current through the Öresund and the Belt Sea into the Kattegat on its way to the ocean. The surface salinity in the Kattegat is therefore low, less than 20‰ in the southeast, whereas the salinity beneath about 15 m depth is from 30 to 34‰. The density difference induces a counter current in the bottom water directed towards the Belt Sea. On an annual basis the requirement for a southgoing deep current in the Belt Sea is fulfilled. Thus the average circulation in the Belt Sea is a two-layered system with an outgoing flow in the surface layer and a bottom layer flowing in towards the Baltic. The deep-water flow penetrates into the Baltic as a thin bottom current over the Darss sill. This constitutes the estuarine component of the water exchange. However, observations show that the flow in the Belt Sea is often a one-layer flow, with the entire water column moving either from or towards the Baltic. This is due to the influence of meteorological conditions, which may support or oppose the estuarine component of the flow. This meteorological force is generated by both the winds and the pressure field, and it can be so strong that the estuarine flow is suppressed completely.

Dietrich (1951) has shown that when there are westerly winds (between SW and NW) over the North Sea and Skagerrak with velocities above about 10 m s^{-1} and a duration of at least several days, the current in the Belt Sea and the Öresund flows towards the Baltic throughout the entire water column. Under other wind directions over the Baltic–North Sea area the current flows out from the Baltic. It should be noted that only the large-scale wind field and not the local winds produces this influence on the flow.

The most favorable conditions for the inflow of saline water into the Baltic occur when there is high pressure over Jutland and low pressure over Sweden; outflow situations occur with high pressure over Scandinavia. Particularly strong inflows occur when an outflow situation is followed by a series of intense west winds. This was the case in November and December 1951 when 200 km^3 of highly saline (22‰) water penetrated into the Baltic over a three-week period (Wyrtki, 1954b). On such occasions the Kattegat becomes filled with water of about 30‰, and the 24 to 25‰ isohalines reach the Gedser–Darss sill. The normal outflow, occurring during stable (anticyclonic) weather conditions, is characterized by weaker currents and longer duration than are the inflow situations, which occur during unstable (cyclonic) weather conditions.

The magnitude of the water exchange can be estimated by several techniques. The most commonly used is the steady-state continuity relationship of Knudsen (1899, 1900). Assuming a fresh-water surplus of 471 km^3 yr^{-1} and annual salinity averages of the incoming and outgoing transports at 17.4 and 8.7‰, respectively, it is found that the inflowing volume is 465 km^3 and the outflowing 935 km^3, taking into account density differences between the water masses. However, the water exchange is not stationary; indeed, an outstanding feature is its variability as governed by the varying forcing conditions.

A relationship between incoming and outgoing flows as a function of a time-dependent fresh-water supply may be derived under the assumption that the Baltic can be treated as a well (or excessively) mixed estuary, as was done by Stommel and Farmer (1953) and in a slightly different way by Kullenberg (1955). Then the outflow at first increases from zero fresh-water supply to a maximum, but falls off sharply to zero when a critical fresh-water supply is reached. Such a course is in fact indicated when the monthly exchange values are used (Svansson, 1972).

Welander (1974), in a theoretical study of the water exchange, used the outer salinity, the fresh-water supply, and the meteorologically driven barotropic transport as forcing functions. He found that a single steady state exists, that total mixing through the interface in the two-layer model must vary with depth (decreasing with increasing depth) to allow a stable steady state, and that the static stability increases with increasing fresh-water supply up to a critical value where the two-layer model breaks down.

Several attempts have been made to relate the water exchange directly to the fresh-water supply, the water level difference between the Baltic and the Kattegat, and surface currents in the Danish Sounds. Jacobsen (1925) initiated this approach. Using current measurements over limited periods of time in sections across the Sound and the Belt Sea, he related monthly mean transports through the sections to the surface current at Drogden. This is obviously a drastic simplification and can give only a crude estimate of the water exchange. Wyrtki (1954a) used the same approach but separated the Sound from the Belt Sea. The transport through the latter was related to the surface current at the light vessel *Halskov Rev* in the Great Belt (Store Baelt). The proportionality constant was obtained by assuming that on a yearly basis the total outflow is equal to the fresh-water supply as given by Brogmus (1952). This implies that the water volume of the Baltic is constant. In reality, the volume necessarily varies, on different time scales, and Wyrtki pointed out that there must be a coupling between these variations, the fresh-water supply, and the water exchange, through a relation of the kind $V = Q_0 + Q_B + Q_Ö + k \cdot v_{\text{Halskov}}$, where V is the volume variation, Q_0 is the fresh-water supply, Q_B and $Q_Ö$ the transport through Great Belt and Öresund, respectively, in km^3 $month^{-1}$, and v_{Halskov} is the summed surface current at *Halskov Rev* light vessel (km $month^{-1}$). The volume variation was determined from water level observations. From this relation k was calculated for the periods 1926–30, 1926–35 and 1946–51 by Wyrtki (1954a). He found $k = 3.4$, 4.1 and 6.8, respectively for these periods. As a possible explanation for the variation of k, Wyrtki considered the variability of the depth of the interface between the out- and ingoing flows and the changes of the position of the light vessel *Halskov Rev* depending upon whether the surface current was in- or outgoing. This suggests that it is necessary to distinguish whether transports are in- or outgoing, as was done by Soskin (1963). We may then split up the relation into one representing inflows giving $Q_i = V - Q_0$ and one representing outflows $Q_u = Q_0 - V$ where Q_i and Q_u are the in- and outgoing transports respectively. Soskin calculated Q_i and Q_u, using the same technique as Wyrtki (1954a), and related these to the mean surface currents at *Halskov Rev* and *Lappegrunden*, lightvessels at the northern entrance to the Öresund. The results of Jacobsen (1925), Wyrtki (1954a), and Soskin (1963) are presented in Fig. 13.3.

The yearly course of the water exchange components reflects a coupling to meteorological factors, which in turn are subject to climatic variations. These intricate relationships are not yet well understood, but it is clear that the water exchange shows a considerable long-term variability.

Major inflows of saline water are particularly important to any description of Baltic bottom water. They occur randomly at intervals of two to five years. Dickson (1971, 1972), studying salinity variations in the European shelf seas, found a clear correlation between positive salinity anomalies and the occurrence of major inflows into the Baltic. He considered the main reason for the salinity variations to be a persistent pressure anomaly pattern over the mid-latitudes of the North Atlantic.

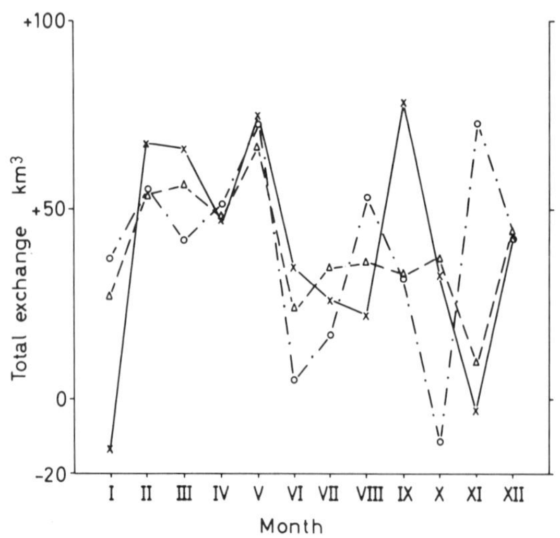

Fig. 13.3. Water exchange after (×) Jacobsen (1925), (o) Wyrtki (1954a) and (Δ) Soskin (1963).

Indeed Dickson (1973) made a successful prediction of a major inflow on the basis of this theory. This demonstrates how large-scale events in the ocean can influence conditions in an enclosed sea like the Baltic.

The yearly averages of water exchange may be used to estimate mean residence time for the Baltic. It is therefore necessary to consider the extensive mixing that occurs in the transition area, primarily in the Belt Sea and the southern Kattegat, producing an increase of the surface salinities in that region. From Falsterbo south of Drogden to Svinbådan in the northern entrance to the Öresund, mean surface salinity increases from 8‰ to 15‰. The corresponding increase in the Belt Sea from Gedser to *Halskov Rev* is about 5‰. From those numbers it can be shown (Kullenberg, 1967) that one third of the water leaving the Baltic returns with the inflow. This implies that 2.9% of the Baltic volume leaves annually, a mean residence time of 35 years.

During the 1970's there have been measurements of currents, water level and meteorological conditions in the transition area using modern equipment. Some preliminary results of these programs have been presented (Belt Project, 1976a). They show a great variability of flow and stratification, frequent periods when there is only a one-layer flow, and a small net exchange relative to the total flow through the Belt Sea.

Salinity and temperature distributions

The Baltic Sea has marked permanent salinity stratification with only a few variations compared to the many variations occurring in the Kattegat. The fresh-water supply generates a brackish surface layer, and incoming deep water generates layers of more saline deep and bottom waters. Occasionally two halocline layers can be defined, of which the shallow primary one is permanent and more pronounced than the secondary deep layer. The fresh-water supply is mixed downwards by a combination of wind-generated mixing and thermohaline convection during fall and early winter. The process generates an almost isohaline surface layer, with no vertical salinity variations during winter and only weak variations during summer.

The thickness of the surface layer varies from basin to basin. It is partly related to the sill depths between basins since deep water flow depends on large cross-sectional areas in the layer beneath the halocline. In the central Arkona basin the depth of the halocline layer varies between 30 and 40 m, and in the Bornholm basin the depth is generally 40 to 50 m. In the Stolpe Channel the depth increases slightly, and in the large deep water reservoir of the Baltic, the central Baltic, it varies between 60 and 70 m, occasionally going down to 80 m in the northern central basin and the Landsort Deep (Fonselius, 1969). In the central basin the depth of the halocline layer has decreased slightly since the beginning of the present century.

The Gulf of Bothnia has not been studied as much as the Baltic proper. The water exchange between the Bothnian Sea and the central basin goes through the Åland Sea, which is deeper and more open than the shallow Archipelago Sea between Åland and Finland. The exchange is relatively continuous and is forced by both the fresh-water supply and meteorological factors which generate the most important currents (Hela, 1977). The incoming water originates from the surface layer of the central basin, which has small salinity and density variations. Due to mixing occurring during the inflow, density variations are further reduced. This implies that stratification in the Gulf of Bothnia is considerably weaker than in the Baltic proper. Stratification is stable during the whole year, but the depth of a weakly developed halocline varies greatly and the halocline can be absent during part of the year, especially in the wintertime. During summer and fall a fairly shallow halocline can occur. The efficiency of water exchange between the Gulf of Bothnia and the central basin further implies that the deep and bottom waters are renewed annually in most of the Gulf although the salinity stratification prevents thermohaline convection to the greatest depths. In the deepest parts of the Gulf the bottom water salinity is about 1.5‰ greater than surface salinity.

The surface salinity in the Bothnian Bay is 1 to 3‰, in the Bothnian Sea 4 to 6‰, in the Gulf of Finland 2 to 7‰, in the Baltic proper 6 to 8‰. The deep-water salinities are 3–7‰ in the Gulf of Bothnia, 5 to 9‰ in the Gulf of Finland, and 10 to 13‰ in the Baltic proper. During particularly strong inflows from the Kattegat bottom water salinity in the Bornholm basin can reach 20‰, and in the eastern and western Gotland basins 14‰ and 11‰, respectively. During such inflows a secondary halocline layer develops at a depth of 110 to 130 m (Voipio and Mälkki, 1972). The passage time of deep water from the Arkona to the Gotland basin varies, depending upon the intensity of the inflow, from four to nine months (Francke and Nehring, 1973). The residence time of deep and bottom waters in the Baltic proper depends primarily upon their densities. The water beneath the sills cannot be replaced except through a new inflow which has enough excess density and kinetic energy to replace it, and the inflowing water advects into the basin at its appropriate density level. After strong inflows of high density water, replacing the old bottom water in the basins, the density of the new bottom water normally must be decreased by vertical mixing before a new inflow can replace it. During these periods, referred to as periods of stagnation, the motion in the bottom water is very weak. The residence time of the bottom water may vary considerably, from less than two years to about five years (Fonselius, 1969, 1976). Only in the Arkona basin is the water renewed every year due to a combined action of inflowing Belt Sea water and thermohaline convection (Krause, 1969).

Temperature distribution and its seasonal variations also display several interesting features. During fall and early winter the isohaline surface layer becomes isothermal through thermohaline convection. Surface water reaches its maximum density at about 2.5°C. However, the isohaline layer is always at a lower temperature. This may be explained by the effective mechanical convection induced by strong winds during winter. Relatively large evaporation may also play a role, as may horizontal salinity variations.

Convection reaches the top of the primary halocline layer, where it is stopped by strong stratification. The surface water is further cooled to slightly less than 0°C, and a thin winter thermocline develops. During the period of spring warming, heat initially is transferred downwards by thermohaline convection. However, normally warming occurs so quickly that a thermocline is developed before the entire isohaline layer has received any heat input. A spring and summer thermocline is developed over most of the Baltic at depths between 15 and 20 m. This implies that the so-called winter water prevails in a layer on top of the halocline where a characteristic temperature minimum is found in most Baltic temperature profiles (Fig. 13.4). It should be noted, however, that heat penetrates to the deep water when strong winds force the thermocline towards the top of the halocline layer, thereby causing a gradual heat input to deep water.

The summer thermocline can be very sharp and effective in suppressing vertical mixing. This implies that nutrients are prevented from being supplied to most of the euphotic zone, which in the Baltic is about 20 m deep. It also implies that during summer the fresh-water supply becomes trapped in the warm surface layer above the thermocline layer, which further increases stratification. The warm surface layer attains its salinity minimum at about the same time as its temperature maximum, towards the end of July. This shows that the minimum is due more to suppression of mixing by the thermocline than to an increase in the fresh-water supply, since this reaches its maximum around May.

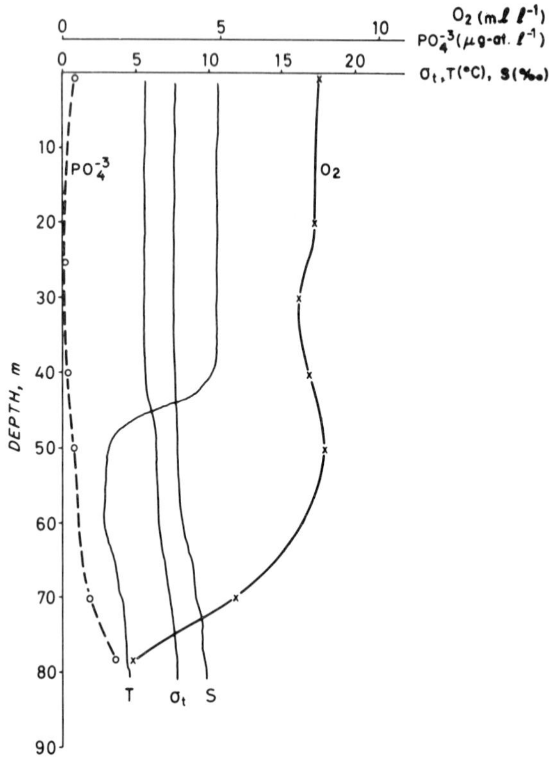

Fig. 13.4. Profiles of salinity (‰), temperature (°C), σ_t, oxygen (×, ml l^{-1}) and orthophosphate (o, μg-at. l^{-1}) from the southern central basin 56°06′N 18°23′E in September, 1977.

In the Gulf of Bothnia a summer thermocline is also developed at 15 to 20 m depth. However, salinity stratification is weak so the heat flux, driven by wind-induced mixing, can gradually penetrate to deeper layers during the fall. This is shown by the considerable delay between the surface and deep-water temperature maxima, which has been used by Hela (1966b) to calculate the vertical eddy diffusivity for heat. In late autumn a pycnocline occurs at about 50 m.

Deep-water temperature in the Bornholm basin is determined by the temperature of the inflowing Kattegat water, which may vary between 2°C and 14°C. High temperature is associated with high salinity. In the central basin deep and bottom water temperatures do not show such large fluctuations. Temperature variations of the inflowing water become moderated by the large volume of the basin as well as by mixing from just above the halocline layer, where the temperature is always low. Below the halocline layer the temperature generally shows an increase towards the bottom. Deep-water temperature normally varies from slightly less than 3°C to slightly less than 6°C.

Both temperature and salinity in the Baltic show important long-term variations, studied in particular by Soskin (1963), Hela (1966a), and Fonselius (1969). Several authors have documented the salinity increases for both surface water and deep water which have occurred since the beginning of this century (Granqvist, 1952; Lindqvist, 1959; Soskin, 1963; Hela, 1966a; and Fonselius, 1969). In the central basin salinity showed a marked decrease during the 1930s, followed by an increase starting about 1938 which probably continued during the 1940s. The great salt inflow during 1951–1952 (Fonselius, 1962) further increased salinity (Fig. 13.5). During the following stagnation period the salinity decreased almost continuously until 1959–1960, when a new major inflow occurred (Fonselius, 1969). The variability as well as the general trend of an increasing salinity is clearly demonstrated (Fig. 13.5). Gaps in the data series around 1915 and 1945 are due to an interruption of observations during World Wars I and II. Fonselius (1969) also demonstrated a general increase of deep-water salinity in the northern central basin and in the Gulf of Bothnia. Matthäus (1977b) studied secular salinity variations for the surface water and demonstrated an increase over the whole

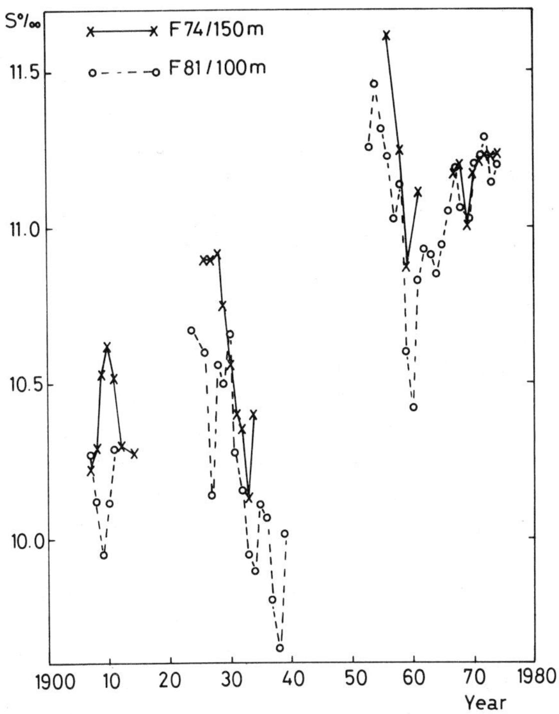

Fig. 13.5. Three-year running average of the mean annual salinity at 150 m at station F74 (59°02′N 21°05′E), and at 100 m at station F81 (57°20′N 19°59′E) in the central Baltic. Data from ICES data bank.

of the open Baltic proper between 0.5 and 0.9‰. Surface observations at Christiansø near Bornholm are shown in Fig. 13.6.

Variations have also influenced the depth of the halocline layer. Hela (1966a) used the method proposed by Tully (1958) to plot salinity versus logarithmic depth to define the upper boundary of the halocline layer and its salinity. In a pure estuary salinities of the permanent halocline layer thus constitute a straight sloping line cut by vertical lines representing the surface and bottom layer salinities. The Baltic is, however, not a pure estuary. Only in the southwestern area in the Gulf of Finland, which is similar to an estuary, and in the northern part of the Bothnian Bay are the estuarine features evident. In the main area of the Baltic proper the salinity of the permanent halocline appears not as a straight line but more as the arc of a circle. Using this technique Hela (1966a) studied one August section from Bornholm to the Gulf of Finland and found the upper limit of the halocline, as defined by the 7.7‰ isohaline, at about 60 m depth, except in the Bornholm basin where it was at about 45 m depth. Fonselius (1969) showed that in the central part of the eastern Gotland basin the top of the halocline in the early part of the century was around 80 m depth. Thus there has been a considerable increase in volume of the deep water.

Hela (1966a) went on to study secular changes of salinity in the surface layer, using representative open sea stations in the central basin, the Bothnian Sea and the Bothnian Bay. These stations, however, had not been occupied frequently and regularly enough to allow computation of annual means. Therefore Hela used the salinity of the upper boundary of the halocline to represent the mean salinity of the upper isohaline layer during the preceding winter. Using this conservative salinity parameter, Hela found a relatively good correlation between variations in the three areas, with the lowest salinities occurring in the 1930s and the highest ones around 1951. The consecutive 10-year running mean values for the central basin station given by Hela are shown in Fig. 13.7. Hela used the results of Jensen (1937) to extend the curve backwards. Jensen had found that on the whole the 5-year running means of surface salinity in the Baltic, at Christiansø, and in the Kattegat were well correlated, with high values in the period 1880–1894, low values in the period 1895–1914, high values between 1915 and 1924, and low ones from 1925 to 1935. Hela concludes that salinity was higher in the years 1880–1895 than around 1905.

Two obvious possible reasons for the secular salinity variations are variations of the fresh-water supply and of the water exchange with the Kattegat and North Sea. Soskin (1963), Fonselius (1969), and Kaleis (1976), among others, have investigated the coupling between fluctuations of river runoff

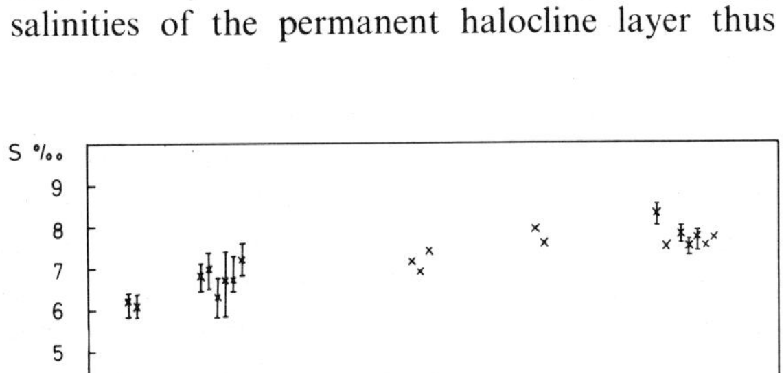

Fig. 13.6. Mean salinity values for the month of April from surface observations at Christiansø, Bornholm basin; bars indicate maxima and minima. Data from Danish Nautical–Meteorological Yearbooks.

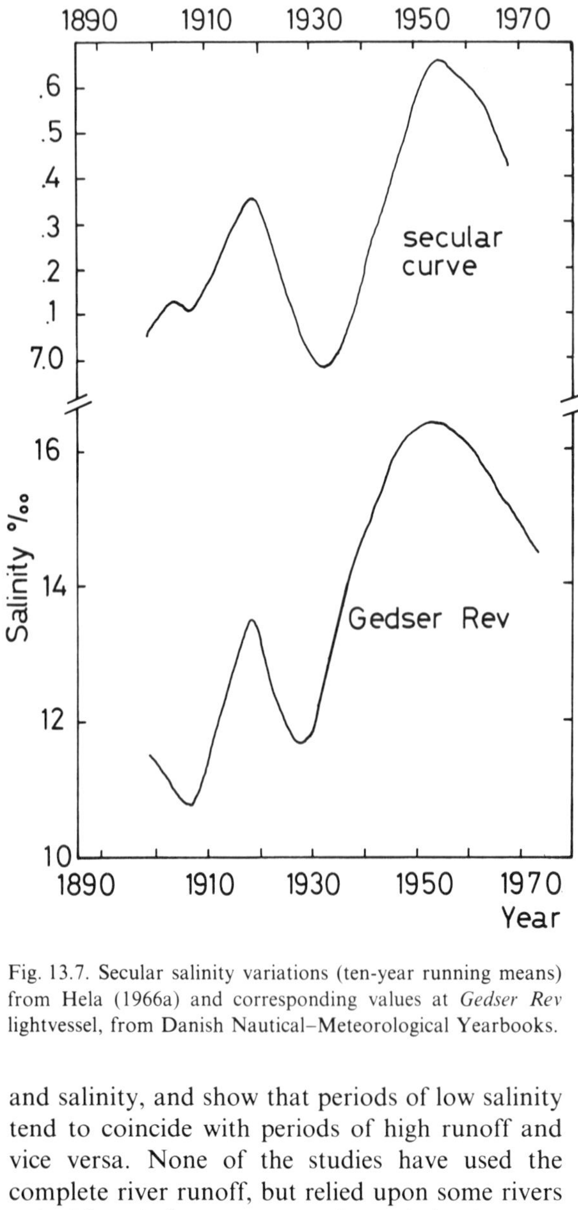

Fig. 13.7. Secular salinity variations (ten-year running means) from Hela (1966a) and corresponding values at *Gedser Rev* lightvessel, from Danish Nautical–Meteorological Yearbooks.

and salinity, and show that periods of low salinity tend to coincide with periods of high runoff and vice versa. None of the studies have used the complete river runoff, but relied upon some rivers only. There is, however, a good correlation between fluctuations in the different rivers which should imply that some representative rivers can be regarded as giving a reliable measure of the total runoff. Kaleis (1976) used three U.S.S.R. rivers representing 28% of the total runoff with an annual average of 120 km^3. He correlated the high and low periods of runoff with the annual average salinities in the water column (0–90 m) of the Bornholm Deep (Fig. 13.2). The results quite clearly show the importance of runoff variations to salinity fluctuations. The low runoff in the period from 1882 to 1898 is in agreement with the salinity found by Jensen (1937) for that period. From the previous discussion it is also clear that salinity variations in the Bornholm Deep are representative of the salinity variations in the Baltic.

Changes of river runoff are in turn probably related to changes in atmospheric circulation, as discussed by Soskin (1963). Cyclonic activity over Europe which produces high precipitation decreased during the 1930s when more anticyclonic conditions developed with less precipitation (Soskin, 1963; Fonselius, 1969). Cyclonic activity was also low in the first years of the century when river runoff, however, was large. There are no long-term studies of changes of the atmospheric circulation which can be correlated directly with runoff fluctuations. Hela (1966a) studied the fluctuations related to continentality over northern Europe, but there is no clear correlation between the fluctuations found by Hela and the river runoff fluctuations presented by Kaleis (1976). It appears that this aspect needs further investigation.

The problem is complicated since the water exchange is also coupled to both runoff and atmospheric circulation. As Hela (1966a) points out, the salinity source is the inflowing Kattegat water. This is demonstrated by the correlation betwen salinity variations at *Gedser Rev* light vessel and Hela's secular curve (Fig. 13.7). It is possible that the Kattegat–Belt Sea undercurrent becomes more developed during periods of low runoff, but on the basis of existing data it does not appear possible to prove this. It can be concluded that the mechanisms governing secular salinity variations in the Baltic are not yet well understood.

The temperature of the Baltic deep water also shows considerable long-term variations (Soskin, 1963; Fonselius, 1969). Fig. 13.8. shows the temperature and salinity variations at 400 m in the Landsort Deep. It is clear that there have been strong temperature variations and a marked increase of almost 1°C.

Both temperature and salinity variations have important implications for conditions in the Baltic. Physical conditions will be considered first, and the chemical and biological aspects will be discussed in later sections. Of major importance is density stratification across the primary halocline, which is influenced primarily by salinity differences between surface water and deep water. Salinity increase

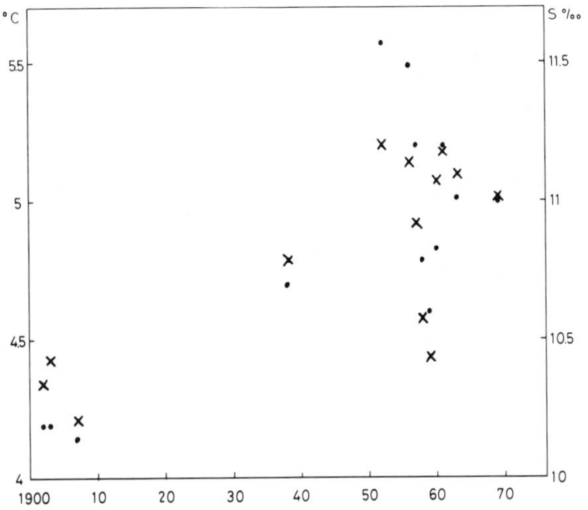

Fig. 13.8. Temperature (×) and salinity (o) variations in Landsort Deep from 1900 to 1970. Data from ICES Data Bank.

during this century has caused an increase of deep-water density in the northern central basin by almost one σ_t unit[1] (Fonselius, 1969). Fonselius investigated variations of the stability across the halocline layer using observations from the 100- and 150-m levels in the Landsort Deep and the Gotland Deep. A general increase of stability was indicated in the Landsort Deep, particularly when only summer values were used, but the trend was not obvious in the Gotland Deep. The major inflows were evident, though, increasing stability markedly in both areas. During the subsequent stagnation period stability decreased until a new inflow occurred.

Instead of using stability over the 100- to 150-m layer, the maximum stability across the halocline layer may be used, although with some caution since it is not always possible to determine the layer of maximum stability. Maximum stability varies according to salinity differences between deep and surface waters (Fig. 13.9). Since long-term changes of salinity appear to occur in both water masses with positive correlation, it is difficult to conclude that there has been either a general increase or decrease of stability. This is important to vertical mixing across the halocline since stability is a major factor influencing mixing.

The rise of the upper limit of the halocline layer,

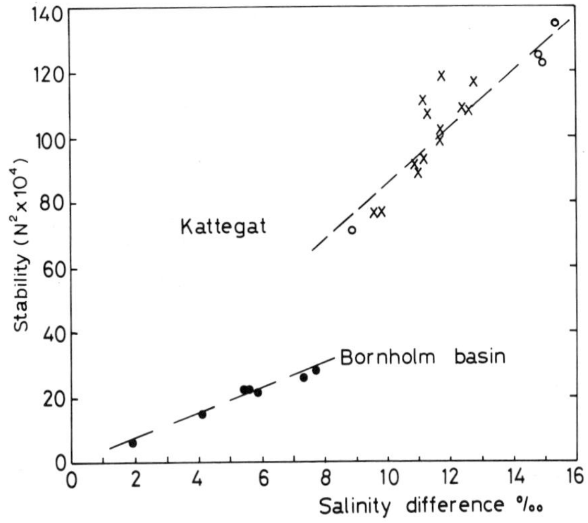

Fig. 13.9. Maximum stability versus salinity differences between deep and surface waters from observations in the Bornholm Basin (●) and the central Kattegat in summer (×) and winter (o). Stability (N^2) is here defined as $\frac{g}{\rho}\cdot\frac{d\rho}{dz}$ where g is the acceleration due to gravity, ρ is density and z is depth. Data from Danish Nautical–Meteorological Yearbooks.

attributed to long-term salinity increase, may have influenced water exchange through the Åland Sea between the Gulf of Bothnia and the Baltic proper (Fonselius, 1969). Due to this rise some deep water from the Baltic proper can penetrate into the bottom layer of the Gulf. Fonselius considers it possible that this has influenced oxygen conditions there since the rise of salinity has increased stability, thereby reducing vertical renewal of the bottom water.

Circulation and mixing

The permanent or mean circulation in the Baltic proper is very weak, with mean velocities of the order of 1 cm s^{-1}. Salinity distribution shows it to be cyclonic, since the isohalines have a NE–SW inclination. In the Gulfs of Finland and Bothnia the mean circulation is likewise cyclonic, with mean velocities of the order of 1 cm s^{-1} or less. The Bothnian Sea and the Bothnian Bay each have one large eddy (Witting, 1912; Palmén, 1930). The mean winds over the Baltic are also weak and the primary forces of the mean circulation appear to be estuarine and thermohaline. The mean flow in the surface layer is slightly better defined along the west

[1]See p. 31 for definition.

coast than along the east coast, doubtless due to outgoing fresh water. There is also a weak vertical shear in the mean circulation.

The fluctuating part of the motion dominates and is induced primarily by varying meteorological conditions, both winds and pressure gradients. Fall and winter storms in the Baltic can be both vigorous and persistent. Tidal effects in the Baltic are very weak.

Witting (1912) and Palmén (1930) investigated currents in the Gulfs of Finland and Bothnia as well as in parts of the northern central basin using observations obtained at light vessels over periods of several years. Palmén demonstrated the influence of the wind on the surface layer current, a finding in general agreement with the Ekman theory. The surface current was found to be almost directly proportional to the local wind, and could reach velocities of about 30 cm s^{-1}.

In the southern part of the Baltic proper current measurements were carried out during the 1920s and 1930s from anchored research vessels. On the basis of such measurements Gustafsson and Kullenberg (1936) discovered wind-induced inertial oscillations. Such oscillations subsequently have been observed in most parts of the Baltic at some distance from the coast (Kielmann et al., 1969, 1973; Mälkki, 1975). Using observations from several parts of the Baltic proper, these authors have demonstrated that the response is mainly barotropic for wind periods longer than about 50 h, and that the main baroclinic response is in the range from 10 to 40 h, where the greatest variability is also found. Periods shorter than 10 h are quickly damped. The inertial peak seems always to be dominant, and the energy of the baroclinic mode about ten times larger than the energy of the barotropic mode (Kielmann et al., 1973). Even for periods of about 100 h the barotropic mode amounts to only 50%, implying that models treating the Baltic as homogeneous are crude approximations at best. Inertial motion has been studied in some detail. Kullenberg and Hela (1942) found a fairly good horizontal coherence in the surface layer, but not at 30 m depth. They also observed a 180° phase shift across the halocline. Kielmann et al. (1969) corroborated these results. The 180° phase shift across the halocline layer may imply that the deep layer motion is compensating the surface layer motion.

The fluctuating motion in the top layer is directly proportional to wind stress, and the spectra look very similar. In deeper layers the motion is related to the divergence and curl of wind stress. It seems clear that the main energy source of motion in the Baltic is the wind. Fluctuating velocities can be of the order of 50 cm s^{-1}. Kielmann et al. (1969) found that the kinetic energy in about 60% of the water column was proportional to $N^{-2/3}$, and in the rest of the water column to N, N being the Brunt–Väisälä frequency. This might imply an essentially turbulent motion in the top layer but motion dominated by internal waves in the bottom layer.

Another feature of Baltic circulation is the various seiches which can occur. Barotropic seiches (Neumann, 1941; Krauss and Magaard, 1962) occur frequently, particularly in connection with fluctuations of the wind field, with periods of about five days (Magaard and Krauss, 1966). Kielmann et al. (1973) found a dominating barotropic seiche with a period of about 27 h and a first baroclinic mode which, they suggested, was topographically induced.

Walin (1972a) made a theoretical investigation of the response of a basin like the Baltic to a varying wind stress. He demonstrated the essentially barotropic response for large periods. In particular he pointed out the importance of the coastal boundary zone. In this zone, about 10 km wide, the response is of a special nature due to the influence of the coast. Important vertical motions can be generated, implying an increased vertical flux compared to the vertical flux across the halocline layer in the interior of the basin.

Mixing in the Baltic is characterized in particular by vertical transfer between the various layers, or across pycnocline layers. The static stability of the permanent halocline layer very effectively suppresses vertical mixing across that layer, which is basic to considerations of the relative importance of the transfer in the open sea and in the coastal zone where the halocline reaches the bottom. Walin (1972b) and Shaffer (1977) have studied the coastal zone and demonstrated that the vertical transfer between the surface layer and the deep water induced there by the wind action can be relatively large.

Vertical diffusion in the halocline and thermocline layers of the open Arkona and Bornholm basins has been investigated by Kullenberg (1974a, 1977) by means of tracer experiments. From these experiments the author determined values of the vertical diffusion coefficient in the range from 0.01 to 0.1 $cm^2\ s^{-1}$ with the lowest values occurring in the halocline layer. Shaffer (1977) found values of the order of 1 $cm^2\ s^{-1}$ in the coastal zone. The results of Kullenberg and Shaffer can be used for a further comparison of the diffusive fluxes. Shaffer gives a flux of about 12 mg salt $m^{-2}\ s^{-1}$ for the coastal zone and Kullenberg finds a flux of 0.6 mg salt $m^{-2}\ s^{-1}$, assumed to be representative for the entire Baltic proper. From independent data Shaffer calculates a mean diffusive flux for the Baltic proper of 0.8 mg $m^{-2}\ s^{-1}$.

Unfortunately other direct studies of vertical mixing across the halocline layer or in deeper layers are not available. Hela (1966) used observations at Finnish coastal stations to study the rate of heat penetration. He calculated vertical mixing coefficients, finding values in the range from 0.5 to 1 $cm^2\ s^{-1}$ in the thermocline at 20 to 25 m depth. The values for the surface layer were in the range from 1.5 to 64 $cm^2\ s^{-1}$.

Recently Matthäus (1977a) used mean annual variations of temperature to calculate average vertical exchange coefficients for heat over 10-m layers for ten stations in the Baltic proper. The calculations were carried to depths of 70 to 80 m at the deep stations. The values were largest in the near-surface layer during winter months, with maximum values above 100 $cm^2\ s^{-1}$. During the summer period the corresponding values were 2 to 10 $cm^2\ s^{-1}$. Below 50 m most of the values were less than 1 $cm^2\ s^{-1}$, with a few reaching 10 $cm^2\ s^{-1}$ during fall or winter. The values found by Matthäus generally conform quite well with those found by Kullenberg (1974a, 1977).

A few experimental studies have been made of the horizontal mixing. Schott et al. (1976) and Kullenberg (1977) carried out simultaneous experiments in the surface and halocline layers, respectively. The value of the diffusion velocity, as defined by Joseph and Sendner (1958), was 10 to 20 m h^{-1} in the surface layer and 3 to 5 m h^{-1} for the halocline layer.

Our present knowledge of the mixing processes and energy sources for mixing is very meager. It does not allow a satisfactory evaluation of the vertical transfer. Regarding energy sources it seems plausible that the input from the fluctuating wind is important. It has been shown that this is the case for the ocean (Denman and Miyake, 1973) and also for the Kattegat (Kullenberg, 1976).

Another source of energy for mixing is the inflows. The inflows which penetrate to the bottom and force the old bottom water or part of it to higher levels play the major role. However, inflows which spread by advection at some level in the deep water are also of importance. These influence the deep water layer just beneath the more active halocline layer (Kaleis, 1976).

Ice conditions

The ice conditions in the Baltic, in particular the northern parts, are of great importance for shipping. Therefore, elaborate warning, forecasting and ice breaking services have been established. The freezing and ice covering depends mainly on meteorological conditions and large fluctuations occur from year to year, depending upon the high pressure situation over the North Sea, Scandinavia and Russia. The number of consecutive days with negative air temperature which is required before freezing can vary considerably and has been used to predict the ice development. Palosuo (1966) has prepared a chart of ice coverage probability based on data from 1931 to 1960.

Freezing generally starts in the open Bothnian Bay in early January, in coastal zones of the Bothnian Sea in middle January and further south by middle February (Palosuo, 1966). The Bothnian Bay is normally completely ice-covered by the end of January, and it is ice-covered 100 to 150 days per year. Complete ice coverage normally also occurs in the coastal zone down to the Åland Sea and along the inner parts of the Gulf of Finland and the Gulf of Riga. The Bothnian Sea becomes ice-covered five winters out of ten. In the central parts of the Baltic proper the probability of ice occurrence is less than 25%.

The ice in the coastal zones is normally fast whereas the open sea ice is in motion influenced by winds and currents. The ice thickness varies considerably in the open sea areas, and open zones as well as ice walls occur frequently. Along the coasts relatively broad open lanes often occur.

Recently attempts have been made to investigate the properties of the ice and the dynamics of the ice drift in some detail using modern observation techniques including satellite remote sensing (e.g., Udin and Omstedt, 1976; Blomquist et al., 1976). One motivation for these studies is the requirement of better data for developing and testing forecasts based on numerical modelling. The observations show a good correlation between wind conditions and ice drift. The main forces determining the motion of the ice were found to be the wind stress and the water stress, but the balance of forces was also influenced by the Coriolis force and the internal ice stress.

CHEMICAL CONDITIONS

In recent decades much attention has been paid to oxygen conditions in the deep water. Fonselius (1962, 1967, 1969, 1972) demonstrated the clear trend of decreasing oxygen content of deep and bottom waters in several of the Baltic basins. The oxygen content of these waters is influenced by weak mixing across the halocline layer, by stagnant periods brought about by major inflows and by the consumption due to oxidation of organic matter. Organic matter sinking from the euphotic zone is oxidized in the deep and bottom water, and in addition oxidation of organic matter in the sediments removes oxygen from the bottom water. Given the very weak supply of oxygen from above in the Baltic proper, this process eventually leads to a consumption of all the dissolved oxygen, producing reducing conditions with the formation of hydrogen sulfide. A lateral supply of oxygen by an inflow can prevent this development; a large part of the inflowing oxygen is consumed in oxidization of the hydrogen sulfide once reducing conditions have developed. This implies that less oxygen is transferred to the next basin in the sequence, and that reducing conditions may develop faster since the new oxygen concentration is already low. Clearly the process is also influenced considerably by the rate of consumption which depends upon the supply of organic matter and the temperature of the water. The supply of organic matter depends, in turn, mainly upon primary production.

Fonselius (1962, 1967, 1969) has described developments during the present century, and the discussion here is based largely on his work. Even in the beginning of the century it was known that there was an oxygen deficit in the deep and bottom waters of the central basin. The development of anoxia is illustrated by sequences taken at station F74 in the northern central basin, and at station F81 in the Gotland Deep (Table 13.2). In Fig. 13.10 an example of a similar development in the Gotland Deep at 100 m depth is given. The oxygen variations clearly show the occurrence of inflows.

During the 1920s oxygen-free water was reported in the bottom water of the Gotland Deep (Granqvist, 1931), and during the 1930s a stagnation period occurred with oxygen-free water at the 200 m level in the Gotland Deep (Kalle, 1943; Fonselius, 1962).

During World War II observations were discontinued. Since the war there have been several periods of oxygen-free water and the frequency of occurrence seems to have increased (Fonselius, 1969). The volume of the oxygen-free water that develops during the stagnation periods has also increased considerably.

Oxygen-free periods have occurred in all the deep basins since 1950. In the Landsort Deep the homogeneous conditions below about 100 m suggest

TABLE 13.2

Changes of salinity, temperature and oxygen at a depth of 150 m at stations F74 and F81

F74, 59°22′N 21°05′E				F81, 57°20′N 19°51′E			
Year	S‰	T°C	O_2 (ml l^{-1})	year	S‰	T°C	O_2 (ml l^{-1})
1906	10.1	4.2	3.15	1910	11.71	3.19	–
1939	11.0	5.2	1.5	1938	11.78	5.69	2.51
1967	11.15	4.9	0.4	1957	12.10	5.50	0.00
				1969	12.19	5.91	0.51
				1972	12.37	5.27	–

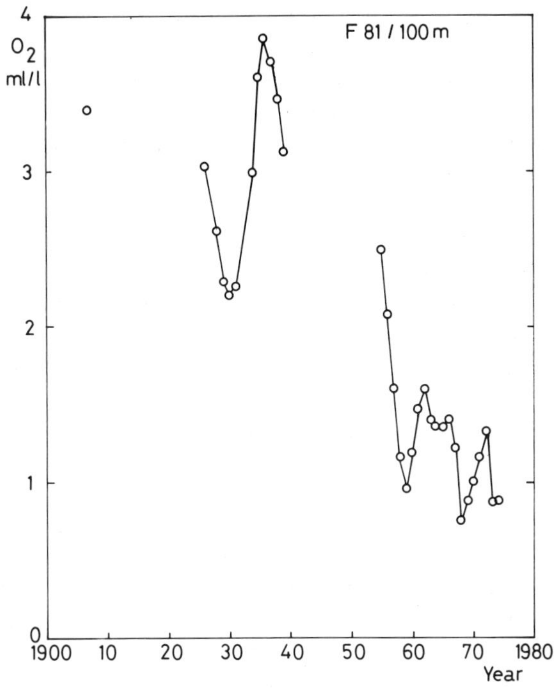

Fig. 13.10. Three-year running averages of annual mean values of dissolved oxygen at 100 m at station F81 (57°20′N 19°51′E) (see also Fig. 13.4). Data from ICES Data Bank.

rather efficient water exchange. Probably for this reason the Landsort Deep has resisted the development of anoxic conditions for a long time (Fonselius, 1969). Hydrogen sulfide was not observed there until the end of 1968 at depths from the bottom to about 300 m. Even in the Arkona basin, oxygen-free bottom waters have been reported.

The Gulf of Finland shows the same trend of generally decreasing oxygen content in deep and bottom waters. Although the supply of deep water from the central basin to the Åland Sea is small (Palosuo, 1964) the oxygen values in the southern Åland Sea have shown a decrease since 1900. In the northern Åland Sea the trend is not so marked. A decrease of the oxygen content in the deep waters of the Bothnian Sea may be related to an increase of the salinity there or to an increased oxidation rate caused by an increasing supply of oxidizable matter. In the Bothnian Bay there is no indication of an oxygen decrease.

Observations during the past decade show that the oxygen-free zone extended most widely during 1968 and 1969, and that subsequently conditions have become better (Jansson, 1978). During the first half of 1976 hydrogen sulfide was observed only in the eastern Gotland basin, whereas in 1969 the hydrogen sulfide zone extended into the Gulf of Finland. The extent of the oxygen-free zone gradually decreased from 1969 to 1976.

Water exchange and the supply of organic material are the two essential factors influencing the oxygen content of the deep water. It seems unlikely that a balance between these two factors should exist; it can be expected that the system will be variable. Sediment cores show that oxic and anoxic conditions in the Baltic deep basins have alternated during historical time. During the present century there have been several changes or perturbations which may have induced a dominance of the oxygen-free state of the system. Several factors which have been suggested are: (1) an increase of static stability across the halocline layer; (2) a decreased depth of the halocline layer; (3) an increase of temperature in deep water; (4) a decrease of oxygen content in the inflowing water; (5) a change of the inflow pattern; and (6) an increase of the supply of oxidizable matter.

There are no clear indications of a general long-term increase in stability, and our present knowledge of the vertical mixing processes does not allow a precise quantification of the effects of a small stability change. It might also be argued that the effective vertical transfer generated in the coastal boundary zone during suitable meteorological conditions is not influenced by the stability. However, stability after a major inflow has a major influence on the time when subsequent bottom-water renewal can occur. Therefore the extreme stability found after major inflows of the mid-century period could have influenced conditions by prolonging the time before subsequent renewal. However, available observations do not really allow this to be proved.

Decreased depth of the halocline layer has probably had an influence on conditions in the southern Åland Sea, as discussed above. However, a general halocline depth between 50 and 60 m also implies that active vertical mixing in the surface layer during fall and winter may more easily reach the top of the halocline. This, together with an increase of the area of the halocline layer, would tend to enhance transfer through the halocline layer.

An increase of the temperature of inflowing water implies both a decreased oxygen content and an increased rate of oxidation. Kullenberg (1970) found a rather good correlation between temperature increase and oxygen decrease in the Landsort Deep, which suggests that the temperature effect is important.

The oxygen content of the Kattegat water varies, and there are indications of some decreases during periods of this century (Herman and Olsen, 1970). This would imply a lower oxygen content of the water flowing into the Baltic proper. The inflow into the successive basins is a cascade process (Fonselius, 1970), which implies that the oxygen content of the incoming water is decreased as the inflow penetrates to the inner basins. A large amount of hydrogen sulfide in the outer parts of the inflow path consumes a substantial amount of the oxygen before the water reaches the inner parts of the central basin. This effect has probably been important in some cases.

There are indications that the inflow pattern has changed, so that since the 1930s the inflows tend to occur during summer and fall rather than during winter and spring. This suggests a higher temperature of the inflowing water and probably a larger content of organic material than before.

The essential source of organic matter is primary production, which will be discussed below. The inflow of oxidizable matter from land has increased during this century, due essentially to human activities (Fonselius, 1972; Jansson, 1978). The possible significance of this will be considered in the discussion of organic matter.

The surface isohaline layer becomes saturated with oxygen during the convection period in fall and winter. During the warm season the absolute content of oxygen above the halocline decreases but the water remains more or less saturated. Beneath the thermocline the oxygen consumed by the oxidation processes is not renewed. In the thermocline layer a minimum of absolute oxygen content develops during late summer, caused by an accumulation of organic material.

The change from oxic to anoxic conditions in the water implies drastic changes in the biological and chemical systems. Fonselius (1972) has discussed the influence of this change on nutrient chemistry. During oxic conditions, regions like the Baltic act as nutrient traps, since both phosphate and nitrate are transferred to the sediments. During reducing conditions, however, inorganic phosphorus is dissolved from the sediments and added to the bottom water where phosphate then accumulates. This process is evident in Baltic profiles, where, during reducing conditions, phosphate concentrations can become very high. When the bottom water is renewed by an inflow, old water is forced to higher levels and becomes partly mixed into the deep water under the halocline layer. From there the phosphate gradually reaches the euphotic zone, about 20 m in depth, by vertical mixing in the whole area and upwelling in the coastal boundary zone. In this way phosphate enrichment takes place not only in deep water but also in the surface layer, as shown by Fonselius (1972).

The significance of this is related to the influence of the phosphate concentration on primary production. Several authors (Sen Gupta, 1968; Francke and Nehring, 1973; Schultz and Kaiser, 1973) have considered nitrogen to be the most likely limiting factor, but Fonselius (1976) argues that it is phosphate. Probably both can act as limiting factors, but during different times of the year.

In the surface waters the atomic ratios of organic C:N:P are approximately 101:19:1 (Voipio, 1969), not very different from values in the open ocean. During wintertime the nitrogen and phosphorus content of the surface waters is generally high and close to the ratio 16:1, whereas during the summer the values decrease to zero or close to it, implying that the nutrients are utilized approximately in the ratio 16:1.

During anoxic conditions the N:P ratio in the intermediate deep water is about 2:1 and in the bottom water 1:1 or slightly less. Fonselius ascribes this fact to the denitrification process. It is conceivable, Fonselius argues, that denitrifying bacteria are mixed upwards to layers just below the halocline so that the process can operate at these depths where the oxygen content may be larger than the starting concentration for denitrification. At these depths the concentration of nitrate is still fairly large. At greater depths where the oxygen concentrations are lower the denitrification process is more effective and the nitrate concentration decreases, whereas the concentration of ammonia increases. On the other hand, the phosphate release from sediments increases the phosphate concentration in the bottom water, an effect also produced by mixing in the deep water. During oxic periods, however, nitrate is found in the deep water, but there are only small amounts of ammonia. The N:P ratio is also larger — about 3:1 — in the bottom water. Almost all of the phosphorus in the deep water is inorganic, whereas there is a considerable amount of organic nitrogen. The ratio C:N (organic nitrogen) is around 16:1, which corresponds better with the composition of humus than of plankton. Fonselius also points out that there are other possible sources of nitrogen, such as rainwater. Nitrogen fixation from the atmosphere occurs due to the blue-green algae that are quite common in the Baltic Sea (Lindahl et al., 1978): urea and amino acids are also present in the waters. Fonselius' conclusion is that it is not nitrogen but phosphate that is the most limiting factor for primary production. If this really is the case, then phosphate enrichment is important to the general conditions in the Baltic Sea. However, it should be recalled that both nitrogen and phosphate may act as the most limiting factor depending upon the area as well as the time of year. Several authors consider that nitrogen is the most limiting factor (Francke and Nehring, 1973). Finally it should be mentioned that silicate is also dissolved from the sediments during anoxic conditions and transferred to the surface layers by the same processes as the phosphate.

PRIMARY AND SECONDARY PRODUCTION

The yearly course of primary production in the central Baltic Sea normally has a spring bloom with maximum values in April or May and a second bloom in July to August. The latter bloom can often yield the highest monthly values. Fonselius (1972) discussed observations of primary production made from the light vessels *Finngrundet* in the Bothnian Sea and *Hävringe* in the northern Baltic proper during the years 1961–1968 and 1963–1967, respectively. The ^{14}C technique was used *in situ* with 24 h incubation time, and the observations were carried out twice every month at five or six depths down to 20 m. The mean annual production was found to be 57 g C m^{-2} at *Finngrundet* and 78 g C m^{-2} at *Hävringe*. The time series at *Finngrundet* showed an increase of production from 1961 to 1968. This was not indicated at *Hävringe*, although the time series there was too short to draw any firm conclusion. Ackefors and Lindahl (1975) presented observations carried out during 1973 and 1974 using the ^{14}C *in situ* method at ten levels down to 20 m and a 4-h incubation time. The average values for the two years were 113 g C m^{-2} in the Bornholm basin and 103 g C m^{-2} in the eastern Gotland basin. These values are higher than those reported by Fonselius as well as those reported by Kaiser and Schultz (1970) for the years 1969–1970. Ackefors (1975) investigated different trophic levels and the energy flow and came to the conclusion that the present annual primary production in the Baltic proper must be about 100 g C m^{-2} in order to support the fish production.

For the Kattegat, Berthelsen and Hansen (1969) found primary production to be in the range 50 to 130 g C m^{-2} yr^{-1} over the period 1966–1969. The trend indicated an increase at some of the sites.

The observations in the period from 1961 to 1974 may indicate an increase of primary production in the Baltic. Since phosphorus concentration has increased as well, this seems to support Fonselius' (1976) argument that phosphate is the most important limiting factor. Fonselius shows that the phosphorus content in both deep water and surface water has increased by a factor of 3 to 4 during the last couple of decades.

Sjöberg and Wilmot (1977) used a combined physical and biological two-dimensional model to qualitatively simulate a spring phytoplankton bloom in the northern Baltic proper (Hobro et al., 1978). The observed time development of the various components of the spring bloom is shown in Fig. 13.11. The results of the simulation suggest that at these latitudes the onset of the bloom is controlled by a rapid increase of insolation during springtime. Vertical mixing and stratification is also important: persistent deep mixing can retard the onset of the bloom. It appears reasonable to conclude that seasonal variations of incoming light have a considerable influence on the timing of production blooms, but clearly other factors, such as the wind conditions with their influence on the mixing, are also important.

The investigation by Ackefors et al. (1977), covering the years 1973–1976, shows that primary production decreases from the Hanö Bight to the Bothnian Sea, and that the spring bloom occurs later in the north than in the south. An increasing trend was indicated only at the southern station, where the annual means varied between 105 and 132 g C m^{-2}. In the Bothnian Sea the correspond-

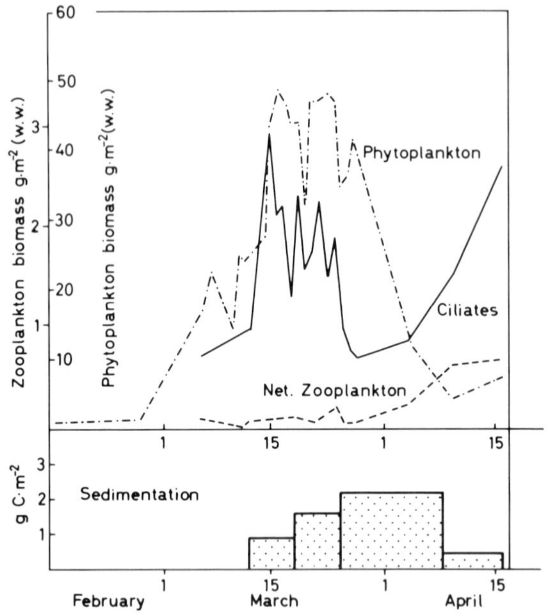

Fig. 13.11. Development of spring bloom in the central Baltic (from Jansson, 1978).

ing variation was between 65 and 71 g C m^{-2} and in the Gotland Sea between 87 and 116 g C m^{-2}. Apart from the light conditions, the authors suggest that the spatial variation is due to a varying nutrient supply which is influenced by upwelling as well as input from various land-based sources.

Ackefors et al. (1977) explained a lack of correlation between chlorophyll *a* and primary production by the fact that the relation between chlorophyll *a*, biomass and primary production changes when insolation, the nutrient supply or the composition of the flora are varying. The dominating phytoplankton species at all stations were blue-green algae, diatoms, dinoflagellates, *Monas* spp. and flagellates. Often the production maxima were dominated by one species with a short bloom. Characteristic features of the phytoplankton biomass were short-time variability and maximum values during the spring bloom. The diatoms were numerous during spring, early summer and autumn, whereas the blue-green algae were abundant during spring and summer.

Zooplankton studies carried out by Ackefors et al. (1977) at seven stations in the Baltic proper that were visited four times a year showed that 40 species normally occur, with copepods dominating, primarily *Pseudocalanus elongatus* and *Temora longicornis* in the open sea and *Acartia bifilosa* in the coastal waters. Salinity and temperature are controlling factors of the zooplankton fauna, with temperature governing seasonal variation and vertical distribution and salinity affecting species composition. The biomass was largest in the southern Baltic with a mean value of 25 g m^{-2} (wet weight); the mean value was 13 g m^{-2} in the northern areas. In the Bothnian Sea only 10 to 12 species were found, with the copepods *Acartia bifilosa* and *Eurytemora* spp. forming about 90% of the biomass, which was smaller than in the Baltic proper. The zooplankton production was about 10 to 15 g C m^{-2} yr^{-1}, showing no marked spatial variations. Most of the production occurred during the period March to November, with the primary maximum during August to September. The authors were of the opinion that phytoplankton production seemed too small to support the zooplankton production. They pointed out, however, that the role of the microzooplankton as food had not been clarified.

Another group which may play an important role in the production cycle in the Baltic is the heterotrophic bacteria, and studies have been initiated to estimate the production of the bacterial biomass (Hagström et al., 1978) as well as the importance of bacteria to the recycling of nutrients (Hagström and Larsson, 1978).

Budget of phosphorus and organic matter

Fonselius (1969, 1976) used a steady-state box model of the surface, deep and bottom water layers to construct a phosphorus balance. The water exchange was first determined by continuity requirements of water and salt for a two-layer model, using long-term annual averages of river inflow (471 km^3), the inflow from the Kattegat (471 km^3) and the outflow through the Danish Sounds (942 km^3). The salinity used for the surface layer was 7‰, for the deep layer 11‰, and for the incoming Kattegat water 17.5‰. By means of these data the volume exchanged vertically between the layers could be calculated. The mean concentrations of phosphorus in the water masses were obtained from hydrographic data, and the input from rivers, sewage and industrial wastes, and the atmosphere were estimated from input studies (ICES, 1977b). The total river input was estimated at 17 000 metric tons P per year, based on Swedish observations. The estimated total content in the surface layer was 132 500 metric tons P, in the deep water 265 700 metric tons, and in the bottom water 78 000 metric tons. The annual exchange across the halocline layer was 80 900 tons downwards and 90 800 tons upwards, and the transfer from the bottom to deep water was 17 700 tons per year. For the surface layer the input was assumed to be balanced by a loss to sediments in shallow water areas, whereas for deep and bottom waters the exchange with the sediments remains unknown. In order to maintain a steady state in deep water, about 18 000 tons phosphorus per year must be lost to the sediments.

Voipio (1969) constructed a phosphorus balance for a three-layer model using estimates of input from various sources together with considerations of chemical processes in the water and sediments as well as the production cycle. He estimated that the exchange between bottom waters and sediments in the Baltic proper was 10 000 metric tons P per year both ways, and that the corresponding exchange across the primary halocline was 40 000 tons. Assuming that a major inflow can completely replace stagnant bottom water, Fonselius (1967) estimated that 110 000 tons P were released from bottom waters in 1961 after such an inflow. This amount had accumulated during about ten years, which implies a release rate from the sediments of about 10 000 tons per year.

Fonselius (1976) also produced an estimate of annual transport and production of organic matter using a two-layer box model. The river and discharge inputs were based on Swedish observations integrated for the entire Baltic Sea area. Fonselius derived an input of about 5×10^6 metric tons C per year excluding the input from benthic and primary production. An estimate of total primary production based on the light vessel measurements cited above gave a total of 26×10^6 tons C per year. He found the benthic production to be negligible. Most of the production is recycled and eaten by higher levels and therefore does not represent a net input to the system.

Fonselius went on to consider the oxygen budget. The total input of oxygen to the deep water by lateral inflow and vertical exchange was calculated by Fonselius (1969), using long-term averages of the water exchange, salinities and the oxygen content of relevant water masses. Using these earlier results, Fonselius (1976) found that only about 10% of the annual production and input of organic matter can be oxidized in deep water.

The implication is that an increased load of organic matter can have serious consequences for oxygen conditions in deep water. The primary production increase over the years indicated by the light vessel observations was about 25% and waste input constitutes about 5% of the total budget. Even with a 30% reduction of the total organic matter budget as given by Fonselius (1976), only about 15% of the organic matter could be oxidized by the yearly input of oxygen. Probably the budgets are as yet not precise enough to allow any firm conclusion on possible implications of an increased discharge of organic matter to the Baltic. Clearly it is important to obtain more accurate budgets and more reliable values on the transfers within the Baltic, in particular the vertical transfer between the various water masses.

Humus is brought to the Baltic in large quantities by rivers. It may be traced through its fluorescence properties, and several such studies have been made throughout the Baltic (Kalle, 1949; Jerlov, 1955; Kullenberg and Nygård, 1971; Karabashev et al., 1971; Højerslev, 1974). Jerlov found a rather good correlation between salinity and the amount of humus, as defined by the content of yellow substance.

OPTICAL PROPERTIES

The optical properties of the Baltic sea waters are influenced by the amounts of humus, yellow substance (with strong absorption in the blue and ultraviolet) and suspended matter (Jerlov, 1955). The distribution of yellow substance conforms with the circulation of outflowing surface water through the Åland Sea and along the Swedish coast. Jerlov (1955) found values of the absorption coefficient at 380 nm due to yellow substance around $2\ m^{-1}$ in the Bothnian Sea, about five times larger than the values in the Kattegat. Typical values of the total light scattering coefficient in the red region of the spectrum are $0.1\ m^{-1}$ to $0.7\ m^{-1}$, with the smallest values at intermediate depths (Kullenberg, 1969; Højerslev, 1974).

In the halocline layer there is often an accumulation of suspended matter. Except in some areas of strong bottom currents, there is generally a turbid bottom layer of about 5 m thickness. A characteristic feature of the Bothnian Sea deep water for depths greater than about 50 m is the gradual increase of suspended matter towards the bottom. The origin of this load of suspended matter is not clear, but it may originate from land and erosion in shallow waters.

The underwater radiant energy distribution is also strongly influenced by humus-like substances; the wavelength of maximum energy penetration is about 550 nm. In parts of the open Baltic proper the 1% level of green (525 nm) light is at 25 to 30 m depth; the 1% quanta (350 to 700 nm) irradiance level is normally found around 20 m depth (Højerslev, 1974). The color of the water is normally green.

The optical properties of water are often good indicators of changes in general water conditions. Jerlov (1975) compared irradiance measurements at practically the same locality in the Baltic obtained during the spring and early summer season in 1937–1938, 1963, and 1973. He found that the penetration of irradiance around 525 nm was slightly better in 1973 than in 1937, and concluded that particulate matter had not increased in the surface layer since 1937–1938.

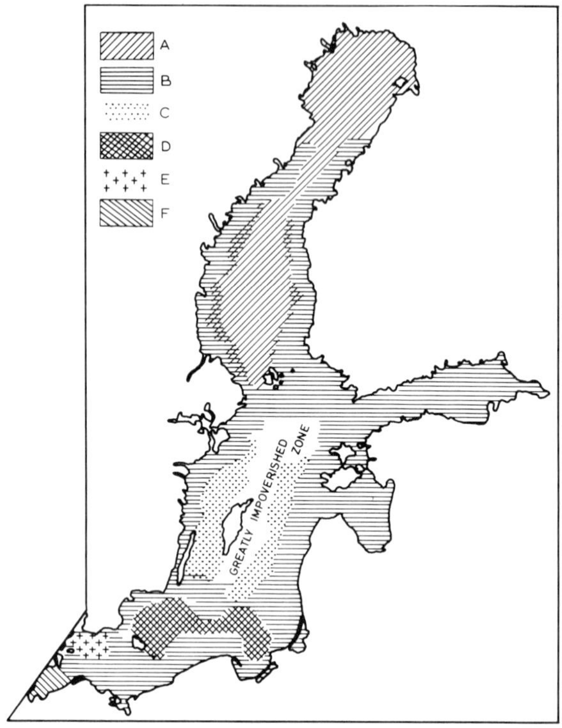

Fig. 13.12. Benthic faunal distribution in the Baltic (Zenkevitch, 1963, as cited by Jansson, 1978).
Legend: A = *Pontoporeia affinis–Mesidothea* community; B = *Macoma baltica* community; C = impoverished community of Polychaeta, Crustacea. *Scoloplos*, *Pontoporeia*, *Femorata*, *Mesidothea*; D = *Macoma calcarea* community; E = *Astarte borealis* community; F = *Cyprina islandica* community.

THE BIOLOGICAL SYSTEM

Jansson (1978) recently published a review of the Baltic biological system in which he pointed out the heterogeneity of the flora. He described the main features of the organisms as: (1) able to withstand low salinities; (2) less specialized in the Baltic and at the same time smaller than members of the same species in the fully marine areas; and (3) low in number of species compared to marine areas (77 macroscopic species in the central Baltic, as compared to 1500 off the Norwegian coast). The faunal distribution is shown in Fig. 13.12 (Zenkevitch, 1963). Jansson also summarized the basic conditions for the ecosystem as: (a) a substratum ranging from hard to soft bottoms, with the rocky archipelagoes as typical features; (b) a strong annual pulse of solar energy input; and (c) brackish waters with marine composition and a physical system of heavy stratification subject to thorough vertical mixing in the surface layer during fall and winter and with intermittent inflows but fairly constant total salinity.

The attached microscopic algae are well developed due to a large area and a long coast line of shallow water with good light conditions. Jansson described the vertical structure of the bottom flora in relation to the solar energy fixing properties of the components and the annual sequence of populations within the community. The belt of blue-green algae is significant for nitrogen fixation and as a nursery ground for crustaceans as well as for the regeneration of nutrients. At the Askö Laboratory in the Landsort region the various components of the system are being studied in detail.

According to Jansson the pelagic subsystem of the Baltic shows an annual sequence similar to that described by Patten (1963) for the North American coast. In winter primary production is dominated by heavy diatoms; in summer green algae and other (dinoflagellates and *Monas* spp.) small and light forms which have high light optima and short generation times predominate. During the fall these are replaced by diatoms and dinoflagellates.

During early spring diatoms (*Skeletonema costatum*, *Thalassiosira baltica*) and dinoflagellates (*Gonyaulax catenata*) may deplete the nitrate within four days with an excess consumption of nutrients and peak production values of 1 g C m^{-2} day^{-1}. The first grazers appear to be unicellular ciliates (Hobro et al., 1978). The larger zooplankton have minimum biomass during early springtime (Ackefors, 1975), although some species such as *Acartia* hatch at the time of the spring phytoplankton bloom. A large part of the primary production is lost to the pelagic system, sinking to the bottom (Kaiser and Schultz, 1975).

During summertime the phytoplankton biomass is low and dominated by small forms (Bagge and Niemi 1971) subject to intense grazing (Kaiser and Schultz, 1975). The larger zooplankton, typically *Temora longicornis* and *Pseudocalanus elongatus*, have increased in biomass (Ackefors, 1975). The low amounts of nutrients are supported by nitrogen-fixing blue-green algae. Blooms of *Nodularia spumigena* have occurred frequently during recent years in large belts during summer (Nyqvist, 1974; Horstmann, 1975). Subsequent to the blue-green bloom, an autumn bloom, mainly of diatoms, dinoflagellates and green algae, develops, during which time the zooplankton biomass reaches its maximum, yielding food for pelagic fish like herring and sprat (Jansson, 1978). During the winter the activity is low and respiration exceeds production for the total system.

The system as described by Jansson (1978) prevails for the northern Baltic proper. In the Belt area the conditions are more variable and several blooms can occur (Von Bodungen, 1975, as cited by Jansson, 1978), whereas in the Gulf of Bothnia the spring and autumn blooms have merged (Lassig and Niemi, 1975). This shows the importance of the north–south extension of the Baltic and illustrates again how the system can differ from region to region.

The total yearly primary production is of the order of 100 g C m^{-2}, in agreement with Ryther's (1970) value for shallow coastal areas of the oceans. Ackefors (1975) estimated the zooplankton production to be about 5 g cm^{-2} yr^{-1}. However, larger values have been observed (Ackefors et al. 1977). About 50% of the sinking organic material is decomposed by bacteria before it reaches deep water (Zsolnay, 1973). Despite this the spring bloom of heavy diatoms provides a large input to the soft bottom system, where a big reservoir of organic matter is built up. Several organisms use this potential energy, building a vertically layered structure with an oxygenated top layer followed by an anoxic layer where hydrogen sulphide producing bacteria dominate.

On sand and gravel bottom the meiofauna is dominated by harpacticoid copepods; nematodes become important on bottom with sediments containing smaller size fractions (Scheibel and Noodt, 1975, cited by Jansson, 1978). The number of species of meiofauna and of macrofauna decreases towards the north. In the macrofauna of the southern Baltic the biomass is dominated by *Cyprina islandica*, *Abra alba*, *Astarte borealis* and *Macoma calcarea*, whereas further north crustaceans like *Saturia (Mesidothea) entomon* and *Pontoporeia affinis* become more important (Jansson, 1978). These species provide important food for the fish in the northern areas, but the fish spawning is hampered there due to critical oxygen conditions. Ankár and Elmgren (1976, cited by Jansson, 1978) have estimated that the total benthic production is reduced by losses within the system until less than half remains for fish food. On the basis of observations in the Askö–Landsort area over several years, Ankár and Elmgren proposed an energy flow model for the benthic ecosystem (Fig. 13.13). Some of the values given in the model are very uncertain, and the flows in and out of the compartments do not balance, probably due to the difficulty of making reliable estimates.

As a general conclusion Jansson (1978) comments on the complexity of the system with its many feedback mechanisms and the necessity of considering the coupling between different levels when assessing ecological rates. Various attempts have been made to study these aspects by means of numerical simulation experiments (Probst, 1975; Sjöberg, 1977).

The Baltic ecosystem is subject to severe natural fluctuations as well as man-made disturbances. At present there are indications of a tendency towards

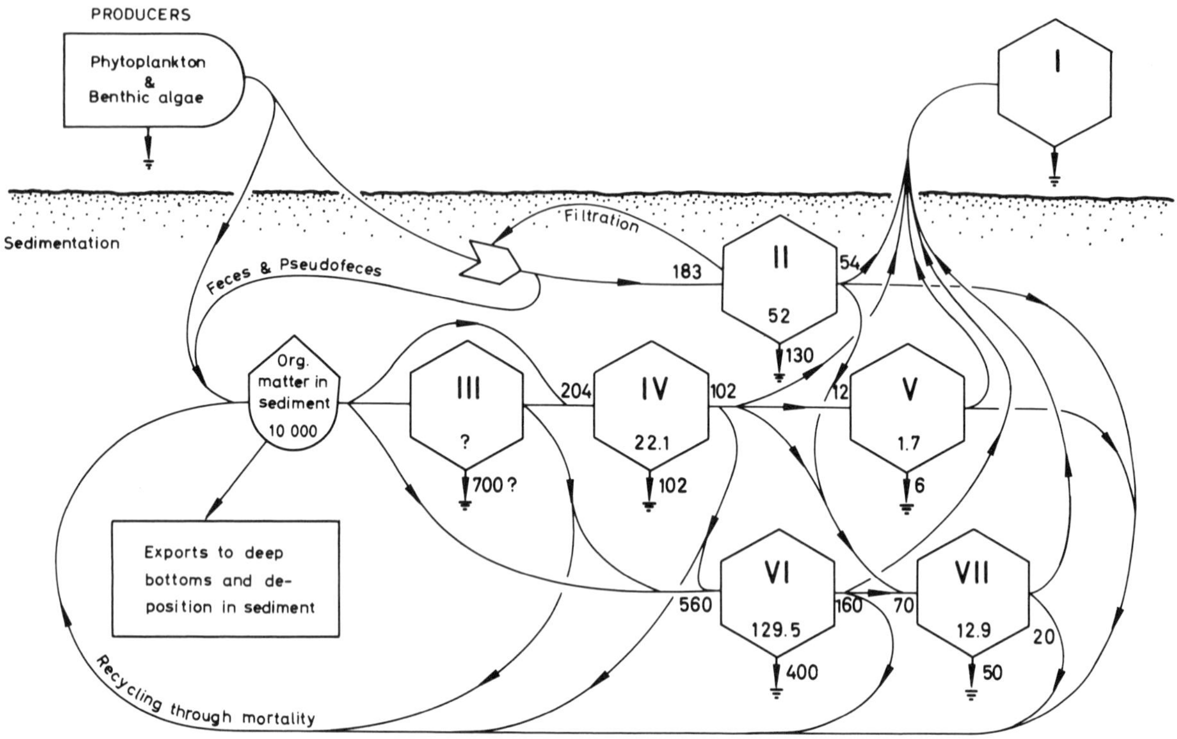

Fig. 13.13. Rough energy flow model of the benthic ecosystem in the Askö–Landsort area. The Roman numerals identify compartments: *I*=fish; *II*=planktivores; *III*=bacteria and ciliates; *IV*=meiofauna, detritivores; *V*=meiofauna, carnivores; *VI*=macrofauna, detritivores; VII=macrofauna, carnivores. The biomass was measured, the production and respiration estimated. Storage are given in kJ m^{-2}; transfer rates as kJ m^{-2} yr^{-1}. Numbers to the left of the hexagons indicate assimilation. Feces and organic excretion are considered as bound to the organic pool in the sediment. (From Ankar and Elmgren, 1976; Jansson, 1978.)

eutrophication of the system which has been considered as normally oligotrophic. The mineral supply has increased, and in certain areas there appears to be a slight increase of primary production. There is also an indication that fish such as pike (*Esox lucius*) and perch (*Perca fluviatilis*), adapted for an oligotrophic system, are being replaced by bream (*Abramis brama*) and roach (*Rutilus rutilus*), which are more suitable to an eutrophic system. In the coastal zones brown algae such as *Fucus* are forced seawards and the green algae *Cladophora* and *Enteromorpha* increase, and in offshore waters there is a tendency towards more frequent blooms of the blue-green alga *Nodularia spumigena*; Jansson (1978) provides a further discussion of these aspects.

FISHING IN THE BALTIC

The Baltic fish stocks are a major resource; the catch of fish has been increasing steadily since the 1930s, due mainly to increased fishing efforts and improved techniques. The catch was about 800 000 t in 1972 (Otterlind, 1976). Most is taken in the Baltic proper, with herring, sprat and cod the most important species, although herring is also fairly abundant in the Gulf of Bothnia. The fish distribution is related mainly to the salinity distribution and oxygen conditions. Only a few ocean fish can become adapted to brackish water; the most important are herring (*Clupea harengus membras*), cod (*Gadus morrhua*), sprat (*Sprattus sprattus*), and flounder (*Platichthys flesus*). Eel (*Anguilla anguilla*) and salmon (*Salmo salar*), depending on adjacent coastal fresh-water environments, are also important species in the Baltic. In the Arkona basin, a great proportion of the fauna is flatfish such as flounder and plaice (*Solea solea*), the latter coming mainly from the west. The varying hydrographic conditions deter migration of mature cod, herring and salmon from the east, and, according to tagging experiments, there exists a distinct boundary at Bornholm (Otterlind, 1976).

The relatively high production and generally good bottom water conditions in the Bornholm basin make this possibly the most important area for the fish fauna, as has been demonstrated by tagging experiments on cod and herring (Otterlind, 1976). In the Bornholm basin many western Baltic fish congregate during summer and autumn for spawning. Important spawning areas are also found in the southern part of the Gotland basin. In the central and northern regions of the Baltic proper variations of salinity and oxygen content have a great influence on the occurrence of cod. This species requires a salinity of at least 10.5‰ for successful spawning. The fish are driven away from the area by an oxygen deficiency below about 2.0 ml l^{-1}. During the 1940s and early 1950s cod were very common down to about 150 m depth east and southeast of Gotland, but in recent years they have been found only occasionally at 100 m depth (Otterlind, 1976). Although the main distribution areas for cod, sprat and flounder are in the Baltic proper, these fish may migrate into the Gulf of Bothnia. Baltic herring, salmon and eels are found regularly in the Bothnian Bay.

Otterlind (1976) describes the results of extensive tagging experiments with herring, cod, plaice and flounder. Generally each species shows distinct migration patterns, both in time and space, often with long migration routes. The plaice move between the Arkona and Bornholm basins, and recruitment seems to be mainly by migration of young fish from the Belt Sea area. Flounder in the central Baltic proper migrate only between the deep and shallow waters in that region. The herring seems to migrate mainly along the whole western coast in the Baltic proper, whereas in the Gulf of Bothnia the migration is more evenly distributed. The migration between these two main areas seems to be insignificant. Cod in the central Baltic proper migrate from the Åland Sea to the southernmost part of the Baltic; cod in the Bornholm basin are more stationary. Migration of cod west of the Bornholm seems to be very small.

Quite clearly the environmental conditions in the Baltic influence the fish stock considerably. Cod were favored at first by the increasing salinity but later were hampered by the increased frequency of oxygen depletion and water containing hydrogen sulfide. To a certain extent the cod have redistributed and have been replaced by an increased abundance of sprat and herring. These species appear to have been favored, at least temporarily, by the slight increase in primary production. However, increased fishing effort has led to exploitation of the Baltic fish stocks and regulations have been imposed by the 1974 Baltic Fishing Convention.

OTHER USES OF THE BALTIC SEA

Marine pollution is a serious threat to the Baltic living resources. Regarding the general pollution situation, including considerations of eutrophication, Zmudzinski (1975) gave an account based on observations made during 1971 to 1973 of the Baltic proper and the transition area. He stressed the importance of the river input of nutrients. Pollution is caused by input of domestic sewage from about 18 million people (1976), industrial and agricultural effluents, atmospheric input and input from marine transportation.

Input studies have been carried out within the framework of ICES (1970, 1977b). It is estimated that domestic sewage discharge amounts to about 2.3×10^6 m^3 day^{-1} directly and 1.2×10^6 m^3 day^{-1} indirectly, of which 40% and 20%, respectively, has not been treated. Information on industrial inputs is very limited. Paper and pulp industries discharge significant amounts of waste containing lignin, which adds a high load of organic matter. The total input of nitrogen from domestic and industrial sources is about 130 000 t yr^{-1} and of phosphorus about 27 000. This is roughly 10% of the total deep water content of phosphorus in the Baltic proper. The information on inputs of metals and organochlorine compounds, in particular PCB (polychlorinated biphenyls), is very meager.

The distribution of pollutants in sediments and living components of the Baltic ecosystem has been studied to some extent (Ackefors, 1971; Jensen et al., 1972: Odén and Ekstedt, 1976; Jensen and Olsson, 1976), but comparable information for the whole area is still limited. Jensen and Olsson (1976) found high levels of PCB (polychlorinated biphenyls) in plankton and plankton feeders like herring, and they also found levels of DDT (dichloro-diphenyl-trichloroethane) about ten times higher in the Baltic than in the Skagerrak. The values were particularly high in the Bornholm to Gotland area.

DDT and PCB in Baltic fish were also studied extensively by Otterlind et al. (1971), who found high levels in muscle tissue of fat fish like herring and salmon and in the liver of fish such as cod. The highest levels of DDT in herring (mean 40 to 50 ppm on fat weight basis) were observed in the area east of Bornholm to southeast of Gotland. The PCB content on the other hand showed an even distribution in the Baltic proper, with mean levels in herring of about 15 ppm in the lipids. Both PCB and DDT content were clearly lower in the Arkona basin than east of Bornholm, and this was also the case for DDT in the Gulf of Bothnia. The authors considered the local state of pollution as the primary cause for the levels found in fish.

Odén and Ekstedt (1976) found that the total DDT and PCB content in the upper 2 cm of the sediments were low in the Gulf of Bothnia, with a slight increase towards the north, but relatively high in the Baltic proper. The PCB content was about 100 times higher there than in the Gulf, whereas the DDT content did not appear to be as large as might be expected from the total input to the Baltic.

An attempt towards a baseline study was made in 1974 and 1975 (ICES, 1977b), covering metals and chlorinated hydrocarbons in fish, but the limited coverage of the sampling does not permit any firm conclusion as to the distribution of pollutants. The values for PCB in herring were in the range 1.7 to 84 mg kg^{-1} fat weight and for DDT in the range 0.6 to 37 mg kg^{-1} of fat. Mercury values in herring muscle tissue ranged from 0.004 to 0.3 mg kg^{-1} wet weight with an even distribution over the Baltic area. Only 4 out of 50 samples, consisting of from 4 to 12 analyses each, had values above 0.1 mg kg^{-1}. With the exception of certain limited areas, like the Öresund where the mercury content in cod and flounder was generally higher than in other areas, the metal contamination in fish such as cod, flounder and herring does not seem to be higher in the Baltic than in the North Sea. The levels for metals as well as for chlorinated hydrocarbons are generally comparable with those found in the North Sea (ICES, 1974, 1977a, b). Recent data on mercury and DDT show a decrease of contamination which undoubtedly is due to precautions taken regarding the use and release of both DDT and mercury since about 1966.

The oil pollution of the Baltic has recently been reviewed by Rudling (1976), who found that almost all soft bottom sediments contained petroleum hydrocarbons. The urbanized areas had the highest concentration in the range from 40 to 400 mg kg^{-1} dry weight. For comparison, sediments from the west coast of Sweden in general had a higher content than those from the Baltic proper.

Seven nations border on the Baltic and the sea is of great importance to all of them. The uses include transportation, fishing, recreation, nonrenewable resources and the use of the sea as a recipient of possible pollutants. In addition, river flows are regulated for power plants as well as for fresh-water supply, and there is a greatly increased demand for fresh water. Long-term observations in the Baltic have revealed both natural and non-natural changes. Since the Baltic is subject to many natural variations on a large scale, an additional small human perturbation may become magnified by natural processes, or it may be hidden for a long time by natural fluctuations. The general concern among the Baltic countries regarding the protection of the Baltic Sea led to the Fishing Convention of 1974 and the Helsinki Convention on the protection of the marine environment of the Baltic Sea Area of 1974, which entered into force in May, 1980.

Cooperative Baltic research is of long standing and is at present very active within groups such as the Baltic Marine Biologists and the Conference of Baltic Oceanographers, as well as more firmly established organizations like ICES. The water balance of the Baltic is currently being studied in a special program within the framework of the International Hydrological Decade and its continuation, the International Hydrological Program.

REFERENCES

Ackefors, H., 1971. Mercury pollution in Sweden with special reference to conditions in the water habitat. *Proc. R. Soc. Lond., B*, 117: 365–387.

Ackefors, H., 1975. Production studies of zooplankton in relation to the primary production in the Baltic proper. *Havsforskningsinst. Skr.*, 239: 123–130.

Ackefors, H. and Lindahl, O., 1975. Studies on the primary

phytoplankton production in 1973–1974 in the Baltic. *ICES/C.M.*, 50: 17.

Ackefors, H., Hernroth, L. and Lindahl, O., 1977. *Studies of the Primary and Secondary Production in the Baltic 1968–1976.* Havsfiskelaboratoriet, Lysekil, 33 pp. (in Swedish).

Ankar, S. and Elmgren, R., 1976. The benthic macro- and meiofauna of the Askö-Landsort area (Northern Baltic proper). *Contrib. Askö Lab. Stockholm*, 11: 115 pp.

Bagge, P. and Niemi, Å., 1971. Dynamics of phytoplankton primary production and biomass in Loviisa Archipelago (Gulf of Finland). *Havsforskningsinst. Skr.*, 239, 19–41.

Belt Project, 1976a. *Fysiske Undersogelser.* Miljøstyrelsen, Copenhagen, 105 pp.

Belt Project, 1976b. *Kemiske og Biologiske Undersøgelser.* Miljøstyrelsen, Copenhagen, 72 pp.

Berthelsen, E. and Hansen, P.M., 1969. Fiskeriundersøgelser i 1969 ved Danmark, Færöerne, og Grönland. H. Plankton. *Skr. Dan. Fisk. Havunders.*, No. 30.

Blomquist, Å., Pilo, C. and Thompson, T., 1976. Sea Ice — 75, Summary report. *Styrelsen för Vintersjöfartsforskning, Forskningsrapp.* 16(9): 1–28.

Brogmus, W., 1952. Eine Revision des Wasserhaushaltes der Ostsee. *Kiel. Meeresforsch.*, 9: 15–42.

Denman, K.L. and Miyake, M., 1973. Upper layer modification at ocean station Papa: observations and simulations. *J. Phys. Oceanogr.*, 3: 185.

Dickson, R.R., 1971. A current and persistent pressure–anomaly pattern as the principal cause of intermediate–scale hydrographic variation in the European shelf seas. *Dtsch. Hydrogr. Z.*, 24: 97.

Dickson, R.R., 1972. The beginnings of a new Baltic inflow? *ICES CM/C*: 10.

Dickson, R.R., 1973. The prediction of major Baltic inflows. *Dtsch. Hydrogr. Z.*, 26: 97.

Dietrich, G., 1951. Oberflächenströmungen im Kattegatt, im Sund und in der Beltsee. *Dtsch. Hydrogr. Z.*, 4: 129–140.

Ehlin, U., Mattisson, I. and Zachrisson, G., 1974. Computer based calculations of volumes of the Baltic area. In: *9th Conf. Baltic Oceanogr. Kiel, April 1974*, Pap. 7.

Fonselius, S.H., 1962. Hydrography of the Baltic deep basins. *Fish. Board Swed., Ser. Hydrogr.*, 13: 40 pp.

Fonselius, S.H., 1967. Hydrography of the Baltic deep basins II. *Fish. Board Swed., Ser. Hydrogr.*, 20: 4–6.

Fonselius, S., 1969. Hydrography of the Baltic deep basins III. *Fish. Board Swed., Ser. Hydrogr.*, 23: 97 pp.

Fonselius, S., 1970. On the stagnation and recent turnover of the water in the Baltic. *Tellus*, 22: 533–544.

Fonselius, S., 1972. On biogenic elements and organic matter in the Baltic. *Ambio, Spec. Rep.*, 1: 29–36.

Fonselius, S., 1976. On the nutrient variability in the Baltic. *Ambio, Spec. Rep.*, 4: 17–26.

Francke, E. and Nehring, D., 1973. Physical and chemical variations in the eastern part of the Gotland basin in 1969/70. *Oikos, Suppl.*, 15: 14–20.

Granqvist, G., 1931. Croisière Thalassologique 1931. *Havsforskningsinst. Skr.*, No. 81: 24.

Granqvist, G., 1952. Harmonic analysis of temperature and salinity in the sea off Finland and changes in salinity. *Havsforskningsinst. Skr.*, 152: 1–29.

Gustafsson, T. and Kullenberg, B., 1936. Untersuchungen von Trägheitsströmungen in der Ostsee. *Sver. Hydrogr.-Biol. Komm. Skr., N. S. Hydrogr.*, 13: 28 pp.

Hagström, Å. and Larsson, U., 1978. *On the Relationships Between Phytoplankton and Bacteria in Primary Production Studies.* Unpublished.

Hagström, Å., Larsson, U., Hörstedt, P. and Normark, S., 1978. *Bacterial Growth Parameters: New Tools in Ecology to Monitor Hetrotrophic Production.* Report to Statens Naturvårdsverk, Sweden (mimeo).

Hallberg, R.O., 1974. Paleoredox conditions in the Eastern Gotland basin during the recent centuries. *Havsforskningsinst. Skr.*, 238: 3–16.

Hela, I., 1944. Über die Schwankungen des Wasserstandes in der Ostsee mit besonderer Berücksichtigung des Wasseraustausches durch die Dänischen Gewässer. *Havsforskningsinst. Skr.*, 134: 108 pp.

Hela, I., 1966a. Secular changes in the salinity of the upper waters of the northern Baltic Sea. *Comm. Phys.-Math. Soc. Sci. Fenn.*, 31(14): 21 pp.

Hela, I., 1966b. Vertical eddy diffusivity of waters in the Baltic Sea. *Geophysica*, 9(3): 219–234.

Hela, I., 1977. Synoptic survey of the hydrography of the Åland Sea. *Havsforskningsinst. Skr.*, 241: 101–110.

Herman, F. and Olsen, V., 1970. Long term fluctuations in oxygen and phosphate concentrations in the Kattegatt, the Belt Sea and western Baltic. *ICES/C.M.*, C: 27.

Hobro, R., Larsson, U. and Wulff, F., 1978. *Dynamics of a Phytoplankton Spring Bloom in a Coastal Area of the Northern Baltic.* Prace Morskiega Inst., Rybacki (Poland).

Højerslev, N.K., 1974. *Inherent and Apparent Optical Properties of the Baltic.* Inst. Phys. Ocean., Copenhagen, Rep. No. 23, 40 pp.

Horstmann, U., 1975. Eutrophication and mass production of blue–green algae in the Baltic. *Havsforskningsinst. Skrift.*, 239: 83–90.

ICES, 1970. *Report on Pollution of the Baltic Sea.* International Council for the Exploration of the Sea, No. 15.

ICES, 1974. *Report of Working Groups for the International Study of the Pollution of the North Sea and its Effects on Living Resources and Their Exploitation.* International Council for the Exploration of the Sea, No. 39.

ICES, 1977a. *The ICES Coordinated Monitoring Programme in the North Sea, 1974.* International Council for the Exploration of the Sea, No. 58.

ICES, 1977b. *Studies of the Pollution of the Baltic Sea.* International Council for the Exploration of the Sea, No. 63.

Jacobsen, J.P., 1925. Die Wasserumsetzung durch den Øresund, den grossen und kleinen Belt. *Medd. Komm. Havunders., Ser. Hydrogr.*, 2(9): 1–72.

Jansson, B-O., 1978. The Baltic — A systems analysis of a semi-enclosed sea. In: H. Charnock and G. Deacon (Editors), *Advances in Oceanography.* Plenum Press, Oxford, pp. 131–184.

Jensen, A.J.C., 1937. Fluctuations in the hydrography of the transition area during 50 years. *ICES Rapp. P.-V.*, 102(1): 3–49.

Jensen, M. and Olsson, M., 1976. DDT and PCB in the Baltic. *Ambio, Spec. Rep.*, 4: 111–124.

Jensen, S., Johnels, A., Olsson, M. and Otterlind, G., 1972. DDT and PCB in herring and cod from the Baltic, the Kattegat and the Skagerrak. *Ambio, Spec. Rep.*, 1: 71–85.

Jerlov, N.G., 1955. Factors influencing the transparency of the Baltic waters. *Göteborg K. Vetensk. Vittehets Samh. Handl., Ser. B.*, 6(14): 19 pp.

Jerlov, N.G., 1975. Long period changes of the optical properties of the Baltic. *J. Cons. Int. Explor. Mer*, 36(2): 188–190.

Joseph, J. and Sendner, H., 1958. Über die Horizontale Diffusion im Meere. *Dtsch. Hydrogr. Z.*, 11: 49.

Kaiser, W. and Schultz, S., 1970. Preliminary results of biological investigations 1969/70 in connection to earlier observations and to the pollution of the Baltic. In: *7th Conf. Baltic Oceanographers, Helsinki, May 1970.*

Kaiser, W. and Schultz, S., 1975. On primary production in the Baltic. *Havsforskningsinst. Skr.*, 239: 29–33.

Kaleis, M.V., 1976. Present hydrographic conditions in the Baltic. *Ambio, Spec. Rep.*, 4: 37–44.

Kalle, K., 1943. Die grosse Wasserumschichtung im Gotlandstief vom Jahre 1933–34. *Ann. Hydrogr.*, 71: 4–6.

Kalle, K., 1949. Fluoreszenz und Gelbstoff in Botnischen und Finnischen Meerbusen. *Dtsch. Hydrogr. Z.*, 2: 117.

Karabashev, G.S., Zangalis, K.P., Solovyev, A.N. and Yakubovich, V.V., 1971. New data on sea water photoluminescence. *Izv. Atmos. Okean. Fis.*, 7(1): 60–68 (in Russian).

Kielmann, J., Krauss, W. and Magaard, L., 1969. Über die Verteilung der kinetischen Energie im Bereich der Trägheits- und Seichesfrequenzen der Ostsee im August 1964. *Kiel. Meeresforsch.*, 25(2): 345–354.

Kielmann, J., Krauss, W. and Keunecke, K.H., 1973. Currents and stratification in the Belt Sea and the Arkona Basin during 1962–1968. *Kiel. Meeresforsch.*, 291(2): 90–111.

Knudsen, M., 1899. De hydrografiske Forhold i de Danske Farvande inden for Skagen i 1894–98. *Beret. Komm. Vidensk. Unders. Dan. Farvande*, 2(2): 19–79.

Knudsen, M., 1900. Ein hydrographischer Lehrsatz. *Ann. Hydrogr. Mar. Meteorol.*, 28: 500–504.

Krause, G., 1969. Ein Beitrag zum Problem der Erneuerung des Tiefenwassers im Arkona-Becken. *Kiel. Meeresforsch.*, 25(2): 268–271.

Krauss, W. and Magaard, L., 1962. Zum System der Eigenschwingungen der Ostsee. *Kiel. Meeresforsch.*, 18: 184–186.

Kullenberg, B., 1955. Restriction of the underflow in a transition. *Tellus*, 7(2): 215–217.

Kullenberg, B., 1967. *Oceanografi før Nordiska Haelsovårdshøgskolan.* Mimeograph, unpublished, 40 pp.

Kullenberg, B. and Hela, I., 1942. Om tröghetssvängningar i Östersjön. *Svet. Hydrogr.-Biol. Komm. Skr., N.S. Hydrogr.*, 16: 1–14.

Kullenberg, G.E.B., 1969. *Light Scattering in the Central Baltic.* Institute of Physical Oceanography, University of Copenhagen, Copenhagen, Rep. No. 5, 12 pp.

Kullenberg, G.E.B., 1970. On the oxygen deficit in the Baltic deep water. *Tellus*, 22: 357.

Kullenberg, G.E.B., 1974a. Some observations of the vertical mixing in the Baltic. In: *9th Conf. Baltic Oceanographers, Kiel, April 1974.*

Kullenberg, G.E.B., 1974b. *An Experimental and Theoretical Investigation of the Turbulent Diffusion in the Upper Layer of the Sea.* Institute of Physical Oceanography, University of Copenhagen, Copenhagen, Rep. 25, 272 pp.

Kullenberg, G.E.B., 1976. On vertical mixing and the energy transfer from the wind to the water. *Tellus*, 28: 159.

Kullenberg, G.E.B., 1977. Observations of the mixing in the Baltic thermo- and halocline layers. *Tellus*, in press.

Kullenberg, G. and Nygård, K., 1971. *Fluorescence Measurements in the Sea.* Institute of Physical Oceanography, University of Copenhagen, Copenhagen, Rep. No. 15, 17 pp.

Lassig, J. and Niemi, Ă., 1975. Parameters of production in the Baltic measured during cruises with R/V Aranda in June and July 1970 and 1971. *Havsforskningsinst. Skr.*, 239: 34–40.

Lindahl, G., Wallström, K. and Brattberg, G., 1978. On nitrogen fixation in a coastal area of the northern Baltic. *Kiel. Meeresforsch., Suppl.* 4, in press.

Lindqvist, A., 1959. *Studien über das Zooplankton der Bottensee II.* Inst. Mar. Res. Lysekil, Ser. Biol. Rep. No. 11, Fishery Board, Sweden.

Magaard, L. and Krauss, W., 1966. Spektren der Wasserstandsschwankungen der Ostsee im Jahre 1958. *Kiel. Meeresforsch.*, 22(2): 155–162.

Mälkki, P., 1975. On the variability of currents in a coastal region of the Baltic Sea. *Havsforskningsinst. Skr.*, 240: 56 pp.

Matthäus, W., 1977a. Mittlere vertikale Wärmeaustauschkoeffizienten in der Ostsee. *Acta Hydrophys., Berlin*, 22(2): 73–92.

Matthäus, W., 1977b. Zur säkularen Veränderlichkeit des Oberflächensalzgehaltes in der offenen Ostsee. *Beitr. Meeresk.*, 39: 37–49.

Mikulski, Z., 1970. Inflow of river water to the Baltic Sea in the period 1951–1960. *Nord. Hydrol.*, 4: 216–227.

Mikulski, Z., 1972. The inflow of river waters to the Baltic Sea in the years 1961–1970. In: *8th Conf. Baltic Oceanographers, Copenhagen, 1972*, 5 pp.

Neumann, G., 1941. Eigenschwingungen der Ostee. *Arch. Dtsch. Seewarte Marineobs.*, 61(4): 59 pp.

Nyqvist, N., 1974. Östersjön blommar. *Forskn. Framsteg*, 6: 1–2.

Odén, S. and Ekstedt, J., 1976. PCB and DDT in Baltic sediments. *Ambio, Spec. Rep.*, 4: 125–130.

Otterlind, G., 1976. Fish stocks and fish migration in the Baltic Sea environment. *Ambio, Spec. Rep.*, 4: 89–102.

Otterlind, G., Jensen, S. and Olsson, M., 1971. DDT and PCB in Baltic fish. *ICES/CM/E*: 31 (mimeograph).

Palmén, E., 1930. Untersuchungen über die Strömungen in den Finnland umgebenden Meeren. *Soc. Sci. Fenn., Comm. Phys.-Math.*, 5(12): 94 pp.

Palmén, E. and Söderman, D., 1966. Computation of the evaporation from the Baltic Sea from the flux of water vapor in the atmosphere. *Geophysica*, 8(4): 261–279.

Palosuo, E., 1964. A description of the seasonal variations of water exchange between the Baltic proper and the Gulf of Bothnia. *Havsforskningsinst. Skr.*, 215: 1–32.

Palosuo, E., 1966. Ice in the Baltic. *Oceanogr. Mar. Biol., Annu. Rev.*, 4: 79–90.

Patten, B.C., 1963. Optimum diversity structure of a summer community. *Science*, 140: 894–98.

Probst, B., 1975. *Ein Modell zur Darstellung des pelagischen*

Kreislaufes in einen marinen Flachwasserökosystem der westlichen Ostsee. Dissertation, Institut für Meereskunde, Kiel, 53 pp (unpublished).

Rudling, L., 1976. *Oil Pollution in the Baltic Sea. A Chemical Analytical Search for Monitoring Methods.* Statens Naturvårdsverk (Swedish Environmental Protection Board), Stockholm, SNV PM 783, 80 pp.

Ryther, J.H., 1970. Photosynthesis and fish production in the sea. *Science*, 166: 72–76.

Scheibel, W. and Noodt, W., 1975. Population densities and characteristics of meiobenthos in different substrates in the Kiel Bay. *Havsforskningsinstitutets Skr.*, 239: 173–178.

Schott, F., Ehlers, M., Hubrich, L.M. and Quadfasd, D., 1976. On measurements of horizontal mass and momentum transfer in small scales during Baltic 75. IFM, Kiel (*J. Phys. Oceanogr.*, 1977).

Schultz, F. and Kaiser, W., 1973. Biological effects of the salt water influx into the Gotland Basin in 1969/70. *Oikos, Suppl.*, 15: 21–27.

Sen Gupta, R., 1968. Inorganic nitrogen compounds in ocean stagnation and nutrient resupply. *Science*, 24: 884–885.

Shaffer, G., 1977. *Calculations of Cross-Isohaline Salt Exchange in a Coastal Region of the Baltic.* Institute of Oceanography, University of Gothenburg, Gothenburg, Rep. No. 24, 26+16 pp.

Simojoki, H., 1949. Niederschlag und Verdunstung auf dem Baltischen Meer. *Fennia*, 71(1): 1.

Sjöberg, S., 1977. Are pelagic systems inherently unstable — a model study. *Ecol. Modell.*, 3: 17–37.

Sjöberg, S. and Wilmot, W., 1977. System analysis of a spring phytoplankton bloom in the Baltic. *Contrib. Askö Lab.*, No. 20: 99 pp.

Soskin, I.M., 1963. *Long Term Changes in the Hydrological Characteristics of the Baltic.* Hydromet. Press, Leningrad, 159 pp. (in Russian).

Stommel, H. and Farmer, G.H., 1953. Control of salinity in an estuary by a transition. *J. Mar. Res.*, 23: 13–23.

Svansson, A., 1972. *Canal Models of Sea Level and Salinity Variations in the Baltic and Adjacent Waters.* Fishery Board Sweden, Ser. Hydr. Rep. 26, 72 pp.

Svansson, A., 1975. *Physical and Chemical Oceanography of the Skagerrak and the Kattegatt. I. Open Sea Conditions.* Fishery Board Sweden, Inst. Mar. Res. Rep. 1, 88 pp.

Tully, J.P., 1958. On structure, entrainment and transport in estuarine embayments. *J. Mar. Res.*, 17: 523–535.

Udin, I. and Omstedt, A., 1976. Sea Ice — 75, Dynamical Report. *Styrelsen för Vintersjöfartsforskning, Forskningsrapp.*, 16(8): 1–64.

Voipio, A., 1969. On the cycle and the balance of phosphorous in the Baltic Sea. *Suomen Kemistil., A.*, 42: 48–53.

Voipio, A. and Mälkki, P., 1972. Variations of the vertical stability in the northern Baltic. *Havsforskningsinst. Skr.*, 23: 3–12.

Von Bodungen, B., 1975. *Die Jahresgang der Nährsalze und der Primärproduktion des Planktons in der Kieler Bucht.* Dissertation. Institut für Meereskunde, Kiel, 166 pp.

Walin, G., 1972a. On the hydrographic response to transient meteorological disturbances. *Tellus*, 241(3): 169–186.

Walin, G., 1972b. Some observations of temperature fluctuations in the coastal region of the Baltic. *Tellus*, 24: 187–198.

Wattenberg, H., 1949. Entwurf einer natürlichen Einteilung der Ostsee. *Kiel. Meeresforsch.*, 6: 10.

Welander, P., 1974. Two-layer exchange in an estuary basin with special reference to the Baltic Sea. *J. Phys. Oceanogr.*, 4: 542–556.

Witting, R., 1912. Zusammenfassende Übersicht der Hydrographie des Bottnischen und Finnischen Meerbusen und der nördlichen Ostsee nach den Untersuchungen bis Ende 1910. *Finnl. Hydrogr. Biol. Unters.*, 7: 1–55.

Witting, R., 1918. Hafsytan, Geoidytan och Landhöjningen utmed Baltiska Hafvet och vid Nordsjön. *Fennia*, 39(5): 1–346.

Wyrtki, K., 1954a. Schwankungen im Wasserhaushalt der Ostsee. *Dtsch. Hydrogr. Z.*, 7: 91–129.

Wyrtki, K., 1954b. Der grosse Salzeinbruch in die Ostsee im November und Dezember 1951. *Kiel. Meeresforsch.*, 10(1): 19.

Zenkevitch, L., 1963. *Biology of the Seas of the U.S.S.R.* Wiley and Sons, New York, N.Y., 955 pp.

Zmudzinski, L., 1975. The Baltic Sea pollution. *Pol. Arch. Hydrobiol.*, 22(4): 601–614.

Zsolnay, A., 1973. Distribution of labile and residual carbon in the Baltic Sea. *Mar. Biol.*, 24: 13–18.

Chapter 14

THE BERING SEA

DONALD W. HOOD

INTRODUCTION

Sandwiched between the northernmost land masses of the North American and Asian continents lies the Bering Sea. It has a mean depth of 1636 m, a surface area of 2.29×10^6 km^2 and a volume of 3.75×10^6 km^3. These dimensions make it the third largest semi-enclosed sea of the world ocean, being surpassed in size only by the Mediterranean and the South China Seas. It derives its name from Vitus Bering, who commanded an extensive Russian expedition during the years 1725 to 1743, which explored the coasts of Kamchatka, the southern side of the Aleutian Islands and the southern mainland of Alaska as far as Prince William Sound (Hood and Kelley, 1974).

From north to south the Bering Sea can be viewed as a sector with a radius of about 1500 km with the Bering Strait located approximately at the apex. The southern arc represented by the Aleutian Islands and the Alaska Peninsula extends between 157°W on the shores of Bristol Bay on the east to the coast of Kamchatka Peninsula at 163°E on the west, a total distance of almost 3000 km.

Included within the Bering Sea boundaries are three major bays: Bristol Bay in the southeast, Norton Sound in the northeast, both bordering Alaska, and the Gulf of Anadyr (Anadyrskiy Zaliv) in the northwest, bordered by Siberia. There are three major rivers, the Kuskokwim and Yukon draining central Alaska and the Anadyr draining western Siberia, which empty into the Bering Sea. All these rivers contain glacial meltwater and runoff from taiga forests and tundra characteristic of the far northern environment.

The Bering Sea is often considered the northern extension of the Pacific Ocean. The exchange of surface water through the Aleutian Island passes occurs relatively freely, and characteristic north Pacific Ocean water is evident within the Bering Sea. However, since the water passage is limited by sill depths of the passes except in the west, and its link with the Arctic Ocean is the narrow (85 km) shallow (45 m) Bering Strait, this sea contains the essential features of a well-defined ocean region and may best be looked upon as a confined sea with characteristics of its own sufficiently different from adjoining oceans to make it a unique ocean region.

Historical setting

The north Pacific Ocean south of 40°N latitude between the coasts of America and Japan was fairly reliably described on maps dating from the late sixteenth century; however, almost total ignorance prevailed for the region north of that transect. It was not until 1643 that a Dutch East India Company expedition commanded by Maerten Gerritszoon Vries discovered a previously much talked about mystery land, Yezo, to the north of Japan, now known as the Kuril Islands (Kuril'skiye Ostrova). In 1639 a barge load of Cossacks made their way down a Siberian river to the Sea of Okhotsk (Okhotskoye More) for a first view of the "Eastern Ocean". By the end of the seventeenth century Russian excursions had found that the Ob, Yenisey and Lena Rivers emptied into the "Icy Sea" to the north while the Amur, Uda, Anadyron and Okhota Rivers flowed south to the "Eastern Sea". Still there was no knowledge of a passage between the two bodies of water.

The question of whether America and Asia were united attracted the attention of Peter I, the Czar of Russia from 1682 to 1725, who initiated a program of westernization by utilizing Swedish prisoners of war in Siberian ports as teachers of shipbuilding and the navigational arts. After the death of Peter I the program was continued by his successor Empress Catherine when Vitus Bering, a Danish explorer, and Lt. Alexei Chirikov were commissioned in their first expedition. In 1728 Vitus Bering with the sister ships *St. Peter* and *St. Paul* sailed through the Bering Strait to settle once and for all that Asia and America were separated. On the second voyage, fatal to Vitus Bering who died of scurvy in 1741, much of the Arctic coast of Asia was charted as well as the southwest Alaskan coast and the Aleutian Island arc. Even though Vitus Bering had remarkable success, his work was doubted and in fact somewhat discarded. The pressing question of a northern passage from the Pacific Ocean led the British Admiralty to issue instructions to their most talented explorer, Captain James Cook, to extend his third voyage in 1778–1780 in further search for a northeast or northwest passage, from the Pacific Ocean into the Atlantic Ocean or the North Sea of Europe by way of Russia. At the helm of the *Resolution* and assisted by Lt. Charles Clarke on the *Discovery*, Cook passed through the Bering Strait into the Arctic Ocean in the summer of 1778. The expedition penetrated to 70°N where Cook viewed the "Chuckchee" Sea and the extremities of both continents, casting aside the possibility of a northwest or northeast passage to India so long sought from the Atlantic Ocean side. The remarkable accuracy of Bering's earlier cartography was demonstrated. The expedition went south to wait out the Arctic winter and visited Norton Sound and many inlets in south-central Alaska. The following summer the expedition again sailed through the Bering Strait but without Captain Cook, who had been murdered during the winter of 1779 by Sandwich Island natives.

Physical dimensions and ice cover

The Bering Sea is divided into a neritic area (0–200 m) to the east and an abyssal area (over 1000 m) of about equal extent to the west. These two areas total 87% of the entire area of the Bering Sea. The continental slope essentially divides the Bering Sea in half in a northwest–southeast direction. It occupies only 13% of the area and generally has a slope of 4 to 5°. The shelf occupying the eastern half continues north through the narrow Bering Strait into the Chukchi Sea. The shallow passage (40 m) restricts all but near-surface water exchange between the Arctic basin and the Bering Sea. The broad continental shelf reaches to nearly 650 km offshore in the northeast part of the sea and is the largest in the world except for the Arctic continental shelves of the Chukchi, East Siberian and Laptev Seas. The shelf region contains several islands which are important to marine mammal and bird ecology and which influence circulation of water and the formation and movement of sea ice. St. Lawrence, the largest of the five major shelf islands, is nearly 200 km long and lies between Norton Sound and the Gulf of Anadyr, south of the Bering Strait. Nunivak Island, the second largest, lies near the coast of Alaska between the mouths of the Kuskokwim and Yukon Rivers. St. Matthew, St. Paul and St. George Islands, the other three larger islands, lie well offshore. St. Matthew is about midshelf, approximately 275 km southwest of St. Lawrence, whereas St. Paul and St. George, the fur seal islands, are near the shelf break some 300 km northeast of Unimak Pass. The deep basin of the Bering Sea, beginning west of Unimak Pass, is separated from the Pacific Ocean by the Aleutian Island arc, which rises from a depth of 7600 m on the Aleutian Trench side and from 4000 m in the Bering Sea basin. Passes through the arc deepen westward, the deepest being 4420 m between Kamchatka and the westernmost Commander (Komandorskiye) Island. North of this passage a sill at 3589 m provides the deepest passage to the basin. Except for a narrow passage at 180°E, the sill depth east of 171°E is much less than 1000 m. The total cross section permitting exchange of water through the southern boundary is only 701 km^2 (Favorite, 1974).

The basin is a vast plain lying at a depth of 3800 to 3900 m with occasional gradual sloping hollows to depths of as much as 4151 m. Two submarine ridges penetrate the basin. The Shirshov Ridge extends south from Kamchatka along 170°E longitude to near the Aleutian arc and separates the eastern and western Aleutian Basin. The North Rat Island Ridge (Bowers Bank) extends 300 km north in a counterclockwise direction. The submarine and physiographic features of the Bering Sea region are shown in Fig. 14.1.

The four major rivers discharging into the Bering

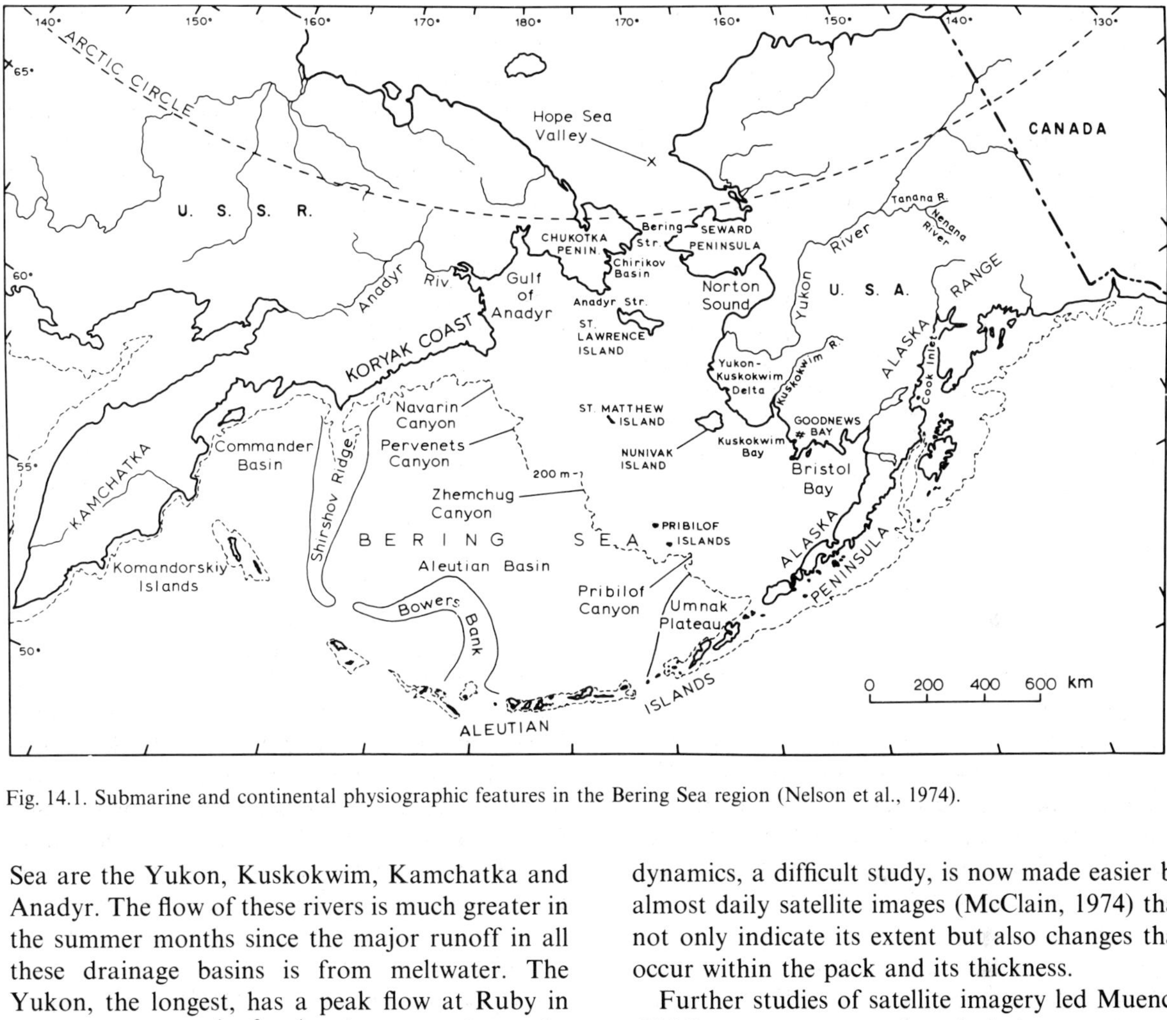

Fig. 14.1. Submarine and continental physiographic features in the Bering Sea region (Nelson et al., 1974).

Sea are the Yukon, Kuskokwim, Kamchatka and Anadyr. The flow of these rivers is much greater in the summer months since the major runoff in all these drainage basins is from meltwater. The Yukon, the longest, has a peak flow at Ruby in August of $4 \times 10^4\ m^3\ s^{-1}$, about equal to the Mississippi, and a mean flow for the year of $5 \times 10^3\ m^3\ s^{-1}$, about two thirds the annual flow of the Columbia.

"The ice of the Bering Sea shelf is seasonal and like all sea ice is not stable, smooth and uniform like that of a freshwater lake, but is an incomplete cover, widely variable in form and structure and typically unstable and dynamic under the influence of surface air and water currents" (Armstrong and Roberts, 1956). The seasonal ice pack occurs over the continental shelf of the northern and eastern portions of the Bering Sea while little ice occurs over the deep southwest portion. The pack persists from some six to eight months of each year, and at its maximum, usually in March or April, may extend southward to or slightly beyond the 500 m isobath at the edge of the continental shelf. Ice dynamics, a difficult study, is now made easier by almost daily satellite images (McClain, 1974) that not only indicate its extent but also changes that occur within the pack and its thickness.

Further studies of satellite imagery led Muench (1975) to comment on the relatively constant position of the southern ice edge throughout most of March and April 1974. Persistent areas of open water existed throughout this period south of St. Lawrence, St. Matthews and Nunivak Islands, while ice appeared to pile up on the north sides. An apparent southward motion existed through the Bering Strait as suggested by ice fracture patterns and flow distribution, the western side having more open water than the eastern due to warmer vertically mixed water. There was a tendency for either open water or thin ice to occur in the northern Gulf of Anadyr, Norton Sound and along the Alaskan Coast south of Norton Sound. A quasi-stationary mass of loose pack ice remained in the region southwest of St. Lawrence Island from April to June, suggesting the presence of a stagnation zone in the horizontal circulation. Southward plumes of ice occurred at times off both sides of St. Lawrence Island as several randomly distributed lines of floes, which may have been due to lines of convergence. Ice floes in this area usually do not exceed 2 m of thickness, although rafting and hummocking can result in thicknesses of 5 to 10 m. In no case is the Bering Sea, even in the Strait, one solid sheet of ice. By July the Bering Sea is usually free of ice, caused largely by melting in place. No icebergs occur since there are no glaciers on either the Siberian or the Alaskan coasts, and passage from the Arctic Ocean is prevented by the circulation of the ice pack and prevailing currents.

Climatology

Mean temperatures in the winter range between −25°C in the Bering Strait and 2°C near the Aleutian Islands. In the summer the mean temperatures are 6 and 10°C, respectively. The mean atmospheric temperature at St. Paul Island for the 12-year period 1960–1971 is given in Fig. 14.2. A consistent two-year cycle prevailed except for 1965 and 1971. Data obtained from the Japanese ship *Shikishima Maru* in the eastern Bering Sea in Fig. 14.3 show air and surface water temperatures to have a cyclic change with intervals of two years for each of the months of March, April and May except 1971, during which time the temperatures were continuously low from the previous year. The two-year cycle was also shown in the pressure pattern as well as by the mean monthly surface pressure between 60°N, 180°W and 60°N, 160°W in

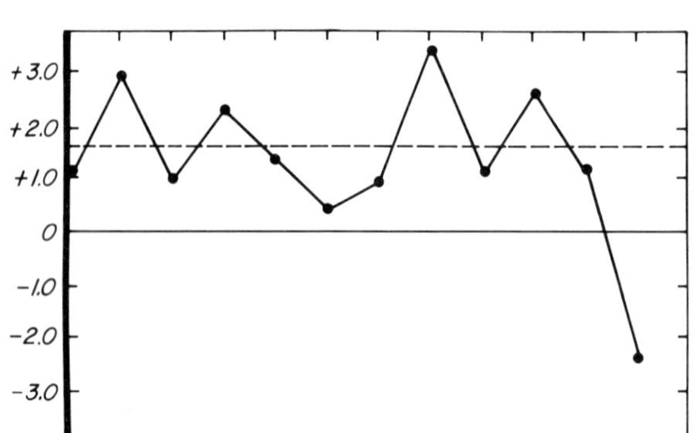

Fig. 14.2. Mean May atmospheric temperature at St. Paul Island for twelve-year period 1960–1971. A two-year cycle is seen in the variation, except for 1965 and 1971 (Konishi and Saito, 1974).

an annual period from March to May from 1967 through 1971. The year 1971 was an anomalously cold year caused by a shift in the low pressure site normally in the Gulf of Alaska to Bristol Bay, causing a northerly wind to prevail over the eastern Bering Sea from April to May 1971.

The Bering Sea lies in one of the regions of maximum gradients in the atmospheric regime. The weather is dominated by the Aleutian low in the winter when there are near sea level pressures of about 1000 mbar, normally located over the south-central sea. The pressures rise to 1011 mbar in the summer when the influence of the eastern Pacific high is felt. Precipitation occurs about 35% of the time, and although fair weather periods occur, usually overcast skies prevail and fog is common.

Resources — non-renewable

The most likely non-renewable resources of the Bering Sea are hydrocarbons and heavy metals. About 75% of the shelf area holds some likelihood of hydrocarbon accumulation, and some areas of the shelf are extremely promising. As seen in Fig. 14.4, the most promising areas are the St. George Basin on the outer margin of the continental shelf, the Bristol Bay Basin and the Norton Basin. These sites have a long geological history of high organic productivity. The outer basins have 1 to 2 km of Oligocene to Recent deposits mainly consisting of diatomite and diatomaceous sand, with some conglomerate. The Bristol Bay Basin contains several kilometers of terrigenous and volcanigenic sediments. The Norton Basin was probably inundated during the Late Miocene and, since the basin was probably receiving continental sediments from the Late Oligocene reservoir, rocks should be common. The Yukon River has been a major source of sediment since the Middle Miocene, and probably large sources of natural gas are present.

The occurrence of gold, platinum and tin placers on the shores of the Bering Sea is well known. Exploration for offshore deposits has thus far been disappointing. Almost all placers found have been in the near-shore zone, generally less than 5 km from shore. The only significant discoveries are rich but small, submerged, stream placer gold off Bluff and much larger, low-grade gold deposits on the sea floor off Nome. Commercial production from the

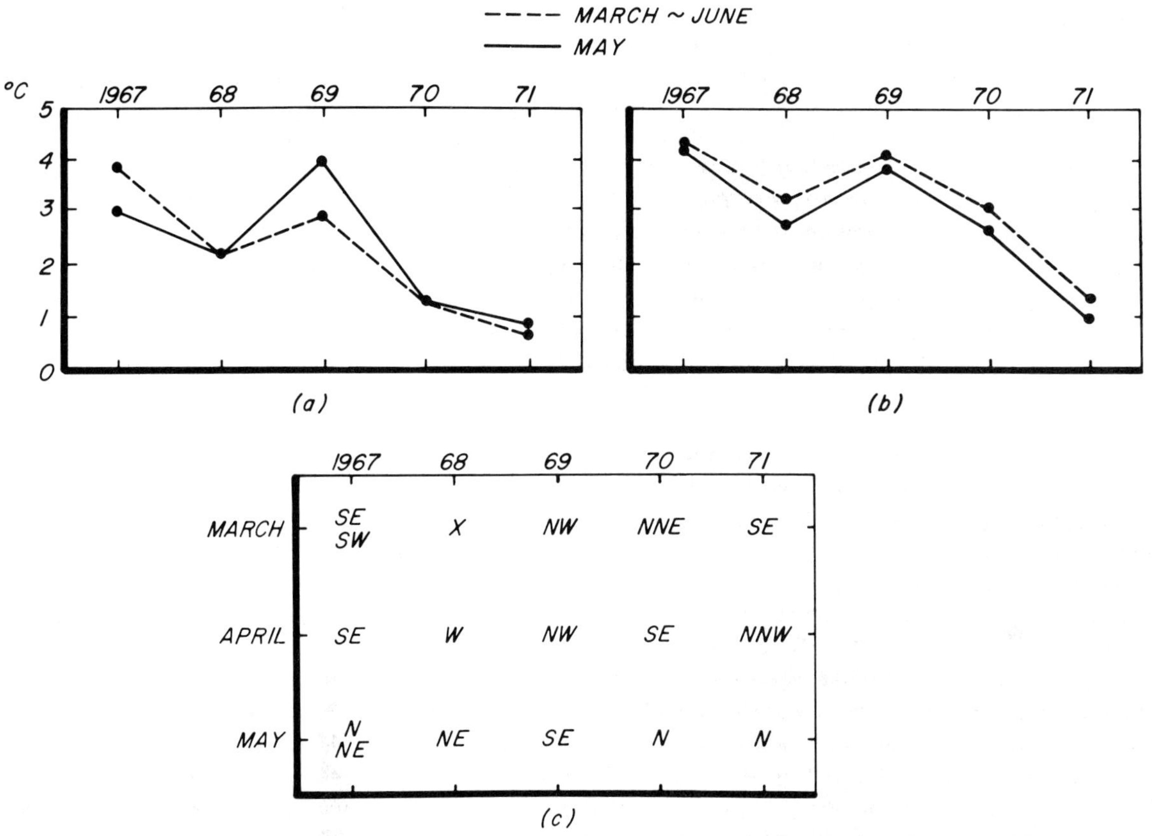

Fig. 14.3. Annual environmental observations made from ship (*Shikishima Maru*) in the eastern Bering Sea from March to June between 1967 and 1971: a. Monthly mean air temperature (°C). b. Monthly mean sea-surface temperature (°C). c. Direction of most frequent winds (Konishi and Saito, 1974).

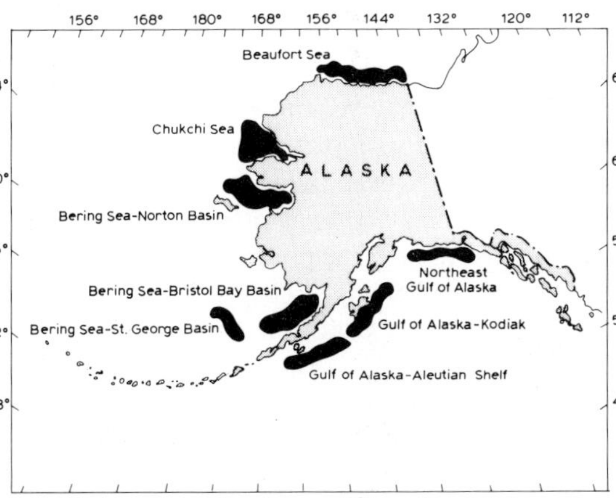

Fig. 14.4. Areas on the continental shelf of Alaska under consideration for leasing (Department of Commerce, 1977).

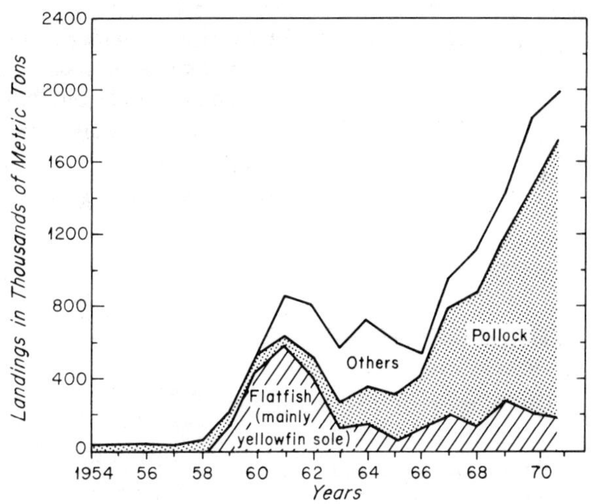

Fig. 14.5. Estimated landings of flatfish, pollock and other species of bottom fish by all nations from the Bering Sea during 1954–1971 (Pruter, 1973).

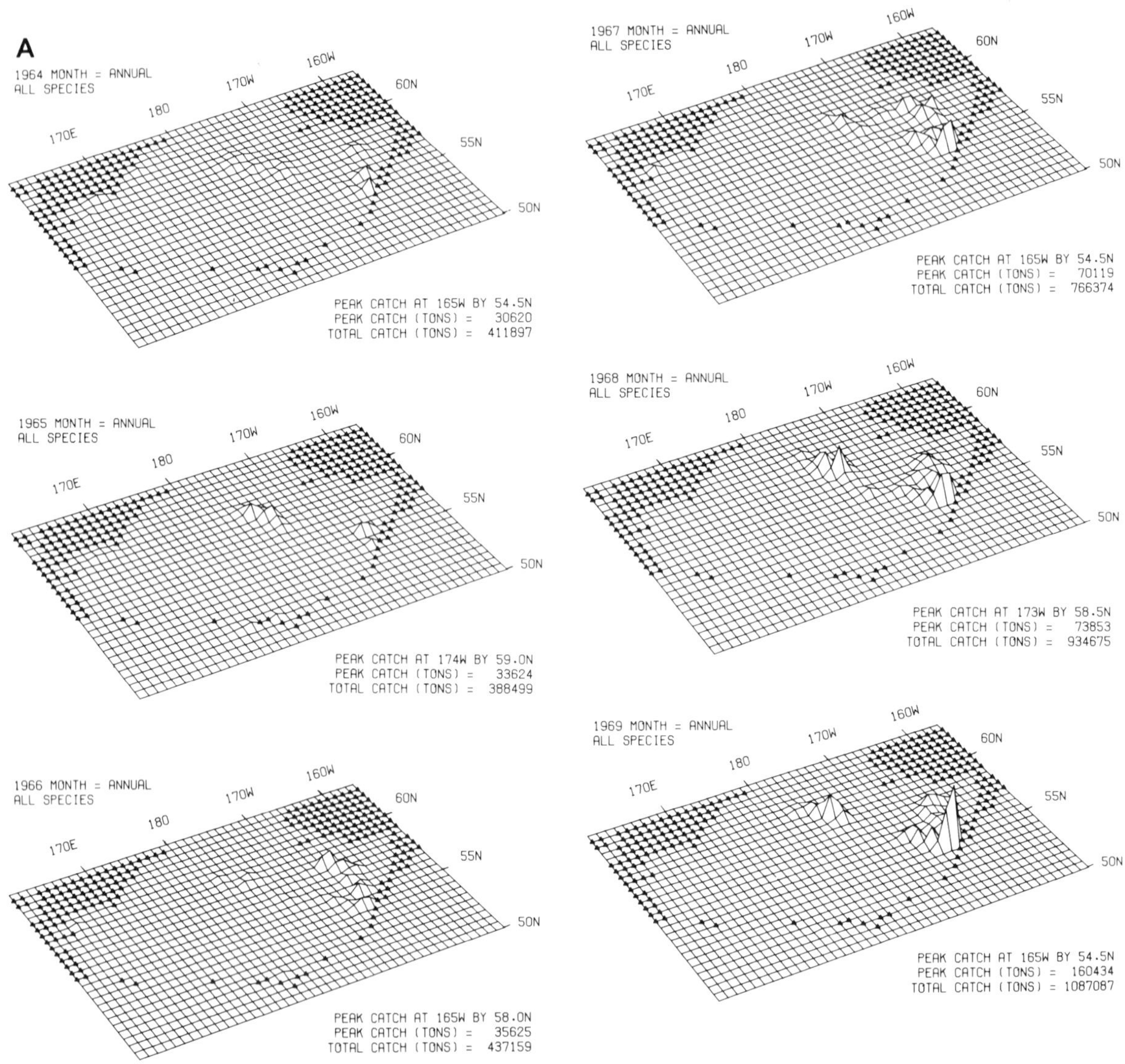

latter deposit began in 1975. The search for tin in and near the Bering Strait and for platinum in and near Goodnews Bay has yet to yield more than minor occurrences.

Resources — renewable

The borders and islands of the Bering Sea have been inhabited by Eskimos in the northern parts and by Aleuts along the Aleutian Arc as long as man has a historic record. As ancient cultures, these natives subsisted on a hunting and fishing economy for centuries in compatibility with an abundance of fish and animals in the area. Not until western man began exploiting the region to supply large populations of people with furs, fish and marine mammals was the natural living wealth of the area affected. As a result, the sea otter, now recovering, was exploited to near the point of extinction, and at least part of the herring's (*Clupea harengus pallasi*) ecological niche appears to have been replaced by the Alaskan pollock (*Theragra chalcogramma*). Another abundant resource, the sockeye salmon (*Oncorhynchus nerka*), once endangered and still under heavy pressure, can probably be restored, but will need international cooperation to avoid further, even accidental, depletion of the stocks. Only in the last fifteen years has the wide diversity of the

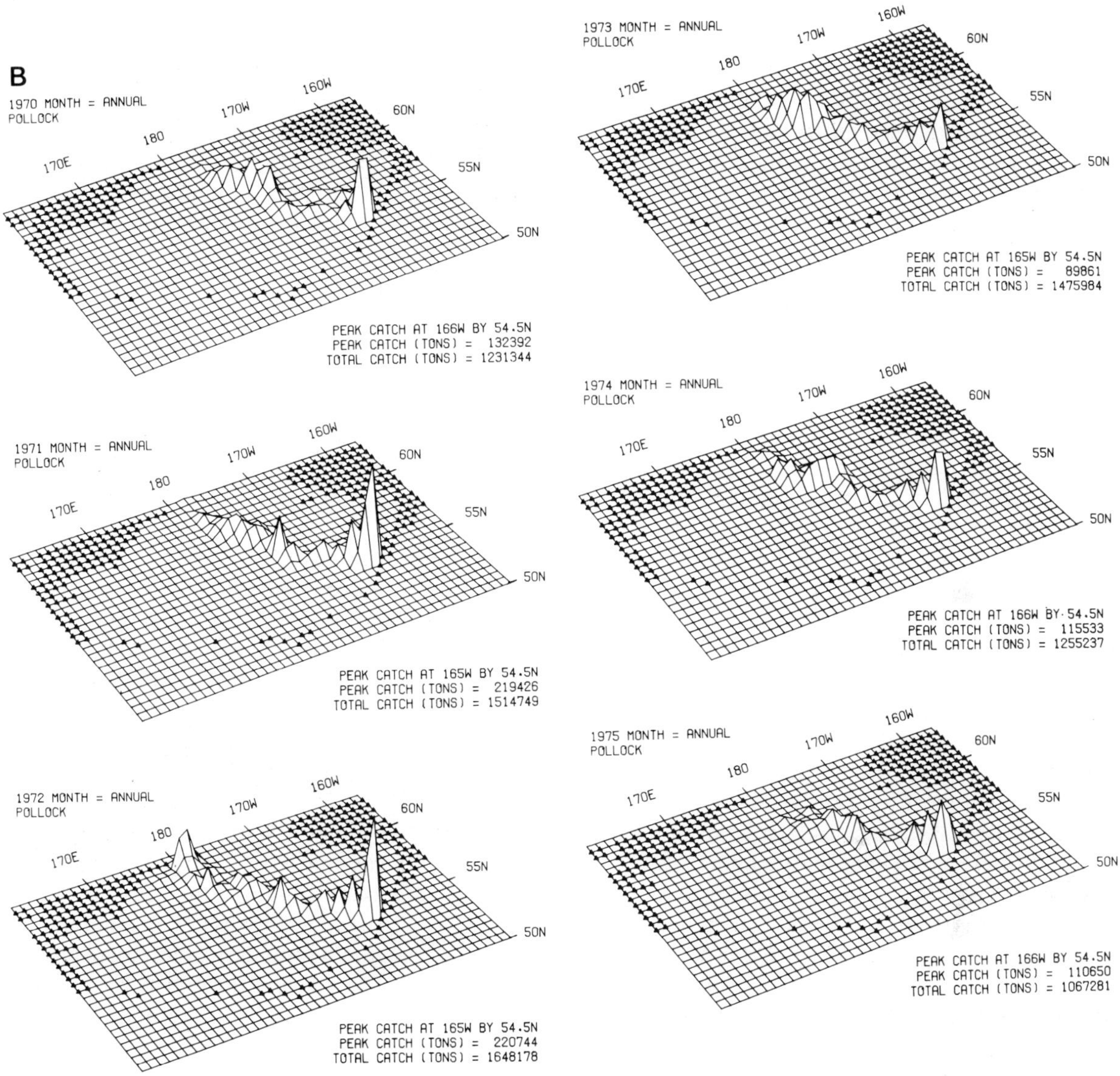

Fig. 14.6. Time and location series graphs showing annual ground fish catch by Japanese mothership, longline-gill net and independent trawl fisheries in the Bering Sea (L.L. Low, pers. comm., 1974). A. All species caught in 1964 to 1969 on eastern continental slope of Bering Sea. B. Pollock caught in 1970 to 1975 on eastern continental slope of Bering Sea.

living resources of this region been fully appreciated. Large harvests of pollock, ocean perch (*Sebastes alutus*, *S. introniger*), yellowfin sole (*Limanda aspera*), black cod (*Anoplopoma fimbria*), halibut (*Hyppoglossus hyppoglossus stenolepis*), rattails (*Nematonurus pectoralis* and *Coryphaenoides cinereus*), tanner and king crab (*Chionoecetes* and *Paralithodes*) and, more recently, shrimp (*Pandalus borealis*) are now being taken by many of the major fishery nations of the world, with Japan harvesting the most, followed by the Soviet Union and the United States. The commercial catch of bottom fish from the Bering Sea between 1959 and 1971 is shown in Fig. 14.5 (Pruter, 1973). L.L. Low (pers. comm., 1974) has produced a time–location plot of all groundfish species caught on the Bering Sea slope by the Japanese mothership fleet and independent trawl fisheries for the years 1964 to 1969. As seen in Fig. 14.5, the catch of Alaskan pollock dominates the total catch after 1968, and Low's data shown in Fig. 14.6 indicate only the pollock catch from 1970 to 1975. The geographical distribution of the groundfish catch indicates an unusual ecological situation in which the catchable groundfish resources of the Bering Sea are limited to a region of only about 300 km along the slope and about 50 km across the eastern slope. Data of Fig. 14.6B also indicate a peak catch of 1.65×10^6 t of pollock in 1972, followed by a rather steady decline in 1973 and 1974 to a catch of 1.07×10^6 t in 1975. The recently imposed 200-mile territorial limit of the United States includes the entire area of the major fishing grounds of the eastern Bering Sea, and an extensive effort is being made to reduce the total catch to a sustainable yield basis. The large annual tonnage of fish harvested by man, however, represents less than 4% of the nekton and benthos required to support the very large population of mammals and birds dependent on this resource (Laevastu and Favorite, 1977).

General

Due to its latitude, the Bering Sea lies in a region with very large annual variations in properties. Incident radiation in the northern region varies annually from almost total daily darkness to total light; the wind torque over the sea is an order of magnitude greater in winter than summer; and the extensive ice cover in the winter is totally absent in the summer. Although warmer seas contain more diverse populations, the colder seas support much larger individual populations. The Bering Sea is a prime example of this general observation. It has one of the largest marine mammal populations in the world, possibly the largest clam population, one of the largest salmon runs, one of the highest densities of birds, the largest eelgrass (*Zostera*) beds and very high yields of pelagic and benthic commercial fisheries.

Today the Bering Sea is the focus of more scientific research than at any time in its history. The base-line studies resulting from the Outer Continental Shelf Environmental Assessment Program of the Bureau of Land Management of the U.S. Department of the Interior; the intensified National Marine Fisheries efforts because of passage recently by the U.S. Senate of the Fisheries Conservation and Management Act of 1977; and the International PROBES (Processes and Resources of the Bering Sea Shelf) study sponsored by the Office of Polar Programs of the National Science Foundation are three major efforts now under way. This review, therefore, might be considered as an interim progress report on the Bering Sea. In the immediate years ahead much more information will doubtless be published than is now available. It is hoped that this document will provide an historical synopsis and indicate some important observations that will aid in the further study of this significant region of the world's oceans.

PHYSICAL OCEANOGRAPHY

Water masses

There are three basic water masses distinguishable in the Bering Sea, commonly named the Alaskan Stream, Western Subarctic and Bering Sea. These waters have in common a dichothermal stratum (a body of cold water at intermediate depths) in the upper layer and a marked halocline. Much of the work on water of the Bering Sea water masses has been reviewed by Takenouti and Ohtani (1974).

The Alaskan Stream water has a complicated structure with varying vertical patterns of temperature and salinity across the current, as shown in Fig. 14.7 for a section along 175°E longitude in winter. The southern boundary of the current has a sharp halocline just below the 100 m level, which disappears toward the north. Entering the Bering Sea, the Alaskan Stream loses its characteristic stratified structure yet still retains the high-temperature thermal layer characteristic of the Subarctic region. The Alaskan Stream represents an extension of the Alaskan Current westward along the Aleutian Islands, providing a source of dilute and relatively warm (4°C) surface water at a depth of 300 m to the northwestern North Pacific Ocean which spreads as far west as 170°E (Favorite, 1967). Most of the transport enters the Bering Sea through Amchitka Pass, while the rest enters through the pass west of Attu Island. A part of the Subarctic Current also joins the northward flow coming from the Alaskan Stream, causing a combined volume transport estimated by Ohtani (1965) at 11×10^6 m^3 s^{-1}. Water thus entering the Bering

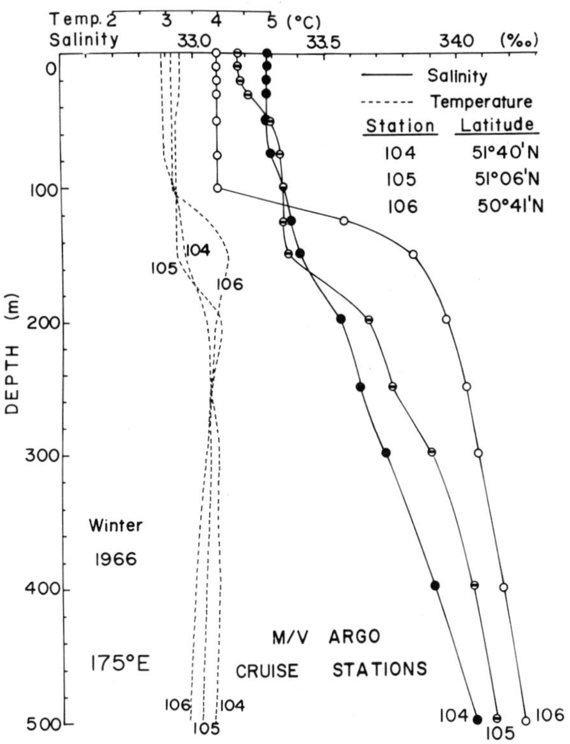

Fig. 14.7. Vertical structures of salinity and temperature of the Alaskan Stream along latitude 175°E in winter (Takenouti and Ohtani, 1974).

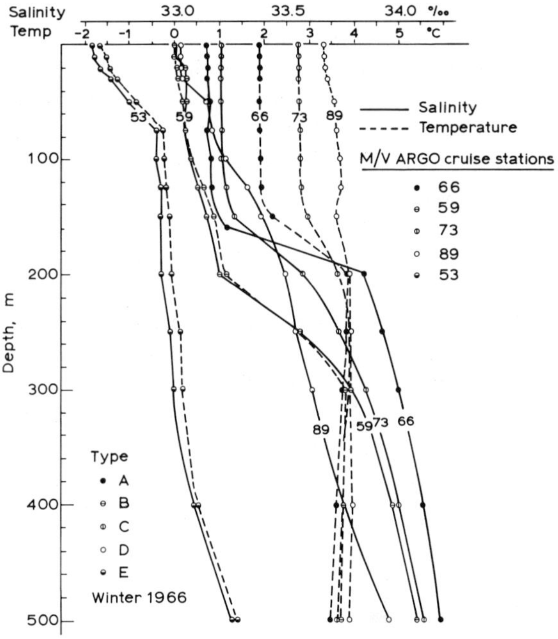

Fig. 14.8. Types of vertical structures in the Bering Sea during winter (Takenouti and Ohtani, 1974).

Sea flows out as the East Kamchatka Current except for a small portion that reaches the Arctic Ocean through the Bering Strait.

The Western Subarctic water, according to Ohtani et al. (1972), is formed in the Bering Sea. Five types of water were characterized as illustrated in Fig. 14.8 for the winter of 1966. Summer deviations occur only in the surface waters above 50 m depth, due to input of fresh water and solar heating. Type B and C waters are formed from D by upwelling of the deep layer associated with the counterclockwise circulation or divergent flow in the surface layer. In the case of B there is less intrusion of low-saline shelf water than with type C. Yet strange winter cooling occurs. Some type B water has a thick surface layer due to depression of surface water by the clockwise eddies. In regions of strong sinking, due to strong currents, the thickness of the upper layer exceeds 500 m with only slight temperature and salinity gradients, thus forming type E water. This type of vertical structure was observed in 1936 and 1959 in areas off the Kamchatka Peninsula, Kamchatka Bay and southeast of Cape Lopatka (Koto and Fujii, 1958; Ohtani, 1965).

Type B water flows southwest from the Bering Sea along Kamchatka Peninsula as the East Kamchatka Current, which passes along the Kuril Islands. The cold and less saline Okhotsk Sea water mixes laterally with the East Kamchatka current near the Kuril Islands, forming the typical Western Subarctic water.

Water over the broad shallow continental shelf away from the immediate coast is cooled during winter from the surface to the bottom by convection currents to give a homogeneous water mass from top to bottom with a temperature of the freezing point. In the summer the surface water temperature rises to 6 to 10°C; however, cold water ranging from −1.7 to 2°C remains under the surface water. Ohtani (1969) discussed the distribution of water temperature, salinity and density from results of summer observations made aboard the *Oshoro Maru* from 1963 to 1966. Along the edge of the shelf there is a zone of discontinuity of temperature and salinity between the oceanic water and the shelf water. The isohalines are nearly parallel to the isobaths and decrease gradually toward the coast. A distinct pycnocline appears between the surface and lower layers at about the depth of 20 to 30 m in the region north of 60°N. This results in very low heat transfer from the surface to the lower layer in the summer.

Near the coast and on the northern part of the continental shelf (north of St. Lawrence Island) as well as in areas of the Gulf of Anadyr and Norton Sound, which are under strong influence of freshwater runoff from land, a very strong halocline develops between the surface and lower layers; thus winter cooling may not reach the bottom in these areas. In general, at depths less than 200 m winter thermohaline convection produces a water column which is almost uniform vertically. The area covered by the cold bottom water on the continental shelf varies considerably from year to year. Maeda (1972) studied the annual variation of the extension of the cold bottom water, using the eastern margins of the 0°C and 2°C isotherms in Bristol Bay as indices. Fig. 14.9 shows these margins for the years 1955 to 1974. The years from 1971 to 1973 were considerably colder than from 1962 to 1967, which correlated well with air temperatures and prevailing winds from the north for these respective years.

Koto and Fujii (1958) studied the characteristics of the dichothermal strata in the Bering Sea region and found that the density (σ_t)[1] of the minimum-temperature water decreased toward the south along the section parallel to the 180° meridian. They concluded that although the dichothermal water appeared to spread southward according to the vertical temperature and salinity distribution observed, this was not the actual case, thus agreeing with Ûda (1935), who maintained that dichothermal strata in the Subarctic Current are not formed as in the Bering Sea Basin but are caused by the horizontal mixing of adjacent waters. The northern basin has two temperature minima in the dichothermal strata. Based on the salinity, the lower temperature minimum was considered as the bottom of the vertical mixing in the winter, while the upper one is formed by horizontal mixing with the cold and less saline water on the shallow shelf region.

A typical temperature cross section in the region of the eastern shelf break north of Zhemchug Canyon presented by Coachman (1975), Fig. 14.10, gives convincing evidence of the intimate association of the temperature minimum with the shelf water. The temperature minimum as shown in Fig. 14.11 is ubiquitously present at the 26.6 σ_t isobar in this region.

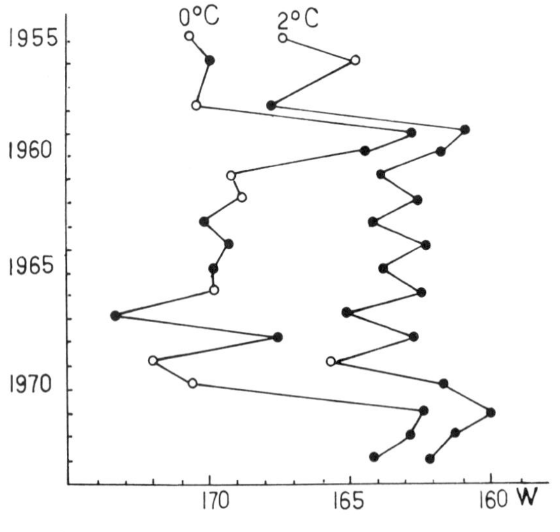

Fig. 14.9. Annual variations of southeastern margins of 0°C and 2°C isotherms in the summer season from 1955 to 1973: closed circles, observed positions; open circles, estimated positions (Takenouti, 1975).

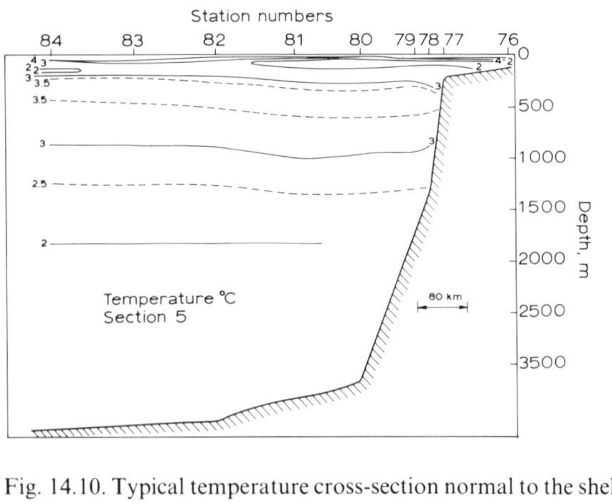

Fig. 14.10. Typical temperature cross-section normal to the shelf break about 100 km northwest of Zhemchug Canyon (Coachman, 1975).

[1]See p. 31 for definition.

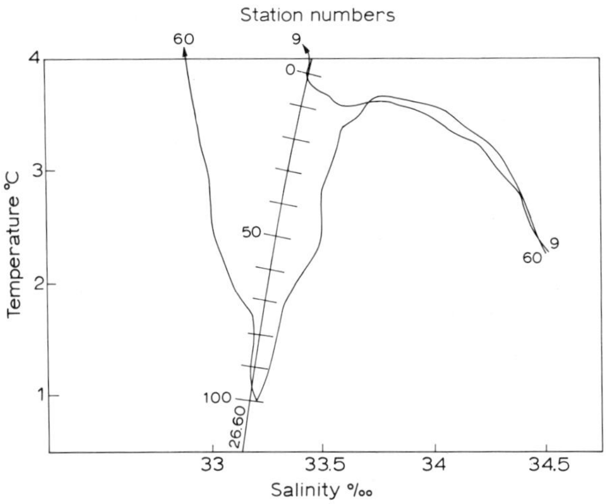

Fig. 14.11. Temperature–salinity diagram of Stations 9 and 60, displaying the extremes of the ubiquitous temperature minimum at $\sigma_t = 26.6$. The scale of 0 to 100 is arbitrary. Station 9 was located at 170°W 54°N and Station 60 at 178°E 61°N, representing the most southern and northern stations along the eastern continental slope (Coachman, 1975).

General circulation

The first circulation scheme for the Bering Sea was proposed by Dall (1881). The first generally accepted modern current chart was produced by Ratmanov (1937), who used dynamic calculations from data obtained on the first oceanographic cruises taken in this region. He concluded that one of the main causes for the Bering Sea circulation was the exchange of water through the passes of the Komandorskiy–Aleutian Island arc. Goodman et al. (1942) enlarged on Ratmanov's scheme by including new observations of currents over the northern and eastern parts of the shelf and in the vicinity of the Bering Strait. A major departure from these schemes was made by Dobrovol'skii and Arsen'ev (1959), who showed a major outflow through the western portion of Near Strait; a major current flowing northwest from the eastern Aleutian Passes parallel to the shelf edges and extending nearly to Cape Navarin; and the presence of numerous gyres in the deep basin. In 1963 Natarov, making observations of the entire area in a single season, was able to compile both a summer and winter circulation pattern, claiming that the main flow takes place through the numerous passes in the eastern part of the Aleutian arc but with a nearly constant input through the eastern portion of Near Strait. The most recent circulation scheme for the whole of the Bering Sea was that of Arsen'ev (1967), shown in Fig. 14.12. Besides incorporating the inflow and outflow regimes of previous workers, he introduces a number of permanent anticyclonic gyres around groups of the Aleutian Islands and on the continental shelf. Hughes et al. (1974), using direct current measurements by means of parachute drogues and indirect measurements by means of Salinity–Temperature–Density recorders (STD) and Nansen bottle data, devised a new circulation scheme for the western Bering Sea for the summer as shown in Fig. 14.13. Taken into account were the direct current measurements, the bathymetry of the region and geostrophic calculations based on all available relevant data.

A number of problems are still unresolved on western Bering Sea circulation. Among the most important are the deep water circulation, quantitative effects of bathymetry, secular and seasonal variations on the flow and the exact role of wind-driven circulation. It is now reasonably well established, however, that the overall circulation in the deep basin is cyclonic in nature. Inflow of surface water is through the eastern passes and the eastern portion of Near Straits, and possibly on the eastern border of other westward straits. Outflow occurs through the passes southwest of Bowers Ridge east of Near Islands (Buldir and Kiska) and through Kamchatka Strait as the Kamchatka Current.

In the eastern Bering Sea, consisting mostly of the broad shallow continental shelf, the circulation remains uncertain. All of the published circulation schemes show an easterly current north of the Aleutian Islands moving across the shelf into Bristol Bay where it flows counterclockwise and eventually moves over the shelf through the Bering Strait. Recent data (L.K. Coachman, unpublished) do not support these conclusions. It has been observed that this easterly current has little transport across the shelf but turns north to join the well-established Bering Slope current which bifurcates near Cape Navarin, part of it flowing northwest of St. Lawrence Island through the Bering Strait. Present work now under way using satellite-tracked drogues, direct-reading current meters and dynamic topography techniques coupled with supplementary chemical data are expected to clarify the complex problems of the current regime of the con-

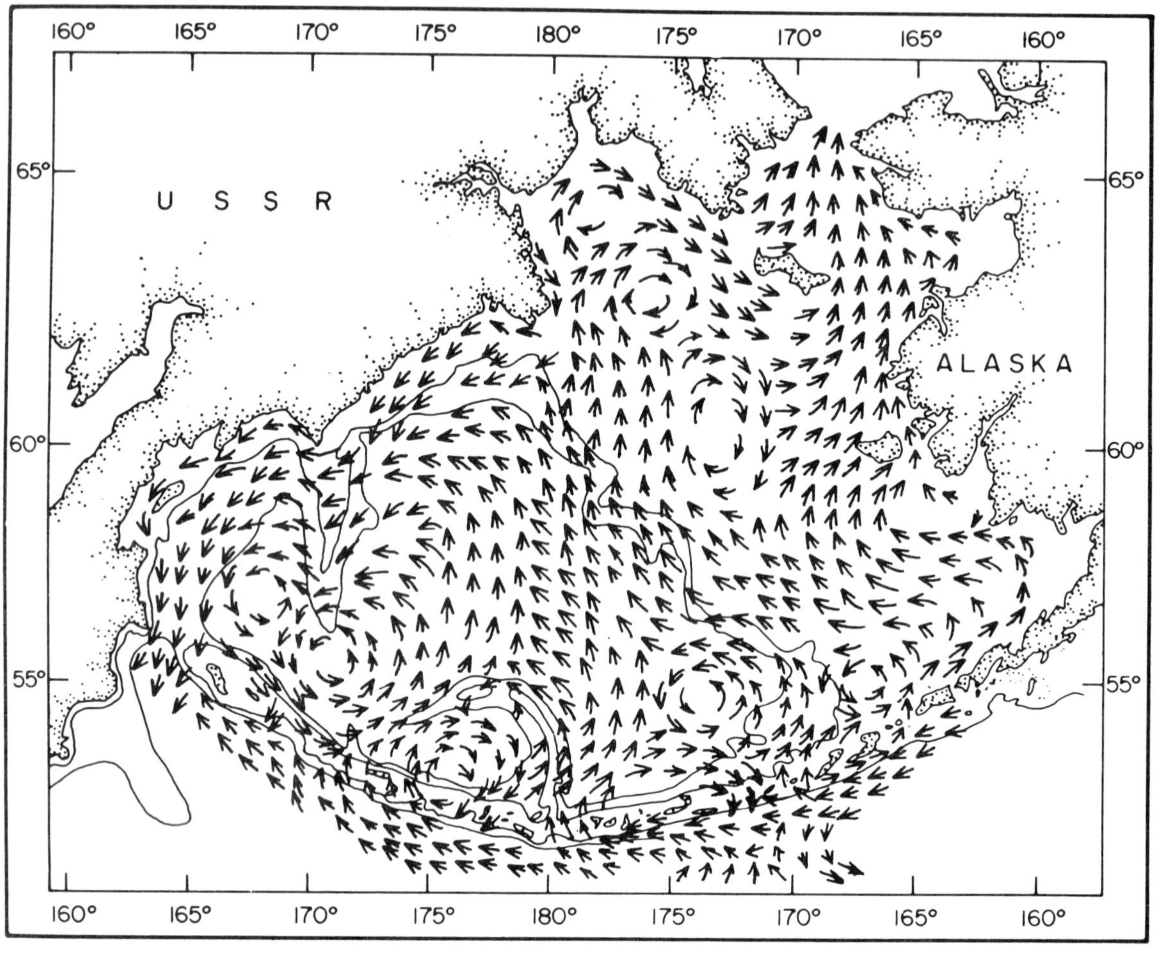

Fig. 14.12. Bering Sea circulation according to Arsen'ev (1967).

tinental shelf. This work is a vital part of the Processes and Resources of the Bering Sea Shelf (PROBES) (Muench et al., 1976), National Science Foundation study and of the Outer Continental Shelf Environment Study (Bruce, 1976) of the National Oceanic and Atmospheric Administration under contract with the U.S. Department of Interior.

All circulation schemes show passage of water from the Bering Sea into the Arctic Ocean (Chukchi Sea) with little evidence of any southward current through the Bering Strait to the Bering Sea.

Exchange with North Pacific and Arctic Ocean

Favorite (1974) lists 39 openings through the Aleutian–Commander (Komandorskiy) Island arc, only 14 of which have an area greater than 1 km^2 or sill depth of at least 200 m. The passes may be classified into five groups: eastern, central, western and the Commander–Near and Kamchatka Straits. These passes, shown in Fig. 14.14, deepen westerward, and their openings constitute 99.9% of the area penetrating the island arc. The depth and area of the major openings of all the passages are shown in Table 14.1.

There have been several conflicting opinions in recent years on the amount and location of flow through the Aleutian passes. The work of Leonov (1960), Natarov (1963), Moiseev (1964) and Batalin (1964) contributed to the understanding which is now fairly well accepted; it was presented by Arsen'ev (1967), who used station data of all workers to deduce the data of Table 14.2. This early

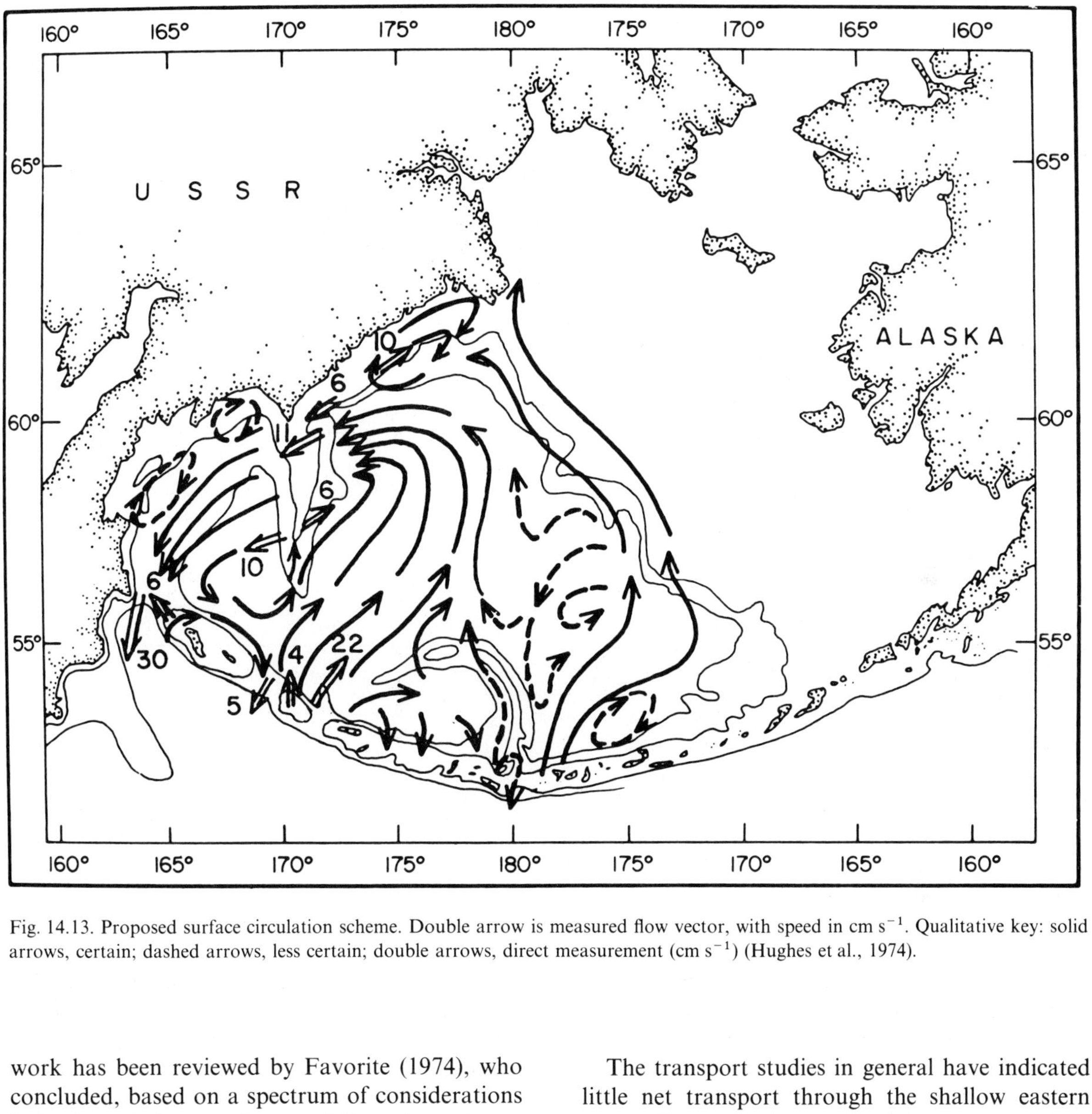

Fig. 14.13. Proposed surface circulation scheme. Double arrow is measured flow vector, with speed in cm s^{-1}. Qualitative key: solid arrows, certain; dashed arrows, less certain; double arrows, direct measurement (cm s^{-1}) (Hughes et al., 1974).

work has been reviewed by Favorite (1974), who concluded, based on a spectrum of considerations including wind stress data, and the volume transport in the Alaskan Stream, that the source water for flow into the Bering Sea is quite variable from year to year and also that the flow through the various passes is highly variable on any time scale. He concluded that north and south flows can occur simultaneously in all major Aleutian passes and that existing data are inadequate to ascertain if a net flow exists. A volume transport of 14×10^6 m^3 s^{-1} into the Bering Sea through the passes east of the Commander Islands was estimated, as compared to the Arsen'ev (1967) estimate of 19.5×10^6 m^3 s^{-1}.

The transport studies in general have indicated little net transport through the shallow eastern passes. However, since most of these computations were completed Kelley and Hood (1971b), Kelley et al. (1971), Hood and Kelley (1976) and Swift and Aagaard (1976) have found extensive areas of upwelled water within and just north of Samalga Pass. A similar situation is indicated for Amukta Pass. The upwelling appears to be caused by the inflow of less dense Alaskan Stream water through the passes where it is buoyed up by the denser surface water of the Bering Sea. The volume of the water transport driving this system has not been measured nor has the extent of upwelling in any

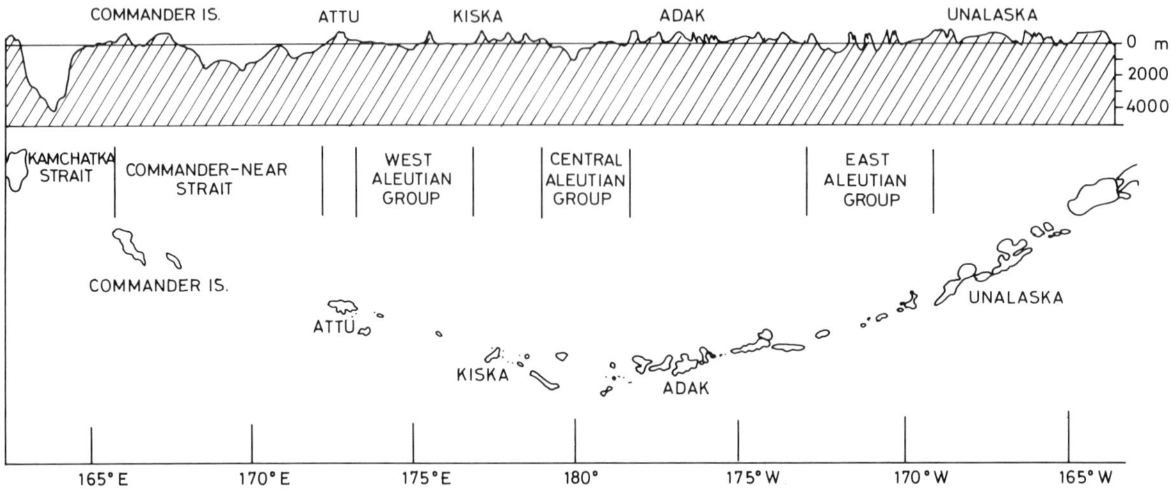

Fig. 14.14. Aleutian–Commander Island arc and a schematic section of island passes and straits (Favorite, 1974).

TABLE 14.1

Depth and area of the major openings in the Aleutian–Commander Island arc (from Favorite, 1974)

General opening[1]	Pass/strait	Depth (m)	Area (km^2)	
East Aleutian group	Samalga	200	3.9	
	Chuginadak	210	1.0	
	Herbert	275	4.8	
	Yunaska	457	6.6	
	Amukta	430	19.3	
	Seguam	165	2.1	
				37.7
Central Aleutian group	Tanaga	235	3.6	
	Amchitka	1155	45.7	
				49.3
West Aleutian group	Kiska	110	6.8	
	Buldir	640	28.0	
	Semichi	105	1.7	
				36.5
Commander–Near Strait	Near	2000	239.0	
	Commander	105	3.5	
				242.5
Kamchatka Strait		4420	335.3	
				335.3
	Total			701.3

[1] The location of these is shown on Fig. 14.14.

region except the area north of Samalga Pass, where a vertical transport rate of 3×10^{-3} cm s^{-1} was measured by Hood and Kelley (1976) using a carbon dioxide gaseous transfer model and of 7×10^{-3} cm s^{-1} by Swift and Aagaard (1976) using a steady-state oxygen model.

Volume transports through the Near and Kamchatka Straits were estimated by Hughes et al. (1974), employing both geostrophic computations and direct measurements by means of parachute

TABLE 14.2

Flow (10^6 m^3 s^{-1}) through Aleutian–Commander Island arc according to Arsen'ev (1967)

	Transport out	Transport in
Kamchatka	21.0[1]	2.6[2]
Commander–Near		14.4
West Aleutian group		0.7
Central Aleutian group		4.4
	21.0[1]	22.1[2]

[1] Above 3000 m; [2] below 3000 m.

drogues. The results of the direct current measurements gave much higher flows than those obtained from the geostrophic computation, which had been the primary method used by earlier investigators. Based on these data, a water balance for the Bering Sea, presented in Table 14.3, was proposed.

Transport through the Bering Strait has been an important question for oceanographers since the first observations by Vitus Bering in 1728. Not until the analysis of all previous work and presentation of much of their own by Coachman et al. (1975) were the physical oceanographic features of the Bering Strait clarified.

There are three water masses in the Bering Strait, based on salinity: Anadyr, Bering Shelf and Alaskan Coastal, indicating their source. Anadyr is characterized by salinities between 32.9 and 33.1‰, Bering Shelf between 32.4 and 32.7‰ and Alaskan Coastal 31.5 to 32.1‰ in the summer. All have seasonal as well as year-to-year fluctuations. Two of the water masses, Anadyr and Bering Shelf, originate south of St. Lawrence Island. Alaskan Coastal Water also has a contribution from the south, but there is also a substantial input from Norton Sound.

The Anadyr water originates in the Bering Sea, probably as a subsurface partial continuation of the Bering Sea Slope current. In passing through the Gulf of Anadyr in summer the water retains, to a considerable degree, its relatively warm temperature (>1°C) and high salinity (>32.9‰). It is modified by mixing on the east and south sides with very cold (< −0.5°C) and less saline (<32.3‰) water occupying the central Gulf. The net effect is slight cooling (by *ca.* 1°C) and freshening (by 0.1 to 0.2‰) during transit. The net result is a mixture of Bering Sea water (80–90%) and of central Gulf water (10–20%). The Bering Shelf water enters the Bering Strait around both ends of the St. Lawrence Island. It is a mixture of water from the Bering Sea with the less saline very cold water resident over the Bering Sea Shelf. The formation of a cold resident water occurs in the winter due to cooling and formation of ice. The upper layer of warm and low salinity water of summer is destroyed during fall by convection and flushing. On the surface where salinities are about 32‰, 1 to 2 m of ice form during the winter, raising the salinity to 33‰ or more. The water is thoroughly mixed top to bottom by the freezing process and isolated in the summer by development at the surface of a warmer, less dense layer. The Alaskan Coastal Water is created from the relatively cold and saline ubiquitous water of the Bering Sea Shelf by dilution with continental runoff, the dilution varying with proximity to the coast. A strong pycnocline is generated in the summer by an abrupt increase in runoff (10-fold increase in 1 to 2 months) of sufficient value to decrease the salinities of the upper layer over the whole eastern portion of the region. In winter the upper layer is destroyed by cessation of runoff and cooling by the atmosphere, which later is followed by extraction of fresh water through ice formation.

Transport through the Bering Strait is almost all to the north, with an average summer flow of 1.6×10^6 m^3 s^{-1}. The major driving force for northward flow is the slope of the sea surface down to the north. The slope was calculated to be 2×10^{-4} cm km^{-1} for 5 to 6 August, 1964 (Coachman and Aagaard, 1966). Continuously acting mechanisms must prevail regionally that tend to add water to the northern Bering Sea and remove water from the Chukchi Sea. Runoff, all being south of the Strait,

TABLE 14.3

Proposed approximate water balance (10^6 m^3 s^{-1}) for the Bering Sea (from Hughes et al., 1974)

	Summer	Winter
Near Strait	+25	+25
Kamchatka Strait	−20	−35
Buldir and Kiska Passes	−5	
Aleutian Passes (including Buldir and Kiska)		+10

may influence the sea level, but runoff is only 10% of the flow. The major cause must be the wind distribution and atmospheric pressure field of the region.

Micro-structure of the shelf

The shelf region of the Bering is the largest in the world except for the very broad shelves of the Chukchi, East Siberian and Laptev Seas, all in the Arctic Ocean. Details of circulation including transport through the Bering Strait are discussed above and by Coachman et al. (1975). The circulation of the southern shelf has never been the subject of a comprehensive survey from which a gross circulation pattern could be interpreted. Until recently the only current measurements available for the southern shelf area were for Bristol Bay (Hebard, 1961). Distribution patterns of water properties have been published from one year's data by Ohtani (1973).

The water boundary conditions, based on information now available in the literature, appear to be: a large, seasonally variable flux on the east of fresh water from Alaska, partially focused as major river flows from the Yukon and Kuskokwim; on the west the Bering Slope current from just west of Unimak Pass in the southeast to Cape Navarin in the northwest; and a wide transition zone from about 100 m depth toward the coast to about 150 m depth on the continental shelf. The overall shelf circulation is unclear, but in general there must be a south to north movement to provide the 1 to 2×10^6 $m^3 s^{-1}$ flux north through the Bering Strait. This may be offset in part by the influx into the southernmost shelf areas of an Alaskan Stream surface layer, which most likely is provided by passage through the eastern Aleutian Island passes.

Recently studies of the shelf region have been accelerated greatly by the advent of the PROBES investigation (Muench et al., 1976) and that of the Outer Continental Shelf Energy Program (Bruce, 1976). Preliminary results of these studies have shown that the motion of water in this region is primarily tidal, with a very slow longer term-flow (1–5 cm s^{-1}) northwestward along the outer shelf, in general following the bottom contours. The mechanism of fluxes across the shelf is lateral diffusion driven mainly by tides (Coachman, 1978). The shelf regime is a broad transition zone between offshore waters of oceanic origin (relatively warm and saline) and the mid-shelf waters (relatively cold and fresh) of outer Bristol Bay. The inner and outer edges of this transition zone are marked by fronts characterized by well-defined horizontal salinity gradients. The outer front lies over the continental slope, and the inner is found at about the 100-m contour. This transition zone is about 100 km wide and some 250 km long, and exhibits a near-zero horizontal salinity gradient. This regime is extremely rich in vertical fine structure, particularly in the low-energy region of the water column between wind mixing at the surface from 20 to 30 m and tidal mixing at the bottom 30 to 40 m. It seems probable that fluxes between fronts are restricted and physically related to the fine structure phenomenon, with multiple layering being alternately influenced by shelf and oceanic source waters.

The earlier suggestion that nutrient-rich water is contributed to the shelf region through upwelling along the continental shelf has no basis in experimental support, at least for the period studied in the spring and summer of 1976–1977 (Coachman, 1978; Hood, 1978).

Ice as a modifier of physical processes

The role that sea ice plays in Bering Sea circulation is poorly understood. The ice cover acts as a complex filter, both mechanically in the transfer of wind stress, and thermodynamically in general air–sea interaction. The Bering Sea ice in particular has an active ice edge boundary (Galt, 1975). There are few on-site studies of the ice edge boundaries because of the inhospitable nature of the region during winter when the processes are active. Ice breakers, which have poor sea-keeping ability in the open water often associated with the edge, find work here difficult. The advent of satellite imagery has greatly increased our knowledge of the physical regime. Muench and Ahlnas (1976) were able to observe the ice distribution from March to June 1974 by NOAA (National Oceanic and Atmospheric Administration) 2, and 3VHRR (Very High Resolution Radiometer image) satellites and Muench and Charnell (1977) from November 1974 to March 1975 by NOAA3 and NOAA4 satellites. These first observations cover the period of late winter ice and break-up, the latter the entire period from inception of ice formation through mid-March with conditions generally conducive to ice formation.

The transition zone or ice edge boundary has a relatively fixed mid-winter position paralleling the shelf break between the Pribilof Islands and Cape Navarin and somewhat north of the shelf break east of the Pribilofs. It is characterized as a region of net ice disintegration, although for limited periods of time ice may actively form there. During the observation period plume-like ice features occurred frequently; they had lengths on the order of tens of kilometers and were spaced about 10 km apart. These features were usually straight and nearly perpendicular to the main ice edge. The long axis was usually oriented from 40 to 90° to the left of the wind, which was always blowing off the ice when these features were evident. At wind speed of less than 5 m s^{-1} the linear band pattern broke down and was replaced by a scalloped ice pattern. The regularity of these spacings between the observed ice bands suggests that wave-like or periodic processes were responsible for their formation. The data suggest that the formation of ice bands may be related to convective processes; they are most probably due to atmospheric roll vortices ubiquitous in the area with about a 10-km spacing. The exact causes of the plume-like features and their effect on the physical processes of the water column are unknown, yet it is clear that the alternate influence of open water and therefore high heat exchange as well as the momentum exchange spatially fixed between plumes of ice, which both insulate and mechanically prevent the transfer of wind energy, must have a profound effect on the oceanographic processes of the shelf region under the influence of ice cover and especially at the active ice edge.

GEOLOGICAL SETTING

The geology of the Bering Sea has been studied since the middle of the 19th century. A thorough review of early work was given by Lisitsyn (1966). Hopkins (1967) reviewed the Soviet and American studies of marine geology and paleontology between 1948 and 1968. Modern geological studies began just after World War II with a series of U.S. Navy Electronic Laboratory cruises from which data the first map of the bottom sediments was prepared by Dietz et al. (1964). The Cenozoic sedimentary and tectonic history was reviewed by Nelson et al. (1974), and the contemporary sedimentary regimes of the eastern shelf region by Sharma (1974).

Sedimentary and tectonic history

The continental shelf of the Bering Sea is an area of continental crust that connects Asia and North America, and the deep basin is flooded by oceanic crust and overlain by a thick sequence of undeformed sediments. Churkin (1972) suggested that the shelf region is a submerged part of a large continental plate that includes all of North America and northeastern Siberia. The abyssal basin is thought to be a northern embayment of the Pacific Ocean that was isolated by the development of the Aleutian ridge prior to the Late Eocene but no earlier than Late Cretaceous time (Scholl et al., 1974).

The tectonic history of the Bering Sea is complicated by the presence of the Kula Plate and the Kula spreading center, which separated the Pacific Plate and the North American–Siberia Plate during Miocene time. The floor of the abyssal basin is probably a remnant of the oceanic Kula Plate, which was destroyed by subduction near the middle of the Tertiary Period (Grow and Atwater, 1970).

Major geologic features

The deep Bering Sea is occupied predominately by two abyssal plains, the Aleutian Basin and the Commander Basin (Fig. 14.1). The two basins are effectively separated by two large, nearly connecting, ridges extending in an S-shape from the Siberian Coast just north of the Kamchatka Peninsula southward to the Aleutian Arc. The Umnak Plateau (also known as the Bering Canyon), a large area of intermediate depth occupying the southeastern corner of the abyssal basin just north of Unimak Pass, may have resulted from foundering of the continental crust as a result of isostatic response to loading of terrigenous material deposited on the continental rise.

The shape of the Bering Sea probably results from the gigantic structure composed of the Alaskan orocline on the east and the Chukatkan orocline on the west, two sharp flexures concave toward the Pacific. The large intervening faulted flexure is submerged on the continental shelf and concave toward the Arctic Ocean. Hopkins and Scholl (1970) believed that large faults, developed mostly during Late Cretaceous and Paleocene times, are strong expressions of the flexure system that probably reflects the bending of Siberia relative to North America as a consequence of rifting in the Atlantic and Arctic basins. The southward shifting of the Alaskan part of the North American continental margin probably resulted from this oroclinical folding.

Cenozoic sediments

From limited data now available, the Late Mesozoic and Early Cenozoic history has been postulated (Nelson et al., 1974). During Cretaceous time, continuing into the Paleocene epoch, the Okhotsk volcanic belt extended through the Bering Sea to the mainland of Alaska. Clastic and volcanic sediments were simultaneously deposited in fault-bounded basins on the Alaskan mainland. Uplift processes soon eliminated sedimentation, and a long erosion period began for continental Alaska, the Chukotka Peninsula and the continental shelves of the Bering and Chukchi Seas. Near the end of Oligocene time, western Alaska, northeastern Siberia and shallow parts of the Bering and Chukchi Seas (known as Beringia) were a surface of marine and subaerial planation. This surface was later warped and the resulting basins filled with Tertiary sediments of the main layer sequence.

The geological history of the main layered sediments in the Bering Sea began after prolonged periods of denudation. The products of erosion from the present Bering continental shelf, southwestern Alaska and the Chukotka Peninsula were probably deposited in the Aleutian and Commander basins. By Late Oligocene time, the sediments from land were being trapped in small basins on the mainlands and in the large shelf basins. Marine sediments were then deposited on the subsiding continental margins and occasionally in the Anadyr Basin. The far north areas were still emergent and receiving continental sediments.

The formation of the Yukon River drainage basin by the uplift of the Alaskan Range during Miocene time resulted in a nearly 50% increase in the land area shedding sediment to the Bering Sea. Biogeographical evidence (Hopkins, 1967) indicates the Bering Strait was inundated and finally formed a continuous marine connection by the end of the Miocene epoch.

Quaternary sediments nearly 1 km thick mantle the deep basins, and thicker deposits may underlie the continental rise (Scholl et al., 1970b). However, these sediments on the continental shelf are rarely thicker than 100 m; in many places they are only a few meters thick, and are absent entirely in some near-shore and insular areas where pure Quaternary bedrock is exposed on the sea floor. Tectonically the Quaternary sediments are relatively undisturbed. Although tectonic activity has occurred north of St. Lawrence Island, near-shore areas of Goodnews Bay (Hopkins, 1959) and worked terraces indicate considerable tectonic activity along the coast of Seward Peninsula (Sainsbury, 1967). Other studies show that lava flows and plugs of Quaternary age are present on the sea floor north of the central part of St. Lawrence Island, near the Pribilof Islands, and on the continental slope along the Aleutian Ridge (Nelson et al., 1974).

During the Pleistocene the Bering Sea was characterized by repeated episodes of subaerial exposure, fluvial sedimentation and massive glacial encroachment accompanied with deposition. Kummer and Creager (1971) give evidence of Siberian glacier encroachment into the Gulf of Anadyr. The Siberian glaciers pushed as far as 150 km from the shore of Chukotka Peninsula and encroached on St. Lawrence Island where they deposited glacial morrainal ridges on the continental shelf (Nelson and Hopkins, 1972). Early to Middle Pleistocene valley glaciers of the Seward Peninsula pushed debris some kilometers seaward of the present coast, and glacial deposits are found well beyond the shore line of Bristol Bay (Coulter et al., 1965). The presence of large partially buried channels suggests that between 10 000 and 16 000 years ago the main Yukon River drained southwest of St. Lawrence Island through the central shelf. The Pribilof and Zhemchug submarine canyons were probably carved by turbidity currents when the Yukon River mouth reached the continental margin (Scholl et al., 1970a).

A very low rate of Holocene sedimentation (2 cm per century, according to C.H. Nelson and M. Rubin, unpublished) in Norton Sound and the Chirikov Basin, but a relatively high rate in the

Chukchi, suggests a bypassing of the Bering epicontinental shelf by Yukon–Kuskokwim sediments. Transport rates of suspended sediments in the western Bering Sea (2–5 g m^{-3}; Lisitsyn, 1966) are much lower than in the eastern Bering Sea (5–25 g m^{-3}; McManus and Smyth, 1970).

Contemporary sedimentation and sedimentary processes on the Bering Sea shelf

Sharma (1974) divided the eastern Bering Sea into five provinces based on their distribution by water currents and provinces of origin. These are: Southeastern Shelf, including Bristol Bay; the Central Shelf, a broad region lying between St. Matthews and Nunivak Island; the Northeastern Shelf, the region bounded by Seward Peninsula, the Yukon Delta, St. Lawrence Island and the Russian coast on the west, generally known as Chirikov Basin and Norton Basin; the Northwestern Shelf, essentially covering the Gulf of Anadyr; and finally, the Outer Shelf, a north–south oriented area parallel to the continental margin (see Fig. 14.15).

The Southeastern Shelf region receives the drainage of numerous rivers and lakes, and the depth increases gradually from the head of the bay to 100 m over a distance of 450 km with a slope of

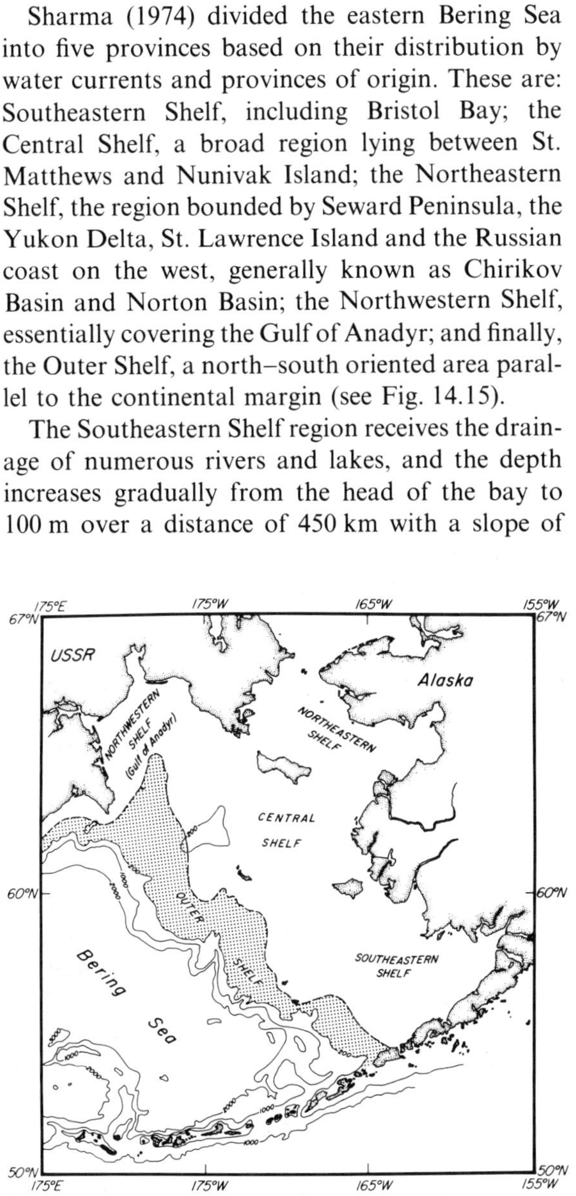

Fig. 14.15. Basinal regimes of the Bering Sea Shelf (Sharma, 1974). Contours in meters.

2.2×10^{-2}‰. Near-shore sediments consist of gravel and coarse sand, and further offshore the sediments become progressively finer. Near-shore coarse sediments are poorly sorted, medium and fine sands deposited on the mid-shelf are moderately well sorted, and offshore the sorting deteriorates with increasing silt and clay. Suspended sediments composed of both biogenic matter and clay-sized matter vary in concentration between 1.6 and 2 mg l^{-1}. The content of organic matter increases with clay content, reaching over 0.5% near the outer boundary, which is associated with a high bioproductive region.

The Central Bering Shelf has two large islands, Nunivak and St. Lawrence. The average depth of this shelf is 50 m and the sediments are 50 to 90% sand. The Kuskokwim River controls the sediment distribution, which varies in composition of silt and sand throughout. The suspended load varies between 5 mg l^{-1} in mid-shelf to 10 mg l^{-1} near the coast of Alaska mainland (McManus and Smyth, 1970).

The Northeastern Shelf region is a shallow basin almost surrounded by land mass. Norton Sound on the eastern part is less than 30 m deep and receives the drainage of the Yukon River in the southeast part. The region has been investigated by Nelson and Hopkins (1972), who find the sediment distribution complex because the basin is covered by both glacially derived relics of the Quaternary and marine contemporary sediments. The marine sediments consist of silt and sand which cover much of the shelf. The Yukon River has built a small delta in the southeastern portion of clayey silt which overlies silty sand. Venkatarathnam (1971) recognized nine heavy-mineral assemblages in this region. In general an area of low heavy-mineral content extends northwest–southeast along a broad shallow valley in the central shelf portion, separating an area of relatively high concentrations on either side.

The Northwest Shelf region or the Gulf of Anadyr is somewhat deeper than the others. The shelf has broad ridges 20 to 30 km at the base and up to 20 m in relief. The thickness of the sediments is between 15 and 20 m, the upper 5 m being of post-glacial origin (Lisitsyn, 1966). The sediments consist of gravels in a 50 to 60 km wide belt extending from shore to depths of 30 to 40 m. Next, in depths to 50 m, are coarse, well-sorted sands. Fine sands, well-sorted, are widespread in depths of

50 to 90 m over an extensive area of 150 to 170 km in width. Most sediments are derived from the Anadyr River, which drains the Anadyr Basin of Siberia.

The Outer Shelf basin parallels the continental margin between 100 and 170 m water depths. It varies in width from about 120 km in the south, becoming quite narrow near the Pribilof Islands, and expanding again to 350 km in the northwest. The slope of the outer shelf is between 4 and 10% except in the Bering Canyon where it is only 0.5%. The sediments are mixtures of sand, silt and clay that are generally poorly sorted, and fine sand. The sediment source varies depending on location, the Kuskokwim River in the south and mid-ridge area and the Anadyr River in the northwest.

Influence of ice

Pack ice and shore-fast ice movements may raft pebbles as far as 100 km, according to Lisitsyn (1966); however, Gershanovich (1967) and Nelson et al. (1974) did not support this, but believed that the present gravel patches originated when the Holocene or earlier transgressions reworked glacial drift of the sea-floor surface. They considered that grounding ice might pick up occasional pebbles and transport it into finer sediments 10 to 20 km from gravel-rich areas.

Sharma (1974) found the sediments in sea ice to be connected with the lower boundary of a middle ice layer. The uniformity of clay minerals and the grain size distribution suggested that the sediments were incorporated in the ice from suspension during the freezing of sea water. The amount of sediment carried by ice varies from year to year and place to place, making a quantitative assessment of this mode of transport difficult to evaluate.

THE BIOTA

Perhaps the most notable observation one makes in oceanographic work in the Bering Sea, aside from the inclement weather and erratic sea state, is the abundance of marine life at all trophic levels. The casual observer would have little reason to fault the statement in the PROBES workshop document (1974):

> "The shelf region of the Bering Sea features one of the largest marine mammal populations in the world, an abundant clam population, the largest bird population per unit area in the world, the world's largest eelgrass beds (*Zostera*), a high catch of pelagic fish per unit area and some of the highest values of daily primary productivity ever measured."

Phytoplankton and primary productivity

The phytoplankton population of the subarctic waters of the Bering Sea has been reviewed by Marumo and Minoda (1976). According to Karohji (1972). 22 species of diatoms and 4 species of dinoflagellates were found to predominate in the Bering Sea (Table 14.4). Motoda and Minoda (1974) reported that the standing stock of surface diatoms ranged from 10^5 to 10^9 cells m^{-3}, with the biomass varying from year to year and showing certain periodical variations with intervals of two to three years.

On one station near Attu Island (53°N, 171°E) in 1967 it was found that diatoms formed 86% of the total phytoplankton biomass and *Denticula seminae* was 79% of these. The nannoplankton formed 93% of the total phytoplankton numbers. There is evidence (Marumo and Minoda, 1976), based on cell counts and chlorophyll *a* distribution, that coccolithophorids contribute as a primary producer at the 30 to 100-m level in these waters. In general the summer diatom communities in the

TABLE 14.4

Dominant phytoplankton species in the Bering Sea (Karohji, 1972)

Bacillariophyceae	
Coscinodiscus curvatulus	*Denticula seminae*
C. narginatus	*Rhizosolenia alata*
C. oculus iridis	*R. hebetata* f. *hiemalis*
C. radiatus	*R. hebetata* f. *semispina*
Chaetoceros atlanticus	*R. styliformis*
C. concavicornis	*Thalassiosira nordenskioldii*
C. convolutus	*Fragilaria oceanica*
C. compressus	*Nitzschia seriata*
C. debilis	*Thaliassiothrix longissima*
C. furcellatus	
C. laciniosus	Dinoflagellata
C. radicans	
C. subsecundus	*Ceratium longipes*
	C. lineatus
	C. furca
	Peridinium spp.

TABLE 14.5

Annual budget of primary production (10^9 kg C) for the Bering Sea (McRoy and Goering, 1976)

Season	Days	Shelf production	Oceanic production	Total production	%Total production
Spring	61	54	65	119	45
Summer	92	23	27	50	18
Fall	61	31	38	69	25
Winter	120	3	3	6	2
Sea ice	61	24	0	24	9
Rivers and lagoons	–	6	0	6	2
Total		141	133	274	100

Bering Sea can be represented by the *Chaetoceros–Hyalochaete* group along the coasts, and by *Thalassiothrix longissima* and the *Chaetoceros–Phaeoceros* group in the offshore regions. In the area of the Bering Sea gyre, the boreal oceanic diatoms predominate (Motoda and Minoda, 1974).

McRoy and Goering (1976) have developed an annual budget for the Bering Sea showing that the increase in productivity begins in February with the development of ice algae on the subsurface layer of sea ice that covers most of the shelf region at that time. This early activity is followed by phytoplankton bloom in the water column in the ice front zone, followed by a conventional spring bloom in the water column as thermal stratification occurs. The primary productivity for the Bering Sea, based on present data, is presented in Table 14.5.

The total primary production of 274×10^{12} $g\,C\,yr^{-1}$ for the Bering Sea, with a total area of 2.3×10^{12} m^2, yields an average productivity of 118 $g\,m^{-2}$ of carbon fixed per year. The unusual feature evidenced in this budget, in addition to the high overall productivity, is the nearly 10% contribution to the total resulting from algae growth in sea ice. Until recently ice was thought to be a deterrent to primary productivity, rather than an asset. An added advantage is also derived from the influence of the spring melt in stratifying the water column, thus allowing a very early annual bloom in the northern part and thereby significantly increasing the productivity season for the Bering Sea.

Zooplankton and secondary production

Gershanovich et al. (1974) states that the border separating the north Bering Sea copepods from those of the south Bering Sea group in the summer lies between Nunivak Island and Cape Navarin. Motoda and Minoda (1974) report 110 species in the Bering Sea, of which 27 were found in the epipelagic zone (0–200 m) on the continental shelf of the eastern Bering Sea. Copepod communities in the epipelagic zone are composed mainly of arctic *Calanus glacialis* and arctic–boreal *Acartia longiremis*, whereas the subarctic oceanic species *Calanus plumchrus*, *C. cristatus*, *Eucalanus bungii* and *Metridia pacifica* dominated the central and western part of the Bering Sea. Along the edge of the continental shelf of the eastern Bering Sea, as well as around Bowers Bank and the Aleutian Islands, mixed communities of eight subarctic oceanic species were found, together with large numbers of *Calanus glacialis* and lesser numbers of *Acartia longiremis*.

Vertical distribution of copepods in the central Bering Sea was studied by Minoda (1971), who found only one species, *Acartia longiremis*, confined to the epipelagic zone (0–200 m) throughout day and night. Another 26 species appearing in the epipelagic zone extend their vertical distribution to the transition zone (200–500 m) or to the bathypelagic zone (below 500 m). Forty-two species do not appear in the epipelagic zone, but always inhabit the transition or bathypelagic zone. This group includes adults of *Calanus plumchrus* and *C. cristatus*, which appear only below 200 m. Fifteen species inhabit only the zone below 1000 m.

The dominant member of the Euphausiacea in the central Bering Sea is *Thysanoessa longipes*, while *T. raschii* is found in the shallow waters of Bristol Bay and the Gulf of Anadyr. *T. inspinata* appears in the area east of Attu Island, probably transported by the Alaskan Stream from the Pacific coast of the Alaskan Peninsula, where it is abundant. *Thysanoessa longipes* in the Bering Sea from June to August is the highest in abundance among euphausids (Takeuchi, 1972).

Sagitta elegaris represents the major part of the Chaetognatha community in the Bering Sea. It is widely distributed on the eastern shelf, and occurs in Bristol Bay in lesser numbers. The Pteropoda are represented by *Limacina helicina helicina* and *Clione limacina limacina*, which are distributed in the western and central Bering Sea as well as the eastern shelf (Takeuchi, 1972).

The day and night variations of zooplankton biomass were examined in detail at one station (54°N, 172°W) by Motoda and Minoda (1974); their results indicated that 61.6% was located in the upper 50 m and 80.3% in the upper 80 m. The average summer zooplankton biomass for the entire Bering Sea on 5° grids for a 15-year period is shown in Fig. 14.16. The maximum values were found in the area of the eastern shelf break. These data show an average of wet weight of 30.7 g m^{-2} or, based on an assumed carbon content of 5% estimated from M. Kotori in Motoda and Minoda (1974), an average summer zooplankton biomass for all the Bering Sea of 1.54 g C m^{-2}.

Additional data for the eastern Bering Sea were presented by Minoda and Marumo (1975) as a result of a 1972 investigation by the Faculty of Fisheries of Hokkaido University; they found considerable variation from place to place but obtained average values comparable to those of Motoda as shown in Fig. 14.16. The exception was a high value of 114.7 g m^{-2} from 0 to 80 m found north of Unimak Pass this particular year.

Secondary production

Cooney (1975) has reviewed the status of secondary productivity measurement for the Bering Sea and concludes that at the present time secondary

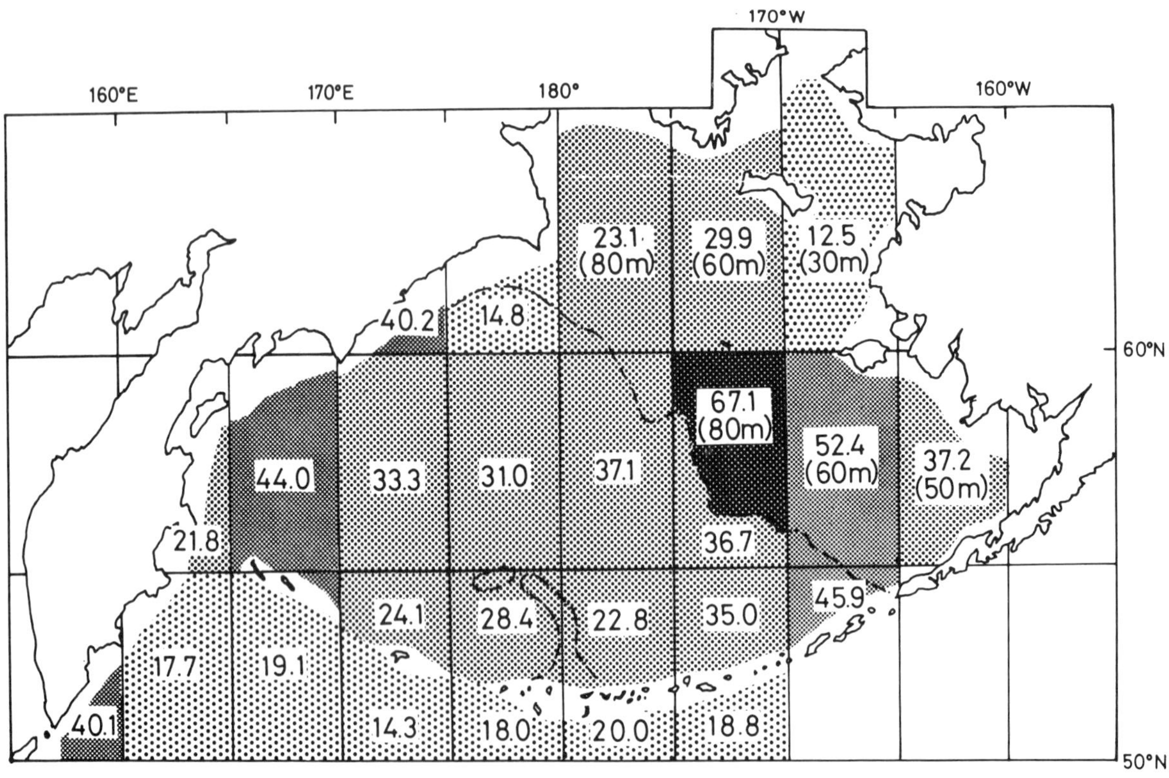

Fig. 14.16. Average summer zooplankton biomass for fifteen years from 1956 to 1970 in 5° grid. Values are expressed in wet wt. g m^{-2} in an 80-m water column, except where otherwise indicated (Motoda and Minoda, 1974).

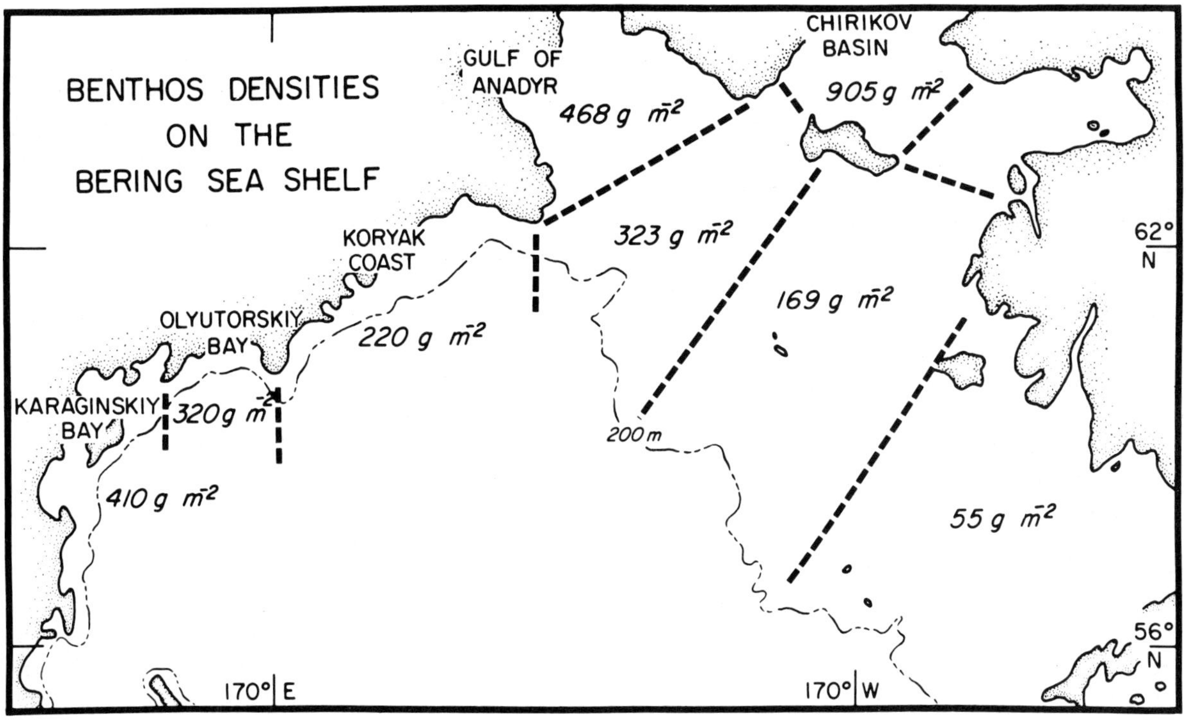

Fig. 14.17. Benthos density in various sectors of the Bering Sea and slope. Values are in wet wt. (Alton, 1974).

rates and food transfer efficiency for this sea are poorly known. Taguchi and Ishii (1972) examined respiration, grazing, excretion and chemical composition of *Calanus cristatus* and *C. plumchrus* and estimated the annual production at between 2.6 and 5.3 g C m^{-2} yr^{-1}. Based on measurements for these two copepods and an estimate of the ecological efficiency of between 10 and 35% (McAllister, 1972) between primary and secondary levels, they estimated a total secondary production of 3.6 to 12.6 g C m^{-2}, of which 50% was due to *C. cristatus* and *C. plumchrus* in the area studied.

The estimate of secondary production by Taguchi and Ishii (1972) is very low compared to that for Station P in the North Pacific (McAllister, 1972). In multiple-year studies McAllister found a secondary production by zooplankton of 3.5 to 17.9 g C m^{-2} yr^{-1} for the years 1960 to 1966, or an average of 11 g C m^{-2} yr^{-1}, with a standing stock of zooplankton of between 0.43 and 1.3 g C m^{-2} yr^{-1} (average 0.74 g C m^{-2} yr^{-1}) and a primary production rate of 34 to 51 g C m^{-2} yr^{-1} (average 40.8 g C m^{-2} yr^{-1}). The average zooplankton standing stock for the summer in the whole of the Bering Sea (Motoda and Minoda, 1974) for a 15-year period was 1.55 g C m^{-2} yr^{-1}, and primary production was estimated at 118 g C m^{-2} yr^{-1} (McRoy and Goering, 1976). The ratio of secondary production to primary production was 0.27 (27%) for Station P, whereas that for the Bering Sea, based on the estimate of Taguchi and Ishii (1972), was between 0.03 (3%) and 0.11 (11%). If, however, it is assumed that the ratio of zooplankton standing stocks between Station P and the Bering Sea are related to secondary production at the two locations, then the Bering Sea should have an average secondary production of about 46 g m^{-2} yr^{-1} (based on the ratio of standing stocks to secondary production at Station P). Thus a ratio of secondary production to primary production of 0.38 would be obtained. It is clear that much more work is needed in the Bering Sea to clarify many questions concerning secondary production in this biologically rich area.

The benthos

Most of the assessment of benthic biomass in the Bering Sea has resulted from Soviet findings which have been reviewed along with other available data by Alton (1974). The data for the northwestern Bering Sea were the result of the *Vitiaz* explorations in 1950–1952 (Belyayev, 1960); those for the northern Bering Sea by voyages of the *Krasnoarmeets* in 1932–1933 (Makarov, 1937); and, those for the eastern shelf slope and Bristol Bay in 1958–1960 (Neiman, 1960, 1963). Animals in these cases were collected in quantitative sediment grabs and separated from the sediment by washing samples through screens having a minimum mesh size of approximately 1 mm. This manner of sample sorting would include organisms larger than smaller macrobenthos but exclude the micro- and meiobenthos, according to the definitions of Horikoshi (1970). The distribution of total biomass of the benthos found by previous workers was summarized by Alton (1974) and is shown in Fig. 14.17. According to a classification made by Horikoshi (1975), the deeper part of the Bering Sea is recognized as poor in benthic biomass, while the shallower parts of the deep-sea portion are considered medium. The southeast region including Bristol Bay is poor to medium, and the extensive northeastern and southwestern region is considered rich.

The estimated benthos of the Bering Sea shelf is 3.3×10^{11} kg, of which only 6.4×10^{10} kg, consisting of small clams, polychaete worms and brittle stars, are considered food for bottom-feeding organisms. Of this only 1.1×10^{10} kg (17%) is accessible to the bottom-feeding fish because of the cold temperatures prevailing in many parts of the Bering Sea (Moiseev, 1964; Maeda, 1972).

Alton (1974) considered the benthos to have a relatively minor role in the economy of the Bering Sea; this is reflected in the relative contribution of benthic organisms to the sustenance of important fish populations. It appears that only 3% of the yield of demersal and semi-demersal fish species, the rock and flat-head sole, are strictly benthophagic forms. Pollock, the main component of the trawl catch, feeds on amphipods, euphausids, copepods, small fish and squid. It has been pointed out, however, by H. Feder (pers. comm., 1976) that the eggs, larvae and young of many benthic organisms spend much of their early life among the pelagic plankton community where they are prey for the juvenile stages of many fish. The king crab (*Paralithodes*) and tanner crab (*Chionoecetes*), in high abundance on the Bering Sea Shelf, are especially good examples of macrobenthic organisms with a pelagic stage in their life history. Neiman (1963) has likened the productivity of the Bering Sea benthos to that of the Barents Sea, where Zenkevitch (1947) estimated benthic production at 25% of the biomass. There is at present no basis for a direct estimate of benthic productivity of the Bering Sea.

Fishes

Historical data on fishes of the Bering Sea come from four concentrated periods of time. First, the exploratory period began with the Vitus Bering Expeditions in 1732–1749, followed by the work of George Wilhelm Stellar on the voyages of Captain James Cook, which formed the first substance of marine biological information for the area; the second, the *Albatross* period, 1890–1900, was stimulated by the fur-seal controversy and increasing fisheries interest (Jordan and Gilbert, 1899); the third period was dominated by Soviet investigations during the late 1930's (Andriashev, 1939); and the fourth investigations, following World War II, by the United States (Ellson et al., 1949), Japan and the Soviet Union (Moiseev, 1964).

The background knowledge of ichthyofauna of the Bering Sea has been carefully reviewed by Andriashev (1939), Schmidt (1950) and Wilimovsky (1974), while Low (1975) has considered the availability of fish catch statistics. Eight families contain nearly three-fourths of the total of 300 fish species supported by the Bering Sea, as shown in Table 14.6.

TABLE 14.6

Proportion of eight predominant families to total species composition of Bering Sea fish fauna (from Wilimovsky, 1974)

Family	% of total fish species
Cottidae	22
Liparidae	15
Stichaeidae	8
Pleuronectidae	8
Zoarcidae	6
Agonidae	5
Scorpaenidae	5
Salmonidae	4
Total	73

The polar Anadyr cold water lying northwest of St. Matthew Island has a strong temperature influence on the distribution of fishes. The bottom morphology further distinguishes the habitat into three main regions: the immediate inshore regions in water depths of less than 50 m, the large eastern shelf region and associated shelf break, and the two ocean basins separated by Bowers Bank and Shirshov Ridge. Superimposed on these are the influences of salinity, circulation and chemical parameters which control the production of food resources.

The cold temperatures ($<0°C$) of the "Anadyr cold" region support only fifteen species, mainly benthic, which are of arctic origin. The benthic fishes consist of two broad groups; one is made up of the wide-ranging species, mostly pleuronectids and lycodids which migrate vertically up and down the banks, and the other consists of very deep water benthic forms represented by several families of a few species each. The pelagic fauna is highly diverse. In the uppermost layers the salmonids predominate; somewhat deeper scorpeaenids, gadids and their relatives and finally the macrouroids are found in the bathypelagic zones. The Bering Sea fish fauna are unique in having remarkably higher proportions of cottids and liparids than any other sea.

Mammals and birds

There are 25 species of marine mammals which inhabit the Bering Sea at some time during the year. A list of these and estimated numbers (Laevastu and Favorite, 1977) is given in Table 14.7. Prorating their occurrence in terms of full-time residence, a conservative estimate of the marine mammal fauna is equivalent to 1.5 million full-time residents having a biomass of about 450 000 metric tons. During residence they consume some 9 to 10 million tons of nekton and benthos annually, converting at least 10% into flesh and redistributing over 1.2 million tons of carrion and feces back into the system (Fay, 1975). Birds are also a significant component of the Bering Sea ecosystem. There are about 40 species of birds considered as "sea birds" that frequent the Bering Sea (Gabrielson and Lincoln, 1959), including 14 of the 23 known species of Alcidae. These birds spend most of their life at sea, usually coming to shore only to breed. In addition, more than 50 species of other birds, not generally regarded as sea birds, frequent the coastal waters, where they utilize marine resources on a seasonal basis. Comparative data on wintering birds (Shuntov, 1966) indicate bird densities of from 27 to 45 birds km^{-2} in the southern portion of the Bering Sea, or about double the density encountered in the Pacific Ocean near Japan, three times greater than in the Sea of Japan and from ten to twenty times greater than that found in both the South China Sea and the Indian Ocean. Summer densities, if comparative information were available, would probably increase the differences. The standing stock of birds is estimated at 27 million, weighing 14 500 metric tons. This biomass is estimated to consume 1 million metric tons of marine fauna and 35 000 metric tons of flora, largely eelgrass, each year.

Phytobenthos

The phytobenthic standing stocks along the coast of Alaska are known to be high (Firth, 1969), but a quantitative estimate in the Bering Sea has not been made. Eelgrass, *Zostera marina* L., has been studied extensively (McRoy, 1966, 1970; McRoy and Goering, 1974; McRoy et al., 1972) in the Bering Sea coastal regions, particularly on the north side of the Alaskan Peninsula in Izembeck Lagoon, where the largest single meadow of eelgrass in the world is thought to exist. This monocotyledon grows as far north as northern Norton Sound on the east and into the Gulf of Anadyr on the west, the density ranging from 62 to 1840 g dry matter m^{-2}. Only near Nova Scotia and in the area of Great South Bay, New York, have comparable densities or productivities been reported. The average eelgrass bed in the Bering Sea is estimated to produce about 1000 g C m^{-2} yr^{-1} to give a total production of 5×10^6 t, of which about' 1×10^6 t are contributed to the Bering Sea as detritus at the end of the growing season.

Eelgrass, being rooted, has the capacity to extract nutrients from the sediment and thus contribute to the production of the entire ocean by a physical process different from those of the floating phytoplankton. It not only extracts nutrients from the sediment, but it appears to contribute phosphorus to the water column through excretions while simultaneously functioning in the transfer of

TABLE 14.7

The marine mammals of the Bering Sea and estimated numbers on a winter–summer basis. From Fay (1974) and Laevastu and Favorite (1977)

Species	Estimated number, eastern Bering Sea	
	summer	winter
Order CARNIVORA		
Suborder FISSIPEDIA		
Ursus maritimus Phipps (polar bear)		
Enhydra lutris L. (sea otter)		
Suborder PINNIPEDIA		
Eumetopias jubata (Schreber) (stellar sea lion)	100 000	50 000
Callorhinus ursinus (L.) (northern fur seal)	900 000	100 000
Odobenus rosmarus (L.) (walrus)	10 000	125 000
Phoca vitulina (L.) (harbor seal)	150 000	200 000
Phoca hispida (Schreber) (ringed seal)	200 000	300 000
Phoca fasciata (Zimmerman) (ribbon seal)	–	–
Erignathus barbatus (Erxleben) (bearded seal)	125 000	250 000
Order CETACEA		
Suborder ODONTOCETI		
Berardius bairoii (Stejneger) (bottled nosed whale)	2000	–
Mesoplodon stejnegeri (True) (Stejneger's beaked whale)		
Ziphius cavirostris (G. Cuvier) (goose beaked whale)		
Physeter catodon (L.) (sperm whale)	20 000	–
Orcinus orca (L.) (killer whale)	800	800
Phocoenoides dalli (True) (dall porpoise)	5000	5000
Phocoena phocoena (L.) (harbor porpoise)		
Monodon monocerus (L.) (narwhal)		
Delphinapterus leucas (Pallas) (belukha)	2000	2000
Suborder MYSTICETI		
Balaena mysticetus (bowhead whale)	2000	2000
Eschrichtius gibbosus (Erxleben) (gray whale)	5000	
Megaptera novaeangliae (Borowski) (humpback whale)	300	
Balaenoptera musculus (L.) (blue whale)		
Balaenoptera borealis (Lesson) (sei whale)		
Balaenoptera physalus (L.) (fin whale)	5000	
Balaenoptera acutorostrata (Lacépède) (minke whale)	2000	

trace metals (copper, lead) from the water to the sediments (McRoy et al., 1972; Barsdate et al., 1974).

CHEMICAL ASPECTS

Much of the interest in the chemistry of the Bering Sea stems from efforts to explain the high biological productivity and effect of wide ranges of climatic conditions on sea-water chemistry. While one can say qualitatively which factors influence productivity, it is not known how these factors interact seasonally nor are there quantitative estimates of their relative importance. The seasonal variations are extreme and possibly trigger major events in the biological cycle which are heavily under climatic control. Ideally a chemist would like to be able to develop a material balance and dynamic model for processes occurring, and thus enlarge his understanding and ability to predict events by knowing the energetics of the system and the rates of reactions or fluxes that are process-limiting. One is far from this level of knowledge in the Bering Sea, as in other highly productive regions of the ocean, but progress is being made, and this section will relate some of the work to the present time.

Chemical inputs

Since most of the sources of water and chemical substances added to the Bering Sea can be identified, it is appropriate to view this sea from the standpoint of inputs. Each input consists of two factors, the amount or rate of supply of water, and the concentration of any given element or ionic species it contains. Both factors must be known for a wide variety of sources. The inputs considered here are rivers, coastal lagoons, circulation, transport of deep water in the Aleutian passes, and contributions from sediment–water mixing on the eastern shelf.

Rivers. Four major rivers enter the Bering Sea — the Anadyr and Kamchatka from the Soviet Union and the Yukon and Kuskokwim from Alaska. Data on the amounts and seasonal distribution of fresh water added by these rivers have been assembled by Roden (1967). The Yukon, the largest entering the Bering Sea, has an annual flow of 16×10^{10} $m^3\ yr^{-1}$. It contains a dissolved solids content of 43×10^{12} g yr^{-1} and a suspended load of 88×10^{12} g yr^{-1}. The total flow of the Yukon River places it 24th in size in the world. Since these rivers are fed largely by ice and snow melt, most of the flow occurs in the summer months in one large seasonal contribution. In mean annual flow, the Yukon has a discharge of only two-thirds that of the Columbia; however, at its peak flow it approaches the size of the Mississippi. The months of May to October account for 90% of the total flow, with 60% occurring in June, July and August. The other rivers have a similar discharge cycle. The area distribution of the river plumes in the Bering Sea is not fully known, but it is clear that the northward movement of water over the shelf transports a large portion of the Anadyr, Yukon and Kuskokwim through the Bering Strait into the Chukchi Sea (Coachman et al., 1975), and much of the sediment load is transported with it (Nelson et al., 1974). Clay minerals derived from the Yukon River have been identified as far north as the western Beaufort Sea (Naidu et al., 1975).

Lagoons. The eastern Bering Sea is bordered with a series of coastal lagoons separated from the open sea by barrier islands and spits. These lagoons change most of their total volume daily through tidal action and contribute water that has been significantly modified by residence. The lagoons harbor dense stands of eelgrass, which attract many kinds of organisms seeking both food and shelter. Results of fairly intensive studies by McRoy (1966, 1970), McRoy and Barsdate (1970), McRoy et al. (1972) and Barsdate et al. (1974) indicate that through biogenic modifications the water flowing from the lagoon is as different from sea water in many important geochemical parameters as is river water entering the sea from terrestrial sources. The eelgrass meadows of Izembeck Lagoon alone produce 166 000 t of particulate carbon (as organic matter containing 7400 t of nitrogen), 1660 t of phosphorus, 3.45 t of copper and 386 t of silica. Only a small portion of the total production appears to be recycled in the lagoon. Avian herbivory accounts for no more than 3% of the total production, and leaching of beached eelgrass less than 0.1%. In addition to the elements exported in the floating eelgrass, it is estimated, based on analysis of exchanging water, that 20 000 t of dissolved organic carbon, 100 t of dissolved inorganic nitrogen and 500 t of dissolved inorganic phosphorus are exported. The phosphorus export can be seen readily in the high phosphorus content of the lagoon effluent plume into the Bering Sea as far as 50 km seaward from the pass to Izembeck Lagoon. At the same time 200 t of copper and 600 t of silica and smaller quantities of lead are lost to the lagoon from Bering Sea water.

Aleutian Island passes. Many studies have been undertaken to estimate the transport of Alaskan Stream water into the Bering Sea. Although questions remain concerning the extent of net transport through the eastern shallow passes, from a chemical point of view it appears that this transport may be the most important. The only point at which transport has an influence on chemical input is when chemical components are added to the receiving system. Inputs of interest to biological productivity are those which add nutrients to the euphotic zone. This occurs in the ocean by some form of vertical transport since bottom waters are enriched in nutrients by that portion of the surface-formed organic matter which settles and decomposes at depth. An upwelling phenomenon was observed by Kelley and Hood (1971a) in the vicinity of Samalga and Amukta Passes, the mixed water being easily identified by its low temperature and oxygen, and high concentrations of nitrate, phosphate, silicate and molecular carbon dioxide. Using the last parameter, Hood and Kelley (1974, 1976) devised a method for direct estimation of the vertical transport rate based on flux of carbon dioxide through the sea surface. These estimates gave an average vertical velocity of 3.1×10^{-3} cm s^{-1} over an area of 3000 km^2 north of Uliaga Island on the north side of Samalga Pass. This rate estimate is probably on the low side, caused by the method of measurement of carbon dioxide transfer rate, which limits the wind effects on water turbulence. The patchy upwelled area extended well beyond the limits of measurement, but the extent was not determined nor was that for Amutka Pass, where the phenomena seem even more intense. Swift and Aagaard (1976) also made an estimate of vertical velocity transport for this area, considering a steady-state oxygen model in which the photosynthetic production of oxygen in the euphotic zone is balanced by an upwelling of low oxygen water. The rate obtained, 7×10^{-3} cm s^{-1}, may be high because the model does not consider the effects of vertical diffusion and oxygen consumption. The actual vertical velocity is probably somewhere between the above two estimates, or about 5×10^{-3} cm s^{-1}. Hood and Kelley (1976) estimated that the enrichment of nitrate to the euphotic zone in this area was at the rate of 4.8×10^6 g km^{-2} day^{-1}. This phenomenon in the passes doubtless provides a large nutrient flux to the rich biota of the Bering Sea.

Sediment–water interactions

The importance of sediment/water exchange as an input and sink for chemical species has been well documented by Berner (1971). While much attention has been paid to the chemistry of *in situ* sediments, few direct measurements have been made of the exchange rates (Berner, 1974; Fanning and Pilson, 1974) between the sediments and water. Reeburgh (1975), using gradient concentration data of Berner (1974) for the Santa Barbara basin, calculated a flux of ammonia from the sediments of 1.5×10^{-4} mol m^{-2} day^{-1}. Assuming like gradients for the Bering Sea, which have not been measured, he calculates a flux of 1.5×10^9 mol day^{-1} contribution of ammonia to the waters of the Bering Sea from this source. This amount, which admittedly is very speculative due to differences in conditions of the two oceanic areas, is about ten-fold that contributed by rivers. This computation clearly points out the possible importance of sediment/water exchange to considerations of chemical budgets in well-mixed systems such as the shallow Bering Sea Shelf.

The vertical distribution of all the carbon dioxide components was measured by Alvarez-Borrego et al. (1972) on a line of stations north of the Aleutian Islands. They found maximum values for the difference between p_{CO_2} (water) and p_{CO_2} (air) of 980 μbar above air at depths of about 800 m. These workers also computed the saturation for calcium carbonate at these stations, and found the water to be supersaturated with respect to both calcite and aragonite down to about 250 m for calcite and about 100 m for aragonite. Below these depths undersaturation occurs to give minimum values of about 55 and 35% for calcite and aragonite, respectively, at 1000 m depth. No water found was warmer than 7°C. They found the ratio of alkalinity to chlorinity to range from 0.128 at the surface to 0.133 near the sea floor. The one GEOSECS station occupied at 53°11′N 177°17′W (Park et al., 1975) showed a higher alkalinity, silica and phosphate and a lower oxygen content from 750 m to the sea floor than for the Aleutian Trench stations at comparable depths. The potential temperature profiles are slightly warmer for the Bering Sea, whereas the salinity comparison shows the same salinity value at the same depth interval.

Nutrients

Probably the most striking feature of the Bering Sea is its remarkably high production under what would casually appear to be a harsh environmental situation. Conversely, however, the system's ability to supply nutrients appears to compensate for the cold temperature, partial annual ice cover and long winter dark periods. Part of this supply results from winter mixing, part from biological remineralization and part from the favorable physical regime that provides sufficient stability in the water column to promote primary production yet not enough to limit some vertical transport from below the euphotic zone during the period of maximum primary productivity. The distribution of nitrogen and phosphorus in the surface waters during the summer clearly indicates that nitrogen, not phosphorus, is generally the principal factor limiting the growth of marine phytoplankton, providing the essential physical structure of the water column is met. The requirement for continuous supply of nutrients is made clear by a simple calculation. The nitrate–nitrogen content during the winter season on the Bering Sea Shelf averages about 25 μg-at. N l^{-1}. This amounts to 10.5 g m^{-2} to a depth of 30 m, which is usually the top of the transition zone during the productive period. Given a ratio of 9.5:1 for carbon to nitrogen in phytoplankton cells (Hattori and Wada, 1974), this amount of nitrogen would produce 100 g C m^{-2} before depletion. Since it is estimated that the production of shelf waters is in excess of 300 g C m^{-2} yr^{-1}, the difference in required nitrogen must come from recycling and convective transport, probably from deeper waters.

Carbon dioxide system

Studies of the carbon dioxide system in the Bering Sea are probably as extensive as those for any part of the world ocean due to the efforts of Ivanenkov (1964), Kelley and Hood (1971a, 1971b, 1971c, 1974), Kelley et al. (1971), Alvarez-Borrego et al. (1972), Gordon et al. (1971, 1973), Park et al. (1975), Kitano (1975), Hood and Kelley (1976) and Hood (1978). A review of much of this work is presented by Park et al. (1974). Most of the problems that have interested these investigators relate to the observations of the extreme variability of the partial pressures of carbon dioxide, both spatially and seasonally in the Bering Sea, thus making it a favorable environment to study the dynamics of the carbon dioxide system. The early observation by Kelley et al. (1971) that partial pressures higher than those in air in the eastern Aleutian Island Passes are associated with values less than air in locations nearby (Fig. 14.18) has led to the discovery of upwelling in these regions. Later Hood and Kelley (1976) reported on an intensive hydrographic and carbon dioxide study in the area near Samalga Pass which showed sea surface p_{CO_2} values as high as 600 μbar (270 μbar higher than air) in upwelled regions. This water was apparently transported without modification from approximately 200 m depth. Hydrographically stratified areas surrounding the upwelling region away from the pass gave values as low as 100 μbar (225 μbar below the overlying air). These studies showed that p_{CO_2} measurements could be used advantageously for detecting upwelled regions as well as highly productive regions in the ocean. This is made possible by the ease of making very precise continuous measurements aboard ship. These authors, and later Hood (1978), who found the lowest p_{CO_2} concentrations ever reported for the open sea (<100 μbar) on the eastern Bering Sea Shelf, have suggested techniques whereby the p_{CO_2} values coupled with carbon dioxide transfer rates through the sea surface may be used to study the rates of some physical and biological processes which influence changes in p_{CO_2} in the sea surface. Vertical transport velocities were estimated in upwelled regions by measuring the flux of carbon dioxide from the surface, and methods for estimating primary and secondary production were suggested for stratified waters where net horizontal convection to the area of measurement can be assumed to be zero and vertical transport is negligible or can be estimated.

Nitrogen

An important consideration in nutrient availability to a growing plant community is the assimilation rate of different forms of nutrients by the community. There are four forms of nitrogen — nitrate, nitrite, ammonia and amino acids — which may be available to plants or are important intermediates to the usable forms. Nitrate in the Bering Sea varies in concentration in the surface waters from zero at some locations in the summer to 25 μg-at. l^{-1} in winter. Nitrite, intermediate between nitrate and ammonia, has a concentration of 0.10 to 0.50 μg-at. l^{-1}. Ammonia varies in concentration between 0.1 and 2.0 μg-at. l^{-1}; the concentration of amino acids, known to be utilized directly in some northern waters (Shell, 1972), has not been determined. Dugdale and Goering (1967) indicate that the ratio of nitrate to total nitrogen uptake varies considerably but tends to decrease directly with latitude. The highest ratio ever recorded was in the Bering Strait (McRoy et al., 1972) where four times more nitrate than ammonia was used preferentially by the phytoplankton. Near the Aleutian Islands a ratio of 2:1 was found in contrast to a

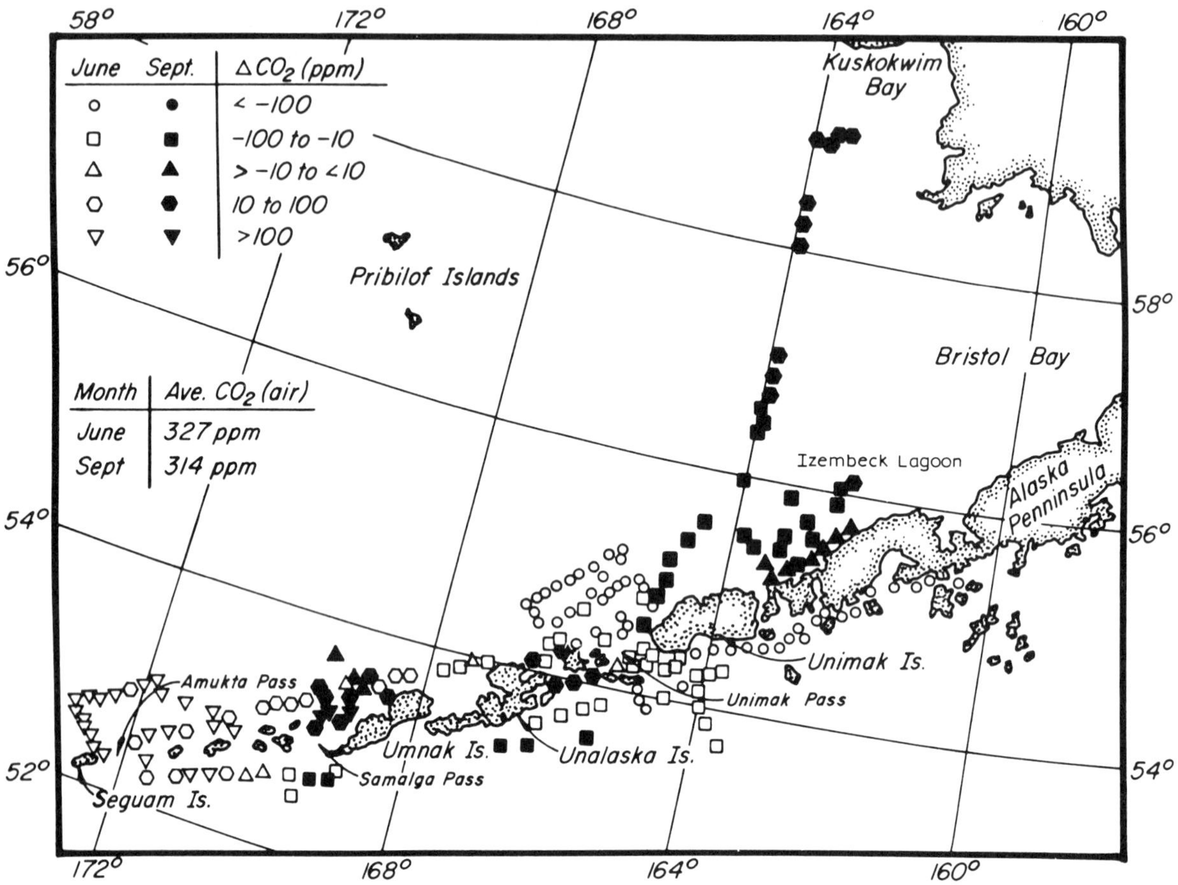

Fig. 14.18. Relative carbon dioxide concentration between air and surface sea water in the vicinity of the Aleutian Island passes. Cruises of the RV *Acona* (Kelley et al., 1971).

southern location at 10°N and 155°W where only ammonia was utilized. This phenomenon is further illustrated by Fig. 14.19, taken from Hattori and Wada (1974). These same authors computed a nitrogen budget for the eastern Bering Sea Shelf based on known inputs and primary production rates, and found that about 50% of the nitrogen required for primary productivity on the shelf must come from recycled nitrogen. This deficit would be even greater (higher percentage recycling) if only summer data were considered. These excessively high requirements for recycling probably indicate that fixed nitrogen from sources other than those presently known is convected to the shelf.

Phosphorus and silica

On the eastern Bering Sea Shelf phosphorus values below 0.2 μg-at. l^{-1}, even when nitrate is zero, are seldom found, indicating that this nutrient is probably not limiting to primary productivity in this region. Suguira (1969, 1974) made extensive studies of both conservative and non-conservative phosphorus in the surface waters of the North Pacific Ocean, including the western Bering Sea. He noted that concentrations of conservative phosphate higher than 1.8 μg-at. l^{-1} are confined to the north of approximately 47°N, decreasing somewhat southward. His explanation for the maintenance of the high surface concentrations is based on vertical mixing around the eastern and central Aleutian Islands. An alternative reason proposed is based on the existence of two types of pycnoclines — one seasonal, the other permanent. The seasonal type, produced by dilution and solar heating, is quite shallow (30–40 m) and is destroyed by winter cooling. Seasonally the highest nutrient concentration in the surface waters occurs in winter, due to

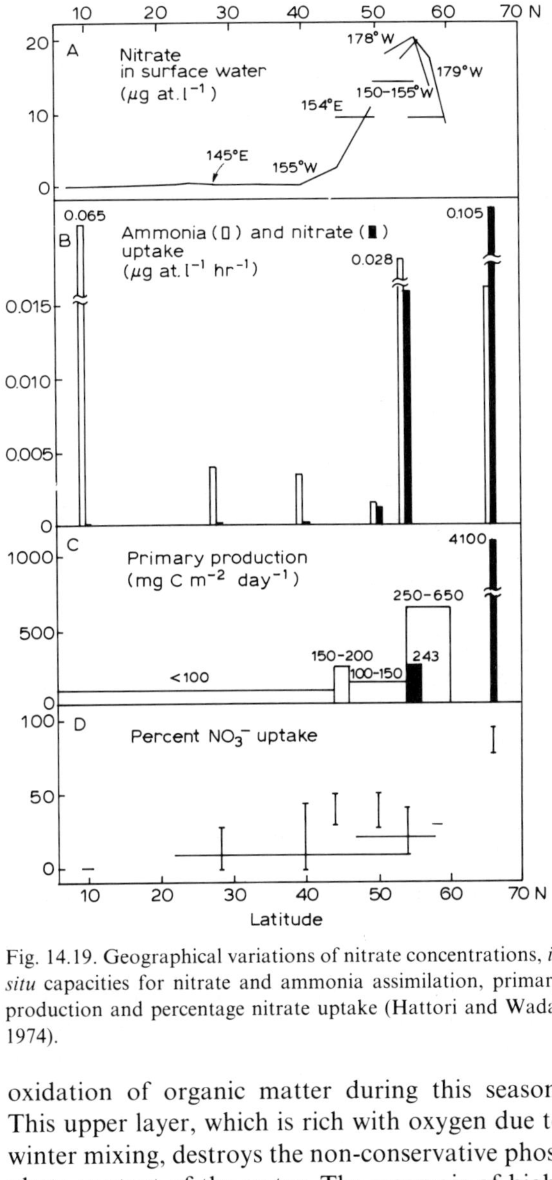

Fig. 14.19. Geographical variations of nitrate concentrations, *in situ* capacities for nitrate and ammonia assimilation, primary production and percentage nitrate uptake (Hattori and Wada, 1974).

oxidation of organic matter during this season. This upper layer, which is rich with oxygen due to winter mixing, destroys the non-conservative phosphate content of the water. The reservoir of high-oxygen, high-conservative-phosphorus water lying between the seasonal pycnoclines then provides the high concentrations of conservative phosphate found.

Silicate in surface waters of the shelf region of the Bering Sea exceeds 10 μg-at. l^{-1} even when nitrate values are near zero, indicating this nutrient is also sufficiently abundant to be non-limiting. Park et al. (1975) in GEOSECS found a silica and oxygen maximum at 3600 m, below which a sharp decrease occurred, reflecting the different water characteristics of the bottom 100 m of deep Bering Sea water. The highest silica concentration observed was 226 μmol kg^{-1}, the highest observed in the North Pacific.

Oxygen

The oxygen concentration in the surface waters of the eastern shelf region during May and June ranges between 8.15 and 10.25 ml l^{-1}, which is supersaturated with respect to air by between 6 and 34%. Moving away from the shelf, the surface water concentration of oxygen drops to near saturation (about 7.65 ml l^{-1} for water at 2.5°C and a salinity of 32.5‰). An oxygen profile by both continuous and discrete sample methods on the GEOSECS Bering Sea Station is shown in Fig. 14.20. Oxygen is seen to reach minimum values of less than 1 ml l^{-1} at depths between 600 and 1800 m, which is a characteristic pattern for the sea off Central America and Peru and in the Arabian Sea and Bay of Bengal in the Indian Ocean. All these areas are characterized by high productivity and a well-defined permanent pycnocline.

Organic matter

Data on the organic matter content of the Bering Sea is largely confined to studies by Loder (1971) and Nishizawa and Tsunogai (1974). Loder examined waters at a series of stations north of Unimak

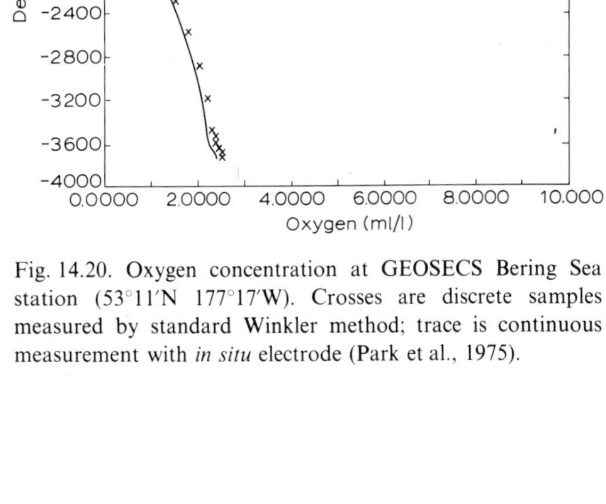

Fig. 14.20. Oxygen concentration at GEOSECS Bering Sea station (53°11′N 177°17′W). Crosses are discrete samples measured by standard Winkler method; trace is continuous measurement with *in situ* electrode (Park et al., 1975).

Pass in the southeastern Bering Sea. The dissolved organic carbon (DOC) was found to vary widely in the top 50 m of the water column, with concentrations between 1.0 and 1.85 mg C l^{-1}. Particulate organic carbon content appeared to correlate well with the light absorbancy of the water, as shown in Fig. 14.21. Nishizawa (1975) observed that particulate carbon in the deeper layers of the Bering Sea never got below 20 μg C l^{-1} and often obtained values above 50 μg C l^{-1}. The general pattern of vertical distribution has much in common with other areas of the Pacific except that the pattern is more easily recognized in the Bering Sea because of the high concentration of particles. A distinct minimum of particulate carbon just below the euphotic zone was observable here where elsewhere it was barely recognizable, if at all. The Bering Sea is apparently one of the most particle-rich areas of the world oceans. Tsunogai et al. (1974), in examining the particulate matter from many stations in the Bering Sea, found that about 33% consisted of ash, which is high in calcium (1–2%), between the depths of 100 and 200 m as compared to above (0.2%) and below (0.5%) this layer. Samples in the Pacific south of the trench were about 20% ash and did not show the calcium maximum between 100 and 200 m but were consistently about 0.4% calcium from top to the bottom depth at which samples were taken (1500 m).

Trace metals

Barsdate and Nebert (1971) and Barsdate et al. (1974) have studied the biologically mediated cycling of copper, lead and zinc in the Bering Sea by anodic stripping voltametry, coupled with electrodialysis and radioactive tracer techniques. Their observations showed that large dissolved organic metal complexes too large to pass through a dialysis membrane are almost absent both in lagoon waters and the open Bering Sea. However, smaller complexes not measurable by existing techniques are expected to be abundant. Experiments in upwelled water in Samalga Pass and at a nearby station without upwelled water definitely showed that a chelating agent (EDTA) must be present for primary production to proceed when the metals are added to upwelled water; they are not necessary in

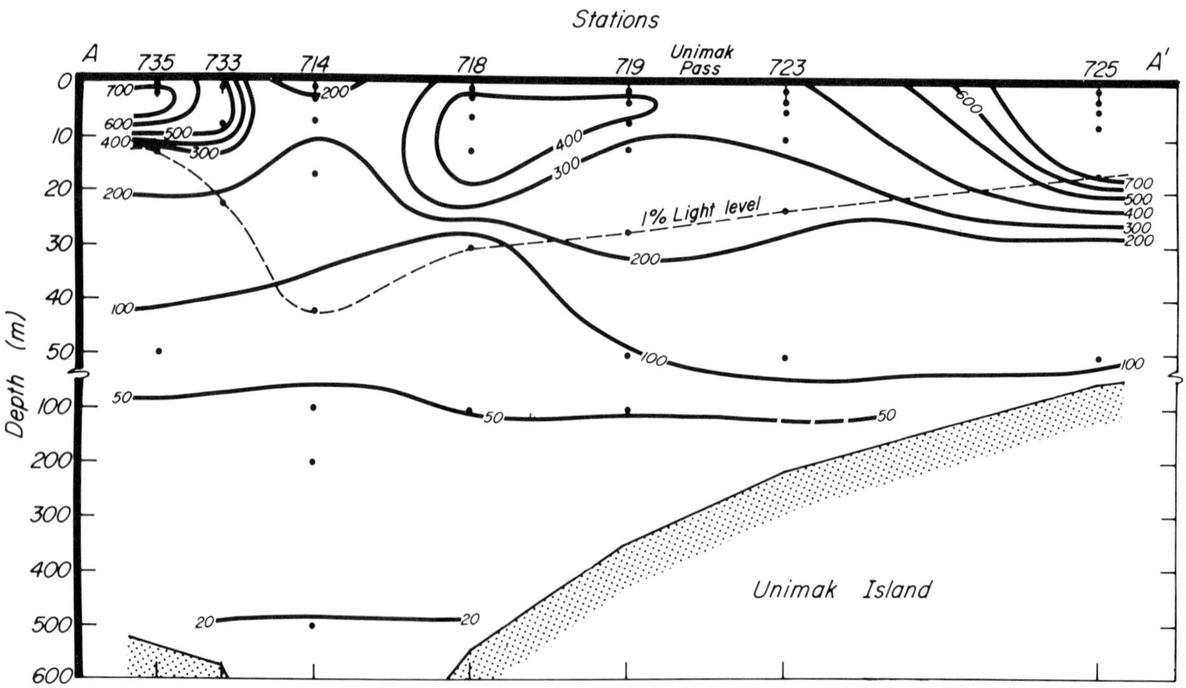

Fig. 14.21. Distribution of particulate organic carbon (POC) (μg C l^{-1}) along east–west section north of Unimak Pass. The depth of the euphotic zone is shown by the dotted line representing the 1% light level (Loder, 1971).

non-upwelled samples. In the case of additions of high concentrations of zinc, copper (20 $\mu g\ l^{-1}$) and lead (5 $\mu g\ l^{-1}$), it required the addition of EDTA (10^{-6} mol l^{-1}) for primary production to occur in either situation.

Nelson et al. (1974) have reported that modern beaches of the southern Seward Peninsula contain anomalies as high as 1.3 p.p.m. mercury, 40 p.p.m. chromium and 75 p.p.m. copper. The normal background content of mercury (0.03–0.05 p.p.m.) immediately offshore from the mercury-rich beaches suggests that the offshore effects of sediment dispersal of naturally and artificially introduced mercury has been minimal in the bottom sediments.

CONCLUDING REMARKS

The high-latitude regions of the world ocean are generally characterized by extremes in light and temperature and a phenological spectrum of biological events that are dependent on seasonal ice and associated low-temperature conditions. Life processes at all trophic levels are greatly complicated by the cold-stressed environment. In addition, the presence of ice limits conventional access to the regions, resulting in a poorly understood ecological domain in the ocean medium.

In the Bering Sea it is apparent that ice conditions exert important influences on the birds and mammals, acting at a gross level as a seasonally variable partition between species which normally reside in ice-covered water and those which are excluded when ice is present. In the spring of the year the primary productivity which occurs in the sea ice is vital to grazing organisms and possibly serves as a source of detritus for benthic biota underlying the ice floes. An understanding of the dynamics which drive this finely tuned system requires detailed information about both the ice regime and the underlying water masses throughout the entire year. The difficult logistical problems of such studies have so far been approached only by the occasional use of icebreakers and of conventional vessels which operate near the ice edge under most favorable conditions.

Until the last few years, with the exception of the U.S.S.R. expeditions of the 1950–1960 era and the multiple-year summer studies by the Faculty of Fisheries of Hokkaido University at Hokadate, Japan, most of the oceanographic research in the Bering Sea has been sporadic. There is a lack of programs planned and executed to respond to specific oceanographic questions, and only a few individual investigators or institutions have maintained a sustained effort in this region of the ocean. The reason for this is clear. The Bering Sea is far from most bases of operation, and the environment is severely harsh for oceanographic study.

More recently the situation has changed rapidly due to a new surge of interest in high-latitude research brought on primarily by the need for new sources of hydrocarbons to meet the critical energy crisis, and the immense pressure on these regions for an increase in marine protein for the increasing world population. These two major areas of need are frequently placed in juxtaposition and to many appear incompatible. The Bering Sea is rich in biological resources and very promising for hydrocarbon extraction. Oceanographic understanding of greater detail than has previously been available for any region in the world is necessary to support any intelligent program which would allow the compatible coexistence of these two extraction industries.

This review comes at the beginning of the most intensive research ever undertaken in the Bering Sea and unfortunately suffers from a lack of material that is as yet unpublished. Despite this inadequacy, there are many observations that indicate the advance that is under way. Especially noteworthy is the recent evidence for the existence of an inner and outer oceanic front in outer Bristol Bay separated by a 50 to 100 km band in which the horizontal salinity gradient is nearly zero; this indicates the existence of a large region (about 25 000 km^2) which is essentially isolated from the rest of the ocean except for diffusive transfer. This region, known to be the most productive of both plants and animals in the Bering Sea, provides a unique opportunity for the study of the environmental events which lead to a sequence of biological events in support of one of the world's largest fisheries. The next review will be eagerly awaited, to learn of the outcome of these and other studies of the Bering Sea, truly one of the world's most interesting ecosystems.

ACKNOWLEDGEMENTS

The author wishes to acknowledge the support of Seoul National University, Department of Oceanography, Seoul, Korea, where he was a visiting scientist for six months in the fall and winter of 1977–1978 and to the PROBES project supported by the Office of Polar Programs of the National Science Foundation. The help of Dr. L.K. Coachman, who allowed the use of some unpublished data, Dr. Shim Jae Heung, who reviewed the biological part of the manuscript, and my wife, Betty, who typed the manuscript is much appreciated.

REFERENCES

Alton, M.S., 1974. Bering Sea benthos as a food resource for demersal fish populations. In: D.W. Hood and E.J. Kelley (Editors), *Oceanography of the Bering Sea. Inst. Mar. Sci., Univ. Alaska, Occas. Publ.*, No. 2: 257–277.

Alvarez-Borrego, S., Gordon, L.I., Jones, L.B., Park, P.K. and Pytkowitz, R.M., 1972. Oxygen–carbon dioxide–nutrients relationships in the southeastern region of the Bering Sea. *J. Oceanogr. Soc. Jap.*, 28: 71–93.

Andriashev, A.P., 1939. *Ocherk Zoogeografii i Proiskhozhdeniya fauny ryb Beringova morya i sopredeinykh vod.* Izd. Leningrad Gos. University, 187 pp.

Armstrong, T. and Roberts, B., 1956. Illustrated ice glossary. *Polar Rec.*, 8: 4–12.

Arsen'ev, V.S., 1967. *Currents and Water Masses of the Bering Sea.* Nauka, Moscow, 135 pp. (in Russian with English summary). [Transl., 1968. Natl. Mar. Fish. Serv., Northwest Fish. Center, Seattle, Wash.]

Barsdate, R.J. and Nebert, M., 1971. Biologically mediated trace metal cycles. *Inst. Mar. Sci., Univ. Alaska, Rep.*, No. 71–2: 58–101.

Barsdate, R.J., Nebert, M. and McRoy, C.P., 1974. Lagoon contributions to sediments and water of the Bering Sea. In: D.W. Hood and E.J. Kelley (Editors), *Oceanography of the Bering Sea. Inst. Mar. Sci., Univ. Alaska, Occas. Publ.*, No. 2: 553–576.

Batalin, A.M., 1964. On the water exchange between the Bering Sea and Pacific Ocean. *Tr. VNIRO*, 49: 7–16 (in Russian). [Transl., 1968. In: *Soviet Fisheries Investigations in the Northwest Pacific, 2.* Natl. Tech. Inf. Serv., Springfield, Va., TT-67-51024, pp. 1–12.]

Belyayev, G.M., 1960. Quantitative distribution of bottom fauna in the northwestern part of the Bering Sea. *Tr. Inst. Okeanol. Akad. Nauk S.S.S.R.*, 34: 85–103 (in Russian). [Transl., 1968. U.S. Naval Oceanogr. Office, No. 337.]

Berner, R.A., 1971. *Principles of Chemical Sedimentation.* McGraw-Hill, New York, N.Y., 240 pp.

Berner, R.A., 1974. Kinetic models for the early diagenesis of nitrogen, sulfur, phosphorus and silicon in anoxic marine sediments. In: E.D. Goldberg (Editor), *The Sea, 5.* Wiley, New York, N.Y., pp. 427–450.

Bruce, H.E., 1976. A study plan for the Alaskan Continental Shelf. In: D.W. Hood and D.C. Burrell (Editors), *Assessment of the Arctic Marine Environment. Inst. Mar. Sci., Univ. Alaska, Occas. Publ.*, No. 4: 61–80.

Churkin, M., 1972. Western boundary of the North American Continental Plate in Asia. *Geol. Soc. Am. Bull.*, 83: 1027–1036.

Coachman, L.K., 1975. Recent studies in Bering Sea Circulation. In: D.W. Hood and Y. Takenouti (Editors), *Bering Sea Oceanography: An Update. Inst. Mar. Sci., Univ. Alaska, Rep.*, No. 75–2: 51–72.

Coachman, L.K., 1978. Circulation on southeastern Bering Sea shelf. In: *PROBES: Progress Report for 1978.* Inst. Mar. Sci., Univ. Alaska, 32 pp.

Coachman, L.K. and Aagaard, K., 1966. On the water exchange through the Bering Strait. *Limnol. Oceanogr.*, 11: 44–59.

Coachman, L.K., Aagaard, K. and Tripp, R.B., 1975. *Bering Strait: The Regional Physical Oceanography.* University of Washington Press, Seattle, Wash., 172 pp.

Cooney, R.T., 1975. Secondary productivity. In: D.W. Hood and Y. Takenouti (Editors), *Bering Sea Oceanography: An Update. Inst. Mar. Sci., Univ. Alaska, Rep.*, No. 75–2: 117–121.

Coulter, H.W. et al., 1965. *Map Showing Extent of Glaciation in Alaska.* U.S. Geol. Survey, Miscellaneous Geologic Investigations, Map I — 415.

Dall, N.H., 1881. Hydrologie des Bering-Meeres und der benachbarten Gewässer. *Petrol. Geogr. Mitt.*, 27: 361–380: 443–448.

Department of Commerce, 1977. Department of Commerce Arctic Research. *Arct. Bull.*, 2: 209.

Dietz, R.S., Carsola, A.J., Buffington, E.C. and Shipek, C.J., 1964. Sediments and topography of the Alaska shelves. In: R.L. Miller (Editor), *Papers in Marine Geology.* MacMillan, New York, N.Y., pp. 241–256.

Dobrovol'skii, A.D. and Arsen'ev, V.S., 1959. On the question of currents in the Bering Sea. *Probl. Severa*, 3: 3–9 (in Russian). [Transl., Natl. Res. Council Can., Ottawa, Ont.]

Dugdale, R.C. and Goering, J.J., 1967. Uptake of new and regenerated forms of nitrogen in primary production. *Limnol. Oceanogr.*, 12: 196–206.

Ellson, J.C., Knake, B. and Dassow, J., 1949. *Report of Alaska exploratory fishing expedition, fall of 1948 to northern Bering Sea. U.S. Fish Wildl. Serv. Fish Leafl.*, 342: 1–25.

Fanning, K.A. and Pilson, M.E.O., 1974. The diffusion of dissolved silica out of deep-sea sediments. *J. Geophys. Res.*, 79: 1293–1297.

Favorite, F., 1967. The Alaskan Stream. *Bull. Int. North Pac. Comm.*, 21: 1–20.

Favorite, F., 1974. Flow into the Bering Sea through Aleutian island passes. In: D.W. Hood and E.J. Kelley (Editors), *Oceanography of the Bering Sea. Inst. Mar. Sci., Univ. Alaska, Occas. Publ.*, No. 2: 3–37.

Fay, F.H., 1974. The role of ice in the ecology of marine mammals of the Bering Sea. In: D.W. Hood and E.J. Kelley (Editors), *Oceanography of the Bering Sea. Inst. Mar. Sci., Univ. Alaska, Occas. Publ.*, No. 2: 383–399.

Fay, F.H., 1975. Mammals and birds in PROBES. In: E.J. Kelley and D.W. Hood (Editors), *Publ. Info. Bull. 74-1.* Institute of Marine Science, University of Alaska, Fairbanks, Alaska, pp. 23–28.

Firth, F.E., 1969. Sea weeds. In: *The Encyclopedia of Marine Resources.* Van Nostrand Reinhold, New York, N.Y., p. 628.

Gabrielson, I.N. and Lincoln, F.C., 1959. *The Birds of Alaska.* Wildlife Management Institute, Washington, D.C., 922 pp.

Galt, J.A., 1975. Discussion of R.D. Muench paper. In: D.W. Hood and Y. Takenouti (Editors), *Bering Sea Oceanography: An Update. Inst. Mar. Sci., Univ. Alaska, Rep.*, No. 75–2: 79–80.

Gershanovich, D.E., 1967. Late Quaternary sediments of the Bering Sea and the Gulf of Alaska. In: D.M. Hopkins (Editor), *The Bering Land Bridge.* Stanford University Press, Stanford, Calif., pp. 32–46.

Gershanovich, D.E., Fadeev, N.C., Liubimova, T.G., Moiseev, P.A. and Natarov, V.V., 1974. Principal results of Soviet oceanographic investigations in the Bering Sea. In: D.W. Hood and E.J. Kelley (Editors), *Oceanography of the Bering Sea. Inst. Mar. Sci., Univ. Alaska, Occas. Publ.*, No. 2: 363–370.

Goodman, J.R., Lincoln, J.A., Thompson, T.C. and Zeusler, F.A., 1942. Physical and chemical investigations: Bering Sea, Bering Strait and Chukchi Sea during summers of 1937 and 1938. *Univ. Wash. Publ. Oceanogr.*, 3: 105–169.

Gordon, L.I., Park, P.K., Hager, I.W. and Parsons, T.R., 1971. Carbon dioxide partial pressures in the North Pacific surface waters — time variations. *J. Oceanogr. Soc. Jap.*, 27: 81–90.

Gordon, L.I., Kelley, J.J. and Park, P.K., 1973. Carbon dioxide partial pressures in the North Pacific surface waters. 2: General late summer distribution. *Mar. Chem.*, 1: 191–198.

Grow, J.A. and Atwater, R., 1970. Mid-Tertiary tectonic transition in the Aleutian Arc. *Geol. Soc. Am. Bull.*, 81: 3715–3722.

Hattori, A. and Wada, E., 1974. Assimilation and oxidation–reduction of inorganic nitrogen in the North Pacific Ocean. In: D.W. Hood and E.J. Kelley (Editors), *Oceanography of the Bering Sea. Inst. Mar. Sci. Univ. Alaska, Occas. Publ.*, No. 2: 149–162.

Hebard, J.F., 1961. Currents in Southeastern Bering Sea. *Bull. Int. North Pac. Fish. Comm.*, 5: 9–16.

Hood, D.W., 1978. pCO_2 Anomalies on Southeastern Bering Sea Shelf. In: *PROBES: Progress Report for 1976–1977. Inst. Mar. Sci., Univ. Alaska, Rep.*, No. 78-x.

Hood, D.W. and Kelley, E.J. (Editors), 1974. *Oceanography of the Bering Sea with Emphasis on Renewable Resources. Inst. Mar. Sci., Univ. Alaska, Occas. Publ.*, No. 2: 623 pp.

Hood, D.W. and Kelley, J.J., 1975. Upwelling phenomena in the eastern Aleutian passes with a suggested method for measuring material transport. In: D.W. Hood and Y. Takenouti (Editors), *Bering Sea Oceanography: An Update. Inst. Mar. Sci., Univ. Alaska, Rep.*, No. 75–2: 245–277.

Hood, D.W. and Kelley, J.J., 1976. Evaluation of mean vertical transports in an upwelling system by CO_2 measurements. *Mar. Sci. Comm.*, 2: 387–411.

Hopkins, D.M., 1959. Cenozoic history of the Bering land bridge. *Science*, 129: 1519–1528.

Hopkins, D.M., 1967. *The Bering Land Bridge.* Stanford University Press, Stanford, Calif., 495 pp.

Hopkins, D.M. and Scholl, D.W., 1970. Tectonic development of Beringia, late Mesozoic to Holocene. *Am. Assoc. Pet. Geol. Bull.*, 54: 2486 (abstract).

Horikoshi, M., 1970. Quantitative studies on the smaller macrobenthos inhabiting various topographical environments around Sagami Bank in the deep sea system of Sagami Bay. *J. Oceanogr. Soc. Jap.*, 26: 37–60.

Horikoshi, M., 1975. Benthological studies in the Bering Sea. In: D.W. Hood and Y. Takenouti (Editors), *Bering Sea Oceanography: An Update. Inst. Mar. Sci., Univ. Alaska, Rep.*, No. 75–2: 123–129.

Hughes, F.W., Coachman, L.K. and Aagaard, K., 1974. Circulation, transport and water exchange in the western Bering Sea. In: D.W. Hood and E.J. Kelley (Editors), *Oceanography of the Bering Sea. Inst. Mar. Sci., Univ. Alaska, Occas. Publ.*, No. 2: 59–98.

Ivanenkov, V.I., 1964. *Hydrochemistry of the Bering Sea.* Akad. Nauk. S.S.S.R., Nauka, Moscow, 137 pp. [As reviewed in Park et al., 1974.]

Jordan, D.S. and Gilbert, C.H., 1899. The fishes of Bering Sea. In: *The Fur-Seals and Fur-Seal Islands of the North Pacific Ocean. Rep. Fur-Seal Inv., 1896–1897*, Part 3, pp. 433–492.

Karohji, K., 1972. Studies on pigmented microorganisms in the Bering Sea and western and northern subarctic regions of the North Pacific Ocean in summer. In: Y. Takenouti et al. (Editors), *Biological Oceanography of the Northern North Pacific Ocean.* Idemitsu Shoten, Tokyo, pp. 99–115.

Kelley, J.J. and Hood, D.W., 1971a. Carbon dioxide in the surface waters of the ice covered Bering Sea. *Nature*, 222: 37–39.

Kelley, J.J. and Hood, D.W., 1971b. Carbon dioxide in the Pacific Ocean and Bering Sea: Upwelling and mixing. *J. Geophys. Res.*, 76: 745–752.

Kelley, J.J. and Hood, D.W., 1971c. *Carbon Dioxide in the Surface Waters of Barents Sea. Southward Expedition — 1970. Inst. Mar. Sci., Univ. Alaska, Rep.*, No. 71–20: 9 pp.

Kelley, J.J. and Hood, D.W., 1974. Upwelling in the Bering Sea near the Aleutian Islands. *Tethys*, 6: 149–156.

Kelley, J.J., Longerich, L.L. and Hood, D.W., 1971. Effect of upwelling, mixing and high primary productivity on CO_2 concentrations in the surface waters of the Bering Sea. *J. Geophys. Res.*, 76: 8687–8693.

Kitano, Y., 1975. Carbonate equilibrium. In: D.W. Hood and Y. Takenouti (Editors), *Bering Sea Oceanography: An Update. Inst. Mar. Sci., Univ. Alaska, Rep.*, No. 75–2: 199–203.

Konishi, R. and Saito, M., 1974. The relationship between ice and weather conditions in the eastern Bering Sea. In: D.W. Hood and E.J. Kelley (Editors), *Oceanography of the Bering Sea. Inst. Mar. Sci., Univ. Alaska, Occas. Publ.*, No. 2: 425–450.

Koto, H. and Fujii, R., 1958. Structure of waters in the Bering Sea and the Aleutian region. *Bull. Fac. Fish., Hokkaido Univ.*, 9: 149–170.

Kummer, J.J. and Creager, J.S., 1971. Marine geology and Cenozoic history of the Gulf of Anadyr. *Mar. Geol.*, 10: 257–280.

Laevastu, T. and Favorite, F., 1977. *Preliminary Report on Dynamical Numerical Marine Ecosystem Model (Dynumes*

II) for Eastern Bering Sea. NWPFC, Provisional Report, 79 pp.

Leonov, A.K., 1960. *Water Balance of the Bering Sea.* Nauka, Moscow. 584 pp. (in Russian). [Transl. available Natl. Tech. Inf. Serv., Springfield, Va., AD627508 and AD689680.]

Lisitsyn, A.P., 1966. *Recent Sedimentation in the Bering Sea.* Inst. Okeanol. Akad. Nauk, Leningrad, 614 pp. (in Russian). [Transl. by Israel Program for Scientific Translations, available from U.S. Dep. Commerce, Clearinghouse for Fed. Sci. Tech. Info., 1969.]

Loder, T.C., 1971. *Distribution of Dissolved and Particulate Organic Carbon in Alaskan, Polar, Sub-polar and Estuarine Waters.* Thesis, University of Alaska, Fairbanks, Alaska.

Low, L.L., 1975. Available fisheries statistics for the Bering Sea. In: D.W. Hood and Y. Takenouti (Editors), *Bering Sea Oceanography: An Update. Inst. Mar. Sci., Univ. Alaska, Rep.*, No. 75–2: 139–151.

McAllister, C.D., 1972. Estimates of the transfer of primary production to secondary production at ocean station P. In: Y. Takenouti (Editor), *Biological Oceanography of the North Pacific Ocean.* Idemitsu Shoten, Tokyo, pp. 575–579.

McClain, E.P., 1974. Environmental earth satellites for oceanographic–meteorological studies of the Bering Sea. In: D.W. Hood and E.J. Kelley (Editors), *Oceanography of the Bering Sea. Inst. Mar. Sci., Univ. Alaska, Occas. Publ.*, No. 2: 579–593.

McManus, D.A. and Smyth, C.S., 1970. Turbid bottom water on the continental shelf of Northern Bering Sea. *J. Sediment. Pet.*, 40: 869–877.

McRoy, C.P., 1966. *Standing Stock and Ecology of Eelgrass* (Zostera marina *L.*) *in Izembeck Lagoon, Alaska.* Thesis, University of Washington, Seattle, Wash., 138 pp.

McRoy, C.P., 1970. Standing stocks and related features of eelgrass populations in Alaska. *J. Fish. Res. Board Can.*, 27: 1811–1821.

McRoy, C.P. and Barsdate, R.J., 1970. Phosphate absorption in eelgrass. *Limnol. Oceanogr.*, 15: 6–13.

McRoy, C.P. and Goering, J.J., 1974. Nutrient transfer between the seagrass *Zostera marina* and its epiphytes. *Nature*, 248: 173–174.

McRoy, C.P. and Goering, J.J., 1976. Annual budget of primary production in the Bering Sea. *Mar. Sci. Comm.*, 2: 255–267.

McRoy, C.P., Barsdate, R.J. and Nebert, M., 1972. Phosphorus cycling in an eelgrass (*Zostera marina* L.) ecosystem. *Limnol. Oceanogr.*, 17: 58–67.

Maeda, T., 1972. On the fishing conditions of Alaska pollock in the Eastern Bering Sea in 1969 to 1970. *Bull. Jap. Soc. Sci. Fish.*, 38: 685–691.

Makarov, V.V., 1937. Data on the quantitative calculations of benthic fauna of the northern Bering Sea and southern Chakotski Sea. *Issled. Morey S.S.S.R.*, 25 (in Russian). [As reviewed by Neiman, 1963.]

Marumo, R. and Minoda, T., 1976. Phytoplankton in the subarctic and tropical waters in the Pacific. *Mar. Sci. Comm.*, 2: 269–282.

Minoda, T., 1971. Pelagic Copepoda in the Bering Sea and the northwestern North Pacific with special reference to their vertical distribution. *Mem. Fac. Fish, Hokkaido Univ.*, 18: 1–74.

Minoda, T. and Marumo, R., 1975. Regional characteristics of distribution of phyto- and zooplankton in the Eastern Bering Sea and Chukchi Sea in June–July, 1972. In: D.W. Hood and Y. Takenouti (Editors), *Bering Sea Oceanography: An Update. Inst. Mar. Sci., Univ. Alaska, Rep.*, No. 75–2: 83–95.

Moiseev, P.A., 1964. Some results of investigations carried out by the Bering Sea Research Expedition. *Tr., VNIRO*, 53: 7–29 (in Russian). [Transl. by Israel Program for Scientific Translations, available from U.S. Dep. Commerce, Clearinghouse for Fed. Sci. Tech. Inf., 1968.]

Motoda, S. and Minoda, T., 1974. Plankton of the Bering Sea. In D.W. Hood and E.J. Kelley (Editors), *Oceanography of the Bering Sea. Inst. Mar. Sci., Univ. Alaska, Occas. Publ.*, No. 2: 207–241.

Muench, R.D., 1975. Satellite observations of Bering Sea ice. In: D.W. Hood and Y. Takenouti (Editors), *Bering Sea Oceanography: An Update. Inst. Mar. Sci., Univ. Alaska, Rep.*, No. 75–2: 191–192.

Muench, R.D. and Ahinas, K., 1976. Ice movement and distribution in the Bering Sea from March to June 1974. *J. Geophys. Res.*, 81: 4467–4476.

Muench, R.D. and Charnell, R.L., 1977. Observations of medium-scale features along the seasonal ice edge in the Bering Sea. *J. Phys. Oceanogr.*, 7(4): 602–606.

Muench, R.D., Elsner, R. and McRoy, C.P., 1976. PROBES: A study of processes and resources on the Bering Sea shelf. In: D.W. Hood and D.C. Burrell (Editors), *Assessment of the Arctic Marine Environment: Selected Topics. Inst. Mar. Sci., Univ. Alaska, Occas. Publ.*, No. 4: 81–85.

Naidu, A.S., Mowatt, T.C., Hawkins, D.E. and Hood, D.W., 1975. Clay mineralogy and geochemistry of some Arctic Ocean sediments: significance on paleoclimate interpretation. In: C. Weller and S.A. Bowling (Editors), *Climate of the Arctic.* Geophysical Institute, University of Alaska, Fairbanks, Alaska, pp. 59–67.

Natarov, V.N., 1963. Water masses and currents of the Bering Sea. *Tr. VNIRO*, 48: 111–133 (in Russian). [Transl. available from Natl. Tech. Inf. Serv., Springfield, Va., TT67-51204.]

Neiman, A.A., 1960. Quantitative distribution of benthos in the eastern Bering Sea. *Zool. Zh.*, 39: 1281–1292 (in Russian). [Transl., 1968. U.S. Naval Oceanogr. Office, No. 402.]

Neiman, A.A., 1963. Quantitative distribution of benthos on the shelf and upper continental slope in the eastern part of the Bering Sea. In: *Soviet Fisheries Investigations in the North-east Pacific*, 1, pp. 143–217 (in Russian). [Transl. by Israel Program for Scientific Translation, available from U.S. Dep. Commerce, Clearinghouse for Fed. Sci. Tech. Inf., 1968.]

Nelson, C.H. and Hopkins, D.M., 1972. Sedimentary processes and distribution of gold in northern Bering Sea. *U.S. Geol. Surv. Prof. Pap.*, 689: 27 pp.

Nelson, C.H., Hopkins, D.M. and Scholl, D.W., 1974. Cenozoic sedimentary and tectonic history of the Bering Sea. In: D.W. Hood and E.J. Kelley (Editors), *Oceanography of the Bering Sea. Inst. Mar. Sci., Univ. Alaska, Occas. Publ.*, No. 2: 485–516.

Nishizawa, S., 1975. Problems concerning particulate material. In: D.W. Hood and Y. Takenouti (Editors), *Bering Sea*

Oceanography: An Update. Inst. Mar. Sci., Univ. Alaska, Rep., No. 75–2: 205.

Nishizawa, S. and Tsunogai, S., 1974. Dynamics of particulate material in the ocean. In: D.W. Hood and E.J. Kelley (Editors), *Oceanography of the Bering Sea. Inst. Mar. Sci., Univ. Alaska, Occas. Publ.*, No. 2: 173–174.

Ohtani, K., 1965. On the Alaskan Stream in summer. *Bull. Fac. Fish., Hokkaido Univ.*, 15: 260–273 (in Japanese).

Ohtani, K., 1969. On the oceanographic structure and the ice formation on the continental shelf in the eastern Bering Sea. *Bull. Fac. Fish., Hokkaido Univ.*, 20: 94–117 (in Japanese).

Ohtani, K., 1973. Oceanographic structure in the Bering Sea. *Mem. Fac. Fish., Hokkaido Univ.*, 21: 65–106 (in Japanese).

Ohtani, K., Akibi, Y. and Takenouti, Y., 1972. Formation of western subarctic water in the Bering Sea. In: Y. Takenouti et al. (Editors), *Biological Oceanography of the Northern North Pacific Ocean.* Idemitsu Shoten, Tokyo, pp. 31–44.

Park, P.K., Gordon, L.I. and Alvarez-Borrego, S., 1974. The carbon dioxide system of the Bering Sea. In: D.W. Hood and E.J. Kelley (Editors), *Oceanography of the Bering Sea. Inst. Mar. Sci., Univ. Alaska, Occas. Publ.*, No. 2: 107–147.

Park, P.K., Broecker, W.S., Takahashi, T. and Reeburgh, W.S., 1975. GEOSECS Bering Sea station. In: D.W. Hood and Y. Takenouti (Editors), *Bering Sea Oceanography: An Update. Inst. Mar. Sci., Univ. Alaska, Rep.*, No. 75–2: 207–238.

PROBES, 1974. *Processes and Resources of the Bering Sea Shelf. Inst. Mar. Sci., Univ. Alaska, Rep.*, No. 75–1: 70 pp.

Pruter, A.T., 1973. *Development and Present Status of Bottom Fish Resources in the Bering Sea.* FAO Technical Conference on Fishery Management and Development, Vancouver, B.C., FI: FMD 673/s-36.

Ratmanov, C.E., 1937. On the hydrology of the Bering and Chukchi Seas. *Issled. Morey S.S.S.R. Gidrometeoizdat (Leningrad)*, 25: 10–118 (in Russian). [Transl. available from Univ. Wash. Libr., Seattle, Wash.]

Reeburgh, W.S., 1975. Recent studies of the Bering Sea. In: D.W. Hood and Y. Takenouti (Editors), *Bering Sea Oceanography: An Update. Inst. Mar. Sci., Univ. Alaska, Rep.*, No. 75–2: 289–292.

Roden, G.I., 1967. On river discharge into the Northeastern Pacific Ocean and the Bering Sea. *J. Geophys. Res.*, 72: 5613–5629.

Sainsbury, C.L., 1967. Quaternary geology of western Seward Peninsula, Alaska. In: D.M. Hopkins (Editor), *The Bering Land Bridge.* Stanford University Press, Stanford, Calif., pp. 121–143.

Schmidt, P., 1950. *Fishes of the Okhotsk Sea.* Akad. Nauk S.S.S.R., Moscow, 370 pp. (in Russian). [Transl., 1965. Israel Program for Scientific Translation, 392 pp.]

Scholl, D.W., Buffington, E.C., Hopkins, D.M. and Alpha, R.T., 1970a. The structure and origin of the large submarine canyons of the Bering Sea. *Mar. Geol.*, 8: 187–210.

Scholl, D.W., Greene, H.C. and Marlow, M.S., 1970b. The Eocene Age of the Adak "Paleozoic?" locality, Aleutian Islands, Alaska. *Geol. Soc. Am. Bull.*, 81: 3583–3592.

Scholl, D.W., Buffington, E.C. and Marlow, M.S., 1974. Plate tectonics and the structural evolution of the Aleutian–Bering Sea region. In: *Contrib. Geol. Bering Sea Basin and Adjacent Regions. Geol. Soc. Am., Spec. Pap.*, 151: 1–31.

Sharma, G.D., 1974. Contemporary depositional environment of the eastern Bering Sea. Part 1. Contemporary sedimentary regimes of the eastern Bering Sea. In: D.W. Hood and E.J. Kelley (Editors), *Oceanography of the Bering Sea. Inst. Mar. Sci., Univ. Alaska, Occas. Publ.*, No. 2: 517–540.

Shell, D., 1972. *Nitrogen Utilization by Phytoplankton in Southeast Alaska.* Thesis, University of Alaska, Fairbanks, Alaska.

Shuntov, V.P., 1972. *Marine Birds and the Biological Structure of the Ocean.* Dalnerostochnoe Knizhnoe Izd., Vladivostok (in Russian). [Transl., 1974. Bur. Sport Fish. Wildl., Natl. Tech. Inf. Serv., Wash., D.C., TT-74-55032, 566 pp.]

Suguira, Y., 1969. Distribution of reserved (preformed) phosphate in the subarctic Pacific region. *Paper. Meteorol. Geophys.*, 15: 208–215.

Suguira, Y., 1974. Phosphate and oxygen problems with reference to water circulation in the subarctic Pacific region. In: D.W. Hood and E.J. Kelley (Editors), *Oceanography of the Bering Sea. Inst. Mar. Sci., Univ. Alaska, Occas. Publ.*, No. 2: 163–171.

Swift, J.H. and Aagaard, K., 1976. Upwelling near Samalga Pass. *Limnol. Oceanogr.*, 21: 399–408.

Taguchi, S. and Ishii, H., 1972. Shipboard experiments on respiration, excretion and grazing of *Calanus cristatus* and *C. plumchrus* (Copepoda) in the northern North Pacific. In: Y. Takenouti et al. (Editors), *Biological Oceanography of the Northern North Pacific Ocean.* Idemitsu Shoten, Tokyo, pp. 419–431.

Takenouti, A.Y., 1975. Recent Japanese physical oceanographic studies of the Bering Sea. In: D.W. Hood and Y. Takenouti (Editors), *Bering Sea Oceanography: An Update. Inst. Mar. Sci., Univ. Alaska, Rep.*, No. 75–2: 45–50.

Takenouti, A.Y. and Ohtani, K., 1974. Currents and water masses in the Bering Sea: A review of Japanese work. In: D.W. Hood and E.J. Kelley (Editors), *Oceanography of the Bering Sea. Inst. Mar. Sci., Univ. Alaska, Occas. Publ.*, No. 2: 39–57.

Takeuchi, I., 1972. Food animals collected from the stomachs of three salmonid fishes and their distribution in the natural environments in the northern North Pacific. *Bull. Hokkaido Reg. Fish Res. Lab., Fish. Agency*, 38: 112 (in Japanese).

Tsunogai, S., Nozaki, Y., Minagawa, M. and Yamamoto, S., 1974. Dynamics of particulate material in the ocean. In: D.W. Hood and E.J. Kelley (Editors), *Oceanography of the Bering Sea. Inst. Mar. Sci., Univ. Alaska, Occas. Publ.*, No. 2: 175–189.

Uda, M., 1935. On the distribution, formation and movement of the dichothermal water in the Northwestern Pacific. *Umi to Sora*, 15: 445–452.

Venkatarathnam, K., 1971. *Heavy Minerals on the Continental Shelf of the Bering Sea.* U.S. Geol. Survey Open file Rep., 93 pp.

Wilimovsky, N.J., 1974. Fishes of the Bering Sea: the state of existing knowledge and requirements for future effective effort. In: D.W. Hood and E.J. Kelley (Editors), *Oceanography of the Bering Sea. Inst. Mar. Sci., Univ. Alaska, Occas. Publ.*, No. 2: 243–256.

Zenkevitch, L.A., 1947. Marine fauna and biological productivity. *Morya*, U.S.S.R., 2 (in Russian). [Transl. by S. Botchrskaya. George Allen and Unwin, London, 955 pp.]

Chapter 15

OKHOTSK SEA, JAPAN SEA, EAST CHINA SEA

SABURO NISHIMURA

INTRODUCTION

The Asiatic continent is fringed on its northeastern border by four marginal seas, the Bering Sea, the Okhotsk Sea, the Japan Sea and the East China Sea, from north to south. Each is enclosed by an island arc which separates it from the Pacific Ocean. The juxtaposition of these enclosed seas within island arcs arranged like a festoon is one of the most prominent features in the geomorphology of the western North Pacific area (Minato and Ijiri, 1966).

The Okhotsk Sea, the Japan Sea and the East China Sea, including the Yellow Sea (Huang Hai) and the Gulf of Bo Hai (Pwok Hai), are considered here; the Bering Sea is treated separately in Chapter 14.

GEOMORPHOLOGY AND HYDROGRAPHY

Each of the three enclosed seas is directly connected to the adjacent sea by narrow and shallow straits. The Okhotsk Sea and the Japan Sea are connected by the Mamiya Strait (Proliv Nevel'skogo) between the continent and Sakhalin, and by the Sôya Strait between Sakhalin and Hokkaido. Connecting the Japan Sea and the East China Sea are the Tsushima Strait between Kyushu and the Tsushima Islands, and the Korea Strait between the Tsushima Islands and Korea (Fig. 15.1). Through these straits the seas are washed by a common current system, the Tsushima Current system, which is a large branch of the Kuroshio (Fig. 15.2).

In some geomorphologic and hydrographic features these enclosed seas differ significantly. Both the Okhotsk Sea and the East China Sea intrude further than the Japan Sea into the continent and they have much wider connections with the Pacific to the southeast through deep straits located within a chain of islands, the Kuril and the Ryukyu Islands, respectively. Consequently, the tidal phenomena in the Okhotsk Sea and the East China Sea depend largely on prominent ocean tides that are accompanied by strong tidal currents. In the Okhotsk Sea the tides are mainly of a mixed character with the diurnal component prevailing. The amplitude varies from 1 m to a maximum value of 13 m in Penzhinskaya Bay (*a* in Fig. 15.1) in the northeasternmost part (Leonov, 1960). In the East China Sea a semidiurnal tide generally prevails, with the exception of the innermost Gulf of Bo Hai where a diurnal tide occurs; the amplitude ranges from 1 to 2 m in the Ryukyus and 1 to 3 m along the coasts of China facing the Yellow Sea and Gulf of Bo Hai up to 4 to 8 m along the west coast of Korea and the coast south of the mouth of the Yangtze (Chang Jiang), with a maximum value of 11 m observed on the mainland coast at Hangzhou Bay (*b*), southwest of Shanghai (*c*) (Ogura, 1933).

The Japan Sea, on the other hand, is more isolated, connected with the Pacific and the neighboring marginal seas only through shallow straits, and it lacks any engulfed shallow areas except for the northernmost Gulf of Tartary (Tatarskiy Proliv) (*d*). It is characterized by tides of an extremely small range — the amplitude is only 0.2 m along most of the Japanese coasts, while it varies from 0.4 to 0.5 m on the Siberian coast, and slightly more than 2 m in the Gulf of Tartary.

The Okhotsk Sea and the East China Sea are also similar to each other in that they receive greater river discharge than does the Japan Sea (Table

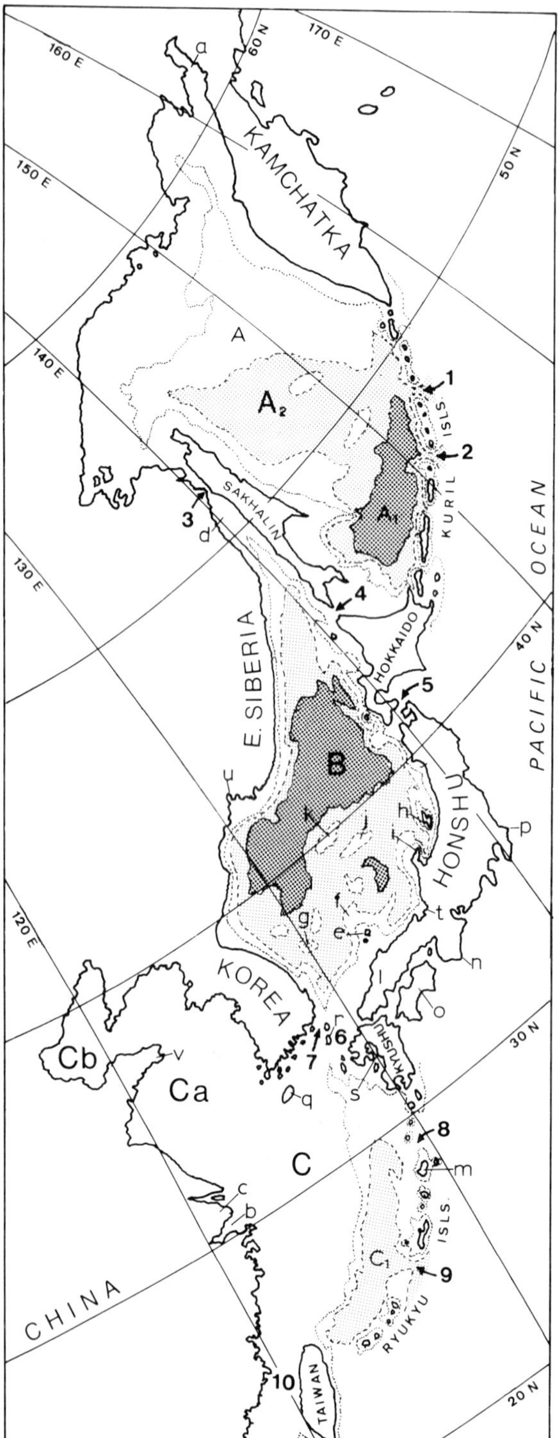

Fig. 15.1. Major bottom topography and the location of main straits in the Okhotsk Sea (A; A_1=Kuril Basin; A_2=Central Basin), Japan Sea (B) and East China Sea [C; Ca=the Yellow Sea; Cb=the Gulf of Bo Hai (Pwok Hai); C_1=the Okinawa Trough]. Areas deeper than 3000 m are stippled thickly; 1000 to 3000 m are stippled thinly; the 200-m contour is shown by the dotted line. (Depths are not shown for the Pacific Ocean.) The sill depths and names of the straits are given in Table 15.2. Localities mentioned in the text are identified by small letters.

15.1). Moreover, the bulk of the discharge into both those seas is supplied by a few big rivers, the Amur into the Okhotsk Sea, and the Yangtze and Huang He into the East China Sea, rather than by many small rivers as is the case of the Japan Sea where no river has an annual discharge larger than 20 $km^3\ yr^{-1}$. Thus, it is natural that a marked estuarine condition prevails in the Okhotsk Sea and the East China Sea, especially in their inner engulfed parts into which most of the big continental rivers are discharged.

Bottom configuration and sill depths

The three enclosed seas are more or less unique in bottom configuration (Fig. 15.1). The Okhotsk Sea and the East China Sea are deepest at the southernmost periphery, i.e., the Kuril Basin (A_1) just inside of the Kuril Islands with a depth of more than 3000 m, and the Okinawa Trough (C_1) inside the Ryukyu Islands with a depth of more than 2000 m. However, the Okhotsk Sea, to the north of the Kuril Basin, has a moderately deep (1000–1700 m) central basin (A_2) of considerable dimensions, surrounded by continental shelves. The East China Sea lacks a central basin, and the Okinawa Trough makes a sharp break through the wide continental shelf which occupies the greater area (>70%) of the East China Sea *sensu stricto* and the entire floor of the Yellow Sea and the Gulf of Bo Hai. Thus, the mean depth is greater in the Okhotsk Sea (777 m) than in the East China Sea (349 m for the sea *sensu stricto*, and 44 m and 21 m for the Yellow Sea and the Gulf of Bo Hai, respectively).

The Japan Sea is characterized by a wide deep basin, the Japan Sea Basin (B), in its northern half, with a very flat bottom at a depth of more than 3000 m. In the southern half the bottom configuration is rather complicated, with scattered shallow areas surrounding the islands of Oki (e),

Take-shima (*f*), Ullung (*g*), Sado (*h*) and Peninsula Noto (*i*), and with two conspicuous isolated banks, the Yamato Bank (*j*; 285 m deep) and the North Yamato Bank (*k*; 418 m deep), separated by deep troughs. The development of the continental shelf is limited, except for the Gulf of Tartary (*d*) and off the San-in District (*l*) of west Honshu; consequently, the mean depth (1350 m) is much greater than in the Okhotsk Sea and the East China Sea.

Table 15.2 shows the maximum depths of the seas and the sill depths of straits and passages connecting them with the ocean or neighboring seas. The deep sills of the Okhotsk Sea and the East China Sea enable the deep water of the Pacific to penetrate, but the deep basin of the Japan Sea is completely isolated by shallow sills and no entry of deep Pacific water is possible.

Climate and its influence on hydrography

The immediate proximity of these enclosed seas to Siberia assures a strong continental climate, especially in their northwestern parts (Gorczynski, 1920). Thus, the summer is very warm and wet, while the winter is bitterly cold and dry. Such climatic features give rise to the following prominent hydrographic characteristics of the enclosed seas: (1) a marked stratification develops in summer when the surface layer is warmed and diluted, with the formation of a distinct seasonal thermohalocline; and (2) a deep chilling and vigorous evaporation takes place in the winter, resulting in the generation of large-scale convectional mixing to a considerable depth with the breakdown of the seasonal thermohalocline and, in the northwestern parts of the seas, in the formation of sea ice.

The seasonal thermohalocline develops at depths of 10 to 75 m throughout the area but, generally speaking, it is shallower and more distinct in the northwestern than in the southeastern part of each

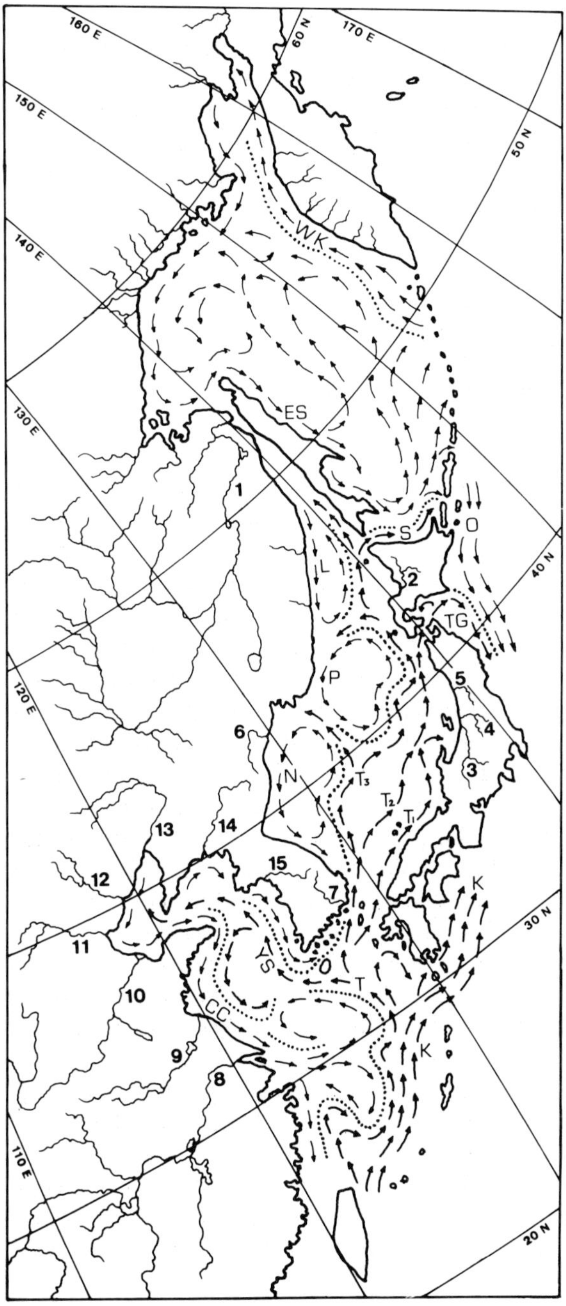

Fig. 15.2. Current systems in the Okhotsk Sea (Zenkevitch, 1963), Japan Sea (Uda, 1934a; Kasahara and Ito, 1972), and East China Sea (Uda, 1934a; Inouye, 1974). Dotted lines indicate boundaries between current systems; those between warm and cold currents are referred to as "polar fronts". The current systems in the Okhotsk Sea are (* identifies warm currents): *ES* = East Sakhalin; *WK* = West Kamchatka; and *S* = *Sôya Currents. In the Japan Sea: *L* = Liman; *P* = Primorskoi; *N* = North Korean (*L*, *P* and *N* are sometimes combined and called the Liman Current); *T* = *Tsushima (T_1 = coastal branch, T_2 and T_3 = offshore branches); and *TG* = *Tsugaru Currents. In the East China Sea: *CC* = Chinese Coastal; and *YS* = *Yellow Sea Currents. In the Pacific Ocean: *O* = Oyashio; *K* = *Kuroshio Currents. The rivers shown are: *1* = Amur; *2* = Ishikari; *3* = Shinano; *4* = Agano; *5* = Mogami; *6* = Tumen; *7* = Naktong; *8* = Chang Jiang (Yangtze); *9* = Huai He (Sha Ho); *10* = Huang He (Hwang Ho); *11* = Hai He; *12* = Luan He (Lwan Ho); *13* = Liao He; *14* = Yalu He; *15* = Hangang.

TABLE 15.1

Surface area, volume and fresh-water gain of the Okhotsk, Japan and East China Seas

Surface area and volume[2]	Annual fresh-water gain		Annual evaporation[3] (km^3)	Annual fresh-water net gain (km^3)
	precipitation[3] (km^3)	river discharge[4] (km^3)		
Okhotsk Sea				
1.58×10^6 km^2 1.23×10^6 km^3	1.11×10^3	0.59×10^3	0.47×10^3	1.23×10^3
Japan Sea				
1.01×10^6 km^2 1.36×10^6 km^3	0.91×10^3	0.21×10^3	0.81×10^3	0.31×10^3
East China Sea[1]				
1.24×10^6 km^2 0.28×10^6 km^3	1.12×10^3	1.30×10^3	1.18×10^3	1.24×10^3

[1] Including the Yellow Sea and the Gulf of Bo Hai (Pwok Hai).
[2] After Udintzev (1957) for the Okhotsk Sea, Fairbridge (1966) and Uda (1966) for the East China Sea. The values for the Japan Sea vary depending upon the delimitation of the sea, e.g., 0.98×10^6 km^2 (Zenkevitch, 1963) up to 1.30×10^6 km^2 (Uda, 1936) in surface area. Here, the values by Tokyo Astronomical Observatory (1978) are adopted.
[3] Calculated based on Samoilenko (1966) and U.S.S.R. Naval Office (1977).
[4] The annual discharges (km^3 yr^{-1}) of the main rivers draining into the seas are: Okhotsk Sea — Amur 371; rivers of northern coast 82; rivers of northwestern coast 57; rivers of Kamchatka 52 (after Bezrukov, 1960). Japan Sea — Shinano 16; Agano 13; Mogami 11; Naktong 11; Ishikari 4 (Takahashi, 1978). East China Sea — Chang Jiang (Yangtze) 1000; Huang He (Hwang Ho) 42; Talu He 38; Hangang 22; Liao He 16 (Yang and Hsu, 1978).

TABLE 15.2

Maximum depth and sill depths of the Okhotsk, Japan and East China Seas

Maximum depth (m)	No.[1]	Straits and passages	Sill depth (m)
Okhotsk Sea			
3374 (Kuril Basin)	*1*	Kruzenshtern Strait	1920
	2	Bussol' (Boussole) Strait	2318
Japan Sea			
3695 (Japan Sea Basin, eastern part)	*3*	Mamiya Strait (Proliv Nevel'skogo)	5
	4	Sôya (La Pérouse) Strait	53
	5	Tsugaru Strait	140
	6	Tsushima Strait (eastern channel)	95
	7	Korea Strait (western channel of Tsushima Strait)	130
East China Sea			
2717 (Okinawa Trough)	*8*	Passage Takara-jima — Amami Ô-shima	800
	9	Passage Kume-jima — Miyako-jima	1200
	10	Taiwan (Formosa) Strait	60

[1] Numbers as shown in Fig. 15.1.

of the seas. The depth to which the winter vertical convection reaches also changes from area to area, ranging from about 150 m to more than 3000 m. It is shallow in the southeastern parts of the seas, especially in the East China Sea and the Japan Sea where warm surface Pacific waters penetrate. In the shelf region of the East China Sea, including all of the Yellow Sea and the Gulf of Bo Hai, and in the northwestern part of the Japan Sea, convectional mixing may develop down to the sea floor where the sinking of the dense water formed by mixing is not obstructed. In the Okhotsk Sea the convection reaches only intermediate depths, usually 100 to 400 m at most (Tabata, 1955; Kitani, 1972) even though the winter climate is very severe. This is because Pacific deep water with a high density penetrates the Okhotsk Sea through the deep straits and fills the entire lower portion of the basin, thus blocking any deeper development of the vertical convection.

Another feature of the climate in this region is the prevalence of monsoon winds. The southeast monsoon prevails in summer and the northwest monsoon in winter, and this seasonal alternation of prevailing winds has a considerable effect on the hydrography of the seas. The effect is particularly prominent in the East China Sea: the surface currents are directed mainly from north to south during the winter northwest monsoon but they are complicated with northward components of variable strength during the summer southeast monsoon (Kwan, 1957). The same pattern of alternation is also suggested for the surface current system of the Japan Sea (Nishimura, 1969).

The notably strong and persistent northwest monsoon may intensify the winter chilling and evaporation of surface water, while the southeast monsoon may contribute to a still higher summer surface temperature by drifting the warmed surface water northward over the seasonal thermohalocline. These processes amplify the large seasonal fluctuation in the northwestern parts of the seas. The seasonal fluctuation in surface temperature occasionally reaches 20°C in the northwestern Japan Sea and more than 25°C in the Yellow Sea and the Gulf of Bo Hai (Table 15.3). Nowhere else in the world is the annual fluctuation larger (Dietrich and Kalle, 1957). The seasonal fluctuation in surface salinity also reflects the monsoon climate. Since precipitation is largely confined to the southeast monsoon, the surface salinity drops in the warm season (Table 15.3). This is particularly obvious for the Okhotsk Sea and the East China Sea where net fresh-water gain is considerably larger than in the Japan Sea (Table 15.1), and naturally the decrease in salinity is most pronounced in the northwest where large rivers discharge.

TABLE 15.3

Seasonal change in the properties of the surface waters of the Okhotsk, Japan and East China Seas

Seas	Summer		Winter	
	temperature (°C)	salinity (‰)	temperature (°C)	salinity (‰)
Okhotsk Sea				
NW region	6–13	29.0–32.8	−1.8[1]–1	32.7–32.9
SE region	7–18	31.0–33.5	−1.8[1]–2	32.8–32.9
Japan Sea				
NW region	16–22	32.0–33.6	−1.8[1]–5	34.0–34.1
SE region	24–27	32.6–33.6	6 –12	34.0–34.5
East China Sea				
NW region	22–28	22.0–32	−1.7[1]–10	30.0–31.0
SE region	27–29	32.0–34	10 –23	31.0–34.5

[1] Freezing point.

After Uda (1934b), Kitani (1972) and U.S.S.R. Naval Office (1977) for the Okhotsk Sea; Uda (1934a, b) for the Japan Sea; Uda (1934b) and Niino and Emery (1961) for the East China Sea.

Currents and water masses

Sixteen discrete water masses are identified in these three enclosed seas. Some of these water masses penetrate the basin from the Pacific Ocean or from adjacent seas; others are formed within the basin by winter convection. The depth ranges and the temperature and salinity characteristics of these water masses are summarized in Table 15.4. A schematic representation of the vertical circulation patterns along longitudinal sections is given in Fig. 15.3.

The East China Sea receives a huge amount of oceanic saline water from the Pacific via the warm current **Kuroshio** (Fig. 15.4); its volume transport through a transverse section on the southern periphery of the sea is calculated to be 20 to 50×10^6 m^3 s^{-1} (Inouye, 1974). While less than 5% of the volume enters the Japan Sea as the **Tsushima Current** and another smaller part enters the Yellow Sea as the **Yellow Sea Current**, the bulk of the transport is returned to the Pacific through the passages between Amami Ô-shima Island (*m*) and Kyushu as the Kuroshio proper. In the Pacific the Kuroshio flows northeastward along the coasts of the Japanese islands to about 35°N, then eventually turns eastward to become the Kuroshio Extension and flows into the central North Pacific. The transport of the Kuroshio proper off Shiono-misaki (central Honshu, *n*) is calculated to fluctuate between 25 and 65×10^6 m^3 s^{-1} (Masuzawa, 1960a); the increase in transport indicates a notable development of the Kuroshio with increasing latitudes.

In the East China Sea the Kuroshio meets low-salinity surface water from the continental coastal area and produces a marked front, together with large-scale eddies and lateral mixing, in the central part of the sea (Fig. 15.4). The Tsushima Current, which originates in the Kuroshio southwest of Kyushu, flows northward along the front and thus undergoes extreme dilution. The temperature and salinity characteristics of the source waters and of the mixed Tsushima Current waters are shown in Fig. 15.5.

The Japan Sea is unique among the enclosed seas under consideration in having strong ocean currents in the form of powerful jet streams. The

TABLE 15.4

Properties of the subsurface water masses in the Okhotsk, Japan and East China Seas (references as in Table 15.3)

Sea basins and water masses	Depth range (m)	Temperature (°C)	Salinity (‰)
Okhotsk Sea			
Sôya Current water[1]	*c*. 5 to 50–150	5–12	33.5–34.0
Intermediate cold water[2]	30 to *c*. 150	−1.6–0	32.5–33.2
Transient layer water[2]	*c*. 150 to 600–750	0–2.0	33.2–33.8
Deep and bottom waters[1]	600–750 to bottom	1.8–2.5	34.0–34.5
Japan Sea			
Tsushima Current water[1]	50–100 to *c*. 150	11–17	34.4–34.7
Central water[2]	*c*. 150 to *c*. 200	9–11	34.1–34.3
Intermediate water[2]	*c*. 50–150 to 200–300	3–4	33.9–34.05
Deep and bottom waters[2]	200–300 to bottom	0.1–1	34.0–34.1
East China Sea (shelf region)			
Yellow Sea Current water[1]	*c*. 20 to 30–70	10–15	33.0–34.6
Bottom water[2]	*c*. 30 to bottom	6–8	32.3–32.4
East China Sea (off-shelf region)			
Kuroshio water[1]	50–100 to *c*. 400	11–20	34.4–34.7
Intermediate water[1]	*c*. 400 to *c*. 800	6–11	34.3–34.4
Deep and bottom waters[1]	*c*. 800 to bottom	4–6	34.3–34.5

[1] Water which penetrates the basin from the adjacent sea or ocean.
[2] Water formed within the basin through winter convection.

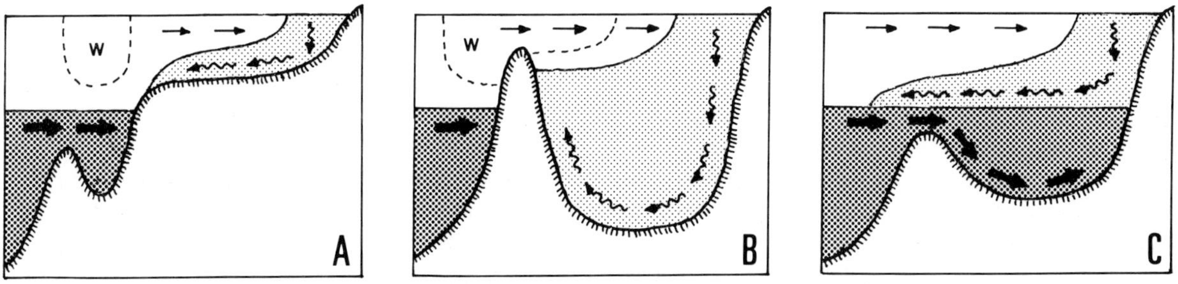

Fig. 15.3. Schematic circulation patterns in longitudinal vertical sections for the East China Sea (A), the Japan Sea (B) and the Okhotsk Sea (C). Thin and thick arrows indicate respectively the flow of the surface and deep waters from the Pacific or the adjacent marginal sea (*w* means the axis of the warm current); waved arrows indicate the sinking and spreading of the moderately dense water formed in the northwestern part of the seas through winter convection.

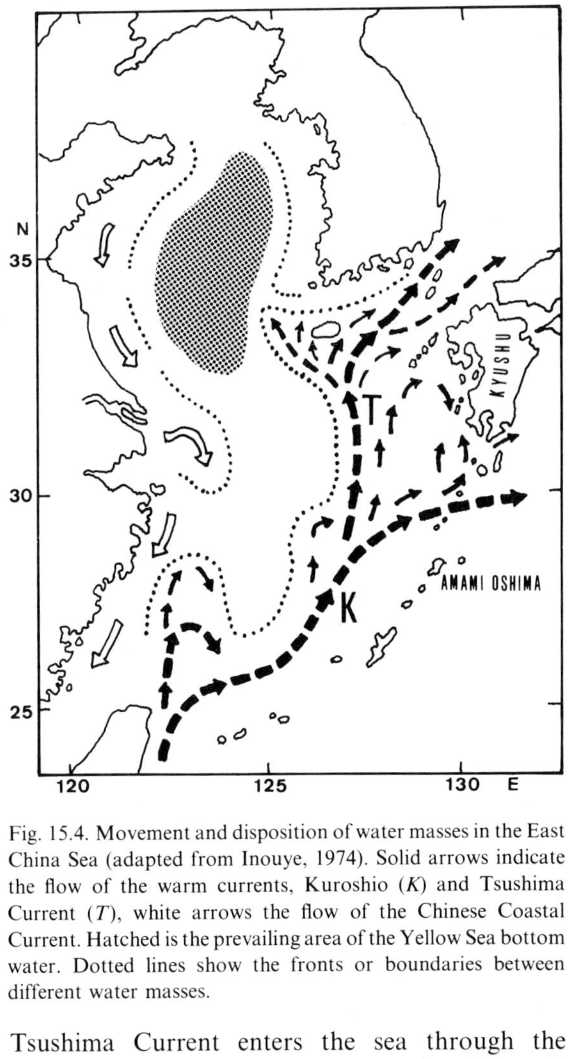

Fig. 15.4. Movement and disposition of water masses in the East China Sea (adapted from Inouye, 1974). Solid arrows indicate the flow of the warm currents, Kuroshio (*K*) and Tsushima Current (*T*), white arrows the flow of the Chinese Coastal Current. Hatched is the prevailing area of the Yellow Sea bottom water. Dotted lines show the fronts or boundaries between different water masses.

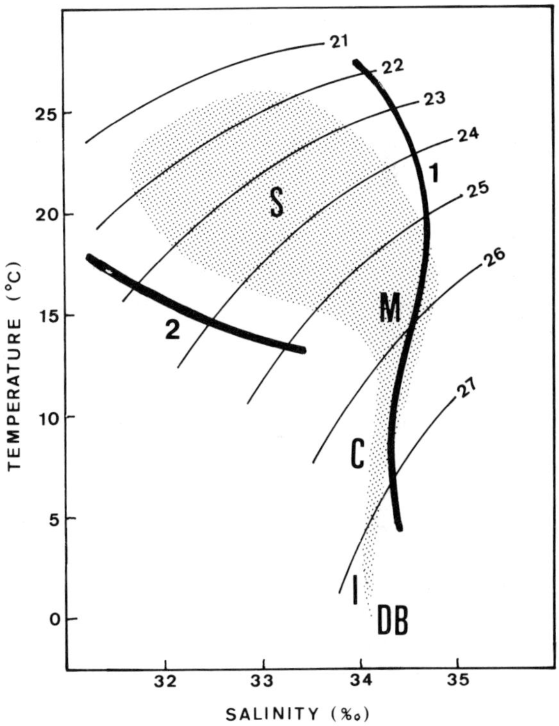

Fig. 15.5. Temperature–salinity relationships in the Tsushima Current area of the Japan Sea (hatched), overlapped on the representative *T–S* curves in the Kuroshio area (*1*) and in the central shelf area (*2*) of the East China Sea. Lines of equal density (σ_t) are superimposed. The waters in the Tsushima Current area are composed of the surface water (*S*), the middle or core water of the current (*M*), the Japan Sea central water (*C*), the intermediate water (*I*), and the deep and bottom waters (*DB*).

Tsushima Current enters the sea through the Tsushima Strait and the Korea Strait, particularly through the latter (Miyazaki and Abe, 1960) and flows northeastward in two or three meandering streams, with the farthest offshore stream being the strongest. Kawai (1974) has given a critical review of these complex current systems. The Tsushima Current eventually flows out into the

Pacific, mostly through the Tsugaru Strait but partly through the Sôya Strait into the Okhotsk Sea (Fig. 15.2).

To the northwest of the warm current streams there are cold currents, generally referred to collectively as the **Liman Current.** The volume transport and the speed of this current are much smaller than those of the Tsushima Current; it can best be described as a counter- or compensation current of the latter (Masuzawa, 1960b). The boundary between the waters of the cold current and the farthest offshore stream of the warm current is generally very sharply defined and is often referred to as the *polar front* (Suda, 1932; Shimomura and Miyata, 1953). In this way, the Japan Sea is divided into two distinct areas: the southeastern warm-water area washed by the Tsushima Current and the northwestern cold-water area under the influence of the Liman Current. This distinction is clearly reflected in the biota and biological production of the Japan Sea.

Some water enters the Okhotsk Sea from the Japan Sea by the **Sôya Current** as the terminal flow of the Tsushima Current, and Pacific waters arrive through passages between the Kuril Islands; all contribute to the formation of a large counterclockwise circulation (Iida, 1961; Kurashima et al., 1966; Watanabe, 1967). However, those waters have almost lost their oceanic nature and the circulation is very weak.

Where winter convection reaches the sea floor, as in the northwestern parts of both the East China Sea and the Japan Sea, the dense water that is formed spreads gradually along the bottom and eventually fills the basins of the Gulf of Bo Hai, the Yellow Sea and the Japan Sea. This water persists in the bottom layer throughout the seasons, little affected by the overlying water masses, and thus forms the *bottom waters* of these enclosed seas (Uda, 1934a, 1966; Inouye, 1974).

The bottom water of the Japan Sea (Suda, 1932; Uda, 1938) is confined to that basin, prevented from spilling out by the shallow sills of the straits; it dissipates little by little into the upper layers through eddy turbulence, and this mixing with upper layer waters gives rise to the formation of the *deep water*. Both the bottom and deep waters are cold water masses of the Japan Sea and together are called the *Japan Sea cold water mass*, sometimes with the inclusion of the *intermediate water*. The bottom water in the Gulf of Bo Hai and the Yellow Sea, on the other hand, flows gradually south along the shelf floor and down over the shelf break into the Okinawa Trough as a kind of cascade (Tsujita, 1963).

The deep and bottom waters of the Japan Sea occupy no less than 90% of the total volume of the basin. There are several theories and calculations to explain the physical properties of these water masses, based on available hydrographic and climatological data (Suda, 1932; Uda, 1934a, 1938; Miyazaki, 1952, 1953). The theories indicate that the distinctive properties of the deep and bottom waters of the Japan Sea are determined by the nature of the water that sinks in the process of winter convection and also by later spreading and mixing.

In the southern part of the Japan Sea and in the Okhotsk Sea the water formed in winter convectional mixing sinks only to intermediate depths and spreads horizontally along isentropic surfaces to form the *intermediate water*. This water in the Tsushima Current region in the southeastern part of the Japan Sea is comparatively high in both temperature and salinity, and is called the *Japan Sea central water* (Miyazaki and Abe, 1960). The intermediate water formed in the transition area between the Tsushima Current and the Liman Current, particularly around the Korean Bight in the western part of the Japan Sea, has lower temperature and salinity and penetrates below the Japan Sea central water; it is the *Japan Sea intermediate water* (Miyazaki, 1952). The intermediate water formed in the Okhotsk Sea is characterized by a low temperature ($<0°C$) near the freezing point. It spreads below the surface water throughout the sea and persists as *intermediate cold water* in all seasons (Uda, 1935, 1955; Kitani, 1972). This is because the surface water is diluted and heated to such an extent during the warm season that vertical stability is established between the surface and subsurface layers and downward heat transfer is inhibited.

The intermediate cold water, often referred to as *dichothermal water*, of the Okhotsk Sea not only spreads within the sea but also flows out into the Pacific through straits in the southern Kuril Islands (Yasuoka, 1967), and contributes to part of the Oyashio. To compensate for this loss, warm and saline Pacific water enters through the Straits of

Boussole (Bussol') and Kruzenshtern in the northern Kuril Islands. The bulk of the penetrating Pacific water sinks to fill the lower portion of the present sea in, for instance, the Kuril (A_1) and central basins (A_2 in Fig. 15.1), as deep and bottom waters (Yasuoka, 1968). Part of the Pacific water, however, mixes vigorously with overlying intermediate cold water, giving rise to the formation of another water mass named the *transient layer water* below the intermediate cold water (Yasuoka, 1967).

The southeastern part of the East China Sea is occupied by Pacific water masses that penetrate the deep passages between the Ryukyu Islands. Below the surface water, the core water of the warm current, or the *Kuroshio water*, is composed of the western North Pacific central water (Miyazaki and Abe, 1960), while the intermediate and deep waters are derived from corresponding water masses in the adjacent area of the Pacific basin. The bottom water, which fills the Okinawa Trough (C_1), apparently is separated from similar water in the Pacific; as yet there is no definite explanation for the origin of this water mass.

BIOTIC STRUCTURE AND DISTRIBUTION

Surface communities

The northwestern parts of the Okhotsk Sea and the East China Sea receive a large amount of fresh water from rivers, and are characterized by a remarkable development of estuarine, brackish, and sometimes of fresh-water populations. In the northwest Okhotsk Sea several species of the circumpolar copepod genera *Eurytemora* and *Acartia* as well as the temperate Far Eastern estuarine copepods *Tortanus derjungi*, *Sinocalanus tenellus* and *Epilabidocera* spp. flourish under the effect of the discharge of the Amur (Brodsky, 1957). Some truly fresh-water animal groups such as Rotifera and Cladocera are also plentiful (Brodsky, 1957; Zenkevitch, 1963). The influence of the discharge of the Amur is also evident in the bottom communities to the extent that a "fresh-water" intertidal zone can be identified in the mouth of the river (Mokievsky, 1949).

Estuarine species are even more abundant in the Yellow Sea and the Gulf of Bo Hai. The percentage of estuarine species of fish in the seas under consideration is given in Table 15.5. It is highest for the Yellow Sea and the Gulf of Bo Hai, followed by the Okhotsk Sea and the northwest Japan Sea, and lowest for the southeast Japan Sea where it is about the same as for the Bering Sea. Estuarine species of other animal groups are also prominent in the East China Sea. Many estuarine copepods belong to the same genera as those in the Okhotsk Sea; other genera of tropical origin such as *Centropages*, *Schmackeria*, *Labidocera* and *Pontella* have also been reported from this area (Burckhardt, 1913; Mashiko, 1954). Additional species are still being

TABLE 15.5

Number and percentage of the estuarine or diadromous species in the fish fauna

Seas	Total number of species	Estuarine or diadromous species			Sources
		number	percentage	Principal families (sp. no.)	
Yellow Sea and Gulf of Bo Hai	280	34	12.1	Gobiidae[a](8) Salangidae[b](5)	Chang et al. (1955), Rass (1965), Chu et al. (1963)
Okhotsk Sea	270	22	8.2	Salmonidae[c](10) Osmeridae[c](4)	Schmidt (1950)
Japan Sea					
NW region	210	15	7.1	Salmonidae[c](6) Osmeridae[c](4)	Lindberg (1947)
SE region	650	30	4.6	Gobiidae[a](10) Theraponidae[a](2)	present author
Bering Sea	297	14	4.7	Salmonidae[c](7) Osmeridae[c](6)	Andriashev (1939)

[a]Tropical to subtropical families. [b]A warm-temperate family. [c]Arctic to cold-temperate families.

discovered and described (Shen and his collaborators, 1955, 1956, 1963, 1964; Chen and Zhang, 1965). Among benthic animal groups, mollusks of the families Assimineidae and Potamididae and crustaceans of Penaeidae, Ocypodidae and Grapsidae are especially noteworthy for their inclusion of many estuarine species in the area (Kamita, 1941; Kamita and Sato, 1941; Miyadi et al., 1953).

The Pacific coasts of Kyushu, Shikoku (*o* in Fig. 15.1) and Honshu as far north as Cape Inubôsaki (35°40′N; *p*), and the islands, peninsulas and capes along these coasts, are noted for a profusion of tropical and subtropical high-salinity shelf species, such as coral reef organisms. The dominant species are found among Kuhliidae, Lutjanidae, Chaetodontidae, Pomacentridae and Scaridae in fishes, Palinuridae in decapod crustaceans, Cypraeidae and Conidae in mollusks, Nephtyidae and various stony corals in coelenterates. Growths of many small green algae such as *Caulerpa*, *Acetabularia* and *Halimeda* are noted. These species, however, are very poorly represented or totally absent along the Japan Sea coasts of Kyushu and Honshu, as exemplified by the sparsity of Cypraeidae (cowries) and Conidae (cone shells) (Table 15.6), in spite of the fact that these coasts are under the influence of warm waters of the Tsushima Current much as the Pacific coasts are washed by the Kuroshio.

Many species of oceanic epipelagic organisms that are common off the Pacific side of the Japanese Archipelago are totally absent or only very rarely found in the Yellow Sea or the Japan Sea — for example, the oceanic tunas of the suborder Scombrina (*Katsuwonus pelamis*, *Thunnus alalunga*, *T. obesus*, *T. albacores*, *Tetrapturus audax*, *T. angustirostris*, *Xiphias gladius*), the oceanic flying fishes (*Exocoetus volitans*, *Prognichthys rondeleti*), and the oceanic whales (*Physeter catodon*, *Balaenoptera borealis*, *B. edeni*). They migrate northward with the warm current up to waters south of Quelpart (Cheju, Saishu) Island (*q*) or the Tsushima Islands (*r*) but never venture into the Yellow Sea or the Japan Sea (Nishimura, 1965a; Okiyama, 1974). They are stenohaline species, and it is very likely that the marked dilution of the surface water in the East China Sea, especially in the northwest, may inhibit their penetration into these seas. The distribution of *Trichodesmium thiebauti*, a pelagic blue-green alga and a typical indicator organism of the Kuroshio, shows a similar pattern (Fig. 15.6). Dense populations of *Trichodesmium* coincide with or are seaward of the axis of the Kuroshio, and none appear in the area northwest of the central part of the East China Sea.

TABLE 15.6

Distribution of tropical coral reef shells in waters adjacent to Japan (after Kawakami and Habe, 1961)

Localities	Number of species	
	Cypraeidae	Conidae
Taiwan (22–25°N)	49	55
Ryukyu Islands (26–27°N)	64	78
Amami Islands (28°N)	48	50
Pacific coast		
Kyushu (32°N, Miyazaki Pref.)	33	22
Shikoku (33°N, Kochi Pref.)	42	51
Honshu (33°30′N, Shiono-misaki)	41	41
Honshu (35°N, Sagami Bay)	30	10
Japan Sea coast		
Kyushu (32°N, Amakusa)	16	3
Kyushu (33°30′N, Fukuoka Pref.)	15	8
Honshu (34°N, Yamaguchi Pref.)	14	4
Honshu (35°30′N, Fukui Pref.)	8	1

The surface salinity around Quelpart and Tsushima Islands drops radically between July and November, the peak period of the northward transport of the warm current (Fig. 15.7) and also of the northward migration of these epipelagic organisms.

During the summer the entire Japan Sea is covered with a surface community dominated by embayment or neritic organisms. These include, among others, plankton animals such as *Noctiluca scintillans*, *Oikopleura dioica*, *Penilia avirostris*, *Sagitta crassa*, and *S. nagae* (= *S. bedoti* of previous authors). The proportion of the chaetognath populations formed by *Sagitta nagae* in Japanese and adjacent waters is shown in Fig. 15.8. It is dominant in the East China Sea, the Yellow Sea, and in offshore areas of the Japan Sea. The abundance of *S. nagae* off the northeast coast of Honshu occurs because of the terminal flow of the Tsugaru Current and the mixing of the Kuroshio and the Oyashio. This chaetognath has been used as an indicator of coastal water by Kuroda (1977).

The Japan Sea lacks the extensive fresh-water

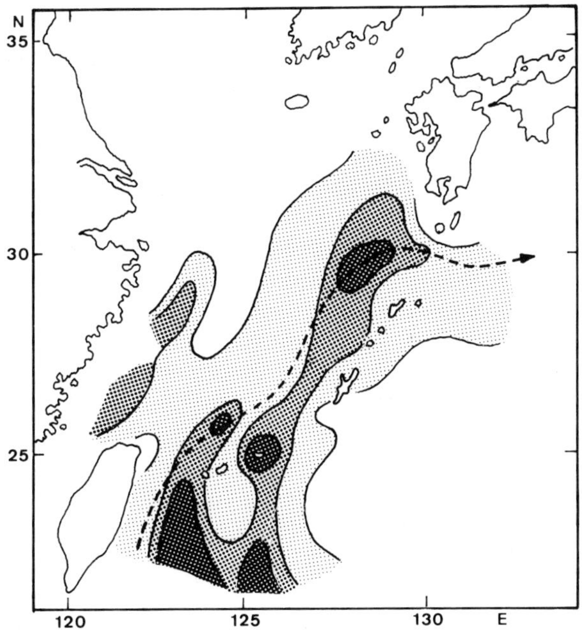

Fig. 15.6. Number of trichomes or filaments per liter of the pelagic colony-forming blue-green alga *Trichodesmium thiebauti* in the surface waters southwest of Japan in July and August 1966 (after Asaoka, 1975). Thickly hatched: more than 2×10^3; moderately hatched: 10^3 to 2×10^3; thinly hatched: less than 10^3; blank area: none. Broken arrow indicates the position of the axis of the Kuroshio.

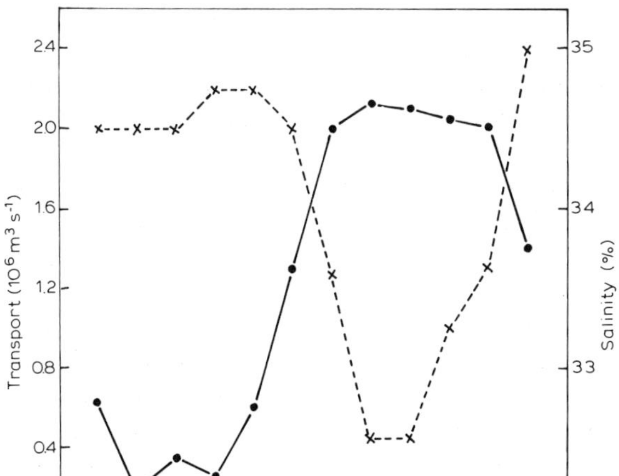

Fig. 15.7. The net northward transport and the surface salinity of the Tsushima Current through a transverse section from west Honshu to southeast Korea (adapted from Moriyasu, 1972). Full line: transport in 10^6 m^3 s^{-1}; broken line: surface salinity in ‰.

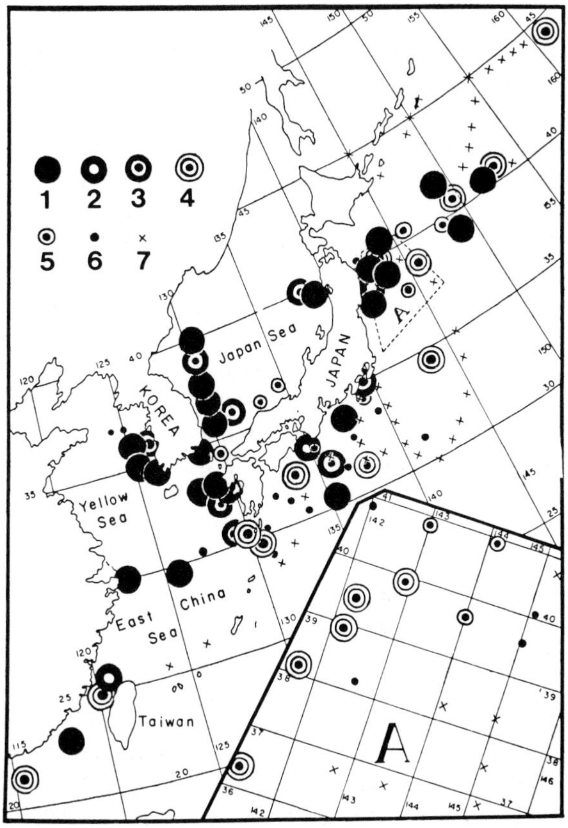

Fig. 15.8. Occurrence of the chaetognath *Sagitta nagae* (=*S. bedoti*) as a percentage of the species of chaetognaths in the Japanese and adjacent waters (after Tokioka, 1959). Legend *1*=81–100%; *2*=61–80%; *3*=41–60%; *4*=21–40%; *5*=11–20%; *6*=10% and less; *7*=absent.

drainage as found in the East China Sea or the Okhotsk Sea. The neritic populations in its surface community are believed to originate in the Yellow Sea or, more probably, in the waters along the southern coasts of the Korean Peninsula. They may have been introduced into the surface community of the Tsushima Current by lateral mixing as the current sweeps past these coastal areas (Nishimura, 1965a). Moreover, the flow of the offshore stream of the Tsushima Current and the northward drift of surface water generated by the southeast monsoon is probably responsible for the wide offshore dispersal of these neritic species within the Japan Sea.

The influence of the warm current is also discernible in the surface communities of the Okhotsk Sea. During the warm season, the surface layer of the sea can be divided into the southern, the central and the northern areas (Aikawa, 1935). The southern area is characterized by the predominance of styliform, chaetoceran or *Nitzschia*-type plankton and many small subtropical copepods, which indicate that the area is under the effect of the Sôya Current, a continuation of the Tsushima Current. The central area is occupied by communities commonly found in the Oyashio waters in the western North Pacific. These waters are characterized by the occurrence of discoid diatoms and large cold-water copepods such as *Calanus plumchrus* and *Eucalanus bungii*. The northern area is populated by local neritic plankton, dominated by cold-water species of *Chaetoceros*. This pelagic distribution agrees well with the circulation pattern of the Okhotsk Sea (Fig. 15.2).

Middle-layer communities

The western North Pacific central water, which has remarkably stable temperatures (14–17°C) and salinities (34.4–34.7‰), flows into the East China Sea and the Japan Sea as the middle layer or core water of the Tsushima Current. In the East China Sea this saline water of oceanic origin is limited to the southeasternmost part around Ryukyu and Kyushu. In the Japan Sea the water, after passing through the Korea Strait (Miyazaki and Abe, 1960), flows both offshore and along the coast and extends over a wide area of the central and southeastern parts of the sea. A distinct middle layer occupied by this water is located at depths of about 50 to 150 m, thicker and deeper in the south and thinner and shallower in the north.

The shelf floor bathed by this saline water is noted for thriving populations of numerous Indo- and West Pacific outer shelf organisms. Among the fishes are found the warm-water families such as Sparidae, Synodontidae, Branchiostegidae, Trichiuridae, Triglidae and Zeidae; the decapod crustaceans include the families Homolidae, Goneplacidae and Leucosidae; the molluscs include Dentaliidae, Nuculidae, Glycymeridae and Limopsidae. These populations clearly identify the shelf floor under the influence of the warm current.

The western North Pacific central water mass often carries certain mesopelagic populations into the East China Sea and the Japan Sea. They include, for instance, various taeniosomous fishes of the families Regalecidae and Trachipteridae, Lamprididae (opah), and Molidae (sunfishes). They penetrate these seas along the middle layer, extending especially deeply into the Japan Sea, without encountering any environmental dilution. Some of them reach northward as far as Hokkaido and the coast of East Siberia. They are, however, unable to colonize these marginal seas, because they cannot survive the severe winter there (Nishimura, 1965a).

The Japan Sea central water mass, prevailing at the depths of 150 to 200 m in the central and southeastern parts of the sea, is the habitat of some oceanic populations that successfully colonize this marginal sea. The populations comprise, for instance, the little gonostomatid fish *Maurolicus japonicus*, little oegopsidean squids *Enoploteuthis chuni* and *Abraliopsis scintillans*, the euphausiid *Euphausia pacifica*, and the hyperiid amphipod *Parathemisto japonica*. They are the only members to be included in the category of primary deep-sea animals, thriving in the Japan Sea. They are mostly equipped with luminous organs and other specialized structures, although they do not inhabit truly deep waters in the Japan Sea basin.

Deep and bottom communities

The deep or bottom communities of the three enclosed seas have the most distinctive populations. The bottom water of the Gulf of Bo Hai and the Yellow Sea has low temperature and low salinity and lies directly below the surface water, which occupies the upper 10 to 30 m (Table 15.4). This cold water mass persists on the bottom from spring through autumn and harbors a community quite different from that of the surface water. The community of the bottom water is characterized by the occurrence of certain cold-temperate and subarctic species that are common to the northern Japan Sea and the Okhotsk Sea. Representative fishes, for example, are *Clupea pallasi*, *Gadus macrocephalus*, and many other species of the families Stichaeidae, Zoarcidae, Hexagrammidae, Cottidae and Liparidae (Rass, 1965), and there are a fair number of invertebrate species of the same climatic category, such as polychaetes (Ushakov, 1955). These cold-water populations are believed to be self-sustaining within the present sea basins, independent and isolated from those in the Japan Sea. Nevertheless, migrations may occur between the populations of the Yellow Sea and the Japan Sea via waters south of the Korean Peninsula during an exceptionally severe winter (Uchida, 1935).

Another interesting feature of the bottom water in the Yellow Sea is the existence of a cloud of floating silt, very fine, less than 10 μm in diameter, that is suspended in the lower 20 to 30 m of the water. Within this silt cloud there is a dense population of the diatom *Melosira sulcata* (= *Paralia sulcata*), reaching more than 10^3 cells l^{-1} (Asaoka, 1975; Fig. 15.9). At the start of the winter convection, the silt and the diatom population begin to be dispersed throughout the water column, bringing about a marked increase in turbidity and a concomitant decrease in the transparency of the water (Fig. 15.10). Minimum transparency occurs during the winter, a murky condition that previously was attributed to discharge from the Yangtze River (Matudaira and Inoue, 1939) despite the apparent discrepancy between the seasons of discharge peak and transparency minimum and the fact that minimum transparency usually was located far from the mouth of the Yangtze, more than 100 km offshore.

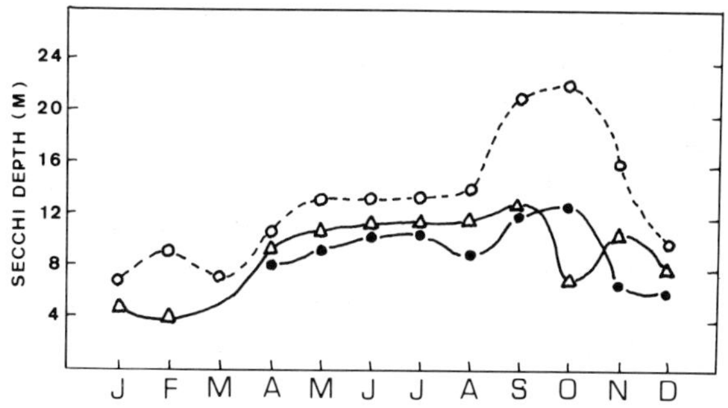

Fig. 15.10. Water transparency (Secchi disc depth) in the Yellow Sea (after Uda, 1934b). The change off west Korea is indicated by dots for 1930 and by circles for 1931, while that off the tip of Shantung Peninsula (*u* in Fig. 15.1) is shown by triangles for the average year.

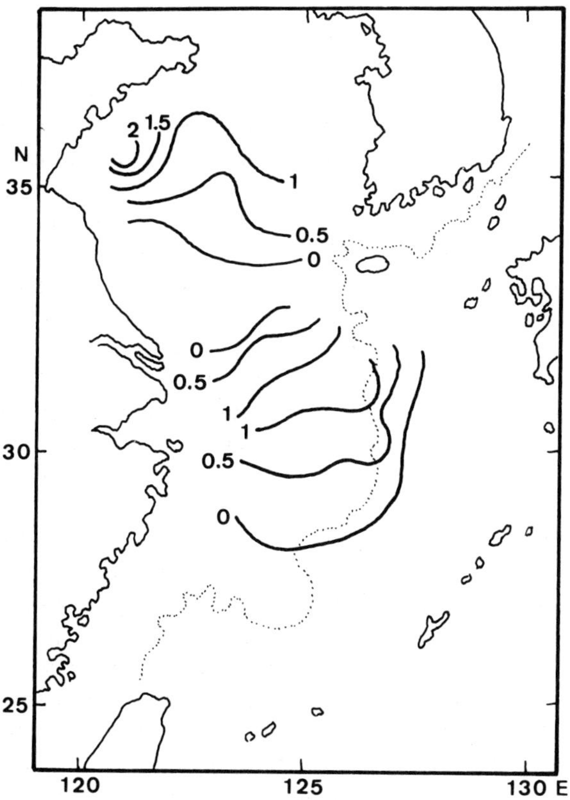

Fig. 15.9. Bottom distribution of the diatom *Melosira sulcata* (10^3 cells l^{-1}) in the East China Sea during the summer seasons, 1960–1963 (after Asaoka, 1975). The 100-m contour is dotted.

The persistence of the cold bottom water throughout the warm seasons seems responsible for the accumulation of fine silt as well as the existence of *Melosira* and other cold-water populations on the bottom. This has been attributed to a possible large-scale counterclockwise circulation that could give rise to a stagnant area in the central part of the shelf region of the Yellow Sea. The validity of this hypothesis has been proved theoretically (Asaoka and Moriyasu, 1966).

The Okhotsk Sea and the off-shelf part of the East China Sea are connected with the Pacific through deep straits. They thus receive both Pacific deep water and its communities, so the deep-water faunas are almost the same as those found in adjacent Pacific depths. They are characterized by a predominance of so-called primary or archaic deep-sea animals — the animals normally occupying a primitive position in phylogenetic relationships and confined entirely as a group to the deep-sea environment (Andriashev, 1935; Schmidt, 1935; Zenkevitch and Birstein, 1960). Representative groups include hyalosponges, polychelid and homarid decapod crustaceans, porcellanasterid sea-stars, elasipod sea-cucumbers, and teleostean fishes such as Bathyclupeiformes, Opisthoproctoidei, Stomiatoidei, Myctophidae and Macrouridae.

The Japan Sea is quite distinct in this respect; most of the typical primary deep-sea animals mentioned above do not exist there. Those that occur are found only in the southernmost part of the sea, possibly transported there accidentally by the Tsushima Current from the East China Sea during their pelagic larval stage. This is typically the case with the Macrouridae, the dominant teleostean group in deep waters of the ocean (Table 15.7). Their occurrence in the Japan Sea is unusual in that they are not found in the deep layers but invariably are confined to the shallower layers (*c.* 50–150 m) in the core water of the warm current.

The deep and bottom waters in the Japan Sea are occupied, instead, by an entirely different fauna characterized by genuinely cold-water groups such as Zoarcidae, Cottidae, Liparidae, Lumpenidae and Pleuronectidae among fishes, Buccinidae among mollusks, Majidae, Crangonidae and Pandalidae among decapod crustaceans (Nishimura, 1968). These are all typical representatives of secondary deep-sea animals (Andriashev, 1935; Schmidt, 1935); they are primarily inhabitants of the shelf or upper bathyal zone in high latitudes and have adapted to low temperature and then penetrated into the deep-sea environment in comparatively recent geological time.

The dominance here of secondary, but not primary or archaic, deep-sea fauna, may be related to the recent origin of the deep basin of the Japan Sea. However, the survival and abundance in the Japan Sea of the secondary deep-sea fauna, sometimes referred to as *pseudo-abyssal fauna*, is probably related to the distinctive properties of the deep and bottom waters filling the basin. As seen from Table 15.4, the deep and bottom waters are significantly colder than those of the East China Sea (Okinawa Trough) and the Okhotsk Sea, and such physical differences seem enough to account for the penetration and flourishing of cold-adapted secondary deep-sea populations in the depths of the Japan Sea, and the failure of archaic deep-sea populations to colonize this basin (Nishimura, 1977). Thus, it appears that climatic as well as geomorphological features enable the present enclosed sea to support its unique deep-sea communities.

In the Okhotsk Sea, intermediate cold water is formed by the same mechanism as the bottom water of the Japan Sea but in the Okhotsk it does not reach to the sea floor because of the penetration of the Pacific deep water. This exerts a profound effect on the vertical distribution of organisms, both pelagic (Vinogradov, 1954b) and benthic (Ushakov, 1953). The populations decrease markedly both qualitatively and quantitatively from

TABLE 15.7

Distribution of macrourid fishes in the Okhotsk, Japan and East China Seas

Macrourid fauna	Remarks
Okhotsk Sea: 4 species	
Coryphaenoides cinereus *Macrurus angustifrons* *Chalinurus pectoralis* *Hemimacrurus acrolepis*	the macrourids form the biggest family among the primary deep-sea fish fauna of the sea, and some of the species seem to be distributed abundantly (Rass, 1963)
East China Sea: 6 species	
Coryphaenoides asper *Coelorhynchus jordani* *C. multispinulosus* *C. japonicus* *Hymenocephalus gracilis* *H. longiceps*	some of the species seem to be abundant (Misu et al., 1955)
Japan Sea: 1 species	
Coelorhynchus multispinulosus	species found only occasionally on the insular shelf in the southwestern region (Nishimura, 1968)

the surface layer to the intermediate cold layer, and then increase again rather significantly in the transient layer below.

The transient layer in this enclosed sea extending at depths between 150 m and 600 to 750 m is biogeographically very important. The populations are qualitatively and quantitatively rich there, and many benthic cold-adapted animals in the Okhotsk Sea fauna appear to be distributed in and around this layer (Schmidt, 1950; Ushakov, 1955).

The Okhotsk Sea may have been the source of the cold-adapted marine populations of the North Pacific or even of the entire Northern Hemisphere (Schmidt, 1950; Ekman, 1953). Populations could have been acclimatized to the cool climate under favorable temperature conditions there during the Late Tertiary to Quaternary epochs, and some might have descended to deep environments to form the so-called secondary deep-sea fauna. In the process of downward migration by these cold-adapted populations, it seems very likely that the transient layer might have afforded a favorable transition or passage route because the layer is occupied by the water formed through mixing of icy-cold convectional water with deep water from the ocean. The layer extends to the sea floor at the junction of the continental shelf and the slope, so that populations could be pre-adapted to existence in the deep environments.

The depth range of the transient layer in the Okhotsk Sea corresponds to that of a zone of the sea floor called by Soviet scientists the "transitional horizon" in marine benthic zonation and is situated between the sublittoral and bathyal zones (Zenkevitch, 1963). This horizon is considered to be not only transitional or intermediate between the sublittoral and bathyal zones with regard to its environmental and faunal elements, but also to have worked as an evolutionary migration route of cold-adapted sublittoral populations to the bathyal environment.

The secondary deep-sea fauna of the Japan Sea bears a strong relationship with that of the transient layer in the Okhotsk Sea; the majority of the faunal elements are identical or differ only at subspecific rank. An immigration might have been made by the cold-adapted populations of the Okhotsk Sea into the Japan Sea during recent geological time — possibly during the Late Pleistocene. The formation and sinking of bottom water may have first occurred by the latest glacial period in the Japan Sea and organisms could then have descended to the deep basin to form the secondary deep-sea fauna (Nishimura, 1977).

In the present-day Okhotsk Sea, however, descent of the cold-adapted populations seems to be inhibited, limiting them to the transient layer, by the influx into the deep basins of warm deep and bottom waters of Pacific origin. The deep basins are only sparsely populated, qualitatively and quantitatively (Ushakov, 1953; Savilov, 1961), and their faunal composition resembles that of the deep waters of the North Pacific, as indicated, for example, by the presence of the primary deep-sea fish group, Macrouridae (Table 15.7).

Paleogeography and faunal development

Description of the geological history of the Far Eastern enclosed seas must await wider sources of material. However, a traditional view (Yabe, 1929) favors the idea that the present-day East China, Japan and Okhotsk Seas once were vast dry lands embracing shallow inland basins or gulfs during the period from the Late Pliocene through most of the Pleistocene (Fig. 15.11). The basins in the East

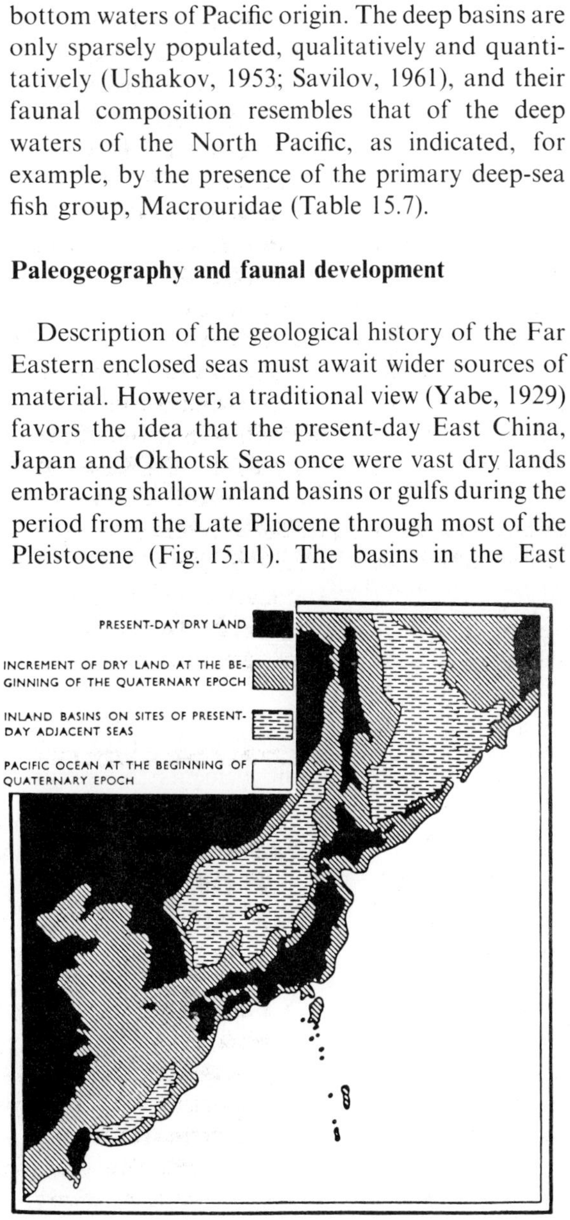

Fig. 15.11. Paleogeography of the far eastern seas at the beginning of the Quaternary epoch according to Yabe's theory (after Zenkevitch, 1963; reproduced by permission of George Allen and Unwin, London).

China Sea and the Okhotsk Sea might have directly bordered the Pacific, while that in the Japan Sea may have been completely isolated or connected with the ocean only by a narrow channel, possibly situated across present-day southern Hokkaido (the so-called Ishikari Depression). Of course many changes may have occurred in the coastal geography due to large-scale fluctuations in sea level during the Pleistocene (Lindberg, 1955, 1972; Hoshino, 1975).

This paleogeographic situation could have had a profound effect on the development of fauna in the Far Eastern seas. First, topographic transformations of enlarged gulfs with wide openings to the Pacific during the interglacial high sea level period into semi- or completely enclosed inland basins during the glacial low sea level period might have been favorable for the development and establishment of many endemic forms, particularly from Indo–West Pacific populations. Numerous Indo–West Pacific populations might then have migrated, in every warm interglacial period, northward to the Far Eastern waters where they are now found, and then been isolated in closed inland basins during the following cold glacial period, eventually to be transformed into phylogenetically as well as climatically different forms. This may have been the mechanism by which the endemic species of the Far Eastern seas, species that are temperate in climatic category and comprise one of the most important biogeographic constituents of the Far Eastern marine fauna, were derived from tropical Indo–West Pacific elements (Nishimura, 1971). In many animal groups the Indo–West Pacific species and the Far Eastern endemic species together constitute more than 90% of the total number of species occurring in the region under consideration (Table 15.8).

The gulfs might also have been significant in the development of many estuarine or brackish-water forms, because presumably they received drainage from ancient big rivers such as the paleo-Huang Ho and paleo-Yangtze, both of which might have discharged into the East China Sea gulf, and the paleo-Amur, which at that time poured into the Japan Sea gulf instead of the Okhotsk Sea (Lindberg, 1955). It is known that the Japan Sea constitutes, together with the lower reaches of the Yangtze, one of the greatest centers of speciation for brackish-water copepods (Mashiko, 1954). The estuarine condition in semi-enclosed gulfs might have been so intense as to force some of the organisms to adapt even to fresh-water life. Thus, the fresh-water biota of the continental area embracing the present-day Yellow Sea and East China Sea frequently includes interesting organisms of marine origin; for instance, *Chinonereis edestus* and *Nephtys oligobranchia* in polychaetes (Chamberlin, 1924; Okuda, 1943), *Eriocheir* in decapod crustaceans, *Fugu* and *Siniperca* in fishes (Nishimura, 1966b, 1967).

Rich occurrences of estuarine species characterize the local fauna in the inner parts of the Yellow Sea and East China Sea, but they are not necessarily confined to these sites on the continental coast. Similar forms are also found in an isolated

TABLE 15.8

Biogeographic elements in the marine fauna of the surrounding seas of Japan, as shown in Brachyura (crabs) and Echinoidea (sea urchins) (after Nishimura, 1971)

Biogeographic elements	Brachyura		Echinoidea	
	number of species	%	number of species	%
Indo–West Pacific	420	62.3	96	50.0
Indo–West Pacific and tropical Atlantic	3	0.4	1	0.5
Circumtropical	1	0.1	1	0.5
Indo–Pan-Pacific	2	0.3	0	0.0
Far–Eastern endemic	241	35.7	81	42.2
Japan–Oregon	2	0.3	1	0.5
Okhotsk–Bering	3	0.4	11	5.7
Arctic–North Atlantic	2	0.3	1	0.5
Austral	1	0.1	0	0.0

population in the inner parts of certain bays in west Kyushu, Japan. Ariake Bay (*s* in Fig. 15.1) is the best known among them; here a number of estuarine organisms with continental affinity are thriving to a remarkable degree, making the area a unique place zoogeographically among the Japanese Islands (Uchida and Tsukahara, 1953). The estuarine organisms in west Kyushu should be considered a relict population because paleontological and archeological evidence suggests that their distribution around Japan was much wider in the Pleistocene and prehistoric Holocene (Miyadi et al., 1953). These bays on west Kyushu might once have constituted, together with those on the continent, parts of a huge interconnecting estuarine system along the entire inner coast of the East China Sea gulf. With the gradual isolation of the Japanese Archipelago from the continent, changes in environmental conditions may have caused a steady diminution in those estuarine populations, which survive only in a few bays on west Kyushu.

BIOLOGICAL PRODUCTION

Primary productivity

An overall view of the geographical distribution of primary productivity in the world ocean is given by Koblentz-Mishke et al. (1970). Within the region under consideration, productivity is shown to range from 100 to more than 500 mg C m^{-2} day^{-1} (Fig. 15.12). An area of high productivity surrounds the continental coasts, notably in the East China Sea. A small area of very low productivity is located in the central part of the Okhotsk Sea. The Japan Sea is divided into a northwestern area of high production and a southeastern area of low production.

The average annual rate of primary production has been compiled and estimated for each of these enclosed seas from the data hitherto published by various authors (Koblentz-Mishke, 1965; Sanger, 1972; Taniguchi, 1972, 1973; Fukataki, 1974). It ranges from about 75 to 200 g C m^{-2} yr^{-1}. Net primary production is obtained by multiplying these values by the area of the respective seas. The results range from about 900 to 3200 million metric tons per year as shown in Table 15.9.

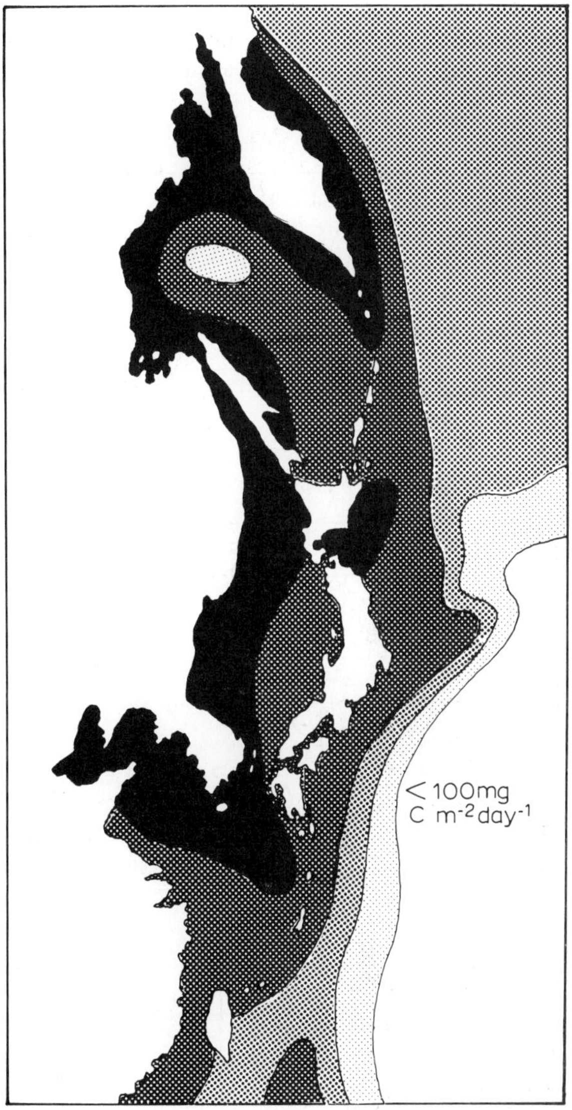

Fig. 15.12. Geographical distribution of primary productivity (mg C m^{-2} day^{-1}) in the Far Eastern seas (adapted from Koblentz-Mishke et al., 1970). Black area: more than 500; thickly hatched: 250–500; moderately hatched: 150–250; thinly hatched: 100–150.

Biomass distribution

Measurements of the abundance and distribution of diatoms, the predominant component of phytoplankton, in the Japan Sea are based on samples collected from the surface by water bottles (Ohwada and Ogawa, 1966). The results for four seasons (Fig. 15.13) indicate that phytoplankton density is always less in the northwestern cold-

TABLE 15.9

Average annual primary production in the Okhotsk, the Japan and the East China Seas (adapted from various authors)

Seas	Productivity ($g\ C\ m^{-2}\ yr^{-1}$)	Net production (wet wt. $10^9\ t\ yr^{-1}$)
Okhotsk Sea	100–150	2.1–3.1
Japan Sea	75–100	0.9–1.3
East China Sea[1]	150–200	2.4–3.2

[1] Including the Yellow Sea and the Gulf of Bo Hai.

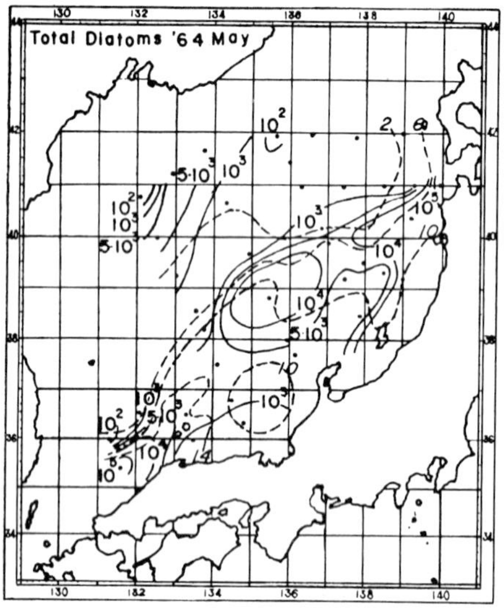

(a) Spring

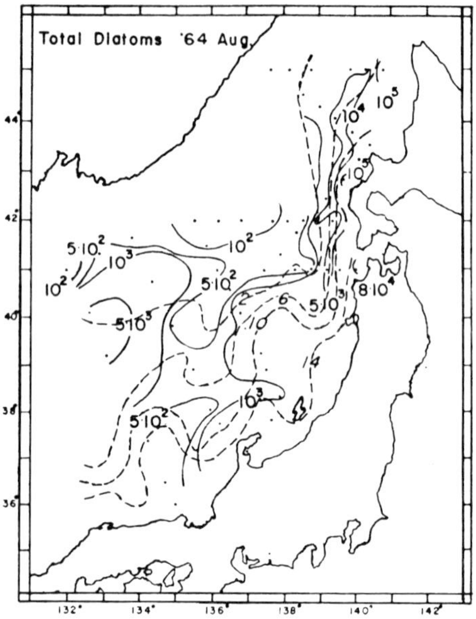

(b) Summer

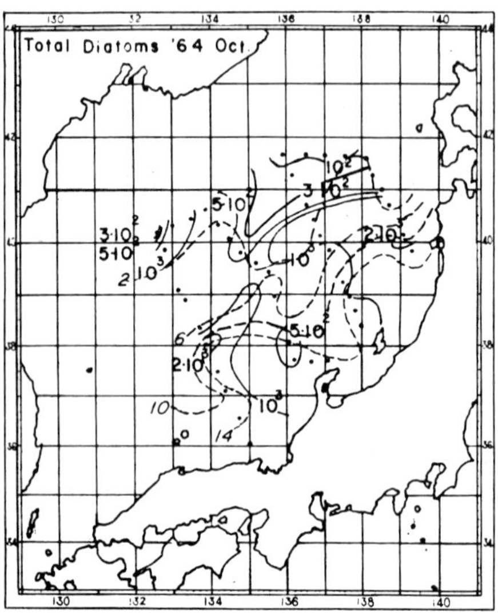

(c) Autumn

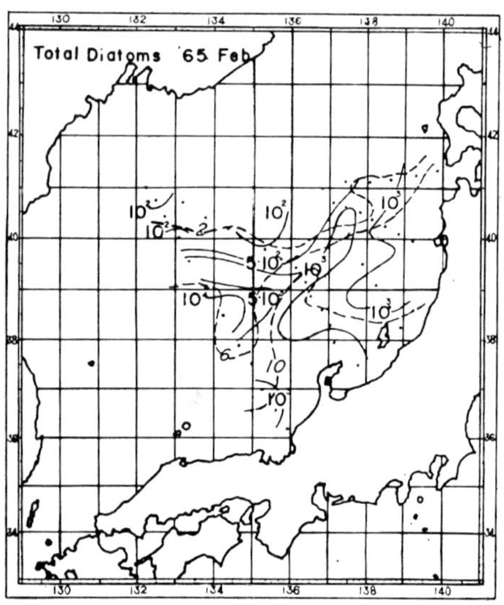

(d) Winter

Fig. 15.13. Diatom distributions (cells l^{-1}) in the Japan Sea (after Ohwada and Ogawa, 1966). Solid lines: diatom abundance; broken lines: water temperature in °C at the 100-m layer.

water area (usually 10^2–10^3 cells l^{-1}) than in the southeastern warm-water area (10^3–10^5 cells l^{-1}). The ratio of the seasonal maximum to the seasonal minimum in population density is also less in the northwestern cold-water area (approximately 50) than in the southeastern warm-water area (approximately 200). The difference between the phytoplankton density in the northwestern cold-water and in the southeastern warm-water area is somewhat larger in spring and summer than it is in autumn and winter.

The lesser occurrence of diatoms in the cold water area appears, at first sight, to contradict the spatial distribution of primary productivity in the Japan Sea (Fig. 15.12). This can be explained, however, by taking into account the grazing effect of herbivorous zooplankton. Zooplankton abundance was estimated from samples collected by vertical net tows in the upper 150 m (Morioka, 1976). The results (Table 15.10) show that throughout the seasons the zooplankton biomass is larger in the northwestern area than in the southeastern area, and that the maximum is found in the spring in both areas. The predominant species constituting the zooplankton biomass in this sea are, in decreasing order of importance, the copepod *Calanus plumchrus*, the euphausiid *Euphausia pacifica*, the amphipod *Parathemisto japonica*, the copepod *Calanus cristatus*, and the euphausiid *Thysanoessa longipes* (Morioka, 1975). *Calanus plumchrus* is by far the most abundant species, forming 65 to 90% of the total zooplankton biomass in the central part of the sea (Nakai, 1942; Meshcheryakova, 1960).

These zooplankters are believed to be bathypelagic normally, but each makes a marked vertical migration to the surface during the young stages. The upward migration takes place in spring and *en masse* — this is one of the most striking biological phenomena in the Japan Sea and in the western North Pacific generally. At shallow depths the young zooplankters feed voraciously on phytoplankton and grow rapidly to the final, remarkably fat, immature stage. They then sink to bathypelagic levels where they spend the longer part of their life and mature sexually. In some of these zooplankters, the young grow so fat that they surpass the adult in body size. This apparently is due to the fact that the adults feed less, or not at all, at bathypelagic depths (Beklemishev, 1954).

If the phytoplankton bloom and zooplankton rise coincide, the phytoplankton could be consumed by these rapidly growing zooplankters and kept at a low level of biomass. Alternatively, if the seasons are out of phase, the phytoplankton could propagate and ultimately attain a considerable biomass without any marked grazing by zooplankton. Evidence suggests that the phytoplankton bloom coincides with the growth of zooplankton herbivores in the northwestern area, but not in the southeastern area of the Japan Sea. The zooplankters rise in middle March to April throughout the sea, and remain in the surface layers until May in the southeastern area and until July in the northwestern area. The vernal bloom of phytoplankton in the southeastern area takes place too early, from February to March, while it occurs during April to May in the northwestern area where it coincides with the rise and growth of zooplankton (Nishimura, 1965b). Thus, production is "unbalanced" (Cushing, 1959; Heinrich, 1962) in the southeastern area, but "balanced" in the northwestern area of this enclosed sea. This conclusion is

TABLE 15.10

Mean biomass of zooplankton (mg wet wt. m^{-3}) in the Japan Sea collected in vertical hauls in the upper 150 m by NORPAC net (0.33-mm mesh); the figures in parentheses are number of hauls (after Morioka, 1976)

Areas	Spring (May)	Summer (July–Aug.)	Autumn (Sept.–Oct.)	Winter (Febr.–Mar.)
Northwestern cold-water area	190 (20)[1]	88 (39)	128 (94)	67 (21)
Southeastern warm-water area	75 (4)[1]	31 (24)	65 (174)	38 (10)

[1] Estimated from figure 1 in Morioka (1976).

compatible with the geographical trends in phytoplankton abundance of the Japan Sea.

Sequence of phytoplankton bloom

One of the most significant factors controlling the initiation of vernal blooms in temperate to high-latitude seas is the relation of the "critical depth" (D_{cr}) to the depth of the "upper mixed layer" (L_m) (Sverdrup, 1953). The critical depth is defined as the depth above which the cumulative amount of organic matter produced through phytoplankton photosynthesis is exactly equal to the cumulative amount of organic matter decomposed through respiration of phytoplankton in the same water column. Phytoplankton growth occurs only when D_{cr} exceeds L_m, other conditions being optimal. A few calculations using a time sequence demonstrate the difference in the timing of the vernal bloom of phytoplankton between the southeastern and northwestern areas of the Japan Sea. It is not possible, however, to calculate the value of D_{cr} directly for the various areas of these enclosed seas because of the lack of pertinent data concerning the extinction coefficient of underwater light and the seasonal and spatial changes of the incident solar energy, so one must rely on an indirect estimate.

An approximate estimate for D_{cr} is derived from the value of the "compensation depth" (D_c), calculated from observed values of the "Secchi disc depth" (D_s). Monthly mean values of D_s for the southeastern area of the Japan Sea are obtained from Uda (1934b), and they are transformed, as shown on Table 15.11, into the approximation of D_{cr} following the relationship given by Enomoto (1959) for the waters west of Kyushu. Values for L_m are estimated from data by Kajiura et al. (1958) on the seasonal sequence in hydrographic sections extending from Cape Tateishi, middle Honshu (*t* in Fig. 15.1), to the central part of the Japan Sea.

The data show that in December and January D_{cr} is equal to or smaller than L_m and therefore no phytoplankton growth can occur. As early as February, D_{cr} is greater than L_m so that phytoplankton growth can occur; this is in fact the time of the start of phytoplankton bloom in the southeastern area of the Japan Sea. The biomass of this bloom is very large, but its specific composition is simple, consisting primarily of a single species, *Coscinodiscus wailesi*, a subtropical diatom. In later stages of the bloom a massive quantity of this diatom, half-disintegrated, gradually sinks to deeper levels; this is another evidence for the unbalanced production in this area (Nishimura, 1965b).

In the northwestern area of the Japan Sea, on the other hand, the critical depth is probably much smaller than the mixed layer depth in winter because the Secchi disc depth is small there, usually 10 to 15 m (Uda, 1934b), while the winter convection develops so markedly that the mixed layer

TABLE 15.11

Monthly sequence in Secchi disc depth (D_s), compensation depth (D_c), critical depth (D_{cr}) and the depth of upper mixed layer (L_m), all in meters, in the southeastern part of the Japan Sea; D_{cr} is unavailable for the summer months, July to September, due to lack of some pertinent data; note that D_{cr} equals or is less than L_m only in December and January (bold-face type)

Months	D_s (m)	D_c (m)	D_{cr} (m)	L_m (m)	Remarks
October	25	75	225	60	seasonal thermocline sinking gradually
November	20	60	180	100	
December	20	60	**120**	**120**	seasonal thermocline merging into permanent thermocline
January	20	60	**120**	**150**	
February	20	60	180	120	
March	25	75	225	150	
April	20	60	240	180	seasonal thermocline developing in
May	15	45	180	10 (200)	upper layers; the values in parentheses indicate the depth of permanent thermocline
June	20	60	240	30 (220)	

is as large as 200 m or more. Under such conditions phytoplankton growth cannot occur. Growth is not possible until April or May when the sea becomes covered by a thin surface layer of diluted water from melting of snow and ice. The phytoplankton bloom in the northwestern area of the Japan Sea occurs in April to May, and this period falls within the time of the rise and vigorous feeding by zooplankters, which keeps the phytoplankton standing crop at a low level. Thus, the biological production in this area may well be described as balanced.

In the Okhotsk Sea, D_{cr} is similarly less than L_m in winter because the incident light intensity is small due to the very low solar altitude, while the winter vertical mixing reaches a considerable depth, possibly to 150 m or occasionally 400 m. Thus, phytoplankton growth cannot occur until June or July when a marked seasonal thermohalocline develops at shallow depths. There is only one phytoplankton maximum — that of the spring to summer bloom — in the Okhotsk Sea, especially in its northern part (Ushakov, 1953), and the bloom seems to be limited to the upper 30 to 50 m above the intermediate cold water (Zenkevitch, 1963). Since the blooming season coincides with the season of the rise and active feeding of bathypelagic zooplankters, as in the northwestern area of the Japan Sea, the biological production is considered to be balanced in this marginal sea, too.

The problem is complicated in the East China Sea. Enomoto (1959) demonstrated that inhibition of phytoplankton growth due to the interaction between D_{cr} and L_m is not feasible even in winter in the waters west of Kyushu. The same situation may prevail for the entire area of the southeastern half of the East China Sea. Phytoplankton growth may be controlled there by the nutrient supply provided by vertical convectional mixing, rather than by the critical depth and upper mixed layer depth relationship as it is in the subtropical seas. Since no important herbivorous zooplankton occurs there in winter, the production may be unbalanced as it is in the southeastern area of the Japan Sea. The facts that phytoplankton growth in this part of the East China Sea starts as early as January (Enomoto, 1959) and the standing crop is generally larger in winter (January to March) than in summer (Asaoka, 1975) appear to confirm this interpretation.

In the northwestern part of the East China Sea the situation is somewhat different. Here the water is shallow and the winter convection is limited in depth, so that L_m cannot be large. Transparency decreases greatly in winter months, with D_s ranging between 5 to 10 m, sometimes dropping to less than 5 m, in this part of the sea (Fig. 15.10) so that D_{cr} can never be large. Since the greatest depths in the central parts of the Yellow Sea and the Gulf of Bo Hai attain respectively 90 m and 60 m, the problem hinges, in fact, upon the subtle difference between these depth values and the critical depth in winter. It is as yet impossible to say which is larger. However, the *Melosira* population of this area is dispersed and appears throughout the layers during the winter, and this suggests that phytoplankton growth is taking place there. In the coastal, shallower parts of these seas, on the other hand, D_{cr} may doubtless exceed the water depth in winter months and can, therefore, bring about an early phytoplankton growth.

The peculiar occurrence pattern of *Melosira sulcata*, the dominant phytoplankter of this area, may be explained in relation to this postulated unbalanced production. This diatom attains a large population size in winter when it is largely free of zooplankton grazing, and it survives in a large quantity and sinks to the sea floor to disintegrate or to spend the warm season in the cool bottom water, embedded within the dusky matrix of floating silt. Upon the start of vertical mixing in the following winter, it may be raised, dispersed throughout the water layers, and begins to propagate under conditions favorable for blooming.

The phytoplankton flora in the Yellow Sea and the Gulf of Bo Hai is very simple; as much as 70 to 90% (by cell number) of the total phytoplankton is composed of *Melosira sulcata* (Kurasige, 1943b). Such a community must be unstable as a system, which again seems to indicate that biological production in this part of the East China Sea might be unbalanced, just as it is in the southeastern part of the Japan Sea where *Coscinodiscus wailesi* predominates in the vernal bloom.

Exploitation of natural products

The natural products from these Far Eastern enclosed seas have been utilized by man since ancient times. Beginning with collecting or hunting in inland waters, people then fished on the seashore of the Japanese Archipelago as early as 5000 years B.C. (Kaneko, 1973). This corresponds to the Atlantic stage in Northern Europe, which was characterized by a prevailing transgression or rise in sea level. A marked increase in marine fishing and hunting occurred during the next subboreal stage when the sea level lowered, leaving a wide coastal plain behind. The people hunted fishes, mainly coastal but sometimes pelagic such as bonito (*Katsuwonus pelamis*) and tuna (*Thunnus thynnus* and *Auxis* spp.), sea mammals and sea birds, and gathered sea urchins and various kinds of shelled mollusks (Naora, 1965). The remains are now found as shell mounds in many places throughout Japan, including the west coast of Kyushu facing the East China Sea, the Japan Sea coast of Honshu, and the Okhotsk Sea coast of Hokkaido. Similar shell mounds have also been found along the East Siberian coast around Vladivostok (*u* in Fig. 15.1) and northward, though the dates seem to be a little later, 2000 to 1000 years B.C. (Okladnikov, 1959).

Fishery techniques, including boat construction and sailing, seem to have been fairly complex in those early days, although improvements have no doubt been made since. By the tenth century A.D. various kinds of fishery products were being manufactured in many districts of Japan as tributes to the central government or for commerce, as gathered, for instance, from a medieval Japanese code "*Engishiki*" compiled in 927 A.D. During the Tokugawa Era (1603–1867), there was considerable development in technical innovations in fishing gear (Yamaguchi, 1957).

Since the Meiji Era (1867–1912) fishing activities have expanded to the high seas in the Japan Sea and in the Okhotsk and East China Seas. This has produced a kind of industrial revolution in Japanese fisheries. The Russians began to share in the natural products of the Far Eastern seas in the 1930s; earlier, their fishing activity was in large part confined to river mouths and coastal zones for migrating salmon and herring (Rass, 1955). After World War II the people of Korea and China entered the fisheries on the high seas.

The natural products from the Far Eastern seas clearly reflect the composition and geographic distribution of the biotic communities. Only some twenty species of fishes are caught commercially in the Okhotsk Sea, but the yields of these fishes are generally high. The most important are herring (*Clupea pallasi*), pink salmon (*Oncorhynchus gorbuscha*), keta (*O. keta*), pollack (*Theragra chalcogramma*), cod (*Gadus macrocephalus*), halibut (*Hippoglossoides elassodon*) and various kinds of plaice (*Clidoderma asperrimum*, *Lepidopsetta bilineata*, *Limanda* spp., *Pleuronectes* spp.). Many of these depend, directly or indirectly, for nutrition on the bathypelagic zooplankters such as *Calanus*, *Metridia*, *Parathemisto* and *Thysanoessa* thriving in the sea. Owing to rich food resources, the fishes grow rapidly to a large size, and attain a fecundity level higher than in, for instance, the North Atlantic (Moiseev, 1957). There are also some large decapod crustaceans that are important commercial items: for example *Paralithodes camtschatica*, *P. brevipes*, *P. platypes* and *Erimacrus isenbecki*.

Most of the shores of the Okhotsk Sea are exposed to strong surf and waves, especially along the northern coasts where the tidal range is extremely large. In addition, species there are subject to killing cold or covered by thick ice during the long-lasting winter. Under such conditions, the littoral biota develops sparsely. Around the Kuril Islands a great wealth of seaweeds, *Laminaria* in particular, develops. This area is under the influence of oceanic warm water from the Pacific, and the tidal range is no more than 0.5 m. Annually great amounts of *Laminaria* are harvested as a food delicacy as well as for the production of alginic acid, iodine and potassium (Miyabe, 1902; Stschapova, 1948).

The northwestern part of the Japan Sea is similar to the Okhotsk Sea with regard to the nature and quantity of living resources; this is quite understandable since the biota and biological production systems are much alike in the two areas. They have many commercial items in common, especially herring, pollack, cod and other cold-water species. The northwestern Japan Sea is distinct, however, in also having some warm-water species of commercial importance. About forty species of fishes are commercially caught, including some temperate forms such as sardine (*Sardinops melanosticta*), anchovy (*Engraulis japonica*), mackerel (*Scomber japonicus*), hairtail (*Trichiurus lepturus*), saury (*Cololabis saira*), and flatfishes of the families Cynoglossidae and Soleidae. The cephalopods *Todarodes pacificus* and *Octopus dofleini* are also of great commercial significance. The pelagic temperate species make a marked seasonal north–south migration over a wide range, occasionally moving from the East China Sea to the northernmost part of the northwestern Japan Sea and vice versa. Among them, sardine and mackerel have undergone a pronounced secular change in population size and migratory habit since the 1930s, possibly due to changes in ocean climate (Kaganovsky et al., 1947; Rumyantzev, 1947).

The southeastern part of the Japan Sea and the off-shelf part of the East China Sea, both under the influence of the Tsushima Current and the Kuroshio, are characterized by rich occurrences of temperate species and of tropical and subtropical species of commercial value in the surface and mid-water levels. Here more than seventy species of fishes are commercially caught, including, in addition to temperate species, tropical and subtropical species such as sea breams (Sparidae), jack mackerel (*Trachurus* and *Decapterus*), yellowtails (*Seriola*), lizard fishes (Synodontidae), *Muraenesox* and its relatives, flying fishes (Exocoetidae), pomfrets (Pampidae) and sillagos (Sillaginidae). These fishes have long been among the food items familiar to and most favored by the Japanese; in particular, the sea bream *Pagrus major* has been highly prized as the "king of edible fish" (Tanaka, 1943).

The cold deeper levels of the Japan Sea yield many cold-water products of the same or closely related species as those fished in the Okhotsk Sea and the northwestern Japan Sea — cod, pollack, halibut (*Hippoglossoides dubius*, a closely related species to *H. elassodon*), plaices, the crabs *Chionoecetes opilio* and *C. japonicus*, and the prawn *Pandalus borealis*. There is thus a marked difference in the specific composition between the yields from the surface to middle layers and from the deep levels, and the transition from one level to the other is sharply defined. Japanese fishermen clearly distinguish between these two levels in their fishing grounds, traditionally calling the shallower ground as *okaba* — the landward ground — where temperate and tropical–subtropical species are caught and the deeper ground *taraba* — the cod ground — where rich yields of cold-water species are harvested.

The littoral biota along the Japan Sea coasts are poorly developed partly because of the small range of tide; many of the genuine intertidal organisms are almost completely absent there. The seasonal fluctuation in sea level is of larger amplitude than the daily tidal range, being higher in summer to autumn than in winter to spring. This plays an important negative role: the organisms that settle on the littoral zone during the high-level season in warm months are then exposed to desiccation and excessive cold in the low-level season from winter to spring, and eventually they perish to be washed away (Nishimura, 1970; Saito, 1970). The rough weather of the northwest monsoon adds to this winter damage of the littoral biota. These situations, together with the lack of many of the stenohaline reef organisms, lead to a marked poverty, qualitative as well as quantitative, in the littoral communities of the Japan Sea as compared with those of the Pacific coasts. Cultures of *nori* (edible *Porphyra*) and a few littoral bivalves are conducted only locally. The sublittoral zone, on the other hand, sustains a fairly rich biota; such products as edible brown alga *Undaria*, agar-agar algae (Gelidiaceae) and abalones (*Haliotis discus* and *H. gigantea*, with local subspecies) are harvested chiefly by diving.

The wide, shallow continental shelf in the Yellow Sea and the Gulf of Bo Hai afford good fishing grounds for trawling. The supposed unbalanced production of phytoplankton together with a huge amount of detritus transported by river discharge may be the causes of the enriched organic content in the bottom sediments of the seas (Hamada and Hosokawa, 1959; Hosokawa et al., 1959). The sea floor supports large populations of commercial species such as croakers (Sciaenidae), gurnards (Triglidae), plaices (*Paralichthys olivaceus*, *Eopsetta grigorjewi*, *Tanakius kitaharai*, *Cynoglossus* spp.), sharks (*Mustelus manazo*, *Squalus mitsukurii*), skates (*Raja kenojei*, *Gymnura japonica*, *Dasyatis* spp.) and shrimps (*Penaeus orientalis*, *Trachypenaeus curvirostris*), most of which are inhabitants of tropical and subtropical estuarine or muddy bottom environments and migrate far into these seas during the warm months for feeding. Vast stretches of sandy or muddy beach with a high detritus content in the sediments, coupled with a large tidal range, favor the development of detritus-feeding infaunal communities within the seas, particularly along their inner shores; the communities are characterized by abundant occurrences of various bivalves and crabs, and many of them are exploited commercially (Kamita, 1941; Kurasige, 1943a).

CONCLUDING REMARKS

Natural products from these enclosed seas have long been important to the people of the Far Eastern countries, especially the Japanese who have had to depend on the products for nutrition and commercial goods. It is no wonder that these seas have had a significant influence on the cultural history of the Japanese people (Inouye et al., 1976). The seas have, in addition, served at times as a barrier against cultural shocks from the outside world and at other times as a route for the influx of foreign cultures.

The significance of these enclosed seas as production sites of valuable natural resources and as communication routes of cultural and economical affairs has not weakened in recent times. It is and will be much augmented in the present and future periods when human activities are greatly increased and nations become more closely connected with one another. It is important to utilize those renewable natural resources on a rational basis and to protect the seas from environmental pollution. Furthermore, it is desirable to augment artificially, if possible, biological production in balance with the natural environments.

Regulation of fishing has already been used for some commercial items (salmon, bottom fishes, crabs and others) within and between nationalities. In recent years, culture methods have been widely introduced into marine fisheries in Japan and other countries for the purpose of increasing the resources and yields. The products include *Porphyra* and *Undaria* among algae, prawns, crabs and some mollusks among invertebrate animals, and yellowtail, sea breams and others among fishes.

Environmental pollution in the Japan Sea poses more serious problems than in the Okhotsk Sea or the East China Sea. This is related to the peculiarities in the topographic features and general circulation patterns. The Japan Sea is more completely enclosed and receives surface water from the East China Sea, not from the open ocean. In addition, the deep and bottom waters of the Japan Sea are isolated and relatively stagnant so that pollutants may accumulate there faster and more easily than in the other marginal seas. It is of vital importance that the nations of the surrounding countries protect these enclosed seas from environmental pollution as far as possible, and to conserve their rich and diversified biota which have been and will continue to be so useful to human welfare in the Far East.

REFERENCES

Aikawa, H., 1935. On the quantitative analysis of the plankton association in the adjacent seas of Japan III. *J. Imp. Fish. Exp. Stn.*, 6: 131–171 (in Japanese).

Andriashev, A.P., 1935. New data on the deep-water fishes of the Bering Sea. *Dokl. Akad. Nauk S.S.S.R.*, 1/2: 105–108 (in Russian).

Andriashev, A.P., 1939. *Outline of the Zoogeography and Origin of the Fish Fauna in the Bering Sea and its Adjacent Waters.* Leningrad University, Leningrad, 187 pp. (in Russian).

Asaoka, O., 1975. Biological oceanography of the East China Sea. *Mar. Sci. Mon., Tokyo*, 7(1): 38–45 (in Japanese).

Asaoka, O. and Moriyasu, S., 1966. On the circulation in the East China Sea and the Yellow Sea in winter (Preliminary report). *Oceanogr. Mag., Tokyo*, 18: 73–81.

Beklemishev, C.V., 1954. The feeding of some common plankton copepods in the Far-Eastern seas. *Zool. Zh.*, 33: 1210–1230 (in Russian).

Bezrukov, P.L., 1960. Bottom sediments of the Okhotsk Sea. *Tr. Inst. Okeanol. Akad. Nauk S.S.S.R.*, 32: 15–95 (in Russian).

Brodsky, K.A., 1957. *Fauna of the Copepoda (Calanoida) and the Zoogeographical Division of the Northern Part of the Pacific Ocean and Adjacent Waters.* Izd. Akad. Nauk S.S.S.R., Leningrad, 222 pp. (in Russian).

Burckhardt, G., 1913. Wissenschaftliche Ergebnisse einer Reise um die Erde von M. Pernod und C. Schröter. III. Zooplancton aus ost- und süd-asiatischen Binnengewässern. *Zool. Jahrb., Abt. Syst. Ökol. Geogr.*, 34: 341–472.

Chamberlin, R., 1924. A new freshwater nereid from China. *Proc. Biol. Soc. Wash.*, 37: 79–82.

Chang, Ch.-L., Cheng, Ch.-T., Cheng, B.-Sh., Li, S.-Ch., Cheng, W.-L. and Wang, W.-B., 1955. *Report of Investigations on the Fishes of the Yellow Sea and the Gulf of Pwok Hai.* Sci. Publ. House, Peking, 353 pp. (in Chinese).

Chen, Q.-Ch. and Zhang, Sh.-Zh., 1965. The planktonic copepods of the Yellow Sea and the East China Sea. I. Calanoida. *Stud. Mar. Sin.*, No. 7: 20–131 (in Chinese).

Chu, Y.T. Chang, Ch.-L. and Cheng, Ch.-T. (Editors), 1963. *Fish Fauna of the East China Sea.* Sci. Publ. House, Peking, 642 pp. (in Chinese).

Cushing, D.H., 1959. The seasonal variation in oceanic production as a problem in population dynamics. *J. Cons., Cons. Int. Explor. Mer*, 24: 455–464.

Dietrich, G. and Kalle, K., 1957. *Allgemeine Meereskunde: eine*

Einführung in die Ozeanographie. Borntraeger, Berlin, 492 pp.

Ekman, S., 1953. *Zoogeography of the Sea*. Sidgwick and Jackson, London, 417 pp.

Enomoto, Y., 1959. Studies on plankton in the west coast of Kyushu II. On conditions for the vernal blooming of phytoplankton. *Bull. Jap. Soc. Sci. Fish.*, 25: 172–182.

Fairbridge, R.W., 1966. East China Sea. In: R.W. Fairbridge (Editor), *Encyclopedia of Oceanography*. Reinhold, New York, N.Y., pp. 238–243.

Fukataki, H., 1974. Fisheries exploitation of the Japan Sea. In: *The Tsushima Warm Current — Oceanographic Structure and Fisheries*. Jap. Soc. Sci. Fish., Kôseisha, Tokyo, pp. 115–128 (in Japanese).

Gorczynski, W., 1920. Sur le calcul du degré de continentalisme et son application dans la climatologie. *Geogr. Ann.*, 2: 324–331.

Hamada, S. and Hosokawa, I., 1959. On the muddy sediments of the East China Sea. *J. Oceanogr. Soc. Jap.*, 15: 39–42 (in Japanese).

Heinrich, A.K., 1962. The life histories of plankton animals and seasonal cycles of plankton communities in the ocean. *J. Cons., Cons. Int. Explor. Mer*, 27: 15–24.

Hidaka, K., 1966. Japan Sea. In: R.W. Fairbridge (Editor), *Encyclopedia of Oceanography*. Reinhold, New York, N.Y., pp. 417–424.

Hoshino, M., 1975. *Eustacy in Relation to Orogenic Stage*. Tokai University Press, Tokyo, 397 pp.

Hosokawa, I., Okabe, S. and Hamada, S., 1959. Oceanographical studies on the sediment in the East China Sea (I). On the contents of organic carbon and total nitrogen. *J. Oceanogr. Soc. Jap.*, 15: 43–47 (in Japanese).

Iida, H., 1961. On the water masses in the coastal region of the south-western Okhotsk Sea. *J. Oceanogr. Soc. Jap.*, 20: 272–278.

Inouye, M., Haga, T. and Hayashiya, T. (Editors), 1976. *Cultural History of the Japanese Archipelago*. Kenkyusha, Tokyo, 328 pp. (in Japanese).

Inouye, S., 1974. The oceanographic aspects of the waters west of Kyushu. In: *The Tsushima Warm Current — Oceanographic Structure and Fisheries*. Jap. Soc. Sci. Fish., Kôseisha, Tokyo, pp. 27–41 (in Japanese).

Kaganovsky, A.G., Starovoitov, P.A. and Kizevetter, I.V., 1947. *The Mackerel*. Primizdat, Vladivostok, 48 pp. (in Russian).

Kajiura, K., 1949. On the hydrography of the Okhotsk Sea in summer. *J. Oceanogr. Soc. Jap.*, 5: 19–26 (in Japanese).

Kajiura, K., Tsuchiya, M. and Hidaka, K., 1958. Analyses of the hydrographic condition of the Japan Sea. *Rep. Expl. Surv. Tsushima Warm Current*. Fish Agency, Tokyo, pp. 158–170 (in Japanese).

Kamita, T., 1941. *Studies on the Decapod Crustaceans of Chosen. I. Crabs*. Fish. Soc. Chosen, Keijô, 289 pp. (in Japanese).

Kamita, T. and Sato, T.N., 1941. Marine fauna of Jinsen Bay, Corea. *J. Chosen Nat. Hist. Soc.*, 8(30): 1–3 (in Japanese).

Kaneko, H., 1973. Development in fishery activity. In: T. Esaka (Editor), *Jômon Earth Wares and Shell Mounds*. Kôdansha, Tokyo, pp. 119–127 (in Japanese).

Kasahara, Sh. and Ito, S., 1972. Distribution and migration of the squid in the offshore area of the Japan Sea. *Proc. Symp. Forest. Fish. Technol.*, 57: 115–143 (in Japanese).

Kawai, H., 1974. Historical change in the interpretation of ocean currents in the Japan Sea. In: *The Tsushima Warm Current — Oceanographic Structure and Fisheries*. Jap. Soc. Sci. Fish., Kôseisha, Tokyo, pp. 7–26 (in Japanese).

Kawakami, I. and Habe, T., 1961. The characteristic aspects of the molluscan fauna in the west coast of Kyushu, Japan. *Rec. Oceanogr. Works Jap., Spec. No.*, 5: 195–197.

Kitani, K., 1972. On the variability of dichothermal water in the Okhotsk Sea. In: A.Y. Takenouti et al. (Editors), *Biological Oceanography of the Northern North Pacific Ocean*. Idemitsu Shoten, Tokyo, pp. 45–62.

Koblentz-Mishke, O.I., 1965. Primary production in the Pacific. *Okeanologiya*, 5: 325–337 (in Russian).

Koblentz-Mishke, O.I., Volkovinsky, V.V. and Kavanova, J.A., 1970. Primary production of the world ocean. In: L. Zenkevitch (Editor), *Program and Method in the Studies of Biogeocoenoses of Aquatic Environments*. Nauka, Moscow, pp. 66–83 (in Russian).

Kurashima, K., Nishida, K. and Nakabayashi, S., 1966. On the open water in the south-eastern part of the frozen Okhotsk Sea and the currents through the Kurile Islands. *J. Oceanogr. Soc. Jap.*, 23: 58–62.

Kurasige, H., 1943a. Tidal flats in Korea. *Kaiyô no Kagaku (Science of the Ocean), Tokyo*, 3(8): 22–24 (in Japanese).

Kurasige, H., 1943b. Quantitative and qualitative characteristics of the marine diatom in the coastal water of Taiyatô at the Yellow Sea side of Tyôsen, in comparison with that of Tataho Bay at the southern coast of the peninsula. *Bull. Fish. Exper. Stn., Gov.-Gen. Chosen*, No. 8: 1–114 (in Japanese).

Kuroda, K., 1977. Chaetognatha in the Kuroshio area south of Japan. II. Revision of indicator species. *Bull. Kobe Mar. Obs.*, No. 194: 26–33.

Kwan, Ping-Hsien, 1957. A preliminary study on the relations between the surface currents and the winds along China coast. *Oceanol. Limnol. Sin.*, 1: 95–122 (in Chinese).

Leonov, A.K., 1960. *Regional Oceanography. Part I*. Gidrometeoizdat, Leningrad, 765 pp. (in Russian).

Lindberg, G.U., 1947. A preliminary list of the fishes of the Japan Sea. *Izv. TINRO*, 25: 125–206 (in Russian).

Lindberg, G.U., 1955. *The Quaternary Period in the Light of Biogeographical Data*. Izd. Akad. Nauk S.S.S.R., Leningrad, 334 pp. (in Russian).

Lindberg, G.U., 1972. *Large-Scale Fluctuations of Sea Level in the Quaternary Period*. Nauka, Moscow, 548 pp. (in Russian).

Mashiko, K., 1954. A note on the brackish-water copepods. *Jap. J. Ecol.*, 4: 13–16 (in Japanese).

Masuzawa, J., 1960a. Kuroshio. In: K. Wadati (Editor), *Encyclopedia of Oceans*. Tokyodô Shuppan, Tokyo, pp. 214–219 (in Japanese).

Masuzawa, J., 1960b. The Japan Sea. In: K. Wadati (Editor), *Encyclopedia of Oceans*. Tokyodô Shuppan, Tokyo, pp. 438–441 (in Japanese).

Matudaira, Y. and Inoue, S., 1939. On the distribution of floating mud, and the state of the sea, off Yangtze River. *J. Oceanogr. Imp. Mar. Obs. Kobe*, 11: 705–730 (in Japanese).

Meshcheryakova, I.M., 1960. Seasonal changes of the plankton in offshore waters of the Japan Sea. *Izv. TINRO*, 46: 95–144 (in Russian).

Minato, M. and Ijiri, S., 1966. *The Japanese Archipelago*. Iwanami Shoten, Tokyo, 2nd ed., 223 pp. (in Japanese).

Misu, H. et al., 1955. Feeding habits of demersal fishes. In: *The Investigations of Demersal Fish Resources in the East China Sea and the Yellow Sea. Bull. Seikai Reg. Fish. Res. Lab.*, No. 2: 96–106 (in Japanese).

Miyabe, K., 1902. Laminariaceae of Hokkaido. *J. Sapporo Agric. Coll.*, 1: 1–50.

Miyadi, D., Kuroda, T. and Habe, T., 1953. On the biogeographic provinces in the Japanese waters. *Seibutsu Kagaku (Biological Science), Tokyo*, 5: 145–148 (in Japanese).

Miyazaki, M., 1952. The heat budget of the Japan Sea. *Bull. Hokkaido Reg. Fish. Res. Lab.*, No. 4: 1–54 (in Japanese).

Miyazaki, M., 1953. On the water masses of the Japan Sea. *Bull. Hokkaido Reg. Fish. Res. Lab.*, 7: 1–65 (in Japanese).

Miyazaki, M. and Abe, S., 1960. On the water masses in the Tsushima Current area. *J. Oceanogr. Soc. Jap.*, 16: 59–68 (in Japanese).

Moiseev, P.A., 1957. Influence of oceanologic regimen of the Far-Eastern seas on commercial fish population. *Proc. UNESCO Symp. Phys. Oceanogr., 1955, Tokyo*, pp. 253–259.

Mokievsky, O.B., 1949. Freshwater intertidal zone of the estuary of Amur. *Dokl. Akad. Nauk S.S.S.R.*, 66: 1187–1190 (in Russian).

Morioka, Y., 1975. Plankton ecology in the Japan Sea. *Nat. Sci. and Mus., Tokyo*, 42(2): 75–79 (in Japanese).

Morioka, Y., 1976. Plankton standing crops in the Japan Sea. *Mar. Sci. Mont., Tokyo*, 8(9): 55–60 (in Japanese).

Moriyasu, S., 1972. The Tsushima Current. In: H. Stommel and K. Yoshida, (Editors), *The Kuroshio*. Tokyo University Press, Tokyo, pp. 353–369.

Nakai, Z., 1942. Plankton resources. *Kaiyô no Kagaku (Science of the Ocean), Tokyo*, 2: 181–188 (in Japanese).

Naora, N., 1965. *The Life and Environment of Ancient Japanese*. Azekurashobo, Tokyo, 270 pp. (in Japanese).

Niino, H. and Emery, K.O., 1961. Sediments of shallow portions of East China Sea and South China Sea. *Bull. Geol. Soc. Am.*, 72: 731–762.

Nishimura, S., 1965a. The zoogeographical aspects of the Japan Sea. Part I. *Publ. Seto Mar. Biol. Lab.*, 13: 35–79.

Nishimura, S., 1965b. The zoogeographical aspects of the Japan Sea. Part II. *Publ. Seto Mar. Biol. Lab.*, 13: 81–101.

Nishimura, S., 1966a. The zoogeographical aspects of the Japan Sea. Part III. *Publ. Seto Mar. Biol. Lab.*, 13: 365–384.

Nishimura, S., 1966b. Origin and history of the teleostean genus *Fugu*. *Nanki Seibutu*, 8: 5–11 (in Japanese).

Nishimura, S., 1967. Origin and history of the Far-Eastern freshwater serranid *Siniperca* and its allied genera (Teleostei:Percida). *Bull. Osaka Mus. Nat. Hist.*, No. 20: 13–30 (in Japanese).

Nishimura, S., 1968. The zoogeographical aspects of the Japan Sea. Part IV. *Publ. Seto Mar. Biol. Lab.*, 15: 329–352.

Nishimura, S., 1969. The zoogeographical aspects of the Japan Sea. Part V. *Publ. Seto. Mar. Biol. Lab.*, 17: 67–142.

Nishimura, S., 1970. A note on the littoral biota of the Japan Sea. *Nanki Seibutu*, 12: 34–38 (in Japanese).

Nishimura., S., 1971. Composition and origin of the marine fauna of Japan. In: S. Nishimura and K. Suzuki, *Common Seashore Animals of Japan in Color*. Hoikusha, Osaka, pp. 170–179 (in Japanese).

Nishimura, S., 1977. *Origin and History of the Japan Sea; an Approach from Biogeographic Standpoint*. Tsukiji Shokan, Tokyo, 2nd ed., 228 pp. (in Japanese).

Ogura, S., 1933. The tides in the seas adjacent to Japan. *Bull. Hydrogr. Dep. Imp. Jap. Navy*, 7: 1–189.

Ohwada, M. and Ogawa, F., 1966. Plankton in the Japan Sea. *Oceanogr. Mag., Tokyo*, 18: 39–42.

Okiyama, M., 1974. Biological features of the Japan Sea — Characteristics in the biota. In: *The Tsushima Warm Current — Oceanographic Structure and Fisheries*. Jap. Soc. Sci. Fish., Kôseisha, Tokyo, pp. 42–55 (in Japanese).

Okladnikov, A.P., 1959. *Primorie in the Far Old Time*. Primizdat, Vladivostok, 292 pp. (in Russian).

Okuda, S., 1943. Occurrence of a freshwater polychaete in Central China. *J. Shanghai Sci. Inst., N.S.*, 2: 99–103.

Rass, T.S., 1955. New regions and new objects for fisheries in the Far-Eastern seas. *Vopr. Ikhtiol.*, 4: 71–81 (in Russian).

Rass, T.S., 1963. Deep-sea rattail fishes (Macrouridae, Pisces) in the Okhotsk Sea. *Tr. Inst. Okeanol. Akad. Nauk S.S.S.R.*, 62: 211–223 (in Russian).

Rass, T.S., 1965. *Fisheries Resources of the European Seas of the U.S.S.R. and Possibilities of Their Augmentation by Acclimatization*. Nauka, Moscow, 106 pp. (in Russian).

Rumyantzev, A.I., 1947. On the changes in the composition of warm-water fish fauna of the Primorie waters in the Japan Sea. *Zool. Zh.*, 26: 47–52 (in Russian).

Saito, Y., 1970. Reason for the poverty in algal vegetation on the Japan Sea coasts, as discussed from ecological viewpoint. *Kagaku (Science), Tokyo*, 40: 561–565 (in Japanese).

Samoilenko, V.S. (Editor), 1966. *The Pacific Ocean. I, Meteorological Conditions above the Pacific Ocean*. Nauka, Moscow, 390 pp. (in Russian).

Sanger, G.A., 1972. Fishery potentials and estimated biological productivity of the Subarctic Pacific region. In: A.Y. Takenouti et al. (Editors), *Biological Oceanography of the Northern North Pacific Ocean*. Idemitsu Shoten, Tokyo, pp. 561–574.

Savilov, A., 1961. Ecologic characteristics of the bottom communities of invertebrates in the Okhotsk Sea. *Tr. Inst. Okeanol. Akad. Nauk S.S.S.R.*, 46: 3–84 (in Russian).

Schmidt, P., 1935. The Okhotsk Sea and its fish fauna. *Vestn. Akad. Nauk S.S.S.R.*, 1935 (5): 29–38 (in Russian).

Schmidt, P., 1950. *Fishes of the Okhotsk Sea*. Izd. Akad. Nauk S.S.S.R., Leningrad, 370 pp. (in Russian).

Shen, C.-J., 1955. A study on the crustaceans in the neighboring seas of Fenghsien, Kiangsu. *Acta Zool. Sin.*, 7: 75–101 (in Chinese).

Shen, C.-J. and Bai, Hs.-E., 1956. A study on the Copepoda in the spawning ground of the mackerel off Yentai. *Acta Zool. Sin.*, 8: 177–234 (in Chinese).

Shen, C.-J. and Li, F.-Hs., 1963. The Copepoda in the mouth of the Chien Kiang, Kwangtung. *Acta Zool. Sin.*, 14: 571–596 (in Chinese).

Shen, C.-J. and Tai, A.-Y., 1964. Descriptions of eight new species of freshwater copepods from the delta of the Chu Kiang. *Acta Zool. Sin.*, 16: 225–246 (in Chinese).

Shimomura, T. and Miyata, K., 1953. On the oceanographic character of the low temperature region off Sado Island I. *Bull. Jap. Soc. Sci. Fish.*, 19: 424–428 (in Japanese).

Stschapova, T., 1948. Geographical distribution of the order Laminariales in the northern part of the Pacific Ocean. *Tr. Inst. Okeanol. Akad. Nauk. S.S.S.R.*, 2: 89–138 (in Russian).

Suda, K., 1932. On the bottom water of the Japan Sea (Preliminary report). *J. Oceanogr. Imp. Mar. Obs. Kobe*, 4: 221–240 (in Japanese).

Suda, K., Hidaka, K., Matudaira, Y., Kurasige, H., Kawasaki, H. and Kubo, T., 1932. The results of the oceanographical observations on board M.S. "Synpu Maru" in the southern part of the Japan Sea in the summer of 1930. *J. Oceanogr. Imp. Mar. Obs. Kobe*, 4: 1–173 (in Japanese).

Sverdrup, H.U., 1953. On conditions for the vernal blooming of phytoplankton. *J. Cons., Cons. Int. Explor. Mer*, 18: 287–295.

Sverdrup, H.U., Johnson, M.W. and Fleming, R.H., 1942. *The Oceans; Their Physics, Chemistry and General Biology*. Prentice-Hall, Englewood Cliffs, N.J., 1087 pp.

Tabata, T., 1955. On the hydrography of the southern side of Kuril Islands and of the southern part of Okhotsk Sea. *Low Temp. Sci., Sapporo*, 9: 159–170 (in Japanese).

Takahashi, Y. (Editor), 1978. *River Hydrology*. Kyoritsu Shuppan, Tokyo, 218 pp. (in Japanese).

Tanaka, S., 1943. *Taste and Nutrition of Edible Fishes*. Jidaisha, Tokyo, 341 pp. (in Japanese).

Taniguchi, A., 1972. Geographical variation of primary production in the western Pacific Ocean and adjacent seas with reference to the interrelations between various parameters of primary production. *Mem. Fac. Fish., Hokkaido Univ.*, 19: 1–33.

Taniguchi, A., 1973. Photosynthetic activity of the marine phytoplankton communities. *Mar. Sci. Mon., Tokyo*, 5(2): 19–25 (in Japanese).

Tokioka, T., 1959. Observations on the taxonomy and distribution of chaetognaths of the North Pacific. *Publ. Seto Mar. Biol. Lab.*, 7: 349–456.

Tokyo Astronomical Observatory, 1978. *Science Almanac*. Maruzen, Tokyo, 855 pp. (in Japanese).

Tsujita, T., 1963. The fishery oceanography of the East China Sea and Tsushima Strait. 2. On the characteristics and fluctuations of the cold water masses in view point of the production oceanography. *Bull. Seikai Reg. Fish. Res. Lab.*, No. 28: 57–68.

Uchida, K., 1935. The marine fish fauna of Korea. *Zool. Mag., Tokyo*, 47: 132–134 (in Japanese).

Uchida, K. and Tsukahara, H., 1953. On the fish fauna of Ariake Bay. *Zool. Mag., Tokyo*, 62: 121–122 (in Japanese).

Uda, M., 1934a. The results of simultaneous oceanographical investigations in the Japan Sea and its adjacent waters in May and June, 1932. *J. Imp. Fish. Exper. Stn.*, No. 5: 57–190 (in Japanese).

Uda, M., 1934b. Hydrographical researches on the normal monthly conditions in the Japan Sea, the Yellow Sea, and the Okhotsk Sea. *J. Imp. Fish. Exper. Stn.*, No 5: 191–236 (in Japanese).

Uda, M., 1935. On the distribution, origin and movement of the dichothermal water in the North-Eastern Sea region adjacent to Japan of the North Pacific Ocean. *Umi to Sora*, 15: 445–452 (in Japanese).

Uda, M., 1936. Results of simultaneous oceanographic investigations in the Japan Sea and its adjacent waters during October and November, 1933. *J. Imp. Fish. Exper. Stn.*, No. 7: 91–151 (in Japanese).

Uda, M., 1938. Notes on the origin of deep water in the seas adjacent to Japan. *Jap. J. Limnol.*, 8: 195–204 (in Japanese).

Uda, M., 1955. Research on the fluctuation of North Pacific circulation. *Rec. Oceanogr. Works Jap.*, 2: 1–13.

Uda, M., 1966. Yellow Sea. In: R.W. Fairbridge (Editor), *Encyclopedia of Oceanography*. Reinhold, New York, N.Y., pp. 994–998.

Udintzev, G.B., 1957. Bottom topography of the Okhotsk Sea, *Tr. Inst. Okeanol. Akad. Nauk S.S.S.R.*, 22: 3–76 (in Russian).

Ushakov, P., 1953. *The Fauna of the Okhotsk Sea and its Life Conditions*. Zool. Inst. Akad. Nauk S.S.S.R., Leningrad, 459 pp. (in Russian).

Ushakov, P., 1955. *Polychaetous Annelids of the Far-Eastern Seas of the U.S.S.R.* Izd. Akad. Nauk S.S.S.R., Moscow and Leningrad, 445 pp. (in Russian).

U.S.S.R. Naval Office, 1977. *Atlas of the Oceans: the Pacific Ocean*. Moscow, 302 maps (in Russian).

Vinogradov, M.E., 1954a. Daily vertical migration of zooplankton in the Far-Eastern seas. *Tr. Inst. Okeanol. Akad. Nauk S.S.S.R.*, 8: 164–199 (in Russian).

Vinogradov, M.E., 1954b. The vertical distribution of zooplankton biomass in the Kuril–Kamchatka trench. *Dokl. Akad. Nauk S.S.S.R.*, 96: 637–640 (in Russian).

Watanabe, K., 1967. Summary of drift ice in the Okhotsk Sea. *Inst. Low Temp. Sci. Publ.*, 1967: 667–686.

Yabe, H., 1929. The latest land connection of the Japanese Islands to the Asiatic continent. *Proc. Imp. Acad. Tokyo*, 5: 167–169.

Yamaguchi, K., 1957. *History of Fisheries in Japan*. Tokyo University Press, Tokyo, repr. ed., 351 pp. (in Japanese).

Yang, L.-H. and Hsu, Ch.-Y., 1978. *The Seas of China*. Guangdong (Kwangtung) People's Publ. House, Guangzhou (Kuangchou), 72 pp. (in Chinese).

Yasuoka, T., 1967. Hydrography in the Okhotsk Sea (1). *Oceanogr. Mag., Tokyo*, 19: 61–72.

Yasuoka, T., 1968. Hydrography in the Okhotsk Sea (2). *Oceanogr. Mag., Tokyo*, 20: 55–63.

Zenkevitch, L., 1963. *Biology of the Seas of the U.S.S.R.* Interscience, New York, N.Y., 955 pp.

Zenkevitch, L. and Birstein, J.A., 1960. On the problem of the antiquity of the deep-sea fauna. *Deep-Sea Res.*, 7: 10–23.

Chapter 16

THE GULF OF ST. LAWRENCE

L.M. DICKIE AND R.W. TRITES

INTRODUCTION

The Gulf of St. Lawrence (Fig. 16.1) is a triangular-shaped semi-enclosed sea on the east coast of Canada. The southern apex of the triangle is at about 46°N latitude, 62°W longitude. The northern base is at about 50°N between 57° and 67°W. It has a surface area of about 226 000 km^2 (Forrester and Vandall, 1968), and is thus about half the area of the Baltic Sea or the Black Sea and one-third to one -quarter the area of the North Sea, Sea of Japan or Gulf of Mexico. The enclosed volume is 34 500 km^3 (Forrester, 1964) and the mean depth is 152 m.

Like the Gulf of Mexico and the Black Sea and unlike the other major semi-enclosed seas, the Gulf receives the discharge from a major river, the St. Lawrence, which enters its northwest corner and

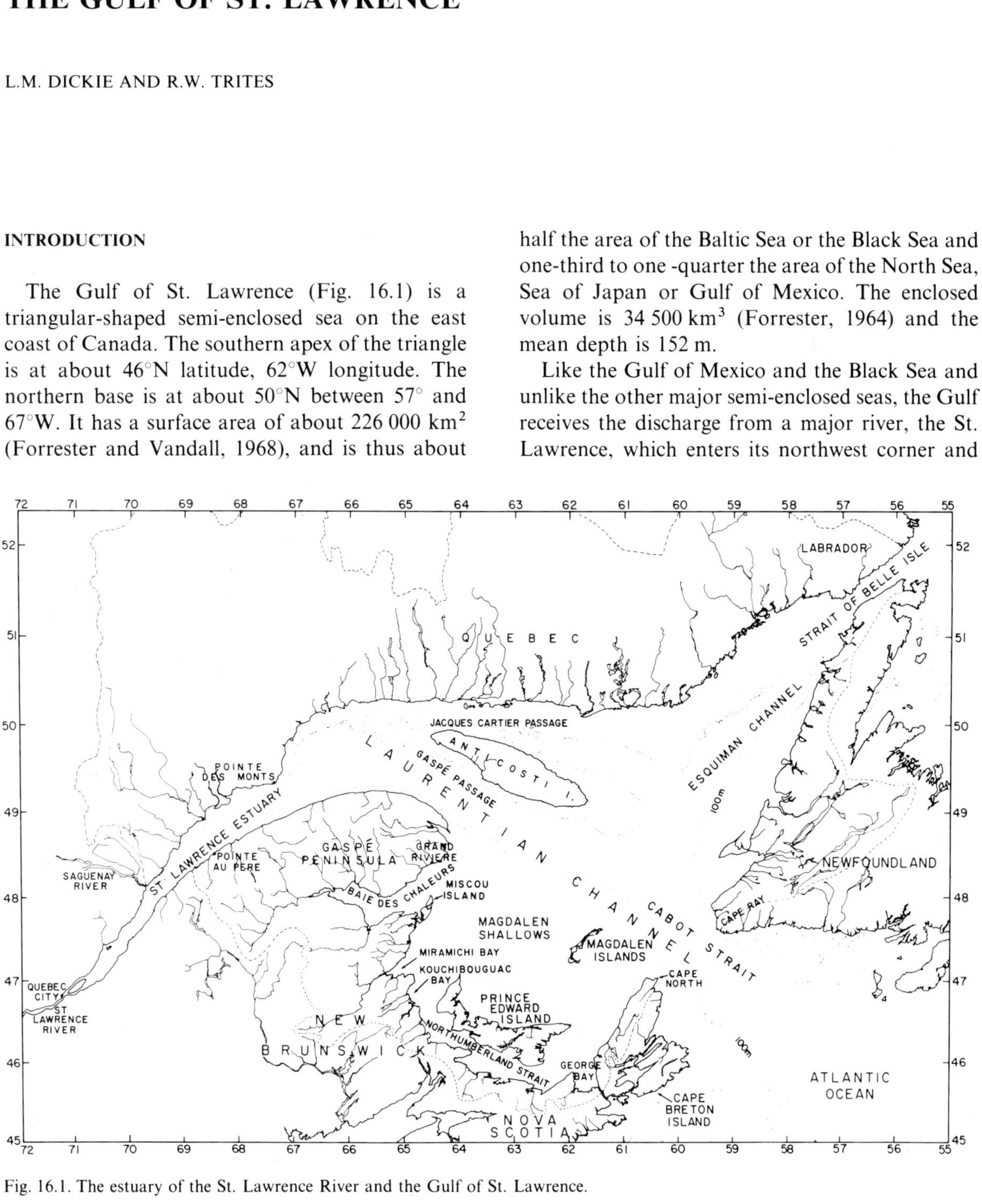

Fig. 16.1. The estuary of the St. Lawrence River and the Gulf of St. Lawrence.

has a very important influence on its dynamics. In addition, the Gulf of St. Lawrence is fed, especially along its northern coast, by a large number of intermediate and small rivers. Its drainage basin is large, some nine times the area of the Gulf proper. The ratio of the drainage basin to the surface area is larger than in any other semi-enclosed sea. There are two connexions with the ocean. The larger, Cabot Strait on the southeast, is 104 km wide with a maximum depth of 480 m. The smaller, Strait of Belle Isle at the northeast corner, is 16 km wide with a sill depth of only 60 m.

The location on the eastern seaboard of North America, coupled with the St. Lawrence River, makes it a major avenue for access to the interior of the continent. It is also the setting for important fisheries, oil exploration activities and recreation, despite the fact that it is frequently almost completely covered by ice for two months of the year. In addition, the many rivers entering it are subject to extensive and progressive alteration, or "regulation", principally for hydroelectric power generation. The present and potential extent of these alterations seems to pose an example of major direct effect by man on the dynamics of a semi-enclosed sea. Under these circumstances, it becomes a matter of particular interest to understand, predict and control the consequences of these man-made modifications.

GEOLOGY AND MORPHOLOGY

Geologically the Gulf of St. Lawrence is a drowned Palaeozoic, mainly Carboniferous, lowland. The original uplift area was probably well eroded in the Palaeozoic when the major channels were established and then refilled to be eroded again into their present form, more or less, by the Early Cretaceous (Loring and Nota, 1973). At this time the original peneplain was probably tilted to the southwest and collected the river drainage along the forerunner of the deep Laurentian Channel, which cuts across the floor of the present Gulf from the middle estuary of the river in the northwest corner through Cabot Strait, thence across the continental shelf for some 250 km to the continental slope. Along much of this channel the depth now exceeds 350 m, dipping to more than 500 m near Cabot Strait, thus opening the Gulf to access from ocean waters of intermediate depth (Fig. 16.2). From the north-central side of the Laurentian Channel a branch (possibly a sunken rift valley) called the Esquiman Channel extends northeastward towards the Strait of Belle Isle where the narrower, shallower channel provides a connexion with the inshore waters of the Labrador continental shelf. The Esquiman Channel itself has a further branch to the northwest into the Jacques Cartier Passage, between the northern coast and Anticosti Island, a large oblong island rising from the deep floor and lying on a northwest–southeast axis in the northwest corner of the Gulf.

If the deeper channels are taken as bounded by the 200-m contour, then about half of the surface area lies over deeper channels. To the north of the Laurentian Channel and on the borders of the Jacques Cartier Passage, the bottom slopes rather sharply from deep water to either the mainland or to Anticosti Island. To the north of Esquiman Channel a secondary trough lies between it and land, thus continuing the steeply sloped marginal topography. The same steep sides characterize the inner half of the south side of the Laurentian Channel from the junction of the St. Lawrence and Saguenay Rivers eastwards towards the tip of the Gaspé Peninsula, which juts into the northwest corner of the Gulf south of the river. The remaining half of the south side of the Laurentian Channel is bordered by a broad shallow sandy plateau which is generally shallower than 75 m. A similar plateau, which is narrower and smaller, borders the east side of Esquiman Channel, between it and the island of Newfoundland. Altogether, about 25% of the area of the Gulf is shallower than 75 m.

The broad southwestern shallows are a prominent feature and appear to play an important role in the realization of the potential for biological production. Sitting on this broad sunken plain are two major islands. To the southwest, Prince Edward Island is a large elongated island of erosion-resistant red sandstone lying parallel to the coastline and separating Northumberland Strait, a strait of water over 200 km in length and 15 to 30 km wide, from the main circulation. Northeast of Prince Edward Island, near the edge of the shallows, occurs a smaller island — really a series of three igneous islands topped with red sandstone and connected by long sandbars — called the Magdalen Islands (Îles de la Madeleine). Current

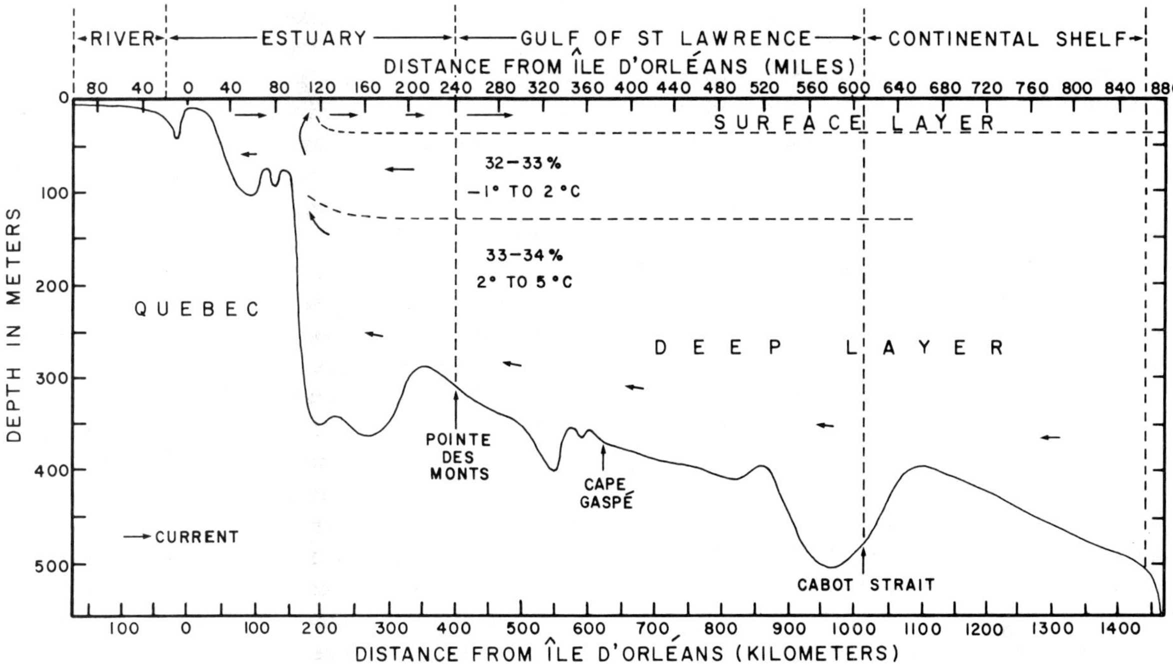

Fig. 16.2. Schematic longitudinal profile along the St. Lawrence River and Laurentian Channel from Île d'Orléans to the edge of the continental shelf.

patterns and sediment compositions suggest that this central group of islands is in dynamic equilibrium with the forces of erosion and deposition. All the islands are prominent topographic features which affect the circulation in the Gulf.

TEMPERATURE AND SALINITY DISTRIBUTIONS

Thermally, the water of the Gulf of St. Lawrence is three-layered in summer (Figs. 16.2 and 16.3A). A warmed surface layer of salinity 27 to 32‰ ranges in thickness from 10 to 15 m in the north and east to between 15 to 30 m over parts of the Magdalen Shallows. It is underlain by an intermediate layer of very cold (−0.5° to 1.0°C) moderately saline (31.5–33‰) water centered at about 80 m but at times extending to as deep as 150 m. Below this, and filling the deeper channels, is a warmer water body with a maximum temperature of 4° to 6°C, and salinity about 34.6‰, which fills the deeper channels. Trites (1972) pointed out that summer–winter differences are insignificant for these deeper waters which comprise about 45% of the total water volume of the system.

The deepest layer is continuous with the deeper water in the channels of the continental shelf which originates in part from the southward flowing Labrador Current (Dunbar, 1955). Lauzier and Trites (1958) found that the maximum temperature of this layer was correlated with its volume, which in turn was related to the strength of the Labrador Current. In the late 1950s the deep layer reached a peak temperature for recent years of about 6°C and has since displayed a slow downward trend in keeping with the general cooling of eastern Arctic and Greenland waters. Temperature reached about 4°C by 1973. El Sabh (1977) analyzed 70 oceanographic sections across Cabot Strait at various times of the year between 1950 and 1970. Mean monthly transport patterns were computed. An example for August is shown in Fig. 16.3B. On average the inward flow of water through this most important exchange passage occupies all the deeper layers of the section and extends throughout the whole water column on the northeastern or Newfoundland side. Inward flow is most rapid at depths of from 50 to 150 m, while the outflow is most rapid at the surface. Net inflow, however, is largely confined to depths greater than 100 m and reaches a maximum at depths of 150 to 300 m depending on the season. In August 1966, for example (Fig. 16.3B), the maximum net inflow was at 250 to 275 m.

The origin and dissipation of the middle, cold layer has been a subject of speculation (Tremblay and Lauzier, 1940; Forrester, 1964; Banks, 1966). This layer is continuous with the superficial waters of the Labrador Shelf through the Strait of Belle Isle and resembles them in its temperature and salinity characteristics. There is certainly some significant exchange through this passage. However, the rates of exchange are small compared with the very large total volume of this layer in the Gulf. It appears, therefore, that the layer is largely formed *in situ* from the surface layers during winter when vertical mixing and convection extends to depths of 100 to 125 m and the two upper layers are thermally indistinguishable.

As the spring freshets and vernal warming establish the strongly differentiated surface layer, the exchange of water through the Strait of Belle Isle can be seen in the outflow of warmed surface water and the inflow of colder, saline Labrador water which is added to the intermediate layer. Banks (1966) calculated that the annual inflow from this source could be as much as 12% of the volume of the intermediate layer.

The intermediate layer is identifiable in cross sections of the Cabot Strait but appears to lie largely below the depth of net outflow. Dissipation appears therefore to be largely by entrainment and upwelling into the surface outflow or, perhaps more important, by mixing and advection at the main estuarial upwelling areas discussed below.

The origin and dynamics of the surface layers are complex and are particularly interesting because of the interaction with human activities which influence the fresh-water inflow. The major input of fresh water originates from the St. Lawrence River. El Sabh (1977) estimated that the river provides 75% of a total mean annual river discharge rate of 19 000 $m^3 s^{-1}$. Trites (1972) estimated a further contribution of 2500 $m^3 s^{-1}$ as the net difference between mean annual precipitation less evaporation over the Gulf itself. The total fresh-water inflow appears to be mixed rapidly with salt water, leading to a total transport of waters into and out of the Gulf which is far in excess of the net fresh-water input. Dawson (1913) estimated that the volume of water involved in the Gaspé current system was more than 100 times the volume of the St. Lawrence River runoff. El Sabh (1977) estimates that at

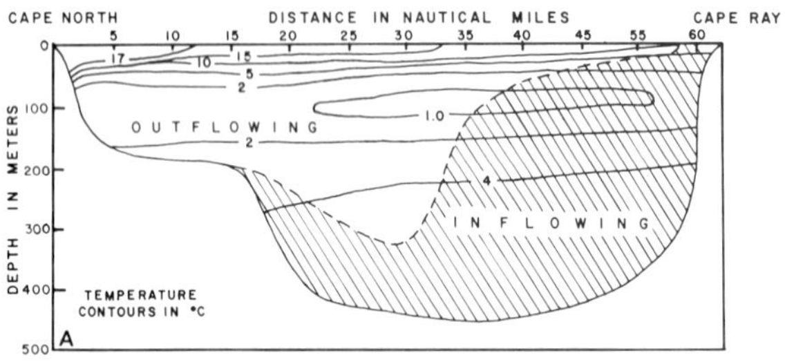

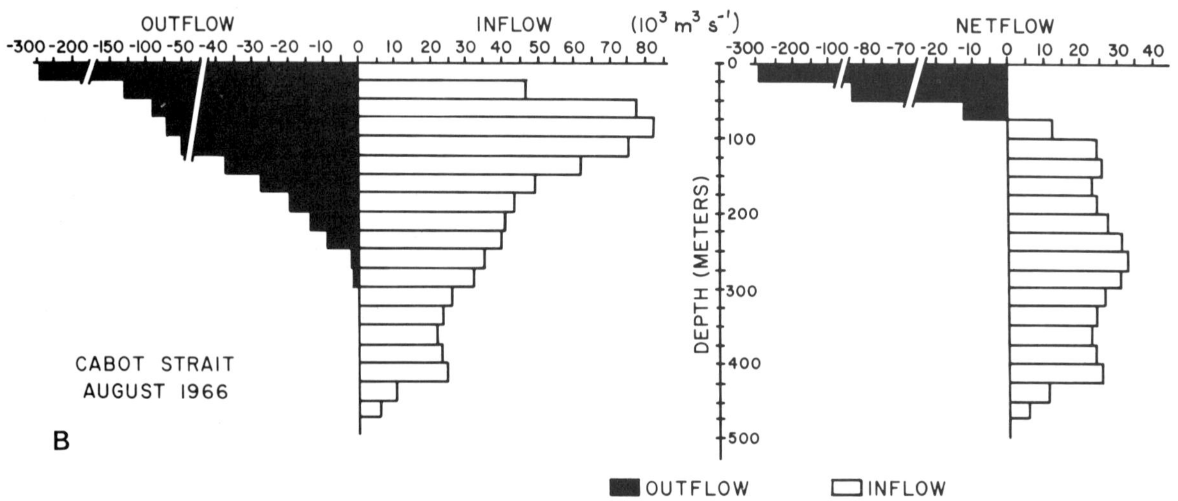

Fig. 16.3A. The temperature profile in Cabot Strait, August 1966, with residual current patterns as measured by moored current meters. B. Computed flow distribution with depth in Cabot Strait, August 1966 (after El Sabh, 1977).

Cabot Strait the total outward volume transport is about 30 times the volume of the net transport. While this outward transport is generally strongest at the surface, on the south side in August when the spring freshet effects are felt in the area, outflow rates in excess of 20 cm s^{-1} may be found more than 50 km from shore at depths of 30 m.

The fresh-water inflow from the St. Lawrence River begins to be mixed with seawater near Île d'Orléans just down-river from Québec City. Salt water is found on the river bed here at times of low volume outflow. The major mixing, however, appears to take place at the point of confluence of the Saguenay and St. Lawrence Rivers (Fig. 16.4). In this region, over a distance of only 20 km, the river bottom drops sharply from less than 40 m to a depth of more than 350 m, marking the head of the Laurentian Channel. In this area the mixing processes are so strong that at certain times the water is virtually homogeneous in a vertical section at 5°C and 28‰, even in July. Temperature and oxygen are low at the bottom of the Laurentian Channel, and salinity, nitrate, phosphate and silicate are all high. However, isopleths of these characteristics

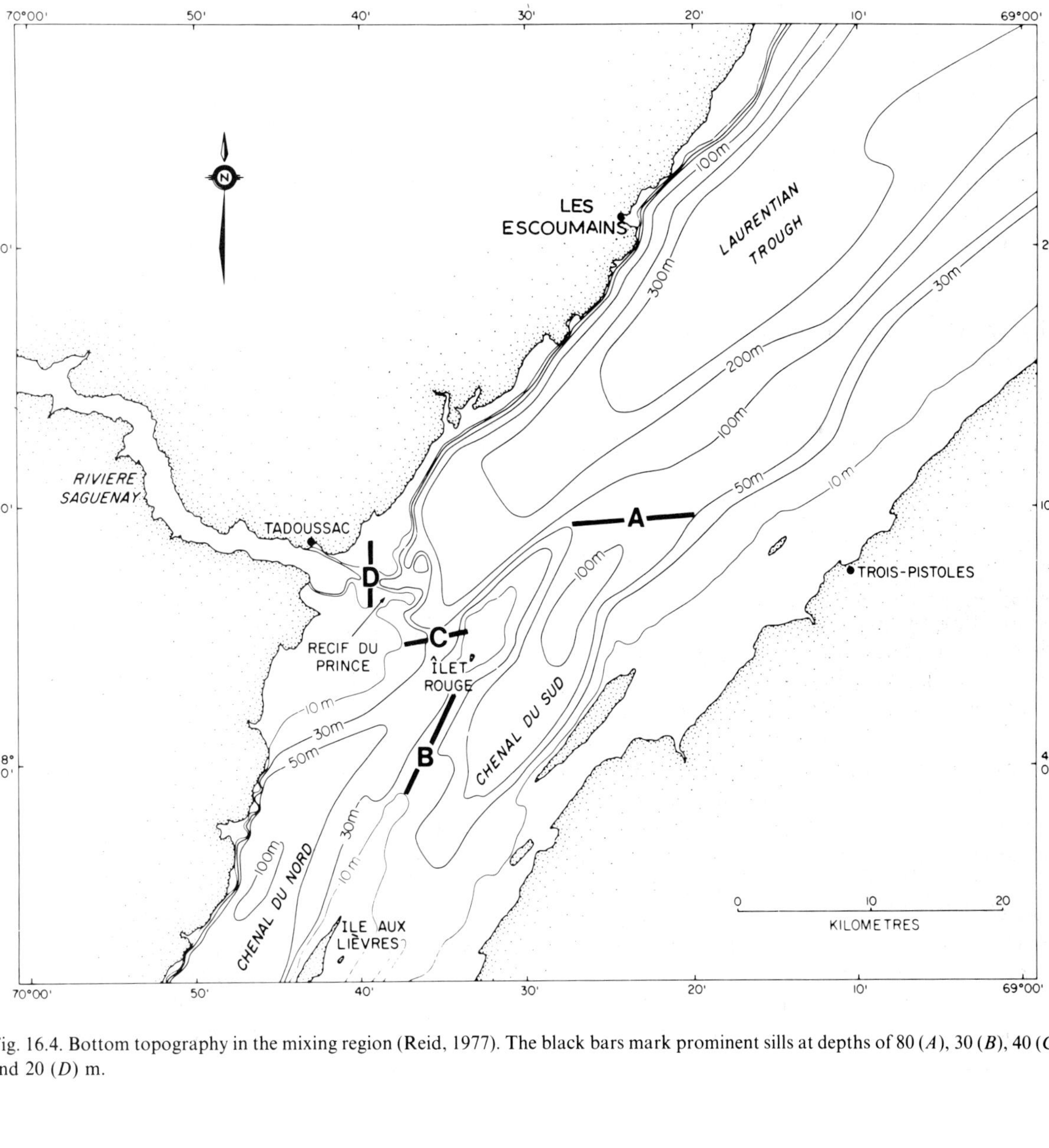

Fig. 16.4. Bottom topography in the mixing region (Reid, 1977). The black bars mark prominent sills at depths of 80 (*A*), 30 (*B*), 40 (*C*), and 20 (*D*) m.

slope rather steeply upward towards the surface as the bottom begins to shoal in the mixing region and tend towards the vertical in the area of intense vertical mixing. In the case of nutrients, downstream concentrations, especially in winter, tend to be higher at the surface than at intermediate depths, strengthening the image of a region of vigorous upwelling, mixing and surface outflow.

At present there is no conclusive demonstration of the relative importance of the factors responsible for the upwelling and mixing of the surface and deeper waters. Neu (1975) argues that entrainment of fresh water in the upper estuary is a primary mechanism. In support of this thesis, the volume of the Gaspé current, which originates with salt and fresh-water mixing in the estuary, varies seasonally with river outflow by far more than the volume of the fresh water itself. However, as pointed out by Neu (1975), the volumes of salt water involved in the Gaspé current are very large indeed. The current reaches a salinity of 27‰ a few kilometres downstream, indicating increases in volume transport of greater than ten times the fresh-water outflow, which suggests that sources of energy other than that from turbulent river flow must be involved.

Huntsman (1923) was apparently the first to suggest that tidal oscillations may provide the main source of energy for this mixing system. The tides in the North Atlantic are propagated into the Gulf of St. Lawrence principally through Cabot Strait. As might be expected, the response in the Gulf to this external forcing is relatively complex (Farquharson, 1962, 1966). However, for much of the area the principal characteristics in the vertical tide can be described in terms of the semidiurnal lunar constituent (M_2). A plot of the cotidal lines reveals the existence of two amphidromic points, the most important one near the Magdalen Islands. The M_2 tide propagates counterclockwise about this amphidromic point, with the cophase lines appearing as "spokes" extending outwards from the amphidromic "hub". There is approximately a 6-h phase lag between Cabot Strait and Gaspé Passage. Tidal range increases away from the amphidromic point reaching values of 3 to 4 m in the region of most vigorous upwelling.

Forrester (1970) pointed out that the rapid dissipation of M_2 tidal energy took place precisely in the region around Îlet Rouge in mid-river off the entrance to the Saguenay. Forrester (1974) and Ingram (1975) found evidence that strong internal tides appear to be propagated from this region, observations confirmed by data on nutrients and chlorophyll distribution collected by Therriault and Lacroix (1976). This evidence supplemented by special cruises in 1973 is summarized by Reid (1977), who offers the most comprehensive view of the mixing mechanism. He found that the combination of vertical oscillations and tidal currents across sills which occur in the area of abrupt shoaling may be responsible. The rising tide lifts intermediate and deep water from the deep northern channel and spills it across mid-river sills to the shallower south channel. On the falling tide the sills hold the water in the area of strongest river outflow on this southern side, where it is vigorously mixed and flushed down-stream by the combination of an ebbing tidal stream and fresh-water runoff. Reid's findings are of particular interest in that they suggest that at high fresh-water discharge levels, such as occur in some years in May, the amount of fresh water in the system may tend to suppress the effects of the mixing mechanism by increasing the stability and deepening the halocline to or below the mid-river sill depths. Reid also found that the nitrogen flux in the region of intense mixing and upwelling was of the order calculated by Steven (1974) to be delivered to the Gulf proper via the Gaspé current. The nutrient entrainment into the current at down-stream locations was insignificant, thus supporting Steven's concept of a nutrient pump in this region of most vigorous mixing.

While there seems now to be overwhelming evidence that upwelling and mixing in the region seaward of the Saguenay is important in determining the composition and dynamics of Gulf of St. Lawrence waters, they are certainly affected by other forces as well. For example, Blackford (1966) using an electrical analogue model, and Murty and Taylor (1970) with a numerical model, were able to reproduce most of the features of the main surface circulation from a knowledge of the winds. Obviously the circulation down-stream from the mixing area is augmented by wind influences. Furthermore, there is evidence for upwelling at various other points in the Gulf proper. Most of these are around the perimeter or around islands and thus probably serve primarily to remix the surface and intermediate water layers that were separated during spring–summer warming. One such area was identified by Steven (1974) north of Anticosti Island over the shallow ridge between the Island and the mainland, although it is questionable how important it is relative to mixing which may be induced by three moderate-sized rivers emptying into the same region (Lauzier et al., 1957). According to aerial photos of ice in winter, a second probably wind-induced upwelling occurs south of Anticosti Island. A third, possibly more important upwelling appears to take place around the Magdalen Islands, since sea-bed drifters released on the Magdalen Shallows show a counterclockwise convergence of the bottom circulation towards the islands (Lauzier, 1967). Bottom drifters also tend to move towards the western shore of the Magdalen Shallows in the region south of Miscou Island, suggesting an upwelling induced by prevailing westerly winds in this area. Blackford (1965) found direct evidence of onshore bottom currents into Kouchibouguac Bay from current meters moored in June and July, 1964. These regions of local upwelling are especially important areas of capture of lobsters, cod and herring, and are spawning grounds for at least the herring and cod, suggesting that the currents may affect both the movements of the animals and the concentrations of their food and their larvae.

CIRCULATION PATTERNS

The picture of the general circulation pattern has emerged from a variety of measurements and computations, including direct current measurements from moored meters at selected sites and times, use of drift bottles, sea-bed drifters and drogues, and geostrophic calculations. A counterclockwise circulation is present at all times, with a normally persisting two-way flow in Cabot Strait and the Strait of Belle Isle. Examination of the surface current patterns on a smaller spatial and temporal scale reveals marked fluctuations from the general pattern. An indication of the spatial complexity is illustrated in Fig. 16.5, taken from El Sabh (1976) who found evidence for eddies of the order of tens of kilometres in diameter. El Sabh's monthly surface geostrophic current patterns show that the location and intensity of most of these eddies vary appreciably except for one in the area between Anticosti and Pointe des Monts which persisted at all times.

Blackford (1965) and Keyte and Trites (1971) obtained evidence of smaller clockwise and counterclockwise gyres, of the order of 10 to 30 km, present in the surface waters of the Magdalen Shallows. These may result primarily from eddies created by currents in the vicinity of headlands such as the northern tip of Prince Edward Island or the eastern extremity of Gaspé. They appear to be more or less embedded in the general southerly and easterly flow across the Magdalen Shallows. Evidence for such structures was also obtained by El Sabh (1976) who suggests that their duration is less than one month. Their general significance cannot be ascertained at present as most attempts at direct observation have been frustrated by equipment difficulties. However, El Sabh found evidence for enhancement of mid-water nutrient values in the centre of the strong persistent counterclockwise gyre west of Anticosti Island and suggested that clockwise gyres in the eastern area might contribute to the relatively low productivity there.

The circulation patterns for the surface waters are best known from measurements taken during the open-water season. The winter situation is less well known, although some ten winter cruises have been made in recent years, and certain deductions can be made from the patterns of ice formation and dissipation (El Sabh and Johannessen, 1972). In addition, drifting ice stations have been maintained twice in recent years in the Gaspé current area (Ingram et al., 1968; Johannessen et al., 1968). In general, observations of ice drift confirm the summer circulation with the stronger currents over the deep channels. This is particularly clear in aerial reconnaissance of the movements of ice originating in early winter from the freeze-up of the St. Lawrence River, or entering from the Labrador coast through the Strait of Belle Isle.

The annual cycle

A number of major hydro-meteorological parameters undergoing cyclic seasonal changes affect the physical system: fresh-water inflow, precipitation, evaporation, air temperature, insolation, cloud cover, and wind. Owing to the inadequacy of the data, very little has been done to determine the

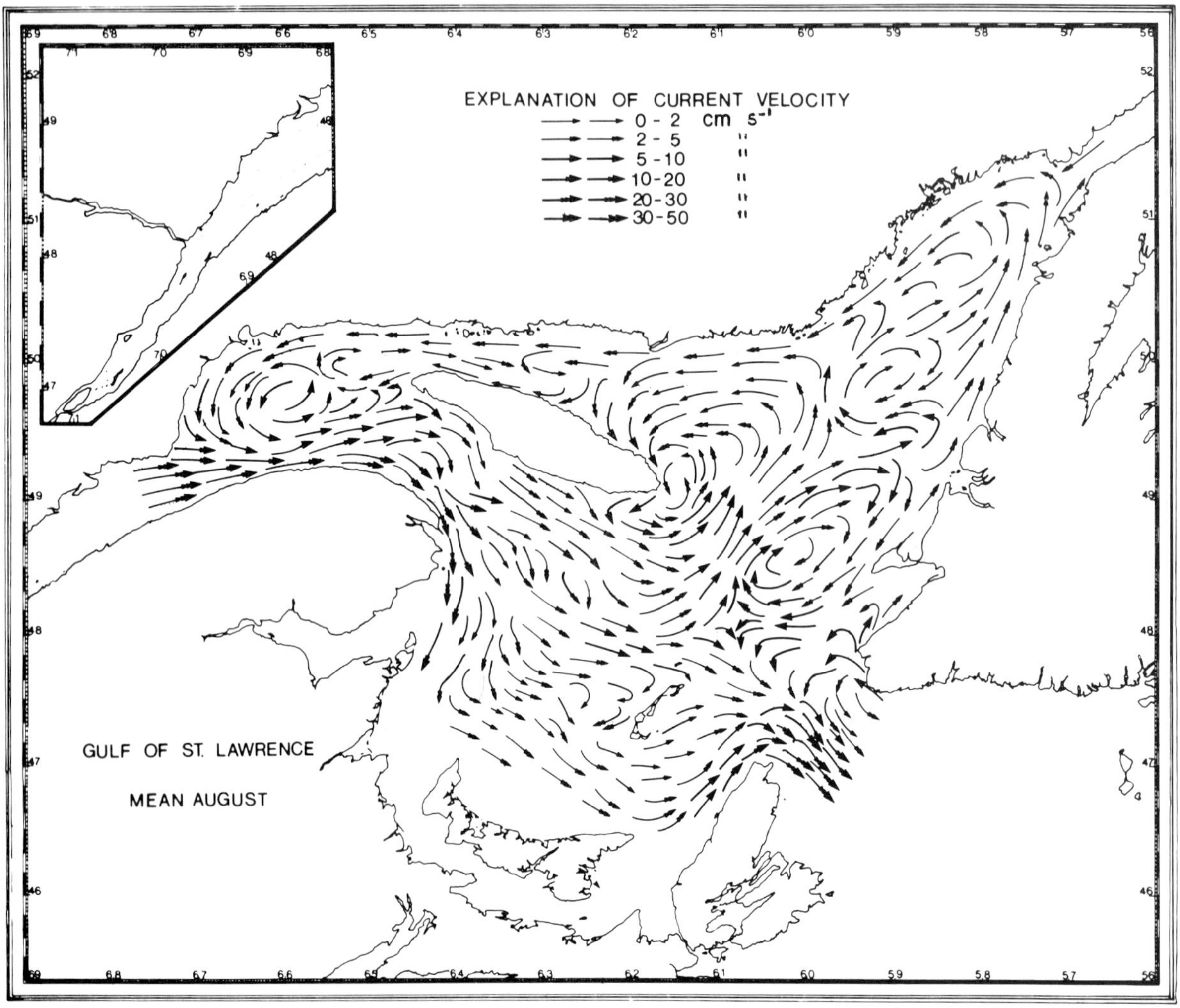

Fig. 16.5. Field of surface geostrophic currents in the Gulf of St. Lawrence in August (after El Sabh, 1976).

complete energy balance in the Gulf. Computation of ice potentials and heat budget was first undertaken by Lauzier and Graham (1958). Subsequently, work by Coombs (1962), Matheson (1967), and Boudreault (1969) shed light on the relative importance of the various terms in the heat budget. For spring and summer months, the radiation terms constitute the most important component in the energy balance of the surface layer; in autumn and winter evaporation and conduction terms dominate the thermal budget. The advection of heat by the currents is an important component, and appears to have significant seasonal fluctuations. However, the relative or absolute magnitudes of heat transport through the Strait of Belle Isle or Cabot Strait so far have not been measured.

Fresh-water inflow has already been identified as a major variable and river discharges have been monitored for a number of years on the major rivers. Trites (1972) used data available for the period 1957–1965 from the St. Lawrence and several other major monitored rivers and, extrapolating from their drainage basins, arrived at monthly mean rates which varied from 8×10^3 m^3 s^{-1} in winter to nearly 30×10^3 m^3 s^{-1} in May. His calculated annual average of only 13.4×10^3 m^3 s^{-1} is to be compared, however, with a mean of 19.1×10^3 m^3 s^{-1} calculated in much the same way by El Sabh (1977) but based on the 21-year period 1950 to 1970. The reasons for the difference are clear from the annual averages plotted in Fig. 16.6A from El Sabh's data. Fig. 16.6B shows the monthly mean river discharges as compiled by El Sabh for the longer period.

In addition to river discharge, Trites (1972) noted the importance of direct precipitation on the Gulf and evaporation from it. His estimates for precipitation are also plotted in Fig. 16.6B. Direct winter precipitation appears to be about half as large as river discharge, but decreases relatively during the spring when the freshets appear from winter snow and ice melt on land. The effect of precipitation must, however, be corrected for evaporation. For this calculation, the total evaporation is estimated to be equivalent to 75 cm of water per year. We have applied this equally throughout the year thus subtracting a uniform value of 6.1×10^3 m^3 s^{-1} from the sum of inputs to calculate a monthly net fresh-water input to the system in Fig. 16.6B.

It is of considerable practical importance to compare variations in the month-to-month discharge with the annual variations. The peak monthly values of May are approximately double the minimum monthly values in March. But the

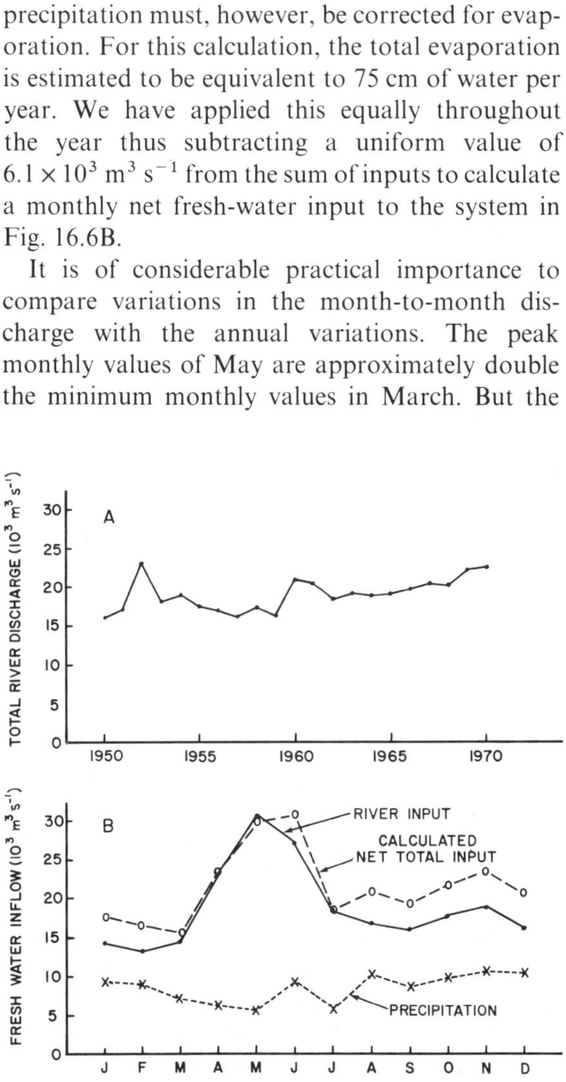

Fig. 16.6A. Mean annual total river discharge into the Gulf of St. Lawrence 1950–1970 (from El Sabh, 1977). B. A monthly fresh-water budget for the Gulf of St. Lawrence. (Net total = river + precipitation − evaporation.)

average flow of the high run-off year of 1952 was 1.4 times the average of the low flow year of 1959. That is, year-to-year variations are generally less than half the within-year variations. Clearly, in assessing the significance of differences in discharges over extended periods of time the within-year distribution pattern must be reckoned as of major importance.

In assessing the importance of seasonal variations, three additional factors must be taken into account. In the first place, our assumption of consistency of the rate of evaporation over the year undoubtedly leads to overestimates in winter when the surface is ice-covered and underestimates in autumn when there are cooler, drier northerly winds. Coombs (1962) recognized this in his attempts to derive an autumn heat budget, but found it impossible to dissociate the effects from the combined influence of sea–air temperature differences and winds. A second major factor must be the ice-cycle itself. Ice is formed *in situ* and the resulting sea ice, having salinities as low as 4.5‰ (Coombs, 1962), might be considered equivalent to about 0.9 times the volume of fresh water resulting from its melting. Forrester and Vandall (1968) have estimated the mean ice volume for 1962–1967 at the beginning of March as approximately 40×10^9 m^3. If this were melted over the two-month period March–April, it would yield a discharge of about 7×10^3 m^3 s^{-1} which is to be compared with the estimated fresh-water river discharge of 14.4×10^3 m^3 s^{-1} for March shown in Fig. 16.6B. A third effect is concerned with time lags in flow of fresh water through different parts of the system. El Sabh (1977) notes that the spring freshets which affect the Gaspé current in May are not felt at Cabot Strait until August. Lag times of this order were also calculated by Trites (1972) and by Sutcliffe et al. (1976). It is clear from these considerations that information permitting one to determine the full effect of fresh-water fluxes on seasonal dynamics of the system is not yet available.

Ice cover of the Gulf arises from three significant sources whose influence varies from time to time. Ice presence usually begins in the north and west (Fig. 16.7) in early January. At about the same time drift ice appears in the northeast in the Strait of Belle Isle from areas outside it. *In situ* freezing throughout the Gulf becomes established in January and by the end of January about half the

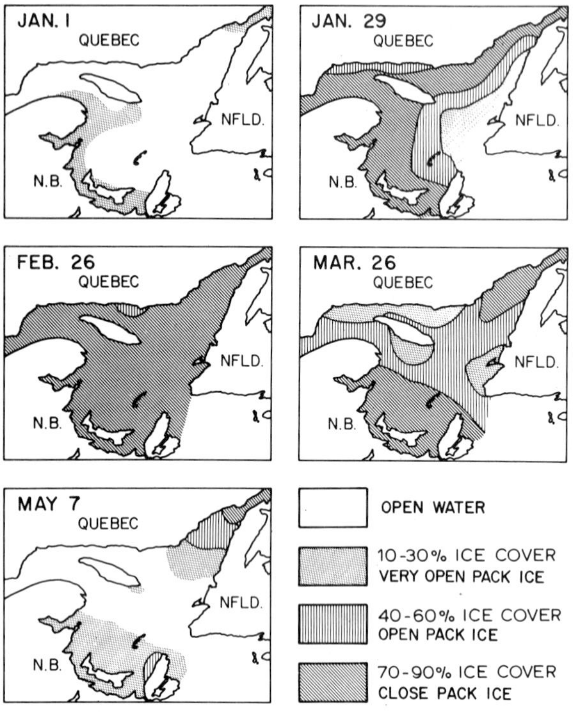

Fig. 16.7. Five-year mean ice concentration in the Gulf of St. Lawrence (after Matheson, 1967).

surface is occupied by close pack ice, with a lighter cover over all but a small area northeast of Cabot Strait where surface inflow occurs. By the end of February the entire surface is typically occupied by close pack ice. The continued circulation in winter is well reflected by ice patterns in March and early April when cover is reduced to less than half over all of the deep-water channels except in the extreme northeast. Ice cover in the Strait of Belle Isle area and over the Magdalen Shallows is still from 10 to 30% in early May but clears rapidly as surface albedo is lowered and insolation increases. However, heavy ice from the drift of arctic ice southward along the Labrador coast is not uncommon in the Strait of Belle Isle as late as June.

Despite the ice cover and low winter temperatures which may persist until early May, the surface waters warm rapidly during early summer. Over the shallows south of the Magdalen Islands open surface water temperatures of about 20°C are encountered in August. Water in the bays and inlets along the coast of Prince Edward Island or in Northumberland Strait are warmed even more. As a result, a number of shallow-water animal and plant species typical of warm temperate regions south of Cape Cod, but which can tolerate the winter cold, flourish in this region. In the open Gulf, particularly over the Magdalen Shallows, pelagic warm-water migrant fishes are an important component of the biomass.

THE BIOLOGICAL PRODUCTION SYSTEM

Biological production depends on the physical and chemical environment, and cannot realistically be considered apart from the cycle of physical events outlined above. In fact, however, basic biological production in the Gulf has not been studied in detail until recently and understanding of the observations is only beginning to emerge as mechanisms underlying the physical dynamics are revealed.

The biological system has practical and scientific importance. Although the southern Gulf comprises less than 15% of the continental shelf area available to east coast Canadian fishermen, it regularly has yielded about 25% of Canadian total commercial catch by weight. Total fishery landings from the area averaged 435 500 m t yr^{-1}, over the ten-year period 1967–1976, more than half of which is estimated to have been produced in the area of the Magdalen Shallows. The corresponding fishery landings per unit area were about 2.0 km^{-2} yr^{-1} in the area as a whole, but about 3.5 t km^{-2} yr^{-1} for the Magdalen Shallows. This can be compared to a fishery landing for the entire Grand Banks and Scotian Shelf area which is of the order of 2.9 t km^{-2} yr^{-1}. Thus, the high production of the southwestern Gulf contrasts with a low production in the northwest, a matter of considerable practical as well as general interest.

The locally high fisheries production in the southwestern Gulf must result in large measure from a combination of vigorous upwelling and mixing near the head of the Laurentian Channel with favourable growing qualities of the broad down-stream expanse of the Magdalen Shallows. This expectation is confirmed by results of the biological program of McGill University, carried out under sponsorship of the International Biological Programme (IBP). The results were summarized by Steven (1974).

Primary and secondary production

According to Steven, the spatial and temporal distribution of biological properties at the base of the trophic system reflected the physical dynamics in considerable detail. Nutrients were high and oxygen, primary production and biomass were low in spring in the immediate upwelling area. Downstream, however, a region of high primary production appeared from about Pointe des Monts in the lower estuary eastward including Gaspé Passage, then extending south deep into the middle of the Magdalen Shallows (Fig. 16.8). From mid-May to mid-June carbon fixation exceeded 200 mg C m^{-2} h^{-1} in this area. As might be expected, nutrient concentrations here were lower than in the middle estuary, but total nitrate remained as high as 100 mg at. m^{-3} in much of the region at least until early June, suggesting that the nutrient replenishment mechanism must be continuously effective. The pattern of production and of high biomass distribution in the lower estuary in June was confirmed in a detailed survey by Platt (1972). The high phytoplankton biomass showed signs of predation by the developing zooplankton biomass in late May.

By mid to late June surface water flows through

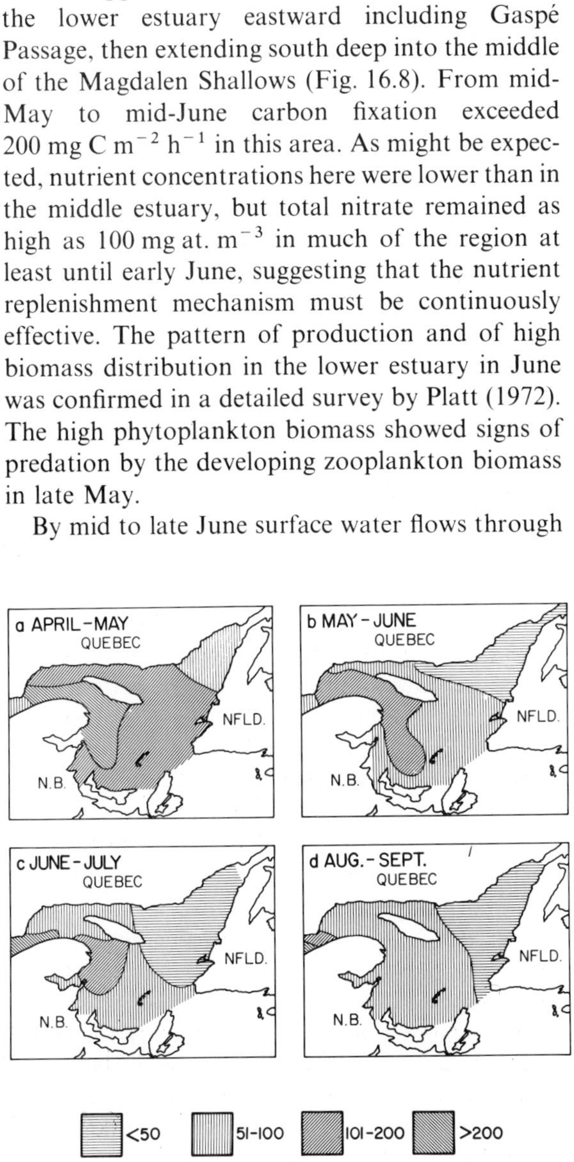

Fig. 16.8. Average carbon fixation rates (mg C m^{-2} h^{-1}) for selected 30-day periods from April through September, 1969 to 1972 (after Steven, 1974).

the middle estuary subsided to near summer levels and simultaneously vigorous primary production began to appear there. Values reaching 200 mg C m^{-2} h^{-1} were found throughout the region of mixing in July. At the same time a strong development of zooplankton was sustained throughout the entire estuary and Magdalen Shallows. In subsequent years Sameoto (1976) found high concentrations of euphausiids (maximum of 4.5 gm wet weight m^{-3}) living in the deeper water of the lower estuary in daytime; they made strong vertical migrations into regions of high chlorophyll during the night. In late June, during the IBP, total zooplankton concentrations reached 50 to 100 mg dry weight m^{-3} in the estuary and northern Magdalen Shallows while concentrations in the southern shallows averaged over 100 mg m^{-3} with August values recorded in excess of 200 mg m^{-3}.

It is notable that at no time did the northeast region of the Gulf share in this rich productivity. Carbon fixation reached rates in excess of 50 mg C m^{-2} h^{-1} for a brief period during the spring bloom immediately following ice break-up. Phytoplankton concentrations reached 20 to 40 mg chl *a* m^{-2}, but by mid-May had already declined to levels <10 mg chl *a* m^{-2}, and remained low throughout the rest of the year. Steven (1974) called the region a biological desert.

The evidence suggests that the early phytoplankton bloom of the Gulf is initiated by high insolation and high nutrients following the ice melt. This is followed by a general increase in zooplankton biomass throughout the whole Gulf, including the northeast. However, in the northeast region phytoplankton and nutrients quickly subside, although the zooplankton biomass continues high, falling off only in July. In the southern shallows area the extremely high primary production and nutrients accompanied by high zooplankton are apparently sustained throughout the summer months as in a large chemostat, supplied principally by the estuarial upwelling and mixing mechanisms. In the northeast sustained high zooplankton relative to phytoplankton appears to result from the dispersal of organisms from the region of high production on the Magdalen Shallows.

The production rates of the Shallows do not seem to subside until the effects of the river freshets have passed through, in late July in the north, with the extreme down-stream areas apparently benefitting until late August. Primary production, chlorophyll and zooplankton biomass have all declined markedly by October, with the areas of local upwelling immediately around the Magdalen Islands and between Gaspé and Anticosti Island in the lower estuary persisting longest.

Lauzier and Marcotte (1965) suggest that the enriched surface jet of water spreads onto the broad shallows where the slower water movement, increasing sunlight over a large surface area and the resultant surface warming afford ample opportunity for vigorous production. This picture is undoubtedly too simple. As noted earlier, upwelling in the Magdalen Shallows, local tidal mixing and various sizes of gyres may all play a role. However, the essential validity of the simple picture is further maintained by Steven's estimates of primary production. Integrated annual production in the St. Lawrence Estuary averaged about 510 g C m^{-2} yr^{-1}, while for the Gaspé current system it averaged 390 g C m^{-2} yr^{-1}; and for the central Gulf and Shallows 210 g C m^{-2} yr^{-1}. These values are in the upper part of the range of published values for coastal waters, despite the fact that the production season appears to end by late October and the system becomes ice-covered by February.

Fisheries production

The relatively high landings of fish per unit area of Magdalen Shallows can be explained by rich primary and secondary production and therefore must also be associated with the estuarial upwelling and mixing mechanisms. However, the linkages of the various parts of this system are complex and details of the responses appear to be significant in the context of management requirements.

More than 145 species of fish and shellfish occur in the Gulf (Srivastava, 1971), and over 25 species have been landed commercially at one time or another. In addition, important populations of several species of whales and seals live there for varying periods of the year (Sargeant and Fisher, 1957; Sargeant et al., 1970) and large sea bird colonies are found around the shorelines and islands (Brown et al., 1975). Among the important commercial species, a few of the pelagic fishes such as tuna (*Thunnus thynnus*) and mackerel (*Scomber scombrus*) are found there only in summer and early autumn. Mackerel, as well as important populations of herring (*Clupea harengus*) and cod (*Gadus morrhua*), both spawn and feed in the Gulf, but all mackerel and the bulk of adult herring and cod move outside it during the period of ice cover. Other important benthic and deeper water species, such as white hake (*Urophycis tenuis*), American plaice (*Hippoglossiodes platessoides*), witch flounder (*Glyptocephalus cynoglossus*), redfish (*Sebastes marinus*) and lobsters (*Homarus americanus*) appear to remain in the Gulf throughout the year.

The life-histories of the common commercial species are linked with the physical and basic biological production cycles described earlier. There appears to be a quite specific succession of spawnings. American plaice is one of the first to reproduce. Spawning begins under the ice in April and is completed by the end of May. A spring run of herring arrives on the Magdalen Shallows from overwintering grounds south of Newfoundland and from the Scotian Shelf in May (Winters, 1976) and spawning begins in the region of the Magdalen Islands, with later appearances progressively further north in the Baie des Chaleurs and Gaspé area. By late May capelin (*Mallotus vilosus*) appear in the warmed fringes of the Gaspé area and begin spawning, later extending well into the estuary. They are followed by cod which move in from deep waters of the Scotian Shelf and Newfoundland Banks in May and begin spawning in the Gaspé area by early June. Their peak spawnings occur in late June, but extend through September (Powles, 1958). All of these species deposit their eggs on the bottom. This succession is augmented in June and July by extensive mackerel spawning in the pelagic zone throughout the Magdalen Shallows.

By August a separate autumn spawning of "fall-run" herring begins along the western edge of the Magdalen Shallows. In recent years this fall herring spawning has produced a much larger biomass of herring larvae than the spring spawning.

The complex of spawning successions appears to be related to the succession of different kinds and probably sizes of zooplankton on which the larvae feed (Ware, 1975, and pers. comm., 1978). In addition to an apparently inflexible order of succession it is noteworthy that although eggs and larvae are fairly widespread following the time of peak spawning, the benthic spawning sites are concentrated in the areas around the Magdalen Islands and the western shore of the Shallows where there is convergence of bottom currents and local upwelling.

All these features of spawning distributions suggest that success in the higher trophic levels cannot depend solely on a rich broth of primary and secondary producers. Additional elements related to local food concentrations seem to be important to the early stages (Conover, 1968; Dickie, 1974), hence aspects of small-scale heterogeneity in the physical and biological environment would have to be identified before variations in spawning success could be explained completely.

Similar considerations appear to apply at the larval stages. Surveys of eggs and larvae have been undertaken and described by a number of authors, who deal with a variety of species: Kohler et al. (1974) for plaice, cod and herring; MacKay (1976) for mackerel; Messieh and Kohler (1972) for herring; Scarratt (1964, 1973) on lobsters. All appear to agree that periods of peak occurrence of different species of larvae are fairly well defined. The results however show markedly aggregated horizontal distributions and some reports indicate significant differences in vertical distribution with larval size or age. Once again it appears that geographic and temporal aggregations on relatively small scales may be involved in the success of year-class production of both fish and invertebrates and may need to be understood in explaining year-to-year differences.

Nevertheless the interplay of small-scale events in no way obviates the usefulness for prediction of large-scale patterns as they relate to the major features of the system outlined earlier. In this connection the recent work of Sutcliffe and his associates (1972, 1973, 1976, 1977) is of particular interest.

In the course of a study of the impact of freshwater outflow on the nitrogen budget of estuarial systems and embayments, Sutcliffe (1972) examined relations between the St. Lawrence River annual discharge and data on commercial landings of several species which appeared to be produced primarily in the system. Strong correlations appeared between the data series when landings were lagged behind annual river outflow by approximately the same time as it took for the species in question to reach average catchable age. Four different lag periods were established for four different species. It appeared likely that river discharge was affecting the productivity of the species through its influence on larval success, which must be the life-history stage at which year-class abundance was established. Such a thesis would agree with the evidence showing the influence of river discharge on nutrients and primary and secondary production which has been compiled in the years since Sutcliffe's original observations. Correlation coefficients of 0.75 to 0.93, which Sutcliffe found between river discharge and lagged catches, suggested that his data were indexing variables which explained well over half of the total variance in annual landings.

Subsequently Sutcliffe (1973) examined the hypothesis that actual larval abundance information, or landings data more representative of adult abundance, and seasonal (monthly) runoff data would all improve the original correlations, which had used highly aggregated data. The first expectation was confirmed by data available from Scarratt (1964) on lobster larval abundance in the immediate vicinity of the Miramichi River and its outflow. In the second, lobster landings data for the estuary and Magdalen Islands exclusively replaced the original cannery data, and higher correlations were obtained when these were associated with runoff data for particular months rather than with annual averages. Sutcliffe (1973) noted that, on the strength of the correlation, the steady decline in Quebec lobster landings from about 1960 to 1972 should be reversed by 1973. It is pointed out by Mann (1977) that this prediction has been fulfilled by the three subsequent years of landings data.

Sutcliffe's initial studies were based on data for four species, all of which are strongly benthic in their habits: soft-shell clams (*Mya arenaria*), lobsters, haddock (*Melanogramus aeglefinus*) and halibut (*Hippoglossus hippoglossus*). The situation appears somewhat more complex for herring and mackerel, the principal species of the pelagic zone. Winters (1976) found that the direct relationship between herring year-class strength and St. Lawrence River discharge was negligible. However, he established that there was a strong relationship between the sizes of mackerel year classes and summer surface temperatures at the Magdalen Islands — an index which Sutcliffe et al. (1976) found to be strongly related to spring river discharge. Corrections for this temperature effect allowed Winters to demonstrate a remarkably clear parabolic relationship between mackerel adult stock size and subsequent mackerel recruitment, showing that the temperature-related effect apparently completely obscured any simple population effect in the uncorrected data. Winters then showed that herring year-class production as well as growth rates were strongly inversely related to the combined biomass of adult mackerel and herring. There is apparently a strong density-dependent interaction among larvae of the two species, and it appears that post-larval mackerel are significant predators on early life-history stages of herring. Thus, while Winter's observations in general agree with the thesis that year-class strength is related to larval success, the expression of the relationship for herring is dominated by biological interactions with both larval and adult mackerel.

The significance of the relation of the mackerel year-class strength with the environmental temperature is hard to assess. MacKay (1976) found that there was a density-dependent response of mackerel growth rates in relation to the very large 1958 year class, suggesting some general sensitivity of production to food abundance. However, Winter's correlation could be an indirect expression of river discharge on larval food abundance. Alternatively it could be a direct physiological effect of temperature on egg and larval survival in a typically southern species (Worley, 1933), or an expression of variations in local upwelling events on either mechanism, or even some other mechanism indexed by water temperature. In any case, it appears that in the long run the direct effects of environment on herring abundance will be obscured by species interactions when mackerel abundance is variable. Mackerel abundance in its turn seems to be mediated more directly by environmental effects.

In a somewhat similar vein Lett et al. (1975) described the production of cod stocks over at least the period 1965 to 1973 as a remarkably complex result of environmental and stock dynamics effects. However, they did not indicate how they treated the (usually high) variances of the short series of plankton larval catches on which the relationship is based, and the standard error of at least the second of their complex temperature relations is also rather high. It is therefore not possible to evaluate their subsequent complex relationship between environment and number of larvae, year-class strengths at recruitment, and estimated adult biomass of cod and related herring and mackerel population sizes. Their conclusion that cod production seemed constrained by an upper limit to the carrying capacity of the pelagic environment for some juvenile age seems still open to question. However, Paloheimo and Kohler (1968) earlier showed a strong inverse relationship between adult stock size of cod and abundance of younger year classes, which in turn showed density-dependent early growth. In addition, the adult cod growth rate was shown by Kohler (1964) to respond strongly to a change in food availability during a mass mortality of herring. It may tentatively be concluded therefore that the expression of environmental effects on cod abundance, as indexed in temperature by Lett et al. (1975) is probably confounded with other factors which affect both food available and predation or competition with associated species.

In view of the possibly complex interactions of the fish populations in the Gulf of St. Lawrence with environmental variables, the conclusion of Sutcliffe et al. (1977) that effects indexed by the river discharge are powerful predictors of fish production outside the Gulf is of particular interest. Sutcliffe et al. (1976) showed that the total fresh-water discharge into the Gulf of St. Lawrence (RIVSUM) plus its large volumes of entrained water, are a dominant factor in determining the character of water masses on the Scotian Shelf and Gulf of Maine. Correlations between river outflows and lagged sea surface temperatures are strong as far away as Boston, indicating that a significant part of the environmental signal comes through the ocean current system. This result does not seem especially surprising when, as they point out, it is recognized that the total annual volume outflow through Cabot Strait (3×10^4 km^3) is nearly the same as the volume of water on the Scotian Shelf between the Laurentian Channel and Georges Basin (3.6×10^4 km^3).

Sutcliffe et al. (1977), in parallel with the 1972 paper, found highly significant correlations between the catches of some twelve species of fish landed from the Gulf of Maine and Georges Bank, lagged for different appropriate intervals behind indices of the environmental signals. In almost every case the correlation was very nearly as high with an appropriately lagged RIVSUM as with sea temperature of nearby shore stations. Species showing significant relationships included cod, herring and redfish. When information on fishing effort along with the environmental signal was included it improved the overall correlation for cod, yellowtail flounder (*Limanda feruginea*) and menhaden (*Brevoortia tyrannus*) and revealed a significant relationship with haddock. The total list included benthic, pelagic and anadromous species of fish and some benthic invertebrates. As they pointed out, the roster of species for which significant effects were found makes up about 84% of the total commercial landings from the Gulf of Maine and in the majority of cases correlations with environment explain more than 50% of the annual variation. The correlations with the environmental signal are positive in some cases and negative in others, and the monthly environmental signals giving the strongest relations vary from species to species, but in all cases the time lag involved implies that the mechanism acts through the larval or early life history stages.

There seems to be no serious doubt that environmental factors play a dominant direct role in the determination of relative year-class success of commercially exploited marine species throughout the entire area influenced by the Gulf of St. Lawrence. The mechanism of the effect is obscure and may involve factors related to distribution, direct physiological effects on survival, or influence through changes in food abundance or availability. The obvious and simple effects of river discharge on nutrients and on primary and secondary production, together with evidence of density dependence of both growth and recruitment are incontrovertible evidence of the action of the physical system on biological production through the trophic chains. However, the final expression of biological production in adult stock size and catch is apparently influenced by other factors including fishing.

In view of this apparently strong environmental influence, the relative importance of the impact of fishing and the efficacy of fishing regulations as a principal device for regulating production need to be evaluated for both the Gulf and outside shelf. It is certainly surprising that the environmental signal has such strong effects in situations where fishing intensity is considered to be very high. Perhaps equally surprising is the fact that in such a situation the primary expenditures in the direction of "conservation" measures for the fishery are still directed almost solely towards control of fishing. It would appear that direct study of the larvae and their fates as they relate to the physics and biology of the water masses is an inescapable requirement for improved prediction of fishing prospects and would provide essential information for a balanced perspective on the results that can be achieved by control of fishing intensity.

THE IMPACT OF HUMAN ACTIVITY

It is clear that statistics of fresh-water input into the Gulf of St. Lawrence provide an index for the dynamics of the entire coastal system from the St. Lawrence River mouth to Cape Cod. The effects on fisheries production are of immediate practical importance. However, of equal significance are the relationships between the surface dynamics and the formation and dissipation of ice cover within the Gulf because of the consequences for transportation, mineral exploration, fishing activity and possible effects on the climate of the surrounding land. It appears that an understanding of the factors affecting the fresh-water input and its relation to the estuarial mixing mechanism is essential to an understanding of the impact of man on the entire system.

Unfortunately an adequate description of the heat budget has not yet been made, so it is uncertain to what extent the general circulation controls the formation and dissipation of ice. However, the inflow of warm deep water must balance the seawater content of the outflow, and this seems to vary from year to year and seasonally more or less in proportion to the freshets. According to Coombs (1962) the advection of heat from the deep water source may be of the order of 10% of surface heat flux. Therefore in addition to the present practice of short-term ice forecasting, it may someday be possible to make forecasts of overall ice conditions by making use of data on fresh-water discharge. Over the past twenty years conditions have varied from light, short ice seasons to heavy prolonged ice cover, suggesting that the entire system must respond significantly and probably rapidly to changes in the external forces.

In addition, the surface outflow through either mixing processes in the estuary or subsequent entrainment must be involved in dissipation of the intermediate cold layer formed during the winter. Thus, year-to-year as well as seasonal variations in the dynamics of the superficial layers must give rise to differences in the total heat content of the system which are greater than would be reflected in the surface layers alone. These total differences must be important in determining the subsequent ice cover.

Given the apparent importance of the dynamics of surface layers to the entire system, it is particularly significant that the fresh-water drainage system of the Gulf is being progressively altered by human activity. Since the mid-nineteenth century the descendants of European settlers have made increasing use of the drainage area bordering the Gulf. Extensive canal systems have been built for transportation, and large dams and reservoirs have been constructed for hydroelectric power. In these constructions the object of the changes has generally been to control at least the seasonal variations — that is, to hold back the large spring freshets and to augment the winter outflows. Neu (1975) analyzed changes in the seasonal pattern for the period 1964 to 1970 and found that in this seven-year period alone fresh-water flow in the estuary down-stream from Pointe des Monts had been artificially increased during mid-winter by an average 3600 $m^4\ s^{-1}$ and decreased in May by an average 7800 $m^3\ s^{-1}$.

Variations in the seasonal flow and the distribution of the seasonal effects at various points are illustrated in Fig. 16.9. Neu (1975) estimated that by 1970 the range of natural seasonal variation in river discharge at Pointe des Monts had been reduced to about 50% of its original value.

The magnitude of the man-made alterations in the annual hydrological cycle is impressive. At present, however, larger reservoirs are being added. One storage basin recently completed on the Manicouagan system (Manic 5) has a storage

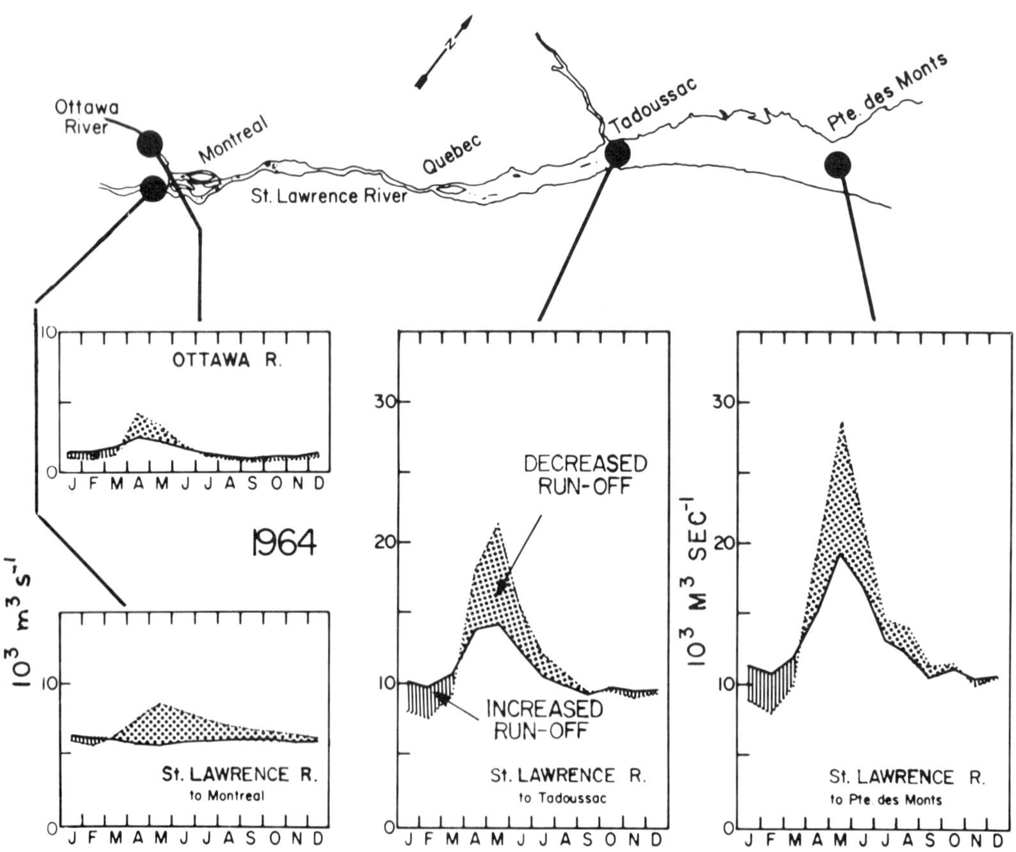

Fig. 16.9. The calculated effect of river regulation on monthly runoff at various points along the St. Lawrence River in 1964 (after Neu, 1975).

capacity which exceeds 10% of the total annual fresh-water discharge from *all* Canadian rivers. When full it will contain a volume of water equivalent to more than 200 days discharge of the river at Montreal. The live storage volume of about 140 km^3 means that the ratio of the discharge during the three-month winter minimum period to that of the remaining nine months, which was formerly at about 1:4, will be altered to about 2:3 (H.J.A. Neu, pers. comm., 1978). When the new storage capacity is fully utilized it will be possible to smooth year-to-year variations in the total river discharge. Neu (1975) argues that these changes both seasonally and from year to year must have profound effects on the entire ecosystem of eastern North America and that the weight of evidence implies that the effects will be deleterious. The evidence certainly supports the view that effects are potentially important and far-reaching, but the direction or magnitude of the consequences of change is not a simple matter to evaluate from the evidence on hand.

Improved prediction of the consequences of further modification of the fresh-water inflow appears to rest on a better understanding of the forces driving the physical system, and of their linkages with biological production, especially at the higher trophic levels. There can be little doubt that a major source of energy for the upwelling of deep water can be found in the tides. Forrester's (1970) and Reid's (1977) calculations imply that this mechanism is of primary importance and Reid's calculations further suggest that the effectiveness of mixing, and the introduction of nutrients and of relatively warm deep water into the surface layer during some six months of the year, is maximal at intermediate flow rates. If this is a correct interpretation, it follows that modification of the seasonal outflow towards an intermediate level might enhance both nutrient enrichment and warming processes and benefit many of those biological production systems which show correlations with RIVSUM. If, on the other hand, mixing is directly related to the rate of fresh-water outflow over a wider range of outflows, the effects of altered variations may have profoundly deleterious effects.

Unfortunately for any such simplistic speculation, the evidence from the biological systems shows that the relations are more complicated. Production of fish eggs and larvae seems rather inflexibly related to the seasonal pattern. The production of different species in different areas seems to be related to outflow during various months and is not necessarily directly linked with the time of spawning. Furthermore, both positive and negative correlations with river outflow have been revealed by the biological studies. The fact that seasonal variations in outflow are of the same magnitude as year-to-year variations compounds the difficulty of *a priori* interpretation or prediction. To add to the difficulties of predicting the fisheries system, we are still uncertain of the factors which index a reduction or increase in average ice cover, but this must also be related to both seasonal and average outflows. Since ice cover is likely to affect the biological production system directly, its interaction with outflow serves to emphasize a need for directed and comprehensive multidisciplinary studies of the dynamics of this natural system as it relates to the artifactual manipulations.

The effect of variations in the Gulf of St. Lawrence system does not stop at the water's edge. The "maritime effect" of the large water body on the climate of the contiguous land area can be traced in its moderating effects on temperature and humidity extremes and its effects on cloud cover, fog and precipitation of rain and snow, particularly over the land area to the east and north. All of these climatic factors are sensitive to the alternation of open-water and ice-cover seasons as well as to the effects of stored heat. It is again difficult to predict the consequences of any alteration or the extent of its effect, but local trends in average air temperatures at settlements along the western borders of the Gulf are different from those of inland settlements. In general, over the past thirty years, air temperatures in towns on the Gaspé coast and in northern New Brunswick show a relative rise in winter and a lowering in spring-time, which H.J.A. Neu (pers. comm., 1978) points out is in keeping with the expectation of the effect of river regulation on the maritime influence.

The foregoing speculation about man's impact on the Gulf is entirely apart from the effects of contaminants in the water or other sources of pollution in the water column or sediments, customarily considered as aspects of environmental impact. Given the importance of the Gulf as a transportation route, there are, of course, the usual probabilities of contamination from shipping. The importance of the river as a drainage system for an extensively industrialized area of the continent from Chicago to Montréal also enhances possibilities of contamination. Finally, the system is on the lee side of the atmospheric circulation of the continent.

In view of these pollution sources, the evidence reviewed by Trites and Walton (1975) shows surprisingly low levels of contamination. In particular, the river seems not to be a major source of contamination by either oil or trace metals, possibly a reflection of the short residence times indicated by the vigorous outflow of surface waters.[1] In general the amounts of most trace metals and oil are not significantly different from values for seawater in the open ocean. Local exceptions occur for petroleum hydrocarbon concentrations in the vicinity of shipwrecks. Also of considerable concern in the estuary region in particular are high concentrations of mercury in the sediments. These were traced to sources in industrial plants on the Saguenay River system, and led to the institution of apparently effective emission control measures. Except for these two instances, Trites and Walton's conclusions confirm those of Loring and Nota (1973) that by far the bulk of sedimentation of material in the Gulf is derived from natural local weathering of the land shield area to the north of the Gulf and is at a relatively low rate, probably affecting on the average only the top 0.5 m over the last 5000 years. Man's activities in contributing constitutents to the water mass thus seems to be of minor significance in the Gulf compared to his apparently direct effects on the dynamics of the water masses.

[1]Dunbar (1979) argues that previous work, including that carried out by the IBP program, underestimated the importance of the river as a nutrient as well as a pollution source. His argument is based on a draft of a paper by Greisman and Ingram (1977) (quoted in Dunbar as 1978) which described the results of the calculation of the relative contribution of the river and the upwelling to the nitrogen content of the surface waters. Unfortunately for Dunbar's argument this draft was substantially changed before publication and the final version supports Steven's thesis of the overwhelming importance of the nutrient pump as well as being in accord with the conclusions of Trites and Walton (1975). Dunbar's mis-statement is the unfortunate result of an unlikely conjunction of publication procedures and dates.

THE MANAGEMENT AND POLITICS OF CONTROL

In 1971 an eminent international group of meteorologists and oceanographers met in Stockholm at the invitation of the Royal Swedish Academy and issued a warning that no voluntary alteration should be permitted to any single area of the earth's surface greater than about 1 000 000 km^2 because of the possibility of unstable positive feedback effects on world climate [Study of Man's Impact on Climate (SMIC), 1971]. It was thought that man-made changes on smaller scales should be safe for experiment, and were perhaps to be encouraged and studied. The Gulf of St. Lawrence by itself is smaller than the maximum safe size of area of modification as it was judged by the scientists. However, the system affected is large and is being altered measurably, hence serious study might be expected to yield important insights into the relationship between environmental cause and effect. It would be profoundly disturbing if the type of impact suggested above, were to be acknowledged without mounting some effort to identify the probable consequences.

Knowledge of the dynamics of the Gulf of St. Lawrence has been slowly accumulated only since the late nineteenth century. Early observations seem to have resulted from the perceptions and efforts of a few individuals. The names of J.F. Whiteaves (1901) on benthic biology, W.B. Dawson (1897, 1913) and J.W. Sandstrom (1919) on hydrodynamics and A.G. Huntsman (1919, 1923) on biological and physical oceanography figure importantly in early reports with later contributions by L.M. Lauzier (1967) and J.L. Tremblay (Tremblay and Lauzier, 1940). The Canadian Fisheries Expedition of 1914–1915 was the first large-scale operational survey, apparently undertaken as a direct result of the persuasiveness of Dr. A.G. Huntsman. It stood alone in Gulf studies until 1944 when the Government of Canada in cooperation with the Provinces of Nova Scotia, New Brunswick, Prince Edward Island and Quebec, and the Colony of Newfoundland formed the Atlantic Herring Investigation Committee (AHIC) as part of a general effort to assess the capacity of the fisheries to provide employment for veterans of World War II (Leim et al., 1957). During the period 1944–1948 AHIC carried out a physical oceanographic study of the Gulf as part of the general survey of the natural history of the herring stocks there. Modest research efforts were subsequently made by the Province of Quebec from a small marine station at Grande Rivière and by the Fisheries Research Board of Canada from stations at St. Andrews, New Brunswick, St. John's, Newfoundland and Grande Rivière. Increased research efforts, especially in physical oceanography, followed the foundation by the federal government of the Bedford Institute of Oceanography in 1962 and the Marine Ecology Laboratory in 1965, but most research was still dictated by the interests of individuals who carried out programs within the area of their disciplines or engaged in short-term collaboration on specific cruises.

This state of affairs continued until 1970 when the Canadian Government extended its jurisdiction over fisheries resources and declared the Gulf of St. Lawrence to be within national waters. That year the Science Council of Canada, in the course of a study of marine science and technology noted as a direct result of the work of Neu (1968) the possibility that the Gulf system might be progressively modified and recognized the lack of knowledge that would be required to predict and assess the consequences. The Council called for a major integrated program of research to be included in new programs (Science Council of Canada, 1970).

The Science Council of Canada is a quasi-independent body set up by the federal government to comment on science and science policy. A board of responsible citizens has a budget and several permanent staff to carry out in-depth studies, primarily by contracts with experts. There is no obligation on the part of any government to take the advice of such agencies, but occasionally either by direct communication with policy makers, or by judicious publicity, attention is directed to the issues it identifies and the Science Council's information and opinions can then influence government action.

Responsibility for the Gulf of St. Lawrence is complex. The pattern of social activity around the Gulf is strongly dependent on ice cover and climate, which control seasonal access for fisheries as well as for transportation of agricultural and forestry products. Factors affecting the coastal communities are clearly the primary responsibility of the five provincial governments whose territories border on the Gulf, namely Newfoundland and Labrador, Nova Scotia, Prince Edward Island, New Brunswick and Quebec. Additionally, however, responsibility for fisheries and most aspects of transport are constitutionally assigned to the federal government. The changes in fresh-water drainage noted above have been largely the result of a combination of hydroelectric power development and a large "seaway" canal system which opened the Great Lakes to direct access by seagoing vessels. In these developments semi-autonomous public utilities in the Provinces of Quebec, Ontario and of several states of the United States are involved, under general agreements by the federal governments of the two countries.

Clearly the responsibility for control of the Gulf as an integrated phenomenon is divided among many different agencies of different governments. It is not surprising if no one agency finds that the Gulf of St. Lawrence is its most important and pressing problem.

Among these scattered jurisdictions the Science Council report on marine sciences did have an identifiable influence. A fledgling federal department, the "Ministry of State for Science and Technology" found in oceanography a cause to be espoused. This interest resulted in 1974 in the framing and official adoption by the cabinet of the federal government of a formal statement of its marine science policy, to the effect that it encouraged many aspects of research and industrial development in relation to marine matters. But policy statements do not necessarily mean the commitment of funds to any specific program; indeed, in Canada, operational program action is generally outside the responsibility of a "ministry of state". Its responsibility is primarily to encourage action by others. The interest of several provincial governments was apparently stimulated by the policy, both in the prospects for continental shelf oil and the general topic of river outflow and climatic effects. But resultant programs were related primarily to the framing of legislation to help enhance provincial claims to the right to control oil exploration activity and to share revenue. There was no explicit or immediate concern for identification or control of local environmental issues or consequent social problems — which were, in any case, largely speculative.

At the federal government level, immediately following the time of the Science Council Study

(1968–1969), a major reorganization began. A Department of Environment was formed, combining a number of responsibilities of formerly separate government departments which were concerned with various aspects of the management of the Gulf system. Agencies concerned with fisheries, atmospheric prediction, wildlife, marine research and environmental protection were amalgamated under a single administration. Interestingly enough, during the early stages of development of this new department, a "Delphi" exercise on project priority was carried out among some 120 directors of its various units, and despite the knowledge that such an exercise would have a significant effect on the overall budget the "Gulf program" was given top priority. However, for a number of reasons, perhaps not the least of which was a diversion of much top administrative and ministerial level effort to preparations for the politically visible Stockholm International Environmental Conference, this priority rating by the directors was ignored in subsequent budgeting.

In the midst of what appeared to be administrative distraction and organizational upheaval, the scientific community at operational levels continued to show concern over the situation in the Gulf. Individuals from nearly every Canadian university from Montreal eastwards, together with scientists from related federal government branches, had convened a workshop in late 1968, which influenced the Science Council study. A second workshop convened in December 1970 attracted additional representatives from federal and provincial government agencies, business and private groups. There was general recognition that the problem was one which was outside the responsibility of any single agency or institution represented, but which clearly required sponsorship by a central government agency if it were to be undertaken as a cooperative interdisciplinary venture. The workshop resulted in a detailed framework and budget proposals for an agreed interdisciplinary natural science program, and recognized the need for a new social sciences program. The package was, of course, once again submitted to the Department of Environment as well as other agencies and committees.

As a result of the various workshops and subsequent initiatives, a number of concrete research programs were formulated and later approved and carried out. The I.B.P. program of McGill University in cooperation with Laval University and the University of Montreal, was developed in direct relation to the Gulf workshops, and was funded by the National Research Council. Provision for parts of the program, amounting to nearly half of the proposed budget, were made in the existing budgets of several of the involved government agencies. When the submission of the total program failed to find a principal sponsor, some of the programs were undertaken by individual agencies, where this was practicable, and the discussions and results of them have led to the initiation of others. However, it is an inescapable conclusion that this attempt by the scientific community in eastern Canada to engage the attention of the public, of the administration or of its own political leadership in a large-scale cooperative interdisciplinary study has failed.

Some discussion of this particular situation seems warranted in relation to investigations of major environmental perturbations in general. The fact is that the oceanographic sciences are frequently dealing, as in this Gulf example, with instances of large ocean–atmosphere systems whose stability is poorly understood. Underlying the attempts to study and comprehend such large systems is the concern that positive feedback effects in them could lead to amplified results from seemingly innocuous causes. It was on this basis that the Stockholm SMIC (1971) conference issued its warning. Subsequent events appear to lend credence at least to the existence of causal interconnections of larger systems. For example, there has been considerable recent success in using measurements of high-altitude pressure systems to forecast long-range weather prospects. The most dramatic illustration is the apparent relation between pressure systems of the southeast Pacific and the onset of El Niño events off the coast of Peru. If cause and effect can be demonstrated to operate on these scales, it would hardly be reassuring to conclude that the only real protection from deleterious consequences of ignorant, man-made environmental perturbation was the possibility of discontinuities in the functioning of large systems.

It is not surprising that administrators or politicians in control of bureaucracies should balk at approving extraordinary short-term expenditures of funds. In general it is their expectation that smaller expenditures sustained over longer periods would achieve the same results, and that this scale of operation is preferable because control rests within ordinary budgets, and study and evolution of control systems can be undertaken more easily. The question posed for the general scientific community by the nature of the history of research applied to the Gulf of St. Lawrence is therefore whether or not the evidence or risk of general system effects is strong enough to justify attempts to develop large-scale programs outside the scales of those allowed by bureaucratic inertia or administrative "common sense". Oceanographic research proposals of the late 1960s and early 1970s were often influenced by a tendency towards the development of "big" science. The subsequent history of the Gulf study shows that significant new knowledge has resulted from a piecemeal approach within the scale of more usual scientific programming, much of which depends on individual initiatives. However, the question of how far this attitude should prevail is still a problem in the management of scientific knowledge which is unresolved but which needs to be vigorously re-examined.

The Gulf of St. Lawrence situation also illustrates another important practical problem in the study of changes in major natural systems. It is generally agreed that complexity alone has not deterred society from undertaking massive or expensive research and development projects. However, it appears that the clarity with which the objectives are formulated, and our ability to identify these objectives with the responsibilities of particular political entities, may be very important factors in program development.

In the foregoing description of the situation, we have pointed out the divisions of responsibilities for various aspects of activity in the Gulf of St. Lawrence. This division of responsibility, and hence of jurisdictional powers, has constantly hampered the abilities of the various workshops or committees to engage in comprehensive considerations of the characteristics of the system with political and administrative authority. Thus, for example, it is agreed by most observers that the evidence is quite strong that past human manipulation of fresh-water outflow may affect the ice cover of the Gulf. There is also a growing consensus with the Science Council recommendation that the objective of research should be to develop our knowledge of the system to the point where the ice cover could be placed under conscious control. But what if scientific study should demonstrate that such control was indeed feasible? Is the prospect for leadership in controlling the consequences of such manipulation likely to sustain the effort that would be required to mount the study and evolve a control system?

With jurisdictional responsibility so greatly divided the answers to these questions are not immediately apparent. From the point of view of potential political leadership, transportation, hydroelectric power and oil-producing companies are likely to welcome reductions in ice cover. So too would inland ports which are ice-bound in winter. On the other hand, there would clearly be economic disruption for present coastal winter ports and associated transport. There may also be severe climatic effects for lee-shore land areas on the east and south margins which may well be subjected to more severe winter snow and fog. The balance of benefits for fisheries between a longer season but possibly altered productivity could not be determined until after a study or even with experience. But industry apprehension would be understandable because of the currently high local productivity.

The issues raised by a consideration of the possible consequences of environmental study are bound to be complex. The prospects of significant change in the welfare of individuals are great, with some gaining where others lose. The solutions to the problems will clearly be political.

In instances where there are clear indications that unintentional environmental effects have been and are being sustained, it seems obvious that it is a simple and responsible act for scientists to promote study and attempt to predict and assess the consequences. However, it is also realistic to recognize that the political task is far more difficult. Those who insist on understanding are inevitably the ones who face the difficulty of trying to ensure the best consequences of intentional acts. It is questionable whether fractioned responsibility of the sort described here is capable of providing the required leadership. The problem is not unique to the Gulf of St. Lawrence, but unfortunately we do not appear to have evolved effective methods of dealing with such large-scale events. The remedy may well require the evolution of new institutions which are less vulnerable to local sectarian interests.

REFERENCES

Banks, R.E., 1966. The cold layer in the Gulf of St. Lawrence. *J. Geophys. Res.*, 71: 1603–1610.

Blackford, B.L., 1965. Some oceanographic observations in the southern Gulf of St. Lawrence — Summer 1964. *Fish. Res. Board Can. Rep. (Oceanogr. and Limnol. Ser.)*, No. 190: 22 pp.

Blackford, B.L., 1966. A simple two-dimensional electrical analog model for wind-driven circulation in the Gulf of St. Lawrence. *J. Fish. Res. Board Can.*, 23: 1411–1438.

Boudreault, F.-R., 1969. Moyennes climatiques du bilan thermique de Grande-Rivière (Baie des Chaleurs) Québec. *Que. Minist. Indus. Comm., Stn. Biol. Mar. Cah. Inf.*, No. 50: 54 pp.

Brown, R.G.B., Nettleship, D.N., Germain, P., Tall, C.E. and Davis, T., 1975. *Atlas of Eastern Canadian Seabirds.* Canadian Wildlife Service, Ottawa, Ont., 220 pp.

Conover, R.J., 1968. Zooplankton — Life in a nutritionally dilute environment. *Am. Zool.*, 8: 107–118.

Coombs, J.A., 1962. A preliminary investigation of the heat budget in the Gulf of St. Lawrence. *Bedford Inst. Oceanogr., Rep.* 62-1: 65 pp.

Dawson, W.B., 1897. *Survey of Tides and Currents in Canadian Waters. Report of Progress (26th January 1897).* Government Printing Bureau, Ottawa, Ont., 49 pp.

Dawson, W.B., 1913. *The Currents in the Gulf of St. Lawrence.* Can. Dep. Naval Serv., Ottawa, Ont., 46 pp.

Dickie, L.M., 1974. Problems of prediction in marine ecosystems. *Proc. Can. Soc. Zool., Annu. Meet.*, June 1974: 111–115.

Dunbar, M.J., 1955. The present status of climate change in the Atlantic sector of northern seas with special reference to Canadian eastern Arctic waters. *Trans. R. Soc. Can., Ser.* 3, 49: 1–7.

Dunbar, M.J., 1979. Biological production in the Gulf of St. Lawrence. In: M.J. Dunbar (Editor), *Marine Production Mechanisms.* International Biological Program 20. Cambridge University Press, Cambridge, pp. 151–171.

El Sabh, M.I., 1976. Surface circulation pattern in the Gulf of St. Lawrence. *J. Fish. Res. Board Can.*, 33: 124–138.

El Sabh, M.I., 1977. Oceanographic features, currents and transport in Cabot Strait. *J. Fish. Res. Board Can.*, 34: 516–528.

El Sabh, M.I. and Johannessen, O.M., 1972. Winter geostrophic circulation in the Gulf of St. Lawrence. *J. Fish. Res. Board Can.*, 29: 595–598.

Farquharson, W.I., 1962. *Tides, Tidal Streams and Currents in the Gulf of St. Lawrence.* Can. Dep. Mines and Tech. Surv., Mar. Sci. Branch, Rep., 15 pp.

Farquharson, W.I., 1966. St. Lawrence Estuary current surveys. *Bedford Inst. Oceanogr., Rep.*, 66–6: 84 pp.

Forrester, W.D., 1964. A quantitative temperature–salinity study of the Gulf of St. Lawrence. *Bedford Inst. Oceanogr., Rep.*, 64-11: 16 pp.

Forrester, W.D., 1970. Geostrophic approximation in the St. Lawrence estuary. *Tellus*, 22: 53–65.

Forrester, W.D., 1974. Internal tides in St. Lawrence estuary. *J. Mar. Res.*, 32: 55–66.

Forrester, W.D. and Vandall, Jr., P.E., 1968. Ice volumes in the Gulf of St. Lawrence. *Bedford Inst. Oceanogr., Rep.*, No. 68-7: 16 pp.

Greisman, P. and Ingram, G., 1977. Nutrient distribution in the St. Lawrence Estuary. *J. Fish. Res. Board Can.*, 34(11): 2117–2123.

Huntsman, A.G., 1919. Some quantitative and qualitative plankton studies of the eastern Canadian plankton. In: *Canadian Fisheries Expedition 1914–15.* Department of Naval Service, Ottawa, Ont., pp. 405–482.

Huntsman, A.G., 1923. The influence of tidal oscillations on vertical circulation in estuaries. *Trans. R. Soc. Can., Ser. 3*, 17: 11–14.

Ingram, R.G., 1975. Influence of tidal-induced vertical mixing on primary productivity in the St. Lawrence estuary. *Mem. Soc. R. Sci. Liège*, 6: 59–74.

Ingram, R.G., Johannessen, O.M. and Pounder, E.R., 1968. A pilot study of ice drift in the Gulf of St. Lawrence. *Mar. Sci. Centre, McGill Univ., Rep.*, No. 5: 17 pp.

Johannessen, O.M., Pounder, E.R., Serson, H., Keys, J., Lindsay, D., Seifert, W. and Banke, E., 1968. Preliminary Report on the "ice drift study" in the Gulf of St. Lawrence, Winter 1968. *Mar. Sci. Centre, McGill Univ., Rep.*, No. 4: 21 pp (9 figs.).

Keyte, F.K. and Trites, R.W., 1971. Circulation patterns in the Gulf of St. Lawrence: June 1968 and Sept. 1969. *Fish. Res. Board Can., Tech. Rep.*, No. 271: 108 pp.

Kohler, A.C., 1964. Variations in the growth of Atlantic cod (*Gadus morhua* L.). *J. Fish. Res. Board Can.*, 21: 57–100.

Kohler, A.C., Faber, D.G. and McFarlane, N.J., 1974. Eggs, larvae, and juveniles of fishes from plankton collections in the Gulf of St. Lawrence. *Fish. Mar. Serv., Res. Dev. Tech. Rep.*, No. 490:105 pp.

Lauzier, L.M., 1967. Bottom residual drift on the continental shelf area of the Canadian Atlantic coast. *J. Fish. Res. Board Can.*, 24: 1845–1859.

Lauzier, L.M. and Graham, R.D., 1958. Computation of ice potentials and heat budget in the Gulf of St. Lawrence. *Fish. Res. Board Can., Rep. (Oceanogr. and Limnol.)* Ser. 3(11): 7 pp.

Lauzier, L.M. and Marcotte, A., 1965. Comparison du climat marin de Grande-Rivière (baie des Chaleurs) avec celui d'autres stations de la côte atlantique. *J. Fish. Res. Board Can.*, 22: 1321–1334.

Lauzier, L.M. and Trites, R.W., 1958. The deep waters in the Laurentian Channel. *J. Fish. Res. Board Can.*, 15: 1247–1257.

Lauzier, L.M., Trites, R.W. and Hachey, H.B., 1957. Some features of the surface layer of the Gulf of St. Lawrence. *Bull. Fish Res. Board Can.*, No. 111: 195–212.

Leim, A.H., Tibbo, S.N., Day, L.R., Lauzier, L., Trites, R.W., Hachey, H.B. and Bailey, W.B., 1957. Report of the Atlantic herring investigation committee. *Bull. Fish. Res. Board Can.*, No. 111: 317 pp.

Lett, P.F., Kohler, A.C. and Fitzgerald, D.N., 1975. The role of stock biomass and temperature in the recruitment of southern Gulf of St. Lawrence Atlantic cod *Gadus morhua. J. Fish. Res. Board Can.*, 32: 1613–1627.

Loring, D.H. and Nota, D.J.G., 1973. Morphology and sediments of the Gulf of St. Lawrence. *Bull. Fish. Res. Board Can*,, No. 182: 147 pp.

MacKay, K.T., 1976. *Population Biology and Aspects of Energy Use of the Northern Population of Atlantic Mackerel*, Scomber scombrus *L*. Thesis, Dalhousie University, Halifax, N.S., 225 pp.

Mann, K.H., 1977. Destruction of kelp-beds by sea-urchins: A cyclical phenomenon or irreversible degradation? *Helgol. Wiss. Meerunters.*, 30: 455–467.

Matheson, K.M., 1967. The meteorological effect of ice in the Gulf of St. Lawrence. *Mar. Sci. Centre, McGill Univ., Rep.*, No. 3: 110 pp.

Messieh, S.N. and Kohler, A.C., 1972. Distribution, relative abundance and growth of larval Herring (*Clupea harengus*) in the southern Gulf of St. Lawrence. *Fish. Res. Board Can. Tech. Rep.*, No. 318: 31 pp.

Murty, T.S. and Taylor, J.D., 1970. A numerical calculation of the wind-driven circulation in the Gulf of St. Lawrence. *J. Oceanogr., Soc. Jap.*, 26: 15–26.

Neu, H.J.A., 1968. Man-made changes — the potential benefits and liabilities. In: *Bedford Institute, Gulf of St. Lawrence Workshop, Dartmouth, N.S.*, pp. 75–79 (unpublished).

Neu, H.J.A., 1975. Runoff regulation for hydropower and its effects on the ocean environment. *Can. J. Civil Eng.*, 2: 583–591.

Paloheimo, J.E. and Kohler, A.C., 1968. Analysis of the southern Gulf of St. Lawrence cod population. *J. Fish. Res. Board Can.*, 25: 555–578.

Platt, T., 1972. The feasibility of mapping the chlorophyll distribution in the Gulf of St. Lawrence. *Fish. Res. Board Can. Tech. Rep.*, No. 332: 28 pp.

Powles, P.M., 1958. Studies of reproduction and feeding of Atlantic cod (*Gadus callarias* L.) in the southwestern Gulf of St. Lawrence. *J. Fish. Res. Board Can.*, 15: 1383–1402.

Reid, S.J., 1977. Circulation and mixing in the St. Lawrence Estuary near Ilet Rouge. *Bedford Inst. Oceanogr., Rep., Ser.* BI-R-77-1: 36 pp.

Sameoto, D.D., 1976. Distribution of sound scattering layers caused by Euphausiids and their relationship to chlorophyll *a* concentrations in the Gulf of St. Lawrence estuary. *J. Fish. Res. Board Can.*, 33: 681–687.

Sandstrom, J.W., 1919. The hydrodynamics of Canadian Atlantic waters. In: *Canadian Fisheries Expedition 1914–15*. Department of Naval Service, Ottawa, Ont., pp. 221–346.

Sargeant, D.E. and Fisher, H.D., 1957. The smaller Cetacea of eastern Canadian waters. *J. Fish. Res. Board Can.*, 14: 83–115.

Sargeant, D.E., Mansfield, A.W. and Beck, B., 1970. Inshore records of Cetacea for eastern Canada. *J. Fish. Res. Board Can.*, 27: 1903–1915.

Scarratt, D.J., 1964. Abundance and distribution of lobster larvae (*Homarus americanus*) in Northumberland Strait. *J. Fish. Res. Board Can.*, 21: 661–680.

Scarratt, D.J., 1973. Abundance, survival, and vertical and diurnal distribution of lobster larvae in Northumberland Strait, 1962–63, and their relationships with commercial stocks. *J. Fish. Res. Board Can.*, 30: 1819–1824.

Science Council of Canada, 1970. Canada, science and the oceans. *Sci. Counc. Can., Rep.*, No. 10: 37 pp.

Srivastava, V.M., 1971. Fish of the Gulf of St. Lawrence; an unabridged bibliography. *Fish. Res. Board Can., Tech. Rep.*, No. 261: 141 pp.

Steven, D.M., 1974. Primary and secondary production in the Gulf of St. Lawrence. *Mar. Sci. Centre, McGill Univ., Rep.*, No. 26: 116 pp.

Study of Man's Impact on Climate. (SMIC), 1971. *Inadvertent Climate Modification*. Mass. Inst. Tech. Press, Cambridge, Mass., 308 pp.

Sutcliffe Jr., W.H., 1972. Some relations of land drainage, nutrients, particulate material, and fish catch in two eastern Canadian bays. *J. Fish. Res. Board Can.*, 29: 357–362.

Sutcliffe Jr., W.H., 1973. Correlations between seasonal river discharge and local landings of American lobster (*Homarus americanus*) and Atlantic halibut (*Hippoglossus hippoglossus*) in the Gulf of St. Lawrence. *J. Fish. Res. Board Can.*, 30: 856–859.

Sutcliffe Jr., W.H., Loucks, R.H. and Drinkwater, K.F., 1976. Coastal circulation and physical oceanography of the Scotian Shelf and the Gulf of Maine. *J. Fish. Res. Board Can.*, 33: 98–115.

Sutcliffe Jr., W.H., Drinkwater, K. and Muir, B.S., 1977. Correlations of fish catch and environmental factors in the Gulf of Maine. *J. Fish. Res. Board Can.*, 34: 19–30.

Therriault, J.-C. and Lacroix, G., 1976. Nutrients, chlorophyll, and internal tides in the St. Lawrence estuary. *J. Fish. Res. Board Can.*, 33: 2747–2757.

Tremblay, J.L. and Lauzier, L.M., 1940. L'origine de nappe d'eau froide du Saint-Laurent. *Nat. Can.*, 67: 5–23.

Trites, R.W., 1972. The Gulf of St. Lawrence from a pollution viewpoint. M. Ruivo (Editor), *Marine Pollution and Sea Life*. Fishing News (Books), London, pp. 59–72.

Trites, R.W. and Walton, A., 1975. A Canadian coastal sea — The Gulf of St. Lawrence. *Bedford Inst. Oceanogr., Rep. Ser. BI-R-75-10*: 29 pp (33 figs.).

Ware, D.M., 1975. Relations between egg size, growth and natural mortality of larval fish. *J. Fish. Res. Board Can.*, 32: 2503–2512.

Whiteaves, J.F., 1901. *Catalogue of the Marine Invertebrata of Eastern Canada*. Geological Survey of Canada, Kings Printer, Ottawa, Ont., 271 pp.

Winters, G.H., 1976. Recruitment mechanisms of southern Gulf of St. Lawrence Atlantic herring (*Clupea harengus harengus*). *J. Fish. Res. Board Can.*, 33: 1751–1763.

Worley, L.G., 1933. Development of the egg of the mackerel at different constant temperatures. *J. Gen. Physiol.*, 16: 841–857.

Chapter 17

GULF OF CALIFORNIA

SAUL ALVAREZ-BORREGO

INTRODUCTION

The peninsula of Baja California was first described by Hernan Cortes, who pioneered the exploration of the Pacific soon after the conquest of Mexico–Tenochtitlan. In his fourth letter to the king of Spain (1524) he wrote about an island — Baja California — his captains had reported to him, and Cortes himself sailed to the area of La Paz bay in 1535 (Leon-Portilla, 1972). Finally, in 1539 Cortes sent Francisco De Ulloa with three ships to explore further north. De Ulloa circumnavigated the Gulf and named it "Mar Bermejo", the Vermilion Sea, because of the reddish color of the muddy waters of the Colorado Estuary, and not, as is often assumed, because of the red plankton blooms of the central region (Van Andel and Shor 1964). The first hydrographic cruise into the Gulf was made by the U.S. Fish Commission Steamer *Albatross* in 1889 (Townsend, 1901). Fifty years later, the modern investigation of the Gulf began with the *E.W. Scripps* cruise in 1939, which made a series of 53 hydrographic stations throughout the Gulf, including measurements of temperature, salinity, oxygen and calcium, besides sampling for phytoplankton and zooplankton (Sverdrup, 1941; Roden and Groves, 1959).

The Gulf of California occupies an oceanographically unique position among the marginal seas of the Pacific Ocean (Fig. 17.1A). Lying between the arid peninsula of Baja California in the west and the almost equally arid states of Sonora and Sinaloa to the east, it includes a large evaporation basin, which is open to the Pacific at its southern end (Roden, 1964). It is approximately 1000 km long and 150 km wide on the average. Topographically it is divided into a series of basins and trenches, deepening to the south and separated from each other by transverse ridges (Shepard, 1950) (Fig. 17.1B). It is separated into two areas by the islands Angel de la Guarda and Tiburon. The northern portion is relatively shallow with the exception of Dolphin Basin. Much of this area is at shelf depth with few shelf breaks. South from Angel de la Guarda and Tiburon the Gulf has well-developed continental shelves. On the western side the shelf is generally rocky and narrow, with a sharp shelf break between 80 and 100 m; the eastern shelf is wider (Van Andel, 1964). The coast of Sinaloa, to the east, has many coastal lagoons, most of them in the area of the mouth of the Gulf.

Theories about the formation of the Gulf have developed with the theories of sea-floor spreading, continental drift and plate tectonics (Reichle, 1975). Shepard (1950) concluded that the topography of the Gulf resulted primarily from strike-slip faulting. According to Rusnak and Fisher (1964), the Gulf may have evolved as fractured plates of crustal material moved northwestward toward the Pacific by gravitational sliding on extremely gentle slopes from the regions of western Mexico uplifted by batholithic intrusions. Sykes (1968) suggested that the tensional faults separating the strike-slip faults may be loci of sea-floor spreading. Larson et al. (1968) analyzed magnetic anomalies generated at the East Pacific Rise near the mouth of the Gulf and concluded that the motion of the tip of Baja California away from mainland Mexico has been an average of 6 cm yr^{-1} for the past four million years. The rates of spreading may be obtained from the magnetic patterns and the dates of the earth's magnetic field reversals (Bullard, 1969). Moore and Buffington (1968) suggested the existence of a proto-Gulf before the present spreading episode

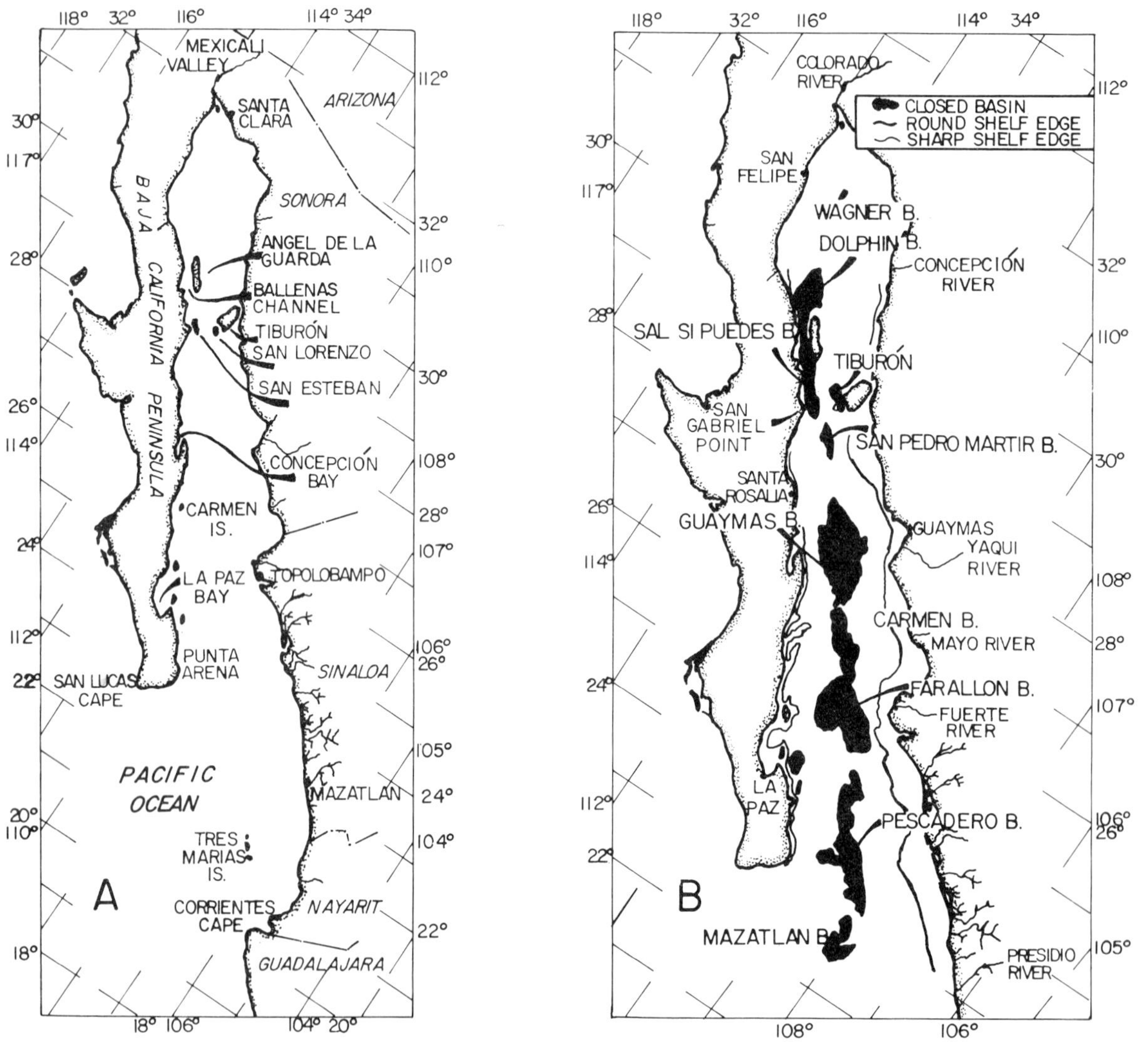

Fig. 17.1. A. Gulf of California. B. Shelves and basins in the Gulf of California. Basins are defined by depth contour at sill depth (after Van Andel, 1964).

began four million years ago. Reichle (1975) has discussed the regional tectonics of the Gulf and presented a general description of a series of transform faults and spreading centers (Fig. 17.2).

CURRENTS, MIXING AND WATER MASSES OF THE GULF

Meteorological aspects

The moderating effect of the Pacific Ocean upon the climate of the Gulf of California is greatly reduced by an almost uninterrupted chain of mountains, 1 to 3 km high, in Baja California. The climate of the Gulf is therefore more continental than oceanic, a fact which contributes to the large annual and diurnal temperature ranges observed there (Hernandez, 1923; cited by Roden, 1964). Roden (1964), using data from Page (1930) and Ward and Brooks (1936), made a summary of the areal distribution of air temperature and rainfall in the Gulf and adjacent regions (Fig. 17.3). In winter the air temperatures decrease toward the interior of the Gulf; there are small air temperature differences between the Gulf and the Pacific coasts of Baja

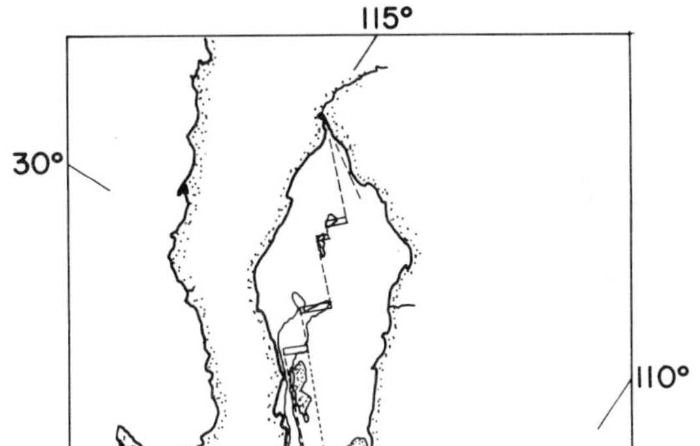

Fig. 17.2. Schematic map of the Gulf of California tectonics. Dashed lines show approximate positions of transform faults; double solid lines show locations of spreading centers as suggested by the bathymetry (after Reichle, 1975).

California, but for the same latitude the air temperatures of the Mexico mainland's coast are higher by about 2°C than those of the Baja California coast. In summer the air temperature increases toward the interior of the Gulf and the temperature differences between the east and west coasts of Baja California are large, sometimes exceeding 10°C. There is more precipitation on the east than on the west side of the Gulf. The northern half of the Gulf is dry and desert-like, with annual rainfall of less than 100 mm. In the southeast, rainfall along the coast increases to about 1000 mm per year. South of Angel de la Guarda and Tiburon Islands most of the rain falls between June and October, but in the north most of the rain falls during winter. The mean annual air temperature range increases from about 6°C at Cape Corrientes to 18°C at the northern end of the Gulf. The number of rainy days per year decreases from about 60 at Cape Corrientes to about 5 along the central Baja California coast.

The year-to-year variation of rainfall in the Gulf is large and depends very much on the incidence of tropical storms and hurricanes. Harris (1969), using data from weather satellites and a review of available literature, indicated that from 1961 (the year the first Tiros satellite was launched) to 1968 there were 93 tropical cyclones in the eastern North Pacific, with wind velocities higher than 34 knots (63 km h^{-1}); 30 of them were hurricanes with wind velocities higher than 74 knots (137 km h^{-1}). According to Harris (1969) the eastern North Pacific is an area of frequent tropical cyclone activity and is second in the world only to the typhoon region of the western Pacific. Hurricanes and tropical storms occur from June to November, with the greatest frequency in September. The general trend of these tropical cyclones is to travel north or northwestward from their origin, which is from 15 to 20°N, and 100 to 110°W. Some of them enter the Gulf of California and dissipate at different places in the states of Sinaloa and Sonora, but a few travel all the way to Arizona. There is a long list of damages that hurricanes cause every year at places such as Manzanillo, Mazatlan, La Paz and San Felipe, including many sunk fishing boats and homeless families. Sometimes there are hundreds of deaths, as there were during hurricane Liza that reached La Paz on October 1, 1976. No systematic studies have been done on the ecological effects of these hurricanes, especially their effects on coastal lagoons and estuaries of the east coast where some commercially important species, crustaceans and molluscs mainly, spend part or all of their lives.

Winds in the gulf are extremely variable. Near the coast a land and sea breeze system prevails and diurnal wind changes are usually larger than annual ones. In the offshore regions of the Gulf northwesterly winds prevail from November to May, and southeasterly ones during the rest of the year. Moderate northwest gales that last two or three days at a time are frequently experienced in the upper Gulf between December and February. These winds are particularly strong in Ballenas channel between the mountainous Baja California

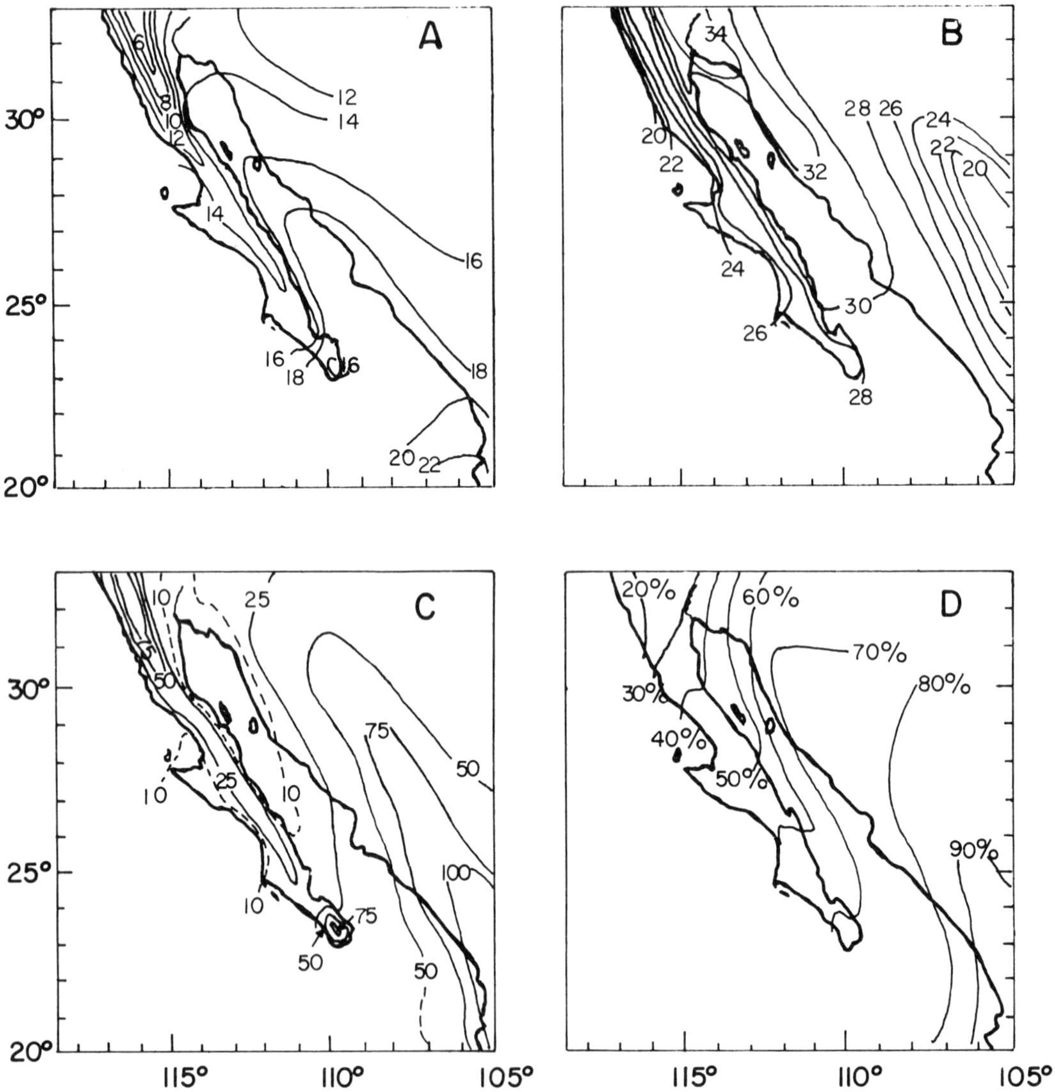

Fig. 17.3. A. Mean air temperature for January (°C). B. Mean air temperature for July (°C). C. Mean annual rainfall (cm). D. Mean percentage of summer rain (May to October) (after Page, 1930; and Ward and Brooks, 1936; cited by Roden, 1964).

coast and the equally high island of Angel de la Guarda; they may on occasion raise such a heavy sea that navigation becomes impossible (Roden, 1964).

Roden (1964) estimated the evaporation from the sea surface of the southern portion of the Gulf to be roughly 130 cm yr^{-1}; he used the equation $E = K\Delta e W$ for this purpose, where E is evaporation, Δe is the vapor pressure difference between the sea surface and the air above it, W is the wind speed and K is a proportionality factor. He did not apply the equation to the northern portion of the Gulf because advection of dry desert air is considerable there. Direct measurements from evaporation pans at coastal stations give figures that fluctuate between 200 and 250 cm yr^{-1}, with minima in winter and maxima in summer (Roden, 1964).

Fresh-water input by rivers has only a local impact, mostly during the rainy season. Most river water is impounded and used for agricultural and urban purposes, and progressively less river water is allowed to reach the Gulf. The northern states of Sonora and Baja California suffer from an acute lack of water. This semitropical desert area could use huge amounts of water if it were available. There is also a great need for fresh water in the southern areas of Arizona and California. During the late 1960s there was a plan to construct a nuclear power and water desalination plant on the northern Sonora coast, with an anticipated fresh-water production capacity of 4×10^9 litres per day (United States–Mexico–International Atomic Energy Agency study team, 1968). It was not constructed because of political reasons. Therefore, we can anticipate that sometime in the near future no fresh river water at all will reach the Gulf.

General aspects of circulation and tides

Circulation in the Gulf is not yet well described. It is complex and changes with time in a way that is far from being fully understood. Currents in the Gulf have rarely been measured directly. According to Roden (1958), the bulk of water in and near the Gulf is the same as in the equatorial Pacific, modified at the surface by evaporation and by admixture with California Current water. Thorade (1909), using temperature charts constructed from ships' logs, deduced the general circulation in the Gulf. He concluded that the low temperatures along the east coast in winter are the result of upwelling produced by northerly winds. He may have been the first to describe the frontal system near Cape San Lucas, between the cold California Current water and the warm Gulf water.

Using 1947 charts of ship-drifting prepared by the U.S. Hydrographic Office, Roden (1958) concluded that in winter the surface circulation of the Gulf is characterized by southward currents north of Cape Corrientes, and during summer a current flows northward along the coast of Mexico and enters the Gulf at the eastern and central regions of the mouth, with a southward flow near Baja California. This is in agreement with the results obtained by Granados-Gallegos and Schwartzlose (1974), who used drift bottles. These authors concluded that there is a general pattern of southward flow in winter throughout the entire Gulf, northward flow in summer south of Tiburon, San Esteban and San Lorenzo islands, and some northward flow in the northern Gulf, with some eddy circulation. During the spring and fall the flow may be in many directions. Counterclockwise gyres have been detected in the northern Gulf by means of satellite photographs (Vonder Haar and Stone, 1973; Granados-Gallegos and Schwartzlose, 1974; Lepley et al., 1975).

Currents in the Gulf have been estimated, using dynamic computations (Sverdrup, 1941; Roden and Groves, 1959; Griffiths, 1968; Stevenson, 1970; Warsh and Warsh, 1971; Roden, 1972; Rosas-Cota, 1976; Alvarez-Sanchez et al., 1978), but only Stevenson (1970) and Rosas-Cota (1976) have presented results for all seasons of the year. Roden (1964) estimated the surface currents at the central portion of the Gulf, near Guaymas, using a linear and stationary model and considering the effects of geostrophic forces, a component due to the horizontal atmospheric pressure gradient and a component produced by wind stress. For February he obtained southward values of 4, 1 and 5 cm s^{-1} respectively for the three components, with a total speed of 10 cm s^{-1}, and for August northward values of 13, 2 and 6 cm s^{-1}, with a total speed of 21 cm s^{-1}. He also estimated vertical currents at both sides of the Gulf for February and found upwelling velocities of roughly 3.2 m day^{-1} near the east coast. Upwelling occurs on the east coast during winter with northwesterly winds, and on the west coast during summer with southeasterly winds.

Results from dynamic computations are in fair agreement with data from drift bottles. Rosas-Cota (1976) reports geostrophic currents to the south in winter and to the north in summer (Fig. 17.4), with more complex reversing conditions and with gyres present during spring and fall. Maximum computed speeds are about 40 to 50 cm s^{-1}. Alvarez-Sanchez et al. (1978) compared geostrophic currents at the mouth of the Gulf with results from drifting drogues and found an agreement for the upper 100 m of better than 70% at three stations and 85% at two stations. Brown (1965) made a series of six drogue stations in the Gulf, from the mouth near La Paz to the area between San Lorenzo Island and Baja California in November 1963. Drogues were situated at 10, 50, 100, 200, 300 and 1000 m. His results show the great complexity of the Gulf circulation, with currents changing with geographic location and depth. Round (1967) concluded that lateral transport in the Gulf of California is not great, basing his observation on the good agreement between the diatom content of surface sediment samples, their position in the Gulf, and the phytoplankton communities above them.

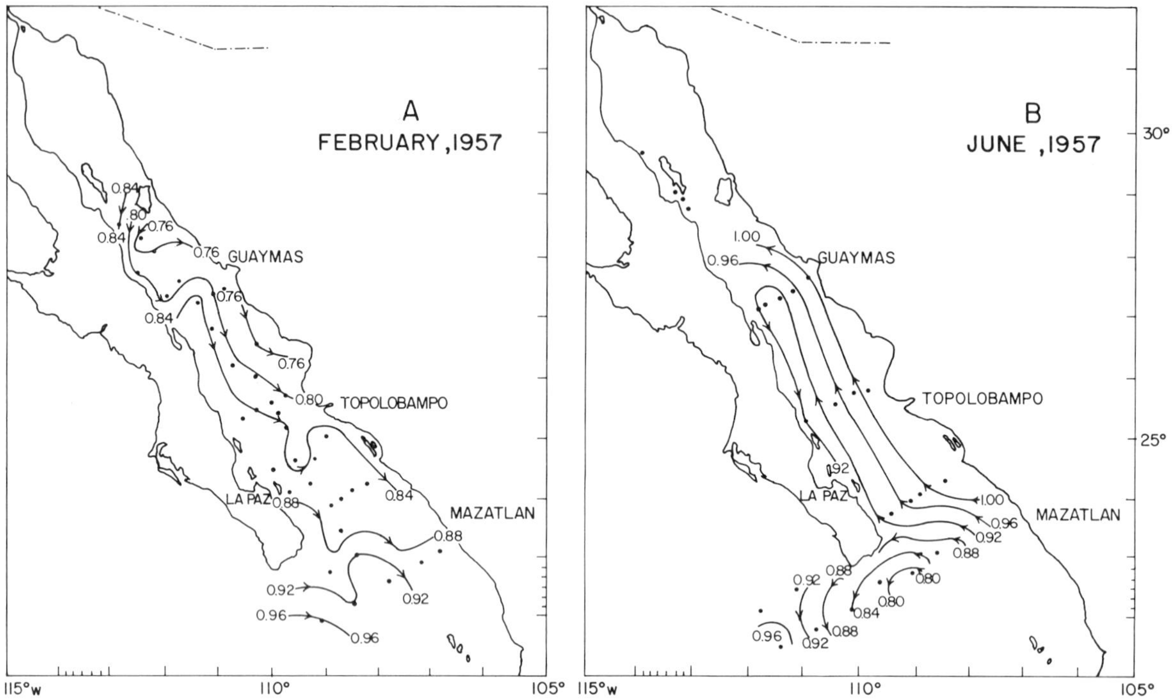

Fig. 17.4. Dynamic topography of the sea surface (meters) referred to the 500-dbar surface (after Rosas-Cota, 1976).

Grijalva-Ortiz (1972) and Stock (1976) computed tidal currents for different points in the Gulf and found that the current ellipses for the principal lunar diurnal component (M_2) have the major axis increasing from the mouth (3 cm s^{-1}) to the head (60 cm s^{-1}) (Fig. 17.5). Stock (1976) compared his computed results for the northern Gulf with current meter data of R.A. Schwartzlose (unpublished results) and found fair agreement for the component along the Gulf axis, but the computed transverse component was significantly larger than the measured data. Tidal currents in the narrows between the islands and the coast and in the passages connecting semi-enclosed lagoons with the Gulf are strong. The speed of these currents is variable and depends upon the stage of the moon and the prevailing winds, but they have been reported up to 3 m s^{-1} (6 knots) (Roden, 1964).

The tides in the Gulf of California are among the most spectacular and dangerous in the world, with reported spring ranges of 10 m at the northern end. The tidal wave is progressive (see Ch. 1, p. 6) so that the time of high or low water is progressively later travelling north up the Gulf, with the time difference between the entrance and the vicinity of the Colorado River being approximately 5.5 h for high water and 6 h for low water. The result is that low water at one end of the Gulf occurs at about the same time as high water at the other end (Roden, 1964). This is shown for the principal lunar semidiurnal component in Fig. 17.6A.

The difference between semidiurnal and diurnal tides is striking. The semidiurnal tide enters the Gulf with a moderate amplitude of 30 cm for the principal lunar component (M_2). The speed of the wave decreases within the Gulf and the amplitude decreases to one third of its initial value near the middle of the Gulf. It then accelerates and its amplitude increases at the head to fifteen times the amplitude in the middle of the Gulf (165 cm for M_2). There it seems to be absorbed by the coastline, suggesting that most of the tidal force is dissipated, an observation that is also indicated by the amphidromic region near the middle of the Gulf (Fig. 17.6B). Hendershott and Speranza (1971) also indicated the existence of a virtual amphidrome for the semidiurnal tides in the Gulf, but their computed results situated it further to the south.

By contrast, the amplitude of the diurnal tide increases very slowly and monotonically to about

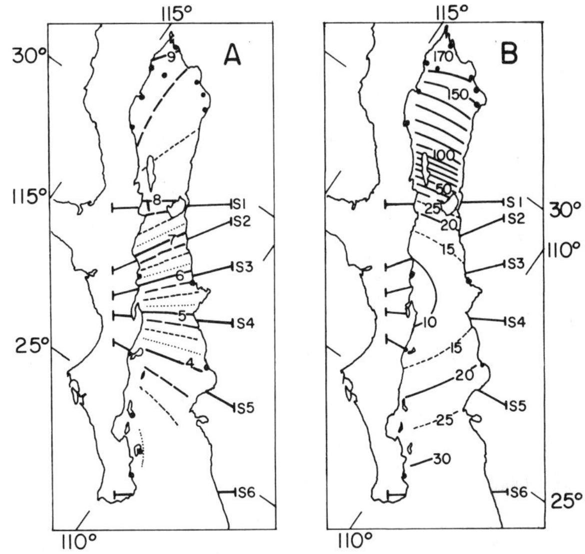

Fig. 17.6. The effect of the principal lunar semidiurnal tidal component (M_2) on A, the phase lag in lunar hours after equilibrium tide at Greenwich; and on B, the tidal amplitude in centimeters (after Filloux, 1973). Cross-sections (*S1–S6*) used in the computations are identified on the right side of each figure.

twice its amplitude at the mouth. Its nearly uniform phase suggests that it is like a standing oscillation, and that most of the energy entering the Gulf does leave again (Filloux, 1973).

Currents and water masses at the entrance

Different authors have defined the entrance to the Gulf of California in different ways. For the purpose of the following discussion, the region of the entrance to the Gulf will be considered roughly as that between a line connecting Cape San Lucas and Cape Corrientes and a line connecting La Paz with Topolobampo (Fig. 17.1). This is a transitional zone that has a very complicated and dynamic oceanographic structure. Sverdrup (1941) studied currents in the outer portions of the Gulf by means of dynamic computations, using data from the February–March 1939 cruise of the *E.W. Scripps*, with discouraging results. With hydrographic lines across the mouth, the dynamic topography of the 500-dbar relative to the 1500-

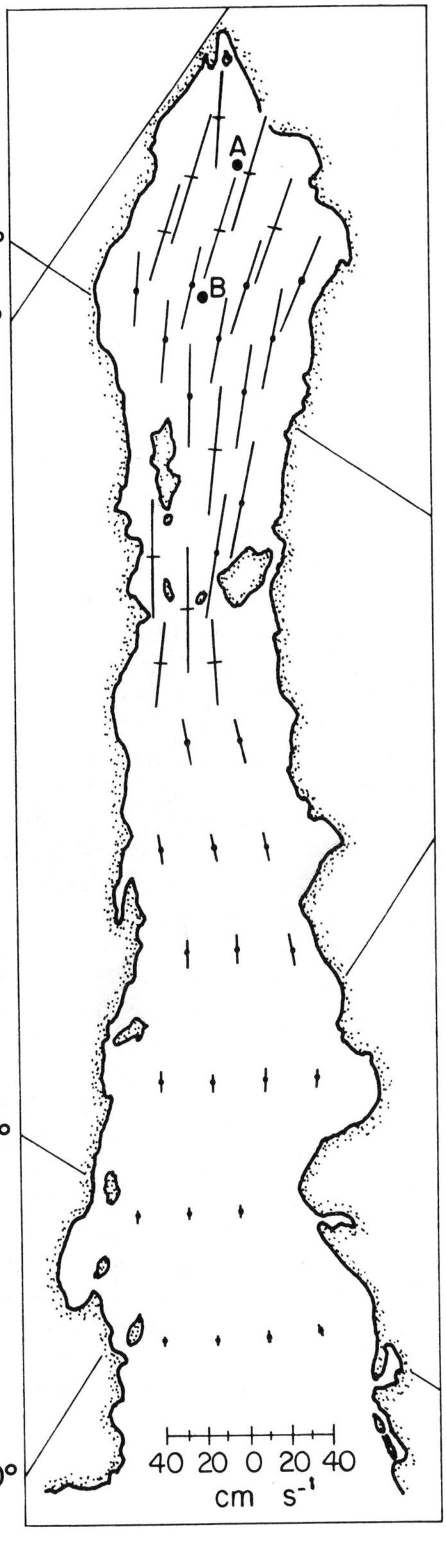

Fig. 17.5. Model current ellipses for the principal lunar semidiurnal tidal component (M_2). *A* and *B* indicate current meter locations (after Stock, 1976).

dbar surface did not indicate to him any regular exchange of water between the Gulf and the adjacent region of the Pacific; instead it showed a sequence of separate highs and lows.

The influence of the Gulf upon the adjacent Pacific Ocean is small. Flux into and out of the Gulf has been estimated as 1.19×10^6 m^3 s^{-1} and 1.17×10^6 m^3 s^{-1}, respectively, the difference being accounted for by evaporation, roughly estimated as 1.7×10^4 m^3 s^{-1} (Roden, 1958). Warsh and Warsh (1971), using a variable reference level to obtain agreement between inflow and outflow, estimated minimum transports from 2.57 to 3.65×10^6 m^3 s^{-1}. However, Roden (1972) reported fluxes of the order of 10 to 12×10^6 m^3 s^{-1}, without discussion on the discrepancy with previous results.

At the entrance to the Gulf of California there are three kinds of surface water: cold California Current water of low salinity ($S‰ \leq 34.60$), which flows southward along the west coast of Baja California; warm eastern tropical Pacific water of intermediate salinity ($34.65 \leq S‰ \leq 34.85$), which flows into the area from the southeast; and warm highly saline ($S‰ \geq 34.90$) Gulf of California water (Roden and Groves, 1959; Griffiths, 1968; Stevenson, 1970).

Beneath these three water masses, successively with depth, are: subtropical subsurface water, with a salinity maximum of about 34.80‰, which is obscured in the Gulf; Antarctic intermediate water, characterized by a deep salinity minimum of about 34.50‰ (Griffiths, 1968); and Pacific bottom water below the intermediate water characterized by an increase in salinity to about 34.68‰.

Roden (1972) made very intensive sampling across the mouth of the Gulf, using a salinity, temperature, depth recorder (STD) with 34 stations spaced approximately 9 km apart, to 1500 m depth where possible (Fig. 17.7). He distinguishes four distinct salinity layers. At the surface, values in excess of 34.8‰ result from evaporation. This water lies above the high stability layer and extends almost to the coast of Sinaloa. In the immediate vicinity of this coast, surface salinities were low at the time of Roden's observations because of the rainy season and the discharge from the Presidio River, 20 km south from Mazatlan. The thickness of the high salinity surface layer varies from 150 m near the Baja California coast to 50 m in the eastern half of the section. Below the saline top layer lies the shallow salinity minimum. It has a typical thickness of 50 m and occurs between 75 and 125 m. The lowest salinities observed within this layer increase from 33.9‰ in the open ocean to 34.3‰ at the Gulf entrance to 34.7‰ halfway up the Gulf (Roden, 1964). Beneath the shallow salinity minimum lies the subsurface salinity maximum. It occupies the depth range between 125 and 400 m and is characterized by salinities between 34.6 and 34.8‰. The deep salinity minimum occupies the depth range between 600 and 900 m (Fig. 17.7). The temperature does not show any special structure; it decreases monotonically with depth from about 25°C at the surface to about 4°C at 1200 m.

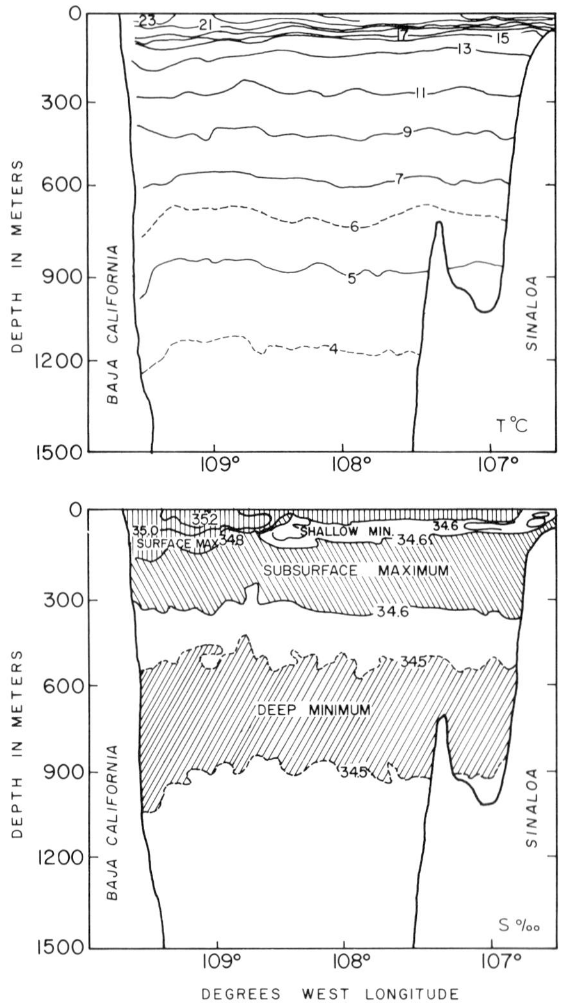

Fig. 17.7. Temperature (°C) and salinity (‰) across the Gulf of California entrance. The section runs along latitude 23°15′N, and was occupied on 5–6 December 1969 (after Roden, 1972).

The formation of fronts at the entrance to the Gulf is a feature that has attracted the attention of those interested in commercial fisheries, such as that for tuna. Griffiths (1968) notes that the most important oceanographic feature of the entrance to the Gulf is a strong front between the California Current and Gulf surface waters. At Cape San Lucas this front is roughly straight, but to the south and west it becomes more sinuous and much weaker and is formed more and more by California Current and subtropical surface waters (Fig. 17.8). Griffiths (1968) used spring 1960 data for his study of this frontal system. According to him, at Cape San Lucas the stronger flow of the California Current water seemed to hold back the Gulf outflow at the surface, and to the south of the Cape it penetrated Gulf surface water at depths between 50 and 100 m, spreading horizontally or affecting in some way the entire Gulf entrance, often in a complicated manner.

Hydrographic conditions inside the Gulf of California

Sverdrup (1941) was the first to present detailed temperature, salinity and oxygen sections across the Gulf. He found that south from Angel de la Guarda and Tiburon Islands temperature decreases regularly with depth, with less than 2°C below 2600 m; but at Ballenas Channel, between Angel de la Guarda and Baja California, and in the northern Gulf, the vertical distribution of temperature is entirely different, with values of about 11°C even in

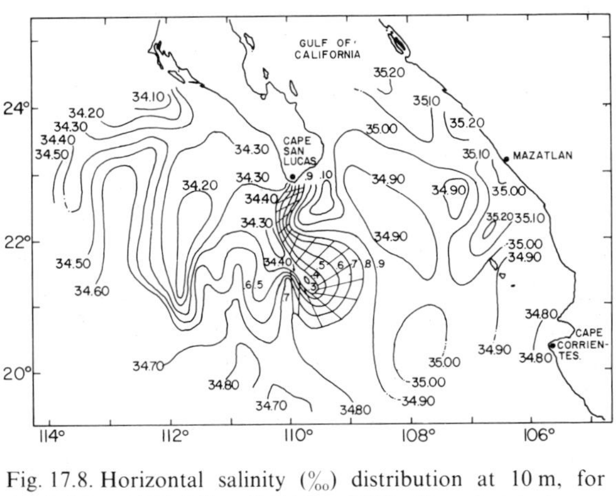

Fig. 17.8. Horizontal salinity (‰) distribution at 10 m, for April–May 1960 (after Griffiths, 1968). Cross-hatched area indicates the front.

the deepest parts (Fig. 17.9B). Temperature and salinity in different sections indicated upwelling at the eastern ends. A salinity minimum is present at depths between 500 and 1100 m, south of Angel de la Guarda and Tiburon, corresponding to Antarctic intermediate water (Fig. 17.9C). The highest surface salinity, 35.5‰, was found by Sverdrup (1941) in the central part of the northern Gulf, but about 70 km from the Colorado River mouth the salinity was nearly as high, 35.12‰. Thus, a direct effect of the Colorado River fresh water was not evident in March–April 1939, due to excessive evaporation even in winter.

Compared to temperatures for similar latitudes in the neighboring Pacific, surface temperatures in the Gulf are warmer from April to September and about equal during the remaining months of the year. The annual range is large and increases from about 9°C near Cape Corrientes to roughly 22°C near the Colorado River mouth. The lowest surface temperatures are persistently found in the vicinity of Angel de la Guarda, where tidal mixing is strong. During October to June, temperatures at the mouth are higher than at the head of the Gulf.

There is no clearcut sequence of change from the winter conditions with warmer water on the west side to summer conditions with warmer water on the east side of the Gulf. December through May, isotherms run mostly east to west south of 28°N, but north of Angel de la Guarda the trend changes and becomes parallel to the coast. Beginning in June and extending through September, the isotherms run parallel to the Gulf over most of its length, with warmer water on the Sonora and Sinaloa coast and upwelling along the Baja California coast. October is a month of change, and the opposite situation develops in the central and southern area, with higher temperatures on the Baja California coast and upwelling on the east coast. Charts show that upwelling is better developed and extends over a greater distance (23–28°N) along the east coast than off the Baja California coast, where it extends from Punta Arena to Concepción Bay (Fig. 17.10) (Roden, 1964; Robinson, 1973). Surface salinities in the northern two-thirds of the Gulf range between 35.0‰ and 35.8‰ and are 1 to 2‰ higher than those at comparable latitudes outside the Gulf. Salinities higher than 36‰ are encountered locally in semi-enclosed and shallow bays in the northern region and along the Baja California coast (Roden, 1964).

In the central Gulf of California, roughly considered as the region between a line connecting La Paz with Topolobampo and a line south of Tiburon Island, depths are great and the water is open to the Pacific Ocean to the southeast. Here the thermocline is well developed between April and October; it reaches its maximum strength in August when the temperature difference between the surface and 150 m is about 16°C. In August the salinity distribution in the upper layers is characterized by a minimum of about 34.8‰ between 25 m and 75 m in the eastern half of the Gulf. Since the only source of low-salinity water is from the southeast, the salinity minimum probably indicates the presence of a current flowing to the northwest between these depths (Roden, 1964). This minimum is related to the shallow salinity minimum found by Roden (1964, 1972) at the mouth of the Gulf. Below the thermocline the water is essentially the same as in the equatorial Pacific. This water is characterized by a salinity minimum of approximately 34.5‰ (Antarctic intermediate water) and a temperature of about 5°C (Sverdrup, 1941) (Fig. 17.9).

At depths between 2500 and 3000 m there is a temperature minimum of slightly less than 1.85°C. From 3000 m to the bottom, temperature increases again at a rate of approximately 0.1°C per 1000 m, due to the adiabatic pressure effect. The salinity distribution between 1000 m and the bottom is surprisingly complex. In general, there is a tendency for salinity to increase from 1000 m towards the bottom (Roden, 1964).

The Ballenas channel is completely isolated from the central part of the Gulf by a submarine ridge connecting Angel de la Guarda Island with San Gabriel Point. The sill depth of this ridge is about 450 m (Rusnak et al., 1964). The topography of the channel (also called Sal Si Puedes basin) is very irregular, with maximum depths exceeding 1500 m. The water in the basin comes from mixing by strong tidal currents between the surface and sill depths.

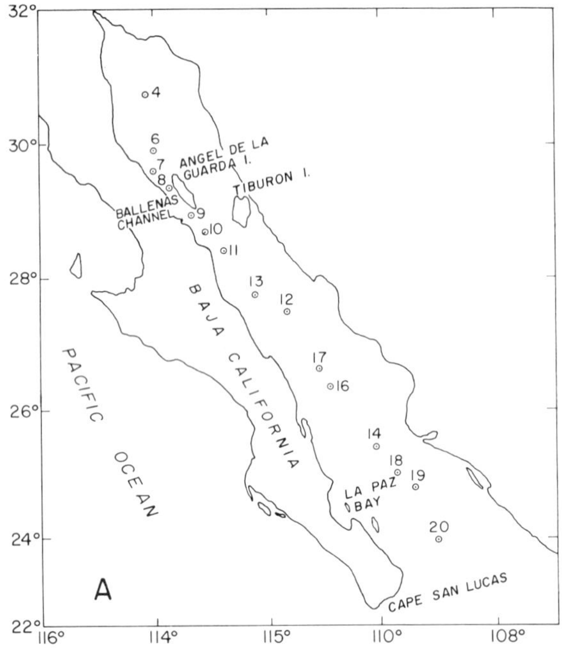

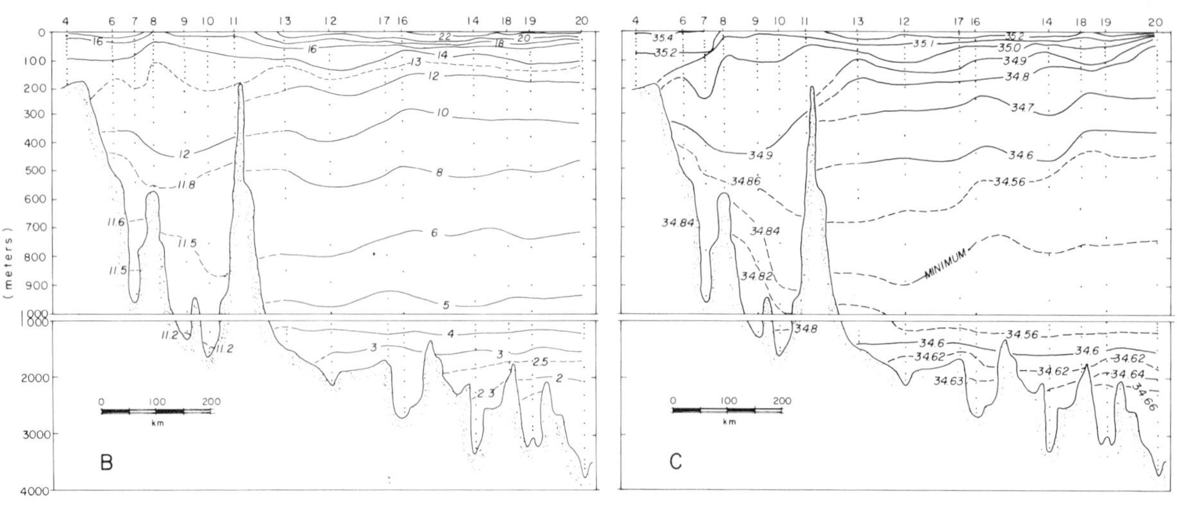

Fig. 17.9. A. Station locations for the April–May 1974 cruise of the *Alexander Agassiz*. B. Vertical temperature (°C) distribution. C. Vertical salinity (‰) distribution (after Alvarez-Borrego et al., 1978).

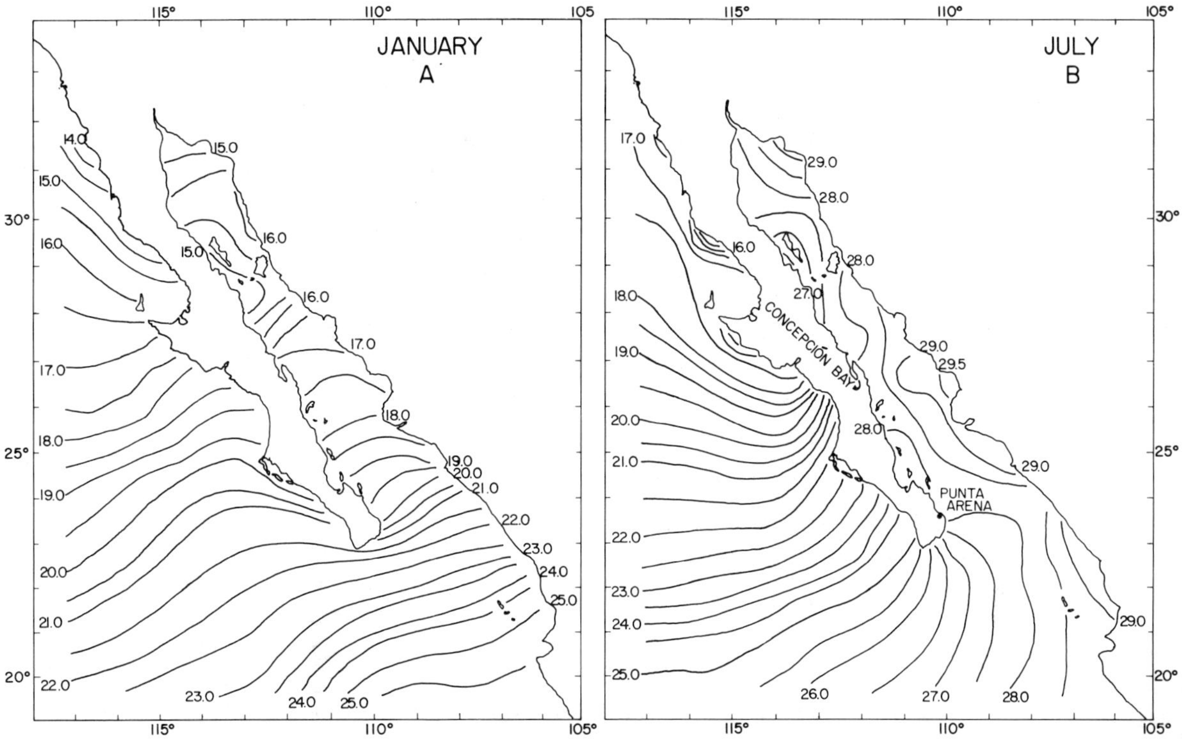

Fig. 17.10. A, January and B, July mean sea surface temperatures (°C) [constructed from graphs that Robinson (1973) presented in °F].

The outstanding hydrographic features of the channel are high temperatures and salinities at great depths. Compared to conditions elsewhere in the Gulf, there is a difference in temperature at 1000 m of 6°C, and of 0.4‰ salinity. The deep salinity minimum, characteristic of the southern half of the Gulf, does not occur here (Fig. 17.9). Above 450 m exchange of water between the northern and central parts of the Gulf is unrestricted. Alvarez-Borrego et al. (1978) compared salinity and temperature data from April and October, 1974. Near the bottom of Ballenas channel ($\sim$1500 m) they found differences of 0.4°C and 0.04‰, with higher values in April. This indicates a greater proportion of sill depth water at the bottom of the basin in October than in April. During winter months stability in the upper layers is weak and limited convection of cold and high salinity water from the bays is likely. During summer the thermocline in Ballenas channel is well developed, as elsewhere in the Gulf (Roden, 1964).

The graphs constructed by Sverdrup (1941) and Granados-Gallegos (1974) indicate that hydrographic conditions at Tiburon basin, between Angel de la Guarda and Tiburon Islands, are very similar to those at Ballenas channel, with very much the same salinity, temperature and oxygen values for the same depths. However, maximum bottom depths for Tiburon basin are only about 550 m, and it has more open communication with the southern Gulf than Ballenas channel. In general tidal mixing plays an important role in the vertical hydrographic structure of the northern Gulf.

In the shallow area north of Angel de la Guarda Island the water is almost isothermal during winter. During late summer the thermocline is strong and the temperature difference between the surface and 150 m amounts to almost 14°C. Salinity in general decreases with depth. In the central northern Gulf it is about 35.2‰ at 100–150 m depth (Roden, 1964). But, with winter conditions, the salinity vertical distribution may show a minimum at about 100 m. Possibly this is due to the influx of colder and more saline water from the head of the Gulf, which moves southward at the bottom (Alvarez-Borrego et al., 1978). The most drastic temperature and salinity changes occur north of 31°N. At the

Colorado River mouth there is no longer a significant fresh-water input. North from Angel de la Guarda and Tiburon Islands, this was the only river with fresh water input throughout the year. The main cause of the change in flow has been the construction of dams on the river, particularly the Hoover dam, but dramatic decrease was introduced by the water storage project begun in Lake Mead in 1934 (Schreiber, 1969). Surface salinities at the Colorado River mouth and adjacent region are about 36.5‰ in winter, and more than 38.5‰ in summer, maintaining in general the same gradient with values increasing northwestward. This clearly indicates that evaporation exceeds precipitation and river input during most of the year. Only during few occasions, caused by local rain and lasting for only a few days, do salinities decrease to about 35‰ here. Surface temperatures range from about 10°C in winter, to about 32°C in summer. The surface temperature gradient reverses at the beginning of spring and fall due to the annual cycle of atmospheric temperature and the shallow depths here, less than 20 m (Alvarez-Borrego and Galindo-Bect, 1974; Alvarez-Borrego et al., 1975).

Roden (1964) explored the possibility that convective overturn was produced by winter cooling in the northern part of the Gulf. For the process to take place, surface density must exceed the density of the subsurface layers. Because density increases with decreasing temperature and increasing salinity, and because the lowest temperatures and highest salinities occur near the coast, the most likely places for convective overturn are the coastal regions. With the data he had, Roden (1964) indicated that the mean minimum temperature in coastal regions is about 13°C, whereas the salinities are generally in excess of 35.5‰. For these values the density at the surface is 1.02680 g cm^{-3}. The same density (*in situ*) is found at a depth of about 100 m, where the temperature is 14°C and the salinity is 35.2‰. Roden (1964) concluded that convective overturn due to winter cooling can therefore be expected to reach roughly 100 m near the coast. Alvarez-Borrego et al. (1975), reported surface temperature minima for December through February at the northernmost part of the Gulf in the range of 8.25 to 12.21°C, producing water heavier than 1.02680. This water could sink deeper than 100 m. However, this region is for the most part shallower than 100 m, so horizontal advection would have to bring this water southward, along the bottom, to move to deeper parts. In the more offshore areas of the Gulf, surface temperatures do not drop much below 15°C, whereas the salinities average about 35.4‰. With these conditions, convection is insignificant (Roden, 1964).

Oxygen distribution

Low oxygen concentrations at intermediate depths are very characteristic of the Gulf waters (Sverdrup, 1941; Roden, 1964) (Fig. 17.11). Sections across the Gulf mouth show that oxygen concentrations are higher than 1 ml l^{-1} in the upper 100 m, and that at a depth of less than 150 m they decrease to less than 0.5 ml l^{-1}. This is the situation for most of the Gulf, with exception of the northern area. At intermediate depths (500–1100 m) the concentration of oxygen occasionally is undetectable by the Winkler method. This plays an important role in the ecology and geology of the Gulf of California, as it relates to organisms and sedimentation patterns. The oxygen minimum at the entrance to the Gulf is more pronounced than in the interior, and covers a larger depth interval. Oxygen increases from a minimum to about 2.4 ml l^{-1} at 3500 m. However, in the central part of the Gulf oxygen concentrations near the bottom are much lower than at comparable depths (2000 m) outside (Roden, 1964).

Another outstanding hydrographic feature of

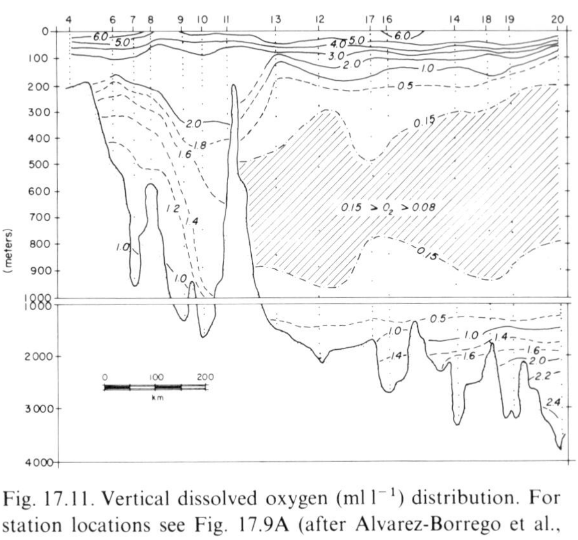

Fig. 17.11. Vertical dissolved oxygen (ml l^{-1}) distribution. For station locations see Fig. 17.9A (after Alvarez-Borrego et al., 1978).

Ballenas channel is high oxygen concentration at great depths. There is no oxygen minimum there (Sverdrup, 1941). Compared to conditions elsewhere in the Gulf, the difference in dissolved oxygen at 1000 m is 1 ml l^{-1} (Roden, 1964), high values being due to tidal mixing in the channel. Oxygen concentration in Ballenas channel was higher in April than in October of 1974, with a difference of 0.45 ml l^{-1} at 1200 m, for the same reason that salinity and temperature were higher in April (Alvarez-Borrego et al., 1978).

North from Angel de la Guarda Island there is no oxygen minimum. Oxygen decreases from about saturation values at the surface (5 to 6 ml l^{-1}) to about 2 ml l^{-1} at 100–150 m depth, and about 1 ml l^{-1} at 300–500 m depth (Sverdrup, 1941; Roden, 1964; Rosenberg, 1969; Granados-Gallegos, 1974; and Villaseñor-Casales, 1974).

Carbon dioxide system

The pH data are very consistent with oxygen data. The *in situ* pH vertical distribution has a minimum of about 7.65 in the central and southern Gulf between 500 and 1000 m. In this region of the Gulf, pH values decrease in spring from about 8.25 at the surface to about 7.80 at 100 m (Fig. 17.12). In the northern Gulf and Ballenas channel there is no pH minimum. In Ballenas channel surface pH readings are the lowest of the whole Gulf, with values of about 8.10, decreasing to 7.95 at 100 m, and decreasing monotonically to 7.70 at 1500 m.

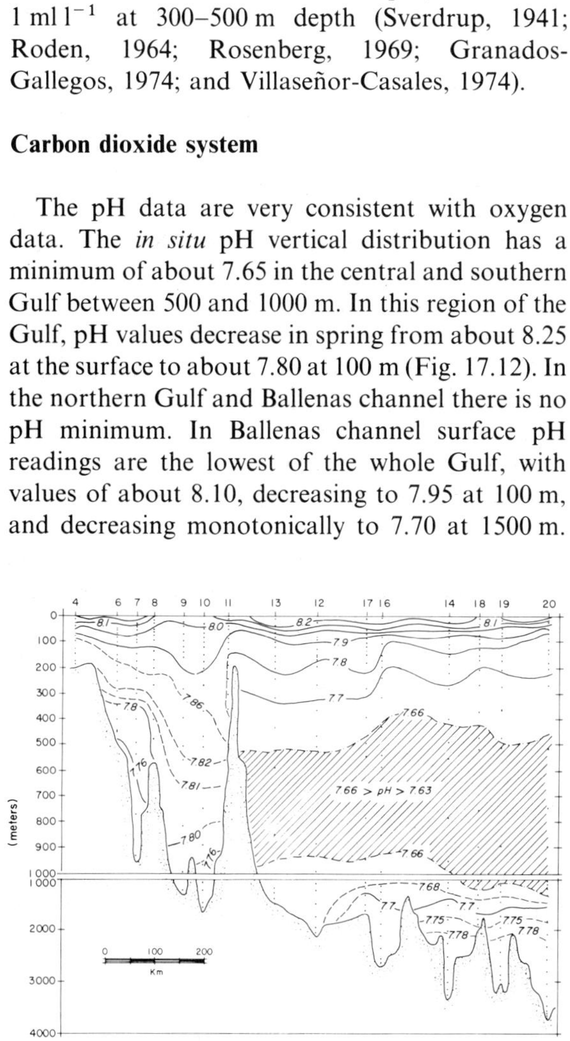

Fig. 17.12. Vertical *in situ* pH distribution. For station locations see Fig. 17.9A (after Gaxiola-Castro et al., 1978).

Surface total inorganic carbon (T_{CO_2}) is maximum in Ballenas channel, with values of about 2.13 mmol kg^{-1} compared to 2.07 mmol kg^{-1} in the southern region of the Gulf and 2.04 mmol kg^{-1} in the northern region. Again, T_{CO_2} has a maximum at intermediate depths in the central and southern regions, which is not present in the northern region and Ballenas channel. Due to the rapid decrease of pH near the surface in the central and southern regions of the Gulf, the percent saturation of calcium carbonate with respect to calcite and aragonite decreases rapidly to undersaturation at about 300 m and 70 m, respectively. Thus, only the surface layer is oversaturated with these minerals south of 28°N. At Ballenas channel the higher temperature and pH at depth causes oversaturation of calcite and aragonite down to 900 m and 130 m, respectively. In the shallow northern Gulf most of the water is oversaturated with these minerals, with the exception of near-bottom waters in Wagner basin, which are not saturated with aragonite (Gaxiola-Castro et al., 1978).

THE GULF AS AN ENVIRONMENT FOR ORGANISMS

Nutrients, organic primary productivity and phytoplankton in the Gulf of California

Although the Gulf of California has been described as an area of great fertility since the time of early explorers, few details are known about the variability in space and time of the concentration of nutrients there. Zeitzschel (1969) reviewed the scarce available data concerning nutrients, and drew the following conclusions. During summer and winter, the surface phosphate concentration is >0.4 μmol l^{-1} in the whole Gulf, even in the southern area. In the northern area surface phosphate concentrations are between 0.9 and 1.9 μmol l^{-1}. The data suggest that phosphate concentrations in the Gulf are far greater than the experimentally established minimum limit of 0.22 μmol l^{-1} for growth of tropical oceanic diatoms (Thomas and Dodson, 1968). Warsh et al. (1973) presented the vertical distribution of phosphate and silicate across the mouth of the Gulf for July 1967. Their graphs show surface phosphate values of about 0.2 μmol l^{-1} increasing rapidly with depth to about 2.3 μmol l^{-1} at 100 m, and to a maximum of about 3.4 μmol l^{-1} at 800 m.

Surface silicate values were about 2 μmol l^{-1}, increasing monotonically with depth to about 100 μmol l^{-1} at 800 to 1000 m. In the upper 50 m, both phosphate and silicate were higher near the west coast, probably due to upwelling there during summer.

In April–May and October, 1974, the R.V. *Alexander Agassiz* of Scripps Institution of Oceanography collected phosphate, nitrate, nitrite and silicate data. Surface concentrations are summarized in Table 17.1. In general, there is a tendency for the surface nutrient concentrations to increase from the mouth to the region of Angel de la Guarda Island, and then decrease again to the central northern Gulf. Highest values were observed at Ballenas channel because of the effect of strong tidal mixing. Values in April and May in general were greater than those of October. Hydrographic stations were occupied at a section along the middle of the Gulf during the April–May cruise; therefore the data are not useful for studying the effect of upwelling. Data from the October cruise do not show gradients across the Gulf that could indicate the effect of upwelling, possibly because the upwelling season had not yet started (Alvarez-Borrego et al., 1978).

During these cruises a subsurface nitrite maximum was detected at 30 to 80 m in most of the locations, with values from 0.2 to 0.6 μmol l^{-1}. In the Ballenas channel there were values greater than 1.0 μmol l^{-1} at this maximum. A second nitrite maximum between 150 and 400 m was observed at the entrance to the Gulf, with concentrations up to 0.7 μmol l^{-1} in April–May, and up to 1.9 μmol l^{-1} in October. With the exception of these maxima, nitrite concentrations were lower than 0.1 μmol l^{-1} at depths greater than 100 m, with values below the detectable level in many cases. The shallow subsurface nitrite maximum is a common characteristic of the oceans and it is caused by bacterial nitrification (Wada and Hattori, 1971), by phytoplankton excretion (Vaccaro, 1965), or both. The deeper nitrite maximum is characteristic of oxygen-deficient zones and it is caused by denitrification (Cline and Richards, 1972). Marine denitrification occurs with oxygen concentrations below 0.1 ml l^{-1} (Goering, 1968). It is not understood why the deeper nitrite maximum was not detected in the interior of the Gulf where the oxygen is almost totally depleted at 400 to 500 m (Alvarez-Borrego et al., 1978).

In the shallow central northern Gulf, very weak phosphate, nitrate and silicate maxima were found at 80 to 125 m, in the April–May cruise, with 2.3 to 2.5 μmol l^{-1} for phosphate, 21 to 23 μmol l^{-1} for nitrate, and 53 to 67 μmol l^{-1} for silicate. These maxima were not detected in October. In April–May these maxima were found with a salinity minimum. They are consistent with the explanation given above that these extremes may be caused by colder, more saline and nutrient-poor water coming from the head of the Gulf and moving southward at the bottom (Alvarez-Borrego et al., 1978). In Ballenas channel, in accordance with the oxygen vertical distribution discussed above, these three nutrients increase monotonically with depth to about 3.0 μmol l^{-1}, 30 μmol l^{-1} and 70 μmol l^{-1}, respectively, at 1500 m. At 1000 m, nutrient concentrations are lower in Ballenas channel than in the central and southern Gulf, with differences of 0.6 μmol l^{-1} for phosphate, 12 μmol l^{-1} for nitrate and 60 μmol l^{-1} for silicate. South from Angel de la Guarda and Tiburon Islands the vertical distribution of these nutrients is very much like that of the open ocean. Silicate increases monotonically with depth to about 170 μmol l^{-1} at 3000 m. Phosphate has a maximum of about 3.5 μmol l^{-1} at 800 to 1200 m, and then decreases to about 2.8 μmol l^{-1} at 3000 m, and

TABLE 17.1

Surface concentrations of nutrients (μmol l^{-1}) for the different regions of the Gulf of California

Southern Gulf	Central Gulf	Ballenas Channel	Northern Gulf
Phosphate			
0.6	1.0	1.7—2.0	0.8–1.0
0.4	0.5	0.9–1.5	0.7–1.0
Nitrate			
0.6	1.9	13.0	0.2–0.4
0.1	0.3	1.0–7.5	0.0–0.2
Nitrite			
0.0	0.09	0.31	0.02–0.20
0.01	0.01	0.13–0.45	0.00–0.09
Silicate			
1.0	0.0–5.0	29.0	11.0–18.0
2.4	2.9	6.6–19.6	6.1–10.2

Upper numbers are from the April–May 1974 cruise of the *Alexander Agassiz*; lower numbers are from the October cruise (after Alvarez-Borrego et al., 1978)

nitrate has a very weak maximum of 43 to 44 μmol l^{-1} at 1000 to 1300 m and decreases to about 39 μmol l^{-1} at 2500 to 3000 m. There is a linear relationship between nitrate and phosphate for the whole Gulf of California with the ratio of the change in nitrate to that in phosphate 16.2 ± 0.7 at the 95% confidence level (Fig. 17.13), which is consistent with the ratio proposed by Redfield et al. (1963) (Alvarez-Borrego et al., 1978).

In the Gulf of California great amounts of silicate are lost from the water column to the sediments through settling and deposition of exoskeletons of microorganisms, mainly diatoms and radiolarians. During the 1939 and 1940 cruises of the *E.W. Scripps* to the Gulf, bottom samples were taken, and revealed that diatomaceous sediments are being laid down over large areas (Sverdrup and Staff, 1940). Gilbert and Allen (1943) studied the vertical distribution of diatoms in the water column, from the surface to 1000 m, in the Gulf. They found that below 60 m the number of empty

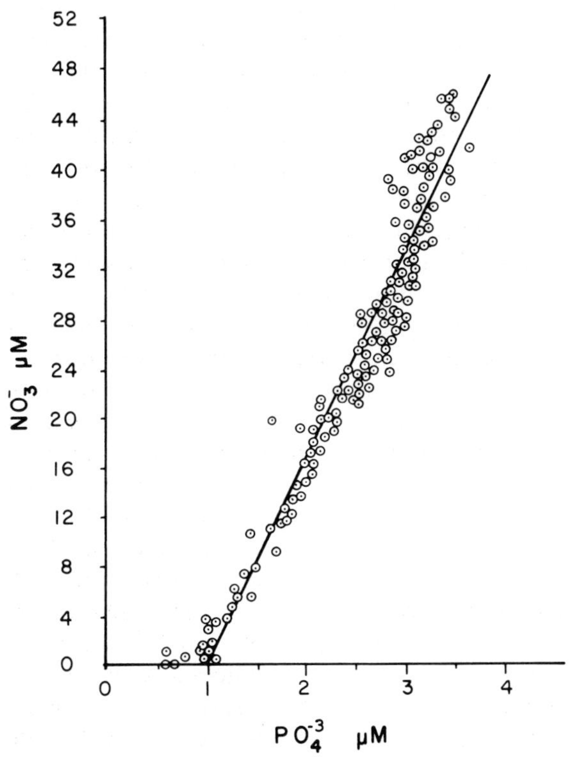

Fig. 17.13. Nitrate versus phosphate diagram, constructed with data from all stations of Fig. 17.9A. The least-squares straight line is shown. The slope interval is 16.2 ± 0.7 at the 95% confidence level (after Alvarez-Borrego et al., 1978).

frustules greatly exceeded that of living cells, usually with a sharp decrease in the number of total frustules between 60 and 300 m, followed by a more gradual decrease below 300 m, so that at 750 m only a few hundred, or at the most one or two thousand, cells per liter were found. The exception was at about 28°20′N, where 15 000 frustules were found at 750 m, whereas populations of only a few thousand cells per liter were found in the upper layers. The specific composition changed considerably with depth. The heavily silicified forms, which were rarely present in large quantities in the surface layers, increased relative to depth and dominated the residue of the diatom population.

Laminated sediments are formed at certain sites where the deposition of silicate has been studied (Calvert, 1964, 1966a, b). Laminated, diatomaceous sediments of the Gulf of California consist of regularly alternating light-colored (diatom-rich) and dark-colored (clay-rich) laminae, approximately 2 mm thick. The mean contents of opal and quartz in the laminae, determined by X-ray diffraction methods, are: light-colored laminae, 52.4% opal and 7.0% quartz; dark-colored laminae, 26.5% opal and 10.0% quartz. Opal is produced biogenetically and quartz is of terrigenous origin. Calvert (1966b) could not distinguish the laminae on the basis of diatom assemblages, but could identify them on the basis of the total content of biogenous and terrigenous materials. Rates of deposition, determined by radiocarbon dating of organic and carbonate carbon in selected core sections, demonstrate that one light-colored lamina and one dark-colored lamina are deposited in a year, and a couplet of laminae constitute a varve (Calvert, 1966b). These varves in the central Gulf of California are confined to the slopes of the basins. They occur where the basin floor intersects the oxygen minimum in the water column. Burrowing organisms do not occur in this poorly oxygenated zone, and the laminations remain undisturbed. Such organisms are present in deeper parts of the basin where oxygen concentrations are higher and their digging activities homogenize the sediments.

In contrast to previously described occurrences of laminated sediments, those of the Gulf of California occur on the slopes of a basin which has free communication with the open ocean (Calvert, 1964). Sediments containing more than 10% by weight opal are confined to the central part of the

Gulf; in the Guaymas basin values exceed 50% by weight. Diatoms constitute the major source of the opal in the sediments. Radiolarians contribute less than 10% by weight of opal, and sponge spicules are an insignificant source. Phytoplankton production is greatest in the central Gulf, the area of richest diatomaceous sediments. The diatom frustules settle to the Gulf floor rather rapidly and are not dispersed uniformly over the entire Gulf (Calvert, 1966a). According to Calvert (1966b), phytoplankton production in the central Gulf is reasonably constant throughout the year. Since river discharge fluctuates greatly during the year as a result of torrential summer rains in coastal Sonora and Sinaloa, the terrigenous sediment supply to the central Gulf is highly variable.

Calculations of accumulation rates of opal and quartz in the laminae using data on lamina compositions and thicknesses demonstrate that the varves originate by a more or less constant rate of accumulation of biogenous material and an annually increased rate of accumulation of terrigenous material as a result of the summer floods. In the central Gulf sediments containing more than 10% opal by weight extend over an area of 3×10^4 km^2, and within this area the mean opal concentration is 25% by weight. The mean sedimentation rate is 2 mm yr^{-1}, and the mean sediment bulk density is 1.0 g cm^{-3}. Thus, the accumulation of biogenous silica in the sediments of the central Gulf proceeds at a rate of approximately 1.5×10^{13} g yr^{-1}. The rivers draining into the Gulf supply approximately 5.5×10^{11} g of dissolved silica per annum; about 4% of the amount deposited (Calvert, 1966a). This river input is now decreasing as more dams are built upstream to divert the water for agricultural uses. Calvert (1966a) indicates that the water exchange between the Gulf and the Pacific Ocean supplies approximately 10^{14} g of dissolved silica per annum to the Gulf. Sufficient silica is available in sea water to account for the accumulation of such richly diatomaceous sediments, given a mechanism — upwelling — which continuously supplies dissolved nutrients to the euphotic zone. Recourse to volcanic sources of silica is then not necessary (Calvert, 1966a).

The Gulf of California represents a subtropical area with exceptionally high rates of primary productivity (Fig. 17.14). From very limited data it can be concluded that rates of primary productivity in

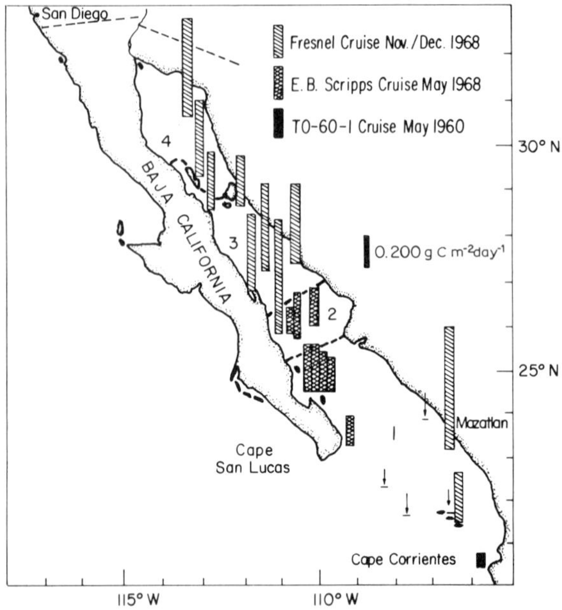

Fig. 17.14. Primary productivity in the Gulf of California integrated over the euphotic zone (after Zeitzschel, 1969).

the Gulf of California are comparable to those of the Bay of Bengal, the upwelling areas off the west coast of Baja California, or North Africa. They are about two to three times greater than that in the open Atlantic or the open Pacific at similar latitudes (Zeitzschel, 1969). Phytoplankton patchiness, and great variability with time, were recognized as important problems in the early studies of organic productivity of the Gulf (Allen, 1938). Differences between localities are striking. In general, diatoms are well represented in the Gulf and dinoflagellates are less abundant (Allen, 1934, 1938). The dinoflagellates are greatly subordinate to the diatoms in spring and even less abundant in autumn. Three dinoflagellate species, *Dinophysis caudata*, *Peridinium oceanicum* and *Ceratium furca* are the most frequently recorded and are widespread, tolerant, oceanic forms (Round, 1967). Small organic particles, 2 to 6 μm in diameter, are abundant throughout the Gulf of California; on an average, 72% of total phytoplankton by number are naked flagellates smaller than 5 μm in diameter, but they are only 10% of the phytoplankton carbon. On the other hand, diatoms make up only about 10% by number on an average, but contribute roughly 51% of the phytoplankton carbon because of their larger size (Zeitzschel, 1970). Gilbert and Allen (1943) give a list of early phytoplankton samplings in the Gulf, and mention that in 1939 the *E.W. Scripps* cruise was the first to sample the Gulf systematically to a depth of 60 m, at intervals of 10 m. They emphasized the need of both continuity of observations and a knowledge of water movements. At any one spot both the biomass and the specific composition may change completely within few hours. Near Guaymas, at noon on December 27, 1940, at an anchor station, there was a mean population of 2000 diatoms per liter with *Pseudoeunotia doliolus* and *Guinardia flaccida* making up almost half of this number; at 07:30 the next morning a second series of samples revealed an increase in numbers of diatoms to 161 400 cells per liter with *Bacteriastrum* sp. and *Chaetoceros* sp., constituting over 90% of this population.

Based on the distribution of phytoplankton, the Gulf has been divided into three geographical regions: the region south of 25°N is the southern region; the region between 25°N and 27°N, the middle section; and the region north of 27°N the northern section (Allen, 1937; Cupp and Allen, 1938; Gilbert and Allen, 1943). Round (1967) distinguished two different zones in the region north of 27°N, one covering the shallow Gulf north of Tiburon, Ballenas channel and Dolphin basin, and the other covering the southern part. The southern region is characterized by the lowest phytoplankton populations of the entire Gulf (Gilbert and Allen, 1943). It supports a large number of *Nitzschia seriata*, *Rhizosolenia alata*, *Chaetoceros atlanticus* and *C. peruvianus*, and many other diatoms. These are predominantly of oceanic character, and are either completely absent or sparse in the other regions of the Gulf. Gilmartin and Revelante (1978) found that these characteristic tropical communities extended into coastal waters immediately adjacent to the southeastern lagoons, indicating that the impact of lagoons on the Gulf in this region is low. These southeastern lagoons had less diversity, a higher degree of dominance, and few elements in common with adjacent waters, reflecting the effect of low flushing rates. Mean numbers of cells per liter for the 60-m water column of the Gulf were 1000 to 10 000 (Gilbert and Allen, 1943; Round, 1967). At the entrance to the Gulf, from April to August 1967, the median concentration of surface chlorophyll *a* increased from 0.65 to 0.97 mg m^{-3}, while the median productivity measured by the ^{14}C method, increased from 5.6 mg C m^{-3} day^{-1} in April to 17.8 mg C m^{-3} day^{-1} in June before returning to 2.6 mg C m^{-3} day^{-1} in August. In this region productivity was highest near Las Tres Marias Islands and second highest near Cape San Lucas, both locations of local upwelling. Chlorophyll *a* and phaeophytin *a* data indicated that the standing crop of phytoplankton was subjected to progressively heavier grazing pressure in the spring and summer by zooplankton (Stevenson, 1970). The average integrated primary productivity value for this region is about 0.27 g C m^{-2} day^{-1} (Zeitzschel, 1969).

The region between 25°N and 27°N has been found to have diatom populations larger than those of the southern section but still relatively poor compared to the more northern regions of the Gulf. In March 1959, Round (1967) found a poor phytoplankton standing crop in the region of Carmen basin, but Gilbert and Allen (1943) reported populations of over 100 000 cells per liter at several stations near Carmen Island taken in 1921, and over 500 000 cells per liter off Topolobampo Bay in 1937. Gilmartin and Revelante (1978) found that the central eastern lagoons had many floristic elements in common with adjacent waters, great diversity and a low degree of dominance, reflecting high flushing indices, which not only tend to export phytoplankton populations but to modify the inshore environment, permitting their survival and growth. The average integrated primary productivity value for this region of the Gulf is about 0.38 g C m^{-2} day^{-1} (Zeitzschel, 1969).

The region between 27°N and 29°N includes the Guaymas area in which samples of over three million cells per liter had been taken in 1937 and 1939. These samples have, as the main constituents, *Asterionella japonica*, *Chaetoceros compressus*, *C. debilis*, *C. radicans* and *Skeletonema costatum*. In 1939 samples from this area were almost a pure culture of *Asterionella japonica*; out of four million diatoms per liter, 3 600 000 were *A. japonica*. In 1940 the dominant species were neritic *Chaetoceros* and *Bacteriastrum* species. No *Asterionella japonica* were found (Gilbert and Allen, 1943). Northwest of Angel de la Guarda Island a tremendous growth of *Coscinodiscus lineatus* and *Asteromphalus heptactis* has been found (Round, 1967). The average integrated productivity value for December is about

0.53 g C m^{-2} day^{-1} in the region north of 27°N up to Ballenas channel (Zeitzschel, 1969).

Gilmartin and Revelante (1978) made a comparative study of the lagoons and the central and southern Gulf during summer of 1972, before the full development of the rainy season. They found a dramatic inshore increase in cell densities, chlorophyll *a* standing crops and rates of primary production. Offshore Gulf stations had mean assimilation numbers of 2.7 mg C (mg Chl)$^{-1}$ h^{-1}, inshore eastern stations a mean of 6.7, and the means of eastern lagoons were between 7.4 and 10.7.

In the northern shallow Gulf Round (1967) found that standing crops of diatoms were very low, in March 1959, with few species from the south extending into this region. In March 1939 Gilbert and Allen (1943) found large numbers of diatoms, with some sampling stations having over 100 000 cells per liter. These authors found enormous numbers of the dinoflagellate *Gonyaulax catenella* or some species closely resembling it. The average integrated primary productivity for this region has been reported as 0.68 g C m^{-2} day^{-1} for December only (Zeitzschel, 1969). During March of 1973 the R.V. *Alexander Agassiz* of Scripps Institution of Oceanography occupied 92 hydrographic stations where chlorophyll *a* samples were taken in the Gulf north of 29°N. Chlorophyll *a* samples were taken down to 50 m where possible, and analysis was made by fluorimeter. Chlorophyll distribution was in general very patchy. Lowest values were found at the central northern Gulf at all depths between the surface and 50 m (1.5 to 2.5 mg m^{-3}). Relatively low surface values were found between Angel de la Guarda and Tiburon Islands, to the north and northwest of Angel de la Guarda (2.5 to 3.5 mg m^{-3}), and at Ballenas channel (~5.0 mg m^{-3}). Highest surface values were found near the northernmost coasts of Baja California and Sonora (30 to 40 mg m^{-3}). There was no clear dependency of chlorophyll concentration as a function of depth, but in general it decreased with depth with the exception of the area around Angel de la Guarda Island, where the concentration was fairly constant down to 30 m, and decreased to about 2.0 to 3.0 mg m^{-3} at 50 m. At 50 m the highest values were found at Ballenas channel and near San Felipe (Gendrop-Funes et al., 1978).

Zooplankton in the Gulf of California

Zooplankton studies of the Gulf of California are very few. Alvariño (1962, 1963) made studies on the systematics of Chaetognatha in the Gulf of California, and gave a description of the different species. Mundhenke (1969) found no clear relationship between the water masses of the Gulf and the geographic distribution of euphausiids. Stevenson (1970) found that plankton abundance was relatively high in the front region near Cape San Lucas, and in the vicinity of Las Tres Marias Islands. At the entrance to the Gulf zooplankton abundance in general increased from April to August. Only in the area north of 31°N had samples been taken systematically throughout a year to study the zooplankton biomass variability as a function of location and time (Farfan, 1973; Cummings, 1977). Zooplankton displacement volumes in the northernmost region of the Gulf vary from the same magnitude to twice as large as those of temperate coastal waters, but they are 3.5 to 7 times greater than values reported for subtropical and tropical seas. Zooplankton volumes in the northern Gulf are maximum during winter (up to 13 to 25 ml of plankton m^{-3} in January), and during this season exceed values for regions of upwelling generally by a factor of two. Copepods are numerically the dominant organisms in all samples, with euphausiids and crab larvae having secondary importance. Copepods also show the greatest variability. Euphausiids have been found only during winter and spring (Cummings, 1977). In the area adjacent to the Colorado River mouth, there is no clear seasonal variation in zooplankton biomass in terms of dry weight. Low and high values were reported for October 1972 and 1973 from 1 to 4 mg m^{-3} and from 3 to 60 mg m^{-3}, respectively. Patchiness is again a problem. High values are reported for January (10–400 mg m^{-3}) and February (10–40 mg m^{-3}), relatively low values for March and April (4–20 mg m^{-3}), intermediate values for May and July (2–40 mg m^{-3}), and high values again for August (25–150 mg m^{-3}) (Farfan, 1973). This area of the Gulf has great turbidity, and encompasses a region where fishing is prohibited throughout the year. It has been reported to be an area of reproduction of many fish and crustacean species, and juveniles are abundant here (Berdegue, 1955; Guevara-Escamilla, 1973). Turbidity is so great that Secchi disk depths are less than 0.5 m near the Baja California coast, and 1 to 2 m near the Sonora coast. Seston concentration values of about 100 mg l^{-1} were found near Baja California at about 31°30′N, in May 1973, using filters of 3 μm pore size. The percentage of organic matter is high, sometimes near 100%, more often about 30% (Garcia-De-Ballesteros and Larroque, 1976).

Fish populations of the Gulf of California

The Gulf of California offers a paradoxical environment for those organisms that would invade its waters and flourish. It has some of the greatest environmental extremes of any of the world's seas, and great productivity (Moser et al., 1974). The fish fauna found here indicate that the Gulf is clearly part of the tropical American or Panamic faunal province. Walker (1960), in his review of the Gulf ichthyofauna, recorded 586 species of fish in the Gulf, of which 526 were shore fishes. Nearly three quarters of the shore fish are tropical or subtropical species with their principal ranges to the south of the Gulf. About 50 species, or 10% of the total, have their principal distributions to the north of the Gulf. The majority of these are found in the northern region of the Gulf as disjunct species from the Southern California or San Diegan fauna. The remaining 92 species, or 17% of the total, are endemic to the Gulf. In a comparison with the Panamic fauna to the south, Walker (1960) has noted that the Gulf has an unusually high number of rocky shore species, a relatively poor representation of muddy bottom forms, and a high percentage of endemic species. The paucity of certain Panamic groups (croakers, catfishes, and anchovies) indicates their inability to adapt to the environmental extremes of the Gulf. These extremes, however, along with the abundance and isolated nature of rocky shore habitats, have permitted an impressive evolution of blennioid fishes. The blennioid families Clinidae, Tripterygiidae, and Chaenopsidae are perhaps the best examples of the profusion of rocky shore fishes in the Gulf. These groups have been reviewed extensively by a number of investigators including Hubbs (1952), Springer (1958), Rosenblatt (1959) and Stephens (1963), and are among the best known of eastern Pacific shore fishes. The mid-water fish fauna, also dominated by warm-water species, is even more depauperate than the shore fish fauna. There seem to be no endemic mid-water fishes in the Gulf. Typically, those mid-water species that are recorded from the Gulf do not penetrate very far north, and most are limited to the southernmost basins. *Benthosema panamense* and *Triphoturus mexicanus* are the only mid-water species that occur in significant numbers in the northern regions of the Gulf. All of the abundant mid-water Gulf species have wide latitudinal ranges outside the Gulf and are equipped to tolerate the seasonal temperature range of the Gulf. Probably more important than this is their ability to tolerate the low oxygen concentrations at the minimum (Moser et al., 1974).

Since 1967 there has been an important sardine fishery of the Gulf of California. A subpopulation of *Sardinops caerulea*, that seems to spend all of its life within the Gulf, has its maximum concentration in the Guaymas area during winter, which is the spawning season. Based on the concentrations of larvae, the mature portion of the subpopulation has been estimated to be between 200 and 250×10^3 t (Sokolov and Wong-Rios, 1973; Sokolov, 1974; De-La-Campa-De-Guzman et al., 1976). A hake population, *Merluccius* sp., has been found principally north of Tiburon Island and along the east coast of the Gulf as far south as Guaymas. The population has been estimated as 28×10^3 t (Mathews et al., 1974). Ichthyoplankton samples were taken in a series of seven cruises to the Gulf of California, during 1956 and 1957, by Scripps Institution of Oceanography and the U.S. National Marine Fisheries Service. In 1956 the northern anchovy, *Engraulis mordax*, was by far the most abundant species, about 38% of the total, followed by the hake, *Merluccius productus*, with about 28%. These were followed by the sardine, *Sardinops sagax*, and the genera *Sebastes* and *Citharichthys*, each of which made up about 4% of the catch. The lanternfish *Triphoturus mexicanus* and the tropical gonostomatid *Vinciguerria lucetia*, ranked sixth and seventh at about 3% of the catch. In 1957, a warmer year, *E. mordax* remained the most abundant, but was followed closely by *V. lucetia*. The hake remained abundant, in third place, followed by *T. mexicanus*, *Citharichthys* sp. and the warm-water lanternfish, *Diogenichthys laternatus*. Larvae of the Pacific mackerel, *Scomber japonicus*, were widespread in the southern region of the Gulf during the February cruises of both years (Moser et al., 1974). The distribution, abundance and ecology of larval tunas at the entrance to the Gulf, were studied during October 1966–August 1967, with samples taken in seven cruises of Direccion General de Pesca e Industrias Conexas of Mexico and the Inter-American Tropical Tuna Commission. *Auxis* sp., the frigate mackerel, was the most abundant with 95.2% of the catch; *Thunnus albacares*, the yellowfin tuna, was second with 3.6%, and *Euthynnus lineatus* was third with 1.2%. Adult specimens of these species, and other tunas, are caught in this region for commercial purposes. Water masses had no influence *per se* on the distribution of larval tunas in this area (Klawe et al., 1970).

During summer 1973 Guevara-Escamilla (1973) collected fish samples with shrimp nets, using commercial shrimp boats, in the region adjacent to the Colorado River mouth, north of San Felipe. He collected a total of 73 species, 18% of which were endemic to the Gulf, 42% were San Diegan and 40% Panamic. One species of special interest in this region is the totoaba, *Cynoscion macdonaldi*, which used to support an important fishery, and now is so scarce that it may even disappear from this area. This region, adjacent to the Colorado River mouth, has an abundance of juveniles and has been declared a sanctuary by the Mexican government, from a line connecting Santa Clara, Sonora, at 114°31′W, with a point 9 km north of San Felipe, Baja California, to the north.

MAN'S IMPACT ON THE GULF OF CALIFORNIA

The waters and organisms of the Gulf of California may still be considered as existing in a natural state with respect to certain types of man's activities. For example, it may still be free of certain pollutants that are already causing problems in other seas of the world. But, unfortunately, man's activities are already altering the Gulf's ecosystems, in some cases dramatically, as in the upper Gulf. Pesticides used in the agricultural areas of the Mexicali valley and Sonora and Sinaloa states may already be entering the Gulf waters in significant quantities. However, there are almost no data on this subject. Guardado-Puentes and Nuñez-Equer (1975) found total concentrations of DDT (dichlorodiphenyl trichloroethane) and its metabolites as high as 0.122 p.p.m. in the sediments of the upper Gulf near the Colorado River mouth. They also found total DDT concentrations up to 0.150 p.p.m. in the clam *Chione californiensis* on the beaches north of San Felipe. Nevertheless, the greatest changes in the upper Gulf have other causes. The decrease of Colorado River fresh-water input has drastically changed the ecological conditions of what used to be an estuarine system, and is now an area of the highest salinities of the whole Gulf. There once was an important fishery of the totoaba, *Cynoscion macdonaldi*, an endemic fish of the northern Gulf. Now this fishery does not exist. There are three possible reasons for this: the lack of estuarine waters for reproduction; the intense fishing of adults near the Colorado River mouth until recent years; and — probably the most important factor and one that is seldom mentioned — the capture of juveniles by shrimp nets (Guevara-Escamilla, 1973). Capturing juveniles by shrimp nets may be a severe problem for many species, especially in the shallow northern Gulf, and the effect may be overlooked simply because there are no important fisheries for all of them. Mathews (1974) has expressed concern about a growing Pacific Mexican shrimp fleet that may already be overexploiting that resource. He estimated that on the average the shrimp nets were passing over each square meter of the Mexican Pacific shrimp grounds about seven times per year. In the upper Gulf the number must be much higher. Many other species caught with the shrimps are considered as "garbage" by the fishermen. Many of these could be useful, but they may already be diminished as a resource. All of them have a role in the Gulf's ecosystem, and man's activities may be changing them without any awareness of the ultimate effects.

REFERENCES

Allen, W.E., 1934. The importance of continuity in marine plankton investigations. *Fifth Pac. Sci. Congr. Calif., 1933. Proc*, 3: 2051–2055.

Allen, W.E., 1937. Plankton diatoms of the Gulf of California obtained by the G. Allan Hancock Expedition of 1936. *Hancock Pac. Exped., Univ. S. Calif. Publ.*, 3: 47–59.

Allen, W.E., 1938. The Templeton Crocker Expedition to the Gulf of California in 1935. The phytoplankton. *Trans. Am. Microsc. Soc.*, 67: 328–335.

Alvarez-Borrego, S. and Galindo-Bect, L.A. 1974. Hidrologia del Alto Golfo de California–I. Condiciones durante otoño. *Cienc. Mar.*, 1(1): 46–64.

Alvarez-Borrego, S., Flores-Baez, B.P. and Galindo-Bect., L.A., 1975. Hidrologia del Alto Golfo de California–II. Condiciones durante invierno, primavera y verano. *Cienc. Mar.*, 2(1): 21–36.

Alvarez-Borrego, S., Rivera, J.A., Gaxiola-Castro, G., Acosta-Ruiz, M.J. and Schwartzlose, R.A., 1978. Nutrientes en el Golfo de California. *Cienc. Mar.*, 5(2): 53–71.

Alvarez-Sanchez, L.G., Wyatt, B. and Stevenson, M.R., 1978. Corrientes en la region de la boca del Golfo de California en la primavera de 1970. *Cienc. Mar.*, 5(1): 105–118.

Alvariño, A., 1962. Zoogeografia de los Quetognatos en el Pacifico especialmente en la region de California. In: *Congreso sobre Biogeografia de los organismos marinos, UNESCO, Mar del Plata, Oct. 1962.*

Alvariño, A., 1963. Quetognatos epiplanctonicos del Mar de Cortes. *Rev. Soc. Mex. Hist. Nat.* 24: 97–203.

Berdegue, J., 1955. La pesqueria de la totoaba (*Cynoscion macdonaldi*, Gilbert) en San Felipe, Baja California. *Rev. Soc. Mex. Hist. Nat.*, 16: 1–4.

Brown, D.M., 1965. *Results of Current Measurements with Drogues, 1963–1964.* Data Rp. Ref. 65-3, Scripps Institution of Oceanography, University of California, La Jolla, Calif., 42 pp.

Bullard, E., 1969. The origin of the oceans. *Sci. Am.*, 221(3): 66–75.

Calvert, S.E., 1964. Factors affecting distribution of laminated diatomaceous sediments in Gulf of California. In: Tj.H. van Andel and G.G. Shor Jr. (Editors), *Marine Geology of the Gulf of California: A Symposium. Am. Assoc. Pet. Geol., Mem.*, 3: 311–330.

Calvert, S.E., 1966a. Accumulation of diatomaceous silica in the sediments of the Gulf of California. *Geol. Soc. Am. Bull.*, 77: 569–596.

Calvert, S.E., 1966b. Origin of diatom-rich, varved sediments from the Gulf of California. *J. Geol.*, 76: 546–565.

Cline, J.D. and Richards, F.A., 1972. Oxygen deficient conditions and nitrate reduction in the eastern tropical North Pacific Ocean. *Limnol. Oceanogr.*, 7: 885–900.

Cummings, J.A., 1977. *Seasonal and Areal Variations of Zooplankton Standing Stocks in the Northern Gulf of California.* Thesis. Dept. of Ecology and Evolutionary Biology, University of Arizona, Tucson, Ariz., 61 pp.

Cupp, E.E. and Allen, W.E., 1938. Plankton diatoms of the Gulf of California obtained by Allan Hancock Pacific Expedition of 1937. *Hancock Pac. Exped., Univ. S. Calif. Publ.*, 3: 61–74.

De-La-Campa-De-Guzman, S., Padilla-Garcia, M.A. and Smith, P.E., 1976. Estimaciones de biomasa de reproductores de sardina Monterrey (*Sardinops sagax*) a traves de censos larvales. Golfo de California. Temporada 1975. In: *Mem. 1er Simp. Nacl. Rec. Pesq. Masivos de Mex.* Instituto Nacional de Pesca. Secretaria de Industria y Comercio, Ensenada, B.C.

Farfan, B.C., 1973. Estimacion de biomasa de zooplancton en la zona norte del Golfo de California. In: *Estudio quimico sobre la contaminacion por insecticidas en la desembocadura del rio Colorado, II.* Universidad Autonoma de Baja California, Rep. a la Dir. de Acuacultura, S.R.H. Ensenada, B.C., pp. 339–364.

Filloux, J.H., 1973. Tidal patterns and energy balance in the Gulf of California. *Nature*, 243: 217–221.

Garcia-De-Ballesteros, G. and Larroque, M., 1976. Elementos sobre la distribucion de turbidez en el Alto Golfo de California. *CalCOFI Rep.*, 18: 81–106.

Gaxiola-Castro, G., Alvarez-Borrego, S. and Schwartzlose, R.A., 1978. Sistema del bioxido de carbono en el Golfo de California. *Cienc. Mar.*, 5(2): 25–40.

Gendrop-Funes, V., Acosta-Ruiz, M.J. and Schwartzlose, R.A., 1978. Distribucion horizontal de clorofila "a" durante la primavera en la parte Norte del Golfo de California. *Cienc. Mar.*, 5(1): 71–89.

Gilbert, J.Y. and Allen, W.E., 1943. The phytoplankton of the Gulf of California obtained by the "E.W. Scripps" in 1939 and 1940. *J. Mar. Res.*, 5(2): 89–110.

Gilmartin, M. and Revelante, N., 1978. The phytoplankton characteristics of the Barrier Island Lagoons of the Gulf of California. *Estuarine Coastal Mar. Sci.*, 7(1): 29–47.

Goering, J.J., 1968. Denitrification in the oxygen minimum layer of the eastern tropical Pacific Ocean. *Deep-Sea Res.*, 15: 151–164.

Granados-Gallegos, J.L. and Schwartzlose, R.A., 1974. Corrientes superficiales en el Golfo de California. In: *Memorias del V Congreso Nacional de Oceanografia.* Escuela de Ciencias Maritimas des Instituto Tecnologico de Monterrey, Guaymas, Son.

Granados-Gallegos, M.L., 1974. *Distribucion de temperatura, salinidad y oxigeno disuelto en el Golfo de California, en Noviembre de 1972.* Thesis, Escuela Superior de Ciencias Marinas, Universidad Autonoma de Baja California, Ensenada, B.C., 56 pp.

Griffiths, R.C., 1968. Physical, chemical, and biological oceanography of the entrance to the Gulf of California, spring of 1960. *U.S. Fish Wildl. Serv. Spec. Sci. Rep. Fish.*, 573: 47 pp.

Grijalva-Ortiz, N., 1972. Tidal computation in the Gulf of California I. *Geofis. Int.*, 12(2): 13–34.

Guardado-Puentes, J. and Nuñez-Esquer, O., 1975. *Concentracion de DDT y sus metabolitos en especies filtro-alimentadoras y sedimentos en el Valle de Mexicali y Alto Golfo de California.* Thesis, Escuela Superior de Ciencias Marinas, Universidad Autonoma de Baja California, Ensenada, B.C., 31 pp.

Guevara-Escamilla, S., 1973. Peces. In: *Estudio quimico sobre la contaminacion por insecticidas en la desembocadura del rio Colorado, II.* Universidad Autonoma de Baja California, Rep. a la Dir. de Acuacultura, S.R.H. Ensenada, B.C., pp. 236–262.

Harris, M.F., 1969. *Effects of Tropical Cyclones upon Southern California.* Thesis, San Fernando Valley State College, San Fernando, Calif., 89 pp.

Hendershott, M.C. and Speranza, A., 1971. Co-oscillating tides in long, narrow bays; the Taylor problem revisited. *Deep-Sea Res.*, 18: 959–980.

Hernandez, J., 1923. The temperature of Mexico. *Mon. Weather Rev. Suppl.*, 23: 24 pp.

Hubbs, C., 1952. A contribution to the classification of the blennioid fishes of the family Clinidae, with a partial revision of the Eastern Pacific forms. *Stanford Ichthyol. Bull.*, 4(2): 41–165.

Klawe, W.L., Pella, J.J. and Leet, W.S., 1970. The distribution, abundance and ecology of larval tunas from the entrance to the Gulf of California. *Inter-Am. Trop. Tuna Comm. Bull.*, 14(4): 507–544.

Larson, R., Menard, W. and Smith, S., 1968. Gulf of California: A result of ocean-floor spreading and transform faulting. *Science*, 161: 781–784.

Leon-Portilla, M., 1972. Descubrimiento en 1540 y primeras noticias de la isla de Cedros. *Calafia (Rev. Univ. Auton. Baja Calif.)*, 2(1): 8–10.

Lepley, L.K., Vonder Haar, S.P., Hendrickson, J.R. and Calderon-Riverol, G., 1975. Circulation in the northern Gulf of California from orbital photographs and ship investigations. *Cienc. Mar.*, 2(2): 86–93.

Mathews, C.P., 1974 ¿Cuanto resistira el camaron? *Cienc. Mar.*, 1(2): 86–91.

Mathews, C.P., Granados-Gallegos, J.L. and Arvizu, J., 1974. Results of the exploratory cruises of the "Alejandro De Humboldt" in the Gulf of California. *CalCOFI Rep.*, 17: 101–111.

Moore, D. and Buffington, E.C., 1968. Transform faulting and growth of the Gulf of California since the late Pliocene. *Science*, 161: 1238–1241.

Moser, H.G., Ahlstrom, E.H., Kramer, D. and Stevens, E.G., 1974. Distribution and abundance of fish eggs and larvae in the Gulf of California. *CalCOFI Rep.*, 17: 112–128.

Mundhenke, D.J., 1969. *The Relationships between Water Masses and Euphausiids in the Gulf of California and the Eastern Tropical Pacific*. Thesis, Naval Post-Graduate School, Monterey, Calif., 114 pp.

Page, J.L., 1930. Climate of Mexico. *Mon. Weather Rev. Suppl.*, 33: 33 pp.

Redfield, A.C., Ketchum, B.H. and Richards, F.A., 1963. The influence of organisms on the composition of seawater. In: M.N. Hill (Editor), *The Sea. Ideas and Observations on Progress in the Study of the Seas, 2.* Interscience, New York, N.Y., pp. 26–77.

Reichle, M.S., 1975. *A Seismological Study of the Gulf of California: Sonobuoy and Teleseismic Observations, and Tectonic Implications*. Thesis, University of California, San Diego, La Jolla, Calif., 249 pp.

Robinson, M.K., 1973. Atlas of monthly mean sea surface and subsurface temperatures in the Gulf of California, Mexico. San Diego Soc. Nat. Hist., Mem., 5: 97 pp.

Roden, G.I., 1958. Oceanographic and meteorological aspects of the Gulf of California. *Pac. Sci.*, 12(1): 21–45.

Roden, G.I., 1964. Oceanographic aspects of the Gulf of California. In: Tj.H. van Andel and G.G. Shor Jr. (Editors), *Marine Geology of the Gulf of California: A Symposium. Am. Assoc. Pet. Geol., Mem.*, 3: 30–58.

Roden, G.I., 1972. Thermohaline structure and baroclinic flow across the Gulf of California entrance and the Revillagigedo islands region. *J. Phys. Oceanogr.*, 2(2): 177–183.

Roden, G.I. and Groves, G.W., 1959. Recent oceanographic investigations in the Gulf of California. *Mar. Res. J.*, 18(1): 10–35.

Rosas-Cota, J.A., 1976. *Corrientes geostroficas en el Golfo de California en la superficie y a 200 metros*. Thesis, Escuela Superior de Ciencias Marinas, Universidad Autonoma de Baja California, Ensenada, B.C., 56 pp.

Rosenberg, D.H., 1969. Distribution of temperature, salinity and oxygen. In: *Environmental Impact of Brine Effluents on Gulf of California. U.S. Dep. Interior Res. Dev. Progr. Rep.*, 387: 16–40.

Rosenblatt, R.H., 1959. *A Revisionary Study of the Blennioid Fish Family Tripterygidae*. Thesis, University of California, Los Angeles, Calif., 376 pp.

Round, F.E., 1967. The phytoplankton of the Gulf of California — part I. Its composition, distribution and contribution to the sediments. *J. Exper. Mar. Biol. Ecol.*, 1: 76–97.

Rusnak, G.A. and Fisher, R.L., 1964. Structural history and evolution of Gulf of California. In: Tj.H. van Andel and G.G. Shor Jr. (Editors), *Marine Geology of the Gulf of California: A Symposium. Am. Assoc. Pet. Geol., Mem.*, 3: 144–156.

Rusnak, G.A., Fisher, R.L. and Shepard, F.P., 1964. Bathymetry and faults of the Gulf of California. In: Tj.H. van Andel and G.G. Shor Jr. (Editors), *Marine Geology of the Gulf of California: A Symposium. Am. Assoc. Pet. Geol. Mem.*, 3: 59–75.

Schreiber Jr., J.F., 1969. Changes in Colorado river flow. In: *Environmental Impact of Brine Effluents on Gulf of California. U.S. Dep. Interior Res. Dev. Progr. Rep.*, 387: 83–87.

Shepard, F.P., 1950. Submarine topography of the Gulf of California, Part 3 of the 1940 *E.W. Scripps* cruise to the Gulf of California. *Geol. Soc. Am. Mem.*, 43: 32 pp.

Sokolov, V.A., 1974. Investigaciones biológico pesqueras de los peces pelágicos del Golfo de California. *CalCOFI Rep.* 17: 92–96.

Sokolov, V.A. and Wong-Rios, M., 1973. Informe cientifico de las investigaciones efectuadas sobre los peces pelagicos del Golfo de California (sardina, crinuda y anchoveta) en 1971. *Progr. Invest. Fom. Pesq. Mex/PNUD/FAO. INP. Inf. Cient. Num. 2. INP/SI*, 12: 1–41.

Springer, V.G., 1958. Systematics and zoogeography of the clinid fishes of the subtribe Labrisomini Hubbs. *Univ. Texas Inst. Mar. Sci. Publ.*, 5: 417–492.

Stephens, Jr., J.S., 1963. A revised classification of the blennioid clinid fishes of the American family Chaenopsidae. *Univ. Calif. Publ. Zool.*, 68: 1–133.

Stephens, Jr. J.S., 1963. A revised classification of the blennioid fishes of the American family Chaenopsidae. *Univ. Calif. Publ. Zool.*, 68: 1–133.

Stevenson, M.R., 1970. On the physical and biological oceanography near the entrance to the Gulf of California, October 1966–August 1967. *Inter-Am. Trop. Tuna Comm. Bull.*, 4(3): 389–504.

Stock, G.G., 1976. *Modeling of Tides and Tidal Dissipation in the Gulf of California*. Thesis, Scripps Institution of Oceanography. University of California, San Diego, La Jolla, Calif., 102 pp.

Sverdrup, H.U., 1941. The Gulf of California; preliminary discussion on the cruise of the E.W. Scripps in February and March 1939. *6th Pac. Sci. Congr. Proc.*, 3: 161–166.

Sverdrup, H.U. and Staff, 1940. Oceanic observations of the Scripps Institution in 1939. University of California, San Diego, La Jolla. *Calif. Rec. Observ.*, 1: 65–160.

Sykes, L., 1968. Seismological evidence for transform faults, sea-floor spreading and continental drift. In: R.A. Phinney

(Editor), *History of the Earth's Crust, A NASA Symposium.* Princeton University Press, Princeton, N.J., pp. 120–150.

Thomas, W.H. and Dodson, A.N., 1968. Effects of phosphate concentration on cell division rates and yield of a tropical oceanic diatom. *Biol. Bull. Mar. Biol. Lab. Woods Hole*, 134: 199–208.

Thorade, H., 1909. Über die Kalifornische Meeresstromung. *Hydrogr. Marit. Meteorol. Ann.*, 37: 17–34; 63–76.

Townsend, C.H., 1901. Dredging and other records of the U.S. Fish Commission steamer *Albatross* with bibliography relative to the work of the vessel. *U.S. Fish Comm. Rep. for 1900*, pp. 387–562.

United States–Mexico–International Atomic Energy Agency study team, 1968. *Nuclear Power and Water Desalting Plants for Southwest United States and Northwest Mexico: A Preliminary Assessment.* Executive summary. TID-24681. Clearinghouse for Federal Scientific and Technical Information, NBS, U.S. Dep. Commerce, Springfield, Va., 22151.

Vaccaro, R.R., 1965. Inorganic nitrogen in sea water. In: J.P. Riley and G. Skirrow (Editors), *Chemical Oceanography*, 1. Academic Press, New York, N.Y., pp. 365–404.

Van Andel, Tj.H., 1964. Recent marine sediments of the Gulf of California. In: Tj.H. van Andel and G.G. Shor Jr. (Editors), *Marine Geology of the Gulf of California: A symposium. Am. Assoc. Pet. Geol., Mem.*, 3: 216–310.

Van Andel, Tj.H. and Shor Jr., G.G., (Editors), 1964. *Marine Geology of the Gulf of California: A Symposium. Am. Assoc. Pet. Geol., Mem.*, 3: 408 pp.

Villaseñor-Casales, A., 1974. *Hidrología de la parte norte del Golfo de California.* Thesis, Escuela Superior de Ciencias Marinas, Universidad Autonoma de Baja California, Ensenada, B.C., 50 pp.

Vonder Haar, S.P. and Stone, R.O., 1973. Oceanographic analysis of orbital photographs of the upper Gulf of California. *Photogrammetria*, 29: 45–61.

Wada, E. and Hattori, A., 1971. Nitrite metabolism in the euphotic layer of the central North Pacific Ocean. *Limnol. Oceanogr.*, 16: 766–772.

Walker, B.W., 1960. The distribution and affinities of the marine fish fauna of the Gulf of California. In: *Symposium: The Biogeography of Baja California and Adjacent Seas. Part 2. Marine Biotas. Syst. Zool.*, 9(3–4): 123–133.

Ward, R. De C., and Brooks, C.F., 1936. *The Climates of North America, I (Mexico, United States, Alaska).* Handbuch der Klimatologie II (j). Borntraeger Verlag, Berlin, 327 pp.

Warsh, C.E. and Warsh, K.L., 1971. Water exchange at the mouth of the Gulf of California. *J. Geophys. Res.*, 76: 8098–8116.

Warsh, C.E., Warsh, K.L. and Staley, R.C., 1973. Nutrients and water masses at the mouth of the Gulf of California. *Deep-Sea Res.*, 20: 561–570.

Zeitzschel, B., 1969. Primary productivity in the Gulf of California. *Mar. Biol.*, 3(3): 201–207.

Zeitzschel, B., 1970. The quantity, composition and distribution of suspended particulate matter in the Gulf of California. *Mar. Biol.*, 7: 305–318.

SYSTEMATIC LIST OF GENERA[1]

MONERA

CYANOBACTERIA (CYANOPHYTA)

MYXOPHYCEAE
Chroococcaceae
Chroococcus
Merismopedia
Microcystis
Nostocaceae
Anabaena
Anabaenopsis
Aphanizomenon
Lyngbya
Nodularia
Oscillatoria
Trichodesmium

EUBACTERIA

Azotobacter
Bacterium
Clostridium
Pseudobacterium
Pseudomonas
Rhodotorula
Spirillum
Treponema
Vibrio

PLANTS

BACILLARIOPHYTA

BACILLARIOPHYCEAE
Achnanthes
Actinocyclus
Asterionella
Asteromphalus
Bacillaria
Bacteriastrum
Biddulphia
Ceratulina
Chaetoceros
Cocconeis
Coscinodiscus
Cyclotella
Denticula
Detonula
Dinophysis
Ditylum
Fragilaria
Guinardia
Lauderia
Leptocylindrus
Mastogloia
Melobesia
Melosira
Navicula
Nitzschia
Paralia
Planktoniella
Poresira
Pseudoeunotia
Rhizosolenia
Skeletonema
Stephanodiscus
Thalassionema
Thalassiosira
Thalassiothrix

CHLOROPHYTA

CHLOROPHYCEAE
Acetabularia
Ankistrodesmus
Bipedinomonas
Caulerpa
Chaetomorpha
Chlamydomonas
Chlorella
Cladophora
Cosmarium
Crucigenia
Dictyosphaerium
Enteromorpha
Entodladia
Euastrum
Halimeda
Micrasterias
Nannochloris
Nephrochloris
Oocystis
Richteriella
Scenedesmus
Spirogyra
Stichococcus
Tetraedron
Ulva

CHRYSOPHYTA

CHRYSOPHYCEAE
Coccolithus
Dictyocha
Distephanus
Monas
Olisthodiscus
Syracosphaera

CRYPTOPHYTA

CRYPTOPHYCEAE
Cryptomonas
Rhodomonas

EUGLENOPHYTA

EUGLENOPHYCEAE
Eutreptia

PHAEOPHYTA

PHAEOPHYCEAE
Corynophlaea
Cystroseira
Dilophus
Ectocarpus
Fucus
Laminaria
Stilophora
Undaria

PYRROPHYTA

DINOPHYCEAE
Amphidinium
Ceratium
Exuviaella
Gonyaulax
Gymnodinium
Gyrodinium
Heterocapsa
Isochrysis
Massartia
Noctiluca
Peridinium
Prorocentrum
Pyrodinium

RHODOPHYTA

RHODOPHYCEAE
Ceramium
Dasya
Gelidium
Kylinia
Laurencia
Phyllophora
Polysiphonia
Porphyra

XANTHOPHYTA

FUNGI

Candida
Fusarium

MAGNOLIOPHYTA (ANGIOSPERMAE)

LILIOPSIDA (MONOCOTYLEDONES)
Lemraceae
Wolffia
Poaceae
Spartina
Potamogetonaceae
Ruppia
Zostera
Hydrocharitaceae
Halophila
Thalassia

ANIMALS

PROTOZOA

CILIATA
Aspidisca
Blepharisma
Chilodonella
Condylostoma
Diophrys
Favella
Frontonia
Lohmaniella
Mesodinium
Metacystis
Prorodon
Sonderia
Spirostomum
Strombidium
Tiarina
Tintinnopsis
Uronema

PORIFERA

DEMOSPONGEA
Cliona

CNIDARIA (COELENTERATA)

ANTHOZOA
Hexacorallia
Diadumene
Metridium
Tealia
Octocorallia

[1]The taxonomic relationships of all genera of plants and animals mentioned in this Volume on *Estuaries and Enclosed Seas* are shown. Genera are listed alphabetically under families, arranged into phyla and subphyla; orders are included in the animal kingdom.

In addition to the contributions of the authors of pertinent parts of the volume, the following collaborators deserve special thanks for their assistance in checking the list and in placing rare or exotic genera in the correct categories: Richard H. Backus, James T. Carlton, Mary Ann Daher, Edward M. Hulburt and John B. Waterbury. Thanks also are extended to Margaret S. Dimmock who typed the final copy of this list.

CNIDARIA *(continued)*

Corallium
Pennatula

HYDROZOA (HYDROMEDUSAE)
Cordylophora
Sarsia

SCYPHOZOA
Aurelia
Chrysaora
Rhizostoma

CTENOPHORA

Beroë
Bolinopsis
Mnemiopsis
Pleurobrachia

PLATYHELMINTHES

TURBELLARIA
Mecynostomum
Procerodes

GNATHOSTOMULIDA

NEMERTINA

Eunemertes
Lineus

ENTOPROCTA

Barentsia

ASCHELMINTHES

GASTROTRICHA
Turbanella

ROTIFERA
Keratella
Trichocerca

NEMATODA

ANNELIDA

OLIGOCHAETA
Marionina
Nais
Paranais

POLYCHAETA
Abarenicola
Arenicola
Boccardia
Capitella
Chaetoparia
Chinonereis
Cistenoides
Clymenella
Ficopomatus
Glycera
Harmothoe
Hesionides
Heteromastus
Lanice
Lepidonotus
Melinna
Mercierella
Nephtys
Nereis
Nerine
Notomastus
Ophelia
Polydora
Protodriloides
Sabella
Scoloplos
Spio
Sternaspis
Streblospio
Stygocapitella

ARTHROPODA

CRUSTACEA

Ostracoda

Branchiopoda
Anostraca
Artemia
Cladocera
Bosmina
Evadne
Penilia
Podon
Copepoda
Acartia
Acartiura
Boeckella
Calamoecia
Calanopia
Calanus
Centropages
Clausocalanus
Corycaeus
Ctenocalanus
Cyclops
Diaptomus
Ectinosoma
Epilabidocera
Eucalanus
Eurytemora
Gippslandia
Gladioferens
Hemiboeckella
Hemicyclops
Isias
Labidocera
Limnocalanus
Mesocyclops
Metridia
Oithona
Paracalanus
Paraleptastacus
Pontella
Pseudocalanus
Pseudodiaptomus
Schmackeria
Scottolana
Sinocalanus
Sulcanus
Temora
Tortanus
Cirripedia
Balanus
Malacostraca
Eucarida
Euphausiacea
Euphausiidae
Euphausia
Thysanoessa
Decapoda
Anomura
Callianassa
Clibanarius
Pagurus
Paralithodes
Upogebia
Astacura
Homarus
Nephrops
Brachyura
Callinectes
Carcinus
Cardisoma
Charybdis
Chionoecetes
Erimacrus
Eriocheir
Gecarcinus
Halicarcinus
Heteropanope
Libinia
Macropipus
Menippe
Ocypode
Panopeus
Portunus
Rhithropanopeus
Sesarma
Uca
Caridea
Alpheus
Crangon
Pandalus
Palinura
Palinurus
Penaeidea
Metapenaeus
Parapenaeus
Penaeus
Trachypenaeus
Hoplocarida
Stomatopoda
Oratosquilla
Peracarida
Amphipoda
Bathyporeia
Gammarellus
Gammarus
Haustorius
Microdeutopus
Orchestia
Parathemisto
Pontogammarus
Pontoporeia
Talitrus
Talorchestia
Cumacea
Isopoda
Cyathura
Eurydice
Idothea
Mesidothea
Paragnathia
Saturia
Sphaeroma
Mysidacea
Diamysis
Gastrosaccus
Mysis
Neomysis
Mystacocarida
Derocheilocaris

INSECTA
Diptera
Clunio
Cricotopus

MOLLUSCA

GASTROPODA
Assiminea
Bithynia
Buccinum
Cerithium
Clione
Crepidula
Haliotis
Hydrobia
Lacuna
Limacina
Littorina
Melampus
Nassa
Nassarius
Ovatella
Rapana
Retusa
Rissoa
Theodoxus
Trochus
Turritella
Urosalpinx

BIVALVIA
Abra
Arctica
Argopecten
Astarte
Barnea
Brachidontes
Cardium
Cerastoderma
Chione
Corbula
Crassostrea
Diala
Divaricella
Donacilla
Laevicardium
Loripes
Macoma
Mactra
Mercenaria
Meretrix
Modiolus
Mya
Mytilus
Nucula
Ostrea
Pandora
Pecten
Pholas
Portlandia
Scrobicularia
Syndesmia
Tapes
Tellina
Teredo
Venus
Yoldia
Zirfaea

SCAPHOPODA
Dentalium

CEPHALOPODA
Abraliopsis
Alloleuthis
Eledone

MOLLUSCA (*continued*)

Enoploteuthis
Loligo
Octopus
Sepia
Sepiola
Todarodes

TENTACULATA

BRYOZOA

Bowerbankia
Conopeum
Lepralia
Membranipora

CHAETOGNATHA

Sagitta

ECHINODERMATA

ASTEROIDEA
Asterias
Astropecten
Hacellia
Marthasterias

CRINOIDEA
Antedon

ECHINOIDEA
Brissopsis
Centrostephanus
Echinocardium
Schizaster
Strongylocentrotus

HOLOTHUROIDEA
Stichopus
Synapta

CHORDATA

TUNICATA

ASCIDIACEA
Rhodosoma

APPENDICULARIA (LARVACEA)
Oikopleura

VERTEBRATA

Pisces

AGNATHA
Cyclostomata
Petromyzontidae

GNATHOSTOMATA
Chondrichthyes
Lamniformes
Carcharhinidae
Mustelus
Rajiformes
Dasyatidae
Dasyatis
Gymnura
Mobulidae
Manta
Rajidae
Raja
Squaliformes
Squalidae
Squalus

Osteichthyes
Actinopterygii
CHONDROSTEI
Acipenseriformes
Acipenseridae
Acipenser
Huso

TELEOSTEI
Anguilliformes
Anguillidae
Anguilla
Muraenesocidae
Muraenesox
Atheriniformes
Atherinidae
Atherina
Menidia
Pranesus
Belonidae
Cyprinodontidae
Aphanius
Fundulus
Lucania
Exocoetidae
Exocoetus
Prognichthys
Poeciliidae
Poecilia
Scomberesocidae
Cololabis
Batrachoidiformes
Batrachoididae
Opsanus
Beryciformes
Berycidae
Holocentridae
Clupeiformes
Clupeidae
Alosa
Brevoortia
Caspialosa
Clupea
Clupeonella
Ethmalosa
Hilsa
Sardina
Sardinella
Sardinops
Sprattus
Engraulidae
Anchoa
Engraulis
Cypriniformes
Cyprinidae
Abramis
Rutilus
Gadiformes
Gadidae
Gadiculus
Gadus
Gaidropsarus
Melanogrammus
Merlangius
Microgadus
Micromesistius
Molva
Theragra
Trisopterus
Urophycis
Macrouridae
Chalinurus
Coelorhynchus
Coryphanoides
Hemimacrurus
Hymenocephalus
Macrurus
Nematonurus
Merlucciidae
Merluccius
Zoarcidae
Macrozoarces
Gasterosteiformes
Gasterosteidae
Gasterosteus
Gonorynchiformes
Chanidae
Chanos
Lampridiformes
Lamprididae
Regalecidae
Trachipteridae
Myctophiformes
Myctophidae
Benthosema
Diogenichthys
Triphoturus
Synodontidae
Saurida
Perciformes
Acanthuridae
Ammodytidae
Bathyclupeidae
Blennidae
Blennius
Branchiostegidae
Carangidae
Decapterus
Seriola
Trachurus
Chaetodontidae
Chaenopsidae
Cichlidae
Tilapia
Clinidae
Emmelichthyidae
Smaris (= *Spicara*)
Ephippidae
Girellidae
Crenidens
Gobiidae
Didogobius
Glossogobius
Lepidogobius
Pomatoschistus
Istiophoridae
Tetrapturus
Kuhliidae
Labridae
Crenilabrus
Tautoga
Tautogolabrus
Lutjanidae
Mugilidae
Chelon
Liza
Mugil
Mullidae
Mullus
Upeneus
Percichthyidae
Morone
Siniperca
Percidae
Perca
Polynemidae
Pomacentridae
Pomatomidae
Pomatomus
Scaridae
Sciaenidae
Corvina
Cynoscion
Menticirrhus
Micropogonias
Scombridae
Auxis
Euthynnus
Katsuwonus
Pelamys (= *Sarda*)
Scomber
Thunnus
Serranidae
Dicentrarchus
Serranus
Siganidae
Siganus
Sillaginidae
Sillago
Sparidae
Lithognathus
Pagrus
Rhabdosargus
Sparus
Stenotomus
Sphyraenidae
Sphyraena
Pampidae (= Stromateidae)
Stichaeidae
Stromateidae
Poronotus (= *Peprilus*)
Theraponidae
Thunnidae (= Scombridae)
Trichiuridae
Trichiurus
Tripterygiidae
Xiphiidae
Xiphias
Zanclidae
Pleuronectiformes
Bothidae
Citharichthys
Lepidorhombus
Paralichthys
Psetta (= *Scophthalmus*)
Rhombus
Cynoglossidae
Cynoglossus
Pleuronectidae
Clidoderma
Eopsetta
Glyptocephalus
Hippoglossoides
Hippoglossus
Lepidopsetta
Limanda
Platichthys
Pleuronectes
Pseudopleuronectes
Rhombosolea
Tanakius
Soleidae
Solea
Salmoniformes
Esocidae
Esox
Galaxiidae
Galaxias
Gonostomatidae
Maurolicus

CHORDATA (*continued*)

Vinciguerria
Osmeridae
Mallotus
Osmerus
Salangidae
Salmonidae
Oncorhynchus
Salmo
Salvelinus
Scorpaeniformes
Agonidae
Anoplopomatidae
Anoplopoma
Cottidae
Cottus
Leptocottus
Myoxocephalus
Hexagrammidae
Ophiodon
Liparidae (= Cyclopteridae)
Scorpaenidae
Sebastes
Triglidae
Prionotus
Siluriformes
Ariidae
Syngnathiformes
Syngnathidae
Syngnathus
Tetraodontiformes
Molidae
Tetraodontidae
Fugu (= *Sphoeroides*)
Zeiformes
Zeidae

Aves

Anseriformes
Anatidae
Mergus
Charadriiformes
Alcidae
Falconiformes
Pandionidae
Pandion
Pelecaniformes
Phalacrocoracidae
Phalacrocorax

Mammalia

Carnivora
Fissipedia
Mustelidae
Enhydra
Ursidae
Ursus
Pinnipedia
Odobenidae
Odobenus
Phocidae
Callorhinus
Erignathus
Eumetopias
Phoca
Cetacea
Odontoceti
Delphinidae
Berardius
Delphinapterus
Delphinus
Orcinus
Tursiops
Monodontidae
Monodon
Phocoenidae
Phocoena
Phocoenoides
Physeteridae
Physeter
Ziphidae
Mesoplodon
Ziphius
Mysticeti
Balaenidae
Balaena
Balaenopteridae
Balaenoptera
Eschrichtius
Megaptera

AUTHOR INDEX[1]

[1] Page references to text are in roman type, to bibliographical entries in italics.

SYSTEMATIC INDEX[1]

[1]In this index, no attempt has been made, for larger taxonomic entities, to list all the pages where subordinate taxa are mentioned. These may be found by use of the Systematic List of Genera (pp. 451–454). For some major groupings, more detailed entries will be found in the General Index.

GENERAL INDEX[1]

[1]Page references to complete chapters are in bold type.

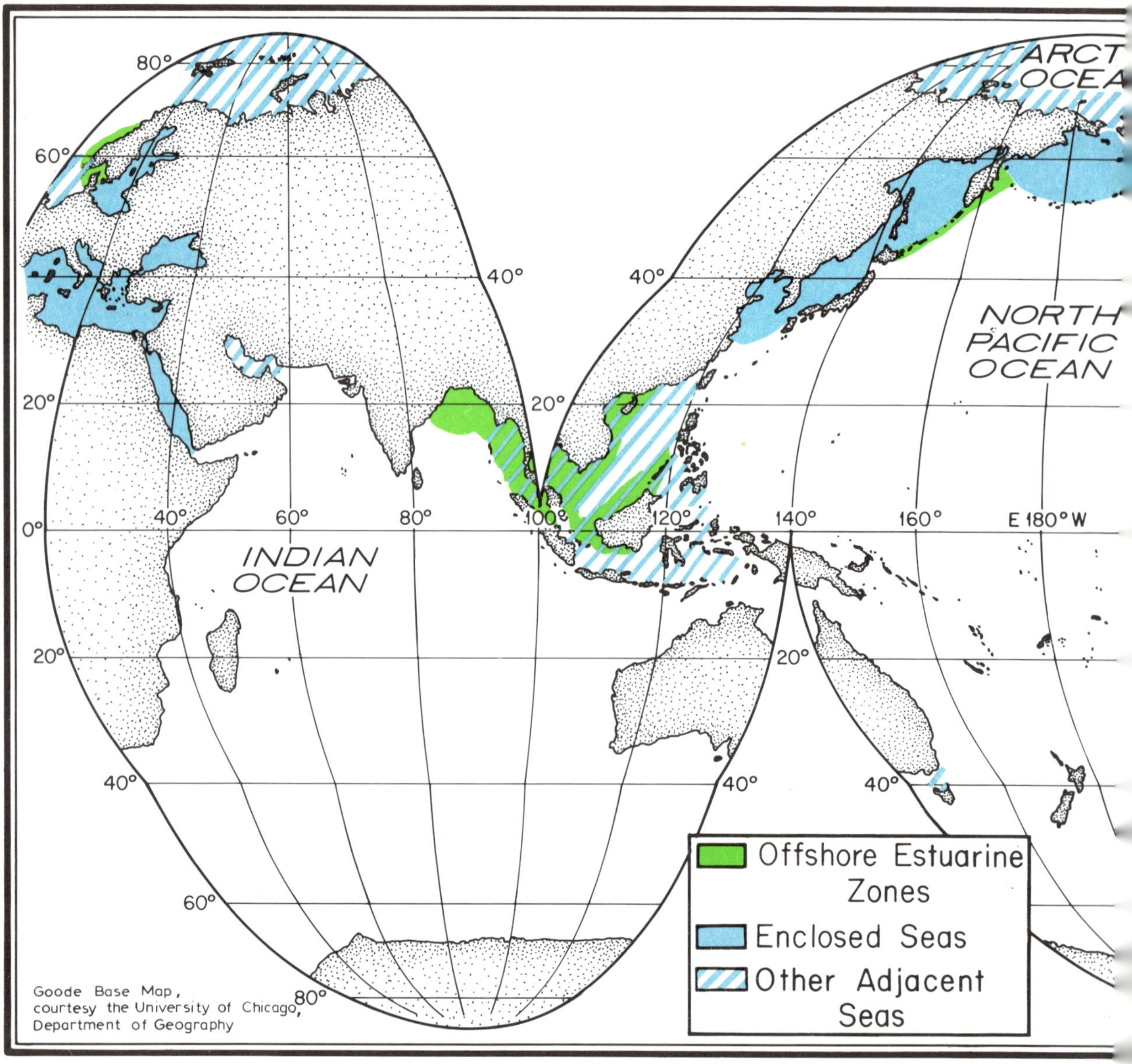

ARCT
OCEA
NORTH
PACIFIC
OCEAN
INDIAN
OCEAN
80°
60°
40°
20°
0°
20°
40°
60°
80°
40°
60°
80°
100°
120°
140°
160°
E 180° W
Offshore Estuarine Zones
Enclosed Seas
Other Adjacent Seas
Goode Base Map,
courtesy the University of Chicago,
Department of Geography